# ANNUAL REVIEW OF NUCLEAR AND PARTICLE SCIENCE

# ANNUAL REVIEW OF NUCLEAR AND PARTICLE SCIENCE

Volume 35, 1985

J. D. JACKSON, *Editor*

University of California, Berkeley

HARRY E. GOVE, *Associate Editor*

University of Rochester

ROY F. SCHWITTERS, *Associate Editor*

Harvard University

ANNUAL REVIEWS INC.     4139 EL CAMINO WAY     PALO ALTO, CALIFORNIA 94306 USA

ANNUAL REVIEWS INC.
Palo Alto, California, USA

*International Standard Serial Number: 0163-8998*
*International Standard Book Number: 0-8243-1535-9*
*Library of Congress Catalog Card Number: 53-995*

TYPESET BY AUP TYPESETTERS (GLASGOW) LTD., SCOTLAND
PRINTED AND BOUND IN THE UNITED STATES OF AMERICA

Annual Review of Nuclear and Particle Science
Volume 35, 1985

# CONTENTS

ANNUAL REVIEWS INC. is a nonprofit scientific publisher established to promote the advancement of the sciences. Beginning in 1932 with the *Annual Review of Biochemistry*, the Company has pursued as its principal function the publication of high quality, reasonably priced *Annual Review* volumes. The volumes are organized by Editors and Editorial Committees who invite qualified authors to contribute critical articles reviewing significant developments within each major discipline. The Editor-in-Chief invites those interested in serving as future Editorial Committee members to communicate directly with him. Annual Reviews Inc. is administered by a Board of Directors, whose members serve without compensation.

ANNUAL REVIEWS OF

| | | |
|---|---|---|
| Anthropology | Medicine | |
| Astronomy and Astrophysics | Microbiology | |
| Biochemistry | Neuroscience | |
| Biophysics and Biophysical Chemistry | Nuclear and Particle Science | |
| Cell Biology | Nutrition | |
| Earth and Planetary Sciences | Pharmacology and Toxicology | |
| Ecology and Systematics | Physical Chemistry | |
| Energy | Physiology | |
| Entomology | Phytopathology | |
| Fluid Mechanics | Plant Physiology | |
| Genetics | Psychology | |
| Immunology | Public Health | |
| Materials Science | Sociology | |

SPECIAL PUBLICATIONS

Annual Reviews Reprints:
  Cell Membranes, 1975–1977
  Cell Membranes, 1978–1980
  Immunology, 1977–1979

Excitement and Fascination of Science, Vols. 1 and 2

History of Entomology

Intelligence and Affectivity, by Jean Piaget

Telescopes for the 1980s

A detachable order form/envelope is bound into the back of this volume.

# SOME RELATED ARTICLES IN OTHER *ANNUAL REVIEWS*

From the *Annual Review of Astronomy and Astrophysics*, Volume 23 (1985):

*Astronomer by Accident*, T. G. Cowling

*Big Bang Nucleosynthesis: Theories and Observations*, Ann Merchant Boesgaard and Gary Steigman

*On Stellar X-Ray Emission*, R. Rosner, L. Golub, and G. S. Vaiana

From the *Annual Review of Earth and Planetary Sciences*, Volume 13 (1985):

*Cosmic Dust: Collection and Research*, D. E. Brownlee

*Downhole Geophysical Logging*, A. Timur and M. N. Toksöz

From the *Annual Review of Energy*, Volume 10 (1985):

*The Three Mile Island Unit 2 Core: A Post-Mortem Examination*, R. S. Denning

*Inherently Safe Reactors*, Irving Spiewak and Alvin M. Weinberg

*Military Sabotage of Nuclear Facilities: The Implications*, Bennett Ramberg

*Decommissioning of Commercial Nuclear Power Plants*, J. T. A. Roberts, R. Shaw, and K. Stahlkopf

*International Cooperation in Magnetic Fusion*, W. M. Stacey, Jr. and M. Roberts

From the *Annual Review of Fluid Mechanics*, Volume 17 (1985):

*Jakob Ackeret and the History of the Mach Number*, N. Rott

From the *Annual Review of Materials Science*, Volume 15 (1985):

*Ferromagnetism and Superconductivity*, H. R. Khan and C. J. Raub

From the *Annual Review of Physical Chemistry*, Volume 36 (1985):

*Molecular Spectroscopy: A Personal History*, Gerhard Herzberg

*Relativistic Effects in Chemical Systems*, Phillip A. Christiansen, Walter C. Ermler, and Kenneth S. Pitzer

Ann. Rev. Nucl. Part. Sci. 1985. 35 : 1–23

# NUCLEAR PROCESSES AT NEUTRON STAR SURFACES

*Ronald E. Taam*

Department of Physics and Astronomy, Northwestern University, Evanston, Illinois 60201

CONTENTS

## 1. INTRODUCTION

During the past two decades, the increasing emphasis on exploring the cosmos in the x-ray and gamma-ray wavelength regime from spacebound observatories has stimulated much study of the properties of neutron stars. For a detailed description of the structure of neutron stars, see the review by Baym & Pethick (2). These compact objects are but one of the end states of a star's never-ending struggle against gravity.

In contrast to their spectacular formation in a supernova event, the final

1

0163–8998/85/1201–0001$02.00

evolution of a neutron star is usually uneventful; its last reserve of thermal energy is radiated into space, as all nuclear fuel has been exhausted. This passive evolution to the stellar graveyard may be temporarily interrupted, however, if a neutron star accumulates matter from either a close binary companion or from the interstellar medium. In fact, much of our knowledge about the properties of neutron stars can be traced to the observations of interacting binary stars in which a neutron star is a member. As a consequence of the mass flow to the neutron star surface, the gravitational binding energy of the matter is released and converted into x-ray radiation. Thus, neutron stars are observed indirectly as matter-accreting x-ray sources, and not directly (unless they are pulsating radio sources), because of their low intrinsic luminosities. Only those x-ray sources with companion stars greater than about 15 $M_\odot$ and less than about 2 $M_\odot$ are expected to be observed since stars in these mass ranges provide the required accretion rates to sustain a strong x-ray source for an appreciable length of time. For the upper mass range, mass transfer is accomplished via a stellar wind, whereas for the low mass range, mass is transferred via Roche lobe overflow (72).

Although the evolution of binaries to the x-ray stage and the physical concepts involved in the accretion process are of great interest in their own right, they are beyond the scope of the present review. Here, we restrict our attention to the physical concepts governing the ultimate fate of the accreted matter. Important in this regard is the fact that potential nuclear fuel, primarily in the form of hydrogen and helium, is accreted. Because of the high pressures exerted by the weight of the newly accreted matter in the surface layers of the neutron star, this matter will fuse to form iron peak nuclei and eventually dissolve into a neutron fluid at densities greater than about several times $10^{11}$ g cm$^{-3}$. The transformation of this matter into iron peak nuclei may take place in either a continuous (62) or explosive (23) manner. Since the gravitational binding energy of accreted matter exceeds its nuclear binding energy by more than an order of magnitude, nuclear burning will be of little observational consequence unless the matter is stored for a period of time and then burned rapidly. It is only in this way that the energy released from the fusion process can dominate that from accretion for a short time. Such transient behavior has, in fact, been detected in the x-ray and gamma-ray wavelength regime in the form of bursts of radiation. There is now strong theoretical evidence supporting the idea that the nuclear fusion occurring in the surface layers of accreting neutron stars can be explosive and that the nuclear energy released in the accreted matter is responsible for the production of cosmic x-ray (34, 43, 54, 93) and, perhaps, cosmic gamma-ray bursts (20, 63, 93, 94).

It is the main purpose of this review to summarize the basic physics that govern the nuclear processing of matter at the neutron star surface. Specifically, we focus our attention on the conditions for which the accretion of matter is explosive and discuss the various nuclear burning regimes in detail. In the next section we briefly summarize the observational facts and describe the nature of the x-ray and gamma-ray burst sources. In Section 3, we turn to a discussion of the physics and the basic character of the nuclear burning development. The results of the model calculations are briefly reviewed in Section 4 and compared with the observational data in Section 5. Finally, we make some concluding remarks and point to future developments in the last section.

## 2. OBSERVATIONS AND THE NATURE OF X-RAY AND GAMMA-RAY BURST SOURCES

Historically speaking, cosmic x-ray bursts were first reported in the literature in 1976 by Grindlay et al (18) and gamma-ray bursts as early as 1973 by Klebesadel et al (40). A review of the observed characteristics of the x-ray and gamma-ray burst sources can be found in the articles by Lewin & Joss (47, 48), Hurley (29–31), and in AIP conference proceedings edited by Lingenfelter et al (52) and Woosley (91). For the x-ray sources the main properties of the burst include rise-time scales between 0.1 and 5 s, durations of 3 to 100 s, and integrated luminosities ranging up to about $10^{38}$ ergs s$^{-1}$ in which a total energy of $10^{39}$ ergs is emitted.

Typical burst profiles are illustrated in Figure 1. The sources are distinguished by the fact that the bursts are recurrent at irregular time intervals ranging from a few hours to several days. The gamma-ray bursts, on the other hand, have a much harder spectrum with most of their energy emitted in the 30–300-keV range. Their time histories are characterized by rise times of less than about 1 s, total durations ranging from about 1–10 s, and significant temporal structure on time scales as short as milliseconds.

Examples of gamma-ray burst light curves are given in Figure 2. Although recurrent bursts from two sources have been observed, bursts from the great majority of the sources are not known to repeat within the last decade. In contrast to their x-ray burst counterparts, the sources are not seen except during their eruptive state. The spectra of such sources, which can be fitted by an exponential with a power-law tail at high energies (6), exhibit high energy emission and low energy absorption features. For a recent review of gamma-ray burst spectroscopy, see the paper by Teegarden (77).

As argued by Lewin & Clark (44) and Hurley (29), the observational evidence strongly suggests that neutron stars are involved in the cosmic x-ray and gamma-ray burst phenomena. The neutron star origin of x-ray bursts is primarily based upon the analysis of Swank et al (66), in which it was demonstrated that the spectrum of an x-ray burst could be fitted by a cooling blackbody. From this study and work by Hoffman et al (26), it was found that the dimensions inferred for the source emitting region ($\sim 10$ km) were comparable to neutron star radii. Van Paradijs (79, 80) further tested the blackbody assumption and found that if the peak burst luminosities are limited by the Eddington luminosity (the maximum luminosity that can be achieved by a star in hydrostatic equilibrium for which the gradient of the radiation pressure balances the force of gravity) for an object of 1.4 $M_\odot$ and if the sources are distributed about the galactic center, then the size of the emitting region is consistent with the neutron star hypothesis. Further support for the neutron star hypothesis was also provided by the preliminary detection of spectral features, possibly gravitationally red-

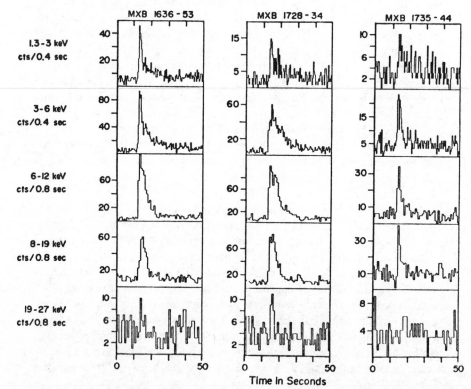

*Figure 1*   Light curves for three different x-ray burst sources, adapted from Lewin & Joss (46). Note that the decay time scale is longer at lower energies, which indicates a cooling of the emitting region. Reprinted by permission from *Nature*, Vol. 270, No. 5634, pp. 211–216, copyright © 1977 Macmillan Journals Limited.

shifted from the neutron star surface (32). Although the interpretation is not definitive, it is consistent with conclusions obtained from arguments concerning the shape of the continuum spectrum.

The neutron star origin of gamma-ray bursts is based upon time histories and spectral features. The short temporal structure in the light curve (<20 ms, but as short as 0.25 ms for the 1979 March 5 event) suggests that a compact object is involved. In addition, the time scales for the oscillatory features in the light curves of the 1979 March 5 event ($\sim 8$ s; 56) and the 1977 October 29 event ($\sim 4.2$ s; 89), an effect probably related to rotation, are similar in kind to the time scales observed in known accreting neutron star x-ray pulsars. The strongest compelling argument for a neutron star is provided by the spectrum of the gamma-ray burst source. Spectroscopic features are seen in emission at high energy ($\sim 420$ and $\sim 720$ keV; 78) and in absorption at low energy ($\sim 40$–$60$ keV; 57). The high energy features

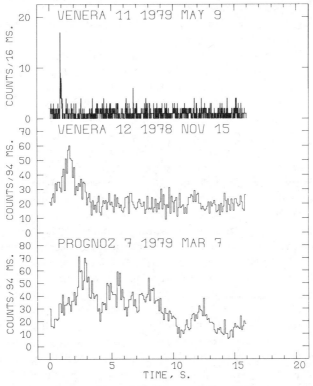

*Figure 2* Temporal development of a gamma-ray burst for three different sources. Note that the time histories are quite diverse. Reproduced from *Advances in Space Research* (30).

may be due to electron-positron pair annihilation at 511 keV redshifted to 420 keV and to the redshifted line corresponding to the first excited state of the $^{56}$Fe nucleus at 847 keV. The redshifts that are deduced are consistent with a neutron star origin. The low energy absorption features, on the other hand, have been tentatively interpreted in terms of transitions between electron Landau levels (cyclotron lines) in a magnetic field of $10^{12}$ gauss, similar to the field strengths inferred for radio and x-ray pulsars. Although these absorption features may be a result of rapid spectral variability (41, 42), such field strengths may be required, on theoretical grounds (7, 63), to confine plasma in order to achieve the observed high spectral temperatures.

Although the neutron star origin of the cosmic x-ray and gamma-ray bursts is generally accepted, the astrophysical circumstances of these phenomena may not be common. Observational evidence bearing on the binary nature of these sources is available for the x-ray burst sources alone. The most direct evidence is provided by the detection of x-ray eclipses every 7.1 hr for the source MXB 1659-29 (8), and by the detection of nearly periodic x-ray absorption dips every 50 min seen in the source MXB 1916-053 (87, 88). Since the companion is not observed, it must be intrinsically faint, and the model that has been advanced for the x-ray burst class consists of a low mass binary system in which matter from one star is transferred via Roche lobe overflow to its neutron star companion (38, 71). It is envisaged that a critical amount of fuel accumulates in the surface layers of the neutron star, whereupon nearly all of the newly added matter is burned in an uncontrolled manner. Support for this picture is provided by the observation that the ratio of the steady x-ray luminosity to the burst luminosity (averaged over the time interval between consecutive x-ray bursts) is comparable to the ratio of the energy yield per gram associated with accretion (i.e. gravitational potential of the neutron star) to that associated with nuclear fusion of hydrogen or helium to iron.

The model for the gamma-ray burst sources is, however, not as secure. There is no direct or indirect observational evidence to distinguish between solitary neutron stars in the interstellar medium or neutron stars in binary systems. It is probably the case, however, that mass accretion is important. The flow will be affected by the presence of the strong magnetic field as matter will tend to be focussed onto the polar caps of the neutron star (10). Once on the surface of the neutron star, the matter may be confined by the magnetic stresses (94) and prevented from spreading over the entire surface. By analogy to the x-ray burst model, it is thought that the emission of gamma rays would arise naturally from a nuclear explosion in the surface layers of a strongly magnetic neutron star. The physics of nuclear processing under the conditions typical in an accreting neutron star is crucial for these models, and it is to these principles that we now turn our attention.

# 3.   NUCLEAR PROCESSES

Since the accreted matter is of cosmic or nearly cosmic abundance, it is rich in hydrogen and helium and contains a small admixture of heavier elements (principally C, N, O, and Fe). As mentioned in Section 1, this matter will be processed to iron peak nuclei and eventually dissolve into neutrons. Depending upon the physical state (i.e. the density and temperature) of the surface layers of the neutron star, the matter may burn stably or unstably. For example, in the layers of the star where the degree of electron degeneracy is low (i.e. where quantum exclusion effects are unimportant in the equation of state) nuclear burning can be stabilized. Consider the consequences of a positive temperature perturbation. For temperature-sensitive nuclear reactions, the rates will proceed faster, the gas will heat up, and, as a result of the increased pressure, expand. This expansion cools the layer, thus rendering it thermally stable.

On the other hand, in the surface layers of the neutron star where the degree of electron degeneracy is high (the pressure depends primarily upon the density and not on the temperature), exothermic burning leads to a potentially thermally unstable situation (58) because the pressure does not increase in response to the temperature perturbation. Hence, there is no tendency for the gas to expand. Such conditions are favorable for a thermal instability provided that the nuclear energy is produced faster than can be removed by energy transport processes (e.g. by conduction, radiative diffusion, convection, or neutrino emission). If the reaction rates are very temperature sensitive, the nuclear burning becomes explosive and thermal stability will not be attained until the degeneracy of the electron gas is lifted (i.e. the equation of state of the gas becomes temperature sensitive). Thus, explosive nuclear burning in the neutron star envelope is virtually assured because (a) the nuclear reaction rates are highly temperature sensitive, (b) the burning occurs in a region where the pressure is primarily temperature independent, and (c) the energy loss processes are inefficient.

The physical state of the matter in the accreted layer of the neutron star is determined by the equations of hydrostatic equilibrium, energy conservation, and energy transport, corrected for general relativistic effects, supplemented by an equation of state. Since the gravity is approximately constant in the surface layers, the pressure at the base of the accreted layer is given by the product of the gravity and the column density of the accreted layer. The temperature in the envelope is not as easily obtained since it depends upon the previous thermal history (67) of the neutron star as well as upon the effect of compressional heating associated with the accretion of matter in the outer layers (here, gravitational potential energy is released as underlying mass layers are compressed by the increasing weight of the overlying layers).

However, if the thermal structure has reached a steady state, the envelope structure can be determined independently of any initial conditions. By assuming such a steady state, Ergma & Tutukov (13), Fujimoto et al (17), and Taam (68) find that nuclear burning in the accreted layer is thermally unstable over a wide range of accretion rates onto a neutron star, $10^{-8} > \dot{M} > 10^{-15}$, where $\dot{M}$ is in units of solar masses per year. In the following, we discuss the various burning regimes.

## 3.1   Hydrogen Burning

Because the rate at which hydrogen burns is limited by the weak interaction associated with the transformation of a proton into a neutron, hydrogen burning by itself is not explosive. The rapid burning on time scales less than 1 s, as required by the temporal variation of the burst light curves, rules out hydrogen as the primary nuclear fuel. Although hydrogen burning cannot be directly responsible for the rapid nuclear energy release expected in the cosmic burst phenomena, it can significantly affect the overall development of nuclear burning because the liberated nuclear energy may be sufficient to heat the neutron star envelope to temperatures where another nuclear fuel can be ignited. If the burning of this other fuel (e.g. helium) is not impeded by weak interactions, then explosive burning may be expected. Hydrogen burning can thus trigger the rapid nuclear burning; its role can be of great importance in the nuclear burning development.

Consider burning at low temperature ($T < 10^6$ K). This is expected if the neutron star is old (i.e. has cooled for $\sim 10^6$ yr) and if it is accreting at a low rate $\dot{M} < 10^{-14}$ $M_\odot$ yr$^{-1}$. At sufficiently high densities $\rho \approx 1.4 \times 10^7$ g cm$^{-3}$ in the accreted matter, the Fermi energy of the electrons exceeds the difference in the rest mass energy between a neutron and a proton (corresponding to 1.29 MeV) and the inverse beta reaction p(e$^-$, $\nu$)n occurs. At these high densities, the reverse reaction is prohibited since all the electron energy levels to which the electron could decay are filled. As discussed by Rosenbluth et al (62), the neutron then would combine with a proton to form deuterium via p(n, $\gamma$)$^2$H. The fusion process continues with the interaction of deuterium with a proton via $^2$H(p, $\gamma$)$^3$He followed by $^3$He(n, $\gamma$)$^4$He, releasing about 7 MeV per proton. As a result, the temperature rises and the character of the nuclear burning development is modified as additional reaction channels (primarily the CNO cycle; see below) become accessible (20, 90).

If the envelope temperatures are higher, corresponding, for example, to higher accretion rates ($\dot{M} > 10^{-14}$ $M_\odot$ yr$^{-1}$), the burning will be initiated at lower densities. Consequently, the pycnonuclear burning (density dependent) described above will not develop, and the burning will initially take place via the pp chain (4, 83). The set of reactions describing this process is slow since it is determined by the rate of the feeble reaction

$p(p, e^+ \nu)^2 H$. Provided that carbon, nitrogen, or oxygen nuclei are present, the energy released from the pp reactions can, however, heat the layer sufficiently to promote the more rapid burning of hydrogen via the CNO cycle (3). In contrast to the pp reaction, the CNO reactions are much more temperature sensitive because the larger Coulomb barriers between the proton and the CNO nuclei are penetrated only by those particles in the tail of the Maxwell–Boltzmann velocity distribution. Although the charged-particle reactions may be rapid, especially for the conditions expected in the surface layers of a neutron star ($\rho > 10^5$–$10^6$ g cm$^{-3}$, $T > 10^7$ K), the weak interactions associated with the positron decays of $^{13}N(e^+ \nu)^{13}C$ (mean decay lifetime of 863 s) and $^{15}O(e^+ \nu)^{15}N$ (mean decay lifetime of 176 s) limit the usual CNO cycle. For the conditions expected in the surface layers, the $^{13}N$ decay is bypassed by the reaction $^{13}N(p, \gamma)^{14}O$ and the rate of the energy production from the cycle is determined by the sum of the positron lifetimes associated with $^{14}O(e^+ \nu)^{14}N$ (mean decay lifetime of 102 s) and $^{15}O(e^+ \nu)^{15}N$. In this case, hydrogen burns to helium via the beta-limited CNO cycle (28). At high densities, $\rho > 10^6$–$10^7$ g cm$^{-3}$, these weak interaction rates are accelerated by electron capture; however, the energy generation rate is still insensitive to temperature. As mentioned above, hydrogen burning cannot by itself be responsible for the thermonuclear runaway, but it can facilitate the ignition of an explosive fuel, the primary one of which we discuss next.

## 3.2   Helium Burning

Although the energy generation rate of hydrogen burning can be temperature sensitive, it is eventually limited by the weak interaction rates, which render the hydrogen-rich layer thermally stable. As a result, a steady state is achieved in which the rate at which the hydrogen is accumulated equals the rate at which it is burned. This equality determines the amount of hydrogen-rich matter that can be accumulated in the surface layers. More matter can be accumulated, however, since the hydrogen is processed into helium. As accretion continues, the mass contained in the helium layer increases, maintained at a temperature governed by energy transport from the hydrogen burning shell. Since the energy transport is dominated by electron conduction (which is efficient because the mean free path of the electron is long in highly degenerate matter), a nearly isothermal distribution is established. After a sufficient amount of helium accumulates on the neutron star surface to reach a critical density, the helium itself will ignite via the triple alpha reaction. In this two-stage process, $^8Be$ is temporarily formed, followed by alpha capture to create $^{12}C$.

For temperatures less than the Debye temperature, $T < 10^8 (\rho/10^9)^{1/2}$ K, the helium nuclei are bound in a Coulomb lattice, and its burning rate is primarily density dependent proportional to $\exp(-A\rho^{-1/6})$, which increases sharply for densities greater than about $2 \times 10^9$ g cm$^{-3}$ (5). If the Coulomb barrier is large compared with the quantum mechanical zero-point energy of nuclei in the lattice, the reaction rate is essentially temperature independent. However, if the excited states of the lattice are populated, the rate becomes temperature sensitive (64). Investigations (13, 90) in this temperature regime indicate that helium burning in the helium layer itself and hydrogen burning (from the overlying hydrogen-rich layer) may release energy sufficiently rapidly to initiate a thermal instability.

For temperatures in excess of the Debye temperature, the burning occurs in the thermonuclear regime. Here, the triple alpha reaction rate is primarily a resonant rate, in contrast to the case at low temperatures. The fact that these reactions are very temperature sensitive in the thermonuclear regime ($\propto T^{30}$) and the fact that the helium layer is characterized by a high degree of electron degeneracy make helium an attractive candidate fuel for explosive burning. In fact, for densities greater than $10^6$ g cm$^{-3}$ and temperatures of $10^9$ K, the burning time scale is less than 1 s. The triple alpha reaction dominates the energy production up to about $10^9$ K at which point its rate saturates, thereby losing its strong temperature dependence. For still higher temperatures, alpha captures onto highly charged nuclei [namely the $^{12}$C$(\alpha, \gamma)^{16}$O$(\alpha, \gamma)^{20}$Ne$(\alpha, \gamma)^{24}$Mg$(\alpha, \gamma)^{28}$Si reaction chain] provide the bulk of the nuclear energy release. The nuclear processing continues primarily via $(\alpha, \gamma)$ reactions supplemented by secondary flows involving $(\alpha, p)$ and $(p, \gamma)$ reactions up to higher atomic masses. The extent of this flow, however, is hindered by the increasing Coulomb barriers, and is determined not only by the initial helium abundance, but also by the pressures involved at the onset of the thermal runaway. The pressure dependence can be understood from the fact that the explosive burning occurs under conditions of constant pressure until the amount of nuclear energy released exceeds the Fermi energy of the electrons, whereupon the gas is lifted out of the electron degenerate regime. As the dominant source of pressure becomes temperature dependent, the layer expands to quench the burning. For sufficiently high pressures ($> 10^{23}$ dynes cm$^{-2}$), the temperatures approach about $4 \times 10^9$ K before the burning is quenched. For such conditions the composition in the accreted layer shifts toward nuclear statistical equilibrium, with about $1.8 \times 10^{18}$ erg g$^{-1}$ released in the process. For lower pressures the peak temperatures achieved in the evolution are reduced and the flow stops at $^{28}$Si for $P \sim 10^{19}$ dynes cm$^{-2}$ and $^{36}$Ar for $10^{21}$ dynes cm$^{-2}$ (24).

## 3.3    Combined Hydrogen and Helium Burning

It was recognized by Taam & Picklum (73, 74) that if hydrogen burns at sufficiently high densities ($\rho \approx 10^6$ g cm$^{-3}$) in the envelope of an accreting neutron star, helium must eventually burn within the same mass layer. This is in contrast to the usual circumstance where hydrogen and helium burning occur in separate mass layers. The existence of this regime reveals that the burning of hydrogen and helium can take place via a new kind of nucleosynthetic process.

Consider the case in which hydrogen burning develops at high densities and temperatures ($\rho > 10^6 \text{ g cm}^{-3}$, $T \geqslant 10^8 \text{ K}$). Such conditions are easily realized in the surface layers of accreting neutron stars ($M = 1.4\ M_\odot$) for $\dot{M} > 10^{-10}\ M_\odot \text{ yr}^{-1}$. At these densities and temperatures, hydrogen burns via the beta-limited CNO cycle, and helium burns via the triple alpha reaction. Although helium burning does not significantly contribute to the energy generation rate at a temperature of $10^8$ K and density of $10^6 \text{ g cm}^{-3}$, it is of paramount importance since it forms $^{12}$C (the catalyst for hydrogen burning). This production of $^{12}$C enhances the energy generation rate from the beta-limited CNO cycle, which thus renders the rate temperature sensitive. Since the electrons are highly degenerate, the mass layer becomes thermally unstable; this promotes higher temperatures and accelerates the triple alpha reaction, which eventually dominates for temperatures greater than about $2\text{–}3 \times 10^8$ K. For temperatures exceeding about $4\text{–}5 \times 10^8$ K, alpha captures onto the most abundant nuclei in the CNO group via $^{14}$O$(\alpha, \text{p})^{17}$F and $^{15}$O$(\alpha, \gamma)^{19}$Ne occur, thus changing the character of the nuclear processing. This is especially important because it opens up new channels for the burning of hydrogen and helium outside of the main chains discussed above.

Based upon published rates and estimated rates obtained from a Hauser–Feshbach statistical calculation for charged-particle reactions, the important reaction chains for the nuclear flow to heavier atomic masses were identified in the paper by Wallace & Woosley (84) and further elaborated by Hanawa et al (22). Under the above physical conditions there is a tendency for hydrogen burning to proceed via the so-called rp process, which is characterized by a series of (p, $\gamma$) reactions interrupted by positron decays to produce nuclei near the proton drip line (further proton capture of these proton-rich nuclei is inhibited by photodisintegration). However, at these high densities and temperatures, ($\alpha$, p) reactions can compete favorably with the positron decays to bridge the waiting points; this accelerates the nuclear flow to higher atomic mass nuclei. Reactions such as $^{14}$O$(\alpha,\text{p})^{17}$F$(\text{p},\gamma)$ $^{18}$Ne$(\alpha,\text{p})^{21}$Na$(\text{p},\gamma)^{22}$Mg followed by $^{22}$Mg$(\alpha,\text{p})^{25}$Al$(\text{p},\gamma)^{26}$Si$(\alpha,\text{p})^{29}$P$(\text{p},\gamma)$ $^{30}$S and $^{15}$O$(\alpha,\gamma)^{19}$Ne$(\text{p},\gamma)^{20}$Na$(\alpha,\text{p})^{23}$Mg$(\text{p},\gamma)^{24}$Al$(\alpha,\text{p})^{27}$Si$(\text{p},\gamma)^{28}$P are important as the operation of these reaction chains provides the rapid deposition of nuclear energy that is required for the explosive phase of evolution.

An example of the dominant nuclear flow during a thermonuclear flash for typical conditions in the accreted layer of a neutron star is shown in Figure 3. The maximum atomic mass that is formed depends upon the thermodynamic history of the accreted layer during the rapid phase of nuclear evolution. As mentioned above, this history is determined by the composition and pressure at the base of the accreted layer at the onset of the thermal runaway. For lower pressures, the helium burning will cease at

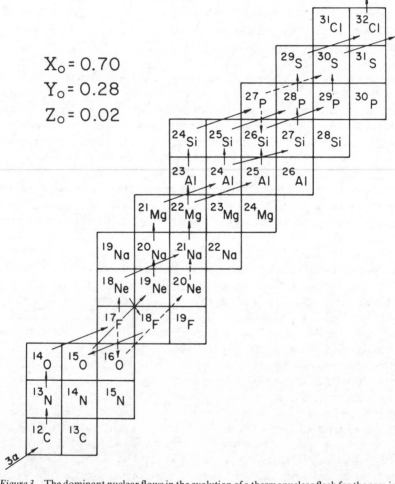

$$X_0 = 0.70$$
$$Y_0 = 0.28$$
$$Z_0 = 0.02$$

*Figure 3*   The dominant nuclear flows in the evolution of a thermonuclear flash for the case in which hydrogen and helium burn together for $T \approx 8 \times 10^8$ K and $\rho \approx 6 \times 10^5$ g cm$^{-3}$. Solid arrows indicate flows that are a factor of 10 larger than the flows corresponding to the dashed arrows. Reproduced from *Astrophysical Journal* (84).

some intermediate nucleus whereupon further processing will involve a series of proton captures followed by positron decays. Among proton-induced reactions, $(p, \gamma)$ reactions are favored over $(p, n)$ and $(p, \alpha)$ reactions since the $(p, n)$ reactions have large endoergic $Q$ values, and the $(p, \alpha)$ reactions are inhibited by the large Coulomb barrier for $\alpha$ particle emission. Although the upward flow to higher atomic masses is inhibited by the increased Coulomb barriers, barrier penetration by proton captures is not as severe as it is for alpha captures. Consequently, the flow can proceed beyond $^{56}$Ni (a pure electron-capture nucleus with a small $Q_{p\gamma}$ value). The extent of this flow is especially important in view of the fact that the flow determines the residual abundance of hydrogen (i.e. some hydrogen may remain after the explosive phase of nuclear evolution). Generally, the extent of the flow depends upon the time scale of proton capture, positron decays, and the rate at which the nuclear flash is quenched. Of particular importance in this nuclear flow is whether the $^{64}$Ge$(p, \gamma)^{65}$As reaction is permitted (85), for, if so, matter can be processed up to $^{96}$Cd with perhaps little hydrogen remaining after the outburst.

## 4.  MODEL CALCULATIONS

The thermonuclear model for high energy transient phenomena has received considerable attention and was recently reviewed (12, 38, 71, 90). From a large number of detailed computations of the thermal, dynamical, and nuclear evolution of accreting neutron stars, it is found that nuclear burning is capable of producing high energy outbursts over a wide range of mass accretion rates. The properties of these outbursts are quite diverse, differing in degree as well as in kind. In particular, the model is capable of producing events characterized by a wide range in energies ($E \approx 10^{38}$–$10^{44}$ ergs) and durations (from $\sim 1$ s to several days).

Nuclear burning is not found to be explosive for all mass accretion rates, however. For very low rates ($\dot{M} < 10^{-15} \ M_\odot \ \mathrm{yr}^{-1}$) the nuclear burning takes place in the purely pycnonuclear regime and the burning can be stable if the energy is conducted to the surface as rapidly as it is generated. Likewise, for very high rates ($\dot{M} > 10^{-8} \ M_\odot \ \mathrm{yr}^{-1}$) the heating associated with compression (see Section 3) leads to burning at such low densities ($\rho < 10^5 \ \mathrm{g \ cm}^{-3}$) that matter burns on time scales longer than 1 s.

In the following, we briefly summarize the main properties as deduced from detailed model computations. In general, each model is characterized by the mass of the neutron star (chosen to be 1.4 $M_\odot$), its initial thermal state, the mass accretion rate, and the composition of the accreted matter. For convenience we divide the discussion of the numerical results into two broad categories: (*a*) models in which the envelope of the neutron star remains in hydrostatic equilibrium throughout the outburst, and (b) models in which hydrodynamical effects are found to be important.

## 4.1   Hydrostatic Models

Hydrostatic models, which closely resemble the x-ray bursters, are characterized by the highest rates of mass accretion ($10^{-10} \lesssim \dot{M} \lesssim 10^{-9} \, M_\odot \, yr^{-1}$). Since there is no observational evidence for a strong magnetic field in these sources, the accreted matter is assumed to be distributed over the entire neutron star surface. The model calculations indicate that nearly all the accreted fuel is burned in the outburst and that most of the nuclear energy released is transported to the surface rather than conducted into the neutron star core. A characteristic theoretical burst light curve is shown in Figure 4. In particular, (a) the surface luminosity rises to about the Eddington value within less than 1 s, and (b) a total energy of about $10^{39}$ ergs is emitted in about 8 s. Calculations by a number of workers [see

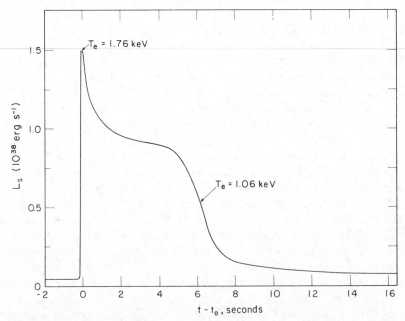

*Figure 4*   Temporal variation of the surface luminosity for a theoretical x-ray burst. Alpha captures onto highly charged nuclei are responsible for the rapid rise time, and proton captures provide the energy for the emission on the shoulder of the burst profile. Here, $t_0$ is the recurrence time scale equal to 11.7 hr. The spectral energy of the burst ($T_e$) is given at the peak of the burst and after an e-folding decay time. Reproduced from *Astrophysical Journal* (70).

Ayasli & Joss (1) and Taam (70) and references therein], although differing somewhat in detail, confirm these general properties. A calculated spectrum of radiation emitted by a neutron star at an effective temperature of 2 keV and a surface gravity of $10^{15}$ cm s$^{-2}$ is shown in Figure 5. Note that the emitted spectrum is shifted to higher frequencies than a blackbody corresponding to the effective temperature (53).

For mass accretion rates in this regime the nuclear outburst can be a pure helium flash (35; see Section 3.2) or a combined hydrogen-helium flash (74; see Section 3.3). For those neutron stars that evolve to a limiting thermal state, the mass accretion rate delineating these two types of flashes is about $2 \times 10^{-10}$ $M_\odot$ yr$^{-1}$ (17, 68) for a 1.4-$M_\odot$ neutron star.

The model computations demonstrate that both types of shell flashes contribute to the x-ray burst phenomena. Based upon the short recurrence time scales of the observed bursts ($< 10^5$ s), the hydrogen-helium shell flash is thought to be more common. On the other hand, the pure helium shell

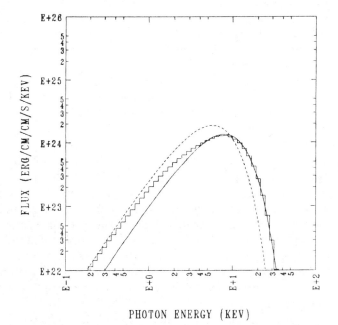

PHOTON ENERGY (KEV)

*Figure 5* Theoretical spectrum of radiation from a neutron star model atmosphere characterized by an effective temperature, $T_{\text{eff}}$, of 2 keV, gravity of $10^{15}$ cm s$^{-2}$, and solar composition. The dashed curve denotes the Planck function at $T_{\text{eff}}$ and the solid curve represents a Planck function ($T_c \approx 2.78$ keV) that best fits the calculated spectrum.

flashes are most likely responsible for those bursts recurring on time scales longer than about a day.

## 4.2   Hydrodynamic Models

Hydrodynamic models are similar in kind to those described above, but are more extreme. For lower mass accretion rates, the steady-state temperatures in the neutron star core are lower and, hence, the amount of fuel that can be accumulated is greater. These differences are further magnified for sufficiently low mass accretion rates ($\dot{M} < 10^{-11}$ $M_\odot$ yr$^{-1}$) that the hydrogen energy generation rate may be affected by the gravitational settling of heavy ions in the strong gravitational field of the neutron star. If elemental diffusion is effective, the energy generation rate from the beta-limited CNO cycle is reduced, which results in lower temperatures than would otherwise be obtained. Thus, the outbursts in the low accretion rate regime are characterized by greater energies and longer durations.

Because the density at the base of the accreted layer is high ($\rho > 10^7$ g cm$^{-3}$) in this regime, the enhancement of the nuclear reaction rates by electron screening is especially important. At such densities, the explosive release of nuclear energy associated with the burning of helium to $^{56}$Ni (see Section 3.2) produces highly super-Eddington luminosities at the base of the accreted layer and these are maintained for a time longer than the energy transport time to the neutron star surface. Consequently, the outer surface layers expand under the action of the radiation pressure gradient and a wind develops. Here, the excess radiative flux (in excess of the surface Eddington luminosity) is efficiently converted into the kinetic and potential energy of the outflowing matter. Mass loss is further facilitated by the fact that the local Eddington luminosity decreases toward the surface as a result of opacity variations related to the temperature and composition dependence of the scattering opacity (21, 61).

The properties of steady-state winds from neutron stars were studied by Ebisuzaki et al (11) and Kato (39). Since the gravitational binding energy of the envelope is much larger than the maximum nuclear energy yield available in the accreted layer, only a small fraction ($< 0.01$) of the accreted envelope is ejected at velocities in the range 0.01–0.1 times the speed of light, corresponding to a mass loss rate of about $10^{18}$ g s$^{-1}$. Since the radiative flux is reduced to the Eddington value as matter is ejected, there is a phase during which the surface luminosity is constant (70, 86).

Although the surface luminosity remains approximately constant at the Eddington value, the effective temperature does not because the location of the photosphere varies. For large photospheric expansions, gravitational redshift corrections become insignificant during the outburst and, as a result, accreting neutron stars exhibiting hydrodynamical behavior are characterized by peak luminosities higher than those that remain in hydrostatic equilibrium. Such hydrodynamical behavior may be especially relevant to those sources that exhibit long x-ray bursts [> 100 s; Swank et al (66)], those that exhibit precursors (49, 76), and, possibly, to the gamma-ray bursters.

In contrast to the x-ray burst phenomenon, the mechanism deemed responsible for producing gamma-ray bursts is still unknown. Because the properties of these bursts are very diverse (see Figure 2) and can be classified into several categories, it is possible that more than one mechanism is involved. However, based upon the success of the thermonuclear flash model in providing an interpretive framework for x-ray bursts, the thermonuclear model has been modified and applied to the gamma-ray burst phenomenon. An essential ingredient in this model is the existence of a strong magnetic field (63, 94). With a strong field the accreted matter is focussed to the poles and confined to a localized region on the neutron star surface by the magnetic stresses. The field also reduces the opacity, which therefore enhances the efficiency of energy transport, but the accreted layer is still expected to be thermally unstable (37). Depending upon the fraction of the surface area to which the matter is localized burst energies can range from about $10^{37}$ to $10^{44}$ ergs. For low mass accretion rates ($\dot{M} < 10^{-12} M_\odot$ yr$^{-1}$) the evolution of the layer is similar to that described above for the hydrodynamic case; however, there is one crucial difference. The ejected matter will interact with the strong magnetic field of the neutron star. Although it is unclear how the energy source is coupled to the gamma-ray-emitting region, it has been suggested that nonthermal radiation associated with synchrotron processes (51) or inverse Compton processes (14, 92) or a combination of both (50) may produce the very high energy photons (> 100 keV) observed.

# 5.    CONFRONTATION OF THEORY WITH OBSERVATION

The thermonuclear flash model for the x-ray burst phenomenon has been developed to the stage where detailed comparisons between theory and observation can be attempted. This statement cannot be made for those sources that emit gamma-ray bursts. Henceforth we concentrate on the transients in the x-ray regime. The results of the detailed numerical calculations reproduce the average properties of the x-ray bursts as summarized in Section 2. In particular, the energetics, temporal structure, and spectral behavior of the average x-ray burster are in quantitative agreement with theory. However, upon examination of individual sources there remain several outstanding theoretical problems. In the following, we list these problems and discuss possible solutions.

## 5.1  Super-Eddington Luminosities

There is some observational evidence suggesting that the Eddington luminosity for a $1.4$-$M_\odot$ neutron star is exceeded for some sources. The evidence is based upon either spectral data or distance determinations. In the former case, the evidence for the super-Eddington luminosity is provided by burst spectra in which the spectral temperatures near the burst maximum (2.5–3 keV) are found to be greater than the maximum Eddington effective temperature (2 keV) for current theoretical neutron star models (55). The spectra thus indicate that the peak burst flux exceeds the Eddington flux by about a factor of 5. Van Paradijs (81) and Czerny & Sztajno (9), however, noted that scattering of photons by electrons in the atmosphere would harden the spectrum. The recent calculations by London et al (53), which include the effects due to Compton scattering, demonstrate that spectral temperatures can exceed the effective temperature by as much as a factor of 1.6, thus resolving the super-Eddington flux problem for these sources. On the other hand, distance estimates based upon the location of a source in a globular cluster (19) or upon statistical arguments of the galactic distribution of sources (33) indicate that the peak burst luminosity (assuming isotropic emission) exceeds the Eddington luminosity by factors ranging from 3 to as large as 10. As a possible resolution to this problem, van Paradijs & Stollman (82) have pointed out the importance of the Doppler shift and aberration of photons produced in a velocity field of an outflowing neutron star wind. Super-Eddington luminosities can, in principle, be observed in the rest frame if the luminosity in the comoving frame of the matter is at the Eddington value. Although the idea is promising, it remains to be demonstrated by detailed calculation that such effects can produce luminosities as great as an order of magnitude larger than the Eddington limit.

## 5.2  Bursts With Short Intervals

The ratio of the energy emitted in a burst to that emitted during the burst-inactive state for most x-ray burst sources is consistent with the idea that the sources of energy for the persistent and burst emissions are the gravitational binding energy and the nuclear binding energy of the accreted matter, respectively. There are important exceptions to this, however, where burst intervals as short as 10 min have been observed (45, 60). Such short intervals are difficult for a theory in which all the accreted fuel is burned in each outburst because there is insufficient time to accumulate the required fuel to initiate a subsequent burst. It is possible that the nuclear flash takes place in localized regions (15); however, one would not expect that successive bursts exhibit similar properties, as is observed in XB1745-24 (60). As another possibility, there are some indications that the accreted matter may not be completely burned in the explosive phase (1, 95; see Section 3.3). Hence, successive bursts of short intervals could be produced if the unburned matter could be reignited. In practice this is difficult to achieve since the unburned matter is hydrogen. Convective mixing of combustible fuel (e.g. $^{12}C$ produced from helium burning of the previous burst) into this region (95) may be important in this regard, but it has not been demonstrated that nearly identical bursts can be produced.

## 5.3   Recurrence Time Correlations

For the asymptotic theory, in which a limit cycle is assumed, the time intervals between successive x-ray bursts decrease (but not necessarily inversely) with increasing mass accretion rate. Although this tendency is exhibited for some sources, it is not exhibited by all. In particular, there exist some bursters for which the burst intervals increase with increasing levels of persistent emission. This may be due to the continually evolving thermal state of the neutron star, especially if the mass accretion rate varies on a time scale shorter than the time required to achieve a thermal steady state ($< 10^5$ yr) in the neutron star core. Scaling arguments relating the burst interval and the level of persistent emission are oversimplifications that do not take into account the effects of thermal inertia (69) or the possibility of shell flashes characterized by a different nuclear burning development (70). Although detailed comparisons are lacking, the effect of thermal inertia is clearly demonstrated in the calculations by Fujimoto et al (16) and Woosley & Weaver (95). In some cases a limit cycle is reached, but in others the effects of thermal inertia are manifested by the irregular recurrence time scales.

## 5.4   Transients With Precursors

There exist x-ray transients similar in kind to the x-ray bursters, but distinguished by longer durations ($\sim 1000$ s) and the appearance of precursors to the main burst. Such transients are probably characterized by a thick layer of helium, which accumulates at a low rate (36). As a result, the outburst is more energetic and of longer duration than that of the x-ray bursters. The precursor may be qualitatively understood in terms of the variation of the photosphere of the neutron star during the outburst (49, 76). As mentioned in Section 4.2, this variation is found in the hydrodynamic models in which the luminosity is close to the Eddington limit. Expansion (contraction) of the photosphere causes the effective temperature, $T_{eff}$ to decrease (increase), and the x-ray luminosity to decline (rise). Thus, the entire burst profile, precursor plus main burst, is interpreted in terms of a very energetic nuclear explosion on the neutron star surface in which the changing location of the photosphere plays a major role. Such behavior is reminiscent of the x-ray bursts that exhibit double peaks (25) in which the photospheric expansion is not so extreme (61) as to shift the energy flux out of the x-ray spectral band. Although the precursor can be qualitatively understood in this way, the detailed calculations (65, 86) fail to reproduce quantitatively the temporal behavior.

## 6.  OUTLOOK

The thermonuclear flash model has provided an interpretive framework for understanding the nature of the x-ray burst phenomenon. As such, it is now generally accepted that the x-ray burst sources are members of close binary systems in which the outbursts are caused by a thermonuclear flash or explosion on the surface of a neutron star member. Thus, x-ray bursters can be viewed as cosmic laboratories in which the structure of matter at high density may be studied. Specifically, a determination of the mass-radius relation of neutron stars from x-ray bursters (or from other compact x-ray sources) would severely constrain hadron interaction models for the equation of state. Such determinations from the observed burst radiation are possible in principle (27, 80). However, in view of the difficulties encountered by the thermonuclear theory in explaining the properties of some individual burst sources (see Section 5), these determinations should be treated with reservation. Further theoretical developments are needed, especially with regard to the construction of more realistic models.

For example, the thermonuclear flash may not occur simultaneously over the entire surface of the neutron star (15, 59, 63). In such cases, the propagation time of the burning front around the star introduces a time delay in the energy production that can modify the burst profile obtained from a spherically symmetric calculation. Thus, a detailed study of the ignition, nuclear evolution, and propagation of the burning front in multidimensional circumstances would be highly desirable. Further complexities are introduced by the presence of a strong magnetic field. In particular, the nuclear fuel is confined by the field to the polar caps of the neutron star, and during the thermonuclear runaway the energy transport by radiation and by convective motions will be complicated by its effects. Detailed studies of the interaction of the outflowing wind with radiation (relevant to the super-Eddington problem) and with a magnetic field are also worthy of attention. These general areas remain largely unexplored and work is needed to test the viability of the thermonuclear flash model for x-ray and gamma-ray bursters.

In addition to the study of these complicating effects, fundamental work is needed to clarify issues related to the basic nuclear physics of the problem. In particular, the nuclear reaction rates and positron decay rates are very poorly known for nuclei near the proton drip line. Since a number of resonances in the compound nucleus can contribute to the reaction rate for each of the many important charged-particle reactions (see Section 3.3), the need to identify the resonances in these nuclei and measure their decay widths cannot be overemphasized. Specifically, for the physical conditions found in the envelopes of accreting neutron stars, the cross sections between protons (and alpha particles) and light to intermediate mass nuclei are determined by the density of these excited states. At low temperature ($< 3 \times 10^8$ K) the cross sections are dominated by a single resonance with the reaction rate determined by energies far out in the wing of the resonance. At higher temperatures many resonances contribute at their respective resonant energies.

Because of the experimental difficulties associated with producing radioactive nuclei of short lifetimes to study the reaction rates of interest, there are few data available on some of these nuclei, and reaction rates have been inferred from the properties of the analogue nucleus (84). This procedure, although reasonable, may be grossly in error as a result of insufficient knowledge of the widths of the resonances. At high temperatures and for heavy nuclei (heavier than Si), the situation improves because the density of excited states in the compound nucleus is high at the reaction energy of interest, and the cross sections can be calculated based upon a statistical approach. However, for light nuclei, the statistical approach may well be in error, and reaction rate studies based on detailed properties of the individual resonances would be advisable.

As stated earlier, the rates associated with weak interactions are also uncertain for the very proton-rich nuclei. Wherever experimental data do not exist, the weak rates have been obtained from calculations based upon the gross theory by Takahashi et al (75). Although these rates are probably adequate for a wide range of physical conditions ($10^8 < T < 10^9$ K; $1 < \rho < 10^6$ g cm$^{-3}$), considerable uncertainty results in the important regime at matter densities greater than $10^6$ g cm$^{-3}$, where electron capture can accelerate the decay process. Finally, there exist no empirical data for very proton-rich nuclei above $^{56}$Ni. Since the hydrogen abundance remaining after the thermonuclear runaway depends critically upon the nuclear flows above $^{56}$Ni, the experimental study of nuclei in this mass regime is urged.

### ACKNOWLEDGMENT

This work was supported in part by the National Science Foundation under grant AST81-09826 A01.

## Literature Cited

1. Ayasli, S., Joss, P. C. *Astrophys. J.* 256: 637–65 (1982)
2. Baym, G., Pethick, C. *Ann. Rev. Astron. Astrophys.* 17:415–43 (1979)
3. Bethe, H. A. *Phys. Rev.* 55:434–56 (1939)
4. Bethe, H. A., Critchfield, C. L. *Phys. Rev.* 54:248–54 (1938)
5. Cameron, A. G. W. *Astrophys. J.* 130: 916–40 (1959)
6. Cline, T. L., Desai, U. D. *Astrophys. J. Lett.* 196:L43–L46 (1975)
7. Colgate, S. A., Petschek, A. G. *Astrophys. J.* 248:771–82 (1981)
8. Cominsky, L. R., Wood, K. S. *Astrophys. J.* 283:765–73 (1984)
9. Czerny, M., Sztajno, M. *Acta Astron.* 33:213–22 (1983)
10. Davidson, K., Ostriker, J. P. *Astrophys. J.* 179:585–98 (1973)
11. Ebisuzaki, T., Hanawa, T., Sugimoto, D. *Publ. Astron. Soc. Jpn.* 35:17–32 (1983)
12. Ergma, E. V. *Astrophys Space Sci. Rev.* 2:163–88 (1983)
13. Ergma, E. V., Tutukov, A. V. *Astron. Astrophys.* 84:123–27 (1980)
14. Fenimore, E. E., Klebesadel, R. W., Laros, J. G., Stockdale, R. E., Kane, S. R. *Nature* 297:665–67 (1982)
15. Fryxell, B. A., Woosley, S. E. *Astrophys. J.* 261:332–36 (1982)
16. Fujimoto, M. Y., Hanawa, T., Iben, I., Richardson, M. B. See Ref. 91, pp. 302–5
17. Fujimoto, M. Y., Hanawa, T., Miyaji, S. *Astrophys. J.* 247:267–78 (1981)
18. Grindlay, J., Gursky, H., Schnopper, H., Parsignault, D. R., Heise, J., et al. *Astrophys. J. Lett.* 205:L127–L130 (1976)
19. Grindlay, J. E., Marshall, H. L., Hertz, P., Soltan, A., Weisskopf, M. C., et al. *Astrophys. J. Lett.* 240:L121–125 (1980)
20. Hameury, J. M., Bonazzola, S., Heyvaerts, J., Ventura, J. *Astron. Astrophys.* 111:242–51 (1982)
21. Hanawa, T., Sugimoto, D. *Publ. Astron. Soc. Jpn.* 34:1–20 (1982)
22. Hanawa, T., Sugimoto, D., Hashimoto, M. *Publ. Astron. Soc. Jpn.* 35:491–506 (1983)
23. Hansen, C. J., Van Horn, H. M. *Astrophys. J.* 195:735–41 (1975)
24. Hashimoto, M., Hanawa, T., Sugimoto, D. *Publ. Astron. Soc. Jpn.* 35:1–15 (1983)
25. Hoffman, J. A., Cominsky, L., Lewin, W. H. G. *Astrophys. J. Lett.* 240:L27–L31 (1980)
26. Hoffman, J. A., Lewin, W. H. G., Doty, J. *Astrophys. J. Lett.* 217:L23–28 (1977)
27. Hoshi, R. *Astrophys. J.* 247:628–31 (1981)
28. Hoyle, F., Fowler, W. A. In *Quasi-Stellar Sources and Gravitational Collapse*, ed. I. Robinson, A. Schild, E. L. Schucking, pp. 17–27. Univ. Chicago Press (1965)
29. Hurley, K. In *Accreting Neutron Stars*, ed. W. Brinkmann, J. Thumper, MPE Report 177, pp. 161–78. Garching: Max Planck Inst. (1982)
30. Hurley, K. *Adv. Space Res.* 3(4):163–74 (1983)
31. Hurley, K. See Ref. 91, pp. 343–51
32. Inoue, H. See Ref. 91, pp. 121–30
33. Inoue, H., Koyama, K., Makishima, K., Matsuoka, M., Murakami, T., et al. *Astrophys. J. Lett.* 250:L71–L75 (1981)
34. Joss, P. C. *Nature* 270:310–14 (1977)
35. Joss, P. C. *Astrophys. J. Lett.* 225:L123–L127 (1978)
36. Joss, P. C. In *Compact Galactic X-Ray Sources*, ed. D. Pines, F. Lamb, pp. 89–96. Urbana: Phys. Dept., Univ. Ill. (1979)
37. Joss, P. C., Li, F. K. *Astrophys. J.* 238: 287–95 (1980)
38. Joss, P. C., Rappaport, S. A. *Ann. Rev. Astron. Astrophys.* 22:537–92 (1984)
39. Kato, M. *Publ. Astron. Soc. Jpn.* 35:33–46 (1983)
40. Klebesadel, R. W., Strong, I. B., Olson, R. A. *Astrophys. J. Lett.* 182:L85–L88 (1973)
41. Lamb, D. Q. See Ref. 52, pp. 249–72
42. Lamb, D. Q. See Ref. 91, pp. 512–47
43. Lamb, D. Q., Lamb, F. K. *Astrophys. J.* 220:291–302 (1978)
44. Lewin, W. H. G., Clark, G. W. *Ann. NY Acad. Sci.* 336:451–78 (1980)
45. Lewin, W. H. G., Hoffman, J. A., Doty, J., Hearn, D. R., Clark, G. W., et al. *Mon. Not. R. Astron. Soc.* 177:83p–92p (1976)
46. Lewin, W. H. G., Joss, P. C. *Nature* 270:211–16 (1977)
47. Lewin, W. H. G., Joss, P. C. *Space Sci. Rev.* 28:3–87 (1981)
48. Lewin, W. H. G., Joss, P. C. In *Accretion Driven Stellar X-Ray Sources*, ed. W. H. G. Lewin, E. P. J. van den Heuvel, pp. 41–115. Cambridge Univ. Press (1983)
49. Lewin, W. H. G., Vacca, W. D., Basinska, E. M. *Astrophys. J. Lett.* 277:L57–L60 (1984)
50. Liang, E. P. T. *Nature* 292:319–21 (1981)
51. Liang, E. P. See Ref. 91, pp. 597–604
52. Lingenfelter, R. E., Hudson, H. S., Worrall, D. M. eds. *Gamma Ray Transients and Related Astrophysical Phenomena*, AIP Conf. Proc. 77. New York: AIP Press. 500 pp. (1982)
53. London, R. A., Taam, R. E., Howard, W. M. *Astrophys. J. Lett.* 287:L27–L30 (1984)
54. Maraschi, L., Cavaliere, A. *Highlights in*

*Astronomy*, ed. E. A. Muller, 4:127–28. Dordrecht: Reidel (1977)
55. Marshall, H. L. *Astrophys. J.* 260:815–20 (1982)
56. Mazets, E. P., Golenetskii, S. V., Ilinskii, V. N., Aptekar, R. L., Guryan, Y. A. *Nature* 282:587–89 (1979)
57. Mazets, E. P., Golenetskii, S. V., Aptekar, R. L., Guryan, Y. A., Ilinskii, V. N. *Nature* 290:378–82 (1981)
58. Mestel, L. *Mon. Not. R. Astron. Soc.* 112:598–605 (1952)
59. Nozakura, T., Ikeuchi, S., Fujimoto, M. Y. *Astrophys. J.* 286:221–31 (1984)
60. Oda, M. See Ref. 52, pp. 319–38
61. Paczynski, B. *Astrophys. J.* 267:315–21 (1983)
62. Rosenbluth, M. N., Ruderman, M., Dyson, F., Bahcall, J. N., Shaham, J., Ostriker, J. *Astrophys. J.* 184:907–10 (1973)
63. Ruderman, M. In *Progress in Particle and Nuclear Physics*, ed. D. Wilkinson, 6:215–34. Oxford: Pergamon (1981)
64. Salpeter, E. E., Van Horn, H. M. *Astrophys. J.* 155:183–202 (1969)
65. Starrfield, S., Kenyon, S., Sparks, W. M., Truran, J. W. *Astrophys. J.* 258:683–95 (1982)
66. Swank, J. H., Becker, R. H., Boldt, E. A., Holt, S. S., Pravdo, S. H., et al. *Astrophys. J. Lett.* 212:L73–L76 (1977)
67. Taam, R. E. *Astrophys. J.* 241:358–66 (1980)
68. Taam, R. E. *Astrophys. Space Sci.* 77:257–65 (1981)
69. Taam, R. E. *Astrophys. J.* 247:257–66 (1981)
70. Taam, R. E. *Astrophys. J.* 258:761–69 (1982)
71. Taam, R. E. See Ref. 91, pp. 263–72
72. Taam, R. E. See Ref. 91, pp. 1–10
73. Taam, R. E., Picklum, R. E. *Astrophys. J.* 224:210–16 (1978)
74. Taam, R. E., Picklum, R. E. *Astrophys. J.* 233:327–33 (1979)
75. Takahashi, K., Yamada, M., Kondoh, T. *At. Data Nucl. Data Tables* 12:101–42 (1973)
76. Tawara, Y., Kii, T., Hayakawa, S. See Ref. 91, pp. 257–62
77. Teegarden, B. J. See Ref. 52, pp. 123–42
78. Teegarden, B. J., Cline, T. L. *Astrophys. J. Lett.* 236:L67–70 (1980)
79. van Paradijs, J. *Nature* 274:650–53 (1978)
80. van Paradijs, J. *Astrophys. J.* 234:609–11 (1979)
81. van Paradijs, J. *Astron. Astrophys.* 107:51–53 (1982)
82. van Paradijs, J., Stollman, G. M. *Astron. Astrophys.* 137:L12–L14 (1984)
83. von Weizsacher, C. F. *Phys. Z.* 38:176–91 (1937)
84. Wallace, R. K., Woosley, S. E. *Astrophys. J. Suppl.* 45:389–420 (1981)
85. Wallace, R. K., Woosley, S. E. See Ref. 91, pp. 319–24
86. Wallace, R. K., Woosley, S. E., Weaver, T. A. *Astrophys. J.* 258:696–715 (1982)
87. Walter, F. M., Bowyer, S., Mason, K. O., Clark, J. T., Henry, J. P., et al. *Astrophys. J. Lett.* 253:L67–L71 (1982)
88. White, N. E., Swank, J. H. *Astrophys. J. Lett.* 253:L61–L66 (1982)
89. Wood, K. S., Byram, E. T., Chubb, T. A., Friedman, H., Meekins, J. F., et al. *Astrophys. J.* 247:632–38 (1981)
90. Woosley, S. E. See Ref. 29, pp. 189–208
91. Woosley, S. E., ed. *High Energy Transients in Astrophysics, AIP Conf. Proc. No. 115.* New York: AIP Press. 714 pp. (1984)
92. Woosley, S. E. See Ref. 91, pp. 485–511
93. Woosley, S. E., Taam, R. E. *Nature* 263:101–3 (1976)
94. Woosley, S. E., Wallace, R. K. *Astrophys. J.* 258:716–32 (1982)
95. Woosley, S. E., Weaver, T. A. See Ref. 91, pp. 273–97 (1984)

*Ann. Rev. Nucl. Part. Sci. 1985. 35 : 25–54*

# FREE ELECTRON LASERS[1]

## W. B. Colson

Berkeley Research Associates, P.O. Box 241, Berkeley, California 94701

## A. M. Sessler

Lawrence Berkeley Laboratory, University of California, Berkeley, California 94720

CONTENTS

## 1. INTRODUCTION

The free electron laser (FEL) uses a high quality relativistic beam of electrons passing through a periodic magnetic field to amplify a copropagating optical wave (1–4). In an oscillator configuration, the light is stored between the mirrors of an open optical resonator, as shown in Figure 1. In an amplifier configuration, the optical wave and an intense electron beam pass through a transversely undulating magnetic field to achieve high gain. In either case, the electrons must spacially overlap the optical mode for good coupling. Typically, the peak electron beam current varies from several amperes to many hundreds of amperes, and the electron energy ranges from a few MeV to a few GeV. The electrons are the power source in a FEL, and provide from a megawatt to more than a gigawatt flowing through the resonator or amplifier system. The undulator resonantly couples the electrons to the transverse electrical field of the optical wave in vacuum.

The basic mechanism of the coherent energy exchange is the bunching of the electrons at optical wavelengths. Since the power source is large, even small coupling can result in a powerful laser. Energy extraction of 5% of the electron beam energy has already been demonstrated. The electron beam quality is crucial in maintaining the coupling over a significant interaction distance, and of central importance to all FEL systems is the magnetic undulator. The peak undulator field strength is usually several kG and can be constructed from coil windings or permanent magnets. In the top part of Figure 2, the Halbach undulator design is shown for one period. The field can be achieved, to a good approximation, using permanent magnets made out of rare earth compounds, a technique developed by K. Halbach (5) and now employed in most undulators. The undulator wavelength is in the range of a few centimeters and the undulator length extends for a few meters, so that there are several hundred periods for the interaction (6–8). The polarization of the undulator can be either linear or circular or a combination (9). The optical wave has the same polarization as the undulator driving it. This is an illustration of the FEL's most important attribute—the flexibility of its design characteristics.

·The transverse undulations of electrons with energy $\gamma mc^2$ generate spontaneous emission in a forward cone of angular width $\gamma^{-1}$. When the undulator fields are strong enough so that the amplitude of the cone's oscillation off-axis is comparable to the cone's width, a detector on-axis at infinity will begin to see several radiation harmonics (10). If the angular

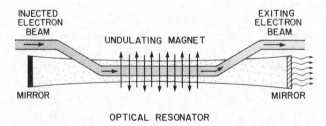

*Figure 1*  The basic elements of a free electron laser (FEL) oscillator are a high quality relativistic electron beam, an undulator magnet that causes the electrons to wiggle, and the resonant optical cavity to provide feedback.

deviations of the cone are larger, then the spectrum becomes broadband like the synchrotron emission from a bending magnet. The total emission energy from a bending magnet and a FEL undulator are similar, but the FEL spectrum is confined to a relatively narrow bandwidth because the electron motion is periodic and the radiation cone stays on the undulator axis. The FEL gain-bandwidth falls within the narrow spontaneous emission spectrum that is determined by the number of undulator periods. The laser linewidth can be much narrower than the spontaneous linewidth as in an atomic laser; the narrow line and long coherence length are established by mode competition.

The laser frequencies driven by the FEL mechanism are much higher than the oscillation frequency of the electrons in the undulator. This is due to a large Lorentz contraction of the undulator wavelength and a large relativistic Doppler shift of the emitted radiation in the forward direction. The relation between the undulator wavelength $\lambda_u$, the optical wavelength $\lambda$, and the electron beam energy is then $\lambda \approx \lambda_u/2\gamma^2$ and the mechanism can be described as stimulated Compton backscattering. It is the relativistic

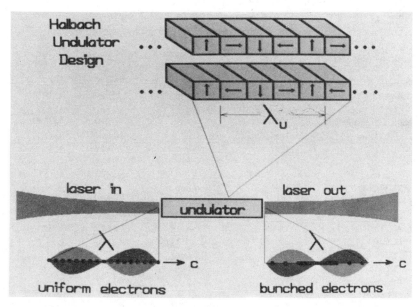

*Figure 2* A practical design for constructing the undulator field is shown at the top where eight permanent magnets are used to form one undulator period. The arrows show the directions of the permanent magnetic field. The interaction of an initially azimuthally uniform electron beam with the radiation in a FEL causes the electron beam to bunch in an optical wavelength. It is this bunching that causes coherent radiation.

factor $2\gamma^2$ that allows the FEL to reach short wavelengths. Low energy beams (5 MeV) are being used to reach wavelengths longer than atomic lasers (500 $\mu$m) and high energy beams (1 GeV) are used for x rays (500 Å), as shown in Table 1 (11–27). The FEL system is also continuously tunable merely by changing the electron energy of the electron source. Figure 3 shows some FEL system configurations, which are explained more fully in Section 2.

Figure 2 illustrates the basic bunching mechanism used to obtain coherent radiation. The electrons leaving the accelerator are randomly positioned over many optical wavelengths. There are typically $10^7$ electrons, or more, in each section of the electron beam one optical wavelength long. As the light and electrons interact at the beginning of the undulator, some electrons gain energy and some lose energy. Those that gain energy move a little faster longitudinally and those that lose energy move a little slower; this creates one bunch in each optical wavelength.

FELs have been described in a number of articles in the general scientific press (28–37). In addition, there is a textbook and a number of review articles on the subject (38–40) and two special issues of *IEEE Journal of*

**Table 1**  Operation of free electron lasers

| Name (Ref.) | Year of first operation | Wavelength | Peak power | Type[a] |
|---|---|---|---|---|
| Stanford (11) | 1976, 1977 | 10 $\mu$m, 3.4 $\mu$m | 130 kW | A, O |
| Columbia (25) | 1977 | 1.5 mm | 8 MW | ASE |
| NRL (26) | 1977 | 400 $\mu$m | 1 MW | ASE |
| NRL/Columbia (23) | 1978 | 400 $\mu$m | 1 MW | ASE, O |
| LANL (13) | 1981, 1982 | 10.6 $\mu$m | 10 MW | A, O |
| NRL (12) | 1981 | 4.6–3.1 mm | 75 MW | ASE |
|  | 1983 | 35 GHz | 17 MW | A |
| Orsay (15) | 1981, 1983 | 6500 Å | 60 mW | A, O |
| MSNW (14) | 1982 | 10.6 $\mu$m | [b] | A |
| Frascati (22) | 1983 | 5145 Å | [c] | A, O |
| TRW (16) | 1983 | 1.57 $\mu$m | 1.2 MW | O |
| NRL (17) | 1984 | 1 cm | 20 MW | ASE |
| MIT (19) | 1984 | 4.3–1.7 cm | 100 kW | A |
| UCSB (20) | 1984 | 0.4 mm | 8 kW | O |
| LLNL (18) | 1984 | 8.6 mm | 80 MW | A |
| Hughes (21) | 1984 | 1 cm | 60 kW | O |
| Erevan (24) | 1984 | 20–40 $\mu$m | 10 W | O |
| Novosibirsk (27) | 1984 | 6000 Å | [d] | A, O |

[a] A = amplifier, O = oscillator, ASE = amplified spontaneous emission.
[b] Output power not measured, but peak loss of electron energy was observed to be 9%.
[c] With an input laser power of 6 W, a gain of $3 \times 10^{-4}$ was measured.
[d] A gain of 1.5% was measured.

# Some FEL Configurations

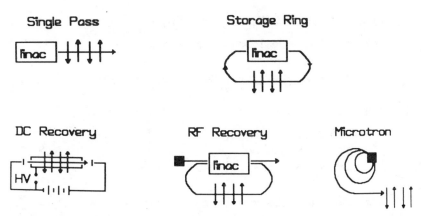

*Figure 3*   Free electron lasers can be in a variety of configurations, as depicted here. In fact, three of these five types have already operated.

*Quantum Electronics* contained many papers on FELs (41, 42). Finally, there are six volumes of conference proceedings that contain hundreds of papers and provide a good introduction to the FEL literature (43–48).

## 2.   GENERALITIES

### 2.1   *History*

The historical development of FELs can be traced back to the microwave tubes, backward wave oscillators, traveling-wave tubes, magnetrons, and klystrons of the 1940s, shown at the top of Figure 4. The traveling-wave tubes were similar in structure to the FEL in that they used mildly relativistic electrons traveling through periodically undulating electric or magnetic fields inside a wave guide. The radiation wavelengths produced were in the centimeter range. A characteristic of all such devices was the closed structure used to store the radiation. These electron tubes were tunable by changing the electron energy and using higher harmonics, and high efficiencies were common. While the Motz (49) tubes used the same configuration as the FEL, the operating mechanism was different. A tube that used the same mechanism as that in a FEL was invented by Phillips (50), but J. M. J. Madey, the inventor of the FEL, was unaware of the Phillips tube, although he did know of Motz's work. Shorter wavelengths could not be reached because electrons did not oscillate fast enough and the closed resonator could not be made small enough.

# GENERAL HISTORY

- Electron Tubes (1930's -> 1960's)
    - free non-relativistic electrons
    - microwave cavity
        long wavelengths, tunable, efficient

- Atomic & Molecular Lasers (1960's -> now)
    - bound electrons
    - optical resonator
        short wavelengths

- Free-Electron Lasers (1976 -> now)
    - free relativistic electrons
    - optical resonator
        short wavelengths, tunable, efficient

*Figure 4*    FELs grew out of the development of electron tubes and atomic lasers. They retain some of the good qualities of both.

Atomic lasers were invented in the 1960s and made use of two new concepts (51): excited electrons in the bound states of atoms or molecules oscillated rapidly to produce optical radiation and this radiation was stored in open optical resonator.

J. M. J. Madey's conception of the FEL (1) came from a mixture of the attributes of microwave tubes and the atomic laser: the Motz undulator and the optical resonator. The relativistic Lorentz contraction and the Doppler shift produced high frequencies from the slower oscillations of the electrons traveling near the speed of light. The FEL is tunable just as the early electron tubes were but it works at short wavelengths.

Independently, R. Palmer, P. Csonka, and K. Robinson were working on the coherent emission of radiation by relativistic electron beams (52).

## 2.2 Basic Concepts

A good theoretical approach to FELs is to solve the relativistic particle dynamics and couple the solutions to the optical wave equation. The more sophisticated analytical methods employed in the analysis of plasmas and lasers are appropriate, but generally not needed. The first classical theory was introduced by M. O. Scully, F. Hopf, et al (53).

The initial electron density has no structure on the scale of the FEL optical wavelength. Individual electrons are only influenced by the radiation field, the undulator magnetic field, and possibly the Coulomb fields of other electrons if the density is large enough. For typical undulator fields and wavelengths, the radiation emitted spontaneously after just one pass is sufficient to define a classical wave. The Lorentz force equations for an electron are

$$\frac{d}{dt}(\gamma\boldsymbol{\beta}) = -\frac{e}{mc}[\mathbf{E_r}+\boldsymbol{\beta}\times(\mathbf{B_u}+\mathbf{B_r})] \qquad \text{1a.}$$

$$\frac{d\gamma}{dt} = -\frac{e}{mc}\boldsymbol{\beta}\cdot\mathbf{E_r} \qquad \text{1b.}$$

$$\gamma^{-2} = 1-\boldsymbol{\beta}\cdot\boldsymbol{\beta}, \qquad \text{1c.}$$

where $\mathbf{E_r}$ and $\mathbf{B_r}$ are the optical electric and magnetic fields, $\mathbf{B_u}$ is the undulator field, $e = |e|$ is the electron charge magnitude, $c$ is the speed of light, $m$ is the electron mass, $\beta c$ is the electron velocity, and $\gamma mc^2$ is the electron energy. Only four of the five equations (Equations 1) are needed to completely specify the problem. The undulator axis is taken along the $z$ axis so that the transverse optical force with contributions from both $\mathbf{E_r}$ and $\mathbf{B_r}$ is proportional to $|\mathbf{E_r}|(1-\beta_z)$. For relativistic electrons $(1-\beta_z) \approx 1/2\gamma^2$, so that the transverse optical force is small; the optical electric and magnetic forces combine almost to cancel when $\gamma \gg 1$.

In order to couple energy out of the electron beam, the time average of $\boldsymbol{\beta}\cdot\mathbf{E_r}$ must be nonzero during the interaction time in the undulator. The role of the undulator is to rotate the transverse electron velocity as the field $\mathbf{E_r}$ passes over it. Note that in Equation 1a the transverse electron motion is determined primarily by the undulator magnet since the transverse optical force is small. However, a randomly distributed electron beam will have $\langle\boldsymbol{\beta}\cdot\mathbf{E_r}\rangle = 0$ with no net energy transfer. But, an energy modulation alters the electron $z$ velocities to cause bunching and coherent emission. While deflections off the mode axis are necessary for coupling, they cannot be too large, since the optical mode has a limited radial extent.

A suitable undulator field (6) around the mode axis is

$$B_x \approx -B\{[1+\tfrac{1}{8}k_u^2(3x^2+y^2)]\cos(k_u z)-\tfrac{1}{4}k_u^2 xy\sin(k_u z)\}$$

$$B_y \approx B\{[1+\tfrac{1}{8}k_u^2(x^2+3y^2)]\sin(k_u z)-\tfrac{1}{4}k_u^2 xy\cos(k_u z)\} \qquad \text{2.}$$

$$B_z \approx -B[1+\tfrac{1}{8}k_u^2(x^2+y^2)][x\sin(k_u z)+y\cos(k_u z)],$$

where $B$ is the peak field strength and $\lambda_u = 2\pi/k_u$ is the undulator wavelength. The electron beams suitable for FELs must be sufficiently aligned so that the transverse excursions are small compared to $\lambda_u$. The average magnetic field strength increases off-axis so that the electrons are focused toward the axis. When electrons are focused back toward the undulator axis, the transverse oscillations are called betatron oscillations. Typical transverse excursions are small enough that $k_u x$ and $k_u y$ are negligible.

With a small, high quality beam, the undulator field sampled by electrons is $(B \cos(k_u z), B \sin(k_u z), 0)$ and the orbits, which are helical, are

$$\boldsymbol{\beta} = [(-K/\gamma) \cos(k_u z), (-K/\gamma) \sin(k_u z), \beta_z] \qquad 3.$$

where $\beta_z \approx 1 - (1 + K^2)/2\gamma^2$ and $K = eB\lambda_u/2\pi mc^2$. Typically $K \approx 1$ and one sees that the transverse oscillations are small.

The optical field polarization that best couples to the above trajectory is given by the vector potential

$$\mathbf{A}(z, t) = \frac{E(t)}{k} \{\sin[kz - \omega t + \phi(t)], \cos[kz - \omega t + \phi(t)], 0\}, \qquad 4.$$

where $E(t)$ is the electric field magnitude, $\lambda = 2\pi c/\omega = 2\pi/k$ is the optical carrier wavelength, and $\phi(t)$ is the optical phase. No $x$ or $y$ dependence is included in $\mathbf{A}$, for now, since we assume the electrons remain well inside the optical mode waist. The optical electric field is $\mathbf{E}_r = -c^{-1}\partial \mathbf{A}/\partial t$. Inserting $\mathbf{E}_r$ and Equation 3 into Equation 1b we have

$$\frac{d\gamma}{dt} = \left(\frac{eKE}{\gamma mc}\right) \cos[(k_u + k)z - \omega t + \phi]. \qquad 5.$$

A particularly useful form of Equation 5 may be obtained in the case where the fractional energy change $\delta\gamma/\gamma \ll 1$. Define the electron phase $\zeta(t) = (k_u + k)z(t) - \omega t$, then eliminate $\dot{\gamma}(t)$ from Equation 5 to get

$$\frac{d^2\zeta}{d\tau^2} = \frac{dv}{d\tau} = |a| \cos(\zeta + \phi), \qquad 6.$$

where $|a| = 4\pi NeKLE/\gamma^2 mc^2$ is the dimensionless optical field strength, $\tau = ct/L$ is the dimensionless time, $L = N\lambda_u$ is the undulator length so that $0 \le \tau \le 1$, and $v = d\zeta/d\tau$ is the electron phase velocity. The electron dynamics have been put in the form of a pendulum equation (54).

The evolution of each electron entering the FEL undulator follows Equation 6. Individual electrons are identified by their initial conditions $\zeta(0) = \zeta_0$ and $v(0) = v_0 = L[(k_u + k)\beta_z(0) - k]$. In weak fields $|a| \ll \pi$, and when $|a| \gg \pi$, the fields are considered strong because the phases evolve significantly in the time $\tau \le 1$. Experiments are usually designed so that the spread in electron velocities does not cause a spread in $v_0$ greater than $\pi$. This can be adjusted by keeping the length $L$ small enough, but a better beam quality allows a greater length $L$ and much more gain.

The optical wave is governed by the wave equation driven by the current $\mathbf{J}_\perp$:

$$\left(\frac{\partial^2}{\partial z^2} - \frac{1}{c^2}\frac{\partial^2}{\partial t^2}\right)\mathbf{A}(z, t) = -\frac{4\pi}{c}\,\mathbf{J}_\perp(z, t), \qquad 7.$$

where the $(x, y)$ dependence has been dropped (see Section 5). The transverse electron current is the sum of all particle currents

$$\mathbf{J} = -ec\sum_m \boldsymbol{\beta}_\perp \delta^{(3)}[\mathbf{x} - \mathbf{r}_m(t)], \qquad 8.$$

where $r_m(t)$ is the trajectory of the $m$th electron and $\boldsymbol{\beta}_\perp = (\beta_x, \beta_y, 0)$. Even the spontaneous emission spectrum in a FEL has a long coherence length so that the field $E(t)$ and phase $\phi(t)$ can be taken to vary slowly over an optical period, $\omega^{-1}$. Then, the terms containing second derivatives in Equation 7 are negligible compared to terms with single derivatives and

$$\frac{d}{d\tau}(Ee^{i\phi}) = \frac{L}{c}\frac{d}{dt}(Ee^{i\phi}) = -\pi eKL\sum_m \frac{e^{-i\zeta}}{\gamma}\delta^{(3)}[\mathbf{x} - \mathbf{r}_m(t)].$$

Then, the wave equation has the simple form

$$\frac{da}{d\tau} = -j\langle e^{-i\zeta}\rangle, \qquad 9.$$

where $a = |a|e^{i\phi}$, the dimensionless current density is $j = 8N(\pi eKL)^2$ $\rho/\gamma^3 mc^2$, $\rho$ is the electron particle density, and the angular brackets represent a normalized average over the electrons. If electrons are bunched at the phase $\pi$, then the optical amplitude is driven with strength $j$ during the time $0 < \tau < 1$ and there is gain. If the phase $\pi/2$ is overpopulated, then the optical phase $\phi$ grows with little gain. Usually, it is a combination of $|a|$ and $\phi$ that is driven because the electron bunching is not perfect.

Figure 5 shows the phase space evolution of a periodic section of the electron beam in the $(\zeta, \nu)$ coordinates. The separatrix path shown is given by $\nu_s^2 = 2|a|[1 - \sin(\zeta_s + \phi)]$; the peak-to-peak height is $4|a|^{1/2}$ and the horizontal position is determined by $\phi$. The "fluid" of electrons starts equally populating all phases and at the phase velocity $\nu_0 = 2.6$ for maximum gain. As the electron fluid evolves in Figure 5 it becomes darker, becoming black at $\tau = 1$. The final bunching is near the phase $\pi$ and the gain and optical phase shift evolution are shown at the right. The initial optical field is weak $a(0) = a_0 = 1$, and the final gain determined numerically is $G \equiv [|a(1)|^2 - a^2(0)]/a^2(0) = 0.135j$.

While we have made a few assumptions, the "pendulum" and wave equations (Equations 6 and 9) form a simple, powerful description of the

FEL (54). They are valid for both weak ($|a| \ll \pi$) and strong ($|a| \gg \pi$) optical fields in either high ($j \gg 1$) or low ($j \ll 1$) gain conditions. It is generally important that both the optical field amplitude and phase are included in the description.

When the optical fields are weak, Equations 6 and 9 can be easily linearized in $a(\tau)$:

$$\dot{a} = ij\langle\exp[-i(\zeta_0 + v_0\tau)]\zeta_1\rangle; \qquad \ddot{\zeta}_1 = |a|\cos(\zeta_0 + v_0\tau + \phi), \qquad 10.$$

where $(\dot{\ }) = d(\ )/d\tau$, $\zeta = \zeta_0 + v_0\tau + \zeta_1$, and $\zeta_1$ is $\zeta$ to lowest order in $|a|$. For a uniform beam distribution, the average of any quantity $f$ is given by

$$\langle f \rangle = \int_0^{2\pi} d\zeta_0(f)/2\pi,$$

the electron coordinates can be removed from Equation 10 and the optical field is determined by the roots to the cubic equation

$$\alpha_r^3 - iv_0\alpha_r^2 - \left(\frac{i}{2}\right)j = 0, \qquad 11.$$

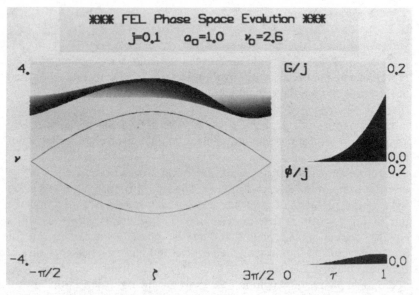

*Figure 5* The electron phase space follows sample electrons through the undulator. The separatrix is shown as a guide to the phase space paths. The electron fluid grows darker as it passes through the undulator. (The same representation is employed in Figures 6 and 8.) Bunching at the phase $\pi$ leads to gain, but also affects the optical phase.

with the field of the form

$$a = a_0 \exp(-iv_0\tau) \sum_{r=1}^{3} c_r \exp(\alpha_r\tau),$$

where the $c_r$ are determined by initial conditions. If $|v_0| \gg \pi$ so that the FEL is far off-resonance, the driving term $j$ is negligible and the trivial uninteresting solution $a \approx a_0$ is obtained, i.e. no gain. If the current density $j$ is large, so that $v_0$ is negligible, the important real root is $\alpha_r = (j/2)^{1/3}(\sqrt{3}/2)$ giving exponential growth. The complex field is then described by $a(\tau) = (a_0/3)\exp[(j/2)^{1/3}(\sqrt{3}+i)\tau/2)]$, and the gain is exponential after an initial bunching time.

Figure 6 shows the phase space evolution in the high gain case where $j = 100$. The electrons are started at $v_0 = 0$ to show how gain is achieved on-resonance. Bunching occurs at the phase $\pi/2$, but in the high gain case a significant optical phase shift changes the position of the separatrix so that, relative to the optical wave, bunching is at phase $\pi$. The resulting exponential growth and phase evolution are shown on the right. The exponential gain only occurs after bunching is established.

In the low gain case, both $v_0$ and $j$ are important in Equation 11. The gain is no longer exponential and all three roots are needed to find the final gain

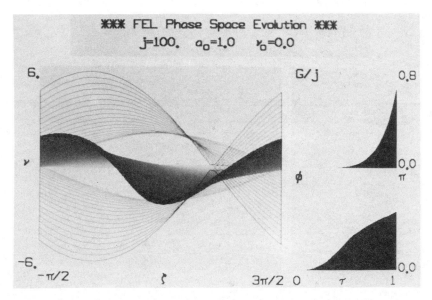

*Figure 6* In the high gain case, there is a substantial optical phase change shift, which then shifts the separatrix. The height of the separatrix is proportional to the $|a|^{1/2}$ and grows with the high gain.

at $\tau = 1$, which is given by

$$G(\nu_0) = j \frac{[2 - 2 \cos \nu_0 - \nu_0 \sin \nu_0]}{\nu_0^3} = -\frac{j}{2} \frac{d}{d\nu_0} \left[ \frac{\sin (\nu_0/2)}{(\nu_0/2)} \right]^2. \qquad 12.$$

The gain is antisymmetric in $\nu_0$ and peaks at $G = 0.135j$ with $\nu_0 = 2.6$. Figure 7 shows the plot of $G(\nu_0)$ above the accompanying optical phase shift $\phi(\nu_0) = j[2 \sin \nu_0 - \nu_0(1 + \cos \nu_0)]/\nu_0^3$. Note that the gain spectrum can be written as the derivative of the spontaneous emission spectrum $[\sin (\nu_0/2)/(\nu_0/2)]^2$. This remains true for a large class of undulator designs and is known as the Madey theorem (55). The theorem states that when an undulator design produces a spectrum $s(\nu_0)$ the gain is proportional to the slope of the spectrum $ds(\nu_0)/d\nu_0$. A second theorem relates the "second moment of the mean electron energy loss evaluated to first order in the optical field strength," $\langle [\delta\gamma^{(1)}]^2 \rangle$, to the "mean energy loss evaluated to second order in the optical field strength," $\langle \delta\gamma^{(2)} \rangle$:

$$\langle \delta\gamma^{(2)} \rangle = \frac{1}{2} \frac{\partial}{\partial\gamma} \langle [\delta\gamma^{(1)}]^2 \rangle.$$

In the FEL oscillator, gain over many passes leads to strong fields. The spontaneous fields either experience exponential growth or the repeated

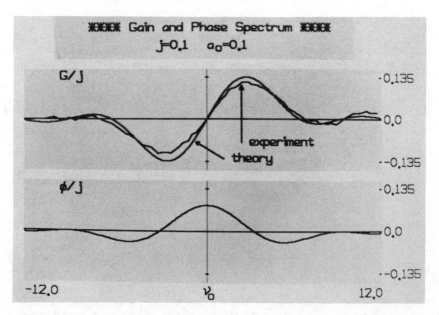

*Figure 7* The final gain and phase of the optical wave are plotted as a function of $\nu_0$. Experimental points are superimposed to show agreement between small amplitude theory and experiment (Orsay) (89).

gain of Equation 12. In stronger fields where $|a| \gtrsim \pi$, the gain process changes and begins to depend on $|a|$. Electron phases now evolve too far in phase space and bunching is difficult to maintain. Figure 8 shows electrons in a strong field $a_0 \equiv |a(0)| = 8$. The separatrix is now large and electrons are trapped in the closed orbit region of phase space. Those near the harmonic circular paths oscillate around the phase $\pi/2$ at a frequency $|a|^{1/2}$; these oscillations are called synchrotron oscillations. There is a decrease in gain, i.e. saturation. When the gain is reduced to equal the FEL system losses, steady-state operation is established.

A method used to extend the saturation limit of FELs was proposed by Kroll, Morton & Rosenbluth (56) and is called the tapered undulator. As electrons lose energy to the optical wave, the undulator properties can be modified to accommodate the new electron energy. As $\gamma$ decreases either the undulator wavelength, $\lambda_u$, or field strength, $B$, can be decreased to maintain resonance. A simple case is that in which both $B$ and $\lambda_u$ change along the undulator so that $K$ is constant. When such a taper is included, the pendulum equation acquires an accelerating term, $\delta = L^2 \, dk_u(z)/dz$,

$$\ddot{\zeta} = \delta + |a| \cos(\zeta + \phi). \qquad 13.$$

In the absence of the field $|a|$, electrons appear to be "accelerated" to higher phase velocities. In strong fields, about half the electron phases are trapped

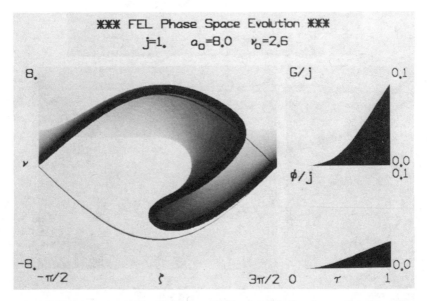

*Figure 8* Phase space evolution in the strong field regime. The "synchrotron" motion of the particles has led to saturation and energy is no longer transferred from the electrons to the optical wave. Even in saturation the phases of the optical wave evolves.

near the phase $\pi$, which drives the optical amplitude and gain. Figure 9 shows the final position of electrons in phase space after trapping has occurred in strong fields $a_0 \approx 40$ and with tapering such that $\delta = 6\pi$. The untrapped electrons are seen at the top of the phase space picture spread over the phase axis randomly. The gain is higher than would be possible at this field strength without tapering. The tapered undulator is a good example of the design flexibility of FELs. The undulator structure [length, polarization, wavelength profile, field profile $B(z)$, etc] are all features that can be modified to enhance performance for a particular FEL application.

An example proposed by Vinokurov & Shrinsky (57) is the klystron FEL (sometimes called a transverse optical klystron FEL, or TOK) where the undulator is split into two sections separated by a drift or dispersive section. The purpose is to achieve higher gain for a given interaction length $L$. The dispersive section acts like the bending magnet of an electron energy analyzer. Small variations in the electron phase velocity $v$ caused by the first undulator section are translated into phase changes $\Delta\zeta = Dv$ at the end of the dispersive magnet and the parameter $D$ measures the strength of the dispersive field. The theoretical description of the field and the electrons uses Equations 6 and 9 with $\Delta\zeta = Dv$ applied to each electron at $\tau = 1/2$. This results in a higher degree of bunching, and therefore greater gain than given by Equation 12.

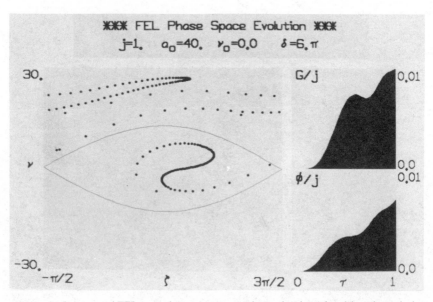

*Figure 9*  In a tapered FEL some electrons are trapped near the phase that drives the optical wave. The untrapped electrons are distributed over many phases and do not drive the wave.

When the undulator is designed to have linear polarization, only the definitions of variables in Equations 6 and 9 change while the form of the equations remains the same. The modifications are $a \rightarrow a[J_0(\xi) - J_1(\xi)]$, $j \rightarrow j[J_0(\xi) - J_1(\xi)]^2$, where $\xi = K^2/2(1+K^2)$, and $B$ becomes the rms undulator field strength.

## 2.3 Transverse Effects

The one-dimensional analysis, which we have employed up to this point, leaves out all transverse effects except the simple periodic undulator motion.

First we discuss electron beam transverse effects. A helical undulator provides focusing of the electrons in both transverse planes. Sometimes a longitudinal, solenoidal field is employed so as to give even more focusing. For some devices the cyclotron resonance in this field coincides, or almost coincides, with the FEL resonance and makes the interpretation of these experiments more complicated (12). On the other hand, this juxtaposition appears to enhance the gain, but is limited to long wavelength applications because of the upper limit on attainable solenoidal field strengths.

For planar undulators there is only "natural" focusing in the plane perpendicular to the sinusoidal motion and the betatron wave number is $k_{\beta y} = eB/\sqrt{2}mc\gamma$ in the non-wiggle plane, where $B$ is the peak field. The resonance condition is maintained as a particle undergoes betatron oscillations. In the wiggle plane, generally some focusing is required (50, 58, 59). Quadrupoles, although they give focusing, seriously degrade FEL performance. A planar undulator field is

$$\mathbf{B} = -B \cosh(k_u y) \cos(k_u z)\mathbf{y} + B \sinh(k_u y) \sin(k_u z)\mathbf{z},$$

so that the motion is

$$x' \equiv \frac{dx}{dz} = \frac{B}{\gamma}\left(1 + \frac{k_w^2 y^2}{2} + \cdots\right) \sin(k_w z),$$

and hence increases as $y$ increases. This increase with $y$ just balances the decrease of $y' \equiv dy/dz$ when $y$ increases and causes $\beta_z$ to be constant. E. T. Scharlemann (60) has shown how shaping the undulator pole faces with a slight parabolic curvature provides horizontal focusing while maintaining $\beta_z$ a constant of the motion. The curvature causes the field to increase off-axis and provides focusing in both $x$ and $y$. If the pole face is given by $y(x) = Y_0(1 - k_u^2 x^2/4)$, then the focusing will be the same in $x$ and $y$ and the electron beam cross section will be round.

It is necessary, in any real FEL, to avoid resonances between the various frequencies to which the particles are subject. For example, one must avoid a resonance between betatron oscillations and integral multiples of $\lambda_u$. Also, one must avoid the usual coupling resonances between the betatron oscillations in $x$ and $y$. There is another kind of resonance that must also be avoided: a synchro-betatron resonance between the "synchrotron motion" of trapped electrons and transverse betatron motion (61, 62).

We turn now to transverse effects of the electromagnetic wave. The simplest effect is the excitation of cavity modes in an oscillator. Figure 10 shows this phenomenon in a computer simulation of the original Stanford experiment where the electron beam has been moved off-axis to excite a combination of higher order modes.

The Rayleigh range is a measure of the effect of diffraction. For a light beam of radius $w$, the Rayleigh range $z_r = \pi w^2/\lambda$ is the propagation distance over which the optical wavefront doubles its area. In a proper FEL design one wants good overlap between the electron beam and the light beam over the whole interaction length so that $z_r$ should be comparable to $L$. However, if the FEL has sufficiently high gain it can provide "guiding" to the light and keep it within the electron beam for many Rayleigh lengths, as in an optical fiber (63, 64). This is seen, dramatically, in Figure 11.

A FEL provides an effective index of refraction, $n$, by changing the optical phase along the interaction length:

$$\mathrm{Re}(n) - 1 \equiv \frac{1}{k}\frac{\mathrm{d}\phi}{\mathrm{d}z} = \frac{j}{|a|Lk}\langle\sin(\zeta+\phi)\rangle,$$

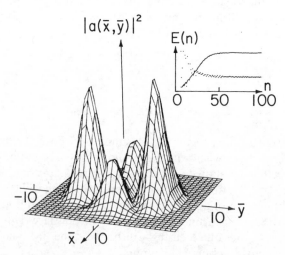

*Figure 10* Typically FELs are made to produce the fundamental mode in an optical resonator with a Gaussian shape in $x$ and $y$. A higher order mode is excited here by moving the electron beam off of the resonator axis. The theoretical calculation employed the parameters of the original Stanford FEL (11).

and

$$\text{Im}(n) \equiv \frac{1}{k|a|} \frac{\mathrm{d}|a|}{\mathrm{d}z} = \frac{-j}{|a|Lk} \langle \cos(\zeta + \phi) \rangle.$$

For an optical fiber, guiding occurs if $\text{Re}(V^2) + \frac{1}{2}\text{Im}(V^2) > 1$, where the (complex) fiber parameter, $V$, is given by $V^2 = (n^2 - 1)b^2k^2$, where $b$ is the electron beam radius. Thus one can readily determine when guiding takes place, provided one can evaluate the averages over particles of $\sin(\zeta + \phi)$ and $\cos(\zeta + \phi)$. When there is gain, we know that the averages of sin and cos are nonzero.

In the exponential growth regime one can evaluate the averages

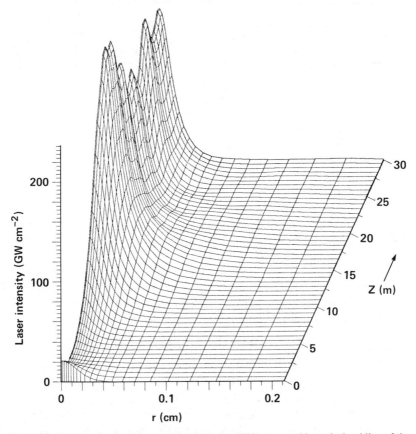

*Figure 11* Intense electron beams going through a FEL can provide optical guiding of the radiation. In the absence of guiding, the radiation would diffract out of the electron beam long before the end of the undulator.

analytically (63–65). One simply augments the wave equation, Equation 7, with $\nabla_\perp^2$ and then approximates this transverse derivative with

$$\nabla_\perp^2 E \approx -\frac{2kE}{z_r}.$$

The result is that Equation 11 becomes, for $v_0 = 0$,

$$\alpha_r^3 + i\alpha_r^2(L/z_r) - \left(\frac{i}{2}\right)j = 0,$$

where the length of the undulator is $L$. Thus the effect of diffraction and optical guiding are included in a one-dimensional theory. Extension to a warm beam and to $v_0 \neq 0$ can be found in the quoted literature (63–65).

## 2.4 *Longitudinal Effects*

The simple pendulum and wave equations (Equations 6 and 9) are valid for a single complex field $a = |a|e^{i\phi}$ with only a single frequency, the carrier frequency $\omega$. A realistic FEL oscillator, or amplifier, produces a spectrum of frequencies surrounding the carrier wave. Usually, the coherence length extends over several optical wavelengths so that the slowly varying amplitude and phase approximation remain valid. To generalize the optical field representation to many modes, the single complex field $a(\tau)$ becomes $a(k, \tau)$ or $a(z, \tau)$.

Driving the carrier phase $\phi$ in the center of the optical wavefront will focus the light along the electron beam path. Even in the low gain, diffraction couples the transverse and longitudinal waves. The phase profile $\phi(z)$ in a low gain oscillator is determined by the resonator mirrors and their Rayleigh length $z_r$. This causes a shift in frequency and a shift in the gain spectrum in an oscillator (66).

Often, the lack of distinct electron energy levels leads to questions about the ultimate coherence capabilities of FELs. In both the FEL and atomic laser, a long coherence length and narrow frequency spectrum is determined by mode competition, not by energy levels. In the low gain case, the weak field gain per pass in each mode is given by Equation 12. The number of modes within the gain bandwidth is about $\gamma^2$ (typically $\gamma^2 \gg 1$). Figure 12 shows the evolution of 100 optical wavelengths, around resonance. The spontaneous emission above resonance experiences gain on every pass, while other wavelengths receive less gain or absorption. The vertical scale follows the photon number $\eta(\lambda)$ over six orders of magnitude in 100 passes. The spectrum clearly narrows as mode competition continues. The photon number evolves as $\exp[G(\lambda)n_p]$, where $n_p$ is the pass number in the low gain oscillator where modes are uncoupled.

Short pulse effects (67) in FELs can also be described by generalizing the field to $a(k)$. An essential concept is "slippage"; this is the distance that light travels over the electron beam while the electrons travel through the undulator. It is given by $L(1 - \beta_z) \approx N\lambda$ using the FEL resonance condition. The ratio of the slippage distance $N\lambda$ to the electron pulse length $\sigma_z$ determines whether or not short effects are important. If $N\lambda \ll \sigma_z$, then the pulse is considered long, and each part of the pulse experiences gain proportional to the local density. If $N\lambda \gg \sigma_z$, then the FEL has short pulses and the modal structure of the pulse is comparable to the gain bandwidth, $\approx N^{-1}$.

Since electrons bunch when they reach the trailing edge of the optical pulse, the optical pulse receives more gain on its trailing edge than on its leading edge and behaves as if it is traveling slower than the speed of light, $c$. This effect is called "lethargy" (68) and must be considered in the oscillator FEL, where the resonator mirror spacing and the electron pulse repetition time must be synchronized (69, 70). The range of mirror positions to achieve synchronism is astonishingly small: only a 4-$\mu$m range was observed in the Stanford experiment. The amount of synchronism within the working range is important in determining the laser linewidth and power.

Other longitudinal effects involve long pulses in the FEL. One is the "trapped particle" instability analyzed by Kroll & Rosenbluth (71). The synchrotron frequency $|a|^{1/2}$ can mix with the carrier wave and produce sideband gain in the FEL. Figure 13 shows the growth of sideband structure in $|a(z)|$ and $\phi(z)$. A window section of a long pulse is four slippage distances long ($-2 < z/N\lambda < 2$). The field $|a(z)|$ is plotted at the top left, with bright regions indicating an intense field and dark regions indicating a low field region. The pass number is plotted along the vertical axis. The "trapped particle" instability starts a modulation in the field magnitude $a(z)$ and the

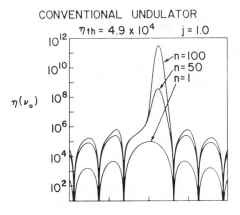

CONVENTIONAL UNDULATOR

$\eta_{th} = 4.9 \times 10^4$    $j = 1.0$

$\eta(\nu_o)$

n=100
n=50
n=1

*Figure 12*   The growth of coherence in the optical wave is shown by following 100 modes from spontaneous emission. The photon density at the wavelengths near peak gain grows more rapidly than the densities at surrounding wavelengths. This narrows the spectrum after only 100 passes. Evidently the laser can become narrowband.

phase $\phi(z)$ with a period equal to the slippage distance. The final spectrum, the Fourier transform of $a(z)$, is shown with its sideband on the bottom right; above is the weak field gain spectrum for reference. The final electron energy spectrum is shown above the gain spectrum. The power and net gain evolution are plotted on the upper right as a function of pass number $n$. The trapped particle instability is expected in nearly all FELs that saturate because of strong fields.

In a linearly polarized undulator, the electron $z$ motion is more complex than in the helical case because there is a periodic oscillation of the electron $z$ velocity even when injected perfectly. The oscillation in $z$, $\Delta z$, is given by $k\Delta z \approx -\xi \sin(2k_u ct)$ where $\xi = K^2/2(1+K^2)$. Since typically $K \approx 1$, the oscillations are a sizeable fraction of the optical carrier wavelength and lead to spontaneous emission and gain in higher optical harmonics (72). To generalize Equations 6 and 9 for a harmonic $hk$, make the replacements $\zeta \to h\zeta$, $v \to hv$,

$$a \to ah[J_{(h-1)/2}(h\xi) - J_{(h+1)/2}(h\xi)], \qquad \text{and}$$

$$j \to jh[J_{(h-1)/2}(h\xi) - J_{(h+1)/2}(h\xi)]^2.$$

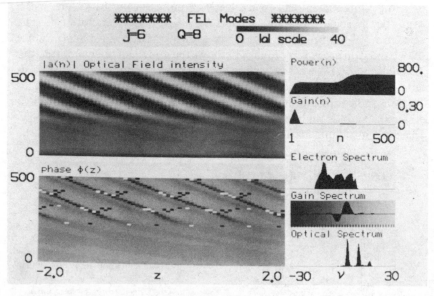

*Figure 13* When the electron synchrotron oscillations mix with the carrier wave, sidebands can be formed. Over many passes the optical wave develops a modulation whose period matches the synchrotron period. The optical power increases with the addition of energy of the sideband.

The form of the equations stays the same, only the couplings are modified. Note that there is gain only in the odd harmonics $h = 1, 3, 5, \dots$ If the undulator field is large enough so that $K \gtrsim 2$, then the coupling to higher harmonics is very strong. Several of the FEL experiments to date have observed coherent emission into higher harmonics, and it should prove to be a useful technique for reaching shorter wavelengths in a FEL.

## 3.    FREE ELECTRON LASER SYSTEMS

FELs can be made in a variety of configurations as is depicted schematically in Figure 3. In Section 3 we describe in more detail a particular linac oscillator, a linac amplifier, and a storage ring oscillator experiment. FEL systems are rapidly evolving and in the future can be expected to be quite different from those described here.

In Table 1 we presented a compendium of those FELs that have operated. Many more FEL devices are under construction and, as one can see from the dates in Table 1, these devices are being brought into operation at an ever-increasing rate. In Table 2 (73–87) we present a representative list of FEL accelerators.

Of great importance to FELs are electron beams of high quality. Two figures of merit of quality, for a given current, are energy spread and brightness. The brightness is defined by $\mathscr{J} = \pi^2 I / \gamma^2 \delta^4 V$, and becomes a measure of "goodness," where $I$ is the current enclosed within the transverse 4-volume $(\delta^4 V = \delta x \delta x' \delta y \delta y')$. For uniform phase space density, the brightness can be approximated by $\mathscr{J} \approx 2I / \gamma^2 \delta x \delta x' \delta y \delta y'$. The quality of a beam depends upon the parameters of the accelerator, the type of accelerator, and, of course, with what care it is aligned, etc. In Table 2 we present brightness and energy spread for a number of accelerators. As one can deduce, the expected performance of FELs far exceeds present achievements.

The development of FELs has been the result of both theoretical advances, which we emphasized in this article, and of experimental advances. In fact, without the latter, we would only have an empty theoretical structure. The experimentalists who have been instrumental in the development of FELs are many in number and, of course, are cited in the references, but special note should be taken of the work of C. A. Brau, D. Prosnitz, D. A. G. Deacon, J. Eckstein, L. Elias, E. Shaw, S. Skrinski, B. Kincaid, C. Pellegrini, J. M. Ortega, M. W. Poole, A. Renieri, P. Elleaume, T. Smith, A. Gover, J. A. Edighoffer, J. M. Slater, G. Dattoli, V. Granatstein, B. Newnam, R. Warren, T. Marshall, J. Walsh, R. Pantell, J. Pasour, and Y. Petroff.

**Table 2**   Selected FEL accelerators

| Accelerator[a] | Beam energy (MeV) | Peak beam current (A) | Pulse length |
|---|---|---|---|
| ETA (73) I | 5 | 1,000 | 15 ns |
| ATA (74) I | 50 | 1,000–10,000 | 50 ns |
| Osaka (75) RF | 20–38 | 1,000–3,000 | 16 ps |
| LANL (76) RF | 20 | 35–65 | 40 ps (micro) 100 $\mu$s (macro) |
| UC Santa Barbara (77) DC | 2.5 | 1.25 | 30 $\mu$s-dc |
| Stanford SCA (78) RF | 80–120 | 4 | 2 ps (micro) 10 ms (macro) |
| LLNL High Brightness Test Stand (79) I | 2 | 20–900 | 70 ns |
| Bell Labs Microtron (80) | 10–20 | 1–5 10 $\mu$s (macro) | 10 ps (micro) |
| UK RF Linac (81) RF | 30–100 | 10.0 | 60 ps (micro) 8.5 $\mu$s (macro) |
| Frascati ENEA Microtron (82) | 20 | 6.5 | 23 ps (micro) 12 $\mu$s (macro) |
| NRL Induction (83) I | 0.55–0.75 | 200 | 2 $\mu$s |
| MIT Pulsed Device (19) | 2.0 | 1,100 | 20 ns |
| Orsay ACO (84) SR | 163 | 3.3 | 0.5–1 ns |
| Stanford SXRC Ring Development (85) SR | 1.2 GeV | 270 | 33 ps |
| Orsay Super ACO Development (86) SR | 400 | 50 | 25–300 ps |
| LBL Design (87) SR | 750 | 327 | 41 ps |

[a] I = induction linac, RF = rf linac, SR = storage ring, DC = DC accelerator.
[b] Edge emittance, i.e. area of $x - x'$ phase space that includes all of the beam.
[c] $1/e$ in $x$ and $1/e$ in $x'$ emittance, or approximately 9 times edge brightness.
[d] $\Delta\gamma/\gamma$ unmeasurably small; variation of $\gamma$ during a pulse.
[e] Estimated.

## 3.1   *The Linac Oscillator*

The experiment of the TRW Group (15) serves to illustrate the linac oscillator. The superconducting accelerator at Stanford has a bunch length 4.3 ps, a peak current of 0.5–2.5 A, and at 66 MeV an energy spread of 0.03% and a beam emittance of $1.5\pi \times 10^{-3}$ cm·rad. The optical cavity had mirrors 12.68 m apart with a 7.5-m radius of curvature. At the optical wavelength of 1.57 $\mu$m, the reflectivity was 99.84%. The undulator consisted of pairs of linear arrays of $SmCO_5$ permanent magnets with wavelength $\lambda_\mu = 3.6$ cm and a peak field of 2.9 kG.

| Accelerator[a] | Pulse rep. rate (Hz) | Beam brightness (A/cm$^2$rad$^2$) | $\Delta\gamma/\gamma$ |
|---|---|---|---|
| ETA (73) I | 1.0 | $2 \times 10^4$ (at 2.5 MeV)[b] | d |
| ATA (74) I | 1.0 | $5$–$6 \times 10^4$ (at 4 MeV)[b] | d |
| Osaka (75) RF | 1.0–720 | $1.8$–$5.4 \times 10^7$ | $7 \times 10^{-3}$ |
| LANL (76) RF | 1.0 | $7.0 \times 10^5$ | $2 \times 10^{-2}$ |
| UC Santa Barbara (77) DC | — | $3.8 \times 10^{6\,b}$ | d |
| Stanford SCA (78) RF | 10 | $8 \times 10^{6\,c}$ | $10^{-4}$ |
| LLNL High Brightness Test Stand (79) I | 1 | $1.5 \times 10^5$ | d |
| Bell Labs Microtron (80) | 100 | $4.2 \times 10^{2\,e}$ | — |
| UK RF Linac (81) RF | 100 | $2.5 \times 10^4$ | $10^{-2}$ |
| Frascati ENEA Microtron (82) | 10 | $4.5 \times 10^4$ | $1.2 \times 10^{-3}$ |
| NRL Induction (83) I | single shot | $6.4 \times 10^3$ | $3 \times 10^{-2}$ |
| MIT Pulsed Device (19) | 0.01 | $1.4 \times 10^6$ | $<0.01$ |
| Orsay ACO (84) SR | 27 MHz | $3.8 \times 10^{6\,c}$ | $10^{-3}$ |
| Stanford SXRC Ring Development (85) SR | 20 MHz | $4 \times 10^{8\,c}$ | $6 \times 10^{-4}$ |
| Orsay Super ACO Development (86) SR | 4.8 MHz | $1.7 \times 10^{8\,c}$ | $3 \times 10^{-4}$ |
| LBL Design (87) SR | 2 MHz | $2.8 \times 10^{9\,c}$ | $2 \times 10^{-3}$ |

The experiment was designed to study the effect of tapering. Furthermore they devised an optical klystron so the multicomponent undulator had the following structure. First, there was a prebuncher section of 15 periods, then a magnetic dispersion section of two periods and a total length of 58.6 cm. Then 90 periods followed that could be tapered and, finally, 15 periods of constant undulator. The tapered part was varied to be a 0, 1, and 2% taper in energy. Beam diagnostics consisted of 14 insertable fluorescent screens so as to be sure the beam was steered properly and the mirrors were aligned using a green light laser.

With a 1% taper, the FEL had an average output laser power of 4 W and the peak power was 1.2 MW. Since the mirror transmission was 0.13% on each end of the cavity, the intracavity optical power was 11 GW cm$^{-2}$. The repetition rate was 10 Hz and the macropulse length 5 ms with the micropulse of 4 ps. The radiation fundamental was at 1.57 $\mu$m and the laser bandwidth was 1.3%.

Above threshold for the laser, the power increased by a factor of $10^{10}$ over that of the spontaneous radiation! The FEL took 305 passages at a gain of 7% per pass to get to 10% of the saturated level. The experimenters also observed coherent radiation at the second and third harmonic of 1.6 $\mu$m.

A study was made of the effect of tapering the undulator. For an untapered case the electron transfer of energy, efficiency, should be $(1/2N)$. The efficiency was measured to be 0.4%, which compares well with the expected value. With a 1% taper the electrons clearly divided into two groups: trapped and untrapped. Most of the electrons (60%) were trapped and decelerated 1–1.8% while the untrapped electrons were unchanged in energy. Thus the beneficial effect of tapering was demonstrated.

## 3.2   The Linac Amplifier

The experiment of the LBL/LLNL is representative of linac amplifier FELs (18, 88). The FEL was run as a single-pass amplifier in the microwave range at 34.6 GHz. The input signal was supplied by a magnetron of peak power 60 kW and a pulse length of 500 ns.

Use was made of the LLNL Experimental Test Accelerator (ETA) (73) to provide a 6-kA, 3.3-MeV beam with an emittance of $0.23\pi$ cm $\cdot$ rad. An emittance filter was used to reduce the beam current to approximately 500 Å with a normalized edge emittance of $0.47\pi$ cm $\cdot$ rad. The highly chromatic transport of the ETA beamline and matching quadrupoles results in a 15-ns, nearly monoenergetic, beam delivered to the interaction region.

The undulator magnet was three meters long, and the undulator period was 9.8 cm. The longitudinal variation of the undulator field provided strong vertical focusing. Horizontally focusing quadrupole magnets, surrounding the undulator, provided horizontal focusing while only slightly reducing the vertical focusing and negligibly affecting the FEL resonance condition.

The interaction waveguide was a rectangular, oversized waveguide immersed in the undulator. The inside dimensions of the waveguide were 9.83 cm wide by 2.91 cm high. The electric field was horizontal and coupled to the $TE_{01}$ waveguide mode, which was excited by the input microwave signal.

The signal gain in the amplified spontaneous emission mode (no microwave input signal) was measured and it was found that the microwave signal grew at a rate of 13.4 dB per meter for a beam current of 450 A. Extrapolating this growth back to the origin, one finds that the effective input noise was 0.35 W.

The amplifier gain was studied both as a function of undulator magnetic field intensity and as a function of undulator length. The peak output power of 80 MW was achieved for both the 2 m and 3 m long undulator. The amplifier went into saturation at 2.2 m; beyond this point, the amplified output power first decreased and then near 3 m started to increase again. The gain as a function of undulator length showed an exponential gain of approximately 15.6 dB m$^{-1}$ up to saturation. This was in close agreement with the small signal gain measurement. The gain curves for the 1-m and 2-m undulators are relatively symmetric about the peak, while the gain curve for the 3 m long wiggler shows a marked asymmetry with a plateau on the long wavelength side of the curve. This asymmetry at saturation is also shown in the numerical simulations.

Study of excitation of other than the $TE_{01}$ mode and study of the effect of varying the undulator parameters (so as to avoid saturation at 80 MW) are to be undertaken in the near future. What has been shown, so far, is that a FEL can be operated in the high gain regime (Gain $\gtrsim$ 2500).

## 3.3    *Storage Rings*

The first, and so far the only, operation of a storage ring FEL oscillator was achieved by the Orsay-Stanford collaboration using the Orsay ring ACO (16, 89). This laser operated in the visible range, at 6500 Å, and produced 75 $\mu$W average power or 60 mW output peak power. The intracavity peak optical power was 2 kW.

The ACO storage ring has a circumference of 22 m and was operated between 160 and 166 MeV. Two bunches were employed, with the average current between 16 and 100 mA. The rms bunch length was (in time units) 0.5 to 1 ns and the energy spread (rms) 0.9 × 10$^{-3}$ to 1.3 × 10$^{-3}$. Because of the strong radiation damping, the transverse size (rms) was 0.3 to 0.5 mm, corresponding to an angular spread of 0.1 to 0.2 mrad.

The optical cavity was 5.5 m long so the round-trip time resonated with the 11 m between electron bunches. The mirror radius was 3 m, the Rayleigh range 1 m. Although the mirror transmission was only 3 × 10$^{-5}$, the round-trip cavity loss was 7 × 10$^{-4}$, primarily because of absorption in the mirror dielectric. In fact, there was mirror degradation due to the radiation harmonics of the undulator, which forced the experimentalists to operate ACO at a reduced energy (originally they had expected to be at 240 MeV) and to operate the undulator at reduced magnetic field ($K = 1.1$ to 1.2), both effects tending, of course, to reduce the flux at higher harmonics.

The permanent magnet undulator had 17 periods with a period of 7.8 cm, and a total length of 1.33 m. It was operated as an optical klystron in order to increase the gain per pass. This increased the gain by about a factor of 2 to 7 so as to reach 2 × 10$^{-4}$ per pass. Lasing with such low gain required careful alignment of the electron beam onto the axis of the optical cavity, high quality mirrors, as well as precise synchronism between the light pulse reflections and the electron bunch revolution frequency. The detuning curve gave only a 1.6-$\mu$m full width at half maximum near laser threshold.

The laser time pulse structure was a series of pulses and showed the electron rf synchrotron frequency (13 kHz), and the 27.2-MHz bunch frequency. The time sequence of pulses is understood as a consequence of theoretical study (90). In frequency space the laser had three lines (near 6500 Å) with the dominant one at 6476 Å. All the lines, corresponding to maximum gains in the klystron FEL, were in the TEM$_{00}$ mode. The width of the lines was 2–4 Å. Tunability was over 150 Å and limited by mirror reflectivity.

The storage ring FEL is the only configuration mentioned where the FEL feeds back on the electron source. On each pass the working FEL "heats" the electron beam by introducing an energy spread. Synchrotron radiation $P_{syn}$ from the bending magnets in the ring damps the excitations. The laser power at saturation is determined by thermodynamic equilibrium, which results in weak fields; this is the Renieri limit (91), $P_{laser} = P_{syn}/2N$. The efficiency of the FEL was only $2.4 \times 10^{-5}$, which is 0.4 of the prediction of Renieri for this case.

## 3.4   *Extensions*

We have seen that FELs can be expected to be efficient, powerful, reliable, tunable sources of radiation in a wide range of wavelengths. In fact, FELs have already been made to operate from the microwave range down to the visible range. It is reasonable to expect that soon we shall have FELs readily available, for many different applications, from microwave wavelengths to soft x-ray wavelengths. When augmented with atomic and molecular lasers and conventional radio tube sources, we can expect to have coherent radiation sources throughout the radiation spectrum (currently, one can see one's way to 300 Å).

Why then should one develop even more devices? Clearly, because they can be designed for special purposes, have special properties, be less expensive, more efficient, etc. The development of FELs is far from completed and really only starting; a number of extensions of FELs appear to be possible. Here, we mention a few of them and refer the interested reader to the appropriate literature.

In the microwave range it is possible to apply a longitudinal magnetic field of sufficient strength that the cyclotron frequency resonates with the radiation frequency. Thus one can arrange a device where there is coincidence between the FEL resonance and the cyclotron resonance as described in Section 2.3 (12, 92, 93).

It is possible to replace the undulator with an electromagnetic field. The attainable magnetic field of an rf wave is less than that of a static or pulsed magnetic field, but the wavelength of the "undulator" can be made less than that of a conventional undulator. Thus, one can get to short wavelengths with a low energy electron beam. The use of an rf wave as an undulator has already been demonstrated (94) and demonstration has been made of an electromagnetic wave undulator FEL by an NRL group (95). This group had the electron beam produce 500 MW of 12.5-GHz radiation through a backward wave oscillator mechanism, and then used this radiation as an undulator for FEL action. In this manner they produced 200-GHz radiation with peak power, not yet optimized, of 0.35 MW. The Santa Barabara group (96) plans to employ the same idea, but use the FEL mechanism to generate the rf field of an "undulator" in a "two-stage FEL."

We have concentrated upon so-called "Compton regime FELs" where there is a strong interaction between the electrons and the optical wave, but where the interaction between electrons is small. In the opposite case, where the electrons interact strongly through Coulomb forces so that a density fluctuation, or plasmon, description of the electron beam is more appropriate, the FEL is said to be in the "Raman regime." An understanding of the collective regime, the Raman regime, is more difficult than that of the Compton regime but offers distinctive features. Experiments (12) have demonstrated 6% conversion efficiency, and large power emission (75 MW) in this regime. One can expect more development of these devices in future years (97).

An interesting extension of the FEL is to operation in a dielectric media (98, 99). Gas loading, for this is the manner proposed to realize the dielectric media, changes the phase-matching condition and so allows a wider parameter space than the vacuum FEL. In fact, this extension can be nontrivial and would appear to allow operation, for example, at smaller undulator magnetic fields than in the conventional FEL. The resonance condition, for relativistic electrons is $n - 1 + \lambda/\lambda_u = (1 + K^2)/2\gamma^2$, for a medium having an index of refraction $n$. Note that the $(n - 1)$ term can easily be comparable to the usual $(1/2\gamma^2)$ FEL term. One can think of this device as being a suitable combination of the Čerenkov effect and the FEL resonance.

Another interesting extension of a conventional FEL is to have an undulator in an isochronous storage ring (100, 101) in which particles with different energies take exactly the same time to go around the ring. Thus bunching at optical wavelengths is preserved around the ring. Most rings do not have this property and thus the electron bunch on entering the undulator is essentially a "new bunch" with random phases. Rings can be made isochronous, to some degree, so that the bunching of a FEL can be preserved. Clearly this is advantageous, and it can be done so as to preserve far-infrared wavelength bunching as has been shown on BESSY (102). A FEL using this concept has not yet been made; it is doubtful that the technique can be extended into the visible, but for the infrared it could make a very interesting device.

Finally, it should be emphasized that "pushing" FELs to shorter and shorter wavelengths, as has been spearheaded by J. M. J. Madey and C. Pellegrini, may require no "new inventions," but nevertheless be difficult and a significant extension. This subject, as one might expect, has received considerable attention (85, 87, 103, 104). Suffice it to say here that it appears possible to construct a FEL oscillator down to about 500 Å, and a single-pass FEL growing from noise to about 300 Å. Just what the limits are remains to be seen, but extending the Orsay achievement by an order of magnitude appears to be possible.

ACKNOWLEDGMENTS

We are grateful for support from the US Air Force Office of Scientific Research Grant No. AFOSR-84-0079 (W.B.C.) and from the Office of Energy Research, US Department of Energy under contract DE-AC03-76SF 00098 (A.M.S.).

*Literature Cited*

1. Madey, J. M. J. *J. Appl. Phys.* 42:1906 (1971)
2. Madey, J. M. J. Stimulated emission of radiation in periodically deflected electron beam. *US Patent 3,822,410* (1974)
3. Elias, L. R., Fairbank, W. M., Madey, J. M. J., Schwettman, H. A., Smith, T. I. *Phys. Rev. Lett.* 36:717 (1976)
4. Deacon, D. A. G., et al. *Phys. Rev. Lett.* 38:892 (1977)
5. Halbach, K. See Ref. 47, p. C1-211
6. Blewett, J. P., Chasman, R. *J. Appl. Phys.* 48:2692 (1977)
7. Elias, L. R., Madey, J. M. J. *Rev. Sci. Instrum.* 50:1339 (1979)
8. Diament, P. *Phys. Rev. Lett. A* 23:2537 (1981)
9. Kim, K.-J. *Nucl. Instrum. Method* 219:425 (1984)
10. Colson, W. B. *Free electron laser theory*, PhD dissertation, Stanford Univ. (1977)
11. Deacon, D. A. G., et al. *Phys. Rev. Lett.* 38:892 (1977)
12. Gold, S. M., et al. See Ref. 48, p. 350
13. Newnam, B. E., et al. *Phys. Rev. Lett.* 38:118 (1977); Newnam, B. E., Hohla, K., Warren, R., Goldstein, J. C. *IEEE J. Quantum Electron.* QE17:1480 (1981)
14. Grossman, W. M., et al. *Phys. Rev. Lett.* 38:52 (1977)
15. Edighoffer, J. A., et al. *Phys. Rev. Lett.* 52:344 (1984)
16. Billardon, M., et al. *Phys. Rev. Lett.* 51:1652 (1983)
17. Pasour, J. A., Lucey, R. F., Kapetanakos, C. A. *Phys. Rev. Lett.* 53:1728 (1984)
18. Orzechowski, T. J., et al. *Phys. Rev. Lett.* 54:889 (1985)
19. Fajans, J., Bekefi, G., Yin, Y. Z., Lax, B. *Phys. Rev. Lett.* 53:246 (1984)
20. Robinson, A. L. Reported in *Science* 226:154 (1984)
21. Dolezal, F., Harvey, R., Palmer, A. Paper presented at *9th Int. Conf. on Infrared and Millimeter Waves*, Osaka (Oct. 1984)
22. Barbini, R., et al. See Ref. 47, p. C1-1
23. McDermott, B. D., et al. *Phys. Rev. Lett.* 41:1368 (1978)
24. Omatuni, T. S., Petrosian, M. L., Petrosian, B. V., Shahbazian, N. T., Oganesian, K. B. *Erevan EPI-727 (42)-84* (unpublished 1984)
25. Marshall, T., et al. *Appl. Phys. Lett.* 31:320 (1977)
26. Granatstein, V., et al. *Appl. Phys. Lett.* 30:384 (1977)

27. Artamonov, et al. *Nucl. Instrum. Methods* 177:247 (1980); Kornyuikhin, G. A., et al. Inst. Nucl. Phys., Novosibirsk, Internal Report (unpublished)
28. Rothenberg, M. S. *Phys. Today* 29:17 (Feb. 1976)
29. "*The Free Electron Laser.*" *Sci. Am.* 236:62 (Jun. 1977)
30. Schwarzchild, B. *Phys. Today* 36:17 (Dec. 1983)
31. *Lasers and Applications* 2:40 (Oct. 1983)
32. Robinson, A. *Science* 221:937 (Sep. 1983)
33. Robinson, A. *Science* 226:153 (Oct. 1984)
34. Schwarzschild, B. *Phys. Today* 37:19 (Nov. 1984)
35. Schwarzschild, B. *Phys. Today* 37:21 (Nov. 1984)
36. Robinson, A. *Science* 226:1300 (Dec. 1984)
37. Thomsen, D. E. *Sci. News* 126:359 (Dec. 1984)
38. Marshall, T. C. *Free Electron Lasers.* New York: MacMillan (1985)
39. Prosnitz, D. "Free electron lasers." In *CRC Handbook of Laser Science and Technology 1*, ed. M. J. Weber, p. 425. Boca Raton: CRC Press (1982)
40. Pellegrini, C. *IEEE Trans. Nucl. Sci.* NS-26:3791 (1979); Morton, P. L. *IEEE Trans. Nucl. Sci.* NS-28:3125 (1981)
41. *IEEE J. Quantum Electron.*, Vol. QE-17, No. 8 (Aug. 1981)
42. *IEEE J. Quantum Electron.*, Vol. QE-19, No. 3 (Mar. 1983)
43. Jacobs, S. F., Sargent, M. III, Scully, M. O., eds. Novel sources of coherent radiation. In *Physics of Quantum Electronics*, Vol. 5. Reading, Mass: Addison-Wesley (1978)
44. Jacobs, S. F., Pilloff, H. S., Sargent, M. III, Scully, M. O., Spitzer, R., eds. Free electron generators of coherent radiation. In *Physics of Quantum Electronics*, Vol. 7. Reading, Mass: Addison-Wesley (1980)
45. Jacobs, S. F., Pilloff, H. S., Sargent, M. III, Scully, M. O., Spitzer, R., eds. Free electron generators of coherent radiation. In *Physics of Quantum Electronics*, Vol. 8. Reading, Mass: Addison-Wesley (1982)
46. Jacobs, S. F., Pilloff, H. S., Sargent, M. III, Scully, M. O., Spitzer, R., eds. Free electron generators of coherent radiation. In *Physics of Quantum Electronics*, Vol. 9. Reading, Mass: Addison-Wesley (1982)
47. *Bendor Free Electron Laser Conference*, *J. Phys. Colloq.* C1, Suppl. 2, p. 44 (Feb. 1983)
48. Brau, C. A., Jacobs, S. F., Scully, M. O., eds. Free electron generators of coherent radiation, *SPIE* (Int. Soc. Opt. Eng.) 453 (1984)
49. Motz, H., Thon, W., Whitehurst, R. N. *J. Appl. Phys.* 24:826 (1953)
50. Phillips, R. M. *IRE Trans. Electron Devices* 17:231 (1960)
51. Schawlow, A. L., Townes, C. H. *Phys. Rev.* 112:1940 (1958); Maiman, T. H. *Phys. Rev. Lett.* 4:564 (1960)
52. Palmer, R. B. *J. Appl. Phys.* 43:3014 (1972); Csonka, P. *Part. Accel.* 8:225 (1978); Robinson, K., unpublished notes
53. Hopf, F. A., Meystre, P., Scully, M. O., Louisell, W. H. *Phys. Rev. Lett.* 37:1342 (1976)
54. Colson, W. B. *Phys. Lett.* 64A:190 (1977); Colson, W. B. Ride, S. K. *Phys. Lett.* 76A:379 (1980); Tang, C.-M., Sprangle, P. *J. Appl. Phys.* 53:831 (1982); Shih, C.-C., Yariv, A. *IEEE J. Quantum Electron.* QE-17:1387 (1981); Dattoli, G., et al. *IEEE J. Quantum Electron.* QE-17:1371 (1981); Bonifacio, R., Pellegrini, C., Narducci, L. M. *Opt. Commun.* 50:373 (1984)
55. Madey, J. M. J. *Nuovo Cimento B* 50:64 (1979)
56. Kroll, N. M., Morton, P. L., Rosenbluth, M. N. *IEEE J. Quantum Electron.* QE-27:1436–68 (1981)
57. Vinokurov, N. A., Shrinsky, A. N. Inst. Nucl. Phys., Novosibirsk, *Rep. No. INP 77-59* (unpublished) (1977)
58. Luccio, A., Krinsky, S. See Ref. 45, p. 181 (1982)
59. Dattoli, G., Renieri, A. Experimental and theoretical aspects of the free electron laser. In *Laser Handbook 4*, ed. M. L. Stitch, M. S. Bass. Amsterdam: North Holland (To be published)
60. Scharlemann, E. T. *Lawrence Livermore National Laboratory ELF Note 105P* (published) (1984)
61. Kroll, N., Rosenbluth, M. See Ref. 44, p. 147
62. Fawley, W. M., Prosnitz, D., Scharlemann, E. T. submitted *Phys. Rev. A* 30:1472 (1984); *Lawrence Livermore Natl. Lab. Rep. UCRL90838* (unpublished) (1984)
63. Scharlemann, E. T., Sessler, A. M., Wurtele, J. S. *Proc. Como Conf., Nucl. Instrum. Methods.* In press; *Lawrence Livermore Natl. Lab. Rep. UCRL-91476* (unpublished) (1984)

64. Scharlmann, E. T., Sessler, A. M., Wurtele, J. S. *Phys. Rev. Lett.* 54:1925 (1985)
65. Moore, G. T. *Opt. Commun.* 52:46 (1984); Moore, G. T. *Proc. Como Conf., Nucl. Instrum. Methods.* In press; *IMO Rep. 84-14* (unpublished) (1984)
66. Colson, W. B., Elleaume, P. *Appl. Phys.* B 29:1–9 (1982)
67. Colson, W. B., Renieri, A. See Ref. 47, pp. C1–11
68. Al-Abawi, H., Hopf, F. A., Moore, G. T., Scully, M. O. *Opt. Commun.* 30:235 (1979)
69. Eckstein, J. N., et al. See Ref. 45, p. 49
70. Benson, S., Deacon, D. A. G., et al. *Phys. Rev. Lett.* 48:235 (1982)
71. Kroll, N. M., Rosenbluth, M. N. See Ref. 44, p. 147
72. Colson, W. B. *IEEE J. Quantum Electron.* QE-17:1417–27 (1981)
73. Fessenden, T. J., et al. In *Proc. 4th Int. Conf. on High Power Electron and Ion Beam Research and Technology*, ed. H. J. Doucet, J. M. Buzzi, p. 813. Palaiseau, France: Ecole Polytech.
74. Briggs, R. J., et al. *IEEE Trans. Nucl. Sci.* NS-28:3360 (1981)
75. Takeda, S. Workshop on Generation of High Fields, Frascati, Sep. 25–Oct. 1, 1984. *Nucl. Instrum. Methods.* In press (1985)
76. Warren, R. M., et al. *Lasers 84 Tech. Dig.*, p. 20 (1984)
77. Elias, L. *Lasers 84 Tech. Dig.*, p. 25 (1984)
78. Boehmer, H., et al. *Phys. Rev. Lett.* 48 141 (1982)
79. Birx, D. L., et al. A multipurpose 5-MeV linear induction accelerator. In *Proc. IEEE Power Modulator Symp.*, June 1984; *Lawrence Livermore Natl. Lab. Rep. UCRL-90554* (unpublished) (1984)
80. Shaw, E. D., Chichester, R. J., Sprenger, W. C. *Lasers 84, Tech. Dig.*, p. 19 (1984)
81. Smith, S. D., et al. See Ref. 45, p. 275
82. Bizzarri, U., et al. *Lasers 84 Tech. Dig.*, p. 21 (1984)
83. Leiss, J., Norris, N. J., Wilson, M. A. *Part. Accel.* 10:223 (1980)
84. Billardon, M., et al. See Ref. 15, p. 1652
85. Deacon, D. A. G. *Lasers 84 Tech. Dig.*, p. 27 (1984)
86. Ballili, J. P., et al. *Report of Prospectives for Super ACO, LURE*, Orsay, Internal Report (unpublished) (April 1981)
87. Peterson, J. M., et al. Contributed paper to the Castelgandolfo Conf., *Nucl. Instrum. Methods in Physics Research.* To be published (1985)
88. Orzechowski, T. J., et al. See Ref. 48, p. 65
89. Billardon, M., et al. See Ref. 47, pp. C1–29
90. Elleaume, P. *J. Phys.* 45:997 (1984)
91. Renieri, A. *Nuovo Cimento* B 59:64 (1979)
92. Freund, H. P. See Ref. 48, p. 361; Bernstein, I. B., Friedland, L. *Phys. Rev.* A 23:816 (1981)
93. Lin, A. T. See Ref. 48, p. 336; Friedland, L., Hirschfield, J. L. *Phys. Rev. Lett.* 44:1456 (1980)
94. Tsumoru, S., et al. *Jpn. J. Appl. Phys.* 22:844 (1983)
95. Granatstein, V. L., Carmel, Y., Gover, A. See Ref. 48, p. 344
96. Elias, L. B. *Phys. Rev. Lett.* 42:977 (1980)
97. Freund, H. P., Ganguly, A. K. See Ref. 48, p. 367
98. Kimura, W., Wong, D., Piestrup, M., Fauchet, A., Edighoffer, J., Pantell, R. *IEEE J. Quantum Electron.* QE-18:239 (1982); Walsh, J., Johnson, B. See Ref. 48, p. 376
99. Fauchet, A. M., et al. See Ref. 48, p. 423
100. Deacon, D. A. G. *Phys. Rev. Lett.* 44:449 (1980)
101. Deacon, D. A. G. *Phys. Rep.* 76:349 (1981)
102. Gaupp, A. See Ref. 47, pp. C1–147
103. Madey, J. M. J., Pellegrini, C., eds. *AIP Conf. Proc.*, Vol. 118, Brookhaven, Sep. 1983, New York (1984)
104. Proc. Conf. on Coherent and Collective Properties in the Interaction of Relativistic Electrons and Electromagnetic Radiation, Como, Sep. 1984. *Nucl. Instrum. Methods.* In press (1985)

*Ann. Rev. Nucl. Part. Sci. 1985. 35 : 55–75*

# HADRON-NUCLEUS COLLISIONS AT VERY HIGH ENERGIES

## K. Zalewski

Institute of Nuclear Physics, 30 055 Kraków, Poland

CONTENTS

## 1.  INTRODUCTION

Thousands of papers have been written on the high energy (multi-GeV region) hadron-nucleus interactions. In this short review we present some results that seem particularly relevant. The selection, necessarily subjective, is biased toward effects where new understanding has been gained and/or

0163–8998/85/1201–0055$02.00

differences from scattering on nucleons are particularly striking. Many references, especially to older papers, have been omitted and, therefore, each reference implies "and references contained therein."

For the very large amount of work invested by many physicists into the study of high energy hadron-nucleus interactions, the following motivations are usually given.

## 1.1    New Physics

In experiments with nuclei, new phenomena are expected. For instance, Burnett et al (1) described two events: $Si + Ag$ or Br and $Ca + C$. The laboratory energies of the incident nuclei were 4 and 100 TeV/nucleon and the multiplicities of charged secondaries were 1015 and 760. The authors' estimate of the energy density in the central region is 4 GeV/fm$^3$. This is a very large number. In the same units the energy density in nuclear matter is 0.15 and in a proton it is 0.47. Estimates of the energy density necessary for the transition from hadronic matter to a quark-gluon plasma range from 0.6 to 1.0 (1). Moreover, the high energy density region can extend over many tens of cubic fermis and live a time of several fm/$c$ (2). This shows that the study of high energy hadron-nucleus interactions gives access to unexplored and potentially interesting fields of physics. The trouble is that the expected experimental signatures for plasma formation, for example, are not very clear (3). On the other hand, cosmic ray physicists report various spectacular effects that no one had expected (4, 5).

## 1.2    New Insights into Old Physics

According to some experts, hadron-nucleus interactions can supply information about elementary interactions, or nuclei, that is not available otherwise. Such analyses are model dependent. Assuming, as is now customary, that the scattering on a nucleus is a series of interactions with single nucleons constituting the nucleus, it makes sense to ask how the objects produced on a nucleon behave *in statu nascendi*, i.e. during the first say $10^{-23}$ sec (laboratory time) after their creation. The classical cascade model, in which they behave like the particles observed in the final state, grossly overestimates the final multiplicity and is considered excluded by the data.

In this branch of physics the inconsistency of a model with experiment is often controversial. The phenomena are complicated and all models are presented as crude first approximations. Thus, it is usually possible to avoid any specific contradiction with experiment by introducing corrections. Demonstrations that a model is wrong are often interpreted by the authors of the model as proof that their main idea was completely misunderstood. Models die slowly. It is only when they become extremely implausible and/or complicated that nobody cares to defend them any more. For one of the last attempts to defend the classical cascade model, see (6).

From the failure of the classical cascade model one concludes that there is a formation time between the interaction and the time when the final hadron with its standard properties appears. In the laboratory, because of time dilatation, this time is long for fast objects, which therefore undergo little or no rescattering in the nucleus, and short for slow objects, which should interact as suggested by the classical cascade model (7). The transition region between these two situations is very diffuse. The uncertainty principle estimate that the momenta in the transition region should be of the order of $p_c = m^2 R$ (where $R$ is the nuclear radius and $m$ is a mass of the order of particle masses and transverse momenta) is reasonable. Estimates of $m$ are model dependent. A typical estimate is $m^2 = \langle m_0^2 + p_T^2 \rangle$, where $m_0$ denotes the mass of the object produced (pion, vector meson, cluster, etc), $p_T$ is its transverse momentum, and the brackets $\langle \ \rangle$ denote averaging. Very crudely, $0.2 \lesssim m \lesssim 1$ GeV. The weakness of estimating $p_c$ is that it is not known whether it should be applied to particles, clusters, partons, or other objects. It has been claimed that a classical cascade model modified only by the assumption that particles cannot interact before a time $m^2/p$ after their creation leads to contradiction with data on hadron-deuteron (h-d) scattering (8). Within specific models it is possible to extract from data good estimates of the formation time (or zone) of hadrons (9).

## 1.3   Conservative Motivation

Hadron-nucleus interactions are interesting on their own merit. One should keep in mind, however, that experiments with nuclear targets are also made in order to extract information about interactions with nucleons. Sometimes it is impossible to use an elementary target, or an elementary projectile. This is the case in most cosmic ray research, where the highest energy interactions are being observed (4, 5), or in the study of hadron-neutron interactions. It may be more economical to use nuclear targets, which compared to hydrogen targets yield larger cross sections for the same target volume and usually do not require cryogenics. Finally, some nuclear targets, such as nuclear emulsions, or more recently "live targets" made of silicon, have valuable detector properties.

Once the practical importance of nuclear targets is recognized, it is necessary to know how to extract hadron-nucleon data from hadron-nucleus data. The standard procedure has been to assume that, whatever cross section is measured, it behaves as $A^\alpha$, where $A$ is the atomic mass number and $\alpha$ is a constant that is either estimated, or obtained by comparing the results for two or more nuclei. The problem is far from trivial. For instance, at 100 GeV/$c$ incident momentum, the inelastic p-A cross sections for $A > 10$ are well fitted by $\sigma_A = 38.2\, A^{0.72}$ mb (10), but for scattering on hydrogen $\sigma_1 = 31.4$ mb and not 38.2 mb. Reference (10) contains more examples of this kind. A direct proof that the nucleus cannot always be approximated as a cluster of free nucleons is supplied by the EMC effect (11): the (isoscalar) structure function of the nucleon obtained in the standard way from $\mu$-Fe scattering is very significantly different from the same function obtained by the same method from $\mu$-d scattering.

## 2.   TOTAL AND ELASTIC CROSS SECTIONS

### 2.1   *Glauber's Approximation*

Glauber's method of calculating total and elastic cross sections works well for high, but not too high, energies (say, tens of GeV) and, in the case of elastic scattering, for not too large scattering angles. At higher energies (11a) and larger angles (12) difficulties appear. The model has been reviewed many times (13, 13a). Here we concentrate on its interpretation. Comparison with experiment at high energies requires the inclusion of inelastic screening and is discussed in Section 2.3.

In this approach a hadron h propagates across the nucleus along a straight line, just like a light ray in geometrical optics. The $S$ matrix can be written as

$$S_A = \int d^2b \, \langle \exp[i\delta_A(\mathbf{b})]\rangle = \int d^2b \langle S_A(\mathbf{b})\rangle, \qquad 1.$$

where the phase shift $\delta_A(\mathbf{b})$ and $S_A(\mathbf{b})$ are calculated for fixed values of the impact parameter $\mathbf{b}$ and of the positions $\mathbf{r}_1,\ldots,\mathbf{r}_A$ of the target nucleons. Thus the argument $\mathbf{b}$ stands for $\mathbf{b}, \mathbf{r}_1,\ldots,\mathbf{r}_A$. The averaging $\langle \ \rangle$ is over $\mathbf{r}_1,\ldots,\mathbf{r}_A$. The phase shift $\delta_A(\mathbf{b})$ may contain a real part, but it is mostly imaginary (absorption). It is the sum of (complex) phase shifts on all the $A$ nucleons in the nucleus. This again is just like in geometrical optics: When a light ray crosses several layers of gray glass, the absorption is the sum of absorptions on the single layers. "Elastic screening," which in Glauber's approach yields $\sigma_{hA} < A\sigma_{hN}$, has its counterpart in optics: Since a ray is partially absorbed in each layer, further layers are reached by weaker beams and consequently absorb less. The phase shifts on single nucleons are obtained from h-N scattering data. Thus the only nuclear input is the distribution of nucleons in the nucleus $\rho(\mathbf{r}_1,\ldots,\mathbf{r}_A)$. Here for simplicity we neglect the difference between protons and neutrons. The formulae are most easily written in terms of the nuclear profiles $\gamma_A(\mathbf{b}) = 1 - S_A(\mathbf{b})$. Their averages are related to the elastic scattering amplitudes by

$$F_A(\mathbf{q}) = i \int d^2b \langle \gamma_A(\mathbf{b})\rangle \, \exp(i\mathbf{q}\cdot\mathbf{b}), \qquad 2.$$

where $\mathbf{q}$ is the momentum transfer. The assumption about the additivity of the phase shifts means that

$$\gamma_A(\mathbf{b}) = 1 - \left\langle \prod_{j=1}^{A} [1 - \gamma_1(\mathbf{b} - \mathbf{b}_j)] \right\rangle, \qquad 3.$$

where $\mathbf{b} - \mathbf{b}_j$ is the impact parameter of the projectile with respect to the center of nucleon $j$, and $\gamma_1(\mathbf{b})$ is the profile of the nucleon, which can be found by inverting Equation 2 and using the experimental h-N elastic amplitude.

From the optical theorem the total cross section is

$$\sigma_{\text{hA}}^{\text{tot}} = 2 \int d^2 b \, \text{Re} \langle \gamma_A(\mathbf{b}) \rangle \qquad 4.$$

and the cross section for inelastic scattering is

$$\sigma_{\text{hA}}^{\text{in}} = \int d^2 b (1 - \langle |S_A(\mathbf{b})|^2 \rangle) = \int d^2 b \langle \sigma_A^{\text{in}}(\mathbf{b}) \rangle. \qquad 5.$$

Using the additivity of phase shifts, we have

$$\int d^2 b \langle \sigma_A^{\text{in}}(\mathbf{b}) \rangle = \int d^2 b \left\{ 1 - \left\langle \prod_{j=1}^{A} [1 - (1 - |S_1(\mathbf{b} - \mathbf{b}_j)|^2)] \right\rangle \right\}. \qquad 6.$$

An analogous formula holds for the inelastic nondiffractive cross section (14).

## 2.2  Interpretational Problems for $\sigma_{\text{hA}}^{\text{in}}$

Equation 6 strongly suggests a probabilistic interpretation of Equation 5 (15). Let us assume for simplicity that the nucleons in the nucleus are independent:

$$\rho(\mathbf{r}_1, \ldots, \mathbf{r}_A) = \prod_{j=1}^{A} \rho(\mathbf{r}_j). \qquad 7.$$

Then, using Equation 5 for A = nucleon, we may rewrite Equation 6 as

$$\int d^2 b \langle \sigma_A^{\text{in}}(\mathbf{b}) \rangle = \int d^2 b \sum_{v=1}^{A} \binom{A}{v} \langle \sigma_{\text{hN}}^{\text{in}}(\mathbf{b}) \rangle^v [1 - \langle \sigma_{\text{hN}}^{\text{in}}(\mathbf{b}) \rangle]^{A-v}. \qquad 8.$$

On the other hand, from a simple modification of Bernoulli's well-known formula, if for given $\mathbf{b}$ the (suitably normalized) probability of hitting a given nucleon is $\sigma_{\text{hN}}^{\text{in}}(\mathbf{b})$, then the probability of hitting exactly $v$ out of $A$ independent nucleons, with the assumption that at least one nucleon must be hit, is (15)

$$P_v = \int d^2 b \binom{A}{v} \langle \sigma_{\text{hN}}^{\text{in}}(\mathbf{b}) \rangle^v (1 - \langle \sigma_{\text{hN}}^{\text{in}}(\mathbf{b}) \rangle)^{A-v} \bigg/ \int d^2 b \langle \sigma_A^{\text{in}}(\mathbf{b}) \rangle. \qquad 9.$$

Multiplying both sides of Equation 9 by $\int d^2 b \langle \sigma_A^{in}(\mathbf{b}) \rangle$ and summing over $v$, one recovers Equation 8. It is therefore tempting to take Equation 9 seriously as the probability that the incident hadron collides inelastically with exactly $v$ target nucleons. Then for instance, the average number of collisions is

$$\langle v \rangle = A \sigma_{hN}^{in}/\sigma_{hA}^{in} = A \sigma_{hN}/\sigma_{hA}, \qquad \qquad 10.$$

where the second equality introduces the more common notation. For experimental data on h-A cross sections, see (16, 17).

Note that Equation 9 as a physical assumption goes beyond the standard Glauber model. A similar "derivation" could be given for total cross sections, when $\gamma$ is real, but since it is well known that, for example, single and double elastic scattering interfere, nobody would believe the result. The most plausible of formulae like Equation 9 refers to inelastic nondiffractive scattering (14). In the literature, however, various interpretations of the cross sections in Equation 10 occur. They may be inelastic cross sections, or cross sections for particle production.

Equations 9 and 10 are widely used, but at least in the framework of the parton model their interpretation requires some care. Glauber's model was derived originally from quantum mechanics (e.g. 13). It is also possible, however, to derive it from a quantum field theory of partons (18). Using the so-called AGK cutting rules (19) and some plausible, though not obviously compelling, simplifying assumptions, it is possible to derive from the same model the probabilistic Equation 9 (20). This derivation exposes two facts not visible in an elementary probabilistic derivation.

First, in spite of the probabilistic form of the final formula, interference effects are crucial to derive it. In the probabilistic derivation the probability of an inelastic interaction with exactly one nucleon is smaller than $A \sigma_{hN}^{in}(\mathbf{b})/\sigma_A^{in}(\mathbf{b})$, because it contains as a factor the probability of no interaction with the remaining nucleons. In the parton picture, the same effect is obtained by destructive interference between the amplitude for inelastic interaction with one nucleon and no interactions with the others and the amplitudes for inelastic interaction with the same nucleon accompanied by elastic interactions with one or more of the remaining nucleons.

Second, all $v$ interactions must take place simultaneously! Thus, the notion that an incident elementary hadron enters the nucleus, interacts with one nucleon, then propagates freely for a while, interacts with another nucleon, etc is excluded. This simultaneity of interactions, known more technically as domination of nonplanar diagrams, has a well-known physical interpretation. A fast incident patron cannot interact directly with a slow target parton. It can, however, develop a ladder of slower and slower partons until the partons at the bottom are sufficiently slow to interact. This takes some time $\tau_0$ in the rest frame of the incident particle. In the laboratory frame, the necessary time is $\gamma\tau_0$, where $\gamma$ is the Lorentz factor of the incident hadron. At high energies this time is much longer than the time necessary to cross the nucleus. Therefore, the probability of developing two ladders consecutively, while flying across the nucleus, tends to zero with increasing energy. A simple formal proof in the framework of the Reggeon model can be found in (21).

Analogues of Equation 9 can also be written when $\rho$ does not factorize (15). This is particularly important for deuterium, where the two nucleons are strongly correlated by momentum conservation. Deuterium is an especially interesting case, because the probability $P_2$ is in this case measurable in experiment (cf Section 5.1).

One more conclusion from the field-theoretical model is of possible interest. This model suggests that the approach discussed here is suitable only at intermediate energies, while at very high energies a completely different picture, corresponding to multiple Pomeron couplings, takes over. There is no reliable estimate of the energy where this transition could take place.

## 2.3   Inelastic Screening

Suppose that an incident hadron scatters inelastically on one of the nucleons in the nucleus: h $\rightarrow$ h*, where h* is something different from h, e.g. a resonance or a group of particles. It may happen that after scattering on another nucleon h* becomes h again. Such processes, known as inelastic screening, reduce not only the h-A inelastic cross section, but also the elastic and the total cross sections. Their importance was pointed out in (18, 22). It is easy to see that the contribution from inelastic screening to the elastic cross section, and via the optical theorem to the total cross section, increases with energy: The momentum transfers to the nucleus in h $\rightarrow$ h* and h* $\rightarrow$ h must be small, because otherwise the nucleus would break up. The energy difference $E_{h*} - E_h$ (laboratory frame) is therefore limited and so is $(m_{h*})^2 \lesssim m_h^2 + 2p_{in}(E_{h*} - E_h)$. According to this formula, however, the mass range allowed for h* increases with increasing momentum $p_{in}$ of the incident particle. Consequently, at higher energies more inelastic screening processes contribute.

A complete calculation of the inelastic screening corrections is not possible for the moment. For anything more complicated than deuterium the technical difficulties are prohibitive. Even for deuterium there are theoretical uncertainties (23). Some input amplitudes refer to scattering on single nucleons but are not measurable there. There is a contribution, the form factor $\varepsilon$ from (23), that is neglected because it is probably small, but that cannot be reliably calculated. Nevertheless, approximate evaluations of the effect of inelastic screening are possible (24) and significantly improve the agreement between theory and experiment.

Applications of such calculations include neutron-nucleus scattering at $p_{in} = 30$–300 GeV/$c$ (25), p-d and d-d scattering at center-of-mass energies $\sqrt{s} = 53$ and 63 GeV respectively (26); $\alpha$-$\alpha$ and p-$\alpha$ scattering at $\sqrt{s} = 125$ and 88 GeV respectively (27); and $\pi$-He and p-He scattering at $p_{in} = 50$–300 GeV (28). Let us stress two particularly instructive cases.

The regeneration amplitude $K_L^0 \to K_S^0$ on nuclei in the energy range 30–130 GeV can be fitted without inelastic screening. It is necessary, however, to choose for the intercept of the $\omega$ Regge trajectory $\alpha_\omega(0) = 0.39 \pm 0.01$ instead of the usual $0.44 \pm 0.01$ (29). Including inelastic screening, one reproduces the standard value of $\alpha_\omega(0)$ (30–32).

When extracting h-n cross sections from h-d cross sections, it is possible to use Glauber's formulae without inelastic screening, e.g. for total cross sections (33):

$$\sigma_{hn} = \sigma_{hd} - \sigma_{hp} + \sigma_{hp}\sigma_{hn} f/(4\pi). \qquad 11.$$

Here $f$ is the inverse square of the distance between the two nucleons in deuterium averaged over a convolution of the deuteron wave function with the elastic h-N profile function. One expects $f \approx 0.027$ mb$^{-1}$, while the fits at high energies use much larger values. Again the difficulty disappears when inelastic screening is included (31, 34). A formula accurate to about 0.3 mb for energies above 50 GeV is (34)

$$\sigma_{hn} = \sigma_{hd} - \sigma_{hp} + \sigma_{hp}\sigma_{hn} f/(4\pi) + [-0.78 \text{ mb} + 0.35 \text{ mb}$$
$$\times \ln (s/M_0^2)]\sigma_{hn}/\sigma_{NN}. \qquad 12.$$

Here $s$ is the square of the center-of-mass energy, $f$ has its expected value, and $M_0$ is the mass of the first diffractive excitation of h, e.g. the mass of resonance $A_1$ for h = $\pi$. Comparing Equations 12 and 11 one sees why $f$ in Equation 11 must be large at high energies.

The last two examples illustrate a more general phenomenon. In the physics of high energy hadron-nucleus interactions it is amazingly easy to conceal existing, or expose nonexisting, effects by small and apparently harmless changes of parameters.

## 3.    COHERENT INELASTIC PROCESSES

Since neutrons are difficult to detect, coherent processes hA → h*A, where the nucleus survives without breaking or getting excited, and semicoherent processes hA → h*A*, where the nucleus does not break but can be excited, are in practice the only processes in which the final state can be completely controlled. In our notation h* may be any particle or group of particles, while A* is the unbroken nucleus.

The best-known coherent scattering processes are those in which h → h* is a diffractive dissociation process. Examples are $\pi \to 3\pi$ (35), $\pi \to 5\pi$ (36), N → N$\pi$ and N → N$\pi\pi$ (37). Sometimes only the recoil nucleus is detected and the cross sections for all h* are summed (38). Cross sections are usually fitted using a simple extension of Glauber's method (39). It is assumed that hadron h at some point **P** of its trajectory across the nucleus changes into h* and then propagates further as h*. An additional averaging of the amplitude over the position of **P** is included.

Besides the input necessary to describe elastic scattering, this approach involves the amplitude for the process hN → h*N, which can be obtained from experiments on hydrogen targets, and the h*N total cross section, which is fitted as a free parameter. The cross sections $\sigma_{h^*N}^{tot}$ come out amazingly small (35–37) and a study of soluble models (40, 40a, 41) strongly suggests that, in spite of the good fits, the method is unreliable and the resulting cross sections $\sigma_{h^*N}^{tot}$ are meaningless. Incidentally, these models suggest also that the cross sections of vector mesons on nucleons obtained by applying a variant of the method used in (39) to photoproduction are reliable. The failure of the approach of (39) in general is due to inelastic screening, which has been completely neglected there.

Coherent inelastic scattering on nuclei can be valuable for the study of resonance production, because the signal-to-background ratio there is often more favorable than in scattering on protons (41a).

## 4. TYPICAL INELASTIC INTERACTIONS

### 4.1 Experimental Picture

In Figure 1 we show the distribution of negative particles produced in proton-xenon interactions at incident proton momentum 200 GeV/c (42). The variable on the horizontal axis is rapidity

$$y = \frac{1}{2} \ln \frac{E + p_z}{E - p_z},$$
13.

where $E$ is the energy of a particle and $p_z$ is the component parallel to the beam direction of its momentum, both in the laboratory frame. Thus the incident proton would have $y = y_{max}^p \approx 6$ and a pion with the same momentum $y = y_{max}^\pi \approx 8$. Under Lorentz transformation to the center-of-mass frame, or any other Lorentz transformation along the beam axis, the laboratory rapidity (Equation 13) transforms into $y' = y + $ const, and the curve in Figure 1 shifts without change of shape. The quantity on the vertical axis is

$$R(y) = \frac{1}{N_{pXe}} \frac{dN(pXe \rightarrow h^{-}X)}{dy} \bigg/ \left( \frac{1}{N_{pp}} \frac{dN(pp \rightarrow h^{-}X)}{dy} \right). \qquad 14.$$

The quantity $N_{pA}^{-1} dN(pA \rightarrow h^{-}X)$ is the ratio of the number of events, where a negative particle has been observed in a small rapidity interval $dy$, to the total number of inelastic p-A interactions. The symbol X is a reminder that whatever else has been produced in the event is irrelevant. The trick of dividing the distribution for $A > 1$ by the corresponding distribution for a proton target is often used because it significantly reduces the energy dependence of almost any quantity plotted.

The distribution in Figure 1 is dominated by typical events. As discussed in Section 5 some rare events are of great interest, but their contributions are not visible in distributions like this one.

In Figure 1 one can distinguish three rapidity regions: the target fragmentation region ($y \lesssim 1$), where the ratio $R(y)$ is large (note that nucleons from target fragmentation do not contribute to the density of

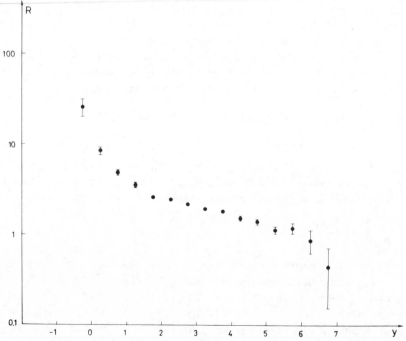

*Figure 1*   The ratio $R(y)$ for negative particles from pXe interactions at $p_{in} = 200$ GeV (data from Ref. 42).

negative particles); the central region ($1 \lesssim y \lesssim 5$), where $R(y)$ is nearly constant; and the projectile fragmentation region ($y \gtrsim 5$), where $R(y)$ drops below one. We discuss these regions in more detail in Sections 4.3 to 4.5. As seen from the figure, there are no sharp limits between the three regions, but this division helps to organize the discussion.

## 4.2  Theoretical Picture

There are many models of inelastic h-A interactions and almost every model has various versions. Let us quote as examples the additive quark model (AQM) (24, 43–47), the dual parton model (DPM) (48–51), and the multichain model (MCM) (52, 53). There are many more. From this mass of papers the outlines of a theoretical picture of h-A interactions at very high energies seem to emerge. Almost every detail is controversial, but the following scenario seems to have become the majority view.

The target nucleus behaves as a swarm of nucleons. Collective models, where multinucleon clusters are basic (for a review and critique, see 24), are losing support. While the projectile crosses the nucleus, there is no particle production but some nucleons, or quarks, get wounded, for example exchange color. The number of such exchanges is usually estimated (in a model-dependent way) using Equation 9. Much later, hadrons appear, beginning with the slowest ones. Hadronization is not localized on the initial hadrons. Most models introduce chains or strings, which later hadronize. Fast hadrons are produced so late that they have no chance to interact with the target nucleus. The slowest hadrons can cascade very much as in the old classical cascade model.

The listing of controversial points is far beyond the scope of this review. Let us mention, however, one problem that seems particularly important. According to some models (e.g. DPM), all the necessary input can be obtained from the study of interactions with single nucleons, and absolute predictions for h-A interactions are possible. According to others (e.g. AQM), scattering on nuclei gives new information not available otherwise (and consequently unpredictable starting from elementary interactions) about the behavior of hadrons (see Section 5.3 for an example).

## 4.3  Projectile Fragmentation

Data on the projectile fragmentation region are usually plotted using the variable $x$—the ratio of the energy of the final particle to the energy of the incident particle. As seen from Equation 13 for ultrarelativistic particles with transverse momenta $p_T \ll p_z$, the range $x_0 \leqslant x \leqslant 1$ corresponds to $y_{max} - y < -\ln x_0$, i.e. to the projectile fragmentation region according to some definition.

High statistics data have only begun to accumulate. A sample of data is shown in Figure 2. Let us note the study of $pA \to hX$ interactions (h = $\Lambda^0, \overline{\Lambda^0}, K_S^0$; A = Be, Cu, Pb) at incident momentum $p_{in} = 300\,\text{GeV}/c$ for $0.2 < x < 1$ (54), $pA \to \Xi^0 X$ (A = Be, Cu, Pb) at $p_{in} = 400\,\text{GeV}/c$ and in the same $x$ range (55, 56), and $h_i A \to h_f X$ scattering ($h_i$ = p, $\bar{p}$, $\pi^+$, $K^+$; $h_f$ = $\pi^\pm, K^\pm, p, \bar{p}$; A = p, C, Al, Cu, Ag, Pb) at $p_{in} = 100$ GeV/$c$ for $0.3 < x < 0.88$ (10). The polarization of hyperons produced by protons on nuclei has also been measured (h = $\Lambda^0, \Xi^0$; A = Be, Cu, Pb) at $p_{in} = 400$ GeV/$c$

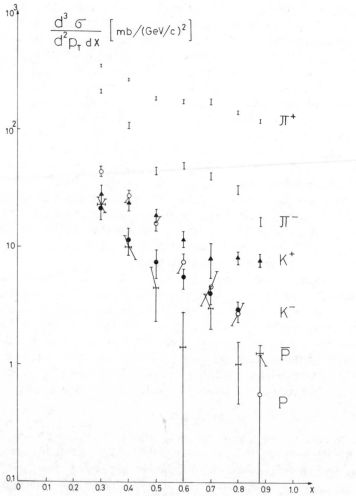

*Figure 2* Differential cross sections for the processes $\pi^+ Pb \to hX$, h = $\pi^\pm, K^\pm, p, \bar{p}$, at incident pion momentum 100 GeV/$c$ and final hadron transverse momentum $p_T = 0.3$ GeV/$c$ (data from Ref. 10).

(57) and (h $= \Lambda^0$; A $=$ W) at $p_{in} = 12$ GeV/$c$ (58). Finally, a study of $h_i A \to h_f X$ summed over all charged $h_f$ of given charge ($h_i = p, \bar{p}$; A $=$ p, Ar, Xe) at $p_{in} = 200$ GeV/$c$ has been reported (42). This work in particular clearly shows the depletion effect [$R(y) < 1$] in the near-forward region.

Models easily fit the rapidity distributions [see (42) for the data of De Marzo et al; (47, 50, 55, 59) for the data from (54); and (55, 60) for the data from (56)], but we have hardly begun the more difficult part of the problem, namely showing that the same parametrization of the input fits many well-measured reactions and taking into account the $p_T$ dependence of the $x$ distributions. Polarization within the large errors is similar to that in pp scattering. The forward depletion comes in most models as the result of energy conservation: The increase of the average multiplicity with $A$ is built into the models. The surplus particles have to take their energy from somewhere, and among other things slow down the fastest particle. For a simple analysis in this spirit, see (61). In the parton model this explanation implies that the fastest partons know how many other partons have interacted. This may seem strange; therefore let us note that other explanations are also being proposed [for example, see Section 6.3 of (24) or (14)].

## 4.4  Central Region

At very high energies the central region of the rapidity distribution $R(y)$ is expected to be a long flat plateau. If the finite intervals in rapidity corresponding to the target and projectile fragmentations are neglected, the length of this plateau should grow with increasing energy as $y_{max} = \ln p_{in}$. Since almost all the particles would be produced in the plateau region, the plateau height would be given by

$$R = \langle n_{pA} \rangle / \langle n_{pN} \rangle, \qquad\qquad 15.$$

where $\langle n_{pA} \rangle$ is the average number of particles of the type considered (e.g. charged particles) produced per one inelastic, nondiffractive p-A interaction.

As seen from Figure 1, we are still very far from this energy range. Both the width and the height of the plateau region are very uncertain and contamination by cascades from the target fragmentation region and by the overlapping projectile fragmentation region is clearly present.

Models agree that at low energies ($E \lesssim 100$ GeV) $R \approx (1 + \langle v \rangle)/2$, where $\langle v \rangle$ was defined in Equation 10 (46, 50). With increasing energy AQM does not predict much change, while the DPM gives $R \approx \langle v \rangle$ asymptotically (46, 50). Models easily reproduce the experimental values of $R$ within their uncertainties (14, 42, 52), though in order to explain the rise of $R(y)$ with decreasing $y$, it is necessary to include cascading into the description. Let us note in particular that the ISR results (62, 63) for particle production in the central region in p-$\alpha$ and $\alpha$-$\alpha$ collisions, at center-of-mass energies per nucleon from 26 to 44 GeV, have also been successfully reproduced (50, 63, 64).

## 4.5   Target Fragmentation

In the old days physicists working with nuclear emulsions classified tracks as black when $\beta = v/c \lesssim 0.3$, gray when $0.3 \lesssim \beta \lesssim 0.7$, and shower or minimum ionizing when $\beta \gtrsim 0.7$. This terminology has survived (e.g. 65). Since $\beta = p/E$, Equation 13 implies that for gray tracks $y \lesssim 0.9$. Thus black and gray tracks belong to the target fragmentation region. Of course many "shower" tracks also belong there.

The black tracks correspond to protons and nuclear fragments. The isotopic composition of the fragments strongly suggests (66) that at least the slowest among them are produced according to a thermodynamic mechanism, on a long time scale. The gray tracks correspond mostly to protons. At high energies their average number $\langle n_g \rangle$ depends little on energy, and its increase with increasing $A$ and/or $\sigma_{hN}$ can be derived from a universal dependence on the average number of inelastic collisions $\langle v \rangle$ (67). Experimentally (67), as a crude approximation $\langle n_g \rangle \approx \langle v \rangle$. This is much more than the expected number of recoil protons, which should be smaller than $\langle v \rangle Z/A \leqslant \langle v \rangle/2$. A simple explanation is that each recoil nucleon interacts with other nucleons (cascades) and thus more protons are ejected. Simple models based on this idea reproduce reasonably well various distributions (68–71), e.g. the joint probability distributions for the numbers of black and shower tracks. This supports the correlation between $n_g$ and $v$ assumed in these models. Not all recoil nucleons end as gray or black tracks. Nucleons with much higher momenta have also been seen (72, 73).

Choosing targets with various atomic numbers $A$, it is possible to scan the region $1 \leqslant \langle v \rangle \leqslant 4$. By selecting samples with given $n_g$, one can extend this range to $1 \leqslant \langle v \rangle \leqslant 6$, and with more events even higher (63, 65, 72, 74). The average $\langle v \rangle$, however, does not provide a complete characteristic of the sample. For example, for samples selected by choosing $A$, the ratio of the dispersion to the average number of shower tracks, $D/\langle n_s \rangle$, depends little on $\langle v \rangle$, while for samples selected by fixing $n_g$, $D/\langle n_s \rangle$ drops from about 0.55 for $n_g = 0$ to about 0.35 for $n_g = 12$ and shows little dependence on $A$ (65).

As seen from Figure 1 not just protons contribute to the rise of particle density in the target fragmentation region. Other contributions come from fragmentation of wounded nucleons (or quarks) in the target nucleus and probably from cascading.

Some groups report the presence of hadrons in regions kinematically forbidden for h-N interactions (references can be traced from 75). This "cumulative" effect might reflect some correlations between nucleons in nuclei.

# 5.   SPECIAL INELASTIC PROCESSES

## 5.1   Double Scattering on Deuterium

Scattering on large nuclei exhibits strong nuclear effects. In scattering on deuterium the nuclear effects are much smaller, but also much easier to analyze. Therefore, experiments with deuteron targets and experiments with heavy targets are often complementary.

In h-d interactions it is experimentally possible to study double scattering (i.e. on both p and n) separately (76, 77). The cross section for inelastic double scattering is usually 15–20% of the total inelastic cross section. A compilation of data and references to earlier work can be found in (78). A simple analysis of these data excludes classical cascade models with (8) or without (76, 77) formation-time corrections. It was soon realized that double scattering provides a valuable check on the Glauber-Gribov theory (79, 80). Some cascading has been found necessary to fit the data in the AQM (81). It is probable that with decreasing uncertainties in both data and theory all the models will reach this conclusion.

## 5.2   Lepton Pair Production

The study of the production of lepton pairs with the invariant mass of the pair $m_{l\bar{l}} \approx m_V$, where $m_V$ is the mass of some vector meson, has been of great interest in the work on vector mesons. The highlights have been the discoveries of $\psi$, $\Upsilon$, and $\Upsilon'$. The production of lepton pairs in the continuum ($m_{l\bar{l}} \not\approx m_V$) for $m_{l\bar{l}} \lesssim 3$ GeV is not well understood. Probably several production mechanism compete there. The production of pairs in the continuum with $m_{l\bar{l}} \gtrsim 4$ GeV is dominated by the Drell-Yan process $q\bar{q} \rightarrow \gamma^* \rightarrow l\bar{l}$ and yields interesting insights into the mechanism of h-A interactions. For a review, see (82), and for some more recent experimental data, see (83–86).

The Drell-Yan picture gives a good description of the data when additivity $[\sigma(hA \rightarrow l\bar{l}X) = A\sigma(hN \rightarrow l\bar{l}X)]$ is assumed. The main theoretical uncertainty is attached to a factor $K \approx 2$, which multiplies the cross section calculated from lowest-order QCD in order to get a normalization in agreement with experiment. This factor has been derived from QCD, but it is not quite clear how reliable this derivation is (83).

As pointed out in (87), additivity and the fact that the cross section for the Drell-Yan process rapidly increases with energy exclude many models. In all models in which at a given incident momentum (and incident hadron) the energy in the $q\bar{q}$ collision is on the average smaller for large target nuclei than for small ones, the increase of the cross section with $A$ must be slower than linear. Examples are models in which a quark interacts more than once, losing energy in each interaction, and models in which a quark gives more energy to other partons when there are more collisions.

Other problems that could be studied using data on the production of high mass $l\bar{l}$ pairs include an analogue of the EMC effect (88, 89) and the contribution of thermal $l\bar{l}$ pairs emitted by a plasma, if this is produced in central heavy-ion collisions (90).

## 5.3   Antiproton Annihilation

The $p\bar{p}$ annihilation cross section decreases with energy according to (91): $\sigma_{\bar{p}p} = 61\ p_{in}^{-0.61}$, where the laboratory momentum of the incident antiproton $p_{in}$ is in GeV/$c$ and $\sigma_{\bar{p}p}$ is in millibarns. Every multiple-collision model of h-A interactions predicts for heavy nuclear targets $\sigma_{\bar{p}A} \gg \sigma_{\bar{p}p}$. One reason is that in $\bar{p}$-A scattering there are many collisions with individual nucleons and many occasions for annihilation. This effect alone yields $\sigma_{\bar{p}A} \approx A\sigma_{\bar{p}p}$ (46). Thus annihilation on heavy nuclei should remain important even at energies where $\sigma_{\bar{p}p}$ is negligible. This fact has been noticed and analyzed in a number of papers (92–96).

Arguments going beyond the standard lore of multiple-scattering models suggest that perhaps $\sigma_{\bar{p}A}$ increases with the target mass number faster than $A$. Early speculations (92, 94) invoked energy degradation of the objects propagating across the nucleus. As explained in the preceding section, this mechanism has lost much of its credibility, though it still can be defended (94). A new idea is put forward in (97): The wave function of an antiproton is completely antisymmetric in the colors of its three valence quarks. According to the additive quark model (or one-gluon exchange mechanism), after a single collision at least one $\bar{q}\bar{q}$ pair remains antisymmetric in color. After two or more collisions the $\bar{q}\bar{q}\bar{q}$ system can become completely symmetric in color. The natural assumption is that it is more difficult to reconstruct a $\bar{p}$ from a totally symmetric $\bar{q}\bar{q}\bar{q}$ than from one of mixed symmetry, which implies that in scattering on nuclei a new mechanism enhancing annihilation operates. A further conclusion is that $\sigma_{\bar{p}A}$ should decrease with increasing energy more slowly than $\sigma_{\bar{p}p}$. This speculation has no experimental support for the moment, but it nicely illustrates the fact that in scattering on nuclei effects unknown from scattering on nucleons could occur.

## 5.4   Large $E_T$ and Large $p_T$ Processes

Let us denote by $I(p_T)\ dp_T$ the cross section for producing a hadron with its transverse momentum in the interval $dp_T$ and by $J(E_T)\ dE_T$ the cross section for producing an event with transverse energy $E_T = \sum |p_T|$ in the

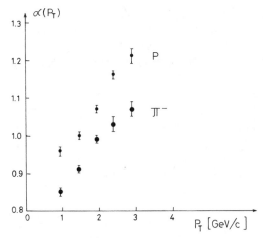

*Figure 3*   Parameter $\alpha(p_T)$ for $\pi^- A \to hX$ interactions for $h = p, \pi^-$, at incident pion momentum 200 GeV/$c$ (data from Ref. 103; data points with errors exceeding 0.02 have been left out).

interval $dE_T$. The $A$ dependence of both $I(p_T)$ and $J(E_T)$ can be parametrized as $A^\alpha$, and in both cases $\alpha$ increases with the argument ($p_T$ or $E_T$) and reaches values bigger than one. [Some data on $\alpha(p_T)$ are shown in Figure 3.] Here, however, the similarity between the two cross sections ends.

Large $E_T$ could mean large multiplicity or large $\langle p_T \rangle$. Data (98) support the first possibility. Since multiplicity distributions are well reproduced by current models, the calculation of $J(E_T)$ should be reliable. Indeed the data (99, 100) are well reproduced by simple model calculations (101). The main deviation is that models somewhat overestimate $J(E_T)$ at large $A$ and/or $E_T$; however, the increase of $\alpha$ with $E_T$ poses no problem.

The fast increase of $\alpha(p_T)$ with $p_T$ (102–105) is more difficult to explain. Many theoretical ideas have been put forward (106, 107), but none have gained general acceptance. Perhaps the most striking experimental result was reported in (104): For $5 < p_T < 6$ GeV/$c$ and center-of-mass energy per nucleon $\sqrt{s}$/nucleon $= 15.5$ GeV the authors report $I_{\alpha\alpha} = (30\text{–}40)I_{pp}$, while at $\sqrt{s}$/nucleon $= 22$ GeV they find $I_{p\alpha} = (3\text{–}4)I_{pp}$. Corresponding data at lower transverse moments are less unexpected (105) and a confirmation of the data from (104) would be of great interest. It seems fair to conclude that the situation concerning $I(p_T)$ is confused and that both new ideas and more good data are needed.

## 5.5   *Central Collisions of Heavy Ions*

Samples of events rich in central collisions are obtained by applying one or more of the following criteria: high multiplicities, many slow tracks (black and/or gray), and in the center of mass system few or no fast baryons produced at angles close to 0° or 180°. The events (1) presented in Section 1.3 are the most impressive specimens, but there are others. Let us mention for example the central $\alpha\alpha$ collisions studied at ISR energies (108). Other cases are quoted in (53, 109–111).

It would be of great interest to find in heavy-ion central collisions some evidence for the formation of quark-gluon plasma. At present the situation is confused. Theoretical estimates (3, 112; for further references see 109) suggest that the energy densities achieved are sufficient to produce the plasma. On the other hand, standard multiple-collision models fit the data well (1, 53, 109–111). Thus, even if the plasma has been produced, it has not been unambiguously identified.

## 6. CONCLUSION

The physics of high energy hadron-nucleus interactions is still in an early stage of development. There is a mass of data, but experiments with large acceptance and particle identification have only begun to appear (42, 113). There are many successful phenomenological models, but the crucial question—is there anything interesting that can be learned from hadron-nucleus interactions and cannot be learned from hadron-nucleon interactions—remains unanswered. Much has been done and progress is clearly visible, but more work is necessary before the programs outlined in the introduction can be fully realized.

ACKNOWLEDGMENT

The author thanks A. Białas, W. Czyż, R. Hołyński, D. Kisielewska, L. Leśniak, and P. Malecki for helpful comments. The hospitality of the Theory Group at the Max Planck Institut für Physik und Astrophysik in Munich, where part of this work was done, is gratefully acknowledged.

*Literature Cited*

1. Burnett, T. H., Dake, S., Fuki, M., Gregory, J. C., Hayashi, T., et al. *Phys. Rev. Lett.* 50: 2062–65 (1983)
2. Kajantie, K., McLerran, L. *Phys. Lett.* 119*B*: 203–6 (1982)
3. Bjorken, J. D. *Phys. Rev. D* 27: 140–51 (1983)
4. Gaisser, T. K., Yodh, G. B. *Ann. Rev. Nucl. Part. Sci.* 30: 475–542 (1980)
5. Bayburina, S. G., Borisov, A. S., Cherdyntseva, K. V., Guseva, Z. M., Denisova, V. G., et al. *Nucl. Phys. B* 191: 1–25 (1981)
6. Artykov, I. Z., Barashenkov, V. S., Eliseev, S. M. *Nucl. Phys.* 87: 241–55 (1966)
7. Kancheli, O. V. *Pisma Zh. Eksp. Teor. Fiz.* 18: 465–69 (1973)
8. Frederiksson, S., Jändel, M. *Z. Phys. C* 10: 185–87 (1981)
9. Białas, A. *Acta Phys. Pol. B* 15: 647–51 (1984)
10. Barton, D. S., Brandenburg, G. W., Busza, W., Dobrowolski, T., Friedman, J. I., et al. *Phys. Rev. D* 27: 2580–99 (1983)

11. Aubert, J. J., Bassompierre, G., Becks, K. H., Best, C., Böhm, E., et al. *Phys. Lett.* 123*B* : 275–78 (1983)
11a. Bujak, A., Devensky, P., Kuznetsov, A., Morozov, B., Nikitin, V., et al. *Phys. Rev. D* 23 : 1895–1910 (1981)
12. Warren, G., Gross, D., Olsen, S. L., Collick, B., Susack, R., et al. *Nucl. Phys.* B 207 : 365–73 (1982)
13. Czyż, W. *Adv. Nucl. Phys.* 4 : 61–131 (1972)
13a. Czyż, W. *Interactions of High Energy Particles with Nuclei*, NBS Monogr. 139. Washington, DC : Natl. Bur. Standards (1975)
14. Białas, A., Czyż, W. *Nucl. Phys. B* 194 : 21–37 (1982)
15. Białas, A., Błeszyński, M., Czyż, W. *Acta Phys. Pol. B* 8 : 389–92 (1977)
16. Denisov, S. P., Donskov, S. V., Gorin, Yu. P., Krasnokutsky, R. N., Petrukhin, A. I., et al. *Nucl. Phys. B* 61 : 62–76 (1973)
17. Carrol, A. S., Chiang, I.-H., Kycia, T. F., Li, K. K., Marx, M. D., et al. *Phys. Lett. B* 80 : 319–22 (1979)
18. Gribov, V. N. *Zh. Eksp. Theor. Phys.* 56 : 892–901 (1969)
19. Abramovskii, V. A., Gribov, V. N., Kancheli, O. V. *Yad. Fiz.* 18 : 595–616 (1973)
20. Jaroszewicz, T., Kwieciński, J., Leśniak, L., Zalewski, K. *Z. Phys. C* 1 : 181–88 (1979)
21. Jaroszewicz, T., Kwieciński, J., Leśniak, L., Zalewski, K. *Phys. Lett. B* 79 : 127–30 (1978)
22. Abers, E. S., Burkhardt, H., Teplitz, V. L., Wilkin, C. *Nuovo Cimento* 42*A* : 365–89 (1966)
23. Kwieciński, J., Leśniak, L., Zalewski, K. *Nucl. Phys. B* 78 : 251–68 (1974)
24. Nikolaev, N. N. *Usp. Fiz. Nauk* 134 : 369–430 (1981)
25. Murthy, P. V. R., Ayre, C. A., Gustafson, H. R., Jones, L. W., Longo, M. J. *Nucl. Phys. B* 92 : 269–308 (1975)
26. Goggi, G., Cavalli-Sforza, M., Conta, C., Fraternali, M., Mantovani, G. C., et al. *Nucl. Phys. B* 149 : 381–412 (1979)
27. Bell, W., Braune, K., Gleasson, G., Drijard, D., Faessler, M. A., et al. *Phys. Lett.* 117*B* : 131–34 (1982)
28. Burg, J. P., Chemarin, M., Chevalier, M., Denisov, A. S., Ekelöf, T., et al. *Nucl. Phys. B* 187 : 205–30 (1981)
29. Roehring, J., Gsponer, A., Molzon, W. R., Rosenberg, E. I., Telegdi, V. L., et al. *Phys. Rev. Lett.* 38 : 1116–19 (1977)
30. Bertocchi, L., Treleani, D. *Nuovo Cimento* 50*A* : 338–48 (1979)
31. Diu, B., Ferraz de Camargo, A. *Z. Phys. C* 4 : 223–29 (1980)
32. Kopeliovich, B. Z., Nikolaev, N. N. *Z.*

*Phys. C* 5 : 333–36 (1980)
33. Biaggi, S. F., Burquin, M., Britten, A. J., Brown, R. M., Burckhart, H., et al. *Nucl. Phys. B* 186 : 1–21 (1981)
34. Jeżabek, M., Zalewski, K. *Acta Phys. Pol. B* 12 : 1021–27 (1981)
35. Bellini, G., Chernenko, L. P., Datsko, V. S., Di Corato, M., Frabetti, P. L., et al. *Nucl. Phys. B* 199 : 1–26 (1982)
36. Bellini, G., Di Corato, M., Menasce, D., Palumbo, F., Sala, A., et al. *Phys. Lett. B* 126 : 140–44 (1983)
37. Goggi, B., Conta, C., Fraternali, M., Livan, M., Mantovani, G. C., et al. *Nucl. Phys. B* 161 : 14–54 (1979)
38. Bujak, A., Devensky, P., Kuznetsov, A., Morozov, B., Nikitin, V., et al. *Phys. Rev. D* 23 : 1911–23 (1981)
39. Kölbig, K. S., Margolis, B. *Nucl. Phys. B* 6 : 85–101 (1968)
40. Czyż, W. *Phys. Rev. D* 8 : 3219–22 (1973)
40a. Czyż, W., Zieliński, M. *Acta Phys. Pol. B* 11 : 615–28 (1980)
41. Miettinen, H. I., Pumplin, J. *Phys. Lett.* 42*B* : 204–8 (1979)
41a. Bellini, G., Frabetti, P. L., Di Corato, M., Palombo, F., Pernegr, J., et al. *Nuovo Cimento* 79*A* : 282–92 (1984)
42. De Marzo, C., De Palma, M., Distante, A., Favuzzi, C., Germinario, G., et al. *Phys. Rev. D* 26 : 1019–35 (1982)
43. Anisovich, V. V. *Phys. Lett.* 57*B* : 87–89 (1975)
44. Białas, A., Czyż, W., Furmański, W. *Acta Phys. Pol. B* 8 : 585–89 (1977)
45. Nikolaev, N. N. *Phys. Lett. B* 70 : 95–98 (1977)
46. Białas, A. *Proc. 13th Int. Symp. on Multiparticle Dynamics*, pp. 328–58. Singapore : World Scientific (1982)
47. Dar, A., Takagi, F. *Phys. Rev. Lett.* 44 : 768–72 (1980)
48. Capella, A., Tran Thanh Van, J. *Phys. Lett. B* 93 : 146–50 (1980)
49. Capella, A., Kwieciński, J., Tran Thanh Van, J. *Phys. Lett.* 108*B* : 347–50 (1982)
50. Capella, A. *Acta Phys. Pol. B* 14 : 359–73 (1983)
51. Ranft, J., Ritter, S. *Z. Phys. C* 20 : 347–55 (1983)
52. Kinoshita, K., Minaka, A., Sumiyoshi, H. *Prog. Theor. Phys.* 61 : 165–75 (1979)
53. Sumiyoshi, H. *Phys. Lett. B* 131 : 241–46 (1983)
54. Skubic, P., Overseth, O. E., Heller, K., Sheaff, M., Pondrom, L., et al. *Phys. Rev. D* 18 : 3115–44 (1978)
55. Takagi, F. *Phys. Rev. D* 27 : 1461–67 (1983)
56. Pondrom, L., Grobel, R., Handler, R., March, R., Sheaff, M., et al. *Proc. 10th Int. Symp. on Multiparticle Dynamics*, pp. 579–86. Bombay : Tata Inst. (1979)

57. Heller, K., Cox, P. T., Dworkin, J., Overseth, O. E., Pondrom, L., et al. *Phys. Rev. Lett.* 51:2025–28 (1983)
58. Abe, F., Hara, K., Kim, N., Kondo, K., Miyashita, S., et al. *Phys. Rev. Lett.* 50:1102–5 (1983)
59. Białas, A., Białas, E. *Phys. Rev. D* 20: 2854–61 (1979)
60. Krasny, W., Zalewski, K. *Acta Phys. Pol. B* 12:973–79 (1981)
61. Białas, A., Stodolsky, L. *Acta Phys. Pol. B* 7:845–49 (1976)
62. Åkesson, T., Albrow, M. G., Almehed, S., Benary, O., Bøggild, H., et al. *Phys. Lett.* 119B:464–70 (1982)
63. Bell, W., Braune, K., Gleasson, G., Drijard, D., Faessler, M. A., et al. *Phys. Lett.* 128B:349–53 (1983)
64. Białas, A., Czyż, W., Leśniak, L. Z. *Phys. C* 13:147–51 (1982)
65. Faessler, H. *Ann. Phys.* 137:44–85 (1981)
66. Minich, R. W., Argawal, S., Bujak, A., Chuang, J., Finn, J. E., et al. *Phys. Lett.* 118B:458–60 (1982)
67. Braune, K., Brücker, W., Faessler, M. A., Frey, R. W., Gugelot, P. C., et al. Z. *Phys. C* 13:191–97 (1982)
68. Babecki, J., Nowak, G. *Acta Phys. Pol. B* 9:401–17 (1978)
69. Nilsson, G., Andersson, B., Otterlund, I. *Nucl. Phys. B* 195:203–21 (1982)
70. Stenlund, E., Otterlund, I. *Nucl. Phys. B* 198:407–26 (1982)
71. Hegab, M. K., Hüfner, J. *Nucl. Phys. A* 384:353–70 (1982)
72. Rees, C. D., Lubatti, H. J., Moriyasu, K., Rock, D., Arnold, R., et al. Z. *Phys. C* 17:95–103 (1983)
73. Azimov, S. A., Inogamov, Sh. V., Kosonowski, E. A., Lipin, V. D., Lutpullaev, S. L., et al. *Phys. Rev. D* 23:2512–21 (1981)
74. Hołyński, R., Krzywdziński, S., Zalewski, K. *Acta Phys. Pol. B* 5:321–26 (1974)
75. Anishin, D. V., Gorenstein, M. I., Zinoviev, G. M. *Phys. Lett.* 108B:47–50 (1982)
76. Dziunikowska, K., Figiel, J., Kisielewska, D., Malecki, P., Rudnicka, H., et al. *Phys. Lett. B* 61:316–20 (1976)
77. Ansorge, R. E., Barlow, R. J., Carter, J. R., Ioannidis, G. S., Neale, W. W., et al. *Nucl. Phys. B* 109:197–206 (1976)
78. Lubatti, H. J., Moriyasu, K., Rees, C. D., Rogers, E., Wróblewski, A. Z. *Phys. C* 7:241–48 (1981)
79. Baker, M., Lubatti, H. J., Rogers, E. O., Weis, J. H. *Phys. Rev. Lett.* 39:375–78 (1977)
80. Baker, M., Lubatti, H. J., Rogers, E. O., Weis, J. H. *Phys. Rev. D* 17:826–38 (1978)
81. Białas, A., Czyż, W., Kisielewska, D. Z. *Phys. C* 12:35–41 (1982)
82. Burgun, G. *Acta Phys. Pol. B* 13:335–55 (1982)
83. Smith, S. R., Childress, S., Mockett, P. M., Rutherford, J. P., Williams, R. W., et al. *Phys. Rev. Lett.* 46:1607–10 (1981)
84. Ito, A. S., Fisk, R. J., Jöstlein, H., Kaplan, D. M., Herb, S. W., et al. *Phys. Rev. D* 23:604–33 (1981)
85. Badier, J., Boucrot, J., Buorotte, J., Burgun, G., Callot, O., et al. *Phys. Lett.* 104B:335–38 (1981)
86. Falciano, S., Freudenreich, K., Juillot, P., Wallace-Hadrill, J. S., Anderson, L., et al. *Phys. Lett.* 104B:416–20 (1981)
87. Jaroszewicz, T., Jeżabek, M. Z. *Phys. C* 4:277–80 (1980)
88. Goodhole, R. M., Sarma, K. V. L. *Phys. Rev. D* 25:120–29 (1982)
89. Chmaj, T., Heller, K. *Acta Phys. Pol. B* 15:473–82 (1984)
90. Kajantie, K., Miettinen, H. I. Z. *Phys. C* 14:257–62 (1982)
91. Rushbrooke, J. G. *Proc. 3rd Eur. Symp. on Antinucleon-Nucleon Interactions, 1976, Stockholm*, ed. G. Ekspong, S. Nilsson, pp. 277–99. Oxford: Pergamon (1977)
92. Jeżabek, M., Zalewski, K. *Acta Phys. Pol. B* 11:425–30 (1980)
93. Pajares, C., Ramallo, A. V. *Phys. Lett. B* 107:238–40 (1981)
93a. Pajares, C., Ramallo, A. V. *Phys. Lett. B* 107:373–76 (1981)
94. Sumiyoshi, H., Kinoshita, K., Minaka, A. Z. *Phys. C* 11:347–52 (1982)
95. Sumiyoshi, H., Kondo, H., Uehara, M. Z. *Phys. C* 19:107–14 (1983)
96. Sumiyoshi, H., Kondo, H., Uehara, M. *Phys. Lett. B* 120:436–40 (1983)
97. Jeżabek, M. *Phys. Lett. B* 126:106–10 (1983)
98. De Marzo, C., De Palma, M., Distante, A., Favuzzi, C., Lavopa, P., et al. *Phys. Rev. D* 29:363–76 (1984)
99. Brown, B., Devensky, P., Gronemeyer, S., Haggarty, H., Malamud, E., et al. *Phys. Rev. Lett.* 50:11–14 (1983)
100. Gordon, H., Kilian, T., Ludlam, T., Stumer, I., Winik, M., et al. *Phys. Rev. D* 28:2736–40 (1983)
101. Brody, H., Frankel, S., Frai, W., Otterlund, I. *Phys. Rev. D* 28:2334–37 (1983)
102. Cronin, J. W., Frish, H. J., Cochet, M. J., Boymond, J. P., Piroué, P. A., et al. *Phys. Rev. D* 11:3105–23 (1975)
103. Frish, H. J., Giokaris, N. D., Green, J. M., Grosso-Pilcher, C., Mestayer, M. D., et al. *Phys. Rev. D* 27:1001–30 (1983)
104. Angelis, A. L. S., Basini, G., Besch, H. J., Breedon, R., Camilleri, L., et al. *Phys. Lett.* 116B:379–82 (1982)

105. Åkesson, T., Albrow, M. G., Almehed, S., Benary, O., Bøggild, H., et al. *Nucl. Phys. B* 209:309–20 (1982)
106. McNeil, J. A. *Phys. Rev. D* 27:292–95 (1983)
107. Sukhatme, U. P., Wilk, G. *Phys. Rev. D* 25:1978–79 (1982)
108. Åkesson, T., Albrow, M. G., Almehed, S., Benary, O., Bøggild, H., et al. *Phys. Lett. B* 110:344–48 (1982)
109. Otterlund, I., Grapman, S., Lund, I. Z.

*Phys. C* 20:281–90 (1983)
110. Pajares, C., Ramallo, A. V. Z. *Phys. C* 20:213–15 (1983)
111. Pajares, C., Ramallo, A. V. *Phys. Lett. B* 120:441–43 (1983)
112. Anishetty, R., Kochler, P., McLerran, L. *Phys. Rev. D* 22:2793–2804 (1980)
113. Brick, D. H., Rudnicka, H., Shapiro, A. M., Smith, W., Widgoff, M., et al. *Nucl. Phys. B* 201:189–96 (1982)

*Ann. Rev. Nucl. Part. Sci. 1985. 35 : 77–105*

# THE INTERACTING
# BOSON MODEL

*A. E. L. Dieperink and G. Wenes*

Kernfysisch Versneller Instituut, 9747 AA Groningen, The Netherlands

CONTENTS

## 1.  INTRODUCTION

Collective models play an important role in nuclear physics. Their main objective is to reduce the complexities of the full nuclear many-body problem by isolating a few essential degrees of freedom. A unique prescription for the definition of collective coordinates cannot easily be given, and in the past a variety of collective models was developed with different starting points. Most of these models can be characterized as geometric approaches (1, 2) in the sense that they picture the nucleus as a liquid drop subject to quadrupole surface vibrations and rotations; the nuclear radius is parametrized in terms of the five quadrupole shape

77

variables $\alpha_\mu$ ($\mu = -2, -1, 0, 1, 2$):

$$R(\theta, \phi) = R_0[1 + \sum_\mu \alpha_\mu Y^*_{2\mu}(\theta, \phi)],$$

where $R_0$ is the average spherical radius. A collective Hamiltonian is then constructed as a polynomial in terms of the $\alpha_\mu$ and their conjugate momenta, $\pi_\mu = -i\hbar(\partial/\partial\alpha_\mu)$. In the harmonic approximation it has the simple form

$$H = \frac{1}{2D} \sum_\mu |\pi_\mu|^2 + \tfrac{1}{2}C \sum_\mu |\alpha_\mu|^2,$$

or equivalently in a second quantized formulation

$$H = \hbar\omega \left( \sum_\mu b_\mu^+ b_\mu + \tfrac{5}{2} \right),$$

where the phonon creation and destruction operators $b_\mu^+, b_\mu$ are linear combinations of the $\alpha_\mu$ and $\pi_\mu$.

To describe actual nuclei a more general phenomenological Hamiltonian is usually constructed as a rotationally invariant analytic function of the variables $\alpha$ and $\pi$; in most applications one takes a kinetic energy part that is quadratic in the momenta and a rather general potential energy function of the conventional invariants $\beta^2 \equiv \sqrt{5}(\alpha\alpha)^{(0)}$ and $\beta^3 \cos 3\gamma \equiv -\sqrt{\frac{2}{35}}(\alpha\alpha\alpha)^{(0)}$, where $\beta$ is the deformation and $\gamma$ the triaxiality parameter that characterize the intrinsic shape (1). Since such a Hamiltonian contains phonon-number-changing terms, the eigenvalue problem is in general infinite dimensional; in practice one usually truncates the Hilbert space, spanned by the states $(b_\mu^+)^n|0\rangle$ ($n = 0, 1, 2, \ldots$), at an arbitrary maximum phonon number $N$, thus effectively reducing the problem to a finite matrix diagonalization (2).

It is, however, possible to formulate this matrix problem in a different framework, in which the number of bosons, $N$, is conserved by embedding the quadrupole phonons in a larger group structure. Such an approach was first proposed by Janssen, Jolos & Dönau (3) in terms of the generators $N$, $b_\mu^+ \sqrt{N - b^+ \cdot b}$, $b_\mu^+ b_{\mu'}$, and $\sqrt{N - b^+ \cdot b}\, b_\mu$, which close under commutation and form a realization of the compact algebra of the group U(6). In this model (which is generally referred to as truncated quadrupole phonon model, TQPM) the "phonon" basis states $(b_\mu^+)^n|0\rangle$ ($n = 0, 1, \ldots, N$) span the finite dimensional representation $[N]$ of the group U(6), and the collective Hamiltonian, which is expressed in terms of the group generators, can easily be diagonalized numerically.

A more elegant formulation of the collective motion in terms of the same group structure was proposed by Arima & Iachello (4–7). They introduced a monopole ($L = 0$) s-boson in addition to the five quadrupole ($L = 2$) bosons and moreover suggested a physical interpretation of the bosons as correlated fermion pairs. This model, the interacting boson model, has received broad attention, which can be attributed on the one hand to its ability to describe experimental data over a wide range of the periodic system (including transitional nuclei) in terms of a few parameters, and on the other hand to the suggested connection with microscopic models. Since the original version of the model (IBA-1) is described in several review articles (e.g. 8, 9) the discussion here is restricted to a short summary of its most characteristic features (Section 2).

More recently a generalized version of the model has been developed (10, 11) in which a distinction is made between the neutron and proton collective variables. This was motivated by the observation that in heavy nuclei, where neutrons and protons occupy different major shells, the effective interaction between the like fermions differs substantially from that between neutrons and protons. A discussion of the properties of this extended IBA model (IBA-2) forms the main part of this article. Its phenomenological aspects are discussed in Section 3 with particular emphasis on the consequences of the neutron-proton degree of freedom for experimental observables. The present status of the microscopic understanding of the model is reviewed in Section 4. Section 5 contains a discussion of the achievements and limitations of the algebraic approach.

## 2.   THE INTERACTING BOSON MODEL–1

We begin with a brief summary of the simplest version of the IBA model, in which no distinction is made between neutron and proton bosons. More details can be found in the review articles (8, 9).

### 2.1   *The Model*

The properties of low-lying collective states in nuclei are dominated by the pairing and quadrupole degrees of freedom. In the IBA-1 model these are incorporated by introducing six bosonic degrees of freedom, divided into a scalar boson with angular momentum $L = 0$ (called an s-boson) and a quadrupole boson with angular momentum $L = 2$ ($d_\mu$-boson, $\mu = -2$, $-1, 0, 1, 2$). The creation $\{s^+, d_\mu^+\}$ and annihilation $\{s, \tilde{d}_\mu \equiv (-1)^\mu d_\mu\}$ operators, which obey the standard boson commutation relations, span a six-dimensional space, and thus provide a basis for the representations of the group U(6). The basis states for an $N$-boson system span the totally symmetric representations $[N]$ of U(6) and can be expressed as $|s^{N-n_d}d^{n_d}[\alpha L]\rangle$, where $n_d$ is the number of $d$ bosons that are coupled to angular momentum $L$ (and additional quantum numbers $\alpha$ necessary to fully specify the states). One then assumes that the properties of low-lying collective states in even-even nuclei can be described by a Hamiltonian that conserves the boson number, is rotationally invariant, and contains at most two-body interactions. A typical Hamiltonian that has frequently been used in empirical studies has the form

$$H = E_0 + \varepsilon \hat{n}_d + \kappa \tilde{Q}^{(2)} \cdot \tilde{Q}^{(2)} + \kappa' L^{(1)} \cdot L^{(1)}.$$   1.

Here $E_0$ contributes to the binding energy only; $\hat{n}_d$ is the d-boson number operator, $\hat{n}_d = d^+ \cdot \tilde{d}$; $\tilde{Q}_\mu^{(2)}$ is the general quadrupole operator

$$\tilde{Q}_\mu = d_\mu^+ s + s^+ \tilde{d}_\mu + \chi (d^+ \tilde{d})_\mu^{(2)},$$   2.

where $\chi$ is a quadrupole structure parameter, and $L_\mu^{(1)}$ the angular momentum operator

$$L_\mu^{(1)} = \sqrt{10}(d^+ \tilde{d})_\mu^{(1)}.$$   3.

The above Hamiltonian describes the spherical region (where the energy difference between s- and d-bosons dominates, $|\kappa| \ll \varepsilon$), the deformed region (where the $\tilde{Q}^{(2)} \cdot \tilde{Q}^{(2)}$ interaction dominates, $|\kappa| \gg \varepsilon$), as well as transitional regions. In general it must be solved numerically, i.e. by matrix diagonalization in a convenient basis. In special limiting cases, however, the eigenvalue problem can be solved in closed form, which has the important advantage that one obtains a direct insight in the nature of the solution. This happens whenever $H$ can be expressed in terms of invariants of a complete chain of subgroups of $G = U(6)$. Arima & Iachello (5–7) showed that in IBA-1 three such situations occur; they are referred to as dynamic symmetries. These cases provide labels as well as closed expressions for the energy eigenvalues:

$$\kappa = 0: \qquad \begin{array}{cccc} U(6) \supset & U(5) \supset & O(5) \supset & O(3) \\ | & | & | & | \\ N & n_d & \tau & L \end{array} \qquad \text{I.}$$

$$\varepsilon = \chi = 0: \qquad \begin{array}{cccc} U(6) \supset & O(6) \supset & O(5) \supset & O(3) \\ | & | & | & | \\ N & \sigma & \tau & L \end{array} \qquad \text{II.}$$

$$\varepsilon = 0, \chi = \pm \tfrac{1}{2}\sqrt{7}: \qquad \begin{array}{ccc} U(6) \supset & SU(3) \supset & O(3). \\ | & | & | \\ N & (\lambda, \mu) & L \end{array} \qquad \text{III.}$$

Here the labels refer to the eigenvalues of the various Casimir invariants, e.g. $n_d$ is the eigenvalue of the linear invariant $\hat{n}_d$ of the group U(5).

As an example, which is of relevance in the discussion of IBA-2, we consider the SU(3) case (III), which corresponds to $\varepsilon = 0$ in Equation 1 and

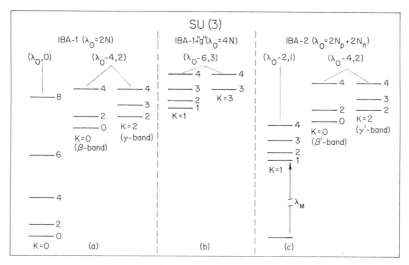

*Figure 1* Low-lying SU(3) representations labelled by $(\lambda, \mu)$: (*a*) for IBA-1, with s- and d-bosons only; (*b*) additional bands that occur if a g-boson is added; (*c*) additional bands in the neutron-proton IBA.

$\chi = \pm\frac{1}{2}\sqrt{7}$ in Equation 2, in some more detail. The energy spectrum is obtained by rewriting the appropriate Hamiltonian, in terms of the Casimir invariants of the subgroups SU(3) and O(3), $H = \frac{1}{2}\kappa C_2[\text{SU}(3)] + (\kappa' - \frac{3}{8}\kappa)C_2[\text{O}(3)]$, with the result (6)

$$E(\lambda, \mu, L) = \frac{1}{2}\kappa[\lambda^2 + \mu^2 + \lambda\mu + 3(\lambda + \mu)] + (\kappa' - \frac{3}{8}\kappa)L(L+1). \qquad 4.$$

The allowed values of the SU(3) representations $(\lambda, \mu)$ follow from the tensor decomposition U(6) $\supset$ SU(3). In each $(\lambda, \mu)$ representation one or more rotational bands [O(3) representations] occur that can be labelled (12) with the Elliott quantum number $K$, $K = \min(\lambda, \mu), \min(\lambda, \mu) - 2, \ldots, 1$ or 0. In Figure 1a the lowest energy states are shown for $\kappa < 0$: $(\lambda, \mu) = (2N, 0)$, the ground-state band, and $(\lambda, \mu) = (2N - 4, 2)$, which contains the so-called $\beta$ band ($K = 0$) and $\gamma$ band ($K = 2$).

## 2.2 *Relationship with Geometric Models*

After the introduction of the IBA model, various groups made detailed studies (13–22) of the relation between IBA and the conventional geometric models. Since a full account of the methods involved and the results is beyond the scope of this article, we restrict ourselves to a few main points that are also of relevance for the later discussion of the IBA-2 version.

The relationship between the algebraic and geometric descriptions can be discussed most conveniently by using the concept of coherent states (22).

In the case of the group U(6), the $N$-boson coherent states (13)

$$|N\alpha\rangle = \mathcal{N} \left( s^+ + \sum_{\mu} \alpha_{\mu}^* d_{\mu}^+ \right)^N |0\rangle \qquad 5.$$

provide a link between the five continuous geometric variables $\alpha_{\mu}$ and the algebraic representation. Two kinds of applications can be distinguished:

1. At the level of quantum mechanics, coherent states are utilized to convert the matrix eigenvalue problem into a differential equation by making use of relations such as $d_{\mu}^+ s|N\alpha\rangle \rightarrow \partial/\partial\alpha_{\mu}|N\alpha\rangle$ etc. In this way an IBA equivalent geometric Hamiltonian can be constructed (16, 18) that is very similar to the one encountered in the conventional geometric models; the important difference is that it must be solved in a compact Hilbert space.

2. By taking the expectation value of the IBA Hamiltonian, one can construct (13, 14, 21) a classical energy functional in terms of the variables $\alpha: E_N(\alpha) = \langle N\alpha|H|N\alpha\rangle$. Minimization of $E_N$ with respect to $\alpha$ yields information on the ground-state equilibrium configuration of the IBA Hamiltonian. From this analysis it follows that the three dynamic symmetries in IBA-1—U(5), SU(3), and O(6)—correspond to an (anharmonic) vibrator, axially symmetric deformed rotor, and $\gamma$-unstable rotor, respectively.

## 2.3   Finite N Effects

As mentioned above a basic ingredient of the IBA model, not present in geometric models, is the dependence on the boson number $N$, and therefore it is of interest to discuss this aspect in more detail. As shown below, the microscopic picture underlying IBA suggests that $N$ should be related to the number of valence nucleon pairs. In the past it was thought that the effects of finite boson number would show up most drastically at high spin states in terms of angular momentum cut-off effects at $L = 2N$ and a reduction of $B(E2)$ values connecting high spin states in the ground-state band (as compared to the rigid rotor values). A number of experimental tests did not unambiguously confirm these expectations. More recently it was realized that at high angular momentum one approaches the limits of applicability of the IBA model since very likely other degrees of freedom (such as quasi-particle excitations and alignment) begin to play a role. On the other hand some investigations for deformed nuclei strongly suggest that finite $N$ effects may have an important bearing on electromagnetic properties of low spin states. Experimentally these effects should appear most clearly in studies of systematic trends of collective properties as a function of $N$, where one does not need to assume that the IBA model works perfectly for *one* nucleus. Here we mention two pieces of evidence that are connected with interband E2 transitions in the deformed region. First, in a phenomenological approach the spin dependence of the interband $\gamma \to$ g E2 transitions can be expressed as (2)

$$B(E2, L_\gamma \to L_g) \approx 2\langle L_\gamma, 2, 2, -2|L_g 0\rangle^2$$

$$\times \{M_1 + M_2[L_\gamma(L_\gamma + 1) - L_g(L_g + 1)]\}. \qquad 6.$$

Here the $M_1$ term describes direct $\gamma \to$ g transitions and the $M_2$ term arises from a mixing between $\gamma$ and g bands. Traditionally this mixing is parameterized (2) in terms of $Z_\gamma = -2M_2/(M_1 + M_2)$. Its empirical value for the rare-earth nuclei when plotted against boson number exhibits a smooth cup-shaped curve with a minimum midshell (see Figure 2). Whereas in the past $Z_\gamma$ has mostly been treated as a purely phenomenological parameter, in the IBA model it is fixed once the Hamiltonian is determined from a fit to energy levels. Warner and Casten (23, 23a) showed that the empirical behavior of $Z_\gamma$ can be nicely described in terms of the IBA model using a special version of the Hamiltonian of Equation 1, with $\varepsilon = \kappa' = 0$. The systematic behavior of $\gamma \to$ g transitions arises primarily from a variation of the structure of the perturbed SU(3) wave functions resulting from the explicit $N$ dependence in IBA.

A second characteristic feature of the IBA model is the prediction of

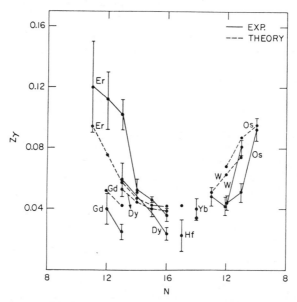

*Figure 2* Empirical and calculated values of the mixing parameter $Z_\gamma$ as a function of the boson number $N$ (from 23).

relatively strong E2 transitions connecting $\beta$ and $\gamma$ bands in deformed nuclei. In a pure rotor limit these transitions would be forbidden in the geometric approach because they would change the vibrational quantum number by two, and therefore they can only occur via bandmixing. On the other hand in the IBA model the $\beta$ and $\gamma$ bands belong to the same SU(3) representation ($\lambda = 2N - 4$, $\mu = 2$) (see Figure 1a), and the direct E2 matrix elements between them do not vanish for finite $N$ (24). More refined IBA calculations in a perturbed SU(3) scheme (23, 23a) predict that for finite $N$ in many cases the $\beta \rightarrow \gamma$ transitions dominate over the $\beta \rightarrow g$ E2 transitions and moreover that it is the direct $\beta \rightarrow \gamma$ matrix element rather than the induced bandmixing mechanism that contributes to the $\beta \rightarrow \gamma$ transitions. These transitions are very difficult to detect experimentally because of their low energy, but a few have been observed and these do indeed follow qualitatively the predicted strength pattern (see Figure 3).

A final remark concerns the possible role of bosons with $L > 2$ such as the $L = 4$ g-boson. Motivated by the evidence from microscopic considerations that in deformed nuclei the $L = 4$ pair plays a non-negligible role, some phenomenological studies (25, 25a) have been made by extending the IBA group structure from U(6) to U(15). The most noticeable effects of the g-boson are (a) an increase of the angular momentum cut-off, and (b) the prediction of additional representations not present in U(6). For example, in the SU(3) limit of U(15) (this extreme limit is probably not very realistic since the neglect of an energy splitting between s-, d-, and g-bosons gives rise to a rather large g-boson percentage in the ground-state band) the lowest representations are $(\lambda, \mu) = (4N, 0)$, $(4N - 4, 2)$, and $(4N - 6, 3)$, the

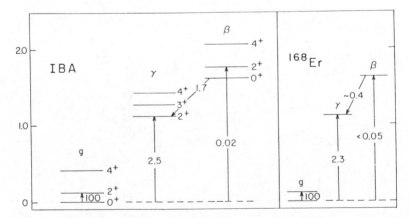

*Figure 3*  Schematic structure of a deformed nucleus in the IBA formalism (*left*) compared with an empirical example (*right*). The numbered arrows give the relative B(E2) strengths between the bands (from 23a).

latter containing $K = 1, 3$ bands that have no IBA-1 counterpart (see Figure 1b). At present it is not clear whether these bands can be identified with experimental ones or not.

# 3. PHENOMENOLOGY OF THE NEUTRON-PROTON IBA MODEL (IBA-2)

## 3.1 *The Model*

The IBA model discussed above can be considered as a purely phenomenological approach without an obvious microscopic basis. Soon after the introduction of IBA-1, in an attempt to establish a link with the shell model, a generalized version was proposed (10, 11) in which a distinction is made between neutrons and protons (IBA-2). Referring for a quantitative discussion of the microscopic foundation to Section 4, in this section we discuss the phenomenological consequences of this degree of freedom.

The basic philosophy underlying the IBA model is that the properties of low-lying collective states are dominated by the nucleons that occupy the valence orbitals, even for nuclei far away from closed shells. As a consequence, the number of valence protons and neutrons plays a central role in the IBA-2 formalism. In the shell model space spanned by the nucleons in the valence orbitals, a collective subspace is defined consisting of strongly correlated pairs coupled to angular momentum $L = 0$ and $L = 2$ ("SD" subspace). On the basis of the observation that in (medium-) heavy nuclei the valence protons and neutrons occupy different major shells, only correlated pairs of neutrons and pairs of protons are introduced; the neutron-proton correlations are treated as a mean field interaction between these pairs. It should be noted that this construction is not at variance with isospin symmetry as long as neutrons and protons occupy different shells, in which case all low-lying states have definite isospin $T = T_0 = \frac{1}{2}(N - Z)$; however, whenever neutrons and protons are filling the same shells it has been shown that a $T = 1$ neutron-proton pair must be added to conserve isospin (26).

The calculational problem is further simplified by treating the nucleon pairs as bosons. As a result, for a given number of $(N_{vn} + N_{vp})$ valence fermions the many-body system is replaced by a system of $N_p = \frac{1}{2}N_{vp}$ proton bosons and $N_n = \frac{1}{2}N_{vn}$ neutron bosons that can occupy two levels, an $L = 0$ s-level and an $L = 2$ d-level. The IBA-2 basis states are denoted by

$$|d_p^{n_{d_p}}(\alpha_p L_p)s_p^{N_p - n_{d_p}} d_n^{n_{d_n}}(\alpha_n L_n)s_n^{N_n - n_{d_n}} L\rangle. \qquad 7.$$

At a purely phenomenological level the most general rotationally invariant, boson-number-conserving, two-body IBA-2 Hamiltonian would contain up to 26 parameters. Since this is an impractical number, one has often used a simplified form suggested by microscopic considerations:

$$H = E_0 + \varepsilon(\hat{n}_{d_p} + \hat{n}_{d_n}) + \kappa \tilde{Q}_p^{(2)} \cdot \tilde{Q}_n^{(2)} + V_{nn} + V_{pp} + \lambda_M \hat{M}. \qquad 8.$$

Here the first term contributes to the ground-state energy only; the second term represents the energy difference between the s- and d-levels (that can be thought to arise from the stronger monopole pairing as compared to the quadrupole pairing). The third term is the quadrupole-quadrupole inter-action between the neutron and proton bosons (which is the image of the corresponding interaction in fermion space). The $\tilde{Q}^{(2)}$ operators have the general form of Equation 2; the structure parameters $\chi_n$, $\chi_p$ are empirically found to be strongly dependent on $N_n$, $N_p$, which can be interpreted as a Pauli effect (see Section 4). The interactions $V_{nn}$ ($V_{pp}$) between the like bosons (that are usually restricted to d-boson-number-conserving inter-actions) appear to be relatively unimportant and are frequently neglected.

The last term in Equation 8 is called the Majorana interaction (10, 11). Its role can illustrated by considering its action on the states in the $N_p = N_n = 1$ system. By taking linear combinations one can construct basis states with a definite symmetry character under p $\leftrightarrow$ n exchange, namely the symmetric states $|s_p s_n\rangle$, $2^{-1/2}|d_p s_n + d_n s_p\rangle$, and $|(d_p d_n)^{L=0,2,4}\rangle$, and the antisymmetric states $2^{-1/2}|d_p s_n - d_n s_n\rangle$, and $|(d_p d_n)^{L=1,3}\rangle$. One sees that the Majorana interaction, given by

$$\hat{M} \equiv (d_p^+ s_n^+ - s_p^+ d_n^+)^{(2)} \cdot (\tilde{d}_p s_n - s_p \tilde{d}_n)^{(2)} - 2 \sum_{\lambda=1,3} (d_p^+ d_n^+)^{(\lambda)} \cdot (\tilde{d}_p \tilde{d}_n)^{(\lambda)}, \qquad 9.$$

by construction acts only on the antisymmetric states, and thus plays the role of a symmetry energy.

The neutron-proton degree of freedom in IBA-2 is mathematically equivalent to that of a particle with spin $\frac{1}{2}$ and therefore one may classify a many-body state by its total "spin," for which the symbol $F$ ($F$-spin) has been introduced (10, 11) with $F_z = \pm\frac{1}{2}$ for p (n). The (anti-) symmetric two-boson np states have $F_z = 0$ and $F = 1$ (0). For general $N_p$, $N_n$ the basis states can be classified according to the irreducible representations of either the group $U^{(p)}(6) \otimes U^{(n)}(6)$, or alternatively the group $U(6) \otimes SU^{(F)}(2)$, where $SU^{(F)}(2)$ is the $F$-spin group. Because of the overall boson symmetry, the $U(6)$ and $SU^{(F)}(2)$ representations are characterized by the same Young diagram with two rows of length $(\frac{1}{2}N + F)$ and $(\frac{1}{2}N - F)$, respectively ($N = N_p + N_n$). The basis states with $F_z = \frac{1}{2}(N_p - N_n)$ and maximum $F$-spin, $F_{Max} = \frac{1}{2}N$, are totally symmetric and have a one-to-one correspondence with those in IBA-1. The other basis states with $F < F_{Max}$ are referred to as mixed symmetry states and their position relative to the $F_{Max}$ states is controlled by the strength of the Majorana term, which has a spectrum $\langle \hat{M} \rangle = (\frac{1}{2}N - F)(\frac{1}{2}N + F + 1)$. Although in realistic cases the Hamiltonian of Equation 8 may contain $F$-spin-breaking terms, in practice one finds that the lowest eigenstates are predominantly symmetric in character.

If the $V_{nn}$ and $V_{pp}$ terms in Equation 8 are neglected there are only four parameters $\varepsilon$, $\kappa$, $\chi_p$, and $\chi_n$ that determine the low-lying spectrum of a particular nucleus. One can distinguish several regimes with different characteristics. For example, for small $N_p$ and $N_n$ the one-body terms dominate and the ground state is mainly a condensate of s-bosons, whereas for larger $N_p$, $N_n$ the two-body interactions become more important and result in more d-boson admixtures in the ground state, the detailed structure of which depends on the values of $\chi$ parameters.

## 3.2  Dynamic Symmetries in IBA-2

To obtain more insight into the nature of the phenomena that can be described in the framework of IBA-2, it is instructive to study special cases of Equation 8 for which the eigenvalue problem can be solved analytically (27) and a simple geometric interpretation can be given (13). A simple inspection shows that in IBA-2 many dynamic symmetries occur. Here we restrict ourselves to a few that appear to be of physical interest, namely the subgroup chains:

$$U^{(p)}(6) \otimes U^{(n)}(6) \supset U^{(p+n)}(6) \supset \begin{cases} U^{(p+n)}(5) \supset O^{(p+n)}(5) \supset O^{(p+n)}(3) & \text{IA.} \\ O^{(p+n)}(6) \supset O^{(p+n)}(5) \supset O^{(p+n)}(3) & \text{IIA.} \\ SU^{(p+n)}(3) \qquad\qquad\qquad \supset O^{(p+n)}(3) & \text{IIIA.} \end{cases}$$

and

$$U^{(p)}(6) \otimes U^{(n)}(6)$$
$$\supset \begin{cases} U^{(p)}(5) \otimes U^{(n)}(5) \supset U^{(p+n)}(5) \supset O^{(p+n)}(5) \supset O^{(p+n)}(3) & \text{IB.} \\ O^{(p)}(6) \otimes O^{(n)}(6) \supset O^{(p+n)}(6) \supset O^{(p+n)}(5) \supset O^{(p+n)}(3) & \text{IIB.} \\ SU^{(p)}(3) \otimes SU^{(n)}(3) \supset SU^{(p+n)}(3) \supset \qquad\qquad O^{(p+n)}(3). & \text{IIIB.} \end{cases}$$

The Hamiltonians corresponding to these two types of group chains differ in the following way. In chains of type (A) the neutron-neutron, proton-proton, and neutron-proton interactions need to be equal since the generators of the group $U^{(p+n)}(6)$ are symmetric combinations of the $U^{(p)}(6)$ and $U^{(n)}(6)$ generators, e.g. $\tilde{Q}^{(2)} = \tilde{Q}_p^{(2)} + \tilde{Q}_n^{(2)}$, such that the Hamiltonian is an $F$-spin scalar. On the other hand this constraint is not present in the chains of type (B), and here the wave functions are characterized by the quantum numbers of the neutron and proton systems separately, rather than $F$-spin.

By writing the IBA-2 Hamiltonian in terms of the Casimir invariants of one of the subgroup chains, one finds analytic expressions for the energy spectrum, just as was the case in IBA-1. For details, see (27).

A new feature of IBA-2 is the following. Physically the bosons can be the image of either particle pairs or hole pairs. With only one type of bosons present the energy spectra are the same for these situations. However, in IBA-2 one must distinguish the situation where both neutrons and protons are particle-like (hole-like) and those where one deals with a particle-like and hole-like combination. In terms of the allowed $N$-boson representations of the group U(6), the particle case can be represented by $[N_p] \equiv [N_p, 0, 0, 0, 0, 0]$ and the hole case by the conjugate representation $[\bar{N}_n] \equiv [N_n, N_n, N_n, N_n, N_n, 0]$. When combining the neutron and proton representations the particle-particle (or hole-hole) and particle-hole cases lead to different results, as is shown below more explicitly for case IIIB.

Some interesting consequences of the neutron-proton degree of freedom can be illustrated by considering a simple schematic Hamiltonian, appropriate for the deformed region

$$H = \kappa \tilde{Q}^{(2)} \cdot \tilde{Q}^{(2)} + \lambda_M \hat{M}, \qquad\qquad 10.$$

where $\tilde{Q}^{(2)} = \tilde{Q}_p^{(2)} + \tilde{Q}_n^{(2)}$ with $\kappa < 0$, $\lambda_M \geqslant 0$. In the following we consider the three special cases $\chi_p = \chi_n = \pm\frac{1}{2}\sqrt{7}$, $\chi_p = -\chi_n = \pm\frac{1}{2}\sqrt{7}$, and $\chi_p = \chi_n = 0$, where analytic solutions can be obtained.

THE SU(3) LIMIT   For $\chi_p = \chi_n = \pm\frac{1}{2}\sqrt{7}$ the Hamiltonian in Equation 10 can be expressed in terms of the Casimir invariants of the subgroups $U^{(p+n)}(6)$, $SU^{(p+n)}(3)$, and $O^{(p+n)}(3)$, and therefore is diagonal in scheme IIIA. The energy spectrum reads

$$E(\lambda, \mu, F, L) = E_0 + \tfrac{1}{2}\kappa[\lambda^2 + \mu^2 + 3(\lambda + \mu) + \lambda\mu]$$
$$- \tfrac{3}{8}\kappa L(L+1) - \lambda_M F(F+1). \qquad 11.$$

The allowed values of $(\lambda, \mu)$ representations of the group $SU^{(p+n)}(3)$ follow from the tensor decomposition of the outer product of the representations $(\lambda_p, \mu_p) \otimes (\lambda_n, \mu_n)$ of $SU^{(p)}(3) \otimes SU^{(n)}(3)$:

$$(\lambda, \mu) = (2N, 0) + (2N-2, 1) + (2N-4, 2)^3 + \dots.$$

Thus in comparison to IBA-1, additional SU(3) representations (such as $\mu = 1$) occur corresponding to mixed symmetry neutron-proton modes, some of which are shown in Figure 1c. Below we discuss the experimental evidence for the $(\lambda = 2N-2, \mu = 1)$, $K^\pi = 1^+$ band.

THE SU*(3) LIMIT   For $\chi_p = -\chi_n = \pm\frac{1}{2}\sqrt{7}$ and $\lambda_M = 0$ the Hamiltonian of Equation 10 is diagonal in the scheme IIIB. In this case one must combine

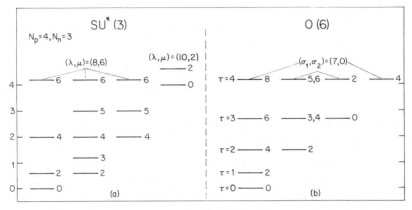

*Figure 4* Comparison of a spectrum in the SU*(3) limit of Equation 10 with that in the O(6) limit.

the $(\lambda_p, \mu_p)$ representations of SU$^{(p)}$(3) with the conjugate representations $(\lambda_n, \mu_n) = (\mu_n, \lambda_n)$ of SU$^{(n)}$(3), with the result

$$(\lambda, \mu) = (2N_p, 2N_n) + (2N_p - 1, 2N_n - 1) + \cdots + (2N_p - 4, 2N_n + 2) + \dots.$$

To distinguish this case from the previous one it will be referred to as the SU*(3) limit (29). Figure 4*a* shows that the angular momentum content of the leading representation $(2N_p, 2N_n)$ is the same as that of the rigid triaxial rotor (1). In practice one expects the degeneracies in this limit to be perturbed by SU*(3) symmetry-breaking terms in Equation 10, e.g. $|\chi_i| \neq \frac{1}{2}\sqrt{7}$ or $\lambda_M \neq 0$. One can show that this leads to a splitting of the angular momentum degeneracy and an ordering of the states into rotational bands connected by strong intraband and weaker interband E2 transitions.

THE O(6) LIMIT    The case of $\chi_p = \chi_n = 0$ in Equation 10 can also be solved in closed form by using the identity

$$(d^+ s + s^+ \tilde{d})^{(2)} \cdot (d^+ s + s^+ \tilde{d})^{(2)} = \frac{1}{2} C_2[O(6)] - \frac{1}{2} C_2[O(5)].\qquad 12.$$

The corresponding spectrum reads

$$E(\sigma, \tau, F, L) = \kappa[\sigma_1(\sigma_1 + 4) - \tau_1(\tau_1 + 3) + \sigma_2(\sigma_2 + 2) - \tau_2(\tau_2 + 1)]$$
$$- \lambda_M F(F + 1).$$

For a comparison with the SU*(3) case the states in the lowest representation with $\sigma_1 = N_p + N_n, \sigma_2 = \tau_2 = 0$ [which coincides with the corresponding $\sigma = N$ multiplet of the O(6) limit of IBA-1] are shown in Figure 4*b*. Two points are worth noticing: (*a*) while in O(6) the first excited $0^+$

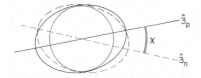

Figure 5  Geometric interpretation of the $(\lambda, \mu) = (2N-2, 1)$ $K = 1$ band in the SU(3) limit of IBA-2 as a small amplitude oscillation between the symmetry axis of the neutron and proton distributions.

state is a member of the ground-state multiplet, in SU*(3) it belongs to a higher representation; and (b) the odd-even staggering of the energy levels in the quasi-$\gamma$-band is reversed.

GEOMETRIC INTERPRETATION OF IBA-2    A geometric interpretation of the various dynamic symmetries of IBA-2 can be obtained (13) using the techniques mentioned in Section 2.2. The appropriate coherent states have a simple product form

$$|N_p\alpha_pN_n\alpha_n\rangle = \prod_{k=p,n} \mathcal{N}_k(s_k^+ + \sum_\mu \alpha_{k,\mu}^* d_{k,\mu}^+)^{N_k}|0\rangle. \qquad 13.$$

The ten quadrupole degrees of freedom $\alpha_n, \alpha_p$ can be expressed in terms of three Euler angles describing the orientation of the matter distribution; four intrinsic deformation $(\beta_{p,n})$ and triaxiality $(\gamma_{p,n})$ variables characterizing the neutron and proton distributions; and three angles that specify the relative orientation of the two systems. Minimization of the energy surface with respect to these variables defines the equilibrium shape. One finds (13, 28) that the $U^{(p+n)}(5)$, $O^{(p+n)}(6)$, and $SU^{(p+n)}(3)$ ground-state matter distributions are symmetric in the neutron-proton degree of freedom and can be characterized by the same geometry as in IBA-1, namely the anharmonic spherical vibrator, the $\gamma$-unstable rotor, and the axially symmetric deformed rotor, respectively.

In addition to totally symmetric excitation modes, in all three cases modes are present that can be interpreted as small amplitude oscillations of the neutrons against the protons $(\Delta F = 1)$. In particular the $(\lambda = 2N-2, \mu = 1)$ $K = 1^+$ band in the SU(3) limit can be depicted as a small amplitude oscillation in terms of the angle $\chi$ between the symmetry axis of the neutron and proton axially symmetric deformed bodies (see Figure 5). The SU*(3) limit can be visualized as a prolate, axially symmetric, deformed proton distribution coupled with an oblate neutron distribution through an attractive quadrupole-quadrupole interaction. The lowest configuration is then one in which the two symmetry axis are perpendicular, corresponding to a triaxial matter distribution.

## 3.3    Mixed Symmetry States in IBA-2

One of the most interesting properties of the IBA-2 model is the prediction of low-lying collective states that are not totally symmetric in the neutron-proton degree of freedom. Whereas in the past such modes were also proposed in geometric models (30, 30a,b) in the IBA-2 approach these states are a direct consequence of the underlying shell model picture, and therefore have a microscopic interpretation. Since strong evidence for the existence of these modes was recently presented, we discuss them in more detail.

While collective mixed symmetry states are predicted to occur in all regions where the IBA model is applicable, they are of primary interest in the SU(3) region where a low-lying $K^\pi = 1^+$ band is predicted with a strong M1 matrix element connecting its $L^\pi = 1^+$ bandhead with the $L^\pi = 0^+$ ground state.

The magnetic dipole operator in the IBA-2 model can in lowest order be expressed as

$$T_\mu(M1) = \sqrt{\frac{3}{4\pi}} [g_p L^{(1)}_{p,\mu} + g_n L^{(1)}_{n,\mu}], \qquad 14.$$

where $g_p$ ($g_n$) are the proton (neutron) boson $g$-factors. (Their microscopic interpretation is discussed in Section 4.) For the following discussion it is convenient to decompose Equation 14 into a part that is diagonal and does not change the $F$-spin symmetry ($\Delta F = 0$) and a part that connects symmetric with mixed symmetric states ($\Delta F = 1$)

$$T_\mu(M1) = \sqrt{\frac{3}{4\pi}} \left\{ \frac{1}{N} (g_p N_p + g_n N_n) L^{(1)}_\mu \right.$$

$$\left. + (g_p - g_n) [N_n L^{(1)}_{p,\mu} - N_p L^{(1)}_{n,\mu}]/N \right\}. \qquad 15.$$

From the first term within the braces, one obtains (31, 32) a simple expression for the $g$-factors of low-lying symmetric states:

$$g_R = \frac{\langle \mu \rangle}{\langle L \rangle} = (g_p N_p + g_n N_n)/N. \qquad 16.$$

This formula has been shown (32) to describe the trend of experimental $g$-factors of $2^+_1$ states in the rare-earth region rather well, with values of $g_p$ ($g_n$) that are smooth functions of $N_p$ ($N_n$) and deviate no more than 25% from their naive values $g_p = 1$, $g_n = 0$ [$\mu_N$].

The general trend for a series of isotopes is a decreasing function of neutron number in the 82–126 shell for $N \leqslant 104$ (where the neutron boson number $N_n$ increases) and an increasing function for $N \geqslant 104$ (where $N_n$ decreases). It has been observed (33) that the application of Equation 16 with the standard values of $N_p$ does not work so well for $Z \leqslant 64$ and

$N \leqslant 88$. This can probably be attributed (33, 34) to a proton subshell closure effect at $Z = 64$, which effectively reduces the number of active correlated valence pairs, and thus $N_p$.

The second term in Equation 15 describes M1 transitions. Whereas in the spherical region this operator vanishes when acting on the ground state (which contains only s-bosons), in the deformed limits of IBA-2 one can easily construct a sum rule for the M1 strength using group theoretical techniques. For example, in the SU(3) limit one finds (13, 32)

$$B(M1, 0^+ \rightarrow 1^+) = \frac{3}{4\pi} \frac{4N_p N_n}{N_p + N_n} (g_p - g_n)^2 \; [\mu_N^2].$$   17.

This formula reflects the fact that only the valence nucleons participate in the collective motion, in contrast to most geometric models that predict a smooth $A^{4/3}$ dependence (35, 36). Using the empirical values of $g_p$ and $g_n$ as obtained from the analysis of $g$-factors (Equation 16), one predicts a $B(M1)$ strength of the order of $2.5\mu_N^2$ for a typical well-deformed nucleus such as $^{156}$Gd. It appears that breaking of the SU(3) symmetry does not change this prediction qualitatively and only leads to a slight redistribution of the strength over several $1^+$ states. It should be noted, however, that up to now the excitation energy of the $K = 1$ band (which strongly depends on the strength of the Majorana interaction) has not been calculated reliably.

Before discussing the experimental situation it is interesting to note that recent calculations in the framework of the Nilsson model, with a conventional pairing plus quadrupole interaction Hamiltonian, lead to qualitatively similar results (36–40). Whereas in the unperturbed Nilsson basis the M1 strength is strongly fragmented, the residual $Q \cdot Q$ interaction tends to concentrate the strength ($\sim 6\mu_N^2$) into one state (with the character of a $0\hbar\omega$ giant resonance) in the region 2.5–5 MeV, depending upon the details of the interaction.

Motivated by the large predicted M1 strength, researchers have performed electron scattering experiments at the Darmstadt linear accelerator to search for it. At present in several rare-earth nuclei at backward angles peaks have been observed (41, 41a) that on the basis of the measured form factor are interpreted as magnetic dipole excitations with an orbital character. As an example Figure 6 shows in $^{156}$Gd a single strong M1 state at $E_x = 3.075$ MeV with an estimated strength of $B(M1, \uparrow) \approx 1.5\mu_N^2$ and possibly smaller pieces of strength present in the background. In a recent $(\gamma, \gamma')$ experiment on the Gd isotopes (42) these states were confirmed and moreover additional information could be obtained. For example, the broad state observed in $^{158}$Gd(e, e') has been resolved in an intriguing doublet, 9 keV apart (see Figure 6). This figure also shows that in the $(\gamma, \gamma')$ work not only the elastically scattered photon was observed but also the decay of the $1^+$ state to the $2_1^+$ level. In this respect it is of interest to note that the ratio $R = B(M1, 1^+ \to 2_1^+)/B(M1, 1^+ \to 0_1^+)$ contains information about the $K$ value of the decaying state. Namely, in the rotational limit $R$ is just a ratio of Clebsch-Gordan coefficients squared, with the value 0.5 or 2.0 depending upon whether $K = 1$ or 0. Experimentally (42) most excited states have $R \approx 0.5$, consistent with the $K = 1$ assignment. Further evidence for the orbital rather than spin-flip type character of this particular M1 mode comes from the fact that in a high-resolution (p, p') experiment (P.

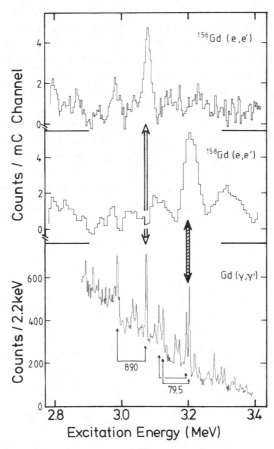

*Figure 6* Comparison of (e, e') spectra on $^{156,158}$Gd [taken at $E_e = 25$ MeV, $\theta_{e'} = 165°$ (41)] with a high-resolution $(\gamma, \gamma')$ spectrum on natural Gd (42). The large arrows point to the elastically scattered photons, and the numbers (in keV) between the lower arrows indicate the positions of the inelastically scattered photons to the first excited $2^+$ state. The peaks at $E_x = 3.075$ MeV in $^{156}$Gd and at 3.20 MeV in $^{158}$Gd have been identified (41, 41a) as $1^+$ states (note that the latter is really a doublet).

von Brentano, personal communication) at $E_p \approx 25$ MeV on $^{156}$Gd the state has not been observed. Namely, at low energies, protons couple predominantly with the spin current and much less with the nuclear convection current (43) (this in contrast to electrons).

### 3.4  Transitional Nuclei and Triaxial Shapes

The form of the Hamiltonian in Equation 8 is very convenient for the study of transitional regions in between the limiting cases. In most applications one tries to describe a series of isotopes by varying the neutron boson number $N_n$ and adjusting the parameters $\varepsilon$, $\kappa$, and $\chi_n$; in practice one usually finds that the first two of these parameters are rather constant and that only the structure parameters $\chi$ vary strongly with boson number. These phenomenological parameters can then be compared with the results of microscopic calculations (discussed in Section 4).

A question of particular interest in IBA-2 is the interplay between the O(6) limit (with $\chi_n = \chi_p = 0$, corresponding to $\gamma$-unstable nuclei) and the SU*(3) limit (with $\chi_p = -\chi_n \mp \pm\frac{1}{2}\sqrt{7}$, corresponding to a triaxial matter distribution in the ground state). Whereas in many respects these two limits give rise to very similar nuclear properties, a more detailed analysis indicates characteristic differences, in particular the odd-even staggering of the energies and certain B(E2) values in the $\gamma$ band. As an illustration we present in Figures 7 and 8 a comparison of the low-lying spectrum and some

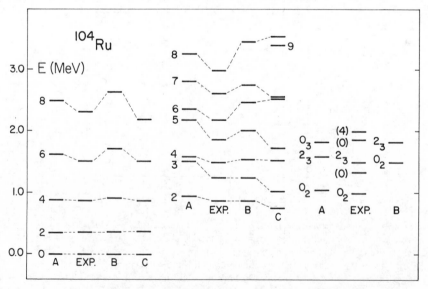

*Figure 7*  The experimental level scheme of $^{104}$Ru compared with a perturbed O(6) spectrum (*A*), a perturbed SU*(3) spectrum (*B*), and a rigid triaxial rotor spectrum (*C*) (from 29).

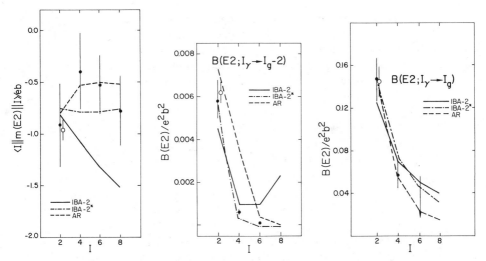

*Figure 8* Experimental (44) quadrupole moments and $B(E2)$ values in $^{104}$Ru as a function of spin $I$ compared with calculated values. AR stands for asymmetric rotor. IBA-2 results are from (45), IBA-2* from (29).

$B(E2)$ values extracted from a recent Coulomb excitation measurement (44) for the transitional nucleus $^{104}$Ru with the results of two IBA-2 calculations. One sees that the perturbed SU*(3) parametrization of (29) describes the experimental data better than the $\gamma$-unstable one (45).

## 4.    MICROSCOPY OF IBA-2

### 4.1    *Introduction*

In most microscopic approaches to collective motion one first determines the ground-state wave function from a variational principle (e.g. by using a quasi-particle or BCS transformation); subsequently, excited states are described in terms of collective particle-hole excitations built upon this ground state. In contrast, in the IBA approach the starting point is the spherical shell model in which the properties of low-lying states in nuclei are described in terms of the valence nucleons that move outside a closed spherical core. Therefore in applications of IBA to nuclei far away from closed shells it is implicitly assumed that the magic numbers remain well defined even in strongly deformed nuclei, where the Nilsson splitting can be rather large compared to the difference in spherical single-particle energies. The validity of this assumption is difficult to assess; however, the empirical indications for finite $N$ effects provide some evidence that the shell structure remains a meaningful concept even in strongly deformed nuclei. Since the full $0\hbar\omega$ shell model problem for a (medium-) heavy nucleus cannot possibly be solved exactly, further approximations must be made. In the IBA-2 approach one proceeds in two steps: (a) a truncation of the full shell model space into a subspace containing only collective $L = 0$ (S) and $L = 2$ (D) pairs (SD subspace); (b) this collective fermionic subspace is mapped onto a boson space (sd space). These two steps are discussed in more detail in the following, where we find it convenient to distinguish between the spherical and deformed regions.

## 4.2   SD Subspace

The question of the optimal definition of the collective subspace has been addressed by several groups (46–52). Whereas in the beginning these studies were restricted to simple systems (single-$j$ shell) recently more realistic cases were considered (nondegenerate multi-$j$ shells including neutron-proton interactions). For a quantitative discussion it is convenient to define a shell model space by a set of quasi-degenerate single-particle orbitals $\{j_1 \ldots j_n\}$. In many cases of practical interest this set comprises the $Z = 50$–82 proton orbitals d5/2, g7/2, h11/2, s1/2, d3/2, and the $N = 82$–126 neutron orbitals h9/2, f7/2, i13/2, f5/2, p3/2, and p1/2. In this space[1] one defines collective pair creation operators with angular momentum $L = 0$ and $L = 2$:

$$S^+ = \tfrac{1}{2} \sum_j \alpha_j (2j+1)^{1/2} [a_j^+ a_j^+]_0^{(0)}, \qquad\qquad 18a.$$

and

$$D_\mu^+ = \tfrac{1}{2} \sum_{jj'} \beta_{jj'} (1 + \delta_{jj'})^{1/2} [a_j^+ a_j^+]_\mu^{(2)}, \qquad\qquad 18b.$$

where $\alpha_j$ and $\beta_{jj'}$ are structure constants. The many-body collective SD subspace is then defined by acting on the closed shell with the operators of Equation 18

$$|\text{"SD"}\rangle = (S^+)^{N-n} \mathscr{P}_0 [(D^+)^n [\alpha L]] |0\rangle,$$

where $\mathscr{P}_0$ is a projection operator that projects out all nucleon pairs coupled to $L = 0$ that are present in $(D^+)^n$; the label $n$ can be interpreted as the number of broken pairs present in the basis state, i.e. such a state has generalized seniority $w = 2n$ (50).

SPHERICAL CASE    In the spherical case, the lowest eigenstates are those with a small number of D-pairs. The structure of the S-pairs (i.e. the coefficients $\alpha_j$) can be obtained by minimizing the shell model Hamiltonian with respect to the state with only S-pairs: $|S_p^{N_p} S_n^{N_n}\rangle$. For the determination of the $\beta_{jj'}$ in the D-pairs, several prescriptions can be given. For example, one may

---

[1] For simplicity we drop the subscripts p, n; the collective space should be understood as a direct product of neutron and proton spaces.

optimize the excitation energy of the D-pair by diagonalizing the shell model Hamiltonian within the space of "one-broken pair," which is equivalent to minimizing the energy with respect to the states (keeping the $\alpha_j$ fixed) $|D_p(S_p)^{N_p-1}S_n^{N_n}\rangle$ and $|D_n(S_n)^{N_n-1}S_p^{N_p}\rangle$. Another possibility is to maximize the strength of E2 transitions within the SD space by defining $D_\mu^+ \approx [Q_\mu^{(2)}, S^+]$, where $Q_\mu^{(2)}$ is the quadrupole operator, $Q_\mu^{(2)} = \sum_i r_i^2 Y_{2\mu}(\hat{r}_i)$. While these two methods lead to D-pairs that are qualitatively similar, it is generally felt that the latter is less realistic because the differences in single-particle energies are neglected.

Although the resulting structure constants $\alpha_j$ and $\beta_{jj'}$ depend somewhat on the details of the shell model Hamiltonian, some general features can be recognized. For example, in the calculation for $_{62}$Sm isotopes ($N_p = 6$) Scholten (50) finds that the $\alpha_{p,j}$ as a function of $N_n$ vary very little (see Figure 9); as a function of $N_p$ there is a stronger but smooth variation. Results for the $\beta$ coefficients are similar.

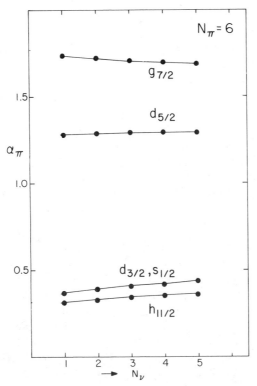

*Figure 9* The structure constants $\alpha_j$ for the protons calculated as a function of the number of neutron pairs, $N_n$, in the Sm isotopes (from 50).

Another important question is how good the SD truncation is, or more particularly how one can deal with the coupling to the noncollective space. In the latter space one may distinguish higher angular momentum pairs such as an $L = 4$ G-pair, and less collective $L = 0, 2$ pairs orthogonal to the S- and D-pairs. The effect of the G-pair, which couples to the SD space mainly through the hexadecapole component in the neutron-proton interaction, can be taken into account most conveniently by explicitly including the $L = 4$ pair in the collective space. After mapping onto the corresponding boson space one can eliminate the g-boson by renormalizing the interactions in the sd space. From practical calculations it appears that its main effect is a lowering of the d-boson energies by about 20%.

The coupling with the low angular momentum states is more difficult to take into account. As an example we consider an $L = 0$ state with two D-pairs that can be expanded as

$$[D^+ D^+]^{(0)} (S^+)^{N-2} |0\rangle = \alpha_0 |S^N\rangle + \sum_i \beta_i |S_i' S^{N-1}\rangle$$

$$+ \gamma |\mathscr{P}_0 [DD]^{(0)} S^{N-2}\rangle, \qquad\qquad 19.$$

where $S_i'$ denote all $L = 0$ pairs orthogonal to the S-pair. Because of the required orthogonality onto zero-broken pairs, one wants to map only the last term in Equation 19 onto the boson space.

It has been argued (46, 46a) that states containing the first excited S'-pair, $|S_{i=1}' S^{N-1}\rangle$ [sometimes referred to as a pair vibrational state (53)], can occur at an unperturbed energy comparable to that of the two-broken-pair state, and therefore are expected to mix appreciably ($\beta_1 \approx \gamma$ in Equation 19), in particular for (almost) semi-magic nuclei. In such cases one is forced to treat the effect of the intruder states explicitly as a new degree of freedom. (It is expected that this problem is less severe if both protons and neutrons are active since the neutron-proton interaction effectively lowers the D-pair energy.)

DEFORMED NUCLEI    Whereas in the spherical region the truncation of the shell model space to S + D-pairs only appears as an intuitively natural approximation scheme, its validity in the deformed region is less obvious. In fact the applicability of the IBA model to deformed nuclei has been questioned (54). For a quantitative discussion of the situation for an axially symmetric deformed nucleus in terms of IBA-2, it is convenient to work in the intrinsic frame, assuming that the ground-state band can be expressed as a product of neutron and proton intrinsic states with projection $K = 0$ on the symmetry axis: $\Phi_0 = \Phi_{p,0} \cdot \Phi_{n,0}$. (We note that this assumption does not imply that the neutrons and protons are uncorrelated; on the contrary, the neutron and proton symmetry axis are assumed to coincide, as a result of the attractive neutron proton interaction.) Either of the intrinsic states $\Phi_{k,0}$ can be approximated by a condensate of $N_k$ Cooper pairs in deformed single-particle orbitals:

$$\Phi_0 = \mathscr{N}(\Lambda_0^+)^N|0\rangle, \qquad\qquad 20.$$

where $\mathscr{N}$ is a normalization constant.[2] To investigate their angular momentum structure the $\Lambda_0^+$ operators are expressed as a sum of multipole pair operators:

$$\Lambda_0^+ = \sum_L X_L \Gamma_{L,0}^+, \qquad\qquad 21.$$

where $\Gamma_{L,0}^+$ creates a pair of fermions with angular momentum $L$ and projection $K = 0$:

$$\Gamma_{L,0}^+ = \tfrac{1}{2}\sum_{jj'} \alpha_{jj'}^L (1+\delta_{jj'})^{1/2}[a_j^+ a_{j'}^+]_0^{(L)}.$$

The structure coefficients, $\alpha_{jj'}^L$, and the multipole amplitudes, $X_L$, can be determined by minimizing the expectation value of the Hamiltonian with respect to Equation 20, so that the validity of the truncation of the sum in Equation 21 to small $L$ values can be investigated by comparing with a complete expansion. A recent calculation of the Tokyo group (51, 51a, 55) yields $X_0^2 + X_2^2 \approx 0.85$ and $X_0^2 + X_2^2 + X_4^2 \approx 0.95$, which indicates an appreciable contribution from the $L = 4$ pair. However, it is more relevant to study the effect of the truncation on observables such as the moment of inertia and quadrupole moments. The latter requires the calculation in the intrinsic frame of $Q_{\text{int}} \equiv \langle\Phi_0|Q_0^{(2)}|\Phi_0\rangle$. In the special case of the aligned Nilsson limit (no pairing) one can show that the $\Lambda_0$ pairs can be regarded as $N$ "independent pairs" (55)

$$Q_{\text{int}} \equiv \langle\Lambda_0^N|Q_0^{(2)}|\Lambda_0^N\rangle = N\langle\Lambda_0|Q_0^{(2)}|\Lambda_0\rangle. \qquad 22.$$

In general, however, blocking effects must be taken into account; these tend to reduce the estimate of Equation 22.

From these studies (55) it emerges that the SD space accounts for only 70% of the value of $Q_{\text{int}}$, while the inclusion of the G-pair results in approximately 96% of the exact value. A similar investigation or the moment of inertia also indicates an appreciable contribution from the G-pair.

### 4.3 Mapping Onto Boson Space

Several methods have been developed for obtaining the boson image of the truncated shell model problem. While some of these methods are formally

---

[2] This state can also be regarded as a number-projected Hartree-Fock-Bogoljubov state.

exact (but difficult to implement in practice since they give rise to non-Hermitean Hamiltonians or to higher-order boson-boson interactions), others are more practical but involve approximations that are difficult to control.

SPHERICAL NUCLEI    In most microscopic IBA calculations in vibrational nuclei, a procedure has been followed first proposed by Otsuka, Arima & Iachello (56) (OAI method). In this method, one assumes a correspondence between boson and fermion states based upon common labels; this amounts to the equivalence of d-boson number $n_d$ and the fermion seniority $v$ in the case of a single-$j$ shell, or generalized seniority in the multi-$j$ shell case,

$$|S^{N-n_d}\mathcal{P}_0[D^{n_d}(\alpha L)]\rangle \to |s^{N-n_d}d^{n_d}(\alpha L)\rangle.$$

To construct operators in the boson space one assumes a simple boson image of the fermion operator and evaluates the coupling constant by equating certain matrix elements in both spaces (in general the states with $n_d = 0, 1$). As an illustration we consider the mapping of the quadrupole operator, $Q_{F,\mu}^{(2)} = \sum_i r_i^2 Y_{2\mu}(\hat{r}_i)$, for which one assumes a boson image of the form

$$Q_{F,\mu}^{(2)} \to \tilde{Q}_{B,\mu}^{(2)} = e_B[d_\mu^+ s + s^+ \tilde{d}_\mu + \chi(d^+\tilde{d})_\mu^{(2)}]. \qquad 23.$$

By equating the matrix elements of Equation 23 for $n_d = 0$ and 1, we obtain

$$\langle S^{N-1}D\|Q_F^{(2)}\|S^N\rangle = e_B\langle s^{N-1}d\|(d^+s)^{(2)}\|s^N\rangle = e_B N\sqrt{5},$$

and

$$\langle S^{N-1}D\|Q_F^{(2)}\|S^{N-1}D\rangle = e_B\chi\langle s^{N-1}d\|(d^+\tilde{d})^{(2)}\|s^{N-1}d\rangle = e_B\chi\sqrt{5}.$$

Evaluating the left-hand side of these equations one obtains the boson couplings $e_B$ and $\chi$ as a function of $N$. In practice one finds that the quadrupole structure parameter $\chi$ absorbs most of the Pauli effects and varies strongly with the boson number, as can also be seen in Figure 10.

With the help of these one-body images one can also construct the image of simple separable Hamiltonians, such as the one of Equation 8. In this way several groups have calculated the IBA parameters for nuclei not too far from closed shells. Comparison of these parameters with the empirical ones shows in general at best a qualitative agreement. In particular both the calculated d-boson energies, $\varepsilon_d$, and the strength $\kappa$ of the $\tilde{Q}_p^{(2)}\cdot\tilde{Q}_n^{(2)}$ interaction are about a factor two too large compared to the empirical ones.

It has been shown (50, 57) that the situation can be improved by including renormalization effects coming from the coupling with states outside the collective SD space, such as S′, D′, and G, which have been estimated either

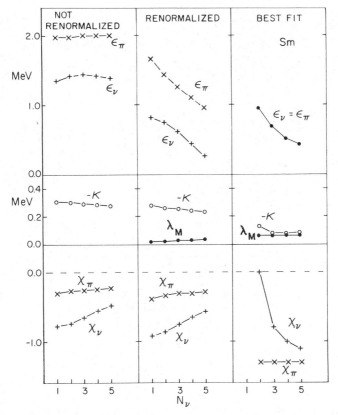

*Figure 10* The IBA-2 parameters $\varepsilon$, $\kappa$, $\chi$, and $\lambda_M$ calculated for the Sm isotopes compared with those obtained from an empirical fit (from 50).

in second-order perturbation theory or by more sophisticated methods. In particular it was found that the lowest order values of the parameters $\varepsilon_d$ and $\kappa$ are reduced appreciably; the renormalized values of $\varepsilon_d$ decrease with increasing $N$ in agreement with empirical trend (see Figure 10). A direct comparison with the experimental spectra for the Sm isotopes shows that the calculated level spacings are still too large (see Figure 11). Clearly, not all effects coming from the coupling to the excluded space can be absorbed in a renormalization of the parameters of Equation 8. In certain cases they give rise to more complicated interactions, such as three-body terms.

DEFORMED NUCLEI    In the deformed region where the number of d-bosons in the ground state is large, the OAI mapping is probably less justified. As an alternative, Otsuka (55) worked out a mapping of intrinsic states, as is suggested by the structure of the states of Equation 20; namely the map-

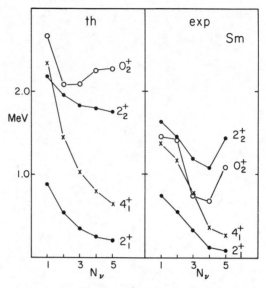

*Figure 11*    Comparison of the lowest states calculated with the renormalized IBA parameters of Figure 10 with the experimental states in the Sm isotopes (from 50).

ping $\Lambda_0^+ = \sum X_L \Gamma_L^+$ onto a boson $\lambda^+ = \sum X_L \gamma_L^+$, and consequently $(\Lambda_0^+)^N \rightarrow (\lambda^+)^N$.

The boson image of one-body operators can be constructed by the method described above with the difference that one now must equate matrix elements for intrinsic states, such as the ground-state, $\beta$, and $\gamma$ bands. The results obtained in this way (55) for the IBA-2 parameters show a good agreement with the empirical ones. It is important to include the $L = 4$ pair in the mapping; the required renormalization of the parameters in the sd-boson space has been carried out using a unitary transformation that transforms a linear combination of d- and g-bosons into a new d-boson.

Several points need further attention. For example, it is not obvious that the boson intrinsic state $(\lambda^+)^N|0\rangle$ with the given $X_L$ is an eigenstate of the boson Hamiltonian. Another question that arises is whether the structure of the $\Lambda_0$ pair for excited bands is the same for the ground-state band, which seems a necessary condition for a successful mapping onto bosons. This can be investigated for the $\beta$ band in axially symmetric deformed nuclei by using a trial wave function of the form $|\beta\rangle = \mathcal{N}_\beta \Lambda_\beta^+ (\Lambda_0^+)^{N-1}|0\rangle$, with $\Lambda_\beta^+ = \frac{1}{2}\sum X_L^\beta \alpha_{jj'}^{\beta,L}[a_j^+ a_{j'}^+]_0^{(L)}$. The multipole coefficients $X_L^\beta$ and structure parameters $\alpha_{jj'}^{\beta,L}$ can be determined by minimizing the shell model Hamiltonian with the constraint of orthogonality onto the ground state of Equation

20. In a somewhat schematic study Maglione et al (58) reported that less than 50% of the $\Lambda_\beta$-pair was contained in the SD space, and moreover that the coefficients $\alpha_{jj'}^{\beta;L}$ showed little resemblance to those for the ground state. This might indicate that the interpretation of the first excited $0^+$-band in deformed nuclei as a $\beta$-vibrational state is not fully justified.

## 5. CONCLUDING REMARKS

In the past decade the field of nuclear collective motion was revitalized by the advent of the interacting boson model. While no fundamentally new degrees of freedom were introduced, the IBA approach has several attractive properties that distinguish it from more conventional models and greatly contribute to its widespread use. First, by emphasizing the symmetry properties of the system an efficient and elegant formulation of the phenomenological aspects of collective motion is obtained, with the help of the powerful machinery of algebraic techniques. Secondly, the introduction of the neutron-proton version of the IBA model, in combination with the interpretation of the bosons as correlated pairs of nucleons, provided a link with the shell model and led to a novel microscopic formulation of nuclear collective motion that is rather different from the conventional approaches, in which the particle-hole rather than particle-particle correlations are emphasized.

At present one cannot claim that a fully satisfactory quantitative microscopic derivation of IBA-2 has been achieved. In particular in the deformed region one lacks a small expansion parameter that could control successive approximation. Also a satisfactory understanding of the so-called Majorana interaction that plays the role of neutron-proton symmetry energy is still missing.

Some limitations of the IBA model have also become apparent. For example, the rigid structure of the pairs independent of the excitation energy prohibits straightforward application of the model to higher excited states, where the effect of Pauli blocking or angular momentum must be taken into account (backbending region). Also, in some cases there seems to be an explicit need to introduce additional degrees of freedom; such as the pair-vibrational mode that appears to play a role for the description of low-lying $0^+$ states (53) near closed shell nuclei.

Another point of future interest is the treatment of the coupling of the $0\hbar\omega$ surface modes with the $2\hbar\omega$ giant resonances that represent the irrotational flow. The IBA description of the low-lying collective states in terms of valence nucleons only is quite successful for many purposes, but it appears too naive if one asks more detailed questions. For example, collective transition densities as measured in electron scattering show a much more surface-peaked behavior than one can obtain in $0\hbar\omega$ model space (59); this indicates the importance of the coupling with giant resonances. It would be of interest to study whether one can incorporate this degree of freedom by embedding the IBA model in a larger (non-compact) group structure. In this way one may make contact with the efforts of the Toronto group (60), which has formulated collective motion in terms of an $Sp(6, R)$ algebra.

ACKNOWLEDGMENT

We wish to thank Dr. O. Scholten for a critical reading of this manuscript. This work is part of the research program of the Stichting FOM, which is supported by the Nederlandse Organisatie voor Zuiver Wetenschappelijk Onderzoek (ZWO).

*Literature Cited*

1. Bohr, A., Mottelson, B. R. *Nuclear Structure*, Vol. II. New York: Benjamin (1975)
2. Gneuss, G., Greiner, W. *Nucl. Phys. A* 171:449 (1971)
3. Janssen, D., Jolos, R. V., Dönau, F. *Nucl. Phys. A* 224:93 (1974)
4. Arima, A., Iachello, F. *Phys. Rev. Lett.* 35: 1069 (1975)
5. Arima, A., Iachello, F. *Ann. Phys. NY* 99: 253 (1976)
6. Arima, A., Iachello, F. *Ann. Phys. NY* 111:201 (1978)
7. Arima, A., Iachello, F. *Ann. Phys. NY* 123:468 (1979)
8. Arima, A., Iachello, F. *Ann. Rev. Nucl. Part. Sci.* 31:75 (1981)
9. Arima, A., Iachello, F. *Adv. Nucl. Phys.* 13:139 (1984)
10. Arima, A., Otsuka, T., Iachello, F., Talmi, I. *Phys. Lett.* 66B:205 (1977)
11. Otsuka, T., Arima, A., Iachello, F., Talmi, I. *Phys. Lett.* 76B:139 (1978)
12. Elliott, J. P. *Proc. R. Soc. Ser. A* 245:128, 562 (1958)
13. Dieperink, A. E. L. *Prog. Part. Nucl. Phys.* 9:121 (1983)
14. Dieperink, A. E. L., Scholten, O., Iachello, F. *Phys. Rev. Lett.* 44:1747 (1980)
15. Van Roosmalen, O. S., Dieperink, A. E. L. *Phys. Lett.* 100B:299 (1981)
16. Ginocchio, J. N., Kirson, M. W. *Nucl. Phys. A* 350:31 (1980)
17. Hatch, R., Levit, S. *Phys. Rev. C* 25:614 (1982)
18. Moshinsky, M. *Nucl. Phys. A* 338:156 (1980)
19. Klein, A., Li, C. T., Vallieres, M. *Phys. Rev. C* 25:2733 (1982)
20. Weiguny, A. *Z. Phys. A* 301:335 (1981)
20a. Assenbaum, H. J., Weiguny, A. *Z. Phys. A* 310:75 (1983)
21. Feng, D. H., Gilmore, R., Deans, S. R. *Phys. Rev. C* 23:1254 (1981)
22. Gilmore, R., Feng, D. H. *Prog. Part. Nucl. Phys.* 9:479 (1983)
23. Casten, R. F., Warner, D. D., Aprahamian, A. *Phys. Rev. C* 28:894 (1983)
23a. Warner, D. D., Casten, R. F. *Phys. Rev. C* 28:1798 (1983)
24. Bijker, R., Dieperink, A. E. L. *Phys. Rev. C* 26:2688 (1982)
25. Wu, H. C., Zhou, X. Q. *Nucl. Phys. A* 417:67 (1984)
25a. Arima, A. *Nucl. Phys. A* 421:63c (1984)
26. Elliott, J. P., White, A. P. *Phys. Lett.* 97B:169 (1980)
27. Iachello, F. *Prog. Part. Nucl. Phys.* 9:5 (1983)
28. Balantekin, A. B., Barrett, B. R., Levit, S. *Phys. Lett.* 129B:153 (1983)
29. Dieperink, A. E. L., Bijker, R. *Phys. Lett.* 116B:77 (1982)
30. Lo Iudice, N., Palumbo, F. *Nucl. Phys. A* 326:193 (1979)
30a. De Franceschi, G., Palumbo, F., Lo Iudice, N. *Phys. Rev. C* 29:1496 (1984)
30b. Suzuki, T., Rowe, D. J. *Nucl. Phys. A* 289:461 (1977)
31. Sambataro, M., Dieperink, A. E. L. *Phys. Lett.* 107B:249 (1981)

32. Sambataro, M., Scholten, O., Dieperink, A. E. L., Piccitto, G. *Nucl. Phys. A* 423 : 333 (1984)
33. Wolf, A., et al. *Phys. Lett.* 123B : 165 (1983)
34. Scholten, O. *Phys. Lett.* 127B : 144 (1983)
35. Lipparini, E., Stringari, S. *Phys. Lett.* 130B : 139 (1983)
36. Bes, D. R., Broglia, R. A. *Phys. Lett.* 137B : 141 (1984)
37. Iwasaki, S., Hara, K. *Phys. Lett.* 144B : 9 (1984)
38. Hamamoto, I., Åberg, S. *Phys. Lett.* 145B : 163 (1984)
39. Hilton, R. R. *Z. Phys. A* 316 : 121 (1984)
40. Kurasawa, H., Suzuki, T. *Phys. Lett.* 144B : 151 (1984)
41. Bohle, D., Richter, A., Steffen, W., Dieperink, A. E. L., Lo Iudice, N., et al. *Phys. Lett.* 137B : 27 (1984)
41a. Bohle, D., Küchler, G., Richter, A., Steffen, W. *Phys. Lett.* 148B : 260 (1984)
42. Berg, U. E. P., et al. *Phys. Lett.* 149B : 59 (1984)
43. Carr, J. A., et al. *Phys. Rev. Lett.* 54 : 881 (1985)
44. Stachel, J., et al. *Nucl. Phys. A* 383 : 429 (1982); 419 : 589 (1984)
45. Van Isacker, P., Puddu, G. *Nucl. Phys. A* 348 : 125 (1980)
46. Allaart, K., Bonsignori, G. *Phys. Lett.* 124B : 1 (1983)
46a. Van Egmond, A., Allaart, K. *Nucl.*

*Phys. A* 425 : 275 (1984)
47. Van Egmond, A., Allaart, K., Bonsignori, G. *Nucl. Phys. A* 436 : 458 (1985)
48. Pittel, S., Duval, P. D., Barrett, B. R. *Ann. Phys. NY* 144 : 168 (1982)
49. Gambhir, Y. U., Ring, P., Schuck, P. *Phys. Rev. C* 25 : 2858 (1982)
50. Scholten, O. *Phys. Rev. C* 28 : 1783 (1983)
51. Otsuka, T., Arima, A., Yoshinaga, N. *Phys. Rev. Lett.* 48 : 387 (1982)
51a. Yoshinaga, N., Arima, A., Otsuka, T. *Phys. Lett.* 143B : 5 (1984)
52. Maglione, E., Vitturi, A., Dasso, C. H., Broglia, R. A. *Nucl. Phys. A* 404 : 333 (1983)
53. Sakata, F., Iwasaki, S., Marumori, T., Takada, K. *Z. Phys. A* 286 : 195 (1978)
54. Bohr, A., Mottelson, B. R. *Phys. Scr.* 22 : 468 (1980); 25 : 915 (1982)
55. Otsuka, T. *Phys. Lett.* 138B : 1 (1984)
56. Otsuka, T., Arima, A., Iachello, F. *Nucl. Phys. A* 309 : 1 (1978)
57. Zirnbauer, M. *Nucl. Phys. A* 419 : 241 (1984)
58. Maglione, E., Vitturi, A., Catara, F., Insolia, A. *Nucl. Phys. A* 411 : 181 (1983)
59. Scholten, O. *Proc. Int. Workshop on Interacting Boson-Boson and Boson-Fermion Systems*, ed. O. Scholten. Singapore : World Scientific (1985)
60. Le Blanc, R., Carvalho, J., Rowe, D. J. *Phys. Lett.* 140B : 155 (1984)

Ann. Rev. Nucl. Part. Sci. 1985. 35 : 107–34

# JET PRODUCTION IN HADRONIC COLLISIONS

## L. DiLella

Experimental Physics Division, CERN, 1211 Geneva 23, Switzerland

CONTENTS

## 1. INTRODUCTION

A typical hadron collision at high energy generally produces a large number of secondary particles. For example at the CERN Proton-Antiproton Collider, with a total center-of-mass energy $\sqrt{s} = 540$ GeV, an average of about 29 charged particles and about 31 photons are emitted (1), with broad distributions around these mean values.

For each secondary particle we can define a transverse momentum variable $p_T = p \sin \theta$, where $p$ is the particle momentum and $\theta$ is the emission angle with respect to the beam axis. It is found experimentally that the $p_T$ distribution falls very rapidly with increasing $p_T$, and the average value measured for charged particles at Collider energies (2, 3) is only $\langle p_T \rangle \approx 0.4$ GeV/$c$.

107

0163–8998/85/1201–0107$02.00

Collisions giving rise to secondaries with small transverse momenta represent the largest fraction of the total inelastic cross section, and are generally referred to as soft collisions. In a very small fraction of collisions, however, high-$p_T$ secondaries are emitted. Such hard collisions were first observed in 1972 (4–7) at the CERN Intersecting Storage Rings (ISR) in proton-proton interactions at $\sqrt{s}$ values between 30 and 62 GeV. They were interpreted in the framework of the parton model (8) as the result of elastic or quasi-elastic scattering of two point-like constituents of the incident protons (9).

In the parton model the two incident hadrons are considered, at any instant, as being composed of independent point-like constituents (partons), each carrying a fraction $x$ of the incident hadron momentum. Today we know that the partons are quarks (q), antiquarks ($\bar{q}$), and gluons (g), all carrying a new quantum number (color), and we have a non-Abelian gauge theory, quantum chromodynamics (QCD) (10), as the best candidate to explain the strong interaction among these elementary constituents. Large-angle scattering of two high-$x$ partons results in two outgoing partons with high $p_T$. At this stage the strong forces among partons, which are presumably responsible for color confinement within the hadrons, induce a final-state interaction among the two high-$p_T$ partons and the other partons; this results in the production of many hadrons (this step is referred to as hadronization, or fragmentation). Since this final-state interaction involves mostly low momentum transfer mechanisms, the final result is the production of two highly collimated systems of hadrons (jets), each having a total four-momentum approximately equal to that of the parent parton. Because the incident partons have low $p_T$, the two jets are approximately coplanar with the beam axis; however, their longitudinal momenta are not equal and opposite, in general, because the initial partons may have different $x$ values.

After high-$p_T$ particle production was first observed at the ISR (4–7), the study of the structure of events containing high-$p_T$ particles in the final state became one of the main lines of research in the 1970s (11). Most of the early experiments used a trigger based on the detection of a single high-$p_T$ particle, which distorted the structure of the jet to which the trigger particle belonged (this effect is known as trigger bias). However, the hadrons observed at azimuthal angles opposite to the trigger particle were indeed found to have a structure consistent with that expected for a jet (12). Further evidence in favor of jet production in hadronic collisions came from the striking similarities between the main features of these jets (average multiplicity, momentum distributions within the jet) and those observed in $e^+e^-$ collisions and lepton-nucleon (eN or $\nu$N) scattering in the deep inelastic region (11).

Following a suggestion by Bjorken (14), experiments were also performed using calorimeters in order to trigger on the whole jet and thereby avoid the trigger bias mentioned above. The first experiments using this technique (15) were performed at Fermilab (16, 17), providing a direct measurement of the cross section for jet production. However, these early experiments used calorimeters with a limited solid angle (of typically 1 sr), and it was conceivable that some effect from trigger bias could still be present in the final state. An extreme possibility was that all the effects observed in high-$p_T$ final states could simply be due to the requirement of one or more high-$p_T$ particles in a limited solid angle, which would distort an otherwise azimuthally symmetric final state.

To overcome this objection, an experiment was performed during 1981–1982 at the CERN Super Proton Synchrotron (SPS) (18) using a calorimeter with full azimuthal coverage and subtending the interval of polar angles $45° < \theta < 135°$ in the center-of-mass frame. This experiment selected hadronic collisions depositing large amounts of energy in the calorimeter, and found that these final states consisted mostly of many low-$p_T$ particles distributed symmetrically in azimuth, in disagreement with the structure expected for high-$p_T$ jets. The same conclusions were reached by a similar experiment performed at Fermilab (19).

These negative results were in sharp contrast with the case of $e^+e^-$ annihilation into hadrons, where jets are obvious above $\sqrt{s} = 10\,\mathrm{GeV}$ (20). The azimuthally symmetric structure of these events was interpreted either as the effect of multiple gluon bremsstrahlung from the initial-state partons (21); or as the effect of the tails of the multiplicity distributions in ordinary soft collisions (22). In either case, they cast doubts on the possibility that jet production in hadronic collisions would ever be observed in an unbiased way.

As we discuss in the next section, this pessimistic view has been contradicted by the dramatic emergence of unambiguous jets at the CERN $p\bar{p}$ Collider. These observations have greatly improved our understanding of the mechanisms responsible for jet production in hadronic collisions at high energy. The purpose of this article is to review the main experimental results obtained recently on this subject, and to discuss their interpretation in the theoretical framework of QCD.

## 2.   EVIDENCE FOR JET PRODUCTION IN HADRONIC COLLISIONS

The first experiment to obtain clear evidence for jet production using a technique free from instrumental bias was the UA2 experiment at the CERN $p\bar{p}$ Collider (23). Figure 1 shows the UA2 detector, which contains a total-absorption calorimeter (the central calorimeter) covering the full azimuth over the polar angle interval $40° < \theta < 140°$. This calorimeter (24) is subdivided into 240 independent cells, each subtending the interval $\Delta\theta \times \Delta\varphi = 10° \times 15°$, so that for each event it is possible to measure the total transverse energy $\sum E_\text{T}$, defined as $\sum E_\text{T} = \sum_i E_i \sin \theta_i$, where $E_i$ is the energy deposited in the $i$th cell, $\theta_i$ is the polar angle of the cell center, and the sum extends to all cells. The observed $\sum E_\text{T}$ distribution (25), given in Figure 2, shows a clear departure from the exponential when $\sum E_\text{T}$ exceeds $\sim 60$ GeV.

In order to investigate the pattern of energy distribution in the events, energy clusters are constructed by joining all the calorimeter cells that share a common side and contain at least 400 MeV. In each event, these clusters are then ranked in order of decreasing transverse energies ($E_\text{T}^1 > E_\text{T}^2 > E_\text{T}^3 > \ldots$). Figure 3a shows the mean values of the fractions $h_1 = E_\text{T}^1/\sum E_\text{T}$ and $h_2 = (E_\text{T}^1 + E_\text{T}^2)/\sum E_\text{T}$ as a function of $\sum E_\text{T}$. Their behavior reveals that, when $\sum E_\text{T}$ is large enough, a very substantial fraction of $\sum E_\text{T}$ is shared on the average by two clusters with roughly equal transverse energies (an event consisting of only two clusters of equal transverse energies would have $h_1 = 0.5$ and $h_2 = 1$).

The azimuthal separation $\Delta\phi_{12}$ between the two largest clusters (26) is shown in Figure 3b for events with $\sum E_\text{T} > 60$ GeV and $E_\text{T}^1, E_\text{T}^2 > 20$ GeV. A

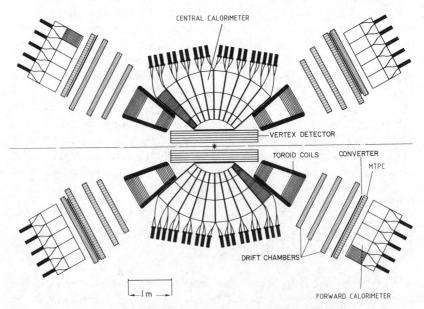

*Figure 1*   View of the UA2 detector at the CERN p̄p Collider.

clear peak at $\Delta\varphi_{12} = 180°$ is observed, which indicates that the two clusters are coplanar with the beam direction.

The emergence of two-cluster structures in events with large $\sum E_T$ is even more dramatically illustrated by inspecting the transverse energy distribution over the calorimeter cells. Figure 4 shows such a distribution for four typical events having $\sum E_T > 100$ GeV. The transverse energy appears to be concentrated within two (or, more rarely, three) small angular regions. These energy clusters appear to be associated with collimated multiparticle systems (jets), as found by reconstructing the charged-particle tracks in these events (see Figure 5). Furthermore, longitudinal shower developments, as measured in the calorimeter, are found to be inconsistent, in general, with those of single particles, but consistent with those of jets containing both charged hadrons and photons (from $\pi^0$ decay).

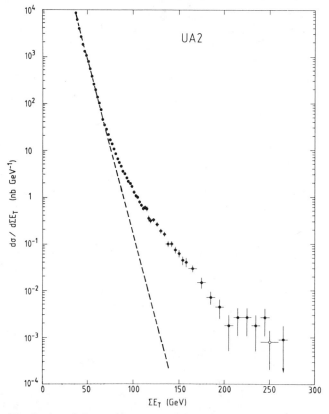

*Figure 2* Distribution of the total transverse energy $\sum E_T$ observed in the UA2 central calorimeter.

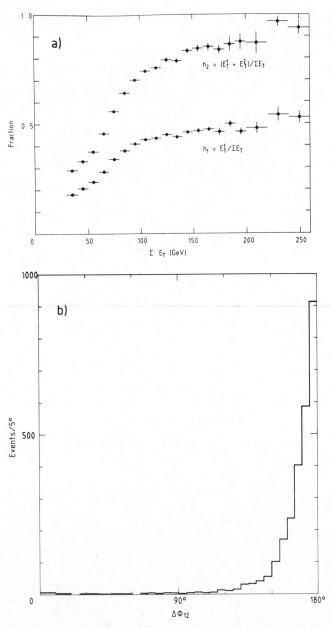

*Figure 3*    (*a*) Mean value of the fraction $h_1$ ($h_2$) of the total transverse energy $\sum E_T$ contained in the cluster (in the two clusters) having the largest $E_T$, as a function of $\sum E_T$. (*b*) Azimuthal separation between the two largest $E_T$ clusters in events with $\sum E_T > 60$ GeV and $E_T^1, E_T^2 > 20$ GeV.

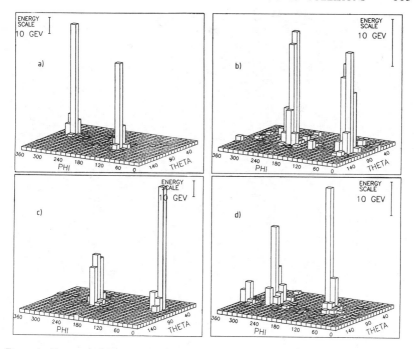

*Figure 4*    Four typical transverse energy distributions for events with $\sum E_T > 100$ GeV in the $\theta$-$\varphi$ plane. Each bin represents a cell of the UA2 calorimeter.

High-$p_T$ jets have also been observed by the UA1 experiment (27) and, in pp collisions, by the Axial Field Spectrometer (AFS) (28) and CMOR (28a) Collaborations at the lower $\sqrt{s}$ values available at the ISR.

## 3.    THEORETICAL INTERPRETATION

As mentioned in the Introduction, jet production in hadronic collisions is interpreted in the framework of the parton model as hard scattering among the constituents of the incident hadrons. Since the initial state contains quarks, antiquarks, and gluons, there are several elementary subprocesses that can contribute to jet production. For each subprocess the scattering cross section, calculated to first order in the strong coupling constant $\alpha_s$ (see the relevant diagrams in Figure 6), is given by the expression

$$\frac{d\sigma}{d \cos \theta^*} = \frac{\pi \alpha_s^2}{2\hat{s}} |M|^2, \qquad 1.$$

where $\theta^*$ is the scattering angle, and $\hat{s}$ is the square of the total energy in the center of mass of the two partons. In QCD the coupling constant $\alpha_s$ is a

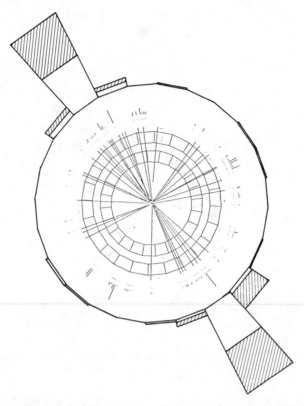

*Figure 5*  View of a typical two-jet event perpendicular to the beams in the UA2 detector. The heights of the trapezoids are proportional to transverse energy. The open and shaded areas represent the energy depositions in the electromagnetic and hadronic sections of the calorimeter, respectively.

function of $Q^2$, the square of the four-momentum transfer in the subprocess. In a model with five quark flavors, $\alpha_s(Q^2)$ is expressed as $\alpha_s(Q^2)$ $= 12\pi/[23 \ln (Q^2/\Lambda^2)]$, where $\Lambda$ is a scale parameter. The property $\alpha_s \to 0$ for $Q^2 \to \infty$, called "asymptotic freedom," allows perturbative calculations of strong processes at high $Q^2$ (10).

Explicit expressions for $|M|^2$ in Equation 1 are given in Table 1 for the various subprocesses (29) as a function of the Mandelstam variables $\hat{s}, t = -\hat{s}(1-\cos \theta^*)/2$, and $u = -\hat{s}(1+\cos \theta^*)/2$ (under the assumption of massless partons). In order to illustrate the relative importance of the subprocesses, Table 1 also displays the numerical values of $|M|^2$ at $\theta^* = 90°$, where $t = u = -\hat{s}/2$. It is evident that terms involving initial gluons, such as gg and qg (or q̄g) scattering, are dominant whenever the gluon density in the incident hadrons is comparable to that of the quarks.

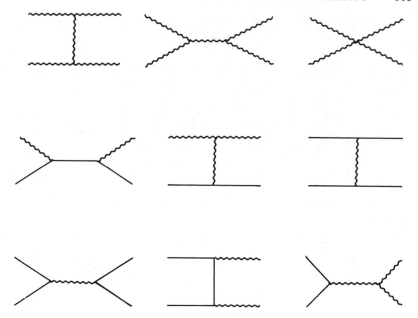

*Figure 6*  First-order diagrams for hard parton scattering. Gluons: wavy line, quarks: straight lines.

## 4.  INCLUSIVE JET YIELD

The cross section for inclusive jet production as a function of the jet $p_T$ and angle of emission $\theta$ can be calculated to leading order in $\alpha_s$ as a sum of convolution integrals (30):

$$\frac{d^2\sigma}{dp_T \, d(\cos\theta)} = \frac{2\pi p_T}{\sin^2\theta} \sum_{A,B} \int dx_1 \, dx_2 \, F_A(x_1, Q^2) F_B(x_2, Q^2)$$

$$\times \, \delta(\hat{s} + t + u)\alpha_s^2 \sum_f \frac{|M|^2_{AB \to f}}{\hat{s}}, \qquad \qquad 2.$$

where $F_A$ and $F_B$ are structure functions describing the densities of partons A and B in the incident hadrons, and the sum extends over all initial parton types A, B, and all possible final states f.

In the case of incident protons or antiprotons, the structure functions $F(x, Q^2)$ are determined experimentally in deep inelastic lepton-nucleon scattering experiments ($Q^2 \lesssim 20 \text{ GeV}^2$) and extrapolated to the $Q^2$ range of interest (up to $\sim 10^4 \text{ GeV}^2$ at Collider energies) according to the predicted QCD evolution (31–33).

**Table 1**  Matrix elements for parton scattering

| Subprocess | $|M|^2$ | $|M|^2$ at $\theta^* = 90°$ |
|---|---|---|
| $qq' \rightarrow qq'^{a}$ <br> $q\bar{q}' \rightarrow q\bar{q}'^{a}$ | $\dfrac{4}{9}\dfrac{\hat{s}^2 + u^2}{t^2}$ | 2.22 |
| $qq \rightarrow qq$ | $\dfrac{4}{9}\left(\dfrac{\hat{s}^2 + u^2}{t^2} + \dfrac{\hat{s}^2 + t^2}{u^2}\right) - \dfrac{8}{27}\dfrac{\hat{s}^2}{ut}$ | 3.26 |
| $q\bar{q} \rightarrow q'\bar{q}'^{a}$ | $\dfrac{4}{9}\dfrac{t^2 + u^2}{\hat{s}^2}$ | 0.22 |
| $q\bar{q} \rightarrow q\bar{q}$ | $\dfrac{4}{9}\left(\dfrac{\hat{s}^2 + u^2}{t^2} + \dfrac{t^2 + u^2}{\hat{s}^2}\right) - \dfrac{8}{27}\dfrac{u^2}{\hat{s}t}$ | 2.59 |
| $q\bar{q} \rightarrow gg$ | $\dfrac{32}{27}\dfrac{u^2 + t^2}{ut} - \dfrac{8}{3}\dfrac{u^2 + t^2}{\hat{s}^2}$ | 1.04 |
| $gg \rightarrow q\bar{q}$ | $\dfrac{1}{6}\dfrac{u^2 + t^2}{ut} - \dfrac{3}{8}\dfrac{u^2 + t^2}{\hat{s}^2}$ | 0.15 |
| $qg \rightarrow qg$ <br> $\bar{q}g \rightarrow \bar{q}g$ | $-\dfrac{4}{9}\dfrac{u^2 + \hat{s}^2}{u\hat{s}} + \dfrac{u^2 + \hat{s}^2}{t^2}$ | 6.11 |
| $gg \rightarrow gg$ | $\dfrac{9}{2}\left(3 - \dfrac{ut}{\hat{s}^2} - \dfrac{u\hat{s}}{t^2} - \dfrac{\hat{s}t}{u^2}\right)$ | 30.38 |

[a] q and q' denote quarks with different flavors.

At Collider energies, jets with $p_T$ around 30 GeV/$c$ produced near 90° arise from hard scattering of partons with relatively small values of $x$ ($x \lesssim 0.1$). In this region gluon jets are expected to dominate, both because there are many gluons in the nucleon at small $x$ and because the subprocesses involving initial gluons have large cross sections (see Table 1). This is in contrast with $e^+e^-$ collisions, where the production of quark jets is the main feature of hadronic final states. Quark jets also dominate at ISR energies ($\sqrt{s} \leq 62$ GeV) where high-$p_T$ jets require initial partons with large $x$ values.

A number of uncertainties affect the comparison between the predicted cross sections and the experimental data. The most obvious one is that Equation 2 predicts the yield of high-$p_T$ massless partons, whereas the experiments measure hadronic jets with a total invariant mass of several GeV. The relation between the parton $p_T$ and the measured cluster transverse energy $E_T$ is usually determined with the help of a QCD-inspired Monte Carlo simulation (34) in which the outgoing partons evolve into jets according to a specific hadronization model, and the detector response to the hadrons is taken into account.

An important uncertainty in the theoretical predictions arises from the $Q^2$ extrapolation of the structure functions, especially those describing the gluons. Another source of theoretical uncertainties is represented by the fact that Equation 2 does not take into account higher-order effects, such as gluon radiation by the initial and outgoing partons. A complete calculation of these corrections is still missing; their effect is usually described by a multiplicative factor $K \lesssim 2$ (35).

Finally, in addition to the statistical errors, the data are also affected by a number of systematic effects, such as uncertainties in the energy calibration of the calorimeters, in the detector acceptance, and in the knowledge of the integrated luminosity. These effects amount typically to an overall uncertainty of $\pm 50\%$ in the measured jet yields. Altogether, a comparison between the theoretical predictions and the experimental results is only possible, at present, to an accuracy not greater than a factor of 2.

Figure 7 shows the inclusive jet production cross section around $\theta = 90°$ as a function of the jet $p_T$, as measured by the UA1 (27) and UA2 (25) Collaborations at the p$\bar{\text{p}}$ Collider, and by the AFS Collaboration (28) at the ISR. The jet yield at the Collider is much larger than that measured at the ISR. This fact was first pointed out by Horgan & Jacob (30) well before these data were available.

Also shown in Figure 7 is a band of QCD predictions (30, 36–38) whose width serves to illustrate the theoretical uncertainties. In spite of the experimental and theoretical uncertainties, the agreement between data and theory is remarkable, especially because the theoretical curves are not a fit to the data but represent absolute predictions.

These data offer the possibility of testing the existence of the three-gluon vertex that results from the non-Abelian structure of QCD. This vertex enters into the QCD calculations of the inclusive jet yield in three different ways: in the elementary subprocesses (see the relevant diagrams of Figure 6); in the $Q^2$ dependence of the gluon structure function; and by determining the variation of $\alpha_s$ as a function of $Q^2$. Such a test can only be performed, however, if the effects of this vertex can be clearly separated from other spurious effects, such as the uncertainties in the theory and the systematic errors in the data. An analysis performed by Furmanski & Kowalski (39) has shown that all these spurious effects can be described to a good approximation by an overall normalization constant; on the contrary, the suppression of the three-gluon vertex changes the $p_T$

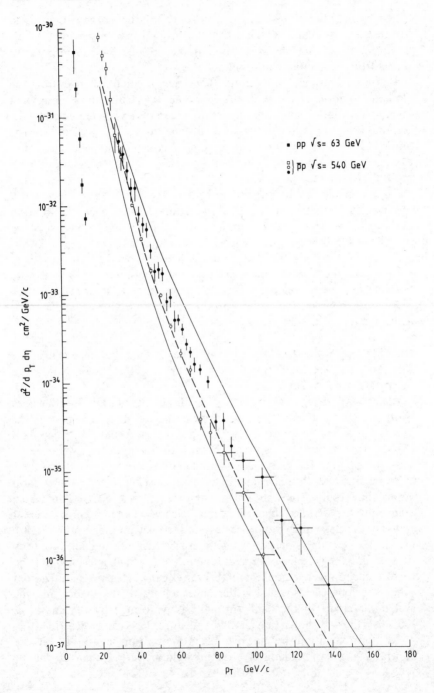

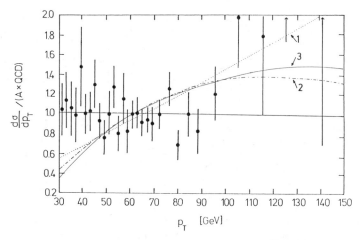

*Figure 8*  Full points: ratio between the inclusive jet yield (25) and a renormalized standard QCD prediction. Curve 1: three-gluon vertex in hard parton scattering removed. Curve 2: three-gluon vertex removed also in the evolution of the structure functions. Curve 3: as for curve 2 with, in addition, $\alpha_s$ = constant (from 39).

dependence of the inclusive jet yield. This is illustrated in Figure 8, which compares the ratio between the UA2 data (25) and the standard re-normalized QCD predictions, with various predictions in which the three-gluon vertex was suppressed, all normalized to the standard QCD result. This comparison provides evidence for the existence of the three-gluon vertex in QCD.

We note finally that, if one ignores the $Q^2$ dependence of the structure functions and of $\alpha_s$, the invariant cross section for jet production at 90° satisfies the scaling law $E\, d^3\sigma/dp^3 = A p_T^n f(x_T)$, where $x_T = 2p_T/\sqrt{s}$, with $n = 4$ (9, 14). Figure 9 shows that the Collider (25, 27) and ISR (28) data, together with recent Fermilab results (40) at $\sqrt{s} = 27.4$ GeV, obey scaling with $n = 5.1 \pm 0.3$ over the range $27 < \sqrt{s} < 540$ GeV. This must only be considered as an approximate behavior, because the $Q^2$ dependence of the structure functions is well established experimentally. The change from $n = 4$ to $n \approx 5$ can be easily accounted for by the expected scale-breaking effects.

---

*Figure 7*  Cross sections for inclusive jet production around $\theta = 90°$, as a function of the jet transverse momentum. Collider data: full circles (UA2, Ref. 25); open circles and squares (UA1, Ref. 27). ISR data: full squares (Ref. 28). The dashed curve represents the original prediction by Horgan & Jacob (30). The two full curves define a band of QCD predictions (30, 36–38).

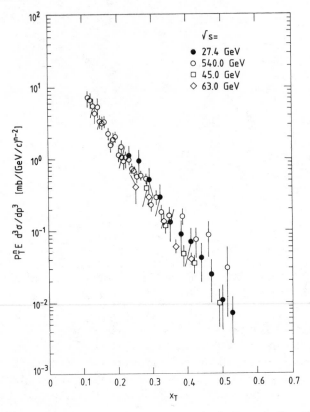

*Figure 9*   Approximate scaling of the invariant cross section for jet production at 90° over the range of $\sqrt{s}$ from 27.4 to 540 GeV (from 40).

## 5. ANGULAR DISTRIBUTION OF PARTON-PARTON SCATTERING

The study of the jet angular distribution in two-jet events provides a way to measure the angular distribution of parton-parton scattering, and can therefore be considered as the analogue of Rutherford's experiment in QCD.

In practice, there are complications arising from the fact that the center of mass of the two partons does not coincide with the center of mass of the colliding hadrons, and, in addition, what is observed experimentally is a mixture of various subprocesses that occur at different center-of-mass energies. We can write

$$\frac{d^3\sigma}{dx_1\,dx_2\,d\,(\cos\theta^*)} = \sum_{A,B} \frac{F_A(x_1)}{x_1}\frac{F_B(x_2)}{x_2}\sum_{C,D}\frac{d\sigma_{AB\to CD}}{d\,(\cos\theta^*)},\qquad 3.$$

where $F_A(x)/x$ $[F_B(x)/x]$ is the structure function describing the density of parton A [B] within the incident hadrons, and the sums extend to all possible subprocesses AB → CD.

If the total transverse momentum $P_T$ of the two-jet system is zero, values of $\theta^*$, $x_1$, and $x_2$ are obtained for each event using the kinematical relations

$$\theta^* = \sin^{-1}(2p_T/M_{jj}); \qquad \begin{aligned} x_1 &= [(P_L^2 + M_{jj}^2)^{1/2} + P_L]/\sqrt{s}, \\ x_2 &= [(P_L^2 + M_{jj}^2)^{1/2} - P_L]/\sqrt{s}, \end{aligned}$$

where $p_T$ is the jet transverse momentum, $M_{jj}$ is the invariant mass, and $P_L$ is the total longitudinal momentum of the two-jet system. For events with $P_T \neq 0$ it is not possible to determine $\theta^*$ unambiguously. In this case only events with $P_T \ll p_T$ are used and $\theta^*$ is determined according to the convention of Collins & Soper (41).

Equation 3 may at first sight appear hopeless in view of the many terms involved. However, in the case of p$\bar{\text{p}}$ collisions the dominating subprocesses are gg → gg, gq → gq (or g$\bar{\text{q}}$ → g$\bar{\text{q}}$), and q$\bar{\text{q}}$ → q$\bar{\text{q}}$, which to a very good approximation have the same cos $\theta^*$ dependence (see Table 1). Equation 3 can then be approximately factorized as

$$\frac{d^3\sigma}{dx_1\, dx_2\, d(\cos\theta^*)} = \left[\frac{1}{x_1}\sum_A F_A(x_1)\right]\left[\frac{1}{x_2}\sum_B F_B(x_2)\right]\frac{d\sigma}{d(\cos\theta^*)}.$$

If $d\sigma/d(\cos\theta^*)$ is taken to be the differential cross section for gluon-gluon elastic scattering, which to leading order in QCD has the form

$$\frac{d\sigma}{d(\cos\theta^*)} = \frac{9}{8}\frac{\pi\alpha_s^2}{2x_1x_2s}\frac{(3+\cos^2\theta^*)^3}{(1-\cos^2\theta^*)^2}, \qquad\qquad 4.$$

then it becomes possible to write

$$\sum_A F_A(x) = g(x) + \frac{4}{9}[q(x) + \bar{q}(x)], \qquad\qquad 5.$$

where $g(x)$, $q(x)$, and $\bar{q}(x)$ are respectively the gluon, quark, and antiquark structure functions of the proton. The factor 4/9 in Equation 5 reflects the relative strengths of the quark-gluon and gluon-gluon couplings in QCD.

The term $d\sigma/d(\cos\theta^*)$ in Equation 4 contains a singularity at $\theta^* = 0$ with the familiar Rutherford form $\sin^{-4}(\theta^*/2)$, which is typical of gauge vector boson exchange. In the subprocesses gg → gg and gq → gq (or g$\bar{\text{q}}$ → g$\bar{\text{q}}$) it arises from the three-gluon vertex. It is also present in the subprocess q$\bar{\text{q}}$ → q$\bar{\text{q}}$, but in this case it would be present in an Abelian theory as well, as for $e^+e^-$ elastic scattering in QED.

Figure 10a shows the cos $\theta^*$ distribution measured in the UA1 experiment (42) for jets with $p_T > 20$ GeV/c produced in the range of pseudorapidity $|\eta| < 2.5$, where $\eta = -\ln\tan\theta/2$. Both the data and the

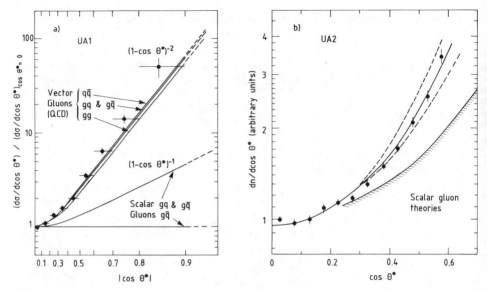

*Figure 10*    (*a*) Distribution of cos $\theta^*$ for hard parton scattering as measured in the UA1 experiment (42). The normalization is defined by setting the value at cos $\theta^* = 0$ equal to 1. (*b*) Distribution of cos $\theta^*$ for hard parton scattering as measured in the UA2 experiment (43). All the different QCD processes (except for q$\bar{\text{q}}$ → q′$\bar{\text{q}}$′), separately normalized to the data, lie in the area between the two dashed curves. The full line is the overall QCD prediction, normalized to the data.

theoretical curves for the three dominant subprocesses are normalized to 1 at cos $\theta^* = 0$. The results of the UA2 experiment (43) are shown in Figure 10*b*, where they are compared with the cos $\theta^*$ distribution predicted by QCD with no approximations. These data cover only the range |cos $\theta^*$| < 0.6 because of the limited polar-angle interval covered by the UA2 calorimeter. Both sets of data agree with QCD expectations, and they clearly show the increase toward the forward direction expected from the $\sin^{-4} (\theta^*/2)$ singularity. The expectations from a theory with scalar gluons are in strong disagreement with the data. A similar analysis had previously been performed at ISR energies using events in which both jets consisted mainly of $\pi^0$'s (44).

## 6.   DETERMINATION OF THE PROTON STRUCTURE FUNCTION

In order to extract the effective structure function $F(x)$, a quantity $S(x_1, x_2)$ is defined in terms of the measured differential cross section $d^2\sigma/dx_1\, dx_2$:

$$S(x_1, x_2) = x_1 x_2 \frac{d^2\sigma}{dx_1\, dx_2} \bigg/ \int_0^{\cos \theta^*_{\min}} K \frac{d\sigma}{d\,(\cos \theta^*)}\, d\,(\cos \theta^*),$$

where $\theta^*_{min}$ is the smallest scattering angle for which both jets fall within the detector acceptance ($\theta^*_{min}$ is itself a function of $x_1$ and $x_2$). The form of $d\sigma/d(\cos\theta^*)$ is taken from Equation 4 and $K$ is the numerical factor describing higher-order QCD corrections (see Section 4). The approximate factorization property $S(x_1, x_2) = F(x_1)F(x_2)$ is found to be verified by the data within errors.

Figure 11 shows the function $F(x)$ obtained in the UA1 (42) and UA2 (43) experiments using $\Lambda = 0.2$ GeV and $K = 1$. In addition to the statistical errors there is a systematic uncertainty of $\sim 50\%$ in the overall normalization, which includes one half of the theoretical uncertainty on $K$ [what is actually determined from the data is $F(x)\sqrt{K}$]. Also shown in Figure 11 are curves representing the function $g(x)+(4/9)[q(x)+\bar{q}(x)]$ as expected from fits to $\nu$ and $\bar{\nu}$ deep inelastic scattering data (45). The Collider results agree with the behavior expected at the large $Q^2$ values appropriate to the collider experiments ($\langle Q^2 \rangle \approx 2000$ GeV$^2$), and they show directly the very large gluon density in the proton at small $x$.

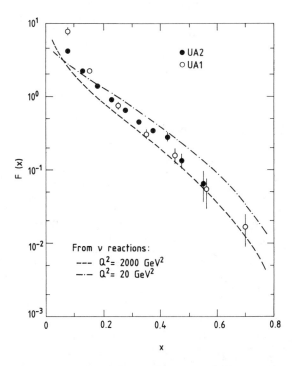

*Figure 11*   Effective structure function measured from two-jet events (42, 43). The dashed lines are obtained from deep inelastic neutrino scattering experiments (45).

# 7. TOTAL TRANSVERSE MOMENTUM OF THE TWO-JET SYSTEM

If the two partons that undergo hard scattering have no initial transverse momentum, the total transverse momentum of the final two-jet system, $P_T$, should be equal to zero. In reality, this does not happen because the incident partons have a small primordial transverse momentum, and, furthermore, both incident and outgoing partons may radiate gluons.

Experimentally, $P_T$ is determined from the sum of two large and approximately opposite two-dimensional vectors $\mathbf{p}_{T1}$ and $\mathbf{p}_{T2}$, and it is therefore sensitive to instrumental effects such as the calorimeter energy resolution and incomplete containment due to edge effects in the detector. These effects can be made small by considering only the component of $P_T, P_\eta$, parallel to the bisector of the angle defined by $\mathbf{p}_{T1}$ and $\mathbf{p}_{T2}$.

Figure 12 shows the distribution of $P_\eta$ as measured in the UA2 experiment (43) for two-jet events having $E_T^1 + E_T^2 > 40$ GeV, $\Delta\varphi > 120°$, and with each jet satisfying the conditions $E_T > 10$ GeV, $|\cos\theta| < 0.6$. For $P_\eta < 15$ GeV/$c$, this distribution corresponds to a $P_T$ distribution of the form $P_T \exp[-AP_T^2]$, with $\langle P_T \rangle = 7 \pm 1$ GeV/$c$, whereas for $P_\eta > 15$ GeV/$c$ the data lie systematically above this simple parameterization.

These results are in good agreement with a QCD prediction (46)

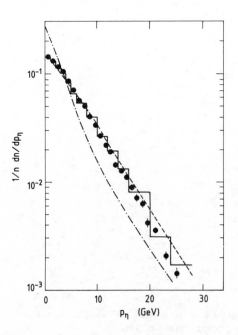

*Figure 12* Distribution of the component $P_\eta$ of the total transverse momentum of the two-jet system, as measured in the UA2 experiment (43). The dashed line is a QCD prediction (46); the dotted-dashed line represents the same prediction, but assumes that gluons radiate as quarks. The histogram represents the standard QCD prediction with the detector effects taken into account.

illustrated by the curve of Figure 12. In QCD, gluon radiation by a gluon (g → gg)—which occurs because of the three-gluon vertex—has a rate 9/4 times higher than that of q → qg, and predictions based on the assumption that gluons radiate like quarks disagree with the data (see Figure 12). Since gluon jets dominate in the $P_T$ range explored at the Collider, we can consider the good agreement between the data of Figure 12 and the theoretical predictions as further evidence in favor of a QCD description of high-$p_T$ jet production.

## 8.  JET FRAGMENTATION

In the preceding sections we discussed the properties of jet production independently of the internal structure of the jet itself. In this section we discuss the properties of the jet internal structure.

High-$p_T$ partons produced far from the mass shell in a hard collision evolve according to two distinct time scales. At early times, which correspond to distances much shorter than the typical size of a hadron, gluon radiation and the creation of q$\bar{q}$ pairs result in the production of a jet of colored partons, which can be described using perturbative QCD (47–49). The following step occurring on a longer time scale involves the long-distance nonperturbative properties of QCD, which lead to the confinement of partons in colorless bound states observed as hadrons. Nonperturbative techniques providing a sufficient understanding of confinement are still missing at present, and two plausible models are commonly used, one dealing with each jet independently (50), the other (referred to as the Lund model) taking into account the configuration of all colored partons in the final state (51).

A jet generally consists of many particles (fragments) whose motion with respect to the jet axis can be described by two variables: (a) the fractional longitudinal momentum $z = \mathbf{p} \cdot \mathbf{P}/P^2$, where $\mathbf{p}$ is the fragment momentum, and $\mathbf{P}$ is the total jet momentum (obviously $0 < z < 1$); and (b) $q_T$, the component of $\mathbf{p}$ perpendicular to $\mathbf{P}$. The $z$ distribution, $D(z)$, is called the jet fragmentation function.

A systematic uncertainty in the scale of $z$ arises from the determination of $\mathbf{P}$, which depends on the criteria used to define a jet. Furthermore, the function $D(z)$ cannot be measured reliably in the low-$z$ region because the final state contains many low-$p_T$ particles that do not belong to the jet but are associated with the partons, which did not take part in the hard collision (these particles, called spectators, are absent, of course, in the hadronic final states resulting from $e^+e^-$ collisions).

Figure 13 shows the function $D(z)$ for charged fragments, as determined by the AFS Collaboration (52) at the ISR ($\langle P \rangle \approx 14$ GeV/$c$), and by the

UA1 Collaboration (53) at the p$\bar{\text{p}}$ Collider ($P_T > 20$ GeV/$c$). In the ISR experiment, all particles with $p_T > 1$ GeV/$c$ are considered; in the UA1 experiment, only those contained in a 35° half-aperture cone around the jet axis are considered. Also shown in Figure 13 is the function $D(z)$ for charged fragments of jets produced in e$^+$e$^-$ collisions at $\sqrt{s} = 34$ GeV (54). While the fragmentation function measured at the ISR is similar to that measured in e$^+$e$^-$ collisions, for $z > 0.2$, the Collider results are systematically lower than the other two sets of data. This is in qualitative agreement with QCD,

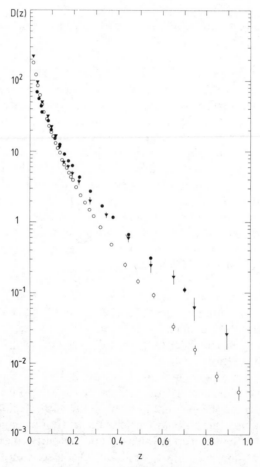

*Figure 13*  Jet fragmentation function for charged particles. Full triangles: jets from pp collisions at ISR energies (52). Open circles: jets from p$\bar{\text{p}}$ collisions at $\sqrt{s} = 540$ GeV (53). Full circles: jets from e$^+$e$^-$ collisions at $\sqrt{s} = 34$ GeV (54).

which suggests that gluon jets (dominating at the $p\bar{p}$ Collider) should be softer than quark jets (dominating in the ISR and $e^+e^-$ data), mainly because gluon radiation by a gluon has a higher probability than gluon radiation by a quark.

Figure 14$a$ shows the average transverse momentum of the charged fragments with respect to the jet axis, $\langle q_T \rangle$, as a function of $z$, for Collider jets (53). The value of $\langle q_T \rangle$ is seen to increase from $\sim 0.5$ GeV/$c$ for $z \approx 0.1$ to $\sim 1$ GeV/$c$ for $z$ values above 0.5. The distribution $dn/dq_T^2$ for $z > 0.1$ is shown in Figure 14$b$, where it is compared with the shape of the $p_T$ spectrum of charged particles in ordinary soft collisions (2).

The average charged-particle multiplicity in a jet, $\langle n_{ch} \rangle$, can be obtained, in principle, by integrating the function $D(z)$ over the range $0 < z < 1$. However, because of the contamination of spectator particles mentioned previously, the uncertainty on $D(z)$ at small $z$ values is too large to make the result meaningful.

An attempt to overcome this difficulty has been made by the UA2 Collaboration (55) at the $p\bar{p}$ Collider. Events are considered in which the two jets with the highest $E_T$ are separated in azimuth by at least $150°$ and have an invariant mass $M_{jj}$ of at least 40 GeV. For these events, Figure 15$a$ shows the distribution $dn/d\varphi$, where $\varphi$ is the azimuthal separation between each charged particle observed over the range $20° < \theta < 160°$, and the axis of the jet with the highest $E_T$. A lower limit to the true jet multiplicity is obtained under the assumption that there are no jet fragments at $\varphi = \pi/2$, but only charged spectators with a uniform $\varphi$ distribution. The average charged-particle multiplicity in a jet is then

$$\langle n_{ch} \rangle = \frac{1}{2} \int_0^\pi \frac{dn}{d\varphi} \, d\varphi - \frac{\pi}{2} \left[ \frac{dn}{d\varphi} \right]_{\varphi = \pi/2}. \qquad 6.$$

Values of $\langle n_{ch} \rangle$ thus obtained are shown in Figure 15$b$ as a function of $M_{jj}$. Also shown for comparison are values of $\langle n_{ch} \rangle$ as a function of $\sqrt{s}$ for hadronic final states from $e^+e^-$ collisions (54), modified according to Equation 6. This procedure allows for a model-independent comparison of jets for $p\bar{p}$ and $e^+e^-$ reactions, suggesting that jets at the $p\bar{p}$ Collider have higher mean multiplicities than would be expected from extrapolation of $e^+e^-$ data. Fragmentation models based on QCD (56), illustrated by the curves shown in Figure 15$b$, predict a higher relative multiplicity of gluon jets with respect to quark jets. At Collider energies one expects that the fraction of gluon jets in the data of Figure 15$b$ varies from $\sim 75$ to $\sim 30\%$ as $M_{jj}$ varies from 40 to 140 GeV; this would explain the behavior of $\langle n_{ch} \rangle$ as a function of $M_{jj}$ observed at the Collider.

The distribution $dn/d\varphi$ shown in Figure 15$a$ gives an idea of the average

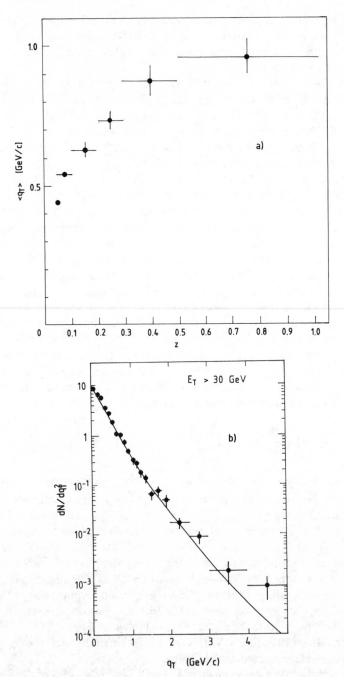

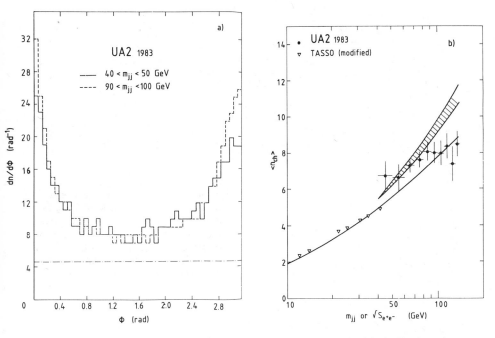

*Figure 15* (*a*) Distribution of the azimuthal separation between the axis of the leading jet and all charged particles, normalized to the total number of events (55). The dashed-dotted horizontal line represents the azimuthal charged-particle density measured in the same experiment for ordinary soft collisions. (*b*) Mean charged-particle multiplicities in jets derived according to Equation 6. Open triangles: data from $e^+e^-$ collisions (54). Full circles: data from UA2 (55). The solid curve is the prediction of the model of Ref. 56 for quark jets; the shaded area represents the predictions for gluon jets.

angular size of a jet. If the average transverse-energy density is plotted as a function of $\varphi$, instead of the charged-particle density, jets appear to be much narrower, as shown in Figure 16. This is because particles with large $\varphi$ separations from the jet axis are mostly soft. Predictions based on QCD (56) reproduce these results rather well: for jet transverse energies $20 < E_T < 30$ GeV, the data are in better agreement with the behavior expected for gluon jets, whereas radiation from quarks is needed in order to explain the behavior of jets with larger $E_T$. Fragmentation models not including the

←

*Figure 14* (*a*) Average transverse momentum of charged particles in a jet with respect to the jet axis, as a function of $z$, for collider jets (53). (*b*) Distribution of the charged-particle transverse momentum with respect to the jet axis (53). The curve represents the shape of the distribution of the transverse momentum (measured with respect to the beam axis) in ordinary soft p$\bar{\text{p}}$ collisions (2).

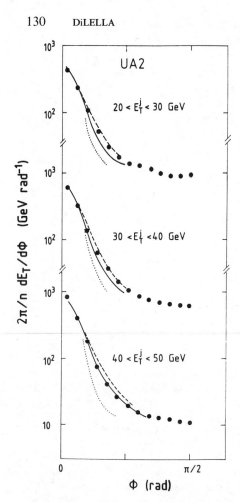

*Figure 16*   Distribution of the azimuthal transverse energy density for three ranges of jet transverse energies. Dotted curves: predictions of the Field & Feynman fragmentation model (57). Dashed (solid) curves: prediction of the QCD model of Ref. 56 for gluon (quark) jets.

effect of gluon radiation from the outgoing partons (57) produce jets that are much narrower than observed.

## 9.   MULTIJET FINAL STATES

It is known that hadronic final states from $e^+e^-$ annihilation at high energy sometimes exhibit a clear three-jet structure. This is interpreted as an effect of gluon bremsstrahlung radiated by one of the two outgoing quarks (54). Such an effect is also expected in the case of hadron collisions, where, however, gluons can be radiated not only by the outgoing high-$p_T$ partons but also by the incident partons and at the parton scattering vertex as well.

We expect such radiation to be much less correlated with the two main jets than in the case of gluons radiated directly by the two outgoing partons.

The first evidence for three-jet events in hadron collisions was obtained in an ISR experiment (58); however, three-jet events are identified much more easily at Collider energy (one such event is shown in Figure 4$d$). Approximately 30% of the events with two jets having $E_T^1 + E_T^2 > 60$ GeV are observed in the UA2 experiment to contain an additional jet with $E_T > 4$ GeV (26); this is the case for only one out of about twenty unbiased final states at Collider energies. The angular correlation between the third jet and the two main ones can be studied by defining the angle $\omega_{13}$ ($\omega_{23}$) between the axis of the first (second) jet and that of the third jet. The distributions of $\cos \omega_{13}$ and $\cos \omega_{23}$, shown in Figure 17, exhibit a clear correlation between the second and third jets, which is to be expected if one assumes that the third jet is due to a gluon radiated by the parton that gave rise to the second jet. The drop near $\cos \omega = \pm 1$ is an instrumental effect resulting from the inability of the detector to resolve jets if they are separated by less than $\sim 30°$.

It should be possible to measure the strong coupling constant $\alpha_s$ by comparing the rates of three-jet and two-jet events. However, unlike three-jet final states from $e^+e^-$ annihilation, which are all of the $q\bar{q}g$ type, final states from hard hadron collisions also contain qgg and ggg types. Since gluon radiation by a gluon is enhanced by a factor of 9/4 with respect to gluon radiation by a quark, precise knowledge of the relative contributions from the various diagrams is needed in order to extract a reliable value of $\alpha_s$ from the data. Such a detailed analysis is under way, and the preliminary

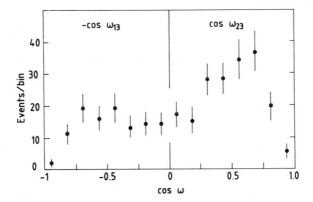

*Figure 17*  Angular correlation between the third jet (with $E_T > 4$ GeV) and the two leading ones (with $E_T^1 + E_T^2 > 60$ GeV) (from 26).

results (26, 27) are in qualitative agreement with QCD predictions (38, 59–61).

## 10.  MULTIJET INVARIANT MASS DISTRIBUTIONS

In the previous sections we have seen that all jet production and fragmentation properties studied so far are in agreement with QCD predictions. It is not surprising, therefore, that two-jet invariant mass distributions are also well described by QCD, and this has indeed been found to be the case at the $p\bar{p}$ Collider (25, 27) for two-jet invariant masses ranging from $\sim 50$ to $\sim 250$ GeV.

The main purpose of studying multijet invariant mass distributions is to search for heavy particles decaying into quarks and gluons. In this case, one expects to observe a peak at the mass of the heavy particle, superimposed on the smooth QCD continuum. A typical example is the production of the weak intermediate bosons, $W^{\pm}$ and $Z^0$, which are expected to decay into $q\bar{q}$ pairs with a branching ratio of $\sim 75\%$ (62).

The energy resolution that can be achieved for high-energy hadrons with conventional calorimeters (15) can be parametrized as $\sigma/E = aE^{-1/2}$ ($E$ in GeV), where $a$ is typically 0.5 to 0.8. A further limitation to the energy resolution for high energy jets arises from two additional effects:

1. In most calorimeters the responses to photons and charged hadrons differ by as much as 30% (15). This effect gives rise to an uncertainty in the jet energy measurement because of the large fluctuations of the jet particle composition.
2. Jets have an intrinsic invariant mass of several GeV, with large jet-to-jet fluctuations because of the jet angular spread. The value of this mass cannot be determined, in general, because the energies of the jet fragments are not individually measured. In practice, neglecting the jet invariant mass is a significant limitation to the attainable mass resolution for multijet systems.

All these effects limit the multijet mass resolution to a value of the order of 10 GeV in the region of the $W^{\pm}$ and $Z^0$ masses. As a consequence, the data sample collected so far at the $p\bar{p}$ Collider is not large enough to provide a statistically significant signal above the QCD continuum.

## 11.  CONCLUSIONS

High-$p_T$ particles from hadronic collisions were observed at the ISR more than ten years ago; however, substantial progress in understanding the mechanisms responsible for this phenomenon has only been achieved in the last three years, thanks to the use of fine-grain hadronic calorimeters with full azimuthal coverage, and especially to the new energy domain opened up by the CERN $p\bar{p}$ Collider.

At a center-of-mass energy of 540 GeV, jets are observed in very clean experimental conditions, and their production properties can be studied, without the need to know their fragmentation, in a way very close to the ideal case in which one would directly observe the scattered partons.

Both jet production and fragmentation properties have been found to be well described by QCD. In particular, the three-gluon vertex, which is characteristic of the non-Abelian structure of QCD, is clearly needed to explain the observations. It is possible for the first time to measure directly the very high gluon density in the proton at low $x$ values. In addition, the data show evidence for higher-order QCD corrections, such as gluon radiation in both the initial and final states. Jets associated with gluon radiation in the initial state have also been observed as an effect of QCD corrections to the production of the weak intermediate vector bosons $W^{\pm}$ and $Z^{0}$ at the $p\bar{p}$ Collider (63, 64).

Jet production will remain the dominant feature of hard hadronic collisions at the next generation of hadron colliders, because the inclusive jet cross section, integrated above a given value of $p_T$, is expected to increase with $\sqrt{s}$. It has been suggested (65) that the increase of the jet yield with energy may be responsible for the rising of the total hadronic cross sections (66).

In the coming years the study of hadronic final states containing high-$p_T$ jets will provide more quantitative tests of QCD; and at the same time the increased statistics will improve the sensitivity in the search for structures in the multijet mass distributions. Such structures, if observed, would be the manifestation of the production of new heavy particles decaying into quarks and gluons.

*Literature Cited*

1. Alpgård, K., et al. *Phys. Lett. B* 121:209–15 (1983)
2. Arnison, G., et al. *Phys. Lett. B* 118:167–72 (1982)
3. Banner, M., et al. *Phys. Lett. B* 122:322–28 (1983)
4. Büsser, F. W., et al. *Proc. 16th Int. Conf. High Energy Phys.*, *Chicago, 1972*, 3:317. Batavia: Fermi Natl. Accel. Lab. (1972)
5. Alper, B., et al. *Phys. Lett. B* 44:521–26 (1973)
6. Banner, M., et al. *Phys. Lett. B* 44:537–40 (1973)
7. Büsser, F. W., et al. *Phys. Lett. B* 46:471–76 (1973)
8. Feynman, R. P. *Phys. Rev. Lett.* 23:1415–17 (1969)
9. Berman, S. M., Bjorken, J. D., Kogut, J. B. *Phys. Rev. D* 4:3388–3418 (1971)
10. Wilczek, F. *Ann. Rev. Nucl. Part. Sci.* 32:177–209 (1982)
11. Darriulat, P. *Ann. Rev. Nucl. Part. Sci.* 30:159–210 (1980)
12. Hansen, K., Hoyer, P., eds. Jets in High Energy Collisions, *Phys. Scr.*, Vol. 19. 202 pp. (1979)
13. DiLella, L. In *Proc. 10th Int. Symp. Multiparticle Dynamics, Goa, India, 1979*, pp. 385–419. Bombay: Tata Inst. Fund. Res. (1980)

134    DiLELLA

14. Bjorken, J. D. *Phys. Rev. D* 8:4098–4106 (1973)
15. Fabjan, C. W., Ludlam, T. *Ann. Rev. Nucl. Part. Sci.* 32:335–89 (1982)
16. Bromberg, C., et al. *Phys. Rev. Lett.* 38:1447–50 (1977); 43:565–68 (1979)
17. Corcoran, M. D., et al. *Phys. Rev. Lett.* 44:514–17 (1980)
18. DeMarzo, C., et al. *Nucl. Phys. B* 211:375–413 (1983)
19. Brown, B., et al. *Phys. Rev. Lett.* 49:711–15 (1982)
20. Söding, P., Wolf, G. *Ann. Rev. Nucl. Part. Sci.* 31:231–93 (1981)
21. Field, R. D., Fox, G. C., Kelly, R. L. *Phys. Lett. B* 119:439–44 (1982)
22. Singer, R., Fields, T., Selove, W. *Phys. Rev. D* 25:2451–54 (1982)
23. Banner, M., et al. *Phys. Lett. B* 118:203–10 (1982)
24. Beer, A., et al. *Nucl. Instrum. Methods A* 224:360–95 (1984)
25. Bagnaia, P., et al. *Phys. Lett. B* 138:430–40 (1984)
26. Bagnaia, P., et al. *Z. Phys. C* 20:117–34 (1983)
27. Arnison, G., et al. *Phys. Lett. B* 123:115–22 (1983); 132:214–22 (1983)
28. Åkesson, T., et al. *Phys. Lett. B* 118:185–98 (1982); 123:133–38 (1983)
28a. Angelis, A. L. S., et al. *Nucl. Phys. B* 244:1–22 (1984)
29. Combridge, B. L., Kripfganz, J., Ranft, J. *Phys. Lett. B* 70:234–38 (1977)
30. Horgan, R., Jacob, M. *Nucl. Phys. B* 179:441–60 (1981)
31. Owens, J. F., Reya, E. *Phys. Rev. D* 17:3003–9 (1979)
32. Baier, R., Engels, J., Petersson, B. *Z. Phys. C* 2:265–77 (1979)
33. Glück, M., Hoffmann, E., Reya, E. *Z. Phys. C* 13:119–30 (1982)
34. Paige, F. E., Protopopescu, S. D. *Proc. 1982 DPF Summer Study on Elem. Part. Phys. and Future Facilities, Snowmass, Colorado*, ed. R. Donaldson, R. Gustafson, F. Paige, pp. 471–77. New York: Am. Inst. Phys. (1982)
35. Furman, M. *Nucl. Phys. B* 197:413–45 (1982)
36. Humpert, B. Two-jet production in p$\bar{\text{p}}$ collisions, *CERN TH 3934* 86 pp. (1984)
37. Antoniou, N. G., Argyres, E. N., Contogouris, A. P., Vlassopulos, S. D. P. *Phys. Lett. B* 128:257–61 (1983)
38. Kunszt, Z., Pietarinen, E. *Phys. Lett. B* 132:453–57 (1983)
39. Furmanski, W., Kowalski, H. *Phys. Lett. B* 148:247–50 (1984)
40. Cormell, L. R., et al. *Phys. Lett. B* 150:322–26 (1985)
41. Collins, J. C., Soper, D. E. *Phys. Rev. D* 16:2219–25 (1977)
42. Arnison, G., et al. *Phys. Lett. B* 136:294–300 (1984)
43. Bagnaia, P., et al. *Phys. Lett. B* 144:283–90 (1984)
44. Angelis, A. L. S., et al. *Nucl. Phys. B* 209:284–300 (1982)
45. Abramowicz, H., et al. *Z. Phys. C* 12:289–95 (1982); 13:199–204 (1982); 17:283–307 (1983). See also Bergsma, F., et al. *Phys. Lett. B* 123:269–74 (1983)
46. Greco, M. *Z. Phys. C* 26:567–68 (1985)
47. Sterman, G., Weinberg, S. *Phys. Rev. Lett.* 39:1436–39 (1977)
48. Einhorn, M. B., Weeks, B. G. *Nucl. Phys. B* 146:445–56 (1978)
49. Shizuya, K., Tye, S.-H. H., *Phys. Rev. D* 20:1101–14 (1979)
50. Bassetto, A., Ciafaloni, M., Marchesini, G. *Phys. Rep.* 100:201–72 (1983)
51. Andersson, B., Gustafson, G., Ingelman, G., Sjöstrand, T., Artru, X. *Phys. Rep.* 97:31–171 (1983)
52. Åkesson, T., et al. *Z. Phys. C* 25:13–20 (1984)
53. Arnison, G., et al. *Phys. Lett. B* 132:223–29 (1983); Ghez, P. Jet fragmentation properties in the UA1 experiment. In *5th Topical Workshop on Proton-Antiproton Collider Physics, Saint-Vincent, Italy, 1985.* Singapore: World Scientific. In press (1985)
54. Althoff, M., et al. *Z. Phys. C* 22:307–40 (1984)
55. Bagnaia, P., et al. *Phys. Lett. B* 144:291–96 (1984)
56. Marchesini, G., Webber, B. R. *Nucl. Phys. B* 238:1–29 (1984); Webber, B. R. *Nucl. Phys. B* 238:492–528 (1984)
57. Field, R. D., Feynman, R. P. *Nucl. Phys. B* 136:1–76 (1978)
58. Angelis, A. L. S., et al. *Phys. Lett. B* 105:233–38 (1981)
59. Kunszt, Z., Pietarinen, E. *Nucl. Phys. B* 164:45–75 (1980)
60. Gottschalk, T., Sivers, D. *Phys. Rev. D* 21:102–30 (1980)
61. Berends, F., Kleiss, R., DeCausmaecker, P., Gastmans, R., Wu, T. T. *Phys. Lett. B* 103:124–28 (1981)
62. Ellis, J., Gaillard, M. K., Girardi, G., Sorba, P. *Ann. Rev. Nucl. Part. Sci.* 32:443–97 (1982)
63. Arnison, G., et al. *Phys. Lett. B* 129:273–82 (1983)
64. Bagnaia, P., et al. *Z. Phys. C* 24:1–17 (1984)
65. Cline, D., Halzen, F., Luthe, J. *Phys. Rev. Lett.* 31:491–94 (1973)
66. Bozzo, M., et al. *Phys. Lett. B* 147:392–98 (1984)

*Ann. Rev. Nucl. Part. Sci. 1985. 35 : 135–94*

# ON THE PRODUCTION OF HEAVY ELEMENTS BY COLD FUSION:
## The Elements 106 to 109

*Peter Armbruster*

Gesellschaft für Schwerionenforschung mbH, D-6100 Darmstadt, West Germany

CONTENTS

## 1. INTRODUCTION—EARLY CHARGED-PARTICLE REACTIONS TO PRODUCE HEAVY ELEMENTS

After a short historical introduction (Section 1), this article presents new insights into the mechanism limiting the fusion of heavy nuclides (Section

135

0163–8998/85/1201–0135$02.00

2). Fusion is finally limited by the increasing Coulomb forces in the formation process of a compound system, as well as in its deexcitation. Moreover, nuclear structure effects in all stages of evaporation residue (EVR) formation are shown to be of importance. The wide field of fusion reaction studies and possible experimental techniques is projected onto the task of element synthesis, and only those aspects that are of relevance here are covered. The better understanding of EVR formation (Section 2) and the new experimental techniques (Section 3) that enabled the production of elements 107–109 (Section 4) are also discussed. In Section 5 ground-state properties and the nuclear structure of the heaviest isotopes, together with their production cross sections, are discussed. Finally, an outlook on how eventually to go beyond $Z = 109$ is given.

Isotopes of elements beyond Md were synthesized by fusion of heavy nuclei. This nuclear reaction was most successful in heavy-element production: The elements 102–109 were made during the last 25 years and about 40 isotopes of these elements were known in 1984. The synthesis of elements 102 to 106 started from the heaviest possible actinide targets fused with light projectiles ranging from $^{10}B$ to $^{26}Mg$. This technique depends on the availability of heavy actinides in microgram quantities, and was widely used with the accelerators of LBL and ORNL in the US and of JINR in the USSR. The Berkeley and Dubna groups developed the pioneering techniques of using heavy-ion accelerators to fuse nuclei and detection systems to identify the new species. The many experiments done between 1955 and 1975 gave us the first five elements, 102–106, produced by fusion reactions and detected first not by chemical methods, as the elements from Pu to Md, but by careful analysis of the radioactive decay of the nuclear species. A gradual shift from chemical methods to physical methods occurred.

In the years 1955–1965, the years of apprenticeship, ingenious new methods were invented and perfected. Many isotopes were discovered, but retrospectively many of them had wrong mass assignments, wrong α energies, decay times, and branchings, and very often the claims of discovery did not withstand rigorous inspection. For elements detected by physical methods, no clear criteria for naming elements exist. Element naming was taken by scientists as a right in itself. Claims have been maintained in spite of experiments proving the contrary; and controversy over the names of elements continues even now. From 1965 to 1975 results were consolidated. The number of new conflicting results became smaller and smaller. The method of fusing actinide target atoms with light projectiles had reached great perfection. Most of the data now accepted were obtained at Berkeley and Dubna during this time. The reactions investigated before 1969 are compiled in Reference (1). The more recent results are reviewed in References (2, 3).

The heaviest element made starting from actinide targets is 106 (4). The experiment performed at Berkeley in 1974 established a preliminary limit to the method. $^{249}$Cf bombarded by $^{18}$O yields $^{263}$106 via a 4n-reaction channel; 87 atoms of $^{263}$106 were identified via mother-daughter correlations. The $\alpha$ lines of $^{259}$104 and $^{255}$102 known previously were correlated to $^{263}$106. The isotope decays via $\alpha$ decay, with a half-life of $0.9 + 0.2$ s; and $\alpha$ energies of 9.06 and 9.25 MeV were measured. The production cross section measured at 95-MeV bombarding energy is 0.3 nb. Dubna experiments (5) assign a fission branch to the isotope. Fission activities were later seen at Berkeley (6) as well, but were not assigned. The cross section obtained for the fission activity was 0.6 nb at Dubna and 9 nb at Berkeley. The fission branch of $^{263}$106 remains questionable, as the fission activity could well come from a target-like transfer product.

Different experiments designed to go beyond element 106 have been performed by bombarding $^{249}$Cf and $^{249}$Bk with $^{20,22}$Ne beams (7, 8). Atoms of elements 107 and 108 could not be detected in the large background of transfer products, which could not be suppressed sufficiently. We return to the actinide target reactions at the end of this article, in Section 5.5, and discuss future possibilities of the technique.

The possibility of using other nuclear reactions has been investigated thoroughly, but the results are discouraging. Successive neutron captures and subsequent $\beta$ decays lead to $^{257}$Fm, but the short half-life of $^{258}$Fm (0.4 ms) prevents further build up. Transfer of nucleons from lighter nuclei to the heaviest targets, such as $^{248}$Cm, $^{249}$Bk, $^{249}$Cf, and $^{254}$Es, produced a number of fission activities, which until now could not be assigned (6, 9, 10), but which might reflect isotopes of elements 102 and 103. The use of deep inelastic reactions for heavy-element production was disappointing, because the transfer of many nucleons is difficult to achieve without heating the highly fissionable heavy isotopes, and thus reducing their survivability to cross sections too small for current detection techniques (11, 12).

## 2. LIMITATIONS OF EVAPORATION RESIDUE FORMATION IN FUSION REACTIONS

We do not consider the disappearance of evaporation residue (EVR) formation at high bombarding energies here; instead we restrict ourselves to limitations at energies near the Coulomb barrier, which are relevant for heavy-element production. Following the assumption of a two-step formation process for EVRs—a first step of amalgamating two nuclei to a compound system, and a second step of deexcitation of the compound system —we discuss limitations in the entrance channel (the compound nucleus formation) and then restrictions in the exit channel (the compound nucleus deexcitation). An adequate presentation of the many facets of the physics of EVR formation goes beyond the scope of this article; only those aspects important for the synthesis of heavy elements are reviewed.

Because the formation cross sections for the heavy elements are smaller than 10 nb, that is $10^{-7}$ of the total reaction cross section, it is impossible to perform directly these investigations within reasonable beam times at EVRs of the elements in question. Therefore, the studies on entrance channel limitations were performed mostly with compound systems in the element range $Z = 80$–$90$. Studies on the deexcitation stop at the compound systems of element 104. Moreover, at a level below $10^{-3}$ of the total cross section, it becomes more and more difficult to make numerical predictions. We must combine experimental results with general theoretical concepts, such as "subbarrier fusion" (13, 14) and "extra-push energy" (15, 16) in order to gain some guidance for the reactions of interest. Subbarrier fusion may increase the fusion cross sections of heavy systems, before increasing Coulomb forces finally limit the formation of a compound system. It is shown below that nuclear structure effects are of importance for EVR formation up to the heaviest systems. Subbarrier fusion and the dynamical limitation of compound system formation are covered in Section 2.1, whereas the deexcitation of the compound system in its relevance to heavy element synthesis is discussed in Section 2.2.

## 2.1    The Entrance Channel Limitations

The phenomenon of increased subbarrier fusion of heavy systems was discovered experimentally (17, 18), whereas the limitation of compound system formation had been foreseen by theory early on (19). It seems appropriate to start the discussion with an experiment demonstrating both effects. Figure 1 shows the $xn$ cross sections for the reaction $^{92}Zr(^{90}Zr, xn)^{182-xn}Hg$ (20) together with the fusion barrier $B_B$ calculated using a standard potential—the Bass potential (21), which allows one to reproduce fusion barriers for all the light systems. The measurement shows that there is fusion far below the barrier, and at the same time still a smooth increase over a large energy range beyond the barrier.

In order to discuss the experiments we introduce a fusion probability $p(E)$ for central collisions (22). This quantity together with a survival probability $w(E+Q, l)$ allows us to separate numerically the phenomena in the entrance and exit channels. The procedure is especially adequate for highly fissionable compound systems, for which only in a small $l$ window near the zero angular momentum will EVRs surviving fission be found. The cross section for EVR formation at the kinetic energy $E$ in the c.m.s. system can be formulated by using the quantities $p(E, l)$ and $w(E+Q, l)$:

$$\sigma(E) = \pi \lambdabar^2 \sum (2l+1)p(E, l)w(E+Q, l). \qquad 1.$$

Here $E + Q$ is the excitation energy $E^*$ in the compound nucleus, $Q$ is the $Q$ value of the reaction, and $\lambda$ is the de Broglie wavelength corresponding to the entrance channel. If the compound nucleus hypothesis is assumed to be valid $w(E^*, l)$ does not depend on the entrance channel. We can now define an angular-momentum-weighted average of the fusion probability $\langle p(E, l) \rangle$:

$$\langle p(E, l) \rangle = \pi \lambda^2 \sum (2l + 1)p(E, l)w(E + Q, l)/\pi \lambda^2 \sum (2l + 1)w(E + Q, l)$$

$$= \sigma(E)/\pi \lambda^2 \sum (2l + 1)w(E + Q, l). \qquad 2.$$

This fusion probability $\langle p(E, l) \rangle$ can be determined from the measured EVR cross section $\sigma(E)$, provided $w(E^*, l)$ is known. The function $w(E^*, l)$ can either be determined from the EVR cross section of a corresponding asymmetric reaction (22), which allows us to determine $w(E^*, l)$ in an energy region undisturbed by entrance channel effects, or it can be determined from an evaporation cascade calculation (23), which has been tested to

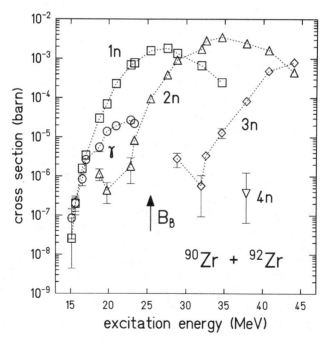

*Figure 1* Excitation functions of different EVR deexcitation channels in the fusion of $^{90}$Zr and $^{92}$Zr (20). Besides the $xn$ channels, the $\gamma$ channel—"radiation-fusion" or "radiation-capture" channel—is seen. As a reference point the Coulomb barrier $B_B$ according to the Bass potential (21) is given. The reaction also shows a subbarrier shift $(B_B - B_a)$ of 6.7 MeV as a dynamical shift ("extra-push") $(B - B_B)$ of 4.2 MeV.

reproduce EVR cross sections adequately (20, 24). The $l$ dependence of $w(E^*, l)$ is determined by the $l$ dependence of the fission barrier, which is assumed to follow the rotating liquid drop model (25). This $l$ dependence restricts the $l$ values contributing to EVR formation to a range much smaller than the broad distribution of angular momenta in the primary fusion process. Because the weighting function $w(E^*, l)(2l + 1)$ used to obtain $\langle p(E, l) \rangle$ is a narrow function peaking at all energies at small $l$ values ($\sim 15\hbar$), $\langle p(E, l) \rangle$ becomes about $p(E, l = 0)$ and the term "fusion probability for central collisions" $p(E)$ is justified.

Figure 2 shows fusion probabilities for three systems: $(^{40}\mathrm{Ar} + {}^{180}\mathrm{Hf})$ (26) showing subbarrier fusion, $(^{90}\mathrm{Zr} + {}^{90}\mathrm{Zr})$ (27) approximating best a one-dimensional tunnelling model (28), and $(^{96}\mathrm{Zr} + {}^{124}\mathrm{Sn})$ showing a strong dynamical hindrance (22). In Figure 3 we give a scheme for the general dependence of $p(E)$. It is characterized by three energies, the adiabatic barrier $B_a$, the Bass potential barrier $B_B$, and the dynamical barrier $B$. The logarithmic slope $(\hbar\omega)^{-1}$ at small energies, and the two parameters, $\sigma$ the barrier fluctuation parameter and $c$ a truncation parameter, are also of importance. The latter can be replaced by the values of $p(E)$ at the energies $B_a$ and $B_B$. The term $B_B$-$B_a$ characterizes subbarrier fusion and $B$-$B_B$ the energy shift due to the dissipation losses in the entrance channel, the extra-push energy. This six-parameter presentation describes both the subbarrier fusion effects and the entrance channel limitations. It replaces the two parameters $B_B$ and $\hbar\omega$ of a one-dimensional tunnelling model (28).

The fusion probability $p(E)$ may be presented as a sum over the tunnelling through a sequence of barriers that are distributed like a Gaussian around a nominal fusion barrier $B$ defined by $p(B) = 0.5$ (29):

$$p(E) = (2\pi\sigma^2)^{-1/2} \int_{-\infty}^{\infty} [\exp -(E' - B)^2/2\sigma^2] T(E - E') \, dE', \qquad 3.$$

with $\sigma$ the width of the barrier distribution and $T$ the transmission coefficient through a one-dimensional barrier calculated using a WKB approximation. As there is always a lowest barrier, $B_a$—the adiabatic barrier—that finally makes $p(E)$ decrease with a slope given by $(\hbar\omega)^{-1}$, the lower integration limit may be truncated and replaced by a truncation parameter $c$. For subbarrier fusion, Equation 3 relates the parameters $\sigma$ and $c$ to the adiabatic barrier $B_a$ and the fusion probability $p(B_a)$, which follow directly from the measured energy dependence of the fusion probability $p(E)$.

The subbarrier fusion phenomena have been analyzed thoroughly. Static deformation of a collision partner increases fusion below the barrier, as the fusion barriers for the different orientations of the collision partners in

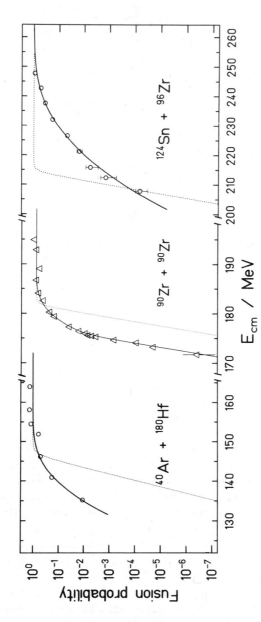

*Figure 2*  Fusion probability $p(E)$, i.e. the probability of fusion in the entrance channel, for three different systems as a function of bombarding energy (22). As reference $p(E)$ for a Hill-Wheeler barrier penetration of the Bass barrier is shown with dotted lines. The system $^{40}Ar + ^{180}Hf$ shows subbarrier fusion $B_B - B_a = 10$ MeV (26), the system $^{90}Zr + ^{90}Zr$ (20) fuses nearly unhindered, whereas the system $^{96}Zr + ^{124}Sn$ (22) shows a strong dynamic hindrance in the entrance channel $B - B_B = 26$ MeV.

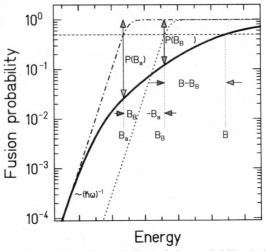

*Figure 3* Six parameters defining schematically the fusion probability. At low energies the slope $(\hbar\omega)^{-1}$ describes the Hill-Wheeler barrier penetration of the lowest possible adiabatic barrier $B_a$. Subbarrier fusion is parametrized by $p(B_a)$ and the subbarrier shift $(B_B - B_a)$, the dynamical hindrance by $p(B_B)$ and the dynamical shift (extra-push) by $(B - B_B)$.

respect to the deformation axis are different. Averaging over all orientations allows us to understand the strong subbarrier fusion of Ar and the deformed nucleus $^{154}$Sm (17). Collective excitations of the collision partners, which are slow compared to the barrier passing time, lead to dynamical deformations that allow fusion below the barrier. Reisdorf et al (30, 31) showed that subbarrier fusion scales with the energies of the lowest vibrational modes. Transfer reactions of positive $Q$ values preceding fusion contribute to subbarrier fusion (32). A coupled-channel calculation taking into account vibrational modes and transfer describes the reactions between different Ni isotopes (33), whereas the data on various Zr on Zr collision systems could not be reproduced by a similar calculation (20). Subbarrier fusion is governed by an interplay of various nuclear structure effects that are still not completely understood. For all systems at very low energies we find a quantum mechanical one-dimensional tunnelling behavior characterized by the logarithmic slope of $p(E)$, which is inversely proportional to a tunneling frequency. The energy $\hbar\omega$ is related to the curvature of the barrier, which depends for a given barrier height on the barrier width and in the inertial mass parameters. Values for $\hbar\omega$ found for the heavy systems are between 2 and 3 MeV.

The Coulomb repulsion between the two collision partners during the fusion process must be small enough to allow the formation of a fused monosystem. Figure 4 demonstrates that nature has set a limit to combining two nuclei with atomic number $Z_p$ and $Z_T$ to form a nucleus with atomic number $Z_p + Z_T$. Systems that were fused successfully are separated from combinations leading to no EVRs. The line of separation is characterized by a well-defined ratio of Coulomb and nuclear forces acting between the two amalgamating collision partners. Beyond that line open points indicate the different failures to produce superheavy elements. For $Z_p + Z_T > 120$ the very first step in the formation of an EVR, the overcoming of the Coulomb barrier, already becomes impossible for any combination of $Z_p$ and $Z_T$. In the range $120 > Z_p + Z_T > 80$ fusion is hindered, not every combination of $Z_p$ and $Z_T$ is successful, e.g. the isotope $^{244}$Fm was made in the reaction $^{206}$Pb($^{40}$Ar, 2n)$^{244}$Fm, but could not be made out of the nearly symmetric partners $^{110}$Pd and $^{136}$Xe (34).

The ratio of disruptive Coulomb forces and attractive surface tension forces governs the amalgamation of two nuclei into one. For a monosystem this ratio is given by the fissility parameter $x$. For a two-touching sphere configuration, Bass (35) defined a corresponding parameter making use of the proximity force. Taking into account that the proton and neutron ratio between the two partners is equilibrated very quickly ($10^{-22}$ s), and that the nuclear system at the Coulomb barrier for those systems of importance here

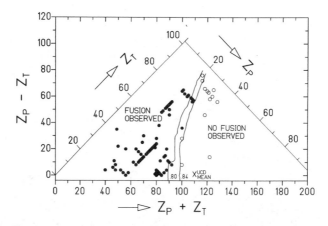

*Figure 4*   Elements $Z_p$ and $Z_T$ fused to element $Z_p + Z_T$. Each combination is a point in the triangle. Successful fusion reactions (in GSI experiments over the years detected by EVRs) are indicated by full points, failures by open points (36). Between $x_{\text{mean}}^{\text{UCD}} = 0.80$ and 0.84, the production goes down below the 100-pb level. Between $Z = 90$ and $Z = 120$, more and more asymmetric combinations are necessary to produce surviving EVRs (34). Beyond $Z > 120$ no fusion is possible.

is more compact than the two-touching sphere configuration, a modified parameter $x_{\text{mean}}^{\text{UCD}}$ describing the ratio of Coulomb and nuclear forces has been defined and applied to organize the vast amount of data (34, 36):

$$x_{\text{mean}}^{\text{UCD}} = 2x(\kappa^2 + \kappa + \kappa^{-1} + \kappa^{-2})^{-1/2} \quad \text{with} \quad x = (Z^2/A)/(Z^2/A)_{\text{crit}};$$

$$(Z^2/A)_{\text{crit}} = 50.88[1 - 1.78(N - Z/N + Z)^2] \quad \text{and} \quad \kappa = (A_1/A_2)^{1/3}. \qquad 4.$$

Passing the Coulomb barrier, an appreciable amount of the radial energy on the long way from a two-touching nuclei configuration to a compound system is transferred into intrinsic excitation energy. The dissipated energy is reflected as a virtual increase of the Coulomb barrier; an extra-push energy is needed to fuse. The extra-push model introduced by Swiatecki (15) predicts a parabolic dependence of the extra-push energy on the scaling parameter $x_{\text{mean}}^{\text{UCD}} - x_{\text{thr}}$:

$$B - B_B = a^2 E_s (x_{\text{mean}}^{\text{UCD}} - x_{\text{thr}})^2 \quad \text{with}$$

$$E_s = 7.6 \times 10^{-4} (Z^2/A)_{\text{crit}}^2 A_2^{1/3} A_1^{1/3} (A_2^{1/3} + A_1^{1/3})^2/(A_1 + A_2), \qquad 5.$$

where $x_{\text{thr}}$ is a threshold value describing the onset of entrance channel limitation and $a$ is a slope parameter. The numerical values for the parameters are calculated or fitted to the data (37). The calculation (38) gives $x_{\text{thr}} = 0.72$ and $a = 22$. The onset of entrance channel limitations is predicted by all friction models (37–40), but it is still open which of the models describes the data best. Accurate measurements of the quantity $B - B_B$ should finally settle which mechanism governs the dissipative losses in the fusion of heavy nuclei.

In Figure 5 are compiled all extra-push values obtained from fits to $p(E)$ for different systems leading to EVR formation (20, 22, 24). The quantity $B - B_B$ with the above parameters follows roughly a quadratic dependence

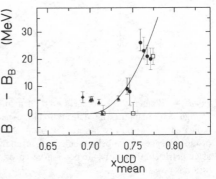

*Figure 5* The dynamical shift in the entrance channel (extra-push), $B - B_B$, as a function of the scaling parameter $x_{\text{mean}}^{\text{UCD}}$ for all systems where EVRs have been detected and $p(E)$ has been analyzed for EVR formation (20, 22, 24). The full curve is the prediction of the extra-push model (15) with the parameters of (38). Triangles are from (20), circles are from (22), and squares are from (24).

as predicted by the one-body dissipation model. It should be mentioned that the analysis of symmetric binary fragmentation data yields smaller extra-push energies and the limitation sets in at higher $x$ values (41, 42). Binary fragmentation includes the cross section of quasi-fission; that is, a contribution leading to a symmetric fragmentation but not necessarily to an equilibrated compound system. For quasi-fission, which needs smaller overlap of the collision partners, we would indeed expect smaller losses by dissipative processes. The data of Figure 5 indicate the importance of nuclear structure effects in addition to the macroscopic dissipation losses. The smallest extra-push energies are observed for spherical, closed shell, collision partners. The $^{48}Ca + ^{208}Pb$ collision system is found to fuse without any hindrance (24); $^{90}Zr$ in the bombardments by $^{124}Sn$ and $^{90}Zr$ beams shows the smallest extra-push energy of all Zr isotopes (20, 22). Systems with several neutrons outside the $N = 50$ shell in Zr isotopes and two protons outside of the $Z = 20$ shell in $N = 28$ isotopes show considerably larger extra-push energies. The nuclear structure effects may contribute to the extra-push energies up to 10 MeV. These nuclear structure contributions to the barrier shifts have to be corrected for in order to obtain the dissipative losses in the macroscopic models. The macroscopic friction models are a first approximation that will describe the gross features, but to make any predictions for an actual collision system nuclear structure effects have to be considered.

Scaling of the extra-push energies may be possible for the macroscopic contribution, but certainly not for the microscopic, nuclear structure part. Whether the large values of the barrier fluctuation parameter observed for entrance-channel-limited systems, $\sigma \approx 10$ MeV (22, 24), reflect subbarrier fusion of microscopic origin or are a macroscopic property of the system is an open question. The analysis shows the barrier fluctuation parameter $\sigma$ to increase with increasing extra-push energy $(B - B_B)$. The shift of the dynamical barrier becomes a quantity that fluctuates with an increasing width around its mean value $B$. The fluctuations getting comparable to the absolute value of the shift make the concept of a well-defined barrier doubtful. For the systems in question the potential energy surface shows no well-defined saddle point, but a large plateau region where small dynamical fluctuations will finally decide into which direction the system develops, to a binary or to a monosystem. The terms $B - B_B$ and $\sigma$ seem to be coupled parameters, defining together the fusion probability at the Bass barrier, $p(B_B)$, a quantity characterizing the hindrance in the entrance channel at the energy of the Bass barrier. The value of $p(B_B)$ may be derived directly from the cross-section value at the barrier energy $B_B$.

Figure 6 shows $p(B_B)$ as a function of $x_{mean}^{UCD}$ for all systems investigated. In this presentation the cross-section values known for heavy-element production beyond $Z = 104$ have been included (Section 5.4). Again, as in Figure 5, nuclear structure effects are evident. They modify $p(B_B)$ by factors larger than 10. A linear fit to all the data points gives for $p(B_B)$ an exponential dependence from $x_{mean}^{UCD} - x_{thr}$ with a slope parameter d $[\ln p(E)]/dx = 71$, equivalent to a factor of two for a change of 0.01 in $x_{mean}^{UCD}$:

$$p(B_B) = 0.5 \exp[-71(x_{mean}^{UCD} - x_{thr})].$$     6.

The entrance channel effects in the fusion of heavy nuclei extracted from the cross sections experimentally via $p(E)$, reduced to the energies $B_B$, $B_B - B_a$, $B - B_B$, the value of the slope parameter $\hbar\omega$, and the values of $p(E)$ at energies $B_a$ and $B_B$ must be understood on the one hand from the macroscopic parameters describing tunnelling, the fusion barrier, and the dissipation of energy of colliding, charged, liquid drops, and on the other hand from nuclear structure effects such as deformation, vibrational modes, and $Q$ values of competing transfer channels. Both theory and experiment are still far from being able to make reliable predictions to support the search for favorable new collision systems, systems that would eventually produce elements beyond the atomic numbers reached so far. The understanding of entrance channel effects is the most challenging task for future investigations, both for theoreticians and for experimentalists. New theoretical microscopic approaches, besides the macroscopic concepts, will have to be applied to the heavy-element synthesis problem, for example the "dissipative diabatic dynamics" of Nörenberg and co-workers (43–45).

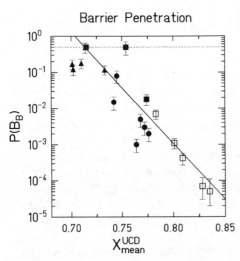

Figure 6 The fusion probability at the Bass barrier $p(B_B)$ characterizing the dynamical hindrance in the entrance channel for all systems where EVRs have been detected (20, 22, 24, 96, 98, 100). The scattering of the experimental values shows the importance of nuclear structure effects, e.g. $^{208}Pb(^{48}Ca, xn)^{256}102$ ($x_{mean}^{UCD} = 0.751$) (24) compared to $^{123}Sb(^{86}Kr, xn)^{209}Fr$ ($x_{mean}^{UCD} = 0.736$) (22). The full line is a fit to Equation 6, giving d $\ln p(E)/dx_{mean}^{UCD} = 71$ and $x_{crit} = 0.71$. The constant slope points to a strongly increasing fluctuation of the dynamical barrier $B$ for systems with a dynamical hindrance (24, 57). Triangles are from (20), circles are from (22), solid squares are from (24, 96, 98, 100), and open squares are from cross-section values (see Section 5.4).

## 2.2   Exit Channel Limitations

Newly born compound systems are hot. The nucleons in the compound system are excited. Excitation energies of 40–50 MeV at the barrier are found for many reactions. As long as the fission barriers are high, this excitation energy is dissipated by a cascade of evaporated particles. However, for small fission barriers the system may be destroyed by fission competing with particle evaporation. The probability of fission during the cooling-down phase of the fused compound system must be small enough to guarantee production rates of at least a few atoms per week. The losses during deexcitation of a fused compound system depend on the stability of nuclei against fission in states far above the ground state. The temperature dependence of the fission barrier is of great importance, as the production cross section of heavy elements is governed by its dependence on intrinsic excitation energy. To minimize the fission losses in the deexcitation phase, the highly fissionable nuclei should be produced with the smallest excitation energy possible, as each step in the competition between neutron emission and fission contributes to the losses. For a given compound system, a high survival probability is obtained if the number of steps in the evaporation cascade is kept small and if for each step the fission barrier has its largest possible value. Systems with the smallest possible excitation energies and angular momenta will survive.

The macroscopic fission barrier $B_f^{macro}$, the barrier protecting a charged liquid drop against spontaneous fission, and the extra-push heating $B - B_B$ are closely related, as both depend on the value of the fissility parameter $x$ (defined in Equation 4). Decreasing fission barriers of liquid drops go together with increasing extra-push heating. Fortunately, the contributions of nuclear structure effects to the fission barrier and to the fusion probability are decoupled. We may try to profit from nuclear structure effects twice: in the entrance channel in order to fuse systems that would not fuse without nuclear structure effects, and in the exit channel in order to produce shell-stabilized heavy nuclei that would not survive deexcitation without shell stabilization.

2.2.1   COLD FUSION   The minimum excitation energy $E_{barr}^*$ of a compound system is the sum of the energy $B$ to pass the Coulomb barrier and the negative reaction $Q$ value. In Figure 7 the minimum excitation for an element $Z_p + Z_T$ made out of a projectile $Z_p$ and a target nucleus $Z_T$ is shown. Those isotopes leading to the smallest excitation energy for a given combination have been chosen. "Extra-push" energy according to the parameters of Figure 5 is added. The size of the squares runs from 15 to

60 MeV. The figure demonstrates the different mechanisms used to reduce the excitation energy, which are explained as follows.

1. There is a trend that heavier compound systems can be produced colder than lighter ones, as long as the extra-push heating (which for symmetric systems becomes of importance for $Z_p + Z_T > 80$) is not yet acting (46).

2. A given compound system produced out of symmetric partners is colder than it would be if produced more asymmetrically. $^{220}$Th as a compound system obtains at the fusion barrier 45 MeV of excitation energy in the reaction $^{40}$Ar $+\,^{180}$Hf, whereas it obtains only 30 MeV in the reaction $^{96}$Zr $+\,^{124}$Sn (22, 26, 47). In the first reaction 3n and 4n channels are observed, whereas in the latter 1n and 2n channels are found. Symmetric fusion cools the system down by at least 15 MeV. This cooling mechanism cannot be applied to heavier systems, as symmetric systems reach the limit of extra-push heating between $Z = 80$ and $Z = 90$, e.g. uranium made out of two $^{110}$Pd nuclei would show the same extra-push heating as element 109 made asymmetrically out of $^{58}$Fe and $^{209}$Bi.

3. Shell stabilization of the ground-state masses of the collision partners increases the $Q$ values and helps to cool the compound system. The breaking of the shells eats up translational energy and leads to colder compound nuclei. This cooling mechanism was first observed in the fusion of $^{40}$Ar and $^{208}$Pb by Oganessian and Demin (48, 49), and it opened a new road to the heaviest elements. The use of Pb and Bi targets together with beams of Ar to Fe allows the production of all elements between Fm and $Z = 109$. A cooling of about 15 MeV, compared to the production using actinide targets, is possible.

The number of evaporated particles in the deexcitation cascade after fusion is a measure of the excitation energy. It has been decreased from 4–5

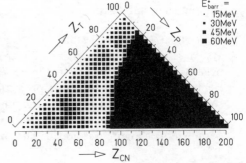

*Figure 7* The excitation energy of the fused systems $(Z_p + Z_T)$ made out of elements $Z_p$ and $Z_T$ taking into account the extra-push energy (36), Equation 5 with parameters as in (38). The squares run from 15–60 MeV. The "cold fusion" valleys at $Z_T = 82$ and the "cold plane" near symmetry for $Z_p + Z_T = 72$–88 is seen.

neutrons in the early fusion studies at Berkeley and Dubna leading to elements $Z = 101-106$ (1-3) via two-neutron reactions found at Dubna in 1974 (49), and one-neutron deexcitation found in 1980 at Darmstadt (50). Deexcitation by pure $\gamma$ emission was found recently in the reactions $^{90}$Zr $+ ^{90}$Zr $\rightarrow ^{180}$Hg at Darmstadt (46) and $^{50}$Ti $+ ^{208}$Pb $\rightarrow ^{258}$104 at Dubna (51). Excitation functions for the reaction $^{90}$Zr $+ ^{90}$Zr have been measured. They show below the barrier at an excitation energy of 17 MeV the one-proton, one-neutron, and the $\gamma$ channels (see also Figure 1). The excitation function of the $\gamma$ channel peaks at 20 MeV and has a width of only 6 MeV. The reaction combines all three mechanisms and leads to a cold fusion reaction. Heavy, symmetric, and shell-stabilized collision partners are fused below the barrier with no hindrance by extra-push heating. The cross section for "radiative fusion" is high, about 50 $\mu$b, as fission competition is still small. The fused system $^{180}$Hg is situated at the triple point of the $(N, Z)$ plane, where proton and neutron separation energies and the fission barrier are about equal. Around $^{180}$Hg we find the nuclei the most resistive to intrinsic excitation. If these are produced cold, they reveal the long-sought "radiative fusion," the ultimate in cold fusion with surprisingly large cross sections. The large cross sections are reproduced by evaporation calculations assuming an E1 strength increased by a factor of five (20, 52, 53).

2.2.2 TEMPERATURE DEPENDENCE OF SHELL STABILIZATION    The fission barrier protects the compound systems against immediate disintegration. The fission barrier is made up of two contributions, a smoothly varying macroscopic liquid drop part and a shell correction, mainly of the ground-state mass. Near element 106 the macroscopic fission barriers fall below 1 MeV. As is shown in Section 5.1 the stability beyond is governed by the shell correction of the ground-state masses alone. The fluctuations of the level density around a continuously increasing Fermi-gas level density determine the shell corrections, which strongly depend on the excitation energy and the deformation of the nuclear system. The study of the temperature dependence of the shell contributions to the fission barrier makes it possible to predict which excitation energy a shell-stabilized compound system may take in order to survive the deexcitation process.

Shell-stabilized, highly fissionable nuclei are found among the actinides. Using these nuclei produced by different nuclear reactions, one can study the contribution of shell effects to EVR formation at excitation energies varying from 10 to 60 MeV. We differentiate between shell-stabilized deformed and spherical nuclei. We choose the energy dependence of $\Gamma_n/\Gamma_f$ for heavy actinide isotopes, as presented in (54, 55), to discuss deformed nuclei. For shell-stabilized spherical nuclei we refer to highly fissionable $N = 126$ nuclei, such as $^{216}$Th (26, 56, 57).

A comparison of EVR cross sections for pairs of reactions leading to two neighboring compound systems gives $\Gamma_n/\Gamma_f$ as a function of excitation energy (54, 55). All steps of the evaporation cascade except the first step are assumed to be equal for an $x$n reaction and $(x+1)$n reaction leading to the same EVR:

$$\Gamma_n/\Gamma_f(E^*_{x+1}) = (\sigma_{x+1}/\sigma_x)_{exp}$$
$$\times P_x(E^*)/P_{x+1}(E^*_{x+1}) \times \sigma_{CN}(E^*_x)/\sigma_{CN}(E^*_{x+1})_{calc}. \quad 7.$$

$P_x$ is the probability of emitting $x$ neutrons at an excitation energy $E^*_x$, $\sigma_{CN}$ is the cross section to produce the compound system at an excitation energy $E^*_x$. $P_x$ and $\sigma_{CN}$ are calculated following standard methods (58, 59). The experimental values for $\Gamma_n/\Gamma_f$ obtained for three nuclei are presented in Figure 8. The values given refer to an angular momentum of zero. The $\Gamma_n/\Gamma_f$ values depend only weakly on the energy $E^*$ in the range investigated. The few points in the range 10–20 MeV are even larger than at higher energies. The results of two calculations are given for comparison. The full line shows the expectation for $\Gamma_n/\Gamma_f$ with shell effects entering into level densities and the fission barriers. An exponential damping of the shell effects with $E_D = 18$ MeV, which follows from microscopic models, is used in the calculation (60, 61). The damping of shell effects leads to the shallow minimum of $\Gamma_n/\Gamma_f(E^*)$, which becomes deeper for systems dominantly shell stabilized. The hatched line is calculated without any shell effects, and obviously the observed increase of $\Gamma_n/\Gamma_f$ in the low energy range cannot be explained. The measured cross sections must include shell effects for all the deformed nuclei analyzed. There would be no heavy elements made by fusion, unless $\Gamma_n/\Gamma_f$ were kept large at low excitation energies by shell effects.

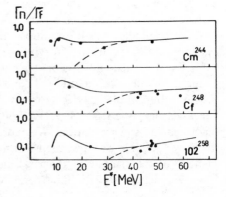

*Figure 8* $\Gamma_n/\Gamma_f(E^*)$ for the three deformed compound systems $^{244}$Cm, $^{248}$Cf, and $^{258}$102. Experimental values are taken from pairs of excitation functions (54, 55). The lines indicate two calculations, full line with shell effects, dotted line without shell effects (58, 60).

A very different behavior was found in experiments near $^{216}$Th. Here nuclei are spherical, highly fissionable, and again their barriers are partly shell stabilized. They may be produced within a range of excitation energies between 20 and 50 MeV. These nuclei are the best approximation to the shell-stabilized spherical superheavy nuclei near $N = 184, Z = 114$. Figure 9 shows measurements of the survival probability for thorium isotopes produced by 4n reactions (57). Two calculations of survival probabilities are shown. The measured cross sections are smaller by about two orders of magnitude compared to an evaporation cascade calculation that included the ground-state shell effects in the fission barriers and a damping of shell effects with $E_D = 18$ MeV (60, 61). A calculation with a reduced value of $E_D = 6$ MeV is given for comparison. It reproduces the trend of the experimental data fairly well.

Figure 10 shows a comparative study of EVR formation cross sections in the excitation energy range of 40–50 MeV. Measured cross sections for nuclei with $B_f < B_n$ are compared to calculations (36, 62). The deformed nuclei in the range $N = 140$–155 have shell corrections of their fission barriers similar to those of the spherical nuclei around $N = 126$ (Figure 10, bottom). The inclusion of shell effects with a damping energy of 18 MeV reproduces well all data for deformed nuclei (Figure 10, top). However, for spherical nuclei the measured values are three orders of magnitude smaller than calculated. Neglecting all shell effects (Figure 10, middle) reproduces the data for spherical nuclei, whereas for deformed nuclei the calculated cross sections would be off by many orders of magnitude, in agreement with the analysis of Figure 8.

We know there is additional ground-state stability of spherical nuclei due to shell effects, which must be observable somewhere in the production

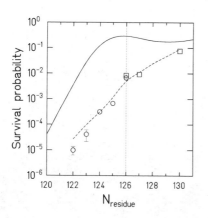

*Figure 9* The survival probability $w(E^*)$ for spherical Th EVRs produced by 4n reactions as a function of the neutron number of the EVR (57). Circles are from ($^{40}$Ar + Hf), squares are from ($^{48}$Ca + Yb) reactions. The lines give two calculations using different damping energies for shell effects (60); full line $E_D = 18$ MeV, dashed line $E_D = 6$ MeV.

cross sections at low energies. The analysis of the data shown in Figure 9 is compatible with shell effects becoming important at energies below 15 MeV ($E_D$ = 6 MeV). At higher energies the spherical compound systems behave as if no shell stabilization exists, whereas in deformed compound systems the shell stabilization is of importance up to 40–50 MeV ($E_D$ = 18 MeV). Introducing the well-known concept of collective enhancement of level densities into the calculation (63), one can explain the small experimental cross sections (64). A deformed nucleus like $^{252}$Fm is not affected by the collective enhancement of level densities, whereas $^{216}$Th shows a temperature-induced deformation. This reduces the influence of the

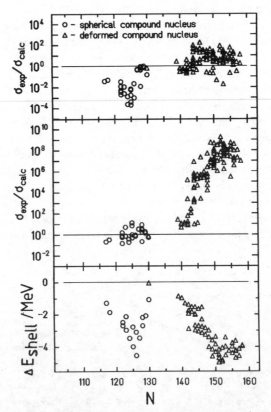

*Figure 10* Comparison of calculated and measured EVRs cross sections at 40–50 MeV. Triangles are deformed nuclei, circles are spherical nuclei. The shell correction $\Delta E_{shell}$ (*bottom*), and calculations with (*top*) and without (*middle*) shell effects compared to experimental values are given (62).

spherical shell effect drastically, because even at moderate excitation energies the nucleus prefers the deformed shapes that are not shell stabilized. Collective enhancement of level densities may help to explain why spherical nuclei like $^{218}U_{126}$ or $^{298}114_{184}$ were not made until now.

The following conclusions for heavy-element production may be drawn from the studies on the limitations of EVR formation.

1. The entrance channel limitation by dissipative losses, as introduced by Swiatecki (15), hinders EVR formation at values of $x_{mean}^{UCD} > 0.70$. Further experiments indicate that there is no well-defined extra-push energy, but large fluctuations of the dissipative losses go together with increasing extra-push energies. The fusion probability at the Bass barrier $p(B_B)$ is proposed to be a more adequate ordering parameter than the extra-push energy for highly hindered systems.

2. Below and above the Bass barrier, nuclear structure effects modify the dissipative losses in EVR formation. Whether or not the same trends as found for subbarrier fusion are of importance when extra-push limitations are acting is an open question. Nuclear structure effects in EVR formation for the region of heavy elements could not until now be predicted.

3. Shell-stabilized deformed EVRs may be produced from compound systems with high excitation energies, as $\Gamma_n/\Gamma_f$ is only weakly dependent on the excitation energy.

4. Shell-stabilized spherical EVRs do not profit from shell stabilization in EVR formation unless their excitation energy is smaller than 15 MeV, a condition eventually to be fulfilled in the reaction $^{110}Pd(^{110}Pd, 2n)^{218}U$, but never for reactions leading into the island of spherical superheavy nuclei.

5. To minimize losses in the evaporation cascade, the excitation energy should be kept as small as possible. The most symmetric combination of shell-stabilized collision partners, not yet limited by extra-push heating, should be chosen in order to fuse as coldly as possible. To produce elements 106 to 109, targets like $^{208}Pb$ and $^{209}Bi$ are bombarded by beams of $^{54}Cr$ and $^{58}Fe$. This method of element production is called cold fusion in contrast to the hotter actinide-based reactions.

# 3.  MODERN EQUIPMENT FOR ELEMENT SYNTHESIS

## 3.1  General Requirements Concerning the Techniques

To fuse two nuclei at the Coulomb barrier and thereby produce the heaviest elements, specific energies between 4.5 and 5.5 MeV/u must be available, corresponding to energies of 80–350 MeV for projectiles between oxygen and nickel. Experimental equipment has to be designed in order to detect isotopes heavier than $Z = 104$ and $A = 250$, the decay modes of which are mainly $\alpha$ decay and fission. Electron capture (EC) branches are expected to be small, and detection of EC is desirable but not of great importance. Half-lives will be smaller than a few seconds. A lower limit on decay times cannot be foreseen. All isotopes are assumed to be made by fusion of heavy nuclei, that is by a reaction with excitation functions as narrow as 10 MeV ($10^{18}$ atoms/cm$^2$). Production cross sections will be below $\sim 10$ nb, the cross sections found to produce isotopes of element 104. The cross-section limit should be pushed to as small a value as possible.

The beam intensities should be as high as target technology allows. In thick Pb and Bi targets, about 3 kW are released by beam intensities of $5 \times 10^{13}$ ions/s. Pb and Bi targets of a few hundred $\mu$g/cm$^2$ at intensities of $4 \times 10^{12}$/s will be heated by about 6 W. Thick targets have been cooled by forced liquid metal cooling at Dubna (65). Thin targets are cooled as rotating targets by black-body radiation cooling (66, 67) or by helium flow cooling at low pressures (68). The beam intensities given are upper limits reached with the most advanced ion sources at the Dubna U-400 or at the Darmstadt UNILAC.

In order to detect primary isotopes of the heaviest elements the separation method should be faster than seconds. In case of the detection of long-lived members of $\alpha$-decay chains, the separation time can be adjusted to the half-lives, which may be as long as days. Chemistry, on-line mass separators, and cluster-loaded He-jet transport systems have separation times of the order of seconds and will be useful only if isotopes in this time range are to be detected. Fast separation systems in the millisecond range are wheels (49, 65), systems in the microsecond range are recoil separators (69, 70).

The luminosity $L$/cm$^2$s $= NI$/cm$^2$s, with $N$ the number of target atoms per cm$^2$ and $I$ the beam current in ions per second, is a quantity related to the reaction rate $A$ by $A = L\sigma$, where $\sigma$ is the production cross section. A luminosity of $10^{31}$/cm$^2$s is equivalent to about 1/pb · d, that is a production rate of one atom per day at $\sigma = 1$ pb. Another important quantity of a separation method is its selectivity against primary beam particles, target recoil atoms, and transfer products. Rotating wheels and He jets do not discriminate radioactive transfer products at all. Recoil separators reduce transfer products by 2–3 orders of magnitude, whereas ISOL systems allow complete suppression of particles with deviating masses. Table 1 compares different methods either used or to be used for heavy-element production. The methods are presented in the following.

**Table 1**  Comparison of specifications of different experimental methods

| Method | Rotating wheels | He-jet | Velocity selector | ISOL system | Gas-filled separator |
|---|---|---|---|---|---|
| Reference | (65) | (68) | (69) | | (70) |
| Application | 104–109 | 104–106 | 104–109 | not used | 102 |
| Decay mode detected | fission | α chain or fission | EVR α chain fission | α chain fission | EVR α chain fission |
| Separation time (s) | $10^{-3}$ | $10^{-1}$ | $10^{-6}$ | 1 | $10^{-6}$ |
| Luminosity $(pb^{-1} d^{-1})$ | 5 | 0.4 | 0.2 | 0.2 | 0.2 |
| Total efficiency | 0.6 | 0.4 | 0.2 | 0.05 | 0.6 |
| Event number detected at 1 pb in 10 days | 30 (only fission) | 2 | 0.4 | 0.05 | 1 |
| Reduction of transfer products | none | none | $10^3$ | excellent | $10^2$ |
| Single-event conclusive | no | no (fission) yes (α chain) | yes | yes | yes |

## 3.2  Techniques

3.2.1  ISOL SYSTEMS  ISOL systems until recently were never used in heavy-element synthesis experiments. They are included in Table 1 because they have some excellent features that in special cases may compensate for their slow separation time and low efficiency (only a few percent).

3.2.2  He-JET TECHNIQUES  As an example of an advanced He-jet facility stands the equipment used by Ghiorso and co-workers (4) for detection of $^{263}106$. The $^{18}O$ beam of $3 \times 10^{12}$/s hits the He-cooled 0.8 mg/cm$^2$ thick $^{249}Cf$ target mounted on a thin substrate. The recoils stopped in helium are transported by a swift flow of helium seeded by a NaCl aerosol through a 5-m long capillary tube. The jet deposits its activity on a wheel, which is moved after a preset collection time in front of surface barrier detectors. These register either $\alpha$ decays or fission events coming from the parent or daughter decays. A decaying parent atom sitting on the wheel either emits its $\alpha$ particle toward the detector or is deposited itself as a recoil on the detector surface. The $\alpha$ decay of the daughters is registered by the detector with a high efficiency. A shuttle system allows one to move the detector in front of a stationary second detector. The decay of a daughter atom transferred as a recoil to the first detector can now be registered either by this detector or by the second one. Three generations of $\alpha$ decays are linked together by the wheel, the shuttling first detector system, and the fixed second detector array. The system has an 80% transport efficiency, a transport time of a second, but no discrimination against the copious transfer products. Nevertheless, a cross section of 0.3 nb has been detected for formation of the EVR $^{263}106$.

Because no correlation to the formation process of the atoms is possible, fission events cannot be correlated. Fission half-lives can be measured by varying the wheel transport time. With decreasing half-lives the restrictions set by the long transportation times become more and more the decisive handicap, e.g. the known isotopes of elements 107 to 109 could not have been detected by any of the advanced He-jet systems installed at the various laboratories.

3.2.3    WHEEL TECHNIQUES    The wheel techniques were pioneered by the Dubna group (49) and later applied by several other groups (6, 34). Their advanced systems to study elements 104–109 are described here (see Figure 11) (65).

The target is a rotating cylinder made of copper, the circumference of which is plated by a 3-mg/cm$^2$ layer of the target substance. The beam hits the wheel tangentially; thus the target is equivalent to a thick target. EVRs are stopped after having passed a layer of a few mg/cm$^2$ in beam direction. Because fission products are emitted isotropically, there is a high probability of detecting at least one of them outside the target layer. Except for a section of about 45° at the irradiation position, the wheel is viewed by mica detectors, which allow for fission product detection. A 60% efficiency for detection of a spontaneous fission event is achieved. The target wheel is cooled and a beam power of 3 kW is safely handled over long irradiation times. The heat generated in the copper wheel is transported via a thin liquid metal layer to the standing water-cooled parts of the system. Beam currents of $5 \times 10^{13}$/s have been used in irradiations with $^{55}$Mn ions. The speed of revolution may be varied between 1 and $10^4$ cycles per min.

Shortest half-lives reported are a few milliseconds. Registration of fission over periods of 100 h in specially selected micas is possible without any disturbing background from spontaneously fissioning natural uranium contaminants. Discrimination of fission products and scattered projectile tracks is possible. The fact that the method can only detect fission is a most serious disadvantage. This is compensated by the excellent luminosity, which is larger by a factor of ten than that of other methods.

The thin layer of target material may be radiochemically treated after the irradiation (72). Long-lived known daughter products of α-decay chains ending in elements Cm to Es have been separated and detected by their α energies. The detection of the heavy actinides is assumed to be related to heavy-element formation. In the case of targets in the lead region these isotopes are not produced as transfer products, nor in other decay channels except for neutron evaporation. The latter is found to contribute predominantly to EVR formation (51). This radiochemical technique is restricted to the special case of reactions using targets in the lead region, and it is difficult to transfer the method to actinide targets.

3.2.4  VELOCITY SEPARATORS    The conservation of mass and momentum in a fusion reaction demands that the fused product of mass $A = A_1 + A_2$ be emitted in beam direction with a well-defined velocity $v$, given by $v = A_1 v_1/$

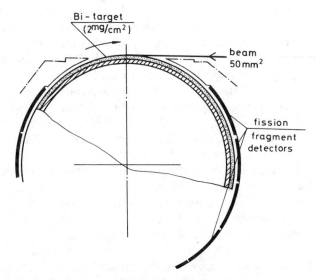

*Figure 11*  Wheel system developed by the Dubna group to synthesize isotopes by "cold fusion" (49, 71).

$A_1 + A_2$. $A_1$ and $A_2$ are the masses of the projectile and the target; $v_1$ is the velocity of the projectile hitting the target nucleus at rest. Particles of velocity $v$ emitted in beam direction must be fusion products. A velocity filter selecting these particles is a kinematic mass separator. Since the projectiles and fusion products are both flying in the beam direction with different velocities, the separation problem is characterized by the suppression factor of primary projectiles in the velocity window of the fusion products. The ratio of projectiles to reaction products at the entrance to the filter is larger than $10^8$. Suppression factors of at least this order of magnitude have to be demanded. Radial and axial acceptance angles of $3°$ guarantee a transport through the velocity filter of $20\%$ for EVRs produced near the Coulomb barrier.

Different versions of velocity filters are discussed and reviewed in (73) in connection with the problem of isotope separation. Several filters are operating but only one system, the Separator for Heavy-Ion Reaction Products (SHIP), is used to search for new elements. It has been operating since 1976 at the UNILAC (69, 74). A collaboration between II. Physikalisches Institut, Giessen, and GSI, Darmstadt—relying on common experiences gained during the construction and operation of the fission product separator LOHENGRIN, Grenoble (75)—completed the SHIP in time to receive the first beams available from the UNILAC (76). Figure 12 shows a sketch of the filter.

The combination of two electric and four magnetic dipole fields, together with two quadrupole triplets, accepts particles radially and axially to $3°$ and focusses all ionic charges within an ionic charge window of $20\%$ (69, 77). The velocity dispersion necessary to separate projectiles and EVRs is maximum in the medium plane of the system. Here the beams are separated. The velocity dispersion is compensated in the second half of the filter, which ion-optically is antisymmetric to the first half. Suppression of the primary beam depends on the velocity difference between projectiles and EVRs. Ar, Kr, and Xe beams producing $A = 180$ EVRs have been suppressed by factors of $10^{12}, 10^{10}$, and $10^8$, respectively. The number of background particles from the primary beam is small enough to use detector systems directly at the focus position without any further stages of beam purification. A time-of-flight (TOF) system behind SHIP allows a redundant velocity measurement. Implantation of the unslowed EVRs with their full energy (30–300 MeV) into surface barrier detectors allows an energy measurement. Together with the TOF measurement, a rough value of the mass can be obtained ($A/\Delta A \approx 10$). These detector systems act as additional stages of beam purification with possible suppression factors up to $10^6$. In subbarrier fusion studies, cross sections were followed over six orders of magnitude and fusion cross sections in the microbarn region could still be measured (30).

## Separator for Heavy Ion Reaction Products    SHIP

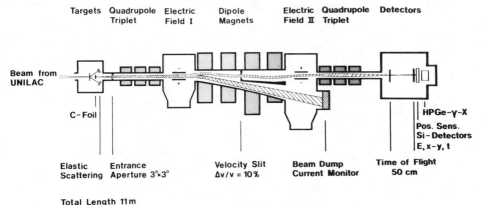

Total Length 11m

*Figure 12*   The velocity filter SHIP (69).

SHIP separates the reaction products spatially, and allows the detection of nuclei produced with sub-microbarn cross sections by exploiting their radioactive decay properties. Thus, it becomes not only a tool of reaction studies as magnetic spectrometers, but also a powerful on-line isotope separator (78). The subsequent $\alpha$ and spontaneous fission decays of the nuclei implanted are correlated among each other and to the signals obtained at the time of implantation into the surface barrier detector (79). Half-lives are determined from a few events by the maximum likelihood method. As an example, we give in Figure 13 the correlation time distribution of the $\alpha$ decay of $^{243}$Fm ($E_\alpha = 8.55$ MeV, $T_{1/2} = 0.18$ s) (80). As the recoil energy from $\alpha$ decay is small, an implanted nucleus does not

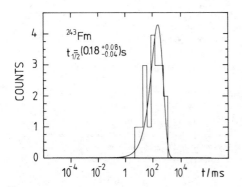

*Figure 13* The maximum likelihood method applied to the recoil $\alpha$ correlations observed in the reaction

$$^{206}Pb(^{40}Ar, 3n)^{243}Fm$$

in order to determine the half-life of $^{243}$Fm (79, 80).

change position. Position-sensitive detectors allow another large reduction of accidental correlations (81). The detection system used at SHIP is shown in Figure 14. An array of seven position-sensitive surface barrier detectors cover an area of 16 cm². The cooled detectors allow for an energy resolution of 25 keV FWHM and position resolution of 0.2 mm FWHM, thus dividing each detector into 100 detector cells.

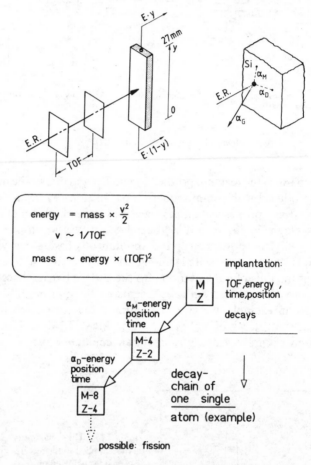

*Figure 14*   The detector system behind SHIP allowing TOF energy measurements of the recoils and the registration of position, time, and decay energy for correlated members of a decay chain (24, 81). In the upper left corner the TOF-energy measuring system to determine the mass of the implanted recoil is shown. The Si detector in the right corner shows an implanted EVR and its decay by three α particles in three subsequent generations. The lower part shows a decay chain with the quantities defining the event.

The detection of single decay chains was made possible by correlated event analysis. The error probability for a chain of correlated events is given as (82)

$$P_{err} = N_0 \prod_{i=1}^{k} \lambda_i \Delta t_{0i}, \qquad\qquad 8.$$

with $N_0$ the number of implanted leading event nuclei during the measuring time, $\lambda_i$ the counting rate in the energy window for the generation $i$ in the decay chain, and $\Delta t_{0i}$ the actually observed time difference between implantation and the $i$th decay. $P_{err}$ is the probability that all $k$ decay signals with their observed energy and time characteristics follow randomly and in any order the same implanted leading event nucleus. For chains with $k$ generations the number of position cells of the position-sensitive detector $F$ enters as $F^{-k}$ into $P_{err}$, e.g. for a three-member chain the SHIP detector reduces $P_{err}$ by a factor of $10^{-6}$.

3.2.5 GAS-FILLED SEPARATORS    Gas-filled separators have never been used for detection of elements beyond 102. But their high potentialities certainly make them a promising instrument in the future. They have been operated as fission product separators (83, 84), as well as fusion EVR separators (85). Facilities in operation now are the Juelich On-line Separator for Fission products, JOSEF (86), and the Small Angle Separating System, SASSY (70). SASSY is a prototype instrument for separation of EVRs and all numbers given in the following refer to this instrument. It has a low atomic number resolution, but a high transport efficiency.

The deflection in a magnetic field depends on the average charge of the ion during its passage through the magnetic field:

$$B\rho/Tm = 0.027(v/v_0)A/\bar{q}, \qquad\qquad 9.$$

with $v_0$ the Bohr velocity $\alpha c$. The ratio $v/\bar{q}$ only weakly depends on the velocity, and $B\rho/A$ becomes a function depending mainly on the atomic number $Z$. A large velocity window and the full ionic charge distribution can be accepted, which gives rise to the high efficiency of the method. The deflection in a gas-filled field, the gas providing the averaging over all possible ionic charges, depends on the mass and atomic number. The resolution is governed by two properties of the ion, the dispersion between neighboring atomic numbers and the fluctuations of the average ionic charges. A gas-filled separator has to be calibrated either with well-defined beams of heavy ions or with reaction products, which are known and can be detected by their radioactivity. For two velocity ranges, $v/v_0 = 2.2$ and $v/v_0 = 4$, the $B\rho/A$ values for different ions in helium have been determined (Figure 15). The dependence is described by a $Z^{-1/3}$ dependence, super-

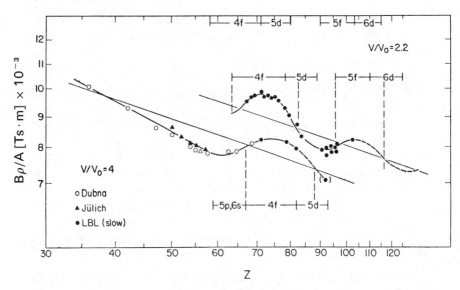

*Figure 15*    The specific magnetic stiffness $B\rho/A$ as a function of the atomic number for helium as a gas filling of the magnet system. Experimental values are presented for two velocities (unit $v_0 = \alpha c$). The straight lines represent a $Z^{-1/3}$ dependence. The lines through the points follow oscillations correlating with the electronic shell structure of the partly stripped recoils passing through the gas (70).

imposed on an oscillating function determined by the electronic shell structure of the ions. These oscillations are most pronounced at small velocity. They govern the resolving power. At atomic numbers for which $B\rho/A$ decreases with $Z$, $Z = 70$–$90$, the difference in $B\rho$ for neighboring elements is small and separation is difficult, whereas for increasing $B\rho/A$, $Z = 90$–$100$, the dispersion between elements is large and separation can be achieved.

Figure 16 shows the principal set-up of SASSY. The system is more compact (4 m) than SHIP and this allows still shorter separation times. The suppression of full energy projectiles in the $^{208}Pb(^{48}Ca, 2n)^{254}102$ reaction was found to be $10^{-15}$. The suppression of target recoils and target-like $\alpha$ emitters is $10^{-3}$. The transmission through the systems was measured to be about 40%. A position-sensitive detector of $5 \times 2\,cm^2$ has registered 15% of all $\alpha$-decaying $^{254}102$ nuclei produced in the target. Use of large area detectors can achieve total efficiencies of more than 50%. Once instrument designers overcome the barrier of mixing dirty atomic physics with clean ion optics, gas-filled separators may become serious competitors to velocity filters as cheap, fast, and efficient separating devices.

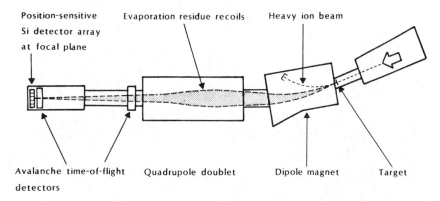

## Small Angle Separatory SYstem SASSY

*Figure 16*    The gas-filled magnetic separator system SASSY (70).

# 4.  PRODUCTION OF ELEMENTS 106–109 BY COLD FUSION REACTIONS

Only one experiment has successfully synthesized isotopes of elements beyond $Z = 105$ starting from actinide targets: the experiment that discovered $^{263}106$ (4). It was described in Sections 1 and 3.2.2. Until recently, all further attempts to synthesize other isotopes of $Z = 106$ or to reach higher elements failed. EVRs from fusion reactions, the cross sections of which are small compared to multinucleon transfer cross sections, could not be detected unambiguously in the large background of similar activities from transfer reactions. Moreover, all searches to synthesize elements around the predicted doubly magic nucleus $^{298}114$ failed (87). These search experiments are not the primary subject of this article, in spite of their interesting fall-out on different reaction aspects. The reactions leading beyond $Z = 106$ all make use of "cold fusion," a method introduced by Oganessian and co-workers (49): Targets around Pb are bombarded by beams of Cr, Mn, and Fe with energies near the Coulomb barrier and isotopes in the chains $N-Z = 46$–$49$ are produced by 1n and 2n channels.

In order to establish the genetics of the isotopes in question, the daughter isotopes of elements $Z < 106$ had to be known. The isotopes of elements $Z < 102$ are well known and their spectroscopic properties are compiled in tables (88). Most of what we know about the isotopes of elements 104 and 105 comes from cold fusion studies and is presented in Table 2. Isotopes

**Table 2** Compilation of data on isotopes of elements 104–109 obtained from cold fusion reactions

### Element 104

| Isotope | $T_{1/2}$ (s) | $E_\alpha$ (MeV) | $b_\alpha$ | Reaction | $E_c^*$ (MeV) | $E_1$ (MeV) | Dose ($10^{17}$) | $\sigma$ (nb) | Comment | Ref. |
|---|---|---|---|---|---|---|---|---|---|---|
| 258 | — | — | 0.13 | $^{208}$Pb($^{50}$Ti,$\gamma$) | 24.0 | 272 | 8.0 | 0.6 | detection of $^{246}$Cf | (51) |
|  |  | 9.01(0.18) |  |  |  |  |  |  |  |  |
|  |  | 8.98(0.29) |  |  |  |  |  |  |  |  |
| 257 | $4.3^{+1.3}_{-0.9}$ | 8.94(0.12) | 1.0 | $^{208}$Pb($^{50}$Ti,n) | 24.0 | 233.5 | — | $6.1\pm1.8$ | decay chain ($\alpha$) | (50, 91) |
|  |  | 8.90(0.06) |  |  |  |  |  |  |  |  |
|  |  | 8.78(0.06) |  |  |  |  |  |  |  |  |
|  |  | 8.71(018) |  |  |  |  |  |  |  |  |
|  |  | 8.60(0.12) |  |  |  |  |  |  |  |  |
|  |  | — |  | $^{208}$Pb($^{50}$Ti,n) | 24.0 | 272 | 8.0 | 5.0 | detection of $^{253}$Fm | (51) |
| 256 | 5 ms |  | — | $^{208}$Pb($^{50}$Ti,2n) | 24.0 | 260 | 0.1 | 6 | detection of of $^{256}$104 (sf) | (90) |
|  | $7.4^{+0.9}_{-0.7}$ ms | 8.81 | 0.02 | $^{207}$Pb($^{50}$Ti,n) | 24.1 | 239 | — | $0.6\pm0.2$ | decay chain (sf) | (50, 91) |
|  |  |  |  | $^{208}$Pb($^{50}$Ti,2n) | 24.0 | 238.5 | — | $5.2\pm0.2$ | decay chain (sf,$\alpha$) |  |
|  | $6.7\pm0.2$ ms | — | 0.01 | $^{208}$Pb($^{49}$Ti,n) | 28.0 | 271 | 8.0 | 0.2 | detection of $^{256}$104 (sf) | (51) |
|  |  |  |  | $^{208}$Pb($^{50}$Ti,2n) | 24.0 | 272 | 8.0 | 6.0 | detection of $^{256}$104 and $^{240}$Cm ($\alpha$) |  |
| 255 | 4 | — | — | $^{207}$Pb($^{50}$Ti,2n) | 24.1 | 260 | 0.2 | 3 | detection of $^{255}$104 (sf) | (90) |
|  |  | 8.77(0.18) |  |  |  |  |  |  |  |  |
|  | $1.4^{+0.3}_{-0.2}$ | 8.71(0.33) | 0.48 | $^{207}$Pb($^{50}$Ti,2n) | 24.1 | 239 | — | $4.8\pm0.4$ | decay chain (sf,$\alpha$) | (50, 91) |
|  |  | 8.63(09) |  |  |  |  |  |  |  |  |
|  | $1.2^{+1.0}_{-0.4}$ | — | — | $^{208}$Pb($^{50}$Ti,3n) | 24.0 | 251 | — | $0.4\pm0.3$ | decay chain (sf) | (50, 91) |
|  |  |  |  | $^{208}$Pb($^{48}$Ti,n) | 30.3 | 259 | 6.0 | 0.2 |  |  |
|  | $1.7\pm0.2$ | — | 0.5 | $^{208}$Pb($^{49}$Ti,2n) | 28.0 | 271 | 8.0 | 1.2 | detection of $^{255}$104 (sf) | (51) |
|  |  |  |  | $^{208}$Pb($^{50}$Ti,3n) | 24.0 | 272 | 8.0 | 0.6 |  |  |

**Element 105**

| Isotope | $T_{1/2}$ (s) | $E_\alpha$ (MeV) | $b_\alpha$ | Reaction | $E_c^*$ (MeV) | $E_1$ (MeV) | Dose ($10^{17}$) | $\sigma$ (nb) | Comment | Ref. |
|---|---|---|---|---|---|---|---|---|---|---|
| 258 |  | 9.30(0.08) 9.17(0.59) |  |  |  |  |  |  | decay chain ($\alpha$) |  |
|  | $4.4^{+0.9}_{-0.6}$ | 9.08(0.28) 9.01(0.05) | 0.67 | $^{209}$Bi($^{50}$Ti, n) | 24.1 | 234 | — | $2.9\pm0.3$ | EC-branch 33% to $^{258}$104 | (24, 50) |
|  | $4.2^{+1.6}_{-0.5}$ | — | 0.75 | $^{209}$Bi($^{50}$Ti, n) | 24.1 | 265 | 2.6 | 2.7 | Detection of $^{258}$104 and $^{246}$Cf | (51) |
| [257] | $5.0^{+1.7}_{-1.1}$ | — | 0.8 | $^{209}$Bi($^{50}$Ti, [2n]) | 24.1 | 270 | 0.95 | 0.83 | $\sigma$ refers to sf only, mass and element number revised in (51) | (92) |
|  |  |  |  | $^{208}$Pb($^{51}$V, [2n]) | 26.3 | 270 | 1.5 | 0.16 |  |  |
|  |  |  |  | $^{205}$Tl($^{54}$Cr, [2n]) | 25.0 | 280 | 1.1 | 0.11 |  |  |
| 257 | $1.4^{+0.6}_{-0.3}$ | 9.16(0.3) 9.07(0.3) 8.97(0.4) | >0.83 | $^{209}$Bi($^{50}$Ti, 2n) | 24.1 | 244 | — | $2.1\pm0.8$ | decay chain ($\alpha$, sf) | (24, 50) |
|  | $1.2^{+0.8}_{-0.5}$ | — | — | $^{207}$Pb($^{55}$Mn, n) | 24.0 | 310 | 23 | 0.03 | sf from $^{257}$105 after $\alpha$ decay of $^{261}$107 | (65) |
| [256] | $1.2^{+0.6}_{-0.3}$ | — | — | $^{207}$Pb($^{51}$V, [2n]) | 26.2 | 270 | 2.2 | 0.3 | $\sigma$ refers to sf only, mass and element number revised in (65) | (92) |
| 256 | $2.6^{+1.4}_{-0.8}$ | — | — | $^{206}$Pb($^{55}$Mn, n) | 24.1 | 310 | 55 | ~0.02 | sf from $^{256}$105 after $\alpha$ decay of $^{260}$107 | (65) |
| [255] | $2.6^{+5.0}_{-1.2}$ | — | — | $^{209}$Bi($^{49}$Ti, 2n) | 27.9 | 270 | 4 | — | EC-branch 10% to $^{256}$104 | (65) |
|  | 1 | — | — | $^{206}$Pb($^{51}$V, [2n]) | 26.3 | 270 | 1.6 | 0.5 | $\sigma$ refers to sf only, mass and element number revised in (65) | (92) |
| 255 | $1.6^{+0.6}_{-0.4}$ | — | >0.5 | $^{209}$Bi($^{48}$Ti, 2n) | 29.9 | 270 | 22 | — | detection of $^{255}$105 (sf) | (65) |

**Table 2** (*continued*)
Element 106

| Isotope | $T_{1/2}$ (s) | $E_\alpha$ (MeV) | $b_\alpha$ | Reaction | $E_c^*$ (MeV) | $E_1$ (MeV) | Dose $(10^{17})$ | $\sigma$ (nb) | Comment | Ref. |
|---|---|---|---|---|---|---|---|---|---|---|
| 263 | $0.9 \pm 0.2$ | 9.25 | 1.0 | $^{249}$Cf($^{18}$O,4n) | 40.2 | 95 | 13.4 | 0.3 | decay chain ($\alpha$) | (4) |
| 261 | $0.26^{+0.11}_{-0.06}$ | 9.56(0.60) 9.51(0.27) 9.47(0.13) | >0.9 | $^{208}$Pb($^{54}$Cr, n) | 22.8 | 264 | 1.6 | 0.5 | decay chain ($\alpha$) | (97) |
| | — | — | >0.5 | $^{208}$Pb($^{54}$Cr, n) | 22.8 | 300 | 5 | 0.3 | detection of $^{253}$Es ($\alpha$) | (98) |
| 260 | $2.5 \pm 1.5$ ms | — | >0.8 | $^{207}$Pb($^{54}$Cr, n) | 22.4 | 290 | — | 0.3 | detection of $^{256}$104 (sf) and built up | (94) |
| | | | | $^{208}$Pb($^{54}$Cr, 2n) | 22.8 | 290 | — | 0.4 | | |
| | $3.6^{+0.9}_{-0.6}$ ms | 9.76 | 0.5 | $^{208}$Pb($^{54}$Cr, 2n) | 22.8 | 268 | 2.5 | 0.3 | decay chain ($\alpha$, sf) | (95) |
| [259] | — | — | >0.2 | $^{208}$Pb($^{54}$Cr, 2n) | 22.8 | 300 | 16 | 0.4 | detection of $^{256}$104 (sf) | (98) |
| | $4$–$10$ ms | — | — | $^{207}$Pb($^{54}$Cr, [2n]) | 22.4 | 280 | 0.3 | 1.0 | mass and element | (93) |
| | | | | $^{208}$Pb($^{54}$Cr, [3n]) | 22.8 | 280 | 0.2 | 1.0 | number revised in (94) | |
| 259 | — | — | ~1.0 | $^{206}$Pb($^{54}$Cr, n) | 23.1 | 290 | — | 0.4 | detection of $^{255}$104 (sf) | (94) |
| | | | | $^{207}$Pb($^{54}$Cr, 2n) | 22.4 | 290 | — | 0.4 | | |
| | $0.48^{+0.28}_{-0.13}$ | 9.63 | >0.8 | $^{207}$Pb($^{54}$Cr, 2n) | 22.4 | 266 | 1.4 | 0.3 | decay chain ($\alpha$) | (95) |
| | — | — | >0.5 | $^{208}$Pb($^{54}$Cr, 3n) | 22.8 | 300 | 5 | 0.02 | detection of $^{255}$104 (sf) | (98) |

**Element 107**

| Isotope | $T_{1/2}$ (ms) | $E_\alpha$ (MeV) | $b_\alpha$ | Reaction | $E_c^*$ (MeV) | $E_1$ (MeV) | Dose ($10^{17}$) | $\sigma$ (nb) | Comment | Ref. |
|---|---|---|---|---|---|---|---|---|---|---|
| 262 | $4.7^{+2.3}_{-1.6}$ | 10.38 | 1.0 | $^{209}$Bi($^{54}$Cr, n) | 22.3 | 262 | 1.2 | 0.2 | decay chain ($\alpha$) | (96) |
|  | — | — | 1.0 | $^{209}$Bi($^{54}$Cr, n) | 22.3 | 290 | 5 | 0.2 | detection of $^{258}$104 (sf) | (51, 65) |
|  |  |  |  | $^{208}$Pb($^{55}$Mn, n) | 24.2 | 310 | 60 | 0.1 | and $^{246}$Cf ($\alpha$) |  |
| [261] | 1–2 | — | [0.8] | $^{209}$Bi($^{54}$Cr, [2n]) | 22.3 | 290 | 4.6 | 0.1 | analysis modified | (92) |
| 261 | — | — | 0.2 | $^{208}$Pb($^{55}$Mn, [2n]) | 24.2 | 290 | 2.2 | 0.05 | later (65) | (65) |
|  |  |  |  | $^{207}$Pb($^{55}$Mn, n) | 24.0 | 310 | 2.3 | 0.03 | detection of $^{257}$105 (sf) | (65) |
| 260 | — | — | 1.0 | $^{206}$Pb($^{55}$Mn, n) | 24.1 | 310 | 55 | ~0.02 | detection of $^{256}$104 (sf) | (65) |

**Element 108**

| Isotope | $T_{1/2}$ (ms) | $E_\alpha$ (MeV) | $b_\alpha$ | Reaction | $E_c^*$ (MeV) | $E_1$ (MeV) | Dose ($10^{17}$) | $\sigma$ (nb) | Comment | Ref. |
|---|---|---|---|---|---|---|---|---|---|---|
| 265 | $1.8^{+2.2}_{-0.7}$ | 10.36 | 1.0 | $^{208}$Pb($^{58}$Fe, n) | 20.4 | 292 | 6 | $0.02^{+0.02}_{-0.01}$ | decay chain ($\alpha$) | (97) |
|  | — | — | 1.0 | $^{208}$Pb($^{58}$Fe, n) | 20.4 | 320 | 30 | 0.004 | detection of $^{253}$Es ($\alpha$) | (98) |
| 264 | — | — | 1.0 { | $^{207}$Pb($^{58}$Fe, n) | 20.2 | 320 | 22 | 0.005 | detection of $^{256}$104 (sf) | (98) |
|  |  |  | { | $^{208}$Pb($^{58}$Fe, 2n) | 20.4 | 320 | 32 | 0.002 |  |  |
| 264 | — | — | 1.0 | $^{209}$Bi($^{55}$Mn, n) | 25.8 | 300 | 130 | 0.002 | detection of $^{255}$104 (sf) | (65, 98) |

**Element 109**

| Isotope | $T_{1/2}$ (ms) | $E_\alpha$ (MeV) | $b_\alpha$ | Reaction | $E_c^*$ (MeV) | $E_1$ (MeV) | Dose ($10^{17}$) | $\sigma$ (nb) | Comment | Ref. |
|---|---|---|---|---|---|---|---|---|---|---|
| 266 | $3.5^{+16.6}_{-1.6}$ | 11.10 | 1.0 | $^{209}$Bi($^{58}$Fe, n) | 19.8 | 299 | 2.8 | $0.015^{+0.035}_{-0.012}$ | decay chain ($\alpha$, sf) | (99, 100) |
|  | — | — | 1.0 | $^{209}$Bi($^{58}$Fe, n) | 19.8 | 320 | 36 | 0.003 | detection of $^{258}$104 (sf) | (51) |
|  |  |  |  |  |  |  |  |  | and $^{246}$Cf ($\alpha$) |  |

$^{258,257}104$ were known from earlier studies (89). The isotopes $^{254,253}103$ are α emitters and were found in the reaction $^{209}\text{Bi}(^{50}\text{Ti}, xn)^{259-xn}105$ as correlated daughter decays of the $Z = 105$ isotopes (50). The half-life of $^{254}103$ is $13^{+3}_{-2}$ s, it emits α particles of 8.46 MeV (64%) and 8.41 MeV (36%); $^{253}103$ has a half-life of $1.3^{+0.6}_{-0.3}$ s and emits α particles of 8.80 MeV (56%) and 8.72 MeV (44%). In Table 2 we give the main references on the isotopes of elements $Z = 104$ and $Z = 105$.

At Dubna the spontaneous fission activities of the isotopes $^{256-253}104$ were investigated (51, 90), whereas at Darmstadt α decay and spontaneous fission activities of $^{256,255}104$ were established (24, 50, 91). Spontaneous fission activities were assigned to $^{257,255}105$ at Dubna (65, 92), α decay of $^{258,257}105$ was found at Darmstadt (24, 50). The production of the isotopes $^{256,255}105$ was established from the occurrence of spontaneous fission (65). The isotopes $^{258,256}105$ are assumed to have an electron capture branch leading to $^{258,256}104$, which gives rise to fission activities with the half-lives of $^{258,256}105$ (50, 65). Delayed fission found after $K_\alpha$ x rays from the EC decay of $^{258}105$ corroborates this interpretation directly (24). As Table 2 shows, there are no conflicting results from the different experiments, neither for the spectroscopic nor for the reaction data for element $Z = 104$ and $Z = 105$ isotopes. The inclusion of 1n reaction channels first found in the $^{208}\text{Pb}(^{50}\text{Ti}, 1n)^{257}104$ reaction at Darmstadt (50) resolved all earlier discrepancies.

## 4.1  The Isotopes of Element 106

Three isotopes, $^{261-259}106$, were produced by $^{54}\text{Cr}$ bombardments of $^{208-206}\text{Pb}$ targets (Table 2). Experiments were performed in 1974 at Dubna (93). A 4–10-ms fission activity observed in $^{208,207}\text{Pb}$ bombardments was then assigned to $^{259}106$, an assignment that could not be confirmed later (94, 95). All irradiations performed since then are compiled in Table 2. Demin and co-workers (94), analyzing the spontaneous fission activities found in $^{54}\text{Cr}$ irradiations of $^{208-206}\text{Pb}$, concluded that the activities seen come from the isotopes $^{256,255}104$ and not from element 106, the isotopes of which are assumed to be α emitters. A value for the $^{260}106$ half-life was given for an activity build-up analysis, and lower limits for the spontaneous fission half-lives of $^{261,259}106$ were inferred. The results presented have been corroborated by experiments at Darmstadt (95, 97).

In an irradiation of $^{208}\text{Pb}$ with $^{54}\text{Cr}$ at 4.85 and 4.92 MeV/u, events correlated to the implanted nucleus with an α energy of 9.56 MeV were found. In the second generation of decays these events were correlated to the isotope $^{257}104$, known to have an α spectrum of several lines, the energies of which were found in the decay chains. It was established (97) that the isotope $^{261}106$ is an α emitter, $E_\alpha = 9.56 + 0.03$ MeV (60%), with a half-life of $0.26^{+0.11}_{-0.06}$ s, in agreement with the conclusions in (94). Its formation cross section in the reaction $^{208}Pb(^{54}Cr, n)^{261}106$ is about 0.5 nb.

The isotope $^{260}106$ has been produced at 4.92 MeV/u in the reaction $^{208}Pb(^{54}Cr, 2n)^{260}106$ (95). It decays by the emission of an $\alpha$ particle of 9.76 MeV into $^{256}104$, which then disintegrates by fission (see Figure 21, Section 5.1). Besides the $\alpha$ branch, a 50% spontaneous fission branch of $^{260}106$ has been detected. The half-life of $^{260}106$ is $3.6^{+0.9}_{-0.6}$ ms, giving a spontaneous fission half-life of about 7 ms. This value is about equal to the half-life of the daughter isotope reached by $\alpha$ decay. The spontaneous fission half-life going from $^{256}104$ to $^{260}106$ does not decrease, a finding again in agreement with the analysis of (94).

In the $^{207}Pb(^{54}Cr, 2n)^{259}106$ reaction, $^{259}106$ was found to have a half-life of $0.48^{+0.28}_{-0.13}$ s, and to decay by $\alpha$ emission with a main $\alpha$ energy of $9.63 + 0.03$ MeV. The spontaneous fission activity assigned to this isotope by the Dubna group in 1974 stems from $^{260}106$ and from $^{256}104$, but not from $^{259}106$. In the reactions used in 1974, isotopes of element 106 were produced, but the isotope upon which the claim of element discovery was built was clearly assigned incorrectly.

## 4.2    The Isotopes of Element 107

Initial experiments designed to synthesize element 107 via the reactions $^{209}Bi(^{54}Cr, xn)$ and $^{208}Pb(^{55}Mn, xn)$ have been reported by the Dubna group (92). Two spontaneous fission activities detected were assigned to $^{261}107$ ($b_\alpha \approx 0.8$) and its daughter $^{257}105$ ($b_\alpha \approx 0.8$). Correlated event chains found in the reaction $^{209}Bi(^{54}Cr, n)$ at Darmstadt (50, 96) showed that the spontaneous fission activity assigned in 1976 at Dubna to $^{257}105$ is mainly due to $^{258}104$, which fissions after EC decay of $^{258}105$ (33%). In the reaction the isotope $^{262}107$ unambiguously was established by seven correlated chains, one of which is given in Figure 17 (96). Both the $\alpha$ decay and the fission branch in the $^{262}107$ chain were confirmed by recent Dubna experiments (51, 65) showing the EC-delayed fission activity of $^{258}104$ as well as the $\alpha$ decay of $^{246}Cf$, a late member of the chain (Figure 18). $^{262}107$ is the one isotope of element 107 established in both laboratories independently and unambiguously.

The 1–2-ms activity assigned to $^{261}107$ (92) was not confirmed directly, but it was argued that the spontaneous fission yield detected in $^{207}Pb(^{55}Mn, xn)$ was compatible with an 80% fission branch of $^{261}107$ necessary to explain the fission yield found for $^{257}105$ (51). A millisecond activity for $^{261}107$ would point to a sharp disappearance of shell stabilization for this isotope or to a sudden loss of the additional stabilization of odd isotopes against spontaneous fission, which has been confirmed for all isotopes of elements $Z = 104$ to $Z = 106$ investigated.

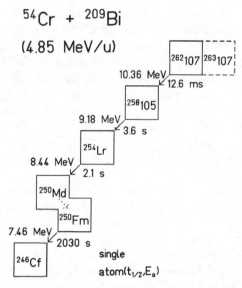

$$^{54}Cr + {}^{209}Bi$$

(4.85 MeV/u)

*Figure 17* Sequence of correlated α decays following the implantation of $^{262}$107 into a position-sensitive Si detector (96).

From the systematics a fission half-life of about 0.1 s should be expected and $^{261}$107 should be an α emitter in the millisecond range. A direct observation of $^{261}$107 by correlated event analysis should be feasible and is highly desirable, as the claim of discovery of element 107 by the Dubna group is built upon the existence of a strong millisecond-fission branch of $^{261}$107. This branch was not found; instead, α decay was observed in a recent experiment at Darmstadt (G. Münzenberg, private communication).

A third isotope, $^{260}$107, was detected at Dubna in the reaction $^{206}$Pb($^{55}$Mn, n) by the EC-delayed fission activity of $^{256}$104 (65). The isotope should be an α emitter as is $^{262}$107. The production cross section was found to be a factor of 10 smaller than for $^{262}$ 107. A confirmation by decay chain analysis is still lacking.

## 4.3  The Isotopes of Element 108

The synthesis of the even elements $Z = 106$ and $Z = 108$ was believed to be more difficult than the finding of the odd ones, $Z = 107$ and $Z = 109$, because their stability against spontaneous fission was expected to be reduced compared to the lifetimes of odd-element isotopes. Very small spontaneous fission half-lives were expected to range from $10^{-6}$ to $10^{-9}$ s. The finding of α decay for $^{266}$109 (99) and the analysis of spontaneous

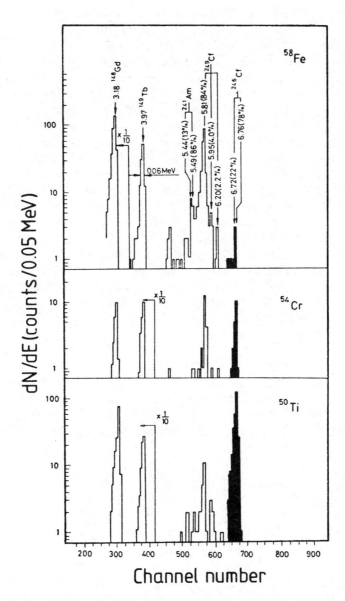

*Figure 18*    The α-particle spectrum for the Cf fractions obtained by bombarding a [209]Bi target with [58]Fe, [54]Cr, and [50]Ti ions (51). The arrows indicate energies and probabilities of the α transitions of nuclei present in the sample. The 3.18-MeV and 3.97-MeV peaks, as well as the complex spectrum having a maximum at 5.81 MeV, are from the tracer activities [148]Gd, [149]Tb, and [249]Cf. The 6.75-MeV α particles are from [246]Cf, a late member of the $N - Z = 48$ decay chain starting at [266]109, [262]107, and [258]105, respectively.

fission rates obtained in $^{54}$Cr bombardments of the isotopes $^{206-208}$Pb (94) led to the conclusion that the isotopes $^{259-261}$106 might be α emitters and encouraged a search for isotopes of elements 106 and 108.

In the reaction $^{208}$Pb($^{55}$Fe, n) at a bombarding energy of 5.02 MeV/u, three correlated events were found; these were assigned to $^{265}$108 (Figure 19) (97). The decay chains connect $^{265}$108 with known isotopes, among which are found $^{257}$104 (1.4 s) and $^{253}$Es (20 d), the latter being used by the Dubna group to prove formation of isotopes in the chain (98). The chains are correlated to $^{261}$106. The second-generation α energy seen in one event and the correlation times measured agree with the previous finding on this isotope. All chains are correlated in the third generation to the isotope $^{257}$104, which once more proves that the mother nucleus for these chains is $^{265}$108. $^{265}$108 decays by α emission, $E_\alpha = 10.36 + 0.03$ MeV, with a half-life of $1.8^{+2.2}_{-0.7}$ ms. In the three events no fission was observed in any generation. Up to element 108 in the sequence $N - Z = 49$ no fission branches have been found. The production cross section obtained is $19^{+18}_{-11}$ pb. The cross section reported by the Dubna group for the same reaction is considerably smaller, which points to either a favorable statistical fluctuation for the events at Darmstadt or to an overestimate of the detection probability in the Dubna measurements (98).

An 8-ms and a 6-ms fission activity were found in the reactions $^{208}$Pb($^{58}$Fe, xn) and $^{207}$Pb($^{58}$Fe, xn), respectively (98). This result was attributed to strong α branches of the isotopes $^{264}$108 and $^{260}$106, which finally lead to $^{256}$104, the fission of which was observed. The even-even isotope $^{264}$108 is concluded to be an α emitter with a negligible fission branch, because the cross section for $^{207}$Pb($^{58}$Fe, n) measured from the fission yield of $^{256}$104 was as large as the cross section for the corresponding

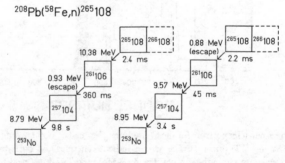

*Figure 19* Decay chains of $^{265}$108 found in the reaction $^{208}$Pb($^{58}$Fe, n)$^{265}$108 at 5.02 MeV/u (97).

reaction leading to $^{265}108$ using $^{208}$Pb targets. A direct observation of $^{264}108$ $\alpha$ decay should be of high priority in order to obtain a mass excess value for an element $Z = 108$ isotope.

Another isotope of element $Z = 108$ with mass number 263 is assumed to exist (65, 98). In the reaction $^{209}$Bi($^{55}$Mn, n), a 1.1-s spontaneous fission activity was seen. This activity was assigned to $^{255}104$, the granddaughter of the $\alpha$-emitting isotope $^{263}108$. Finding the same pattern of fission yields for the $^{58}$Fe irradiations of Pb targets as for $^{54}$Cr irradiations makes an interpretation as proposed (98) and confirmed for the $Z = 106$ isotopes (94) very plausible also for the $Z = 108$ isotopes. A final confirmation of the Dubna analysis by decay chain analysis, as was done for the $Z = 106$ isotopes, is necessary, but even now the existing evidence makes a highly increased stability against spontaneous fission very probable not only for $^{265}108$ but also for the isotopes $^{264,263}108$.

## 4.4    $^{266}109$—The Isotope with the Highest Mass and Atomic Number

In the synthesis of $^{262}107$, the experimental set-up used at Darmstadt was shown to be capable of detecting single event chains with a high degree of significance (96). This encouraged the attempt to synthesize element 109 using the fusion of $^{58}$Fe with $^{209}$Bi as a consequent next step. The isotopes reached with Bi targets and beams of $^{50}$Ti, $^{54}$Cr, and $^{58}$Fe via 1n channels are odd-odd isotopes. Their highly increased stability against spontaneous fission, about a factor $10^5$ in spontaneous fission half-life, makes them best candidates for $\alpha$-decaying isotopes. The $\alpha$ decays connect, via the $A = (4n + 2)$-decay chain, a sequence of odd-odd isotopes. As in previous studies the new isotopes $^{254}$Lr, $^{258}105$, and $^{262}107$ had been found and their $\alpha$ energies measured; it was possible to identify a new member in the chain from one decay chain observed. From a simple logarithmic extrapolation of the production cross sections of the elements 104, 105, and 107, the cross section for element 109 was expected to be about one order of magnitude lower than that for element 107 (Figure 26, top).

In the course of 250 hours of irradiating $^{209}$Bi with $^{58}$Fe beams with specific energies of 4.95, 5.05, and 5.15 MeV/u, totalling a dose of $7 \times 10^{17}$ particles, one correlated decay chain (Figure 20) produced at the highest projectile energy could be observed (99, 100). The particles selected are all those leaving SHIP that have been mass-identified by TOF and energy measurements and that are followed at the same detector position ($\pm 0.4$ mm) by an $\alpha$ decay within 40 ms in the beam pulse, or within 14 s between beam pulses. Of the 529 events registered, there is one event with a mass value $264 \pm 13$ and followed by an $11.10 \pm 0.04$-MeV $\alpha$ particle that is certainly an EVR. All other events can be attributed to α decays from transfer products in the mass range $A = 210–216$.

The event identified as EVR is followed by a further α decay within 22.3 ms. This α particle escaped from the detector; a rest energy of only 1.14 MeV was deposited in the detector. A γ ray of 382.9 keV was found in coincidence. The correlation time found is compatible with the half-life determined for $^{262}107$. The next step observed in the decay chain is a spontaneous fission decay. An energy of $232 \pm 10$ MeV was registered with a time delay of 12.9 s. This is the only fission event found throughout the whole 250-hour irradiation. Via the 30% EC branch of $^{258}105$, the even-even isotope $^{258}104$ is reached and it undergoes fission within ms. The two combined decays explain the virtually long correlation time of the fission event. Using Equation 8 and analyzing the background in the different decay steps, we find a probability of the event being random of $2 \times 10^{-18}$.

All pieces observed in the decay chain agree with an assignment of $^{266}109$ as the mother isotope. All other decay channels were discussed thoroughly (100), and it was concluded from all possible deexcitation channels of the compound nucleus $^{267}109$ that the most probable assignment of the observed decays was the isotope with mass 266 of the new element 109. The isotope $^{266}109$ has a half-life between 2 and 20 ms; it decays by α emission with an energy of $11.10 \pm 0.04$ MeV. The α energy is within the predictions of different mass formulae (101, 102). The energy given might not correspond to the ground-state $Q$ value, as found for many other odd-odd isotopes. The actual $Q$ value may be higher by up to 0.5 MeV. The half-life is too long for the energy given, which suggests a hindrance that again is within the values known for other odd-odd isotopes. The isotope is produced via the 1n channel with an excitation energy in the range 20–26 MeV. The production cross section amounts to 3–50 pb.

The analysis of the one-event assignment used by Münzenberg and co-workers (100) to prove the existence of element 109 is strongly supported by

*Figure 20*  Decay chain of $^{266}109$ found in the reaction $^{209}Bi(^{58}Fe, n)^{266}109$ at 5.15 MeV/u (99, 100).

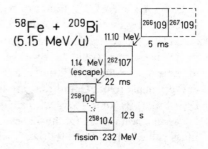

a Dubna experiment using the same reaction (51). At a dose 14 times larger than that at Darmstadt, wheel techniques combined with chemical separation (see Section 3.2.3) gave one fission event and seven $\alpha$ decays of $^{246}$Cf (Figure 18). The production cross section obtained in the experiment is five times smaller than at Darmstadt, a discrepancy observed for the synthesis of $^{265}108$ as well. An increase of the Darmstadt detection probability by a factor of five to a value of 100% is impossible. A corresponding decrease of the detection probability for the Dubna experiments to a value of about 12% would be at variance with the value of 60% given in (65). The production of the isotopes $^{266}109$ and $^{265}108$ by two independent groups and methods makes the existence of elements 108 and 109 as assured as the existence of all the previous man-made elements.

## 4.5   The Naming of Elements

Not hiding a certain self-interest, we want to help settle existing controversies on the naming of elements (103, 104). As a rule we propose the following: Element synthesis becomes production of a given isotope, and a name should be accepted only if the experiment claiming the discovery is reproducible. An isotope is defined by its mass and atomic number, its fingerprints are its decay modes and its half-life. Decay modes of the heavy isotopes in question are electron capture and spontaneous fission, which sometimes are difficult to assign to a specific isotope, and $\alpha$ decay, which (with its decay energies measured with an accuracy of some parts in a thousand) gives a very reliable mode of assignment. The time correlation of subsequent decays is a further method of definite isotope identification. The proposed rule should be applied retrospectively for all elements discovered by isotope identification, that is elements 102–109. A joint commission of physicists and nuclear chemists urgently is needed to solve the long-standing controversies.

## 5.   GENERAL TRENDS IN HEAVIEST ISOTOPES SYSTEMATICS

The measured values of $\alpha$ energies, $\alpha$-decay half-lives, and fission half-lives can be used to establish trends of these quantities up to element 109 and to extrapolate the trends to higher proton numbers. Comparison with theoretical predictions of mass excesses, shell corrections, and fission barriers becomes possible. Besides ground-state properties, production cross sections may be analyzed in order to understand the limitations of EVR formation.

## 5.1 Mass Excesses, Shell Correction Energies, and Fission Barriers

For the two even elements, $Z = 104$ and $Z = 106$, α energies of the even-even isotopes with $N - Z = 48$ have been determined. Figure 21 displays decay chains showing the α decay of $^{256}104$ (24, 91) and $^{260}106$ (95, 106). Only one chain was found for the $^{256}104$ α decay with $E_\alpha = 8.81 + 0.02$ MeV giving an α branch of smaller than 3%; the existence of this α branch was detected via α decay of $^{240}$Cm (27 d) at Dubna as well (98). The α energy of 9.76 MeV for $^{260}106$ is based on the detection of 11 events. The two α energies assuming ground-state to ground-state decay allow us to determine from the experimentally known mass excess of $^{252}$No (105) the mass excesses of $^{256}104$ and $^{260}106$. The mass excesses of $^{256}104$ and $^{260}106$ are $94.2 \pm 0.1$ and $106.6 \pm 0.1$ MeV, respectively (95, 106). These values are compared in Table 3 to different mass excess predictions (102, 107–110). Best agreement is obtained with the prediction of Liran & Zeldes (110); all other approaches fail to reproduce the experimental values by about 1 MeV, the nuclei being more bound by this amount than predicted.

Except for $^{264}108$, all isotopes of the $N - Z = 48$ sequence are known up to $^{266}109$. Taking for the odd elements the highest α energies observed, mass excesses were determined as for the even elements. For odd elements in case of transitions to excited states, the mass excesses obtained would be too large. Table 4 gives the new α energies, mass excesses, and shell correction energies, $\Delta E^{shell} = M_{exp} - M_{macro}$, for the $N - Z = 48$ isotopes beyond $Z = 102$. The macroscopic mass excesses are taken from the tables

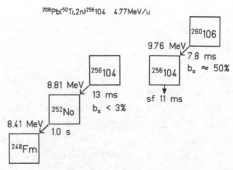

*Figure 21*   Decay chains of the even-even nuclei $^{256}104$ and $^{260}106$ showing α decay (24, 91, 106). From the α energies, mass excesses of the isotopes are determined.

**Table 3**    Mass excess values for $^{256}104$ and $^{260}106$

| Nucleus | Ref. (107) | (108) | (109) | (110) | (102) | (106) |
|---------|-----------|-------|-------|-------|-------|-------|
| $^{260}106$ | 108.27 | 107.29 | 107.7 | 106.94 | 108.13 | $106.62 \pm 0.06$ |
| $^{256}104$ | 95.90 | 94.84 | 95.6 | 94.37 | 95.77 | $94.23 \pm 0.05$ |

published by Møller & Nix (102). Figure 22 presents the shell correction energies for all $N - Z = 48$ isotopes between $Z = 91$ and 109 as given in Table 4 or taken from the latest Wapstra tables (105). There is a small systematic odd-even difference of about 0.2 MeV between odd and even elements all over the range of elements concerned. The new masses do not deviate more than the previously compiled ones. The shell correction increases up to $^{250}105$ and then stays about constant up to $^{266}109$ at a level of $-5.2$ MeV. The calculated shell corrections (102, 111) shown in Figure 22 (top) are systematically larger than the experimental ones, all nuclei are more stabilized than predicted. For the experimental shell correction of $^{276}114$, following the trend of the calculated shell corrections, a value of about $-5.5$ MeV is extrapolated. The shell correction going from $Z = 109$ to $Z = 114$ along $N - Z = 48$ stays constant or even becomes slightly more stabilizing. Figure 23 gives all known experimental shell corrections for

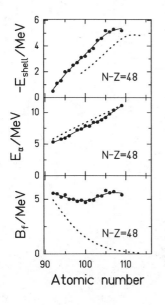

*Figure 22* For isotopes of the sequence $N - Z = 48$ from $Z = 92-109$ shell correction energies (*top*), $\alpha$ energies (*middle*), and fission barriers (*bottom*) are given. Dotted lines represent the calculated shell correction energies (102, 111), and the macroscopic $\alpha$ energies (102) and fission barriers (112) are given.

**Table 4** The $\alpha$ energies, mass excesses, shell corrections, and fission barriers for $N - Z = 48$ isotopes between $Z = 102$ and $Z = 109$

| Isotope | $^{252}102$ | $^{254}103$ | $^{256}104$ | $^{258}105$ | $^{260}106$ | $^{262}107$ | $^{264}108$ | $^{266}109$ | Ref. |
|---|---|---|---|---|---|---|---|---|---|
| $E_\alpha^{max}$ (MeV) | 8.42 | 8.46 | 8.81 | 9.30 | 9.76 | 10.38 | — | 11.10 | (95, 105) |
| $M^{exp}$ (MeV) | 82.87 | 89.66 | 94.23 | 101.52 | 106.62 | 114.48 | — | 128.17 | |
| $M^{macro}$ (MeV) | 86.67 | 94.11 | 98.84 | 106.58 | 111.63 | 119.65 | 125.04 | 133.45 | (102) |
| $\Delta E_{shell}^{exp}$ (MeV) | −3.8 | −4.4 | −4.6 | −5.1 | −5.0 | −5.2 | — | −5.3 | (95) |
| $\Delta E_{shell}^{cal}$ (MeV) | −2.4 | −3.0 | −3.1 | −3.5 | −3.5 | −4.1 | −4.1 | −4.8 | (102, 111) |
| $B_f^{macro}$ (MeV) | 1.1 | 0.9 | 0.7 | 0.6 | 0.5 | 0.4 | — | 0.0 | (112) |
| $B_f^{exp}$ (MeV) | 4.9 | 5.3 | 5.3 | 5.7 | 5.5 | 5.6 | — | 5.3 | (95) |
| $B_f^{cal}$ (MeV) | 5.46 | 5.88 | 5.42 | 5.80 | 5.29 | 5.61 | 5.04 | 5.25 | (111) |

isotopes $109 > Z > 50$ and $157 > N > 82$ (106). For the heaviest isotopes, all available new $\alpha$ energies were used as described for the $N - Z = 48$ isotopes to obtain shell correction energies (95). In addition, calculated shell corrections for the heavier nuclei from Møller and co-workers are shown (102, 111).

The fission barrier $B_f$ is the sum of the macroscopic liquid drop surface and the shell corrections of the ground-state and saddle point masses. The latter is difficult to measure and calculate, but it is known to be a small contribution and is neglected. The macroscopic fission barrier could be taken from different calculations (102, 107) or from an analysis of experimental fission barriers (112). The experimental macroscopic barriers from (112) are used in the following and are given for the $N - Z$ $= 48$ isotopes in Table 4. Figure 22 (bottom) shows the experimental fission barriers, $B_f^{exp} = B_f^{macro} - \Delta E^{shell}$, and the experimental macroscopic barriers for the $N - Z = 48$ isotopes of element $Z = 92$ to $Z = 109$. Surprisingly the barriers stay almost constant, in spite of the strongly decreasing macroscopic barrier, e.g. for $^{260}106$ only 0.5 MeV stem from the macroscopic contribution, whereas 5.1 MeV are due to the shell correction energy of the ground-state mass. For $^{232}U$ with a similar fission barrier the contributions are inverse, almost no shell corrections but a macroscopic barrier giving the main contribution.

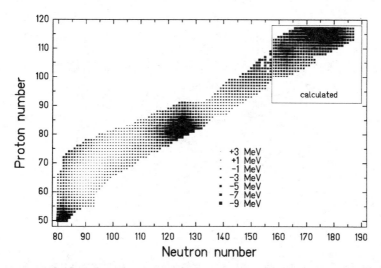

*Figure 23*  Experimental shell correction energies for nuclei with $Z > 50$, $N > 80$ (105, 106) together with the prediction of (102, 111) for nuclei $N > 157$.

The fission barrier protecting the isotopes of the new elements is high because of the strong ground-state stabilization of these nuclides. The stabilization shows an increasing trend in the entire range of the heaviest elements, as is demonstrated in Figure 23. Figure 22 (middle) compares the measured α energies with the α energies expected for nuclei having no shell corrections. For the calculation, the macroscopic masses of (102, 111) are used. The difference in α energies is always smaller than 0.4 MeV, which demonstrates again that the surface as a whole is shell stabilized independently of the individual proton and neutron numbers. The shell stabilization drastically changes the fission barriers, but only negligibly changes the α energies. The α energies reflect small local deviations of the mass surface, but they do not reveal the slowly changing trends toward higher shell correction energies. Only mass excess measurements disclose the increasing shell stabilization of the entire range of nuclides.

## 5.2   Half-lives for α Decay and Spontaneous Fission

The measured partial α lifetimes of the even-even $N - Z = 48$ isotopes were compared to lifetimes corresponding to the macroscopic α energies, Figure 24 (106). The α half-lives were calculated according to a prescription given in (113). All reduced α widths were set equal to one. Although a variation of 1 MeV in the $E_\alpha$ energies changes the lifetimes by three orders of magnitude, only a slight deviation (less than an order of magnitude from the macroscopic expectation) is observed experimentally. Once more this fact demonstrates that the shell corrections vary smoothly and the α lifetimes are affected only little by nuclear structure effects in this mass region. For a given element, α lifetimes are getting shorter and shorter with decreasing neutron number, a trend well described by macroscopic models and observed for many elements beyond $Z = 52$. As isotopes approach the

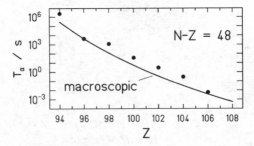

*Figure 24*   The measured α half-lives for even-even $N - Z = 48$ isotopes between $Z = 94$ and 106. The curve is the expectation obtained from macroscopic α energies (102) and the energy half-life conversion formula of (113).

proton drip line, their half-lives drop to the millisecond range. The short $\alpha$ half-lives observed are governed by this macroscopic trend, and nuclear structure is of minor importance only.

The contrary holds for the spontaneous fission lifetimes. To separate nuclear structure effects from macroscopic fission properties, we follow a slightly modified procedure (24), as originally introduced by Swiatecki (114). The fission half-life is given by

$$T_{sf} \text{ (in seconds)} = 3 \times 10^{-21}/P, \qquad\qquad 10.$$

with $P$ a barrier transmission factor, and the numerical factor the barrier knock-on time. Assuming a Hill-Wheeler-type transmission through the fission barrier with curvature $\hbar\omega_f$, we obtain

$$P = [1 + \exp(2\pi B_f/\hbar\omega_f)]^{-1} \approx \exp(-2\pi B_f/\hbar\omega_f). \qquad\qquad 11.$$

With $B_f = B_f^{\text{macro}} - \Delta E^{\text{shell}}$ and the shell corrections at the saddle point neglected, we obtain from Equations 10 and 11 an expression separating the nuclear structure effects from the macroscopic contribution:

$$\log(T_{sf/s}) = 2.73 B_f^{\text{macro}}/\hbar\omega_f - 2.73\Delta E^{\text{shell}}/\hbar\omega_f - 20.5. \qquad\qquad 12.$$

The experimental shell correction energies are obtained from mass excess values as described in Section 5.1, whereas the macroscopic fission barriers are calculated according to the semiempirical description given in (112). For isotopes with small shell correction energies and large values of $B_f^{\text{macro}}$, the barrier curvature parameter $\hbar\omega_f$ may be fitted to the spontaneous fission half-lives. With this value of $\hbar\omega_f$ kept constant the macroscopic expectation may be calculated as a function of the fissility parameter $x$ (Equation 4). This macroscopic spontaneous fission half-life is presented, together with the experimental half-lives, in Figure 25 (top). For $^{260}106$ the macroscopic fission half-life is about $10^{-17}$ s, which compares with 7 ms found experimentally. A stabilization by nuclear structure effects of 15 orders of magnitude is observed. Taking Equation 12 as a presentation with one adjustable parameter, values for $\hbar\omega_f$ as a function of $x$ may be obtained. Figure 25 (bottom) shows $\hbar\omega_f$ for all even-even spontaneous fission emitters. There is a smooth trend of $\hbar\omega_f$ for all isotopes between U and Cf. For Fm and $Z = 102$ some isotopes still follow the trend, whereas the isotopes of elements 104 and 106 show values of $\hbar\omega_f$ that are twice as large as the values for uranium.

Assuming no drastic changes in the inertia parameters, the increased values of $\hbar\omega_f$ point to a narrowing of the fission barrier by a factor of two. In Figure 25 (bottom) those isotopes for which a disappearance of the second fission barrier is predicted are indicated by points (115). Their positions support the hypothesis (48, 116, 117) that the change of half-life systematics observed for $Z = 104$ is caused by such a disappearance. Moreover, these isotopes are shell stabilized. Their macroscopic barriers are smaller than 1 MeV. The values of $\hbar\omega_f$ were fitted to a linear dependence on the fissility parameter, giving for the $x$ dependence of the spontaneous fission half-life

$$\log (T_{sf/s}) = (-3.55x + 6.62)B_f - 20.5. \qquad\qquad 13.$$

The smaller width of the barrier leads to a weaker dependence of the spontaneous fission half-life on the barrier height than that observed for lighter elements. A proportionality factor of 3.55 replaces the factor 6.88 found in (118). The fission through the narrow single-humped barrier most

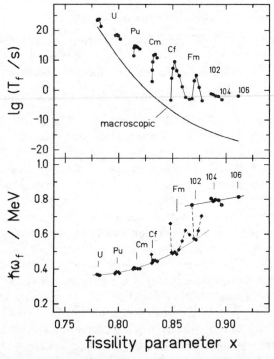

*Figure 25* The upper part shows the fission half-lives of all known even-even isotopes together with the half-lives expected for nuclei with macroscopic fission barriers (112). The barrier curvature parameter is fitted to $^{232}$U and kept constant for all isotopes. The lower part of the figure (24) shows the curvature parameter $\hbar\omega_f$, obtained according to Equation 12 from fission half-lives, experimental shell correction energies (105, 106), and the macroscopic barriers of (112). Diamonds belong to nuclei with a double-humped barrier and points are nuclei with a predicted single-humped barrier (115). The straight line is a fit to the $\hbar\omega_f$ values of the latter.

probably gives a symmetric mass distribution of fission fragments. Until now only the mass distributions of two of these nuclei ($^{260}$104 and $^{258}$102) have been measured and found to be symmetric (119). The total kinetic energy released in the fission of $^{260,258,256}$104 has now been measured (24, 119) and found to follow the systematics established for lighter elements (120).

The odd isotopes $^{255}$104 and $^{257}$105 have partial fission half-lives of 2.7 and 8.2 s, respectively. Comparison to $^{256}$104 allows us to determine the hindrance factors of spontaneous fission for odd protons and neutrons. Hindrance factors of 480 and 610 are obtained (24). Odd-odd isotopes may be hindered by the product of the individual hindrance factors, that is by a factor of $2 \times 10^5$. The $\alpha$ half-lives of the $Z = 107$ and $Z = 109$ isotopes and of the odd isotopes of $Z = 106$ are all shorter than the fission half-lives if we assume these hindrance factors and the fission half-lives of 7.2 ms ($^{260}$106) (95) and $>5$ ms ($^{264}$108) (98) for the neighboring even-even isotopes as a reference. The unconfirmed spontaneous fission half-life of 1–2 ms obtained for $^{261}$107 by the Dubna group (92) was until recently the only exception.

Neither especially large hindrance factors nor nuclear structure effects in $\alpha$ half-lives are responsible for the absence of fission. The occurrence of $\alpha$ decay for the isotopes of elements 106 to 109 is mainly a consequence of a strong ground-state shell stabilization giving rise to increased fission half-lives.

## 5.3  Recent Theoretical Predictions of Shell Corrections and Half-lives

An island of macroscopically unstable, but shell-stabilized, spherical nuclei around $^{298}$114$_{184}$ was predicted as early as 1966 by Myers & Swiatecki (121). Many experiments sought to find these superheavy nuclei in nature or to produce them by nuclear reactions but failed (87). Predictions of shell corrections for nuclei between the heaviest isotopes known and the "superheavy" island are rare. Two calculations using the best macroscopic-microscopic models were published recently (111, 122). The shell corrections of the calculation of (111) are shown as an insert in Figure 23. Besides the strong shell effects at $^{298}$114$_{184}$ ($-7.08$ MeV) already known, another island at $^{272}$109$_{163}$ ($-6.93$ MeV) and $^{270}$108$_{162}$ ($-7.97$ MeV) has been predicted by (111) and (122), respectively. Nuclei not with $N = 184$ but with $N = 178$, 177 are found to have the largest shell corrections, e.g. $-8.97$ MeV for $^{291}$114 and $-8.38$ MeV for $^{294}$116. Following the successful path along $N - Z = 48$ leading to $^{266}$109 and further up to $^{276}$114, nearly constant shell corrections between $-4.6$ and $-5.0$ MeV are predicted. A continuous increase of the shell correction by another $-4$ MeV is expected going into the center of the island. For the isotopes of $Z = 114$ at $N = 163$, a change from deformed nuclei to spherical nuclei is predicted (111). The $N - Z = 48$, 49 paths lead to $Z = 114$ via a chain of deformed nuclei, whereas for $N - Z > 50$ spherical nuclei would be reached.

The new island around $^{272}109$ consists of nuclei that are deformed ($\varepsilon_2 = 0.21$, $\varepsilon_4 = 0.09$) (111). The most interesting feature of these nuclei are the large $\varepsilon_4$ deformations (111, 122). Positive values of $\varepsilon_4$ correspond to a sausage-like shape of the nucleus. These sausage-like nuclei were seen before around $^{188}$Ta. But a microscopic explanation of why positive $\varepsilon_4$ deformation helps to stabilize nuclei was not given until recently. Whether the sausage shape at $N = 164 = 2 \times 82$ has to do with an underlying molecular cluster configuration of two $N = 82$ clusters is open to speculation.

The $N - Z = 48$, 49 isotopes of elements 112–114 are predicted to have $\alpha$ half-lives of about 0.2 $\mu$s and fission barriers of about 5 MeV (111). Experimentally for the elements $Z = 100$–106 the $\alpha$ half-lives are found to be larger than postulated; thus $\alpha$ half-lives may be around 1 $\mu$s for elements 112–114. Experimental shell corrections are stronger by about 1 MeV than calculated; thus fission barriers may be at least 5.5 MeV for the isotopes in question. Fission half-lives of about 20 ms can be extrapolated from Equation 13, and for all isotopes negligible fission branches can be expected. The $\alpha$ chains starting at elements 112–114 will end by fission in $^{257}105$ or $^{256,258}104$, or will continue to still lighter elements. Isotopes $^{277}114$ and $^{273}112$ decay by a chain that never shows fission and finally ends in $^{209}$Bi. Even for the light isotopes of elements 112–114 near $N < 163$, the ground-state stability may still be sufficient to allow detection with present techniques. The stability proceeding to heavier isotopes of element 114 will increase and the expectation of a maximum stability near $N = 184$ remains essentially unchanged. The new aspect that superheavy elements may not decay by fission, but could be detected by $\alpha$ chains and thereby directly give their atomic and mass number, is surprising indeed.

## 5.4   Production Cross Sections

Isotopes of elements $Z = 100$–109 were produced by cold fusion reactions between beams of $Z = 18$–26 and targets of $Z = 81$–83. Trends in production cross sections are discussed in the following. The first experiments producing heavy elements by cold fusion were performed at Dubna in 1974 (49). $^{40}$Ar projectiles bombarded Pb targets and spontaneous fission events were observed, which could be assigned to Fm isotopes. First cross sections and excitation functions were determined (49, 67). The experiments were later repeated at Berkeley (68) and Darmstadt (80). The Darmstadt experiment corroborated the earlier Dubna finding of 2n-reaction channels and ended a controversy between Dubna and Berkeley on the existence of this cold fusion channel (123).

A compilation of the cross sections found in Ar bombardments of Pb targets for all the different reaction channels was published recently (34). The 2n and 3n channels have about equal cross sections, 4n channels are seen with a much smaller cross section. The existing data are reproduced for all reaction channels by modern evaporation codes (23) and do not indicate any dynamical hindrance in the entrance channel. However, subbarrier fusion has to be considered by a barrier fluctuation of 3% in order to reproduce the 2n cross sections, and the damping of shell effects at high temperature has to be taken into account by a damping energy $E_D = 18$ MeV in agreement with microscopic predictions (24, 34).

There are no cross sections known for the production of Md and element 103 by cold fusion. The production of isotopes of element 102 by fusion of Ca and Pb isotopes was studied extensively (68, 124). The high (several $\mu$b) production cross section for $^{254}102$ in the reaction $^{48}$Ca($^{208}$Pb, 2n) found in experiments at Dubna (124) and Berkeley (68) could not be reproduced by the recoil spectrometer SHIP (24). The discrepancy at SHIP of nearly a factor of 10 in the cross section may be due to a reduced efficiency of the recoil spectrometer; indeed, in the case of a 100-ns isomer in $^{254}102$, the efficiency might be considerably decreased by a conversion-induced Auger cascade (24). For element 102 isotopes, the 2n-reaction channel in the $^{48}$Ca reactions has become the dominant channel and all other channels are weaker by at least an order of magnitude. Again an evaporation calculation reproduces the data without assuming any hindrance in the entrance channel (24). More data on the weak channels would definitely help to determine whether or not a small hindrance might already be present. To describe the subbarrier fusion part of the excitation function, a small barrier fluctuation of 1.3% as expected for stiff nuclei is needed. The doubly closed-shell nuclei $^{48}$Ca and $^{208}$Pb fuse without any hindrance at a well-defined barrier, as already found for the symmetric fusion of the two closed-shell $^{90}$Zr nuclei (Figure 5; Section 2.1).

For elements 104 and 105 some excitation functions were measured using reactions with $^{50}$Ti beams (50, 91). The 1n- and 2n-reaction channels have about equal yields. The small cross sections (only a few nb) for the 2n channel and the observation of 1n channels are reproduced by the evaporation calculations only, if a dynamical hindrance at the barrier of about a factor of 25–30 is assumed. Such a hindrance is equivalent to a barrier shift of $21 \pm 3$ MeV and an increase of the barrier fluctuation by a factor of two to about 6%. This extra-push value is in agreement with a $x_{mean}^{UCD}$ scaling (Figure 5) and the large fluctuation was observed for equivalent symmetric systems as well (22).

It was shown that the same parameters in the evaporation calculation reproduce well the excitation function in the $^{249}$Cf($^{12}$C, 4n)$^{257}$104 reaction without assuming any dynamical hindrance, which again corroborates the extra-push scaling that predicts for this reaction no entrance channel limitation. The sudden onset of a dynamical hindrance for the $^{50}$Ti reaction and its absence in $^{48}$Ca reactions points to a nuclear structure effect besides the macroscopic dissipation mechanism underlying the extra-push concept. The addition of two protons to $^{48}$Ca or of two neutrons to $^{90}$Zr in both cases leads to a considerable shift of the fission barrier and an increase in the barrier fluctuation parameter. The $\Gamma_n/\Gamma_f$ values at the low excitation energies present in cold fusion are of the order of a few percent in agreement with the analysis presented in Figure 8, and the high fission barriers as deduced from the analysis of ground-state shell corrections (Section 5.1). A damping of shell effects with temperature as expected from microscopic theory ($E_D = 18$ MeV) is compatible with the data.

Figure 26 (top) gives the cross sections as measured for reactions leading

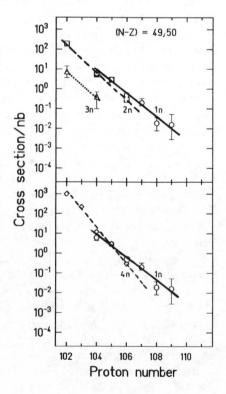

*Figure 26 Upper part:* Experimental cross sections for different xn channels observed in cold fusion reactions leading to compound systems with $N - Z = 49$ and 50 (Table 2). *Lower part:* The 1n cross sections for cold fusion reactions compared to 4n cross sections for $^{249}$Cf- and $^{249}$Bk-based reactions (2, 4). The lines are fits to the data.

to isotopes of elements 102 to 109 via compound systems with $N - Z = 50$ and 49 for even and odd elements, respectively. All values are found in Table 2, except $\sigma_{2n} = 190$ nb and $\sigma_{3n} = 7.3$ nb obtained from the reaction $^{206}$Pb($^{48}$Ca, $xn$) (24). The slope of the cross sections for 3n and 2n reactions is larger than for 1n reactions. On the average the 1n cross section decreases by a factor of 3.5 for each element. For elements 104 and 105, the cross sections for 1n reactions become comparable to 2n cross sections and finally prevail for elements 107 to 109. Comparison of cross sections for pairs of nuclei like $^{207}$Pb/$^{208}$Pb or $^{49}$Ti/$^{50}$Ti shows that more neutrons may increase the yields, e.g. the 2n production cross sections of $^{256,255}$104 in reactions of $^{50,49}$Ti with $^{208}$Pb differ by a factor of four, whereas no difference is found for reactions of the pair $^{208}$Pb/$^{207}$Pb with $^{50}$Ti leading to the same isotopes (Table 2). The difference of one neutron in the lead isotopes for pairs of 2n and 1n reactions leading to a given isotope changes the cross section in favor of $^{208}$Pb by factors of 9, 1.3, and 2.5 for $^{256}$104, $^{260}$106, and $^{264}$108, respectively. $Q$-value arguments favor neutron-rich collision partners. On the average, the available scarce data show an additional neutron to increase the cross section by a factor of two.

The excitation energy for all systems showing 1n cross sections between elements 104 and 109 decreases from 24 to 20 MeV, the fission barriers stay nearly constant (Figure 22, bottom) and still the cross section decreases by nearly three orders of magnitude. If we assume for $B_f = $ constant approximately constant $\Gamma_n/\Gamma_f$ values for all reactions, this decrease can be attributed to an increase of the entrance channel limitation. Figure 27 shows the 1n cross sections as a function of the scaling parameter $x_{mean}^{UCD}$. The

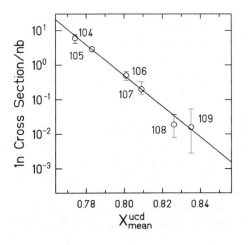

*Figure 27* The 1n cross sections observed for cold fusion reactions as a function of the $x_{mean}^{UCD}$ scaling parameter.

odd-even effect clearly seen in Figure 26 (top) apparently disappears. As all EVRs are found near the Bass barrier, the fusion probability at the barrier $p(B_B)$ is obtained from the cross section values, if the $p(B_B)$ values are normalized to the $p(B_B)$ value obtained for the $^{208}$Pb($^{50}$Ti, xn) reaction and if we assume the decrease in cross section is caused by a decrease of the fusion probability alone. The values of $p(B_B)$ thus obtained (see Figure 6) are lower values because $\Gamma_n/\Gamma_f$ is assumed to stay constant between $Z = 104$ and 109. The value of $p(B_B)$ for production of $^{266}$109 is $5.5 \times 10^{-5}$. An extra-push energy of 85.4 MeV is calculated from Equation 5, with the parameters of Figure 5.

The dynamical barrier is shifted to energies where EVRs cannot survive. Their actual detection at the Bass barrier shows that fluctuations of the order of the shift exist. This corroborates the finding in Section 2.1 that the concept of a well-defined barrier breaks down for systems with large dynamical hindrance. Not only the barrier concept, but also the entrance-exit channel concept, may be questioned for small fusion probabilities and washed-out dynamical barriers. Single high energy neutrons could be emitted in an early stage of the fusion process and cool the system below the fission barrier faster than fission may occur (125). Such a mechanism may be of special importance for small 1n cross sections.

## 5.5   On the Production of Still Heavier Elements

In the experiments performed to produce highly fissionable isotopes, a three-fold limitation of the production process has been observed.

1. The limitation by fission losses in the evaporation cascade—a thermal limitation in the exit channel (Figure 26, bottom).
2. The "extra-push" limitation—a dynamical limitation in the entrance channel (Figures 5 and 6).
3. The nuclear structure–dependent limitation of the shell stabilization with excitation energy—a structural limitation (Figures 9 and 10).

The production of still heavier elements between $Z = 110$ and 114 is, as shown in the analysis of Sections 5.1 and 5.2, not a question of the ground-state stability of these elements, but a question of navigating between the three-fold limitation of the production process.

The first limitation has to be overcome if the production of heavy elements using actinide targets is to be continued. Figure 26 (bottom) compares the largest 4n cross sections using $^{249}$Bk and $^{249}$Cf targets to produce isotopes of elements 100–106 with 1n cross sections for cold fusion reactions. For actinide-based reactions, the two other limitations do not apply as long as the reaction aims at deformed isotopes for elements 110–114 ($N < 165$). As the analysis of mass excesses and fission half-lives points to nearly constant fission barriers, the average $\Gamma_n/\Gamma_f$ values in the evaporation cascade may, in contrast to the trend observed up to element 106, stay fairly large and the cross sections may decrease more slowly than observed for $Z < 106$. This speculation makes reactions such as $^{254}\text{Es}(^{23}\text{Na}, 4n)^{273}110$, $^{249}\text{Cf}(^{26}\text{Mg}, 4n)^{271}110$, or $^{235}\text{U}(^{40}\text{Ar}, 4n)^{271}110$ possible candidates for proceeding beyond 109. The production cross sections will certainly be small. Estimates are uncertain and range between 100 and 0.1 pb. Only the detection of decay chains will allow us to identify the EVRs, since transfer reactions with actinide targets produce fission activities with much larger cross sections than do Pb- and Bi-based cold fusion reactions. The thermal limitation that since 1974 has prevented the actinide-based production of elements heavier than $Z = 106$ may be overcome by using recoil spectrometers and improved detection methods. Compared to the Berkeley 106 experiment (4), the sensitivity of the techniques used to detect $^{266}109$ was improved by at least two orders of magnitude (100).

The entrance channel limitation reduces the cross section for production of element 110 by cold fusion to the level of a few pb (Figure 27). If the $^{58}\text{Fe}$ beams are replaced by $^{64}\text{Ni}$ beams, the cross sections may decrease only marginally because the decrease may be partly compensated by the higher number of neutrons in $^{64}\text{Ni}$, which will lead into a more stabilized $N - Z = 52$ compound system. The best cold fusion reaction for discovering element 110 will be $^{208}\text{Pb}(^{64}\text{Ni}, 1n)^{271}110$ at an energy near the unshifted fusion barrier.

During the years, many attempts to detect superheavy elements have been undertaken (87). One of the more recent failures to produce isotopes of element 116 in the reaction $^{248}\text{Ca}(^{48}\text{Ca}, xn)^{296-xn}116$ gave a lower limit of 100 pb for a time range of 14 orders of magnitude, starting from the microsecond range and ending at half-lives of several years (126). The excitation energy at the barrier in the above reaction is at least 30 MeV. The main reaction channels to be expected are 2n and 3n reactions leading to the isotopes of element 116 with 178 and 177 neutrons. According to the most recent calculations (111), these are the neutron numbers with the highest shell corrections ($\sim -8.4$ MeV). The time range covered in the experiment, following Equation 13, is equivalent to a change of 4 MeV in the shell correction, which is much larger than the uncertainty of the calculation. The spherical nuclei around $N = 178$ very probably are stable enough to be detected in the time range covered. They would have been seen if they were produced. The three-fold limitation in the production process explains at least qualitatively why we failed to produce them.

With $x_{\text{mean}}^{\text{UCD}} = 0.805$ we have have to expect about the same dynamical hindrance as for the production of $^{262}107$, about $10^{-3}$ (Figure 6). With 2n channels at the barrier ($E^* = 30$ MeV), a thermal hindrance of the order of $\Gamma_n/\Gamma_f \approx 0.03$ compared to a 1n channel in a cold fusion reaction is to be expected. Moreover, the production of the spherical nuclei is structurally limited. Taking as a reference the cross section for the production of $^{216}$Th (Figure 9), which is reduced by a factor of 50, a structural limitation of at least the same order of magnitude may be expected. Compared to the production of $^{262}$107 ($\sigma = 0.2$ nb), the additional hindrance of at least a factor of $10^3$ explains why at a limit of 100 pb no atoms of element 116 were found in the above experiments. With cross sections of at most 0.1 pb it is difficult to imagine how the structural limitation blocking the access to the spherical superheavy nuclei could ever be overcome.

The outcome of other experiments attempting to go beyond $Z = 109$ is open. Only experiments can decide which of the limitations, the dynamical of cold fusion reactions or the thermal of actinide-based reactions, may be more easily overcome. The three-fold limitation will stop our search for man-made elements either at $Z = 109$ or at a few atomic numbers more. In any case, the game is limited and we should accept the study of the limitations to be a message in itself. The idea of superheavy elements— elements shell stabilized, not existing in a world of macroscopic nuclear drops—has already been realized. The isotope $^{266}$109 is already completely shell stabilized. For $^{260}$106, the spontaneous fission half-life is increased by 15 orders, and without nuclear structure effects the isotope would never have been detected. Entering the island of $\varepsilon_4$-stabilized $\alpha$ emitters at $Z = 106$ we produced elements that correspond to what superheavy elements are to be. They are not spherical nuclei, but deformed sausage-like entities; they live not billions of years, but milliseconds. We are reluctant to accept the idea that our dream of superheavy elements has already been fulfilled. The detection of one atom of $^{266}$109—the discovery of an element via one atom—may be reward enough, and from a more distant point of view it is difficult to imagine a more beautiful end of the element game. The birth, life, and death cycle of the one atom of element 109, which was born out of two, lived for a short time as the heaviest atom ever made, and disintegrated again into two, reminds us of a Greek saying: "εν το παν—The One that may stand for All."

ACKNOWLEDGMENTS

This article describes the efforts of many physicists, engineers, and technicians during a period of several years. I am happy and thankful to have collaborated with many of them, and without their steady involvement not much of what was reported would have been achieved. Special thanks go to my colleagues S. Hofmann, G. Münzenberg, W. Reisdorf, and

K.-H. Schmidt, who over the years have carried with great enthusiasm the main burden of the GSI experiments, to the UNILAC crew providing the beam—the main tool of our experiments—and to Ch. Schmelzer, the father of GSI, without whom the UNILAC would not exist.

*Literature Cited*

1. Herrmann, G., Seyb, K. E. *Naturwissenschaften* 56 : 590 (1969)
2. Alonso, J. R. *Gmelin Handbuch* Bd. 7b : 28 (1974)
3. Ghiorso, A. In *Actinides and Perspectives*, ed. by M. Edelstein, p. 23. Pacific Grove, Calif: Pergamon (1982)
4. Ghiorso, A., Hulet, K., Nitschke, J. M., Alonso, J. R., Lougheed, R. W., et al. *Phys. Rev. Lett.* 33 : 1490 (1974)
5. Druin, V. A., Bochev, B., Lobanov, Yu. V., Sagajdak, R. N., Kharitonov, Yu. P., et al. *JINR P 7-12056-Dubna* (1978)
6. Somerville, L. P. PhD thesis, Berkeley LBL-14050 (1982)
7. Lee, D., von Gunten, H., Jacak, B., Nurmia, M. J., Liu, Y.-F., et al. *Phys. Rev. C* 25 : 286 (1982)
8. Buklanov, G. V., Demin, A. G., Rubinskaya, L. A., Sagajak, R. N., Utenkov, V. K., Shirokovskij, I. V. *JINR P 7-83-91-Dubna* (1983)
9. Sagajak, R. N., Demin, A. G., Druin, V. A., Lobanov, Yu. V., Utenkov, V. K., Hübener, S. *JINR P 7-82-890-Dubna* (1982)
10. Schädel, M., Lougheed, R. W., Landrum, J. H., Wild, J. F., Dougan, R. J., et al. In *Proc. Int. Workshop on Gross Properties of Nuclei and Nuclear Excitations XI*, ed. H. Feldmeier, p. 16. Darmstadt (1983)
11. Schädel, M., Brüchle, W., Gäggeler, H., Kratz, J. V., Sümmerer, K., et al. *Phys. Rev. Lett.* 48 : 852 (1982)
12. Schädel, M., Kratz, J. V., Ahrens, H., Brüchle, W., Franz, G., et al. *Phys. Rev. Lett.* 41 : 469 (1978)
13. Broglia, R. A., Dasso, C. H., Landowne, S., Pollarolo, G. *Phys. Lett.* 133B : 34 (1983)
14. Jacobs, P. M., Smilansky, V. *Phys. Lett.* 127B : 313 (1983)
15. Swiatecki, W. J. *Phys. Scr.* 24 : 113 (1981)
16. Swiatecki, W. J. *Nucl. Phys.* A376 : 275 (1982)
17. Stokstad, R. G., Reisdorf, W., Hildenbrand, K. D., Kratz, J. V., Wirth, G., et al. *Z. Phys.* A295 : 269 (1980)
18. Beckerman, M., Salomaa, M., Sperduto, A., Enge, H., Ball, J., et al. *Phys. Rev. Lett.* 45 : 1472 (1980)
19. Nix, J. R., Sierk, A. J. *Phys. Rev.* C15 : 2072 (1977)
20. Keller, J. G. Thesis, TH-Darmstadt (1984)
21. Bass, R. *Proc. Symp. on Deep-Inelastic and Fusion Reactions with Heavy Ions, Lect. Notes Phys.* 117 : 281 (1980)
22. Sahm, C.-C., Clerc, H.-G., Schmidt, K.-H., Reisdorf, W., Armbruster, P., et al. *Z. Phys.* A319 : 113 (1984); Sahm, C.-C. Thesis, TH-Darmstadt (1984)
23. Reisdorf, W. *Z. Phys.* A300 : 227 (1981)
24. Hessberger, F. P. Thesis, TH-Darmstadt (1984)
25. Cohen, S., Plasil, F., Swiatecki, W. J. *Ann. Phys.* 82 : 557 (1974)
26. Vermeulen, D., Clerc, H.-G., Sahm, C.-C., Schmidt, K.-H., Keller, J. G., et al. *Z. Phys.* A318 : 157 (1984); Clerc, H.-G., Keller, J. G., Sahm, C.-C., Schmidt, K.-H., Schulte, H., Vermeulen, D. *Nucl. Phys. A* 419 : 571 (1984)
27. Keller, J. G., Schmidt, K.-H., Stelzer, H., Reisdorf, W., Agarwal, Y. K., et al. *Phys. Rev.* C29 : 1569 (1984)
28. Hill, D. L., Wheeler, J. A. *Phys. Rev.* 89 : 1102 (1953)
29. Esbensen, H. *Nucl. Phys.* A352 : 147 (1981)
30. Reisdorf, W., Hessberger, F. P., Hildenbrand, K. D., Hofmann, S., Münzenberg, S., et al. *Phys. Rev. Lett.* 49 : 1811 (1982)
31. Reisdorf, W., Hessberger, F. P., Hildenbrand, K. D., Hofmann, S., Münzenberg, G., et al. *Nucl. Phys. A* 438 : 212 (1985)
32. Beckerman, M., Ball, J., Enge, H., Sperduto, A., Gaes, S., et al. *Phys. Rev.* C23 : 1581 (1981)
33. Beckerman, M., Salomaa, M., Sperduto, A., Molitoris, J. D., DiRienzo, A. *Phys. Rev.* C25 : 837 (1982)
34. Gäggeler, H., Sikkeland, T., Wirth, G., Brüchle, W., Bögl, W., et al. *Z. Phys.* A316 : 291 (1984)
35. Bass, R. *Nucl. Phys.* A231 : 45 (1974)
36. Armbruster, P. *Proc. Int. Conf. on*

*Nuclear Physics*, Florence, Italy, ed. by P. Blasi, R. A. Ricci, p. 343. Bologna: Tipografia Compositori (1983)

37. Bjørnholm, S., Swiatecki, W. J. *Nucl. Phys.* A391:471 (1982)

38. Blocki, J. Private communication, quoted in Ref. 37 (1982)

39. Gross, D. H. E., Nayak, R. C., Satpathy, L. Z. *Phys.* A299:63 (1981); Fröbrich, P. *Phys. Rep.* 116:337 (1984)

40. Davies, K. T. R., Sierk, A. J., Nix, J. R. *Phys. Rev.* C28:679 (1983)

41. Sann, H., Bock, R., Chu, Y. T., Gobbi, A., Olmi, A., et al. *Phys. Rev. Lett.* 47:1248 (1981)

42. Bock, R., Chu, T., Dakowski, M., Gobbi, A., Grosse, E., et al. *Nucl. Phys.* A388:334 (1982)

43. Nörenberg, W. *Phys. Lett.* 104B:107 (1981)

44. Cassing, W., Nörenberg, W. *Nucl. Phys.* A401:467 (1983)

45. Lukasiak, A., Cassing, W., Nörenberg, W. *Nucl. Phys.* A426:181 (1984)

46. Keller, J. G., Clerc, H.-G., Schmidt, K.-H., Agarwal, Y. K., Hessberger, F. P., et al. *Z. Phys.* A311:243 (1983)

47. Schmidt, K.-H., Armbruster, P., Hessberger, F. P., Münzenberg, G., Reisdorf, W., et al. *Z. Phys.* A301:21 (1981)

48. Oganessian, Yu. Ts. *Lect. Notes Phys.* 33:221 (1974)

49. Oganessian, Yu. Ts., Iljinov, A. S., Demin, A. G., Tretyakova, S. P. *Nucl. Phys.* A239:353 (1975)

50. Münzenberg, G., Armbruster, P., Faust, W., Güttner, K., Hessberger, F. P., et al. See Ref. 3, p. 223 (1982)

51. Oganessian, Yu. Ts., Hussonnois, M., Demin, A. G., Kharitonov, Yu. P., Bruchertseifer, H., et al. *JINR E2-84-651* (1984)

52. Schmidt, K.-H., Simon, R. S., Keller, J. G., Reisdorf, W., Hessberger, F. P., et al. *Lect. Notes Phys.* 219:244 (1984)

53. Clerc, H.-G., Sahm, C.-C., Tschöp, E., Schwab, W., Schmidt, K.-H., et al. In *Capture Gamma-Ray Spectroscopy and Related Topics—1984*, ed. S. Raman. New York: Am. Inst. Phys. 636 pp. (1985)

54. Iljinov, A. S., Cherepanov, E. A. *JINR P-7-84-68-Dubna* (1984)

55. Delagrange, H., Lin, S. Y., Fleury, A., Alexander, J. M. *Phys. Rev. Lett.* 39:867 (1977)

56. Schmidt, K.-H., Faust, W., Münzenberg, G., Reisdorf, W., Clerc, H.-G., et al. *Proc. Symp. Phys. Chem. of Fission, Jülich 1979*, 1:409. Vienna: IAEA (1980)

57. Sahm, C.-C., Clerc, H.-G., Schmidt, K.-

58. Iljinov, A. S., Mebel, M. V., Cherepanov, E. A. *JINR P 7-82-638-Dubna* (1982)

59. Iljinov, A. S., Oganessian, Yu. Ts., Cherepanov, E. A. *JINR P 7-81-549-Dubna* (1981)

60. Ignatyuk, A. V., Smirenkin, G. N., Tishin, A. S. *Sov. J. Nucl. Phys.* 21:255 (1975)

61. Schmidt, K.-H., Delagrange, H., Dufour, J. P., Carjan, N., Fleury, A. Z. *Phys.* A308:215 (1982)

62. Keller, J. G., Diplomarbeit, TH-Darmstadt IKDA-Rep-81-7 (1981)

63. Bjørnholm, S., Bohr, A., Mottelson, B. R. *Proc. Int. Conf. Phys. Chem. of Fission, Rochester 1973*, 1:367 Vienna: IAEA (1974)

64. Schmidt, K.-H., Keller, J. G., Vermeulen, D. *Z. Phys.* A315:159 (1984)

65. Oganessian, Yu. Ts. *Int. School-Seminar on Heavy Ion Physics*, Alushta, *JINR D 7-83-644*, Dubna, p. 55 (1983)

66. Marx, D., Nickel, F., Münzenberg, G., Güttner, K., Ewald, H., et al. *Nucl. Instrum. Methods* 163:15 (1979)

67. Gäggeler, H., Iljinov, A. S., Popeko, G. S., Seidel, W., Ter Akopian, G. M., Tretyakova, S. P. *Z. Phys.* A289:415 (1979)

68. Nitschke, J. M., Leber, R. E., Nurmia, M. J., Ghiorso, A. *Nucl. Phys.* A313:235 (1979)

69. Münzenberg, G., Faust, W., Hofmann, S., Armbruster, P., Güttner, K., Ewald, H. *Nucl. Instrum. Methods* 161:65 (1979)

70. Ghiorso, A., Leino, M. E., Yashita, S., Frank, L., Armbruster, P., et al. *LBL Rep. 15955* (1984); and to be published

71. Flerov, G. N. See Ref. 36, p. 365 (1983)

72. Kharitonov, Yu. P., Rykhlyuk, A. V., Hussonnois, M., Bruchertseifer, H., Constantinescu, O., et al. See Ref. 65, *JINR D 7-83-604 Dubna*, p. 589 (1983)

73. Enge, H. A. *Nucl. Instrum. Methods* 186:413 (1981)

74. Faust, W., Münzenberg, G., Hofmann, S. Reisdorf, W., Schmidt, K.-H., et al. *Nucl. Instrum. Methods* 166:397 (1979)

75. Moll, E., Schrader, H., Siegert, G., Ashgar, M., Bocquet, J. P., et al. *Nucl. Instrum. Methods* 123:615 (1975)

76. Ewald, H., Güttner, K., Münzenberg, G., Armbruster, P., Faust, W., et al. *Nucl. Instrum. Methods* 139:223 (1976)

77. Münzenberg, G. *Int. J. Mass Spectrosc. Ion Phys.* 14:363 (1974)

78. Armbruster, P. *Proc. 3rd Int. Conf.*

*Nuclear far from Stability*, Cargèse, *CERN 76-13 : 1.* Geneva: CERN (1976)
79. Schmidt, K.-H., Faust, W., Münzenberg, G., Clerc, H.-G., Lang, W., et al. *Nucl. Phys.* A318: 253 (1979)
80. Münzenberg, G., Hofmann, S., Faust, W., Hessberger, F. P., Reisdorf, W., et al. *Z. Phys.* A302: 7 (1981)
81. Hofmann, S., Faust, W., Münzenberg, G., Reisdorf, W., Armbruster, P., et al. *Z. Phys.* A291: 53 (1979)
82. Schmidt, K.-H., Sahm, C.-C., Pielenz, K., Clerc, H.-G. *Z. Phys.* A316: 19 (1984)
83. Cohen, B. L., Fulmer, C. B. *Nucl. Phys.* 6: 547 (1958)
84. Armbruster, P. *Nucleonik* 3: 183 (1961)
85. Baccho, I., Bogdanov, D. D., Daroczy, S., Karnaukhov, V. A., Petrov, L. A., Ter Akopian, G. M. *JINR P 13-4453 Dubna* (1969)
86. Lawin, H., Eidens, F., Borgs, F. W., Fabbri, R., Gruter, F. W., et al. *Nucl. Instrum. Methods* 137: 103 (1976)
87. Kratz, J. V. *Radiochim. Acta* 32: 25 (1983)
88. Lederer, C. M., Shirley, V. S., eds. *Table of Isotopes*. New York: Wiley. 7th ed. (1978)
89. Ghiorso, A., Nurmia, M., Harris, J., Eskola, K., Eskola, P. *Phys. Rev. Lett.* 22: 1317 (1969)
90. Oganessian, Yu. Ts., Demin, A. G., Iljinov, A. S., Tretyakova, S. P., Pleve, A. A., et al. *Nucl. Phys.* A239: 157 (1975)
91. Hessberger, F. P., Münzenberg, G., Hofmann, S., Reisdorf, W., Schmidt, K.-H., Armbruster, P. *Z. Phys. A* 321: 317 (1985)
92. Oganessian, Yu. Ts., Demin, A. G., Danilov, N. A., Flerov, G. N., Ivanov, M. P., et al. *Nucl. Phys.* A273: 505 (1976)
93. Oganessian, Yu. Ts., Tretyakov, Yu. P., Iljinov, A. S., Demin, A. G., Pleve, A. A., et al. *Pis'ma Zh. Eksp. Theor. Fiz.* 20: 580 (1974)
94. Demin, A. G., Tretyakova, S. P., Utyonkov, V. K., Shirokovsky, I. V. *Z. Phys.* A315: 197 (1984)
95. Armbruster, P. *Lect. Notes Int. Sch. Phys. Enrico Fermi, Varenna, June 1984, GSI-Preprint 84-47* (1984)
96. Münzenberg, G., Hofmann, S., Hessberger, F. P., Reisdorf, W., Schmidt, K.-H., et al. *Z. Phys.* A300: 107 (1981)
97. Münzenberg, G., Armbruster, P., Folger, H., Hessberger, F. P., Hofmann, S., Keller, J. G., et al. *Z. Phys.* A317: 235 (1984)
98. Oganessian, Yu. Ts., Demin, A. G., Huss-

onois, M., Tretyakova, S. P., Kharitonov, Yu. P., et al. *Z. Phys.* A319: 215 (1984)
99. Münzenberg, G., Armbruster, P., Hessberger, F. P., Hofmann, S., Poppensieker, K., et al. *Z. Phys.* A309: 89 (1982)
100. Münzenberg, G., Reisdorf, W., Hofmann, S., Agarwal, Y. K., Hessberger, F. P., et al. *Z. Phys.* A315: 145 (1984)
101. Liran, S., Zeldes, N. *The Hebrew Univ. Jerusalem, Int. Rep.* (1976)
102. Møller, P., Nix, R. J. *At. Data Nucl. Data Tables* 26: 165 (1981)
103. Harvey, B. G., Herrmann, G., Hoff, R. W., Hoffman, D. C., Hyde, E. K., et al. *Science* 193: 1271 (1976)
104. Zvara, I. *Pure Appl. Chem.* 53: 979 (1981)
105. Wapstra, A. H., Bos, K. *At. Data Nucl. Data Tables*, Vol. 19, No. 3 (1977); Wapstra, A. H., Audi, G. *Nucl. Phys. A* 432: 1 (1985)
106. Armbruster, P., Folger, H., Hessberger, F. P., Keller, J. G., Münzenberg, G., et al. *Proc. Int. Conf. AMCO-7*, ed. O. Klepper, p. 284. TH-Darmstadt (1984)
107. Myers, W. D. *Droplet Model of Atomic Nuclei*, New York: IFI/Plenum (1973)
108. von Groote, H., Hilf, E. R., Takahashi, K. *At. Data Nucl. Data Tables* 17: 418 (1976)
109. Seeger, P. A., Howard, W. M. *Nucl. Phys.* A238: 49 (1975)
110. Liran, S., Zeldes, N. *At. Data Nucl. Data Tables* 17: 431 (1976)
111. Leander, G. A., Møller, P., Nix, J. R., Howard, W. M. See ref. 106, p. 466; Møller, P. Private communication
112. Dahlinger, M., Vermeulen, D., Schmidt, K.-H. *Nucl. Phys.* A376: 94 (1982)
113. Rasmussen, J. O. *Phys. Rev.* 113: 1593 (1959)
114. Swiatecki, W. J. *Phys. Rev.* 100: 937 (1955)
115. Randrup, J., Larsson, S. E., Møller, P., Nilsson, S. G., Pomorski, K., Sobiczewski, A. *Phys. Rev.* C13: 229 (1976)
116. Baran, A., Pomorski, K., Lukasiak, A., Sobiczewski, A. *Nucl. Phys.* A361: 83 (1981)
117. Randrup, J., Tsang, C. F., Møller, P., Nilsson, S. G., Larson, S. E. *Nucl. Phys.* A217: 221 (1973)
118. Viola, V. E., Wilkins, B. D. *Nucl. Phys.* 82: 65 (1966)
119. Hulet, E. K., Baisden, P. A., Dougan, R. J., Schädel, M., Wild, J. F., et al. *Nucl. Chem. Div. Ann. Rep.* FY82. Livermore, Calif: Lawrence Livermore Lab; *UCAR-10062-82*, p. 98 (1982)
120. Viola, V. E. *Nucl. Data* 1: 391 (1966);

Unik, J. P., Gindler, J. E., Glendenin, L. E., Flynn, K. F., Gorski, A., Sjoblom, R. K. See Ref. 63, 2:19 (1974)
121. Myers, W. D., Swiatecki, W. J. *Nucl. Phys.* 81:1 (1966)
122. Cwiok, S., Pashkevich, V. V., Dudek, E., Nazarevich, V. *Nucl. Phys.* A410:254 (1983)
123. Viola, V. E., Migney, A. C., Breuer, H., Wolf, K. L., Glagola, B. G., et al. *Phys. Rev.* C22:122 (1980)

124. Orlova, O. A., Bruchertseifer, H., Muzycka, Yu. A., Oganessian, Yu. Ts., Pustylnik, B. I., et al. *JINR P 7-12061-Dubna* (1978)
125. Grange, P., Weidenmüller, H. A. *Phys. Lett.* 96B(1,2):26 (1980); Grange, P., Li, J.-Q., Weidenmüller, H. A. *Phys. Rev.* C27:2063 (1983)
126. Armbruster, P., Agarwal, Y. K., Brüchle, W., Brügger, M., Dufour, J. P., et al. *Phys. Rev. Lett.* 54:406 (1985)

*Ann. Rev. Nucl. Part. Sci. 1985. 35 : 195–243*

# PRODUCTION AND DECAY OF THE b QUARK

*Edward H. Thorndike*

Department of Physics & Astronomy, University of Rochester, Rochester, New York 14627

CONTENTS

0163–8998/85/1201–0195$02.00

# 1. INTRODUCTION

In 1977, the upsilon particle was discovered (1), and was promptly interpreted as a quark-antiquark bound state made up of a new, fifth flavor of quark—the beauty quark, bottom quark, or, more concisely, the b quark. The b quark and its assumed partner, the t quark (truth, top) fitted nicely into a preexisting theoretical framework. In 1970, Glashow, Iliopoulos & Maiani (GIM) (2) had shown that the observed suppression of strangeness-changing neutral current decays of kaons could be understood if, in addition to the then-accepted three flavors of quarks (u, d, s), a fourth flavor (c, charm) existed and formed a doublet with s. The two quark doublets (u, d) and (c, s) would match the two lepton doublets $(v_e, e^-)$ and $(v_\mu, \mu^-)$, canceling triangle anomalies. The charmed quark was discovered in 1974 (3). Soon thereafter, a third charged lepton, the tau lepton, was discovered (4), and the balance between lepton and quark doublets was again lost. In 1973, prior to the discovery of $\tau$ and charm, Kobayashi & Maskawa (5) had shown that with six quark flavors arranged in three doublets, $CP$ violation could be attributed to a phase in the matrix that mixes the quarks prior to their weak decay (the K-M matrix). The b quark, then, was a welcome discovery, because it restored the balance between quark and lepton doublets, and because it allowed a natural framework for description of $CP$ violation.

The b quark nonetheless is accompanied by some challenges. The most profound of these goes by the name of the "family problem," and it is an old problem. Some decades ago it was phrased "Why the muon?" It has since expanded into: "Why three families of quark and lepton doublets?", "Are the three families identical?", "How many more are there?", "How are the families related?" While the discovery of the b quark did not create the family problem, it broadened the problem and at the same time provided more avenues for exploring it.

The existence of the b quark requires one either to find the t quark or to develop a viable theory without a t quark (a "topless" model). As discussed later, data from b-decay studies rule out all topless models so far imagined. Should the t quark not exist, present-day theory will require fundamental changes. [Preliminary data from the CERN pp̄ collider (6) suggest that the t quark does exist, with a mass near 40 GeV.]

While the existence of three quark families provides a natural explanation of $CP$ violation in neutral kaon decay, this explanation is subject to experimental test. The mixing angles and phase of the quark mixing matrix (K-M matrix) can be determined by studying b decay. With these parameters, the $CP$-violating quantities in $K^0$ decay can be calculated and compared with observation. As we shall see, the currently calculated limits on the CP-violating parameters are consistent with measurement. (CP violation should also occur in b decay. No evidence has been seen so far, and the effects expected from the phase of the K-M matrix are quite small.)

The plan of this review is as follows: Section 2 briefly discusses the production of the b quark, both in its hidden form (bound $b\bar{b}$ systems), and in its bare form. Sections 3 to 8 deal with the decay of the b quark. In Section 3, the theoretical framework is developed, both for the Standard Model of b decay and for nonstandard models. In Section 4 the nonstandard models of b decay are confronted with experimental data and found to be wanting. Section 5 presents the direct evidence for B mesons, and gives their masses. In Section 6, experimental information on (b → u)/(b → c), the B semileptonic decay branching ratio, and the B lifetime is presented and used to place tight constraints on the K-M angles. Those effects that arise from the active participation of the light quark q (u or d) in the B meson ($\bar{b}$q) are discussed in Section 7. Section 8 gives experimental information on low-multiplicity hadronic decay branching ratios and inclusive decay properties of B mesons. Section 9 summarizes the emerging picture of b decay, considers the implications for CP violation in the neutral kaon system of the constraints on the K-M angles, and lists open problems.

In discussing neutral b-flavored mesons, $b\bar{d}$ and $\bar{b}d$, there is an arbitrary convention to be chosen as to which is the $B^0$ and which the $\bar{B}^0$. All theoretical papers prior to 1983 followed the natural convention $B^0 = b\bar{d}$, i.e. b-flavored mesons are made from b quarks. In 1983, in its paper (7) showing reconstruction of charged and neutral B mesons, CLEO adopted the opposite convention, based on the perceived need to repeat the pattern set by $K^0$, $\bar{K}^0$, even though we now know better. The Particle Data Group has followed CLEO's convention: $B^0 = \bar{b}d$, $\bar{B}^0 = b\bar{d}$; I reluctantly will do the same. The only place where serious confusion is apt to occur is in CP violation in neutral B decay. There, changing the convention reverses the sign of CP-violating asymmetries.

## 2.  PRODUCTION

### 2.1  Hidden Beauty

The b quark was discovered in its hidden form (bound $b\bar{b}$) at Fermilab in 1977, via the reaction p + nucleus → $\Upsilon$ + X, $\Upsilon \to \mu^+\mu^-$ (1). The dimuon mass spectrum showed a broad peak at 10 GeV, which was suggestive of three poorly resolved lines, at 9.4, 10.0, and 10.4 GeV, each with a true width less than the experimental mass resolution of 210 MeV. The states were named $\Upsilon$ (upsilon), $\Upsilon'$, and $\Upsilon''$ and were interpreted as $J^{PC} = 1^{--}$ $b\bar{b}$ bound states. [I will use the alternate notation $\Upsilon(nS)$, $n = 1, 2, 3$, where $n$ labels the radial excitation of the $^3S_1$ $b\bar{b}$ state.]

With $J^{PC} = 1^{--}$, the upsilon states are expected to be produced by $e^+e^-$ colliding beams, and to appear as sharp spikes in the plot of total hadronic cross section vs center-of-mass energy. Such measurements were made with the DORIS storage ring at DESY (8) and with the CESR storage ring at Cornell (9), and the anticipated narrow spikes were observed. The area of the peaks determined the leptonic partial widths $\Gamma_{ee}$ of the three resonances to be $1.22 \pm 0.06$, $0.52 \pm 0.03$, and $0.38 \pm 0.03$ KeV, respectively (10), which demonstrated a charge 1/3 for the quark. Subsequent measurements (10, 11) showed the leptonic decay branching ratios to be $3.0 \pm 0.3\%$, $1.9 \pm 0.5\%$, and $2.9 \pm 1.0\%$, which determined the total widths of the resonances to be $41 \pm 5$, $27 \pm 7$, and $13 \pm 5$ keV, much narrower than the few-MeV experimental resolution.

There now exists an extensive spectroscopy of the bound $b\bar{b}$ states; it was reviewed by Franzini & Lee-Franzini (12) and by Berkelman (10).

While the discovery of hidden beauty demonstrates the existence of the b quark, hidden beauty provides no information on the b quark's weak decay. The bound $b\bar{b}$ states decay strongly, with b and $\bar{b}$ annihilating each other. The b must be produced in its bare form in order for its weak decay to be studied.

## 2.2  Bare Beauty

Quark-antiquark pairs ($b\bar{b}$ or any other flavor) are produced in $e^+e^-$ colliding beams via the annihilation diagram, Figure 1. For energies below the threshold for producing two b-flavored hadrons, the narrow spikes from hidden beauty appear in the plot of cross section vs energy. At energies far above threshold, unbound $b\bar{b}$ production is expected, with a cross section 1/3 that for $\mu$ pairs (1/3 unit of $R$). The b and $\bar{b}$ will be in oppositely directed jets. At energies just above threshold, structure is expected in the cross section for production of the unbound $b\bar{b}$ systems.

2.2.1  DISCOVERY—$\Upsilon$(4S)  The plot of hardonic cross section vs center-of-mass energy obtained by the CLEO collaboration (13) at CESR is shown in Figure 2. [Similar results were obtained at CESR by the CUSB collaboration (14).] The three narrow lines are the $\Upsilon$(1S, 2S, 3S) states discussed above. Their apparent width shows the experimental resolution. In

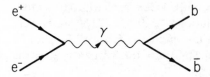

*Figure 1*  A diagram of $e^+e^-$ annihilation leading to b-quark production.

addition, there is a broader peak at 10.578 GeV, labeled Υ(4S). The intrinsic width of that peak is $20 \pm 2 \pm 4$ MeV fwhm (15).

Two pieces of evidence established (16) that bare beauty is produced at Υ(4S): (a) The resonance is broad (20 MeV), indicative of fast decay of Υ(4S) into b-flavored hadrons. Were the $b\bar{b}$ system bound, the decay would proceed through $b\bar{b}$ annihilation, an inhibited strong process, resulting in a narrow (25 keV) line such as Υ(1S)–Υ(3S). (b) There is a substantial yield of high-momentum leptons in Υ(4S) decay, establishing that some of the decay products from Υ(4S) (the b-flavored hadrons) decay weakly.

The Υ(4S) is a bare-b "factory," providing b-flavored hadrons for decay studies. It sits atop a substantial continuum background of lighter quark production. This background can be measured by taking data at energies just below Υ(4S). Most of the studies of the weak decay of the b quark described in Sections 4–8 make use of Υ(4S).

Were it energetically allowed, Υ(4S) would decay to $B\bar{B}$, $B^*\bar{B}$, and $B^*\bar{B}^*$, where B (B*) is a spin-0 (spin-1) b-flavored meson. CUSB (17) has set an upper limit of 0.1 on the B*/B ratio from Υ(4S) decays by searching for, and not finding, the 50-MeV photon from $B^* \to B\gamma$. Using a technique that depends on the dominance of the decay Υ(4S) → $B\bar{B}$, CLEO (7) has reconstructed B mesons and determined the mass difference $M[\Upsilon(4S)] - 2M(B)$ to be 32 MeV. Since this is less than the expected $B^* - B$ mass difference, Υ(4S) can decay *only* to $B\bar{B}$.

If, as expected, charged and neutral B's have different masses, the decays

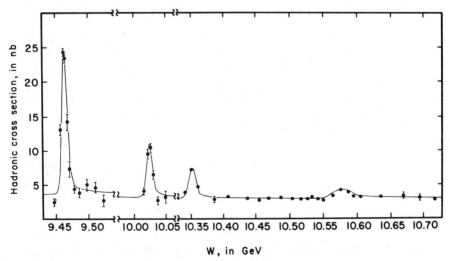

*Figure 2* Hadronic cross section vs center-of-mass energy, showing three narrow resonances Υ(1S, 2S, 3S), and one broad resonance Υ(4S). Results are from CLEO (13).

$\Upsilon(4S) \to B^+B^-$ and $\Upsilon(4S) \to B^0\bar{B}^0$ will have different decay momenta, and therefore may have different branching ratios. The $B^0 - B^\pm$ mass difference is predicted to be near 4 MeV, and is not well measured. CLEO (7) has assumed the threshold approximation for the $\Upsilon(4S) \to B\bar{B}$ decay matrix, which gives decay widths proportional to $P^{2l+1}$ ($P^3$). Using this in conjunction with a predicted $B^0 - B^\pm$ mass difference of 4.4 MeV and the measured $M[\Upsilon(4S)] - 2M(B)$ mass difference, they compute a 60%/40% mix for $B^+B^-/B^0\bar{B}^0$ from $\Upsilon(4S)$ decay. The threshold approximation is assuredly not valid, as the coupled channel analysis of Eichten (18) demonstrates. The momentum dependence, rather, is oscillatory, with three nodes, and the momentum of the $\Upsilon(4S) \to B\bar{B}$ decay places it near the first node. The precise location of the nodes is model dependent, and so different models predict different values for the relative proportions of $B^+B^-$ and $B^0\bar{B}^0$, and also for the $\Upsilon(4S)$ total width. By constraining a coupled-channel analysis to give the correct width, one will remove much of the model dependence and it may be possible to make a reasonably accurate prediction of the proportions of $B^+B^-$ and $B^0\bar{B}^0$. Until such a calculation is performed, however, the proportion of charged to neutral B's will remain essentially unknown. Direct experimental determination of the proportions is very difficult.

2.2.2    OTHER RESONANCES—STRUCTURE ABOVE $\Upsilon(4S)$    With $B\bar{B}$ threshold 32 MeV below $\Upsilon(4S)$, one expects $B^*\bar{B}$ threshold to be about 20 MeV above $\Upsilon(4S)$, $B^*\bar{B}^*$ threshold 70 MeV above, and $B_s\bar{B}_s$ threshold perhaps 200 MeV above. ($B_s$ is the pseudoscalar meson with quark composition $\bar{b}s$.) Both CLEO and CUSB (15, 19) have scanned the region from $\Upsilon(4S)$ up to 11.2 GeV. The two experiments are in good agreement; CUSB hadronic cross-section results are shown in Figure 3. One notes a shoulder on the high side of $\Upsilon(4S)$, a peak labeled $\Upsilon(5S)$, and a second peak or a step (or both) labeled $\Upsilon(6S)$. Because of the complicated coupled-channel nature of the situation, the locations of the 5S and 6S peaks are not directly comparable with potential model predictions. CUSB (20) has presented evidence for $B^*$ production in this region, 50-MeV photons from $B^* \to B\gamma$ decay. No evidence for $B_s$ has been found.

The region just above $\Upsilon(4S)$ can be characterized as follows: No surprises have been encountered. As a source of bare b, the region is inferior to $\Upsilon(4S)$ because of the smaller cross-section enhancement. The $\Upsilon(5S)$ and $\Upsilon(6S)$ are candidates for a $B_s$ "factory" but as yet are unproven.

2.2.3    CONTINUUM PRODUCTION    The absolute value of the hadronic cross section, i.e. the increase in $R$ relative to the region above charm threshold, suggests that b production is present at energies well above $\Upsilon(4S)$. Several

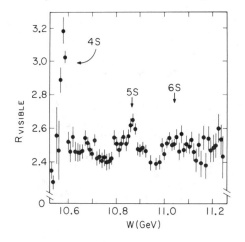

*Figure 3*    Plot of $R = \sigma_{\text{hadron}}/\sigma_{\mu\mu}$ vs center-of-mass energy, showing the region above $\Upsilon(4S)$. Results are from CUSB (19).

PEP (21–24) and PETRA (25–27) experiments demonstrated b production by a more powerful technique that takes advantage of the differences between b-quark jets and light-quark jets. Events containing leptons are selected. For each event, the jet axis is determined, and the longitudinal and transverse momenta of the lepton, relative to the jet axis, are computed. The b-quark jets, with the b decaying semileptonically, are characterized by leptons with large transverse momentum because of the substantial mass of the b-flavored hadron; c-quark jets have leptons with significantly smaller transverse momentum. Having identified the b-quark jet on the basis of the lepton's transverse momentum, one obtains a measure of the longitudinal momentum of the b-flavored hadron from the lepton's longitudinal momentum. Several experiments have used this technique to study the yield of b quarks and the b-quark fragmentation process.

The production cross sections are in agreement with theoretical expectations. The information obtained on the fragmentation process is the mean value of the fragmentation variable $z^1$, and (equivalently) the parameter $\varepsilon$ in

---

[1] Three different fragmentation variables have been used by the various groups. The one preferred from a theoretical viewpoint is $z \equiv (E + P_{\parallel})_{\text{hadron}}/(E + P)_{\text{quark}}$, where $P_{\parallel}$ is the component of hadron momentum parallel to the quark directions. The variable $x \equiv 2E_{\text{hadron}}/\sqrt{s}$ is experimentally more accessible. The two expressions differ because the quark energy is often less than $\sqrt{s}/2$, as a result of initial-state photon radiation and gluon bremsstrahlung. A third variable, $x'$, is calculated by subtracting from $\sqrt{s}$ that energy which is lost to initial-state photon radiation. For b-quark fragmentation, it is estimated that $\langle z \rangle = 1.02\langle x' \rangle = 1.05\langle x \rangle$. I use this relation in combining the results from different experiments to obtain a world average for $\langle z \rangle$.

**Table 1**   Selected b-quark fragmentation information

| Group | Ref. | Particle | Variable | $\langle "z" \rangle$ | $\varepsilon$ |
|---|---|---|---|---|---|
| Mark II | 21 | e | z | $0.79 \pm 0.06 \pm 0.06$ | $0.015^{+0.022+0.023}_{-0.011-0.011}$ |
| | | $\mu$ | z | $0.73 \pm 0.15 \pm 0.10$ | $0.042^{+0.218+0.120}_{-0.041-0.035}$ |
| MAC | 22 | $\mu$ | z | $0.8 \pm 0.1$ | $0.008^{+0.037}_{-0.008}$ |
| DELCO | 23 | e | $x'$ | $0.78 \pm 0.05$ | $0.018^{+0.024}_{-0.011}$ |
| TPC | 24 | e | $x'$ | $0.74 \pm 0.05 \pm 0.03$ | $0.033^{+0.037+0.019}_{-0.019-0.012}$ |
| | | $\mu$ | $x'$ | $0.80 \pm 0.05 \pm 0.05$ | $0.011^{+0.015+0.011}_{-0.007-0.007}$ |
| Mark J | 25 | $\mu$ | z | $0.75 \pm 0.03 \pm 0.06$ | $(0.15 \pm 0.03 \pm 0.05)^2$ |
| TASSO | 26 | e | z | $0.84^{+0.15+0.15}_{-0.10-0.11}$ | $0.005^{+.022+0.020}_{-0.005-0.005}$ |
| | | $\mu$ | z | $0.85^{+0.10+0.02}_{-0.12-0.07}$ | $0.003^{+0.029+0.011}_{-0.003-0.001}$ |
| Combined | | | z | $0.79 \pm 0.03$ | |

the Peterson form for the fragmentation function,

$$D_q(z) = \frac{N}{z[1 - 1/z - \varepsilon/(1-z)]^2}. \qquad \qquad 1.$$

Results are listed in Table 1. The b fragmentation function is quite hard, as expected for heavy quarks, with the b-flavored hadron retaining, on average, 80% of the b quark's energy.

2.2.4   HADRONIC PRODUCTION    As of this writing, there is no convincing evidence for hadronic production of bare b. The CBFP collaboration (28) working at the ISR reported a positive result for $\Lambda_B^0$ production in the $pD^0\pi^-$ decay channel, but this result has not been confirmed (29) and remains controversial because of the surprisingly large $\Lambda_B^0$ cross section, $120^{+82}_{-50}$ $\mu$b. Other bare-b searches have concentrated on multilepton or $J/\psi$ final states but suffer from low sensitivity and large model-dependent acceptance corrections. They have yielded upper limits in the range of 7 to 180 nb. See the work of Kernan & Van Dalen (30) for a recent review.

## 3.   b DECAY—THE THEORETICAL FRAMEWORK

### 3.1   Classification of Decay Reactions

The "Standard Model" of electroweak decays is the obvious, natural framework for analyzing the weak decay of the b quark. Before describing that model, however, I want to take a broader viewpoint (31), make only a few general assumptions, and see what decay reactions are allowed.

**Table 2**    Classification of decay reactions of b

| | | |
|---|---|---|
| 1. | $b \to qq\bar{q}$ | hadronic |
| 2. | $b \to q\ell_i\bar{\ell}_i$ | semileptonic |
| 3. | $b \to q\ell\ell$ | |
| | or $q\bar{\ell}\bar{\ell}$ | double-leptonic |
| | or $q\ell_i\bar{\ell}_j \quad j \neq i$ | |
| 4. | $b \to \bar{q}\bar{q}\ell$ | antibaryonic |
| | or $\bar{q}\bar{q}\bar{\ell}$ | |
| 5. | $b \to 5, 7, 9, \ldots$ quarks and/or leptons | |

(Reactions 3 and 4 are bracketed as **exotic decays**.)

$(\ell_1 = e \text{ or } v_e, \ell_2 = \mu \text{ or } v_\mu, \ell_3 = \tau \text{ or } v_\tau)$

The first assumption is that it is reasonable to talk about b decay, i.e. that decay of the B meson occurs through a b decay vertex. The second assumption is that b decays exclusively into quarks and/or leptons. (If direct photons or stable Higgs come from b decay, this assumption is invalid; otherwise it is fine.) The final assumption is that b decay conserves color. With these three assumptions, there are only four types of decay leading to three quarks and/or leptons, as listed in Table 2. [There are also decays leading to five or more final-state fermions (reaction 5). These add little beyond what is contained in reactions 1 through 4, and are not considered further.]

The hadronic and semileptonic decays can be mediated by $W^{\pm}$ and/or $Z^0$, and are expected within the framework of $SU(2) \times U(1)$ electroweak theory. The double-leptonic decays yield two leptons, or two antileptons, or a lepton and an antilepton of a different flavor. These decays violate lepton number. The antibaryonic decays, with two antiquarks from the decay process and one light antiquark from the initial $\bar{B}$ meson, will of necessity have an excess of one antibaryon. These decays violate both lepton number and baryon number.

I use the term "exotic" to refer to double-leptonic and antibaryonic decays. Exotic decays are not allowed within the context of $SU(2) \times U(1)$ theory, that is they cannot be mediated by $W^{\pm}$ or $Z^0$. They are present in some theories that relate the three quark doublets by a symmetry: the horizontal S(3) model of Derman (32); the SU(7) model of Georgi & Machacek (33); and Georgi & Glashow's (34) E6, for examples. The most natural way for b decay to tell us something profound would be through exotic decays.

Semileptonic decays can be broken down further, according to the final-state quark:

2a.    $b \to c\, e^- v$ (or $\mu^- v$ or $\tau^- v$)

2b.    $b \to u\, e^- v$ (or $\mu^- v$ or $\tau^- v$)

2c. $b \to s\, e^+ e^-$ (or $\mu^+ \mu^-$ or $\tau^+ \tau^-$ or $\nu \bar{\nu}$)

2d. $b \to d\, e^+ e^-$ (or $\mu^+ \mu^-$ or $\tau^+ \tau^-$ or $\nu \bar{\nu}$).

The last two decays involve flavor-changing neutral currents, and are forbidden by the GIM mechanism in the standard K-M model. They must be present in certain topless models. One can similarly break the hadronic decays down further, but it leads to a large number of subreactions, and is not particularly informative.

## 3.2 The Standard Model

The Standard Model (2, 5, 35), as it applies to the decay of the b quark, is illustrated in Figure 4. The quarks are grouped into three left-handed doublets and six right-handed singlets in weak isospin. The leptons form three left-handed doublets and the charged leptons three right-handed singlets. Transitions within the left-handed doublets are mediated by the charged vector boson $W^\pm$. Flavor-changing neutral currents caused by the $Z^0$ cancel because of the GIM mechanism (2).

The quark states for which the weak interaction is diagonal are not eigenstates of the quark mass matrix. The former states $(d', s', b')$ are linear combinations of the latter $(d, s, b)$ with the mixing of quark mass eigenstates into weak interaction eigenstates given by a $3 \times 3$ unitary matrix, the Kobayashi-Maskawa (K-M) matrix (5). This matrix is parametrized by three angles and one phase. The existence of the phase allows the K-M matrix to be complex, and admits the possibility of $CP$ violation in weak interactions. There are several different conventions for representing the K-

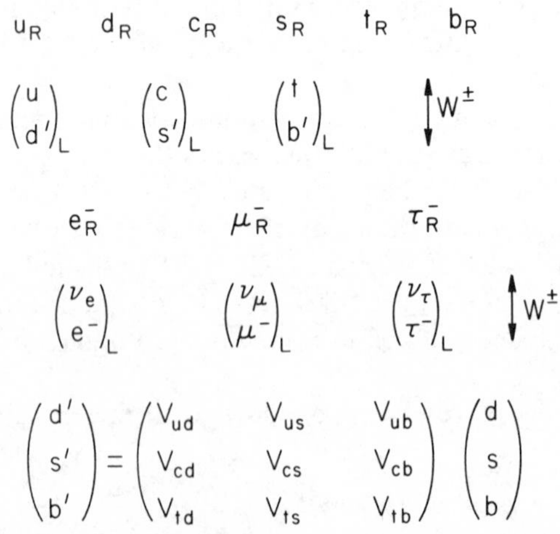

$$
u_R \qquad d_R \qquad c_R \qquad s_R \qquad t_R \qquad b_R
$$

$$
\begin{pmatrix} u \\ d' \end{pmatrix}_L \qquad \begin{pmatrix} c \\ s' \end{pmatrix}_L \qquad \begin{pmatrix} t \\ b' \end{pmatrix}_L \qquad \updownarrow W^\pm
$$

$$
e_R^- \qquad\qquad \mu_R^- \qquad\qquad \tau_R^-
$$

$$
\begin{pmatrix} \nu_e \\ e^- \end{pmatrix}_L \qquad \begin{pmatrix} \nu_\mu \\ \mu^- \end{pmatrix}_L \qquad \begin{pmatrix} \nu_\tau \\ \tau^- \end{pmatrix}_L \qquad \updownarrow W^\pm
$$

$$
\begin{pmatrix} d' \\ s' \\ b' \end{pmatrix} = \begin{pmatrix} V_{ud} & V_{us} & V_{ub} \\ V_{cd} & V_{cs} & V_{cb} \\ V_{td} & V_{ts} & V_{tb} \end{pmatrix} \begin{pmatrix} d \\ s \\ b \end{pmatrix}
$$

Figure 4   Features of the Standard Model, as it applies to decay of the b quark.

M matrix; a frequently used one is shown in Equation 2. The angle $\theta_1$ is essentially the Cabibbo angle.

$$V_{ij} = \begin{pmatrix} c_1 & s_1c_3 & s_1s_3 \\ -s_1c_2 & c_1c_2c_3+s_2s_3e^{i\delta} & c_1c_2s_3-s_2c_3e^{i\delta} \\ -s_1s_2 & c_1s_2c_3-c_2s_3e^{i\delta} & c_1s_2s_3+c_2c_3e^{i\delta} \end{pmatrix}, \qquad 2.$$

where $c_i = \cos \theta_i$ and $s_i = \sin \theta_i$.

Within the framework of the Standard Model, b decay can be written schematically as $V_{cb}(b \to cW_V^-) + V_{ub}(b \to uW_V^-)$, with $W_V^-$ indicating a virtual $W^-$. Thus b decay is determined by the K-M matrix elements $V_{cb}$ and $V_{ub}$. A prime experimental goal of b decay studies is to measure these matrix elements.

### 3.3   QCD and Nonspectator Effects

The model of B decay in which the light antiquark plays no role (is a "spectator") is dubbed the spectator model. Diagrams for hadronic and semileptonic B decay, ignoring QCD, are shown in Figures 5a,b. QCD

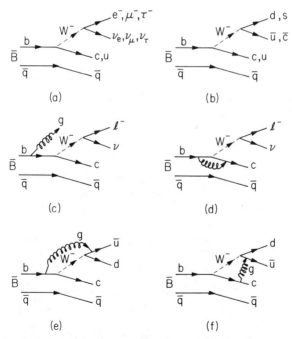

*Figure 5*   Spectator model diagrams of B decay. (*a,b*) Diagrams ignoring QCD; (*c,d*) QCD corrections due to gluon emission and gluon loops; (*e,f*) short-distance enhancement diagrams.

corrections to these diagrams due to gluon emission and gluon loops (Figure 5c,d) reduce the semileptonic and hadronic decay widths by 10 to 15%. Of more importance are QCD corrections of the variety shown in Figures 5e,f, gluon exchange between the quark line containing the b and the quarks emanating from the emitted W. Referred to as short-distance enhancements or hadronic enhancements, these processes alter the color flow. They can be summed in leading logarithms (36, 37); one finds they transform the weak Hamiltonian, e.g. for b → cūd, from

$$H_W = \frac{G_F}{\sqrt{2}} V_{cb}V_{ud}^*[\bar{c}_i\gamma^u(1-\gamma^5)b_i \cdot \bar{d}_j\gamma^u(1-\gamma^5)u_j]$$    3.

into

$$H'_W = \frac{G_F}{\sqrt{2}} V_{cb}V_{ud}^* \left\{ \frac{f_+ + f_-}{2} [\bar{c}_i\gamma^u(1-\gamma^5)b_i \cdot \bar{d}_j\gamma^u(1-\gamma^5)u_j] \right.$$
$$\left. + \frac{f_+ - f_-}{2} [\bar{d}_i\gamma^u(1-\gamma^5)b_i \cdot \bar{c}_j\gamma^u(1-\gamma^5)u_j] \right\}.$$    4.

The coefficients $f_+$ and $f_-$ are given (38) by

$$f_- = \frac{1}{f_+^2} = \left[\frac{\alpha_s(m_t)}{\alpha_s(m_W)}\right]^{12/21} \left[\frac{\alpha_s(m_b)}{\alpha_s(m_t)}\right]^{12/23},$$    5.

where $\alpha_s$ is the strong coupling constant. The total width for hadronic decay is increased by a factor $(2f_+^2 + f_-^2)/3$. For b decay, $f_+ \approx 0.85$, $f_- \approx 1.4$, and hadronic decays are enhanced by 10 to 15%, for $\Lambda_{QCD}$ in the range 150 to 300 MeV.

Decay processes in which the light antiquark takes an active role are called nonspectator processes; diagrams are shown in Figure 6. Leveille (37) gives a useful discussion. In the absence of gluon radiation, nonspectator contributions are suppressed by two powers of the mass of the heavier final-state quark or lepton. This helicity suppression is absent for diagrams with gluon radiation (Figure 6c,d).

All nonspectator processes, whether mediated by W exchange or annihilation, require the b and the light antiquark to coincide spatially, and hence these processes are proportional to the square of the B-meson decay constant $f_B$, defined by $\langle 0|\bar{b}\gamma^\mu\gamma_5 u|B^+\rangle = if_B P_B^\mu/\sqrt{2M_B}$. In a nonrelativistic approximation, $f_B^2 = 12|\psi(0)|^2/M_B$. Estimates (39) for $f_B$ range from 100 to 500 MeV.

The leptonic decay process $B^\pm \to \tau^\pm \nu$ proceeds by an annihilation diagram, and can be cleanly calculated:

$$\Gamma(B^\pm \to \tau^\pm \nu) = 24\pi^2 \left(\frac{f_B}{m_B}\right)^2 \Gamma_0^B |V_{ub}|^2 \left(\frac{m_\tau}{m_B}\right)^2 \left(1 - \frac{m_\tau^2}{m_B^2}\right)^2,$$    6.

where $\Gamma_0^B = G_F^2 M_B^5/192\pi^3$. This reaction provides a theoretically appealing method for determining $f_B^2$ once $|V_{ub}|^2$ is known.

Charged B mesons receive nonspectator contributions only from annihilation diagrams, which are of necessity proportional to $|V_{ub}|^2$. Since $|V_{ub}/V_{cb}|$ is known to be very small (see Section 6.4), it is expected that the hadronic decay width of charged B mesons will be correctly given by the spectator model. Neutral B mesons receive nonspectator contributions only from exchange diagrams, some of which are proportional to $V_{cb}$. Consequently it is expected that the hadronic decay width of neutral B mesons will be increased above the spectator model value, causing it to have a shorter lifetime than the charged B. Since there are no nonspectator contributions to B semileptonic decays, charged and neutral B's will have the same semileptonic decay widths, and the semileptonic decay branching ratios will differ by the same factor as the lifetimes:

$$\frac{\text{Br}(B^0 \rightarrow X\ell v)}{\text{Br}(B^\pm \rightarrow X\ell v)} = \frac{\tau_{B^0}}{\tau_{B^\pm}}. \qquad 7.$$

The nonspectator diagrams contributing to hadronic decays can be calculated in perturbative QCD. The results, however, are viewed as having only qualitative validity, because of the neglect of long-range, nonperturbative effects. Proceeding in this fashion, Leveille (37) suggests

$$\frac{\tau_{B^0}}{\tau_{B^\pm}} \approx \frac{1}{1+0.6(f_B/0.5)^2}, \qquad 8.$$

where $f_B$ is in GeV. If $f_B$ were at the high end of the range estimated (i.e. $\sim 0.5$ GeV), then $\tau_{B^0}/\tau_{B^\pm}$ would be approximately 0.6.

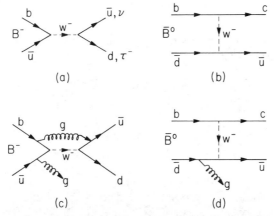

*Figure 6* Nonspectator diagrams for B decay: (*a*) annihilation; (*b*) exchange; (*c*) annihilation with gluon emission; (*d*) exchange with gluon emission.

## 3.4 $B^0 \bar{B}^0$ Mixing

The $B^0$ and $\bar{B}^0$ states can mix, as $K^0$ and $\bar{K}^0$ do, and their decays similarly can violate $CP$. The formalism for describing these processes is much like that for the neutral kaon system. The development given here follows that of J. S. Hagelin (40).

The phenomenological Hamilton can be written

$$
\begin{pmatrix}
M - \dfrac{i}{2}\Gamma & M_{12}^* - \dfrac{i}{2}\Gamma_{12}^* \\[2mm]
M_{12} - \dfrac{i}{2}\Gamma_{12} & M - \dfrac{i}{2}\Gamma
\end{pmatrix}
\begin{pmatrix}
\bar{B}^0 \\[1mm]
B^0
\end{pmatrix},
\qquad 9.
$$

where $\bar{B}^0$ has the quark content $b\bar{d}$, and $B^0$, $\bar{b}d$. The mass eigenstates are

$$
B_{1,2} = \frac{1}{\sqrt{2(1+|\varepsilon|^2)}}\,[(1+\varepsilon)\bar{B}^0 \pm (1-\varepsilon)B^0],
\qquad 10.
$$

with eigenvalues

$$
M_{1,2} = M \pm \mathrm{Re}\sqrt{\left(M_{12} - \frac{i}{2}\Gamma_{12}\right)\left(M_{12}^* - \frac{i}{2}\Gamma_{12}^*\right)},
$$

$$
\Gamma_{1,2} = \Gamma \mp \mathrm{Im}\sqrt{\left(M_{12} - \frac{i}{2}\Gamma_{12}\right)\left(M_{12}^* - \frac{i}{2}\Gamma_{12}^*\right)}.
\qquad 11.
$$

The $CP$ impurity parameter $\varepsilon$ is given by

$$
\varepsilon = \frac{-\mathrm{Re}\,M_{12} + \dfrac{i}{2}\,\mathrm{Re}\,\Gamma_{12} + \sqrt{\left(M_{12} - \dfrac{i}{2}\Gamma_{12}\right)\left(M_{12}^* - \dfrac{i}{2}\Gamma_{12}^*\right)}}{i\left(\mathrm{Im}\,M_{12} - \dfrac{i}{2}\,\mathrm{Im}\,\Gamma_{12}\right)}.
\qquad 12.
$$

The experimental parameters for weak mixing and $CP$ violation can be expressed in terms of $r$ and $\bar{r}$,

$$
r \equiv \frac{\Gamma(\bar{B}^0 \to XW_v^+)}{\Gamma(\bar{B}^0 \to XW_v^-)} = \left|\frac{1-\varepsilon}{1+\varepsilon}\right|^2 \frac{(\Delta M)^2 + (\Delta\Gamma/2)^2}{2\Gamma^2 + (\Delta M)^2 - (\Delta\Gamma/2)^2}
$$

$$
\bar{r} \equiv \frac{\Gamma(B^0 \to XW_v^-)}{\Gamma(B^0 \to XW_v^+)} = \left|\frac{1+\varepsilon}{1-\varepsilon}\right|^2 \frac{(\Delta M)^2 + (\Delta\Gamma/2)^2}{2\Gamma^2 + (\Delta M)^2 - (\Delta\Gamma/2)^2},
$$

$$
\qquad 13.
$$

where $\Delta M = M_1 - M_2$ and $\Delta\Gamma = \Gamma_1 - \Gamma_2$.

If a strong production process has led to a $B^0\bar{B}^0$ initial state, mixing manifests itself as weak decays of $B^0 B^0$ or $\bar{B}^0\bar{B}^0$, in addition to decays of

$B^0\bar{B}^0$. One defines a mixing parameter $y$ by

$$y \equiv \frac{N(B^0 B^0) + N(\bar{B}^0 \bar{B}^0)}{N(B^0 \bar{B}^0)},$$ 14.

where $N(B^0 B^0)$ is the number of produced $B^0 \bar{B}^0$'s that decay as $B^0 B^0$, and similarly for $N(\bar{B}^0 \bar{B}^0)$ and $N(B^0 \bar{B}^0)$. If mixing is complete, $y = 1$, while if there is no mixing, $y = 0$.

The expression for $y$ in terms of $r$ and $\bar{r}$ depends on the relative angular momentum of the initial $B^0$ and $\bar{B}^0$, because of Bose-Einstein statistics. Since $\Upsilon(4S)$ has spin 1, the $B^0 \bar{B}^0$ system into which it decays is in a p state. For this case

$$y = y(\text{odd}) = \frac{r + \bar{r}}{2}.$$ 15.

$CP$ violation can cause the rate for the mixing transition $B^0 \rightarrow \bar{B}^0$ to differ from that for $\bar{B}^0 \rightarrow B^0$. For an initial state consisting of $B^0 \bar{B}^0$, asymmetries in the number of $B^0 B^0$ vs $\bar{B}^0 \bar{B}^0$ decays, and in the number of $B^0$ vs $\bar{B}^0$ decays, might be observed. Any tag of $B^0$ vs $\bar{B}^0$ or $B^0 B^0$ vs $\bar{B}^0 \bar{B}^0$ may be used. The one common in the literature is the charge of the lepton in semileptonic $B^0/\bar{B}^0$ decay. Thus we have

$$a \equiv \frac{N(B^0 B^0) - N(\bar{B}^0 \bar{B}^0)}{N(B^0 B^0) + N(\bar{B}^0 \bar{B}^0)} = \frac{N_{++} - N_{--}}{N_{++} + N_{--}}$$ 16.

and

$$l^{\pm} \equiv \frac{N(B^0) - N(\bar{B}^0)}{N(B^0) + N(\bar{B}^0)} = \frac{N_+ - N_-}{N_+ + N_-} \qquad \text{(from } B^0/\bar{B}^0 \text{ decay).}$$ 17.

The two asymmetries are not independent, being related by the mixing parameter $y$:

$$l^{\pm} = a \frac{y}{1 + y}.$$ 18.

The asymmetries $a$ and $l^{\pm}$ can be expressed in terms of $r$ and $\bar{r}$. The expression for $l^{\pm}$ depends (as does that for $y$) on whether the initial $B^0 \bar{B}^0$ state is odd [e.g. $\Upsilon(4S)$] or even. The expression for $a$ does not depend on the initial state, since it only involves those events that have mixed. Equation 18, relating $a$ and $l^{\pm}$, remains valid, providing one uses $y(\text{odd})$ or $y(\text{even})$ as appropriate. The expressions are

$$a = \frac{r - \bar{r}}{r + \bar{r}},$$ 19.

and

$$l^{\pm} = \frac{r - \bar{r}}{2 + r + \bar{r}} \quad \text{(initial state odd)}. \qquad 20.$$

The asymmetry $a$ measures $CP$ violation in the events that have mixed, while $l^{\pm}$ is a combined measure of mixing and $CP$ violation. It is possible for $a$ to be large, but unmeasurable, because there is no mixing. The predicted size of $l^{\pm}$ is a better indication of the accessibility to detection of $CP$ violation.

The $B^0 \leftrightarrow \bar{B}^0$ transitions are induced by box diagrams with two $W^{\pm}$'s exchanged, as shown in Figure 7. Calculations of $M_{12}$ and $\Gamma_{12}$ yield

$$M_{12} = \frac{G_F^2 f_B^2 B_B m_B}{12\pi^2} \left[ \xi_t^2 \eta \left( m_t^2 + \frac{1}{3} m_b^2 + \frac{3}{4} m_b^2 \ln \frac{m_t^2}{m_b^2} \right) \right.$$

$$\left. + O\left( m_c^2, \frac{m_b^4}{m_t^2}, \frac{m_t^4}{M_W^2} \right) \right],$$

$$\Gamma_{12} = -\frac{G_F^2 f_B^2 B_B m_B}{8\pi} \left[ \frac{(2f_+^2 + f_-^2)}{3} \xi_t^2 m_b^2 \right.$$

$$\left. + \frac{8}{3} \left( \frac{3}{2} f_+^2 - f_+ f_- + \frac{1}{2} f_-^2 \right) \xi_t \xi_c m_c^2 + O\left( \frac{m_c^4}{m_b^2} \right) \right], \qquad 21.$$

where $\xi_i = V_{ib} V_{id}^*$, $i = t$ or c. The terms $f_+$ and $f_-$ are the QCD correction factors given in Section 3.3, and $\eta$ is another QCD correction factor, with value $\sim 0.8$. $B_B$ is the "bag parameter," defined by

$$B_B = \frac{\langle \bar{B}^0 | [\bar{b}\gamma_\mu(1 - \gamma_5)d]^2 | B^0 \rangle}{\langle \bar{B}^0 | \bar{b}\gamma_\mu(1 - \gamma_5)d | 0 \rangle \langle 0 | \bar{b}\gamma_\mu(1 - \gamma_5)d | B^0 \rangle}; \qquad 22.$$

it measures the extent to which the B meson spends its time in a state other than pure b$\bar{q}$. Anticipated values for $B_B$ range from 0.3 to 1.5.

More complete expressions, which take into consideration the momentum dependence in the W propagators, are given by Buras et al (38). The above expressions give $M_{12}$ to 20% accuracy or better for top-quark masses below 60 GeV and give $\Gamma_{12}$ to better than 1%; they are quite sufficient for our purposes.

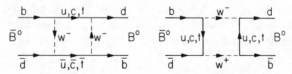

*Figure 7* Box diagrams leading to $B^0\bar{B}^0$ mixing.

## 3.5  CP Violation in B Partial Decay Rates

While the TPC theorem requires that the total decay rate for particle and antiparticle be equal, its requirements on partial decay rates are considerably weaker. There have been several theoretical investigations (41–44) of the possibility for $CP$ violation to reveal itself by an inequality of a partial decay rate $B^+$ vs $B^-$, or $B^0$ vs $\bar{B}^0$. For this to happen with tree-level decay diagrams, there must be two (or more) diagrams contributing to a given decay mode, each with differing weak interaction phases and differing strong interaction phases. Thus if

$$\langle f|T|i\rangle = g_1 M_1 e^{i\alpha_1} + g_2 M_2 e^{i\alpha_2}, \qquad 23.$$

where $g_i$ are the weak couplings, $\alpha_i$ are the strong interaction phases, and $M_i$ are real, we have

$$\langle \bar{f}'|T|\bar{i}'\rangle = g_1^* M_1 e^{i\alpha_1} + g_2^* M_2 e^{i\alpha_2}, \qquad 24.$$

where $\bar{i}'$ is the antiparticle of i with its spin reversed; this leads to

$$\Gamma - \bar{\Gamma} \approx \mathrm{Im}\,(g_1 g_2^*)\sin(\alpha_1 - \alpha_2). \qquad 25.$$

Bernabeau & Jarlskog (41) consider interference between the spectator diagram and the annihilation diagram, Figure 8. They note that effects will be largest for Cabibbo-suppressed decays. They examine the decay $B^- \to D^- D^{*0}$ and its charge conjugate and also a few two-body decays of $B_c$. Their estimate for the asymmetry in the rates for $B^- \to D^- D^{*0}$ vs $B^+ \to D^+ \bar{D}^{*0}$ ranges from a few percent to a few tens of percent, depending on assumptions about the strong interaction phases.

Bander, Silverman & Soni (42) consider the single-loop diagram with gluon emission (Figure 9), obtaining their phase from the loop integration. Interference occurs between the amplitudes from this diagram with

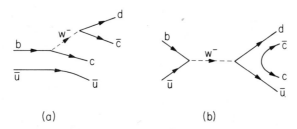

(a)                                    (b)

*Figure 8*  Spectator diagram (*a*) and annihilation diagram (*b*) for the decay process $B^- \to D^- D^{*0}$. Interference between these diagrams can cause the partial decay rate for $B^- \to D^- D^{*0}$ to differ from that for $B^+ \to D^+ \bar{D}^{*0}$ (see 41).

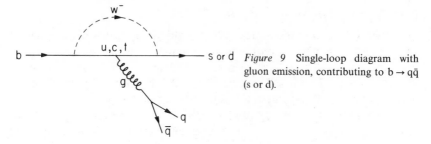

*Figure 9* Single-loop diagram with gluon emission, contributing to b → qq̄ (s or d).

different internal quark lines (u, c, t), and between this diagram and the free-quark diagram (Figure 5b). Asymmetries are estimated for b → duū, suū, scc̄, for which the free-quark diagram contributes, and for b → ddd̄, sss̄, from the single-loop diagram alone. Asymmetries range from a few percent to a few tens of percent.

Carter & Sanda (43) consider the cascade decays of B. They note that $D^0$ and $\bar{D}^0$ have common decay modes, e.g. $K_s^0 Y$, and consequently the cascade decays $B^- \to K_s X D^0 \to K_s X K_s Y$ and $B^- \to K_s X \bar{D}^0 \to K_s X K_s Y$ can interfere, giving a possible difference in the decay rates for $B^- \to K_s^0 K_s^0 XY$ and $B^+ \to K_s^0 K_s^0 \overline{XY}$. This occurs with spectator diagrams, suppressed by one power of a K-M angle—see Figure 10. They also note that the dominant decays $B^0 \to \bar{D}^0 X$ and $\bar{B}^0 \to D^0 X$ can cascade to identical final states $XYK_s$, resulting in possible interference between an initially produced $B^0$ and the $\bar{B}^0$ into which it mixes. The combination of mixing followed by *CP* violation in the decay leads to a different partial decay rate for initially produced $B^0$'s and $\bar{B}^0$'s.

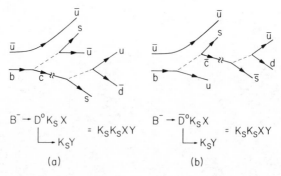

*Figure 10*   Spectator model diagrams for cascade decays of $B^-$, through $D^0$ or $\bar{D}^0$, to a common final state $K_s K_s XY$. Interference between these diagrams can cause the partial decay rate for $B^- \to (D) \to K_s K_s XY$ to differ from that for its charge conjugate (see 43).

## 4. EVIDENCE AGAINST NONSTANDARD DECAYS

Figure 11 gives a classification of all ways the b quark might decay. To decay by emitting a conventional boson, the b must either be in a doublet with a lighter charge $+2/3$ quark (e.g. $b_R$ with $c_R$), or mix with s and/or d, thereby getting a piece of itself into a doublet with a lighter charge $2/3$ quark ($c_L$ or $u_L$). If neither of these occur, then either the b is stable, or it decays exotically. If b mixes with s and/or d, then it can decay by a light charged Higgs (should one exist), or by $W^\pm$ and $Z^0$. In this last case, $b_L$ and $t_L$ either form a doublet or they do not. This list is believed to be complete; should the b decay in any other way, theory would require rather basic changes. The possibility that the b is stable has been eliminated by the observation of $\Upsilon(4S)$, which decays into multihadron final states.

Exotic decays of a double-leptonic variety are predicted by the models of Derman (32), and Georgi & Glashow (34). Since each decay contains two leptons, events from double-leptonic B decays could not avoid a high yield

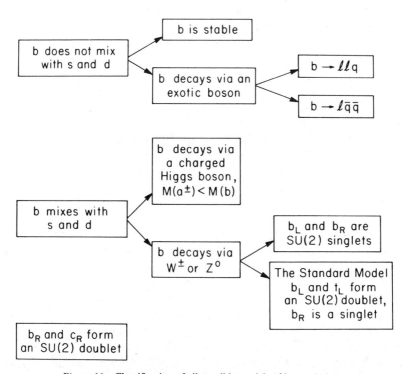

*Figure 11* Classification of all possible models of b-quark decay.

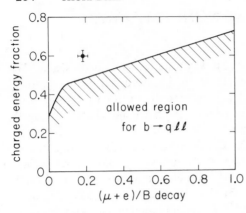

Figure 12   Charged energy fraction in B decay vs lepton yield in B decay, as allowed by double-leptonic decays and as measured. Results are from CLEO (45).

of either e, $\mu$, or $\nu$, or some combination of the three ($\tau$ if present, contributes $\nu$, e, and $\mu$). A high yield of neutrinos implies a large fraction of missing energy, and hence a low fraction of energy appearing in charged particles. One can lower the predicted yield of $\mu$ or e, but only by increasing the yield of $\nu$, thereby lowering the charged energy fraction. Figure 12 shows the region allowed by double-leptonic decays, and also CLEO measurements of muon and electron yields and mean charged energy fraction from $\Upsilon(4S)$ decays. The possibility that b decays double-leptonically 100% of the time is excluded at 99.7% confidence level (45).

Exotic decays of an antibaryonic variety are predicted by the model of Georgi & Machacek (33). Since each antibaryonic decay of a $\bar{B}$ contains one lepton and one antibaryon (ultimately a proton or a neutron), events from antibaryonic B decays must be rich both in $\mu$ or e or $\nu$, and in p or n. The neutrons contribute, along with the neutrinos, to a high missing energy, and consequently a low charged energy fraction. CLEO $\Upsilon(4S)$ measurements of mean charged energy fraction and yields of electrons, muons, protons, and (for good measure) lambdas rule out the possibility that b decays antibaryonically 100% of the time with a confidence level greater than 99.9% (45).

If the b mixes with s and/or d, and there is a charged Higgs (or technipion) $a^{\pm}$ with mass lighter than the b, then the dominant decay of the b would be $b \to a^{-} c$ (or u). The $a^{-}$ would decay into heavy fermion pairs, $\bar{c}s$ or $\tau \nu_{\tau}$. The former mode produces very few $\mu$ or e, far fewer than are observed by CLEO. If one introduces enough of the $\tau \nu$ mode to give the observed rate of $\mu$ and e, then it also gives many $\nu$, with a resulting charged energy fraction well below CLEO's observation. One cannot adjust the mix to get simultaneously a high enough $\mu$ and e yield and a high enough charged energy fraction. The possibility that b decays to a charged Higgs 100% of the time is excluded with 99.5% confidence (45).

The most conventional of the nonstandard models of b-quark decay is the Standard Model minus the top quark. The b quark is in a left-handed singlet (instead of the left-handed doublet with t). It mixes with s and/or d, and decays via $W^{\pm}$ and $Z^0$. Because b is in a singlet, the GIM mechanism is inoperative, and flavor-changing neutral currents will be present. The decays $b \to (s/d) \, \mu^+\mu^-$ and $b \to (s/d) \, e^+e^-$ cannot be avoided. Kane & Peskin (46) have shown that for this class of models $\text{Br}(B \to \ell^+\ell^- X)/\text{Br}(B \to \ell\nu X)$ must exceed 0.12. CLEO's (47) present upper limit on this ratio is 0.029 at 90% confidence level, 0.046 at 99.9% confidence level, convincingly excluding this class of models. Mark J and JADE (48) have reported upper limits of 0.07 at 95% confidence level.

The final nonstandard possibility in Figure 11 is for b, without mixing, to be in a weak isospin doublet with a lighter charge $+2/3$ quark. Since $u_L$, $c_L$ already have partners, the possibilities as pointed out by Peskin & Tye (49) are $u_R$ and $c_R$. Consider a mixture $c_R'' = \alpha u_R + \beta c_R$: $u_R$ cannot be the dominant component in $c_R''$, since this would spoil the agreement of the Standard Model $\bar{u}u$ neutral current with experiment. The u content of $c_R''$ must then be extremely small to suppress the $\bar{u}c$ neutral current coupling, whose presence would induce substantial $D^0\bar{D}^0$ mixing, contrary to experiment. Thus $c_R'' \approx c_R$. In its simplest form, the right-handed doublet model predicts a lifetime of order $10^{-15}$ s for B, and is excluded by the B lifetime measurements discussed in Section 6.3. There is, however, a variant of the model, with a fourth, heavier, charge $-1/3$ b''. The b and b'' mix, and it is a linear combination that forms the right-handed doublet with $c_R$. With an arbitrary mixing angle, a long lifetime is easily obtained.

The shape of the lepton momentum spectrum from semileptonic b decay predicted for $(b_R, c_R)$, i.e. $V + A$, differs from that for $(b_L, c_L)$ i.e. $V - A$. The two theoretical spectra are compared with CLEO data (50) in Figure 14. The Standard Model $V - A$ curve fits the data quite well. The $V + A$ curve does not fit the data. The possibility that b decays 100% of the time by a $V + A$ current is excluded (CLEO, unpublished) with a confidence greater than 99.9%.

One is forced by CLEO data to the conclusions that the Standard Model is the dominant piece of b decay and that the top quark must exist. Nonstandard decays could occur at reduced levels. They are worth searching for because of the illumination they would shed on the family problem.

# 5.   RECONSTRUCTED B MESONS, B MASSES

The bare b quarks produced at $\Upsilon(4S)$ fragment into b-flavored hadrons. CLEO (7, 51) has reconstructed a handful of B mesons that decay by the low-multiplicity hadronic modes $B^- \to D^0\pi^-$, $\bar{B}^0 \to D^0\pi^+\pi^-$, $\bar{B}^0 \to D^{*+}\pi^-$, and $B^- \to D^{*+}\pi^-\pi^-$ (and charge conjugate reactions). For the decays to $D^0$, the subsequent decay $D^0 \to K^-\pi^+$ was used, with $K^-$ identified by time of flight or specific ionization. For the decays to $D^{*+}$, the subsequent decay $D^{*+} \to D^0\pi^+$, $D^0 \to K^-\pi^+$ or $K^-\pi^+\pi^+\pi^-$ was used. Kaons were not identified; rather the trick of cutting on the $D^{*+} - D^0$ mass difference was utilized to suppress background.

The mass spectrum is shown in Figure 13; a peak at 5.273 GeV is evident. Background has been estimated by considering wrong charge combinations, by analyzing data obtained below $\Upsilon(4S)$, and by displacing the $D^0$ mass cut $\pm 200$ MeV from the correct $D^0$ mass. All methods are in agreement. The $D^0$ sideband estimate is shown in Figure 13.

To obtain the best value for masses, CLEO used only decays to $D^{*\pm}$, because apparent decays to $D^0$ include contamination from decays to $D^{*0}$ with subsequent decays $D^{*0} \to D^0\pi^0$, $D^0\gamma$. This contamination shifts the measured mass by a few MeV. The CLEO analysis constrained the B energy to the beam energy, and thus measured the difference between the $\Upsilon(4S)$

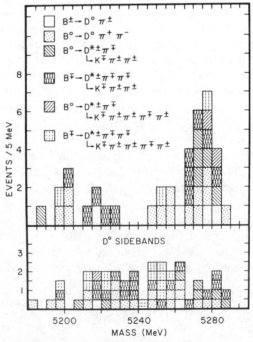

*Figure 13*  Mass spectrum of reconstructed B mesons. Background as estimated by the $D^0$ sideband method is also shown. Results are from CLEO (51).

**Table 3** B-meson masses, as measured by
CLEO (51)

| Particle | Mass (MeV) |
|---|---|
| $\langle B \rangle$ | $5273.0 \pm 1.3 \pm 2.0$ |
| $B^-$ | $5271.2 \pm 2.2 \pm 2.0$ |
| $\bar{B}^0$ | $5275.2 \pm 1.9 \pm 2.0$ |
| $\bar{B}^0 B^-$ difference | $4.0 \pm 2.7 \pm 2.0$ |

mass and twice the B meson mass to be $31.7 \pm 2.9 \pm 4.0$ MeV. The mass of
$\Upsilon$(4S) has been determined by making use of the energy scale provided by
beam depolarization measurements made at VEPP-4, DORIS II, and
CESR (52). The resulting B masses are given in Table 3.

Theoretical predictions (18, 53) for the $\bar{B}^0 - B^-$ mass difference range
from 2.3 to 5.8 MeV. The CLEO data are not precise enough to test these
predictions. As discussed in Section 2.2.1, the measured difference between
twice the B meson mass and $M[\Upsilon(4S)]$ can be used, in conjunction with a
theoretical prediction for the $\bar{B}^0 - B^-$ mass difference, to infer the relative
proportion of decays of $\Upsilon$(4S) to $B^0\bar{B}^0$ compared to $B^+B^-$.

# 6. DETERMINATION OF STANDARD MODEL PARAMETERS

According to the Standard Model, the b quark mixes with s and/or d, and
decays by $b \rightarrow cW_V^-$ and/or $b \rightarrow uW_V^-$, with the virtual $W^-$ being par-
titioned among $e^- \nu_e$, $\mu^- \nu_\mu$, $\tau^- \nu_\tau$, $d\bar{u}$, $s\bar{c}$, $d\bar{c}$, and $s\bar{u}$.

The most basic quantities to determine experimentally are the rates for
$b \rightarrow cW_V^-$ and $b \rightarrow uW_V^-$. These rates cannot be calculated from the Stand-
ard Model alone, as they are respectively proportional to $|V_{cb}|^2$ and $|V_{ub}|^2$,
both functions of the K-M angles, which are free parameters of the
theory. Rather, measurement of these rates helps determine the K-M
angles.

Of secondary importance, one would like to determine the partitioning of
the virtual $W^-$ among its various decay modes. This partitioning is
calculable from the Standard Model (except for the Cabibbo-suppressed
modes $d\bar{c}$ and $s\bar{u}$, which require a knowledge of the Cabibbo angle $\theta_1$), and
in principle would provide a check on the theory. In practice, the
partitioning depends in part on QCD corrections and nonspectator effects,
and so it is these aspects of the theory that would be tested.

To date there have been measurements of the B semileptonic decay
branching ratios $Br(B \rightarrow X\mu\nu)$ and $Br(B \rightarrow Xe\nu)$, and the B lifetime $\tau_B$.

Additionally, upper limits on the ratio of the rates for $b \to uW_V^-$ and $b \to cW_V^-$ [hereafter, the $(b \to u)/(b \to c)$ ratio] have been obtained.

## 6.1   $(b \to u)/(b \to c)$

A variety of methods have been attempted to measure $(b \to u)/(b \to c)$. These include (1.) the kaon yield, (2.) the charmed-meson yield, (3.) a search for exclusive B decays not containing charm, (4.) the lepton momentum spectrum in semileptonic b decays, and (5.) lepton-hadron correlations in semileptonic b decays. All have been carried out at CESR, using B's from $\Upsilon(4S)$.

### 6.1.1   KAON YIELD

Kaons are a tag for charm, and consequently a high kaon yield in B decay indicates a dominance of $b \to c$ over $b \to u$. Kaons will also arise from $W_V^- \to \bar{c}s$, and from $s\bar{s}$ pairs popped from the ocean as the primary quarks from B decay hadronize. From model calculations, one expects 0.7 to 0.8 kaons per B decay from $b \to u \; W_V^-$, and 1.44 to 1.55 kaons per B decay from $b \to c \; W_V^-$. CLEO measures (54) $1.45 \pm 0.095 \pm 0.14$, from which one infers $(b \to u)/(b \to c) = 0.05 \pm 0.20$.

This method is of historical importance, in that it was the high kaon yield observed by CLEO that first signalled the dominance of $b \to c$ over $b \to u$ (16, 55). However, the accuracy of the method is limited by uncertainties in the model calculation, and it can no longer compete with methods (4.) and (5.) discussed below.

### 6.1.2   CHARMED-MESON YIELD

CLEO (56; submitted for publication) has measured the yields of $D^0/\bar{D}^0$ and $D^{*\pm}$ mesons from B decay, using the decay modes $D^0 \to K^-\pi^+$ and $D^{*+} \to \pi^+D^0$, $D^0 \to K^-\pi^+$. The results, $0.56 \pm 0.14$ $D^0/\bar{D}^0$ per B decay and $0.23 \pm 0.05$ $D^{*\pm}$ per B decay, are based on assumed branching ratios for $D^{*+} \to \pi^+D^0$ and $D^0 \to K^-\pi^+$ of 60% and 4.5%, and scale inversely with these values. As just given, the results on D's per B decay do *not* include in their quoted errors the contribution from the uncertainties in the D branching ratios. These branching ratios currently have relative errors of $\pm 25$ and $\pm 20\%$, respectively.

Without a measurement of the $D^\pm$ yield from B decay, one must make some plausible assumptions in order to obtain $(b \to u)/(b \to c)$. If one assumes that the ratio $r$ of $D^+$ to $D^0$ from direct production (as distinct from $D^+$ or $D^0$ from $D^*$ decay) equals the ratio of $D^{*+}$ to $D^{*0}$, then the charm yield equals $(1 + r)(D^0 - b_+ D^{*+})$, where $D^0$ and $D^{*+}$ are the yields of $D^0$ and $D^{*+}$, and $b_+$ is the branching ratio for $D^{*+} \to \pi^+D^0$. The charge of the B and the charge of the D to which it decays are negatively correlated, i.e. $B^\pm$ tends to decay to $D^0/\bar{D}^0$, $B^0/\bar{B}^0$ tends to decay to $D^\pm$. As the correlation between B and D charge varies from 0 to $-1$, $r$ will vary from 1 to the $B^0\bar{B}^0/B^+B^-$ production ratio. Using $r = 0.85 \pm 0.15$, one finds $0.8 \pm 0.3$ D + D* per B decay. To get b → $c W_V^-$, one must add F and charmed-baryon production, and subtract all charm production arising from $W_V^- → \bar{c}s$. These are both likely to be small. Ignoring them, one has (b → u)/(b → c) = $0.25 \pm 0.38$.

Again, we see the dominance of b → c over b → u, but again there are rather large errors. Because this technique, like the kaon yield, focuses on detecting the b → c decay mode, it will not be useful in measuring a small b → u component. We now consider methods based on signals of b → u.

6.1.3 EXCLUSIVE B DECAYS NOT CONTAINING CHARM    CLEO (51) has searched for the exclusive B decays $B^0 → \pi^+\pi^-$ and $B^\pm → \rho^0\pi^\pm$, found evidence of neither, and set upper limits on the branching ratios of 0.05% for the former and 0.06% for the latter.

It is not obvious how to convert these limits to limits on b → u. Formally, one can assume that the fraction of b → u decays that form two-body final states is related to the fraction of b → c decays that form two-body final states:

$$\frac{Br(B^0 → \pi^+\pi^-)}{Br(b → u W_V^-)} = X_1 \frac{Br(B^- → D^0\pi^-)}{Br(b → c W_V^-)},$$

and

$$\frac{Br(B^- → \rho^0\pi^-)}{Br(b → u W_V^-)} = X_2 \frac{Br(\bar{B}^0 → D^{*+}\pi^-)}{Br(b → c W_V^-)}, \qquad 26.$$

thus hiding one's ignorance in $X_1$ and $X_2$. As shown in Section 8.1, the two-body final states containing D or D* typically have 1% branching ratios, so (b → u)/(b → c) is less than $0.05/X$. Because light quark fragmentation is softer than heavy quark fragmentation, $X_1$ and $X_2$ are assuredly less than 1, perhaps much less than 1. Without knowing more about $X$, one cannot get a limit on (b → u)/(b → c).

6.1.4 LEPTON MOMENTUM SPECTRUM    The method of measuring (b → u)/(b → c) that has so far proved most successful is based on the shape of the lepton momentum spectrum in semileptonic B decay. Here the distinguishing property between u and c is the large mass difference. Because the mass of the u quark is negligible compared to that of the b quark while the mass of the c quark is not, b → $u\ell v$ has a higher end-point momentum than b → $c\ell v$.

The dressing of the quarks complicates this simple picture, and it is not known with precision what the effective mass of the hadronic systems will be. Poling (57) has studied the problem with a spectator-quark model in which the initial-state b and light spectator antiquark are in relative motion, with momentum distributed according to a Gaussian. The mass of the b quark is not fixed, but is allowed to vary to conserve momentum and energy. Spectator antiquark and final-state quark have fixed masses. Altarelli et al (58) use the same spectator-quark model, and also investigate QCD corrections. Poling and Thorndike (unpublished) have used a model in which the b-quark mass is fixed.

Conclusions from these studies are:

1. QCD corrections are not very important for b → cℓν. For b → uℓν their effect is well approximated by assigning a mass of 500 MeV to the u quark.
2. The mass of the charmed hadron system $X_c$ is in the range of the D and D* masses.
3. The mass of the hadron system formed by the u quark ($X_u$) is sensitive to the assumptions about the b-quark mass. For the variable b-quark mass, or (equivalently) for a fixed mass of 5.0 GeV, $M(X_u)$ averages 0.9 GeV,

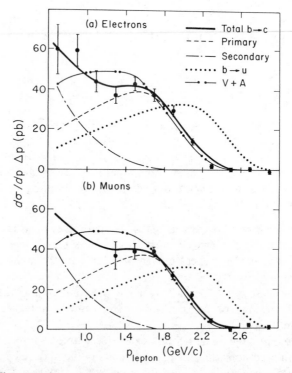

*Figure 14* Electron and muon momentum spectra from B mesons at Υ(4S). The heavy solid curve (total b → c) is the sum of leptons from b → cℓν (primary), and from b → cX, c → sℓν (secondary). Results are from CLEO (50).

while for a fixed b-quark mass of 4.7 GeV, $M(X_u)$ averages 1.3 GeV, and for a b-quark mass of 4.4 GeV, $M(X_u)$ averages 1.7 GeV, no longer small compared to $X_c$.

CLEO (50) has measured both electron and muon spectra in semileptonic B decay, and CUSB (59) has measured the electron spectrum. CLEO results are shown in Figure 14. The solid curve assumes 100% b → c, and includes leptons from the decay of D's produced in B decay. The dotted curve is 100% b → u, following Altarelli et al. There is no evidence for an excess of events above 2.2 GeV, the indication expected for b → u. CLEO has obtained upper limits using the model of Altarelli et al and also using a model in which B → $X_u\ell v$ with $M(X_u)$ fixed. Results are shown in Figure 15. The upper limit following Altarelli et al is (b → u)/(b → c) < 0.04, at 90% confidence level. Using the fixed $M(X_u)$ model, a comparable limit is obtained for $M(X_u)$ less than 0.9 GeV; for $M(X_u)$ = 1.3 GeV, the limit deteriorates to 0.09.

The CUSB group interpret their electron spectrum using the model of Altarelli et al. They find no evidence for b → u, and obtain an upper limit for (b → u)/(b → c) of 0.055.

6.1.5 LEPTON-HADRON CORRELATIONS    If the hadronic system resulting from semileptonic B decay via b → $u\ell v$ is light, then the decay modes B → $\pi\ell v$ and B → $\rho\ell v$ will be important contributors. Indeed, the calculated lepton spectrum resulting from a mix of 1/3 B → $\pi\ell v$, 1/3 B → $\rho\ell v$,

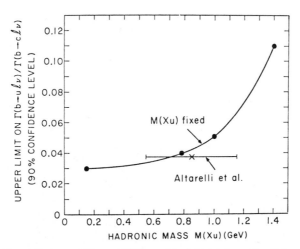

*Figure 15*  Upper limits on $\Gamma(b → u\ell v)/\Gamma(b → c\ell v)$ obtained from fits of the electron and muon momentum spectra with model of Altarelli et al (58), and from a model in which B → $Xu\ell v$ with $M(X_u)$ fixed.

and 1/3 B → $A_1 \ell v$ agrees well with the b → u spectrum calculated according to Altarelli et al.

CLEO is searching for the decay $B^0/\bar{B}^0 \to \pi^{\mp} \ell^{\pm} v$, by studying $\pi^{\mp} \ell^{\pm}$ correlations, and similarly searching for $B^{\pm} \to \rho^0 \ell^{\pm} v$ by studying $\rho^0 \ell^{\pm}$ correlations. The technique is to examine the three-dimensional space defined by the lepton and hadron momenta and their opening angle, looking for regions lightly populated by B → $D/D^* \ell v$, but substantially populated by the sought-after reactions. So far, no evidence for B → $\pi \ell v$ or $\rho \ell v$ has been found. Preliminary upper limits on (b → u)/(b → c) obtained in this way are weaker than those obtained from the lepton spectrum, and are equally dependent on a knowledge of the mass distribution of $X_u$.

In summary, there is no evidence from B decay studies that (b → u)/(b → c) is different from zero. The best upper limit comes from the lepton moment spectrum, and is normally quoted as 0.04. However, this limit is sensitive to the value of the b quark mass used in the spectator model calculation of the b → u lepton spectrum, and deteriorates a factor of two if a quark mass value of 4.7 GeV is used instead of 5.0 GeV.

## 6.2 B Semileptonic Decay Branching Ratio

Both CLEO (50) and CUSB (59, 60) have measured the branching ratios for B → Xev and B → Xμv, using leptons identified in Υ(4S) events. The most recent results are given in Table 4. The systematic errors of the four measurements are largely independent, and are combined in quadrature in obtaining the weighted average given in the table.

Table 4  B semileptonic decay branching ratio measurements

| | | Branching ratio (%) | |
|---|---|---|---|
| Group | Ref. | B → Xev | B → Xμv |
| CLEO | 50 | $12.0 \pm 0.7 \pm 0.5$ | $10.8 \pm 0.6 \pm 1.0$ |
| CUSB | 58, 59 | $13.2 \pm 0.8 \pm 1.4$ | $11.2 \pm 0.9 \pm 1.0$ |
| Υ(4S) combined | | $11.7 \pm 0.6$ | |
| Mark II | 21 | $13.5 \pm 2.6 \pm 2.0$ | $12.6 \pm 5.2 \pm 3.0$ |
| MAC | 61 | $11.3 \pm 1.9 \pm 3.1$ | $12.4 \pm 1.8 \pm 2.2$ |
| DELCO | 23 | $14.6 \pm 2.8$ | |
| TPC | 24 | $11.0 \pm 1.8 \pm 1.0$ | $15.2 \pm 1.9 \pm 1.2$ |
| Mark J | 25 | | $10.5 \pm 1.5 \pm 1.3$ |
| TASSO | 26 | $11.1 \pm 3.4 \pm 4.0$ | $11.7 \pm 2.8 \pm 1.0$ |
| CELLO | 27 | $14.1 \pm 5.8 \pm 3.0$ | $8.8 \pm 3.4 \pm 3.5$ |
| Continuum, combined | | $12.3 \pm 0.9$ | |

Branching ratios have also been obtained by PEP and PETRA groups (21–27, 61), following the method described in Section 2.2.3. To obtain a branching ratio, these experiments must assume the theoretical value for the b production cross section (i.e. 1/3 unit of $R$). Their results, given in Table 4, are consistent with the $\Upsilon(4S)$ determinations and provide evidence that the continuum production cross section is in rough accord with theory. Alternatively, one may assume theory, and conclude from the near equality of the mean branching ratios that either all varieties of b-flavored hadrons have approximately the same branching ratio, or the mix of varieties at $\Upsilon(4S)$ is not too different from the mix in the continuum. [For example, if one neglects $B_s$ and b-flavored baryon production from the continuum, and further assumes that the semileptonic decay branching ratios for charged and neutral B's are in the ratio of 2:1 (see Section 7.2), then the measured ratio of $\Upsilon(4S)$ and continuum values for the mean semileptonic decay branching ratio implies that the $B^0\bar{B}^0/B^+B^-$ production ratio at $\Upsilon(4S)$ lies between 45%/55% and 70%/30%.]

## 6.3    B Lifetime

In the summer of 1983, the MAC and Mark II groups at PEP reported measurements of the mean lifetime of b-flavored hadrons in excess of $10^{-12}$ seconds (62). Since that time, those groups have refined their measurements (63; private communication from W. T. Ford) and three additional groups—DELCO (64) at PEP and TASSO (65) and JADE (66) at PETRA—have also obtained results.

The b-flavored hadrons used in these experiments are produced in the reaction $e^+e^- \to b\bar{b}$ at center-of-mass energies ranging from 29 to 43 GeV. Each experiment first selects a data sample enriched in $b\bar{b}$ relative to $c\bar{c}$, $u\bar{u}$, $d\bar{d}$, and $s\bar{s}$, and then obtains a measure of the decay path.

In all but one case, an impact-parameter analysis was used to measure the decay path. The impact parameter of a track is the distance of closest approach of that track to the beam line. Figure 16 shows a projection of a B decay onto the plane perpendicular to the beam line. The impact parameter $d$ equals $l \sin \psi$, where $l$ is the projected decay path and $\psi$ is the projected angle between the initial B direction and the direction of the track. As the B momentum increases, $\langle l \rangle$ increases as $\beta\gamma$ and $\langle \psi \rangle$ decreases as $1/\gamma$ so that to a first approximation $\langle d \rangle$ is independent of the momentum distribution of the B.

If one knows the axis of the initial B direction, one can assign a sign to the impact parameter, positive (negative) if a forward (backward) laboratory decay angle would be inferred for the track, assuming the parent particle traveled along the axis toward its intersection with that track. Since most B

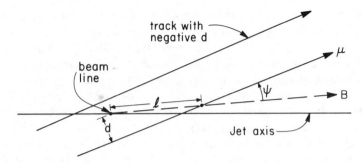

*Figure 16*  Projection of B decay onto plane perpendicular to beam line, showing relation between impact parameter $d$, projected decay path $l$, and projected decay angle $\psi$. A track with negative impact parameter is also shown.

decay products will have forward-going decay angles, the true impact parameter for most tracks from B decays will be positive.

For all experiments performed so far, the experimental error on an individual impact-parameter measurement is large compared to the mean impact parameter expected from the B lifetime. The finite lifetime shows up as a small positive shift in the impact-parameter distribution, resulting in a slight excess in tracks with positive impact parameters over those with negative impact parameters (see Figure 17). It is absolutely essential to assign a sign to the impact parameter; with present experimental resolutions, no useful result can be obtained from the distribution in the magnitude of the impact parameter.

The jet axis of the event (thrust axis or sphericity axis) is used as an approximation to the axis for the B direction. When the angle of the track lies between the jet axis and the true B-direction axis, a positive impact parameter will incorrectly be called negative, which will dilute the desired effect.

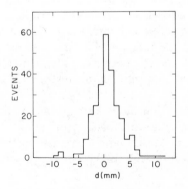

*Figure 17*  Impact-parameter distribution obtained by Mark II (63).

Four of the experiments, Mark II, MAC, DELCO, and JADE, obtain their b-enriched data sample by the procedure described in Section 2.2.3, i.e. they select hadronic events that contain leptons, and further require that the lepton have a large transverse momentum relative to the jet axis. If the lepton instead has a small transverse momentum but a large longitudinal momentum, a c-enriched, b-depleted sample is obtained.

These four groups use the impact parameter of the lepton track as their measure of the decay path. The impact-parameter distribution obtained by Mark II is shown in Figure 17; the other groups obtain similar distributions, all characterized by a positive shift that is small compared to the width of the distribution. One must stare at the distribution for a while and compare symmetrically located positive and negative bins to convince oneself that a statistically compelling effect has been seen.

The JADE group performs a second analysis, as follows. Each hadronic event containing a muon is assigned a probability that the event is $b\bar{b}$. This probability is based both on the transverse momentum of the muon, and on the event shape—$b\bar{b}$ jets are broad, producing a sizeable aplanarity. Two probability-weighted impact-parameter distributions are formed, one weighted with the probability that the event is $b\bar{b}$ (the signal distribution), and the other weighted with the probability that the event is not $b\bar{b}$ (the noise distribution). In one variant of the analysis, the impact parameters of all tracks in the event are used, rather than just the lepton track impact parameter. The quoted JADE result is an average over their various methods; it is preliminary.

TASSO did not use lepton identification to obtain their b-enriched sample, but rather used the event shape, obtaining a b-enriched sample that is 32% $b\bar{b}$ and a b-depleted sample that is 6% $b\bar{b}$. They used all well-fitted high-momentum tracks for their impact-parameter distribution. Since the charged multiplicity in B decay is more than twice that of D decay, using all tracks enhances the sensitivity to the B lifetime, relative to the D lifetime. As with all the experiments, Monte Carlo simulation was used to relate the impact-parameter distribution to the B lifetime.

Mark II has obtained a second measurement of the b lifetime by using a jet vertex technique in place of the impact-parameter analysis. They use the same b-enriched data sample as for the impact-parameter analysis. Each event is then divided into two jets by a plane perpendicular to the jet axis. After discarding tracks from $K^0$ and $\Lambda$, a vertex is formed from all good high-momentum tracks from each jet. The signed distance from the interaction point to the jet vertex *along the jet axis* is determined, and the distribution in this variable is compared against Monte Carlo distributions to obtain the B lifetime.

The six measurements just described are given in Figure 18. Each is an

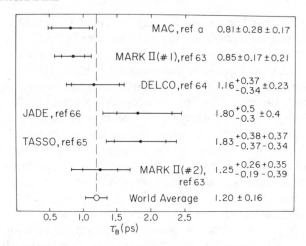

*Figure 18*   Lifetime of b-flavored hadrons. Reference a indicates private communication from W. T. Ford.

average of lifetimes of the various b-flavored hadrons present in $e^+e^- \to b\bar{b}$, with weighting determined both by production cross section and analysis procedure. In particular, the analysis procedures used by MAC, Mark II (#1), DELCO, and JADE give a weighting proportional to the semi-leptonic decay branching ratio, that used by TASSO gives relatively constant weighting factors, and that used by Mark II (#2) gives weighting factors midway between these two. The six measurements agree with each other within errors. They are combined[2] into a world average in Figure 18.

## 6.4   Kobayashi-Maskawa Mixing Angle Determinations

To obtain information on the K-M mixing matrix from the measurements just described, one combines the B lifetime and semileptonic decay branching ratio measurements to obtain the B semileptonic decay partial width, bypassing the need to understand anything about hadronic decays of B.[3]

---

[2] The systematic errors on the lifetime measurements are in part percentage errors, and as such are understated for the smaller measured lifetimes and overstated for the larger measured lifetimes. A straight weighting by combined statistical and stated systematic errors gives too heavy a weighting to the smaller measured lifetimes. I have corrected for this bias in computing the world average.

[3] There is a swindle here, in that the weighting of b-flavored hadrons that is used for the lifetime measurements, i.e. the mix found in continuum production as weighted by the analysis procedure, differs from the mix used for the semileptonic decay branching ratio measurements, i.e. the mix found in $\Upsilon(4S)$ decay. I shift the lifetime measurements downward by $0.05 \pm 0.05$ ps to correct for this weighting.

The B meson, in its semileptonic decay, is treated as a free, point-like b quark. The formulas for $\Gamma_{B \to X\ell\nu}$ and $(b \to u)/(b \to c)$ are:

$$\Gamma_{B \to X\ell\nu} = \frac{G_F^2 m_b^5}{192\pi^3} (|V_{ub}|^2 + f_{ps}|V_{cb}|^2),$$    27.

and

$$\frac{b \to u}{b \to c} = \frac{Br(b \to u\ell\nu)}{Br(b \to c\ell\nu)} = |V_{ub}|^2/(f_{ps}|V_{cb}|^2).$$    28.

The mass and phase space factor in these formulas hide our considerable ignorance of semileptonic B decay. It is unclear whether the $m_b$ in Equation 27 should be a constituent-quark mass or a running-quark mass $m_b(\mu)$, and, if the latter, what the appropriate energy scale $\mu$ should be. Since the constituent mass is above 5.0 GeV and the running mass $m_b(m_b) = 4.25$ GeV (67), and since the mass is raised to the fifth power, the choice makes a considerable difference. It is also unclear what masses to use for $m_b$ and $m_c$ when calculating the phase space suppression factor $f_{ps}$. The theoretical literature contains little discussion of these issues.

In what follows, I use $m_b = 4.8 \pm 0.2$ GeV in Equation 27. This value for $m_b$ corresponds to a running mass at an energy scale $\mu$ of 2 GeV, typical of momentum transfers in B decay. The range allowed by the error includes the mass values chosen by many (but not all) other workers. The phase space factor is computed using physical meson masses in the reaction $B \to D\ell\nu$. Finally, I take the upper limit on $(b \to u)/(b \to c)$ as 0.09, appropriate for an effective b-quark mass in $b \to u\ell\nu$ of 4.7 GeV.

The error on $m_b^5$ is $\pm 21\%$, larger than the $\pm 15\%$ error in the B lifetime. Further progress in determining K-M angles requires improvements in theory as well as in experiment.

Using the values $\tau_B = 1.15 \pm 0.17$ ps and $Br(B \to X\ell\nu) = 0.117 \pm 0.006$, and the upper limit $(b \to u)/(b \to c) < 0.09$ in Equations 27 and 28, one obtains the restrictions on the $V_{ub}$-$V_{cb}$ plane shown in Figure 19. For scale, recall that $V_{us} = 0.23$. The allowed region of the $V_{ub}$-$V_{cb}$ plane is very small indeed. While neither $\theta_2$ nor $\theta_3$ is well determined, both are shown to be substantially smaller than $\theta_1$.

Specifically, one has

$$|V_{ub}| = s_1 s_3 < 0.012,$$

$$|V_{cb}| = |s_3 - s_2 e^{i\delta}| = 0.053 \pm 0.010,$$

$$s_3 < 0.05,$$

$$s_2 < 0.12.$$

The implications of these limits are discussed in Section 9.

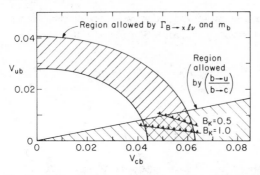

*Figure 19* $V_{ub}$-$V_{cb}$ plane, showing regions allowed by B lifetime measurement and by (b → u)/(b → c) upper limit, and showing their intersection (*cross-hatched*). Also indicated (*lines with points*) are those regions allowed by Re($\varepsilon$), for two values of $B_K$, assuming $m_t$ = 40 GeV. (See text, Section 9.2.)

# 7.   EFFECTS DUE TO THE LIGHT QUARK IN B = b̄q

There are several effects in B decay that are made possible by the active participation of the light quark bound to the b̄ quark in the B meson. These include (1.) the decay $B^{\pm} \to \tau^{\pm}\nu$; (2.) a difference in the lifetimes of charged and neutral B mesons; (3.) $B^0\bar{B}^0$ mixing; and (4.) *CP* violation in B meson decays.[4]

## 7.1   $B^{\pm} \to \tau^{\pm}\nu$

Equation 6, an expression for the partial decay width of the mode $B^{\pm} \to \tau^{\pm}\nu$, can be combined with Equation 27, an expression for the semileptonic decay width, and evaluated numerically, to give the convenient ratio

$$\frac{\text{Br}(B^{\pm} \to \tau^{\pm}\nu)}{\text{Br}(B \to X\ell\nu)} = 0.19\left(\frac{m_B}{m_b}\right)^5\left(\frac{f_B}{0.5}\right)^2\left(\frac{b \to u}{b \to \text{all}}\right), \qquad 29.$$

where $f_B$ is in GeV. Since $f_B$ is believed on theoretical grounds to be 0.5 GeV or less, (b → u)/(b → all) is believed on experimental grounds to be 0.09 or less, and Br(B → X$\ell\nu$) is measured to be 11.7%, one sees that Br($B^{\pm} \to \tau^{\pm}\nu$) will be less than 0.3%. Present experimental upper limits, from CLEO (68), are two orders of magnitude larger, and thus not interesting. Although a measurement of the $B^{\pm} \to \tau^{\pm}\nu$ branching ratio, combined with an in-

---

[4] Some manifestations of *CP* violation in B decay do not involve the active participation of the light quark—see Bander et al (42).

dependent measurement of (b → u)/(b → c) would provide a theoretically clean determination of $f_B$, the experimental difficulties are appreciable and make it unlikely that such a measurement will be forthcoming soon.

## 7.2  $B^\pm/B^0$ Lifetime Ratio

As noted in Section 3.3, the ratio of the lifetimes of the charged and neutral B mesons is equal to the ratio of their semileptonic decay branching ratios. Attempts at measurements have so far been limited to the ratio of branching ratios, and are of three varieties.

7.2.1  MEAN SEMILEPTONIC DECAY BRANCHING RATIO    If one could reliably calculate the b semileptonic decay branching ratio (i.e. the B semileptonic decay branching ratio ignoring nonspectator effects), then one could interpret the difference between that calculation and the measured charge averaged semileptonic decay branching ratio as indicative of nonspectator effects. These effects will lower the neutral B semileptonic decay branching ratio, but will not affect the charged B. Davies & Tye (69) have performed such a calculation, using three different sets of quark masses and two values of $\Lambda_{QCD}$. Their results, both for B decay and D decay, are given in Table 5. Since there are no nonspectator contributions to $D^\pm$ decay, they choose among the six sets of computations by requiring agreement with the measured (70) charged D semileptonic decay branching ratio, $19^{+4}_{-3}\%$. They obtain a semileptonic decay branching ratio for $B^\pm$ of $14-15\frac{1}{2}\%$. Since the measured average of charged and neutral B semileptonic decay branching ratios is $11.7\pm0.6\%$, this calculation strongly suggests substantial nonspectator effects, a ratio of charged to neutral branching ratios as large as 2.0.

Table 5  Calculated B semileptonic decay branching ratios, ignoring nonspectator effects, for three sets of quark masses, and two values of $\Lambda_{QCD}$ (69)

|  | I | II | III |
|---|---|---|---|
| $m_b$ (GeV) | 4.8 | 5.0 | 5.2 |
| $m_c$ (GeV) | 1.35 | 1.6 | 1.7 |
| $m_s$ (GeV) | 0.15 | 0.3 | 0.5 |
| $m_{u,d}$ (GeV) | 0 | 0.15 | 0.35 |
| Br(D → X$\ell\nu$) |  |  |  |
| $\Lambda = 0.2$ GeV | 0.12 | 0.16 | 0.26 |
| $\Lambda = 0.4$ GeV | 0.08 | 0.11 | 0.21 |
| Br(B → X$\ell\nu$) |  |  |  |
| $\Lambda = 0.2$ GeV | 0.13 | 0.14 | 0.16 |
| $\Lambda = 0.4$ GeV | 0.12 | 0.13 | 0.15 |

7.2.2  DILEPTONS    The yield of leptons from $B\bar{B}$ decay at $\Upsilon(4S)$ is proportional to the mean semileptonic decay branching ratio $\bar{b} = f_\pm b_\pm + f_0 b_0$, where $f_\pm$ ($f_0$) is the fraction of charged (neutral) B's produced, and $b_\pm$ ($b_0$) is the semileptonic decay branching ratio of the charged (neutral) B meson. The yield of dileptons is proportional to the mean squared branching ratio $\bar{b^2} = f_\pm b_\pm^2 + f_0 b_0^2$. If $b_\pm$ and $b_0$ are different, $\bar{b^2}$ will exceed $(\bar{b})^2$. By measuring the yield of both leptons and lepton pairs, CLEO (47) determined $\bar{b^2}/(\bar{b})^2$, and from this obtained information on $b_\pm/b_0$. They find $\bar{b^2}/(\bar{b})^2$ to be consistent with 1.0, and less than 1.21 at 90% confidence level. Assuming $\Upsilon(4S)$ decays 50% into $B^+B^-$ and 50% into $B^0\bar{B}^0$ (i.e. $f_\pm = f_0$ = 0.5), then we find $0.31 < b_\pm/b_0 < 3.2$ at 90% confidence level. If the $\Upsilon(4S)$ decay proportions are 60% into $B^+B^-$, 40% into $B^0\bar{B}^0$, then the limits are $0.34 < b_\pm/b_0 < 4.0$. The limits are very weak, consistent either with the ratio near 2 hinted at by the discussion in Section 7.2.1 above, or with a ratio near 1.

7.2.3  LEPTON-HADRON CORRELATIONS    The B mesons produced in $\Upsilon(4S)$ decay have the relatively small momentum of 0.4 GeV. When one of the decay products from one of the B's has a high momentum, then the other decay products of that same B will, on average, recoil against the high-momentum particle, and show a pronounced peaking in their angular distribution with respect to the axis defined by the high-momentum particle. Decay products from the other B in the $\Upsilon(4S)$ event will have little knowledge of the direction of the high-momentum particle, and will be distributed very nearly isotropically with respect to the axis it defines. The angular distribution thus has a peak component (from particles from the same B), and an isotropic component (from particles from the other B).

CLEO is using this kinematic effect to measure $b_\pm/b_0$. In events containing a high-momentum lepton, angular distributions are plotted separately for particles with the same charge as the lepton and for those with the opposite charge from the lepton. $B^+B^-$ events show equal peaking of same and opposite charges, and an excess opposite charge in the isotropic component. $B^0\bar{B}^0$ events, in contrast, show an excess of opposite-charge tracks in the peaked component, and equal numbers of tracks of each charge in the isotropic component. By measuring the net charge (relative to the lepton charge) of the peaked component and the isotropic component, one measures $b_\pm/b_0$. Preliminary analysis (CLEO, unpublished) yields $b_\pm/b_0 < 2.7$.

## 7.3   $B^0\bar{B}^0$ Mixing

The formalism of Section 3.4 expresses the mixing parameter $y$ in terms of $f_B^2 B_B$, $m_t$, and the K-M angles and phase. An approximate expression, valid when $y$ is small, is

$$y \approx 0.7 \left( \frac{f_B^2 B_B m_t^2}{m_b^4} \right)^2 \left| \frac{s_2}{s_3 - s_2 e^{i\delta}} \right|^4 . \qquad 30.$$

One notes that $y$ depends on the K-M angles in a manner different from $(b \to u)/(b \to c)$ or $\tau_B$. It also depends on the poorly known combination $B_B f_B^2 m_t^2$.

The mixing parameter $y$ is plotted as a function of the K-M phase $\delta$ in Figure 20. Here the more exact formulas of Equations 11–13, 15, and 21 are used. The K-M angles are adjusted as $\delta$ is varied so that $(b \to u)/(b \to c)$ and $\tau_B$ remain constant. The curves are independent of the value of $\tau_B$ chosen, but depend on $(b \to u)/(b \to c)$ and on $B_B f_B^2 m_t^2$ as shown. If $B_B f_B^2 m_t^2$ were known, a measurement of $y$ in conjunction with measurements of $(b \to u)/(b \to c)$ and $\tau_B$ would uniquely determine $\theta_2$ and $\theta_3$, and would determine $\delta$ to within a two-fold ambiguity, $\delta$ or $2\pi - \delta$.

A measurement of the $B^0 \bar{B}^0$ mixing parameter requires a tag of b flavor (vs antiflavor). The most commonly suggested possibility is the sign of the lepton in semileptonic B decay. Thus, from $B\bar{B}$ events from $\Upsilon(4S)$ decay, one has

$$y = \frac{N_{++} + N_{--}}{N_{+-} \text{ from } B^0 \bar{B}^0} , \qquad 31.$$

where $N_{++}$ ($N_{--}$) is the number of events with positive (negative) like-sign dileptons, and the denominator includes opposite-sign dilepton events from neutral $B\bar{B}$ decay, but excludes those from $B^+ B^-$ decay. CLEO (47) has attempted such a measurement. After accounting for those like-sign

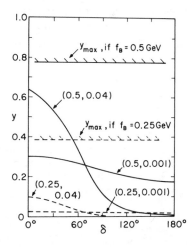

Figure 20   Mixing parameter $y$ calculated as a function of the K-M phase $\delta$, for $f_B = 0.5$ or $0.25$ GeV, and for $(b \to u)/(b \to c) = 0.04$ or $0.001$. For these calculations, $m_t = 40$ GeV, $B_B = 1$. CLEO's measured upper limits on $y$ for the two assumed values of $f_B$ are also shown.

dilepton events that arise from secondary processes (e.g. $B \to X\ell^+\nu$, $\bar{B}$ $\to XD \to \bar{K}\ell^+\nu$), and from hadrons misidentified as leptons, they find no excess attributable to $B^0\bar{B}^0$ mixing. Their upper limit on $y$ is given in Figure 21. Because one must subtract from the detected opposite-sign dilepton events those due to $B^+B^-$ decay, the upper limit obtained is a function of $(f_0/f_\pm)\,[\mathrm{Br}(B^0 \to X\ell\nu)/\mathrm{Br}(B^\pm \to X\ell\nu)]^2$.

JADE (71) has obtained information on mixing by a quite different technique. They measured the forward-backward charge asymmetry for the process $e^+e^- \to b\bar{b}$, detecting b quarks from the continuum as described in Section 2.2.3. They measure an asymmetry of $-0.228 \pm 0.060 \pm 0.025$, in agreement with the Glashow-Salam-Weinberg model prediction of $-0.252$. The $b\bar{b}$ mixing will reduce asymmetry below the GSW prediction, and the good agreement is taken as evidence that mixing is not large. To be quantitative, one needs to know the fractions of the various b-flavored hadrons in continuum production and (since they are detected with a lepton tag) their semileptonic decay branching ratios. One expects that, in addition to $B^\pm$ and $B^0$, $B_s^0$ and b-flavored baryons are present. The former are expected to mix more strongly than $B^0$, while the latter will not mix. For illustrative purposes, I have ignored $B_s$ and baryons, and taken equal production rates of $B^0$ and $B^+$, to calculate an upper limit. This is plotted in Figure 21, as a function of the ratio of neutral and charged semileptonic decay branching ratios. The limit is weaker than CLEO's.

If one takes seriously the relation between $\tau_B \pm/\tau_{B^0}$ and $f_B$ given by Equation 8, then (e.g. assuming $B_B = 1$, $m_t = 40\,\mathrm{GeV}$) one can confront the plots of $y$ vs $\delta$ of Figure 20 with the CLEO upper limits, also indicated on Figure 20. If $f_B$ is near the high end of the range predicted (i.e. near 500 MeV), then the present upper limit is on the verge of being interesting, and a factor-of-three improvement would either reveal an effect or rule out a substantial range of $\delta$. However (as is more likely), if $f_B$ is in the middle of the range predicted (i.e. near 250 MeV), then very substantial experimental

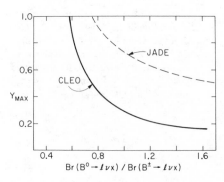

*Figure 21* Upper limits on the mixing parameter $y$, as obtained by CLEO (47) and by JADE (71). The CLEO limit depends on $f_0/f_\pm$, as indicated in the text. A 40%/60% mixture has been assumed.

improvements are required before $B^0\bar{B}^0$ mixing measurements will provide information about the K-M matrix.

## 7.4  *CP Violation in B Decay*

*CP* violation can manifest itself in B decay in two different ways. (*a*) The rate for the mixing transition $B^0 \to \bar{B}^0$ can differ from that for $\bar{B}^0 \to B^0$. (*b*) A partial decay rate for $B^+$ (or $B^0$) can differ from the charge conjugate decay rate for $B^-$ (or $\bar{B}^0$). Both manifestations must occur at some level if the phase of the K-M matrix differs from zero.

*CP* violation in $B^0\bar{B}^0$ mixing implies $r \neq \bar{r}$, which requires (Equation 13) Re $(\varepsilon) \neq 0$. This condition in turn requires that $M_{12}$ and $\Gamma_{12}$ differ in phase (Equation 12). An inspection of Equations 21 shows that the first term in $\Gamma_{12}$ has the same phase as $M_{12}$, and therefore will not contribute to *CP* violation. The second term, proportional to $m_c^2$, will contribute if $\xi_c$ and $\xi_t$ differ in phase.

An approximate expression for *a*, valid when *a* is small, follows from a few pages of algebra:

$$a \approx 4\pi \frac{m_c^2}{m_t^2} \frac{s_3 s_\delta}{s_2}. \qquad\qquad 32.$$

If one further assumes that $y_{\text{odd}}$ is small, one obtains

$$l^\pm = \frac{9f_B^4 B_B^2 m_t^2 m_c^2}{m_b^8} \frac{s_2^3 s_3 s_\delta}{|s_3 - s_2 e^{i\delta}|^4}. \qquad\qquad 33.$$

For the range of K-M angles allowed by the experimental data on $(b \to u)/(b \to c)$ and B lifetime, and for $f_B\sqrt{B_B} \lesssim 0.5$ GeV and $30 \lesssim m_t \lesssim 50$ GeV, one finds (using the more exact expressions, Equations 11–13, 19–21) that *a* does not exceed 0.04, and $l^\pm$ does not exceed 0.003. These are very unattractive targets, well below foreseeable experimental capabilities. It nonetheless is desirable to attempt measurements of the *CP*-violating asymmetries, since a positive effect at a level above that just indicated would demonstrate the existence of a source of *CP* violation other than the phase in the K-M matrix.

CLEO (72) has measured the single-lepton charge asymmetry $l^\pm$, using both muons and electrons. The combined asymmetry, $0.00 \pm 0.03$, includes leptons from $B^\pm$ decay, as well as from $B^0/\bar{B}^0$. Since those from $B^\pm$ must of necessity have no asymmetry, if one assumes a 40%/60% $B^0\bar{B}^0/B^+B^-$ production mix, one has $l^\pm = 0.00 \pm 0.08$. A reduction in error of two orders of magnitude is required to see Standard Model *CP* violation.

At present, there has been no experimental work on searches for *CP* violation in partial decay rates of B mesons.

**Table 6**  B hadronic decay branching ratio measurements by CLEO: three-body and $D^*\rho$

| Method | Ref. | Reaction | Branching ratio (%) |
|---|---|---|---|
| Reconstructed[a,b] | 51 | $B^- \to D^{*+}\pi^-\pi^-$ | $3.2 \pm 2.0$ |
| Reconstructed[b] | 51 | $\bar{B}^0 \to D^0\pi^+\pi^-$ | $9 \pm 6$ |
| Partially reconstructed[a] | 73 | $\bar{B}^0 \to D^{*+}\rho^-$ | $6 \pm 4$ |

[a] Result scales inversely with branching ratio assumed for $D^{*+} \to D^0\pi^+$; 60% used here.
[b] Result scales inversely with branching ratio assumed for $D^0 \to K^-\pi^+$; 4.5% used here.

# 8.   HADRONIC DECAY BRANCHING RATIOS AND INCLUSIVE PROPERTIES

## 8.1   *Low-Multiplicity Hadronic Decay Branching Ratios*

Hadronic decay branching ratios can be obtained from the CLEO (7, 51) reconstructed B mesons mentioned in Section 5. Because of the small number of events, the accuracy is poor. The branching ratios for $B^- \to D^{*+}\pi^-\pi^-$ and $\bar{B}^0 \to D^0\pi^+\pi^-$ are given in Table 6. The second reaction will also contain a contribution from $\bar{B}^0 \to D^{*0}\pi^+\pi^-$. The branching ratio for $\bar{B}^0 \to D^{*+}\pi^-$ is given on the first line of Table 7.

For the decay mode $B^- \to D^0\pi^-$, $D^0 \to K^-\pi^+$, the statistics can be improved (51) by dropping the requirement that the kaon be identified. Apparent $B^- \to D^0\pi^-$ candidates will also consist of $B^- \to D^{*0}\pi^-$, $D^{*0} \to D^0\pi^0$ or $D^0\gamma$, and $\bar{B}^0 \to D^{*+}\pi^-$, $D^{*+} \to D^0\pi^+$ (both with somewhat reduced detection efficiency), since omitting the light decay products from

**Table 7**  B hadronic decay branching ratio measurements by CLEO (51, and unpublished): $D\pi^-$ and $D^*\pi^-$

| Method | | Branching ratio (%) | | | |
|---|---|---|---|---|---|
| | | $D^+\pi^-$ | $D^0\pi^-$ | $D^{*0}\pi^-$ | $D^{*+}\pi^-$ |
| $D^{*+}\pi^-$ reconstructed[a,b] | | — | — | — | $1.7 \pm 1.3$ |
| | $D = D^*$ | — | ←———— $1.3 \pm 0.7$ ————→ | | |
| "$D^0\pi^-$" reconstructed[b] | $D \ll D^*$ | — | 0 | ←———— $2.7 \pm 1.3$ ———— | → |
| | $D = D^*$ | ←———— $0.5 \pm 0.2$ ————→ | | | |
| $\pi^-$ momentum spectrum | $D \ll D^*$ | 0 | 0 | ←——— $1.0 \pm 0.4$ ———→ | |
| $D^{*+}\pi^-$ partially reconstructed[a] | | — | — | — | $1.0 \pm 0.7$ |

[a] Result scales inversely with branching ratio assumed for $D^{*+} \to D^0\pi^+$; 60% used here.
[b] Result scales inversely with branching ratio assumed for $D^0 \to K^-\pi^+$; 4.5% used here.

the D* changes the reconstructed B mass only very slightly. If one assumes $B^- \to D^0\pi^-$, $B^- D^{*0}\pi^-$, and $\bar{B}^0 \to D^{*+}\pi^-$ have equal branching ratios, then that branching ratio is measured to be $1.3 \pm 0.7\%$, as indicated on line 2 of Table 7. If one instead assumes that the branching ratios to D* are large compared to those to D, then the branching ratios are as given on line 3.

CLEO (51) has obtained two-body decay branching ratios by two techniques other than B reconstruction. The first is a measurement of the $\pi^\pm$ momentum spectrum. The B's produced from $\Upsilon(4S)$ decay are moving very slowly. Were they at rest, the two-body decays $B^- \to D^0\pi^-$, $B^- \to D^{*0}\pi^-$, $\bar{B}^0 \to D^{*+}\pi^-$ and $\bar{B}^0 \to D^{*+}\pi^-$ would provide a monoenergetic pion peak at the high end of the spectrum. The motion of the B's broadens the peak into a rectangle. Fitting the spectrum, one obtains an average branching ratio for the four decay modes of $0.5 \pm 0.2\%$, as shown on line 4 of Table 7. If one assumes that the branching ratios to D* are large compared to those to D, then the branching ratios are as given on line 5.

The second technique, dubbed "partial reconstruction," also takes advantage of the slowness of the B, and in addition utilizes the smallness of the energy release in the $D^{*+} \to D^0\pi^+$ decay. In the decay $\bar{B}^0 \to D^{*+}\pi^-$, the $D^{*+}$ recoils approximately $180°$ away from the $\pi^-$, which is very energetic. The soft $\pi^+$ from $D^{*+} \to D^0\pi^+$ very nearly preserves the $D^{*+}$ direction. Measuring only the $\pi^-$ and $\pi^+$ momenta and their included angle, one has a decay process that is once underconstrained. An artificial constraint is introduced, namely that the $D^0$ momentum vector lie in the plane defined by the $\pi^+$ and $\pi^-$ momenta. This procedure allows a pseudomass to be calculated, which differs only slightly from the correct mass and gives a broadened mass peak; this enables the decay mode to be identified and the branching ratio obtained. The result (CLEO, unpublished) is given in the last line of Table 7. The same procedure has been applied (73) to the decay mode $\bar{B}^0 \to D^{*+}\rho^-$, with the result given in Table 6.

Several of the measured branching ratios scale inversely with the branching ratios assumed for $D^{*+} \to D^0\pi^+$ and for $D^0 \to K^-\pi^+$. Here I have used 60% and 4.5%, and indicated in Tables 6 and 7 which results scale.

Table 7 reveals no serious discrepancies among the different methods. The results from the $\pi^\pm$ momentum spectrum agree better with reconstruction and partial reconstruction results if one assumes the branching ratios to $\pi D^*$ are large compared to those to $\pi D$. No experimental evidence contradicts that assumption.

Avery (internal CLEO report) has modified Tsai's calculations (74) of $\tau^\pm$ decay branching ratios to apply to B decay. He obtains $\mathrm{Br}(B \to \pi^-D) + \mathrm{Br}(B \to \pi^-D^*) \approx 2\%$, and $\mathrm{Br}(B \to \rho^-D) + \mathrm{Br}(B \to \rho^-D^*) \approx 5\%$, in agreement with the CLEO data. A calculation by Ali et al (75) gives similar results.

## 8.2 *Inclusive Decay Properties*

Kaon yields were discussed in Section 6.1.1. There we saw that the observed kaon yield was readily explained as resulting in part from D decays and in part from s$\bar{\text{s}}$ pairs popped from the ocean.

CLEO (56; submitted for publication) has measured the yields of $D^0$ and $D^{*\pm}$ mesons. As discussed in Section 6.1.2, the rates are in accord with a dominant b $\rightarrow$ c transition. The momentum spectra are shown in Figure 22. The curve is a V $-$ A, point-quark calculation, which assumes the charmed meson acquires all of the momentum of the charmed quark. The data are only slightly softer than this curve, which implies that the c quark and spectator quark usually combine directly to form the D or D* without the creation of additional pions along that color string.

CLEO (51) measured the charged particle multiplicity for B$\bar{\text{B}}$ events from $\Upsilon$(4S) decay, finding a mean of 11.0 $\pm$ 0.3 and a dispersion (rms spread of the distribution) of 3.1 $\pm$ 0.2. For B$\bar{\text{B}}$ events with one or more identified leptons,

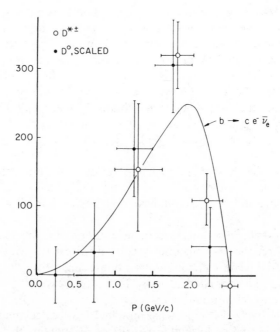

*Figure 22*    Momentum spectra of $D^0$ and $D^{*\pm}$ mesons from B decay at $\Upsilon$(4S). The curve is a V $-$ A point-quark calculation, which assumes the charmed meson acquires all the momentum of the charmed quark. Results are from CLEO (56; and submitted for publication).

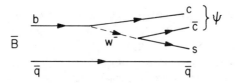

*Figure 23*   Color-suppressed spectator diagram for B → ψX.

the mean is $9.3 \pm 0.3$ and the dispersion is $2.2 \pm 0.2$. From these results, CLEO obtains a mean charged multiplicity in hadronic B decay of $6.0 \pm 0.3$, and a multiplicity of $3.8 \pm 0.4$ in semileptonic B decays.

Since b decays to c very nearly 100% of the time, one expects each B decay to contain a D or D*. Subtracting the mean charged multiplicity for such a particle ($\sim 2.5$, Ref. 76), one sees that in its hadronic decays, B decays to D or D* plus three to four charged pions (and presumably half as many $\pi^0$). Semileptonic B decay leads to a D or D* and $0.3 \pm 0.4$ charged pions.

As pointed out by Fritzsch (77), ψ mesons will be produced in B decay by the color-suppressed spectator diagram shown in Figure 23. Estimates (77, 78) of the branching ratio for B → ψX vary from several percent down to a few tenths of a percent. CLEO (51) has searched for ψ via the $\mu^+\mu^-$ and $e^+e^-$ decay modes. They find a hint of a signal, corresponding to a branching ratio of $1.0 + 0.5 - 0.4\%$ or, alternatively, an upper limit of 1.6% at 90% confidence level.

CLEO (79) observes baryons, both protons and lambdas, from B decay. The momentum spectra are shown in Figure 24. The inclusive branch-

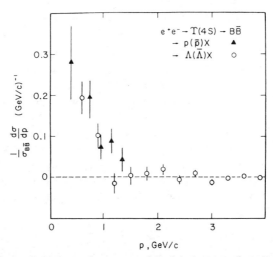

*Figure 24*   Momentum spectra of protons and lambda hyperons from B decay at Υ(4S). Results are from CLEO (79).

ing ratio $1/2[Br(B \rightarrow \Lambda X) + Br(B \rightarrow \bar{\Lambda}X)]$ is $3 \pm 1\%$, and the comparable branching ratio to $p/\bar{p}$, after subtracting those $p/\bar{p}$ from $\Lambda$ decay, is also $3 \pm 1\%$. No evidence suggests that these baryons arise from exotic decays. Indeed the softness of the momentum spectra suggests they do not. If the neutron yield is comparable to the proton yield, then some 10% of B decays contain a baryon-antibaryon pair. This result is consistent with Bigi's (80) estimate of 2 to 13%, which assumes that in the quark decay $b \rightarrow cd\bar{u}$, if the invariant mass of the cd diquark is less than some value characterizing the baryon mass scale, then a baryon-antibaryon pair will form.

# 9. CONCLUSIONS

## 9.1 Emerging Picture of B Decay

The only surprise in the emerging picture of B decay is that there are absolutely no surprises. All has developed as predicted by the simplest theoretical picture, the Standard Model. Using very general considerations, CLEO has ruled out the possibility that any imaginable nonstandard model contributes a major portion to b decay. Further, there is nothing so far seen to suggest even minor contributions from processes outside the Standard Model. Lepton spectra, D spectra, two-body hadronic decay branching ratios, baryon yields, kaon and D yields, charged-particle multiplicities, all are what one would expect from a decay $b_L \rightarrow c_L + W_V^-$. It appears that the third quark doublet is just like the first two.

The K-M matrix, as determined from b decay studies, is "approximately diagonal." The diagonal elements are very close to 1.0, the elements once-removed from the diagonal are 0.05 to 0.2, and $V_{ub}$, twice removed from the diagonal, is less than 0.012. Further, $V_{cb}/V_{us}$ is small, $\sim 1/4$, which means that the third quark doublet is substantially decoupled from the other two. This decoupling manifests itself in the long B lifetime.

The Standard Model treats the K-M angles as free parameters, and can accommodate any values. The approximate diagonality and decoupling with increasing generation number are features that must be explained by a more complete theory, one that deals with the family problem.

A theory that fully explains the family problem should predict both the K-M angles and the quark masses. That theory is yet to be developed. Efforts of a more modest nature derive relations among the K-M angles and the quark masses. For example, by placing restrictions on the form of the 3 $\times$ 3 mass matrix for quark weak interaction eigenstates, Fritzsch (81) arrives at the relationships

$$s_1 = |\sqrt{m_d/m_s} + e^{i\delta}\sqrt{m_u/m_c}|, \qquad 34.$$

$$V_{ub}/V_{cb} = \sqrt{m_u/m_c}. \qquad 35.$$

The former is well satisfied if $\delta$ is near $90°$, while the latter predicts $(b \rightarrow u)/(b \rightarrow c) = 0.008$, well below the present experimental upper limit. Proceeding along lines such as these, one may hope that a more precise determination of the K-M matrix will lead to an understanding of the form of the quark mass matrix, which in turn will supply clues for a solution to the family problem.

## 9.2   K-M Matrix Description of CP Violation in $K^0$ Decays

The quantities $\varepsilon$ and $\varepsilon'$ describing $CP$ violation in the neutral kaon system are calculable (82) in the terms of the K-M angles and phase:

$$\varepsilon = \left[ \frac{B_K G_F^2 f_K^2 m_K}{12\sqrt{2}\pi^2 \Delta m_K} \mathrm{Im}\left( \eta_1 \lambda_c^2 m_c^2 + \eta_2 \lambda_t^2 m_t^2 + 2\eta_3 \lambda_c \lambda_t m_c^2 \ln \frac{m_t^2}{m_c^2} \right) \right.$$

$$\left. + \sqrt{2}\xi \frac{\mathrm{Re}\, M_{12}^{sd}}{\Delta m_K} \right] e^{i\pi/4}, \qquad\qquad 36.$$

$$\frac{\varepsilon'}{\varepsilon} = -\frac{1}{\sqrt{2}} \frac{\xi}{\varepsilon} \left| \frac{A_2}{A_0} \right| \approx 8.4 s_2 s_3 s_\delta. \qquad\qquad 37.$$

Here $\lambda_q = V_{qs}^* V_{qd}$; $B_K$ is the kaon bag parameter defined analogously to $B_B$ in Equation 22; $\eta_1$, $\eta_2$, $\eta_3$ are QCD correction factors with values (83) 0.7, 0.6, and 0.4 respectively, and $A_0$ $(A_2)$ is the isospin zero (two) amplitude for $K^0 \rightarrow \pi\pi$. (The last term in Equation 36 arises from shifting from a quark basis to a basis where $A_0$ is real. It is small and is ignored in the evaluations that follow.)

Re($\varepsilon$) has been measured (70) to be $(1.621 \pm 0.088) \times 10^{-3}$. The value of $\varepsilon'/\varepsilon$ is consistent with zero, with a recent measurement (84) giving $-0.0046 \pm 0.0058$.

$B_K$, needed to evaluate Equation 36, is poorly known. Shrock & Treiman (85) obtained $B_K = 0.4$ using the MIT bag model. Donoghue et al (86) obtained $B_K = 0.33$ using SU(3) and PCAC, but Bijnens et al (87) have shown that there are substantial corrections to this calculation, and values as high as 1.5 cannot be excluded. The top-quark mass is also poorly known. With all this freedom, it is easy to obtain large values of Re($\varepsilon$) from Equation 36. For illustrative purposes, I have assumed $m_t = 40$ GeV, and for $B_K = 0.5$ and 1.0 indicated on Figure 19 that portion of the allowed region of the $V_{ub}$-$V_{cb}$ plane that can give values of Re($\varepsilon$) as large as the measured value.

It is also easy to obtain simultaneously a large value for $\varepsilon$ from Equation

36 and a small value for $\varepsilon'/\varepsilon$ from Equation 37. If $m_t = 40$ GeV, then $\varepsilon'/\varepsilon \gtrsim 0.006/B_K$. With increasing $m_t$, this limit falls.

The hypothesis, that $CP$ violation in neutral kaon decay results from the phase in the K-M matrix, cannot at present be quantitatively tested. Experimental and theoretical uncertainties permit too much flexibility. To rectify this situation, one needs improvements in the experimental determinations of $\varepsilon'/\varepsilon$ and $m_t$, and in the theoretical calculations of $B_K$. A better knowledge of the K-M angles [particularly an improved limit on—or measurement of—(b → u)/(b → c)] is also highly desirable.

## 9.3   Open Problems in B Decay

### 9.3.1   DETERMINATION OF K-M MATRIX

Our present restrictions on the K-M matrix come from measurements of the B lifetime and upper limits on (b → u)/(b → c). The experimental uncertainty in the lifetime measurement is already smaller than the theoretical uncertainty in how to relate the measurement to the K-M matrix—the choice of quark masses discussed in Section 6.4. The extraction of (b → u)/(b → c) from the lepton spectrum also suffers from weakness in the theory, again phrased as a choice of quark masses—see Section 6.1. Efforts to improve the experimental results are under way. It is hoped that they will be matched by improvements on the theoretical front.

To determine $\theta_2$, $\theta_3$, and $\delta$ requires a third measurement, in addition to $\tau_B$ and (b → u)/(b → c). In Section 7.3, it was shown how a measurement of the $B^0\bar{B}^0$ mixing parameter $y$ would serve this purpose. It was also noted that one must know $m_t$, $B_B$, and $f_B$ to use $y$ in this way.

### 9.3.2   NONSTANDARD DECAYS

Some new physics, outside the Standard Model, is responsible for the existence of the three (or more) families and implies relationships among the families (e.g. horizontal symmetries). It is likely that this new physics induces interactions, and consequently decays. Since the physics is outside the Standard Model, the decays will necessarily be nonstandard. This line of thinking has motivated searches for rare decays of $\mu$, K, and $\tau$.

The search for nonstandard b decays is more difficult and cannot compete in terms of the smallness of the branching ratio : b decay does have the advantage of higher mass, and consequently more open channels. For example, antibaryonic decays of kaons are energetically forbidden. The equivalent for $\tau$ decay, $\tau \to \bar{q}\bar{q}\bar{q}$, is allowed to u,d,s quarks (e.g. $\tau \to \pi^0\bar{p}$) but not to a c quark. A further advantage of b decay is that the standard decays are suppressed, by the factor $|V_{cb}|^2 \approx 1/400$, a factor not present in $\tau$ decays. Searches for exotic b decays and for flavor-changing neutral currents in b decays are in progress.

ACKNOWLEDGMENTS

I wish to thank Karl Berkelman, Ashok Das, Vishnu Mathur, Ronald Poling, Carl Rosenfeld, and Henry Tye for useful discussions and constructive criticism. The National Science Foundation provided financial support for the preparation of this review.

*Literature Cited*

1. Herb, S. W., et al. *Phys. Rev. Lett.* 39:252–55 (1977); Innes, W. R., et al. *Phys. Rev. Lett.* 39:1240–42 (1977); Ueno, K., et al. *Phys. Rev. Lett.* 42:486–89 (1979)
2. Glashow, S. L., Iliopoulos, J., Maiani, L. *Phys. Rev. D* 2:1285–92 (1970)
3. Aubert, J. J., et al. *Phys. Rev. Lett.* 33:1404–6 (1974); Augustin, J.-E., et al. (Mark I). *Phys. Rev. Lett.* 33:1406–7 (1974)
4. Perl, M. L., et al. (Mark I). *Phys. Rev. Lett.* 35:1489–92 (1975)
5. Kobayashi, M., Maskawa, T. *Prog. Theor. Phys.* 49:652–57 (1973)
6. Arnison, G., et al, (UA1). *Phys. Lett.* 147B:493–508 (1984)
7. Behrends, S., et al, (CLEO). *Phys. Rev. Lett.* 50:881–84 (1983)
8. Berger, C., et al, (PLUTO). *Phys. Lett.* 76B:243–45 (1978); Darden, C. W., et al, (DASP II). *Phys. Lett.* 76B:246–48 (1978); Bienlein, J. K., et al, (DESY-Heid.). *Phys. Lett.* 78B:360–63 (1978); Darden, C. W., et al, (DASP II). *Phys. Lett.* 78B:364–65 (1978)
9. Andrews, D., et al, (CLEO). *Phys. Rev. Lett.* 44:1108–11 (1980); Bohringer, T., et al, (CUSB). *Phys. Rev. Lett.* 44:1111–14 (1980)
10. Berkelman, K. *Phys. Rep.* 98:145–87 (1983)
11. Haas, P., et al, (CLEO). *Phys. Rev. D* 30:1996–97 (1984)
12. Franzini, P., Lee-Franzini, J. *Ann. Rev. Nucl. Part. Sci.* 33:1–29 (1983)
13. Andrews, D., et al, (CLEO). *Phys. Rev. Lett.* 45:219–21 (1980)
14. Finocchiaro, G., et al, (CUSB). *Phys. Rev. Lett.* 45:222–25 (1980)
15. Besson, D., et al, (CLEO). *Phys. Rev. Lett.* 54:381–84 (1985)
16. Thorndike, E. H., (CLEO). *High Energy Physics—1980, Proc. 20th Int. Conf., Madison, Wisc.*, ed. L. Durand, L. G. Pondrom, pp. 705–11. New York: Am. Inst. Phys. (1981)
17. Schamberger, R. D., et al, (CUSB). *Phys. Rev. D* 30:1985–87 (1984)
18. Eichten, E. *Phys. Rev. D* 22:1819–23 (1980)
19. Lovelock, D. M. J., et al, (CUSB). *Phys. Rev. Lett.* 54:377–80 (1985)
20. Han, K., et al, *Phys. Rev. Lett.* 54: In press (1985)
21. Nelson, M. E., et al, (Mark II). *Phys. Rev. Lett.* 50:1542–45 (1983); Hanson, G. G., (Mark II). *Proc. Int. Europhysics Conf. High-Energy Physics, Brighton*, pp. 147–51. Didcot: Rutherford Appleton Lab. (1983)
22. Fernandez, E., et al, (MAC). *Phys. Rev. Lett.* 50:2054–57 (1983)
23. Koop, D. E., et al, (DELCO). *Phys. Rev. Lett.* 52:970–73 (1984)
24. Aihara, H., et al, (TPC). *LBL Rep. 17545* (1984); *John Hopkins Preprint JHU-8501* (1985)
25. Adeva, B., et al, (MARK J). *Phys. Rev. Lett.* 51:443–46 (1983)
26. Althoff, M., et al, (TASSO). *Z. Phys. C* 22:219–29 (1984); *Phys. Lett.* 146B:443–49 (1984)
27. Behrend, H.-J., et al, (CELLO). *Z. Phys. C* 19:291–99 (1983)
28. Basile, M., et al, (CBFP Collab.). *Nuovo Cimento Lett.* 31:97–111 (1981)
29. Drijard, D., et al, (ACCDHW Collab.). *Phys. Lett.* 108B:361–66 (1982)
30. Kernan, A., Van Dalen, G. *Phys. Rep.* 106:297–398 (1984)
31. Thorndike, E. H. *Proc. V.P.I. Workshop on Weak Interactions as Probes of Unification, Virginia Polytechnic Institute—1980*, ed. G. B. Collins, L. N. Chang, J. R. Ficenec, pp. 614–20. New York: Am. Inst. Phys. (1981)
32. Derman, E. *Phys. Rev. D* 19:317–29 (1979)
33. Georgi, H., Machacek, M. *Phys. Rev. Lett.* 43:1639–41 (1979)
34. Georgi, H., Glashow, S. L. *Nucl. Phys. B* 167:173–80 (1980)
35. Glashow, S. L. *Nucl. Phys.* 22:579–88 (1961); Salam, A., Ward, J. C. *Phys. Lett.* 13:168–71 (1964); Weinberg, S. *Phys. Rev. Lett.* 19:1264–66 (1967)
36. Gaillard, M. K., Lee, B. W. *Phys. Rev.*

*Lett.* 33:108–11 (1974); Altarelli, G., Maiani, L. *Phys. Lett.* 52B:351–54 (1974)

37. Leveille, J. P. *Proc. CLEO Collab. Workshop on B Meson Decay*, pp. 37–67. Available as *CLEO Preprint 81/05*, Univ. Rochester (1981)

38. Buras, A. J., Slominski, W., Steger, H. *Nucl. Phys. B* 245:369–98 (1984)

39. Mathur, V. S., Yamawaki, M. T. *Phys. Rev. D* 29:2057–66 (1984); Reinders, L. J., Yazaki, S., Rubinstein, H. R. *Phys. Lett.* 104B:305–14 (1981); Shuryak, E. V. *Nucl. Phys. B* 198:83–101 (1982); Krasemann, H. *Phys. Lett.* 96B:397–401 (1980); Golowich, E. *Phys. Lett.* 91B:271–74 (1980)

40. Hagelin, J. S. *Nucl. Phys. B* 193:123–49 (1981)

41. Bernabeau, C., Jarlskog, C. *Z. Phys. C* 8:233–38 (1981)

42. Bander, M., Silverman, D., Soni, A. *Phys. Rev. Lett.* 43:242–45 (1979)

43. Carter, A. B., Sanda, A. I. *Phys. Rev. Lett.* 45:952–54 (1980); *Phys. Rev. D* 23:1567–79 (1981); Bigi, I. I., Sanda, A. I. *Nucl. Phys. B* 193:85–108 (1981)

44. Chau (Wang), L.-L. See Ref. 31, pp. 419–30; Chau, L.-L. *Phys. Rep.* 95:1–94 (1983); Chau, L.-L., Cheng, H.-Y. *Phys. Rev. Lett.* 53:1037–40 (1984)

45. Chen, A., et al, (CLEO). *Phys. Lett.* 122B:317–21 (1983)

46. Kane, G. L., Peskin, M. *Nucl. Phys. B* 195:29–38 (1982)

47. Avery, P., et al, (CLEO). *Phys. Rev. Lett.* 53:1309–13 (1984)

48. Adeva, B., et al, (Mark J). *Phys. Rev. Lett.* 50:799–802 (1983); Bartel, W., et al, (JADE). *Phys. Lett.* 132B:241–45 (1983)

49. Peskin, M. E., Tye, S.-H. H. In *Proc. Cornell $Z^0$ Theory Workshop*, ed. M. E. Peskin, S.-H. H. Tye, pp. 405–10. Available as *CLNS Preprint 81-485*, Cornell Univ. (1981); see also Barger, V., Keung, W. Y., Phillips, R. J. N. *Phys. Rev. D* 24:1328–42 (1981)

50. Chen, A., et al, (CLEO). *Phys. Rev. Lett.* 52:1084–88 (1984)

51. Giles, R., et al, (CLEO). *Phys. Rev. D* 30:2279–94 (1984)

52. Artamonou, A. S., et al, (VEPP-4). *Phys. Lett.* 118B:225–29 (1982); Barber, D. P., et al, (DORIS II). *Phys. Lett.* 135B:498–504 (1984); MacKay, W. W., et al, (CESR). *Phys. Rev. D* 29:2483–90 (1984)

53. Mathur, V. S., Yamawaki, M. T. See Ref. 39; Chan, L.-H. *Phys. Rev. Lett.* 52:253–56 (1983)

54. Thorndike, E. H., *Proc. DESY Theory Workshop on Electroweak Interactions and Particle Structure*, pp. 1–40. DESY internal report T-84-02 (1984)

55. Brody, A., et al, (CLEO). *Phys. Rev. Lett.* 48:1070–74 (1982)

56. Green, J., et al, (CLEO). *Phys. Rev. Lett.* 51:347–50 (1983)

57. Poling, R. A. *A Study of the Muon Decays of b-Flavored Hadrons*. PhD thesis, Univ. Rochester (1981)

58. Altarelli, G., Cabibbo, N., Corbo, G., Maiani, L., Martinelli, G. *Nucl. Phys. B* 208:365–80 (1982)

59. Klopfenstein, C., et al, (CUSB). *Phys. Lett.* 130B:444–48 (1983)

60. Levman, G., et al, (CUSB). *Phys. Lett.* 141B:271–75 (1984)

61. Stone, S. *Proc. 1983 Int. Symp. on Lepton and Photon Interactions at High Energies*, ed. D. G. Cassel, D. L. Kreinick, pp. 203–43. Cornell: Newman Lab. Nucl. Stud. (1983)

62. Fernandez, E., et al, (MAC). *Phys. Rev. Lett.* 51:1022–25 (1983); Lockyer, N. S., et al, (Mark II). *Phys. Rev. Lett.* 51:1316–19 (1983)

63. Hayes, K. See Ref. 54, pp. 41–96

64. Klem, D. E., et al, (DELCO). *Phys. Rev. Lett.* 53:1873–76 (1984)

65. Althoff, M., et al, (TASSO). *Phys. Lett.* 149B:524 (1984)

66. Wolf, G., *SLAC-PUB-3446* Stanford, Calif: SLAC (1984)

67. Gasser, J., Leutwyler, H. *Phys. Rep.* 87:77–169 (1982)

68. Rucinski, G. J. *Evidence Against Exotic Decays of b-Flavored Hadrons*. PhD thesis, Univ. Rochester (1983)

69. Davies, C. T. H., Tye, S.-H. H. *Cornell Preprint CLNS-85/635* (1985)

70. Particle Data Group. *Review of Particle Properties. Rev. Mod. Phys.* 56:S1–S304 (1984)

71. Bartel, W., et al, (JADE). *Phys. Lett.* 146B:437–42 (1984)

72. Kagan, H., MacKay, W. W., Thorndike, E. H., (CLEO). *Proc. 18th Rencontre de Moriond*, ed. J. Tran Thanh Van, 2:105–21. Gif-sur-Yvette: Editions Frontieres (1983)

73. Chen, A., et al, (CLEO). *Phys. Rev. D*. In press (1985)

74. Tsai, Y. S. *Phys. Rev. D* 4:2821–37 (1971)

75. Ali, A., Korner, J. G., Kramer, G., Willrodt, J. Z. *Phys. C* 1:269–77 (1979)

76. Schindler, R. H., et al, (Mark II). *Phys. Rev. D* 24:78–97 (1981)

77. Fritzsch, H. *Phys. Lett.* 86B:164–66, 343–46 (1979)

78. Kuhn, J. H., Nussinov, S., Ruckl, R. *Z. Phys. C* 5:117–20 (1980); DeGrand, T. A., Toussaint, D. *Phys. Lett.* 89B:256–58 (1980); Wise, M. B. *Phys. Lett.* 89B:229–31 (1980)

79. Alam, M. S., et al, (CLEO). *Phys. Rev. Lett.* 51:1143–46 (1983)

80. Bigi, I. I. *Phys. Lett.* 106B: 510–12 (1981)
81. Fritzsch, H. *Phys. Lett.* 73B: 317–22 (1978); *Nucl. Phys.* B155: 189–207 (1979)
82. Gaillard, M. K., Lee, B. W. *Phys. Rev. D* 10: 897–916 (1974); Ellis, J., Gaillard, M. K., Nanopoulos, D. V. *Nucl. Phys.* B 109: 213–43 (1976); Gilman, F. J., Hagelin, J. S. *Phys. Lett.* 133B: 443–48 (1983)
83. Gilman, F. J., Wise, M. B. *Phys. Lett.* 93B: 129–33 (1980); *Phys. Rev. D* 27: 1128–41 (1983)
84. Bernstein, R. H., et al. *Phys. Rev. Lett.* In press (1985)
85. Shrock, R. S., Treiman, S. B. *Phys. Rev. D* 19: 2148–57 (1979)
86. Donoghue, J. F., Golowich, E., Holstein, B. *Phys. Lett.* 119B: 412–14 (1982)
87. Bijnens, J., Sonoda, H., Wise, M. B. *Phys. Rev. Lett.* 53: 2367–70 (1984)

*Ann. Rev. Nucl. Part. Sci. 1985. 35 : 245–70*

# THE TRANSITION FROM HADRON MATTER TO QUARK-GLUON PLASMA

## H. Satz

Fakultät für Physik, Universität Bielefeld, D-4800 Bielefeld, West Germany

## CONTENTS

## INTRODUCTION

During the past two decades, our concept of an elementary particle has undergone a fundamental change. We now understand hadrons as bound states of quarks, and thus as composite. In strong interaction physics, quarks have become the smallest building blocks of nature. But the binding force between quarks increases with the distance of separation, making it impossible—as far as we know today—to split a given hadron into its quark constituents. If we insist on individual existence, the hadron remains elementary.

This modification of our hadron picture has led to remarkable consequences in strong interaction thermodynamics: at high density, nuclear matter must become a quark plasma. In return, strong interaction

245

0163–8998/85/1201–0245$02.00

thermodynamics has shown us the limits of quark confinement: in sufficiently dense matter, quarks can become free.

Such high densities prevailed in the very early universe, until about $10^{-6}$ seconds after the big bang; only then were quarks confined to form hadrons. To create and study such a primordial plasma in the laboratory is one of the great challenges for current experimental physics. Various estimates (e.g. 1) indicate that the collision of heavy nuclei at very high energies may indeed produce a terrestrial "little bang," providing short-lived bubbles of the quark-gluon plasma. First experiments toward this ultimate goal are expected to start in the summer of 1986, using existing accelerators at Brookhaven National Laboratory and at CERN. A dedicated large-scale machine for this purpose was recently proposed (2).

Phenomenological indications for critical behavior in strong interaction thermodynamics were first seen quite early (3). However, these considerations were based on the dynamics of elementary hadrons and thus could only indicate at what point the resulting statistical mechanics was expected to break down; they said nothing about what might happen beyond that point. The advent of the quark model suggested a new state of matter, and quantum chromodynamics (QCD) supplied the theoretical basis for a two-phase picture of strongly interacting systems.

The perturbative evaluation of QCD at small coupling leads to an equation of state for the quark-gluon plasma in the high-density limit (e.g. 4). It breaks down at sufficiently low densities—a result that has also been taken as an indication for critical behavior (5, 6). In this approach, it is the low-density regime that remains unattainable.

To cover the entire density range of strong interaction thermodynamics, a nonperturbative evaluation method for QCD is necessary. The lattice formulation (7) provides the basis for such a method—so far, the only one we have. It leads to a partition function whose form is that of a generalized spin system and which can therefore be dealt with by methods developed in statistical physics.

In lattice QCD, the existence of a deconfined phase is now rigorously established (8). First indications for a deconfinement transition had in this context been obtained in the strong coupling approximation (9a,b). The real breakthrough occurred, however, when it became clear that the lattice formulation of QCD could be evaluated by computer simulation (10). After initial studies of the confining potential, this method was quickly extended and applied to gauge field thermodynamics (11–13). Today, it is used extensively in studying the phase structure and the features of strongly interacting matter as they are predicted by QCD.

The first part of this article is an introduction to the conceptual basis of critical phenomena in strongly interacting matter and to the formulation of statistical quantum chromodynamics. In the second part, we summarize the results so far obtained in the lattice evaluation of QCD thermodynamics and attempt to assess their reliability. The final part presents some particularly interesting open questions as well as a few comments on the present status of the experimental attempts to study strong interaction thermodynamics.

# STATISTICAL QUANTUM CHROMODYNAMICS

## The Gauge Field Theory of Strong Interactions

Quantum chromodynamics (QCD) describes the interaction of quarks and gluons in the form of a gauge field theory, very similar to quantum electrodynamics (QED) of electrons and photons. In both cases we have spinor matter fields interacting through massless vector gauge fields. In QCD, however, the intrinsic color charge is associated with the non-Abelian gauge group SU(3), in place of the Abelian group U(1) for the electric charge in QED. The quarks thus carry three color charges, and the gluons, transforming according to the adjoint representation, carry eight. The intrinsic charge of the gauge field is the decisive modification in comparison to QED; it makes the pure gluon system directly self-interactive, in contrast to the ideal gas of photons. As a result, the three-dimensional Laplace equation, which in nonrelativistic QED leads to the Coulomb potential $V \sim 1/r$, becomes effectively one-dimensional for massive quarks, with the confining potential $V \sim r$ as the solution.

The Lagrangian density of QCD is given by

$$\mathscr{L} = -\tfrac{1}{4}F^a_{\mu\nu}F^{\mu\nu}_a - \sum_f \bar{\psi}^f_\alpha(i\partial\!\!\!/ - g\!\!\!A\!\!\!/)^{\alpha\beta}\psi^f_\beta, \qquad 1.$$

with

$$F^a_{\mu\nu} = (\partial_\mu A^a_\nu - \partial_\nu A^a_\mu - gf^a_{bc}A^b_\mu A^c_\nu). \qquad 2.$$

Here $A^a$ denotes the gluon field of color $a$ ($a = 1,\dots,8$) and $\psi^f_\alpha$ the quark field of color $\alpha$ ($\alpha = 1,2,3$) and flavor $f$. We restrict ourselves here to the effectively massless u and d quarks, which suffice to form all nonstrange mesons and baryons. Strange and exotic quarks are much more massive and hence thermodynamically suppressed at finite temperatures. The inclusion of quark masses would add a term

$$\mathscr{L}_m = \sum_f m_f \bar{\psi}^f_\alpha \psi^{\alpha,f} \qquad 3.$$

in Equation 1. The structure functions $f^a_{bc}$ are fixed by the color gauge group, whose generators we denote by $\lambda_a$; with them, we define $A = A^a\lambda_a/2$

in Equation 1. The generators satisfy

$$[\lambda_a, \lambda_b] = if_{ab}^c \lambda_c. \qquad\qquad 4.$$

If we would set $f = 0$, the Lagrangian density (Equation 1) would simply reduce to that of QED, as there would then be no self-interaction among the gluons.

Equation 1 contains one dimensionless coupling constant, $g$, and hence provides no scale. The resulting invariance under scale transformations implies that QCD predicts only the ratios of physical quantities, not absolute values in terms of physical units.

In QCD, hadrons are color-neutral bound states of quarks or of quark-antiquark pairs; they are thus the chromodynamic analogue of atoms of positronium as the electrically neutral states in QED. In both cases is the binding radius determined as the point at which the attractive potential just balances the kinetic energy required by the momentum uncertainty at that spatial localization. The difference between the two theories becomes most significant at large distances: while a finite ionization energy $\Delta E$ suffices to break the electrodynamic bond, this is not possible in the case of quark binding.

As the fundamental theory of strong interactions, QCD must then predict the ratios of all hadron masses as well as describe hadronic scattering processes. The successful application of perturbative QCD to scattering at large momentum transfer was decisive in establishing it as the basic theory (e.g. 14). The hadronic mass spectrum is presently under intensive investigation, and calculations seem to be well on the way toward definitive results (e.g. 15). Here we take QCD as the dynamical input needed for the statistical description of strongly interacting matter.

## The Physical Basis for Deconfinement

For composite hadrons of nonvanishing spatial extension, the concept of hadronic matter appears to lose its meaning at sufficiently high density. Once we have a system of mutually interpenetrating hadrons, each quark finds in its immediate vicinity, at a distance of less than a hadron radius, many other quarks. There does not seem to be a way for a given quark to identify those other quarks which at lower density were its partners in some specific hadron, and we should therefore now consider the system as quark matter.

The mechanism for the deconfinement of quarks in dense matter is provided by the screening of their color charge (9b, 16). In dense atomic matter, the long-range Coulomb potential that binds ions and electrons into electrically neutral units is partially screened by the other charges present and thus becomes of much shorter range,

$$e_0^2/r \to (e_0^2/r) \exp(-r/r_D). \qquad 5.$$

Here $r$ denotes the distance from the probe to the charge $e_0$. The Debye screening radius $r_D$ is inversely proportional to the charge density $n$,

$$r_D \sim n^{-1/3}. \qquad 6.$$

Hence, at sufficiently high density, $r_D$ will become smaller than the atomic binding radius $r_A$. A given electron can now no longer feel the binding force of "its" ion and it is therefore set free; at this point, insulating matter becomes electrically conductive (17). We expect deconfinement to be the chromodynamic analogue of this Mott transition. Since screening is a phenomenon occurring at high density and hence at short range, the difference between electrodynamic and chromodynamic forces at large $r$ here is not important. Moreover, the decrease of the color charge with increasing density, resulting from asymptotic freedom (e.g. 18), further enhances the deconfinement.

The color conductivity of strongly interacting matter thus constitutes a rather natural signal for the deconfinement transition: it should vanish for normal hadronic matter as the color insulating state and become nonzero when the system turns into a plasma and hence a color conductor. In insulating solids, however, the electric conductivity $\sigma_e$ for $T > 0$ is not strictly zero, but only exponentially small (17),

$$\sigma_e \sim \exp(-\Delta E/T), \qquad 7.$$

with $\Delta E$ denoting the ionization energy. Above the Mott transition temperature, $\sigma_e$ is significantly nonzero because Debye screening has globally dissolved the Coulomb binding between ions and electrons, but even below this point, thermal ionization can locally produce some few free electrons, making $\sigma_e$ small but nonzero. The corresponding phenomenon in QCD is the production of a quark-antiquark pair in form of a hadron (19). If we try to remove a quark from a given hadron, the confining potential will rise with the distance of separation until it reaches the value $m_H$ of the lowest $q\bar{q}$ state; at this point, an additional hadron will form, whose antiquark neutralizes the quark we were trying to remove, and the separation thus becomes possible. Local hadron production therefore plays the role of ionization, and we expect that the color conductivity $\sigma_c$ will not vanish in the confinement regime, but instead be given by

$$\sigma_c \sim \exp(-m_H/2T), \qquad 8.$$

where $m_H$ is the mass of the lowest $q\bar{q}$ state. Both electric and color conductivity should vanish identically at $T = 0$. In the chromodynamic case, however, we can let $m_H \to \infty$ and consider the thermodynamics of a pure gauge field system. In this case, we expect from Equation 8

$$\sigma_c \begin{cases} = 0 & T \le T_c \\ > 0 & T > T_c \end{cases},$$ 
9.

so that here $\sigma_c$ should vanish in the entire confinement regime and thus form a true order parameter for the deconfinement transition.

## Lattice QCD at Finite Temperature

With the Lagrangian density (Equation 1) provided, the formulation of statistical QCD becomes, at least in principle, a well-defined problem. We have to calculate the partition function

$$Z(\beta, V) = \mathrm{Tr}\{e^{-\beta H}\}.$$ 
10.

In the trace we have to sum over all physical states accessible to a system in a spatial volume $V$; $\beta^{-1} = T$ denotes the physical temperature. Once $Z(\beta, V)$ is obtained, we can calculate all thermodynamic observables in the canonical fashion; thus

$$\varepsilon = (-1/V)(\partial \ln Z/\partial \beta)_V$$ 
11.

gives us the energy density, and

$$P = (1/\beta)(\partial \ln Z/\partial V)_\beta$$ 
12.

give us the pressure.

In practice, the evaluation of statistical QCD encounters two main obstacles. Perturbative calculations lead to the usual divergences of quantum field theory; we thus have to renormalize to obtain finite results. Moreover, we want to study the entire range of behavior of the system, from confinement to asymptotic freedom—i.e. for all values of the coupling. That is not possible perturbatively; we need a new approach for the solution of a relativistic quantum field theory. It is provided by the lattice regularization (7). Evaluating the partition function on a large but finite lattice whose points are separated by multiples of some spacing $a$, we have $1/a$ as largest and $1/(Na)$ as smallest possible momentum; here $Na$ is the linear lattice size. Hence neither ultraviolet nor infrared divergences can occur at finite $N$ and nonzero $a$. We are left with two questions, however: how can we ensure that physical observables are independent of this regularization, and how can we actually carry out calculations? Renormalization group theory answers, as we discuss below, the first of these questions. The Monte Carlo simulation of the lattice form of statistical QCD then allows us to carry out calculations of thermodynamic observables at any coupling—thus answering the second question.

The lattice formulation of statistical QCD is obtained in three steps. First we replace the Hamiltonian form (Equation 10) of the partition function by the corresponding Euclidean functional integral (20)

$$Z_{E}(\beta, V) = \int (dA \ d\psi \ d\bar{\psi}) \exp\left[ -\int_V d^3x \int_0^\beta d\tau \ \mathscr{L}(A, \psi, \bar{\psi}) \right]. \qquad 13.$$

This form involves directly the Lagrangian density, and by integrating over field configurations, we avoid having to project onto the allowed physical states in the trace (Equation 10). The spatial integration over $\mathscr{L}$ is performed over the entire volume of the system, which in the thermodynamic limit becomes infinite, while in the imaginary time $\tau \equiv ix_0$, the integration runs over a finite slice determined by the temperature. The finite temperature behavior of the partition function thus becomes a finite size effect in the integration over $\tau$. Equation 13 is obtained from the trace form in Equation 10. As a consequence, the vector (spinor) fields have to be periodic (antiperiodic)

$$A(\mathbf{x}, \tau = 0) = A(\mathbf{x}, \tau = \beta),$$

$$\psi(\mathbf{x}, \tau = 0) = -\psi(\mathbf{x}, \tau = \beta),$$

$$\bar{\psi}(\mathbf{x}, \tau = 0) = -\bar{\psi}(\mathbf{x}, \tau = \beta), \qquad 14.$$

at the boundaries of the imaginary time integration.

Next, the Euclidean $\mathbf{x} - \tau$ manifold is replaced by a discrete lattice, with $N_\sigma$ points and lattice spacing $a_\sigma$ in each space direction, and $N_\tau$ points and spacing $a_\tau$ for the $\tau$ axis. The overall space volume thus becomes $V = (N_\sigma a_\sigma)^3$, the inverse temperature $\beta^{-1} = N_\tau a_\tau$. The spin quark fields $\psi$ and $\bar{\psi}$ are now defined on each of the $N_\sigma^3 N_\tau$ lattice sites. To ensure the gauge invariance of the formulation, the gauge fields $A$ must, however, be defined on the links connecting each pair of adjacent sites (21).

In the final step, the integration over the gluon fields is replaced by one over the corresponding gauge group variables

$$U_{ij} = \exp\left[ -ig(x_i - x_j)^\mu A_\mu\left(\frac{x_i + x_j}{2}\right) \right], \qquad 15.$$

with $x_i$ and $x_j$ denoting two adjacent lattice sites; thus $U_{ij}$ is an SU(3) matrix associated to the link between these two sites.

The QCD partition function thus becomes on the lattice

$$Z(N_\sigma, N_\tau; g^2) = \int \prod_{\text{sites}} d\psi_i \ d\bar{\psi}_i \prod_{\text{links}} dU_{ij} \exp\left[ -S(\bar{\psi}, \psi, U) \right], \qquad 16.$$

a form somewhat reminiscent of the partition function of a spin system, with $\psi$, $\bar{\psi}$, and $U$ in place of the spin configurations. The QCD action $S$ in the Wilson formulation (7) has the form

$$S = S_G + S_Q, \tag{17.}$$

with (13)

$$S_G = \frac{6}{g_\sigma^2} \frac{a_\tau}{a_\sigma} \sum_{P_\sigma} (1 - \tfrac{1}{3}\text{Re Tr } UUUU)$$

$$+ \frac{6}{g_\tau^2} \frac{a_\sigma}{a_\tau} \sum_{P_\tau} (1 - \tfrac{1}{3}\text{Re Tr } UUUU) \tag{18.}$$

for the action corresponding to the pure gauge field term $F_{\mu\nu}^a F_a^{\mu\nu}$ of the Lagrangian density (Equation 1). It contains two distinct coupling parameters, $g_\sigma$ and $g_\tau$; these are necessary as long as we consider the spatial and temporal lattice spacings $a_\sigma$ and $a_\tau$ as independent variables (22). If we set $a_\sigma = a_\tau \equiv a$, then, we recover one "isotropic" coupling

$$g_\sigma(a) = g_\tau(a) \equiv g. \tag{19.}$$

The actual form of the action $S_G$ is that of a generalized, gauge-invariant Ising model (21). For the usual Ising model, we have a lattice sum over the interactions of next-neighbor spins,

$$S_I \sim \sum_{\substack{\text{next neighbors} \\ (i,j)}} (1 - s_i s_j). \tag{20.}$$

In Equation 18, the spin variables $s_i$ are generalized to the SU(3) color group matrices $U$, and in order to maintain gauge invariance, the product of next-neighbor spins is replaced by that over the four "spins" around the smallest possible closed path in the lattice ("plaquette"). The two terms of Equation 18 thus denote summations over space-space and space-time plaquettes, respectively.

The quark action $S_Q$ in the Wilson formulation of Equation 1 is given by

$$S_Q = \sum_f \bar{\psi}_f (1 - \kappa M)\psi_f, \tag{21.}$$

where the interaction matrix $M$ depends on direction:

$$M_{\mu,nm} = (1 - \gamma_\mu)U_{nm}\delta_{n,m-\hat{\mu}} + (1 + \gamma_\mu)U_{mn}^+ \delta_{n,m+\hat{\mu}}. \tag{22.}$$

Here $\hat{\mu}$ is a unit vector along the lattice link in the $\mu$ direction. The quark coupling strength, the "hopping parameter" $\kappa$, at finite temperature also depends on the link direction, just as $g$ does. The scalar product $\kappa M$ is

$$\kappa M \equiv \kappa_\tau M_0 + \kappa_\sigma \sum_{\mu=1}^{3} M_\mu. \qquad 23.$$

The hopping parameter also reduces to one variable for $a_\sigma = a_\tau \equiv a$:

$$\kappa_\tau(a) = \kappa_\sigma(a) = \kappa. \qquad 24.$$

Since the basic Lagrangian in Equation 1 contains only one coupling, the introduction of a separate quark coupling is a lattice artifact; $\kappa$ must in principle be expressible in terms of $g$. For massless quarks we have in fact (23)

$$\kappa(g) = \tfrac{1}{8}[1 + 0.11g^2 + O(g^4)] \qquad 25.$$

for sufficiently small $g^2$.

With Equations 16–18, 21, and 22, we have a completely defined lattice formulation for the QCD partition function. It gives us $Z[N_\sigma, N_\tau, g = (g_\tau, g_\sigma)]$. To obtain the desired physical partition function $Z(\beta, V)$ and the resulting thermodynamic observables, we choose $a_\sigma = a_\tau$ (of course, after carrying out differentiations such as needed in Equations 11 or 12). The coupling $g$ can then be related to the lattice spacing $a$ by the asymptotic renormalization group relation

$$a\Lambda_\mathrm{L} = \exp\left\{ -\frac{4\pi^2}{(33-2N_f)}\left(\frac{6}{g^2}\right) + \frac{459-57N_f}{(33-2N_f)^2}\log\left[\frac{8\pi^2}{(33-2N_f)}\left(\frac{6}{g^2}\right)\right]\right\}, \qquad 26.$$

with the following reasoning. We want our lattice formulation to provide results independent of the specific lattice used in the evaluation. Renormalization group considerations assure us that this is the case in the vicinity of the fixed point $g = 0$; if coupling $g$ and lattice spacing $a$ are related through the equation

$$a \, dg(a)/da = B(g), \qquad 27.$$

for $g \to 0$, we then recover the continuum theory. Here $B(g)$ is a function of $g$ only; for small $g$ it can be determined in a perturbation expansion, leading to Equation 26, with $\Lambda_\mathrm{L}$ as a dimensional integration constant. In quantitative studies, we must ascertain that at the coupling values used, this solution is indeed valid; although some deviations occur, this seems to be the case for the larger lattices presently used in numerical work (24). With $a(g)$ given by Equation 26, we then have $V = (N_\sigma a)^3$ and $\beta = (N_\tau a)$; this yields $Z(\beta, V)$ from Equation 16 for given $N_\sigma$, $N_\tau$, and $g$.

All physical quantities are through Equation 26 measured in units of the lattice scale $\Lambda_L$. As mentioned, the Lagrangian (Equation 1) contains no dimensional parameter, and hence $\Lambda_L$ is arbitrary. We can thus either consider dimensionless ratios of observables, or calculate a specific quantity, such as the proton or $\rho$ meson mass, to fix $\Lambda_L$ in physical units.

To what extent now is this lattice formulation of statistical QCD equivalent to the continuum form of Equation 13? By letting $a = (x_i - x_j)$ in Equation 15 and subsequently go to zero, we recover the continuum formulation, Equation 13. The converse is not true, however: neither the gluon action (Equation 18) nor the quark action (Equation 22) are unique; various other forms have been considered, which give the same continuum limit (25–31). All physical results should, of course, be independent of the specific choice of action, and finite temperature thermodynamics provides a particularly sensitive test of this "universality." So far, it appears to be quite well satisfied (32).

The quark action leads to some additional problems. If we simply put fermions on the lattice by associating a spinor field with each lattice site, then the derivative in the Lagrangian (Equation 1) leads to the appearance of sixteen degenerate fermions per flavor (7, 30). To avoid this species doubling, Equation 22 gives fifteen of these quarks a mass $m$, with $m \to \infty$ in the continuum limit. Such a procedure, however, also has its difficulties. The continuum Lagrangian (Equation 1) for massless quarks is invariant under chiral transformations (18): massless fermions decompose into independent left-handed and right-handed particles. This invariance is broken in Wilson's lattice form Equations 21 and 22; it is recovered only in the continuum limit. It can in fact be shown (33) that the lattice formulation for massless fermions leads to species doubling, to chiral symmetry breaking, or to nonlocal derivatives. The choice of action thus is to some extent dictated here by the problem under investigation. And it is all the more important to check if all formulations lead to the same results.

## THE COMPUTER SIMULATION OF STATISTICAL QCD

In the preceding section we saw that the lattice formulation of statistical QCD provides a partition function quite similar to that of a generalized spin system. Because even simpler spin systems, such as the three-dimensional Ising model, so far have not been solved analytically, it is not surprising that we also have to take recourse to the standard evaluation method for statistical systems—computer simulation. The advent of modern supercomputers has made the simulation of statistical QCD possible for quite large lattices. Here, as in many other areas of statistical physics (e.g. 34), computer experiments constitute a viable method of obtaining quantitative predictions for systems with many degrees of freedom.

We noted above that the interaction decisive for confinement is contained in the pure gauge field part of the Lagrangian (Equation 1). Gauge field thermodynamics, without quarks, therefore provides a meaningful model for studying the deconfinement transition. As it also imposes less severe computer requirements and avoids the difficulties encountered in the lattice formulation of quarks, it was the first case to be taken up. We follow this order of development and begin with the thermodynamics of SU($N$) gauge fields; to compare the critical behavior of spin and gauge systems, it is of interest to keep $N$ general here. After that, we go on to include dynamical quarks.

## Gauge Field Thermodynamics

The partition function for the SU($N$) gauge field system on the lattice is given by (13)

$$Z(N_\sigma, N_\tau, g^2) = \int \prod_{\text{links}} dU \, \exp[-S_G(U)], \qquad 28.$$

with

$$S_G(U) = \frac{2N}{g_\sigma^2} \xi^{-1} \sum_{P_\sigma} \left(1 - \frac{1}{N} \text{Re Tr } UUUU\right)$$
$$+ \frac{2N}{g_\tau^2} \xi \sum_{P_\tau} \left(1 - \frac{1}{N} \text{Re Tr } UUUU\right). \qquad 29.$$

Here $\xi \equiv a_\sigma/a_\tau$; for $N = 3$, we recover Equation 18. For the energy density (Equation 11), we obtain

$$\varepsilon/T^4 = 6NN_\tau^4[g^{-2}(\bar{P}_\sigma - \bar{P}_\tau) + c_\sigma'(\bar{P} - \bar{P}_\sigma) + c_\tau'(\bar{P} - \bar{P}_\tau)], \qquad 30.$$

with $\bar{P}_\sigma$ and $\bar{P}_\tau$ denoting the lattice average of space-space and space-time plaquettes, respectively:

$$\bar{P}_\sigma = (3N_\sigma^3 N_\tau Z)^{-1} \int \prod_{\text{links}} dU \, \exp[S_G(U)] \left[\sum_{P_\sigma} \left(1 - \frac{1}{N} \text{Re Tr } UUUU\right)\right],$$
$$31.$$

and similarly for $\bar{P}_\tau$. The Euclidean form of the partition function differs from the Hamiltonian version by a normalizing factor corresponding to the $T = 0$ contribution (20). To take this into account, we have subtracted in Equation 30 from $\bar{P}_\sigma$ and $\bar{P}_\tau$ the plaquette average $\bar{P}$ calculated on a large symmetric lattice, of size $N_\sigma^4$ or larger; for the values of $N_\sigma$ usually considered, it gives a good approximation of the zero-point contribution.

The constants $c'_\sigma$ and $c'_\tau$ in Equation 30 arise from the differentiation of the couplings $g_\sigma$ and $g_\tau$ with respect to the temperature; they have been calculated explicitly (35).

Returning for a moment to the structure of Equation 30, we note that the energy density of the gauge field system is obtained from plaquette averages, i.e. from the lattice average of the product of four adjacent "spins" around the smallest closed loop in the lattice. This is the gauge theory equivalent of the conventional Ising model, where the energy would involve the lattice average of the product of two adjacent spins.

For the actual evaluation we now simulate on a computer the $N_\sigma^3 \times N_\tau$ lattice, choosing for convenience $\xi = a_\sigma/a_\tau = 1$. For a given $g$, we place on each link a specific SU($N$) matrix $U$, taking for example all $U = 1$ (ordered or "cold" start), or randomly distributed $U$'s (disordered or "hot" start). Proceeding from this initial configuration, we then assign to every link of the lattice step by step a new matrix $U'$, randomly chosen with the weight $S_G(U)$. After sufficiently many sweeps through the lattice ("iterations"), the plaquette averages stabilize to give $\bar{P}_\sigma$ and $\bar{P}_\tau$ at the chosen value of $g$. Using the renormalization group relation (Equation 26), we have the corresponding temperature—if the chosen $g$ is sufficiently small for this relation to be applicable.

In Figure 1, we show the energy density thus obtained for the SU(2) system, based on calculations using a $10^3 \times 3$ lattice (13); here $\varepsilon$ is normalized to the ideal gas limit

$$\varepsilon_{SB} = (3\pi^2/15)T^4. \qquad\qquad 32.$$

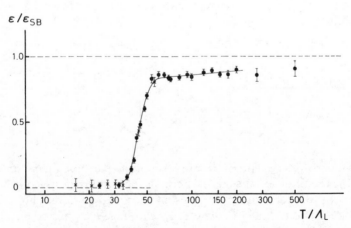

*Figure 1*  The energy density $\varepsilon$ of the SU(2) gauge field system, normalized to its ideal gas limit $\varepsilon_{SB}$; from (13), on a $10^3 \times 3$ lattice.

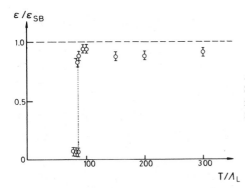

*Figure 2*  The energy density $\varepsilon$ of the SU(3) gauge field system, normalized to its ideal gas limit $\varepsilon_{SB}$; from (37), on an $8^3 \times 3$ lattice.

At high temperatures, the system is seen to behave essentially like a gas of noninteracting gluons. Around $T = 40\Lambda_L$, there is a sudden drop in $\varepsilon$, and below this temperature, the behavior of the system agrees reasonably well with that of an ideal gas of gluonium states of mass $m_G \approx 1$ GeV (36). In Figure 2, the corresponding result is shown for the physically more interesting SU(3) case, evaluated on an $8^3 \times 3$ lattice (37). The behavior is quite similar, but the transition now occurs at $T \approx 80\Lambda_L$ and is discontinuous.

We noted above that the lattice scale $\Lambda_L$ is arbitrary. If we take in each case results for the string tension $\sigma$ at a comparable value of $g$ (10, 38) and use the physical value $\sqrt{\sigma} \approx 400$ MeV, then both the SU(2) and the SU(3) system undergo the transition around

$$T_c \approx 150\text{–}200 \text{ MeV}. \qquad\qquad 33.$$

To assure that the change of behavior exhibited by $\varepsilon$ is indeed due to the deconfinement transition, a corresponding order parameter is needed. It is obtained by noting that the lattice action (Equation 29) possesses a global symmetry under the center $Z_N$ of the SU($N$) gauge group (9a,b). The specific state, in which the system finds itself, may spontaneously break this symmetry, just as the ordered phase of the Ising model breaks the global $Z_2$ symmetry of its Hamiltonian. We are thus looking for the gauge theory analogue of the spontaneous magnetization. It is given (11, 12) by the average of the Polyakov loop

$$L_x(U) = \frac{1}{N} \text{Tr} \prod_{\tau=1}^{N} U_{x;\tau,\tau+1}, \qquad\qquad 34.$$

consisting of the product of all the $U$'s in the temperature direction taken at a given spatial site x. This product becomes a closed loop and thus gauge invariant by the periodicity condition (Equation 14). While $S_G$ is invariant

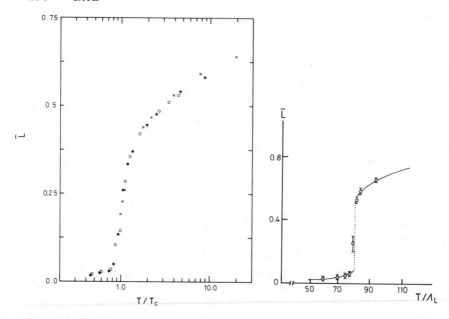

*Figure 3 (Left)*   The deconfinement order parameter $\bar{L}$ of the SU(2) gauge field system, as function of $T/T_c$, with $T_c = 43\ \Lambda_L$; from (32), on a $10^3 \times 3$ lattice, using different actions.

*Figure 4 (Right)*   The deconfinement order parameter $\bar{L}$ of the SU(3) gauge field system; from (39), on an $8^3 \times 2$ lattice.

under global $Z_N$ transformations, the average $\bar{L}$, taken over the lattice and over configurations, acquires a factor $\exp(2r\pi i/N)$, with $0 \leq r \leq N-1$, and thus serves as an indicator for spontaneous symmetry breaking: it is zero for $Z_N$-symmetric states and nonzero if this symmetry is broken. The physical content of the change in symmetry becomes evident by noting that $\bar{L}$ measures the free energy $F$ of a static quark (11, 12):

$$\bar{L} \sim e^{-F/T}.$$    35.

In the confinement regime, $F = \infty$ and hence $\bar{L} = 0$. Once color screening becomes effective, $F$ becomes finite and hence $\bar{L}$ nonzero.[1] The transition from the confined to the deconfined state of the SU($N$) gauge system is thus characterized by the spontaneous breaking of the global center $Z_N$ symmetry.

In Figures 3 and 4 we show lattice results for $\bar{L}$ (32, 39). It thus is clear that

---

[1] This is true provided we remain in the same $Z_N$ phase throughout, which we choose to be the one with $r = 0$.

the transition seen in the energy density behavior is indeed due to deconfinement. The nonzero values of $\bar{L}$ in the confinement regime observed in Figures 3 and 4 are finite lattice size effects (40). The abrupt change of $\varepsilon$ in Figure 2 and that of $\bar{L}$ in Figure 4 suggest that the transition is of first order for the SU(3) system; more detailed studies confirm this (39, 41). The continuous behavior of both $\varepsilon$ and $\bar{L}$ for the SU(2) system, on the other hand, points toward a higher order transition.

This difference in critical behavior for SU(2) and SU(3) gauge systems is in accord with a universality conjecture relating spin and gauge systems (42, 43). By integrating out all degrees of freedom in the partition function (Equation 28) except for the Polyakov loops (Equation 34) on all sites, one obtains from SU($N$) gauge theory a system structurally equivalent to a $Z_N$ spin theory of the same spatial dimensionality. As a result, we expect these theories to show the same critical behavior, and in fact $Z_3$ spin theory (the Potts model) has a discontinuous order/disorder transition (44), while that of the $Z_2$ spin theory (the Ising model) is continuous. More specifically, the critical exponents for the deconfinement transition in SU(2) gauge theory and for the order/disorder transition of the Ising model appear to agree (45), as would be expected if the two theories belong to the same universality class.

Before concluding our survey of gauge field thermodynamics, we want to consider the reliability of the lattice evaluation. The lattice regularization cuts out low and high momenta, which affects even a noninteracting system. This problem can, however, be controlled by comparing gauge theory results with those for an ideal gas evaluated on a lattice of the same size (46). The crucial question is whether the coupling values $g$ used in the actual evaluation are sufficiently small to permit the use of the renormalization group relation (Equation 26). This would assure us that the results obtained correspond to the continuum limit of the theory. In Figure 5 we see a compilation of transition temperature and string tension data for the SU(3) gauge systems (24), shown as function of $6/g^2$. There are clear deviations from scaling, which requires that $T_c$ and $\sigma$ not depend on $g$; it appears possible that the asymptotic form (Equation 26) is valid for $g^2 \lesssim 1$. Even at larger couplings, however, one finds that dimensionless ratios, such as $T_c/\sqrt{\sigma}$ in Figure 5, are independent of $g$. We thus conclude that the lattice results obtained so far are a very promising beginning, but that calculations on larger lattices, combined with finite-size-scaling considerations (47), are certainly necessary.

## QCD Thermodynamics with Quarks

The extension of SU($N$) gauge field thermodynamics to include matter fields has two fundamental consequences.

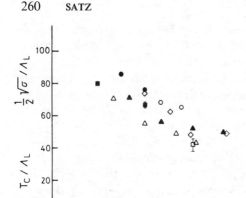

Figure 5 Present lattice results for the deconfinement temperature (*solid and open circles, diamonds*) and the string tension (*solid and open squares, triangles*), both as function of the inverse coupling $6/g^2$; from the compilation in (24).

The presence of the quark term in the Lagrangian (Equation 1) breaks the global invariance under the center of the gauge group (e.g. 48), which holds for the pure gluon system. The physical basis for this effect was already noted in a previous section; for $T > 0$, the color conductivity, just like the electric conductivity, does not vanish identically but is only exponentially small. With the global center symmetry broken, the average Polyakov loop $\bar{L}$ will also remain finite for all $T > 0$. The distinction between the confinement and deconfinement regimes thus becomes somewhat more qualitative in nature (9b, 16, 49).

The second new feature to appear is chiral symmetry. The Lagrangian (Equation 1) is invariant under the chiral transformation

$$\psi \to \psi' = \gamma_5 \psi, \qquad\qquad 36.$$

where $\gamma_5$ is the pseudoscalar Dirac matrix, because it contains only the chirally invariant spinor forms $\bar{\psi}\gamma_\mu\psi$. This invariance would be broken by adding the quark mass term (Equation 3), since $\bar{\psi}\psi$ is not invariant under the transformation in Equation 36. Even for $m_f = 0$, however, chiral invariance may be broken spontaneously; this would correspond to the spontaneous generation of an "effective" quark mass. Since the conduction electrons in a metal acquire an effective mass different from their mass value in vacuum, we may expect such a mass shift here as well. In the confinement regime, chiral symmetry is indeed broken, which leads to constituent quarks with an effective mass of about 300 MeV (for u and d quarks). In the deconfined plasma at high temperature, the quarks become massless again and hence chiral symmetry is restored. In the presence of massless matter fields, statistical QCD thus leads to another transition, from broken to restored chiral symmetry.

What is the relation between deconfinement and chiral symmetry restoration? At present, this question has no really satisfactory answer. There are good arguments that if the two phenomena are distinct, then deconfinement must occur at the lower temperature (50, 51): in general, confining potentials will break chiral symmetry. It may well be, however, that with the global $Z_N$ symmetry broken by the introduction of light quarks, the chiral transition becomes the basic mechanism making deconfinement a genuine phase transition (52).

To obtain for the full partition function (Equation 16) a form suitable for computer simulation, one must integrate over the anticommuting spinor fields. This gives (53)

$$Z(N_\sigma, N_\tau, g) = \int \prod_{\text{links}} dU \, \exp\left[-S_G(U)\right] \left[\det Q(U)\right]^{N_f}, \qquad 37.$$

where

$$Q(U) = [1 - \kappa M(U)] \qquad 38.$$

denotes the fermion matrix in Equation 21. As it connects $\bar\psi$ and $\psi$ over the entire lattice, it is of dimension $(12N_\sigma^3 N_\tau) \times (12N_\sigma^3 N_\tau)$. The evaluation of the determinant of this very large matrix poses at present the main technical problem in the numerical evaluation. For reasonable lattice sizes, it has up to now been carried out only in several approximation schemes:

1. In the hopping-parameter expansion (54, 55)

$$\ln \det(1 - \kappa M) = -\text{Tr} \sum_{l=1}^{\infty} \frac{\kappa^l}{l} M^l, \qquad 39.$$

only the first few leading terms are generally retained for the actual evaluation.

2. In the pseudo-fermion method (56), the quark determinant is written as the integral over complex scalar fields $\Phi$ and $\bar\Phi$

$$(\det Q)^{-1} = \int \prod_{\text{sites}} d\Phi \, d\bar\Phi \, \exp\left[-(\bar\Phi Q \Phi)\right], \qquad 40.$$

which must then be evaluated by a separate Monte Carlo simulation for each configuration of $U$'s.

3. In the microcanonical approach (57), one considers an artificial classical system that is integrable and leads to the partition function (Equation 37). Solving the equations of motion for this system by methods from molecular dynamics, one can then replace the ensemble average (Equation 37) by the corresponding time average, provided ergodicity holds.

The next generation of supercomputers, both faster and with a much larger memory than those employed today, may perhaps bring a precise evaluation within reach. In any case, it is reassuring that all the results obtained so far agree quite well both qualitatively and quantitatively, even though the evaluation schemes differ considerably. In Table 1 we list the calculations carried out up to the end of 1984, indicating both the type of quark action and the fermion determinant scheme used. All find a sudden change in the energy density $\varepsilon$, in $L$ as a measure of confinement, and in the average of $\bar{\psi}\psi$ as a measure of chiral symmetry. For a representative and perhaps most transparent case, we look at the results obtained using Wilson's form Equations 21 and 22 of the quark action, evaluated in the hopping-parameter expansion (Equation 39) (19).

For lattices of small temporal extension $(N_\tau \approx 3\text{--}5)$, the partition function (Equation 37) then has the form

$$Z(N_\sigma, N_\tau, g) = \int \prod_{\text{links}} dU \exp\left[-S_{\text{eff}}(U)\right], \qquad 41.$$

with

$$S_{\text{eff}}(U) \approx S_G(U) - 4N_f(2\kappa)^{N_\tau} \sum_{\text{sites}} \text{Re } L \qquad 42.$$

denoting the effective action in lowest order hopping-parameter expansion. Only closed loops contribute to $S_{\text{eff}}$, and for small $N_\tau$ the dominant contributions are those from Polyakov loops. We note that the action (Equation 41) has the structure of a gauge-invariant spin system in an effective external field (9b, 49, 61), whose strength is $4N_f(2\kappa)^{N_\tau}$. For infinitely heavy ("static") quarks, $\kappa$ vanishes and we recover the pure gauge theory, with $L = 0$ in the confinement regime. For finite quark masses ("dynamic" quarks), $\kappa$ is finite and the external field nonzero; it breaks the global $Z_N$ symmetry and hence makes $L \neq 0$ even when we have confinement. This is how the "ionization" mechanism of local hadron production, discussed above, enters the generalized spin picture.

**Table 1**  Critical parameters in different fermion schemes

| Scheme | $N_\tau$ | $N_f$ | $6/g_c^2$ | $T_c/\Lambda_L^0$ | Ref. |
|---|---|---|---|---|---|
| Wilson $\kappa^4$ | 3 | 2 | $5.30 \pm 0.05$ | $131 \pm 8$ | 19 |
| Wilson $\kappa^5$ | 3 | 2 | $5.25 \pm 0.05$ | $123 \pm 8$ | 19 |
| Kogut-Susskind, canonical | 4 | 3 | 5.3 | 100 | 58 |
| Kogut-Susskind, canonical | 2 | 2 | 4.6 | 89 | 59 |
| Kogut-Susskind, microcanonical | 4 | 4 | 5.1 | 106 | 60 |

The energy density of the quark-gluon system has the form

$$\varepsilon = \varepsilon_G + \varepsilon_Q; \qquad\qquad 43.$$

the gluonic contribution $\varepsilon_G$ is again given by Equation 30, but the plaquette averages are now calculated with the effective action (Equation 41) instead of the pure gluon form (Equation 29). The quark contribution in the lowest order hopping-parameter expansion is given by

$$\varepsilon_Q/T^4 \approx 3N_\tau^3 N_f (2\kappa)^{N_\tau} \, \mathrm{Re} \, \bar{L}. \qquad\qquad 44.$$

The presence of dynamic quarks thus leads to a partial gauge field alignment in an effective external field, and $\varepsilon_Q$ measures the degree of this alignment.

In Figure 6, we show the resulting overall energy density calculated on an $8^3 \times 3$ lattice (19), including terms up to $\kappa^4$ in the hopping-parameter expansion. The relation between $\kappa$ and $g$ is given by Equation 25, that between $g$ and $a$ by Equation 26; we consider the case of two quark flavors. The energy density of the interacting system is again compared to the noninteracting gas limit $\varepsilon_{SB}$ on a lattice of the same size; also shown is the behavior of an ideal gas of $\pi$, $\rho$, and $\omega$ mesons in all their possible charge and spin states. We see that the system at low temperature behaves like a meson gas; around $T/\Lambda_L \approx 150$ it undergoes a sudden transition, and above that

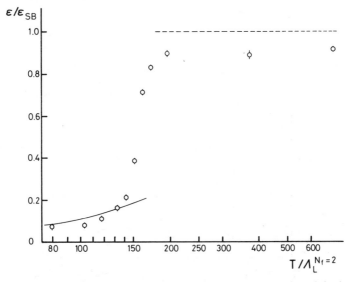

*Figure 6* The energy density in QCD with dynamic quarks, evaluated in hopping-parameter expansion; from (19), on an $8^3 \times 3$ lattice. The solid curve corresponds to an ideal meson gas ($\pi$, $\rho$, $\omega$).

temperature it approaches the ideal gas limit. The deconfinment measure $\bar{L}$ parallels this behavior, as seen in Figure 7; note that with Equation 26, $6/g^2 \approx 5.3$ corresponds to $T/\Lambda_L \approx 150$. We can thus again associate the transition with deconfinement. It should be noted that while $\bar{L}$ is nonvanishing in the confined regime, in accord with the broken $Z_N$ symmetry due to the presence of dynamic quarks, it is, however, quite small there. This agrees with the arguments (9b, 19) presented above, suggesting at most a small symmetry breaking.

As $\bar{\psi}\psi$ is not invariant under the chiral transformation (Equation 36), it provides an order parameter for chiral symmetry restoration. In lowest order hopping-parameter expansion, the lattice average of $\bar{\psi}\psi$ takes the form

$$\langle \bar{\psi}\psi \rangle / T^3 \sim (1 - \mathrm{Re}\,\bar{L}), \qquad\qquad 45.$$

so that chiral symmetry is broken in the deconfinement region, where $\bar{L}$ is small; it is restored as $\bar{L}$ approaches the ideal gas limit of unity. From Equation 42, it is clear that in this approximation the transition points for deconfinement and chiral symmetry restoration coincide. In Wilson's formulation for quarks on the lattice, however, the study of chiral symmetry restoration is not without problems. As was already mentioned, the spurious mass states are removed by making them very massive. It is therefore particularly important to check the transition points for the two

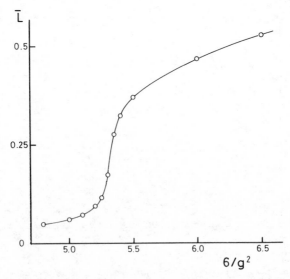

*Figure 7*  The deconfinement measure $\bar{L}$ in QCD with dynamic quarks, evaluated in hopping-parameter expansion; from (19), on an $8^3 \times 3$ lattice.

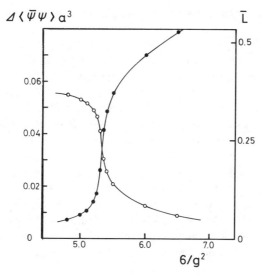

*Figure 8* Deconfinement measure $\bar{L}$ and chiral symmetry measure $\langle\bar{\psi}\psi\rangle$, evaluated in hopping-parameter expansion; from (19), on an $8^3 \times 3$ lattice.

transition phenomena in other formulations as well. We show in Figures 8 and 9 a comparison of $\bar{L}$ and $\langle\bar{\psi}\psi\rangle/T^3$ for Wilson quarks in the hopping-parameter expansion (19) and for Kogut-Susskind quarks in the micro-canonical approach (60): they agree very well, and still further calculations (58, 59) confirm this. It thus appears that deconfinement and chiral symmetry restoration in statistical QCD at vanishing baryon number density occur indeed at the same temperature. A further clarification of the

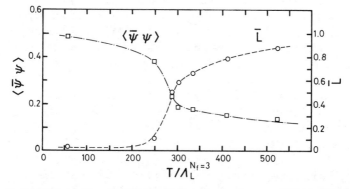

*Figure 9* Deconfinement measure $\bar{L}$ and chiral symmetry measure $\langle\bar{\psi}\psi\rangle$, evaluated in the microcanonical approach; from (60), on an $8^3 \times 4$ lattice.

relation between these phenomena is an interesting challenge in statistical QCD.

To conclude this section, we turn to the question of the actual value of the deconfinement temperature. In Table 1, we list the values of $T_c$ obtained in the different approaches; for purposes of comparison, they are all converted into the units of $\Lambda_L$ for the zero-flavor case. The results of the hopping-parameter expansion are expected to be slightly higher, as the quark masses are not yet very small, when we use Equation 25. Bearing that in mind, we see that deconfinement occurs roughly at

$$T_c/\Lambda_L^0 \approx 100, \qquad\qquad 46.$$

to be compared with the value $T_c/\Lambda_L^0 \approx 80$ for pure gauge field thermo-dynamics on lattices of similar size. For a reliable conversion of this into physical units, we must wait until the evaluation of the hadron spectrum with dynamic quarks is completed (15). At present, the use of available data from comparable lattices gives

$$T_c \approx (200-250) \text{ MeV} \qquad\qquad 47.$$

as the most reasonable value.

## OUTLOOK

### Dense Baryonic Matter

Statistical quantum chromodynamics predicts, as we have seen, a de-confinement transition for strongly interacting matter of vanishing overall baryon number density. In a meson gas, color screening dissolves with increasing temperature the binding between quarks and antiquarks, transforming the system into a chromoplasma of unconfined colored constituents. The decisive element in the screening mechanism is the increase of color charge density, which is here achieved by an increase of temperature resulting in particle production. Matter at high density can, however, also be formed by compressing a system of many nucleons at low temperature; this leads to a high density of baryons and hence also of quarks.

The complete phase diagram of strongly interacting matter must thus describe the phase structure as a function of the temperature $T$ *and* the baryon number density $n_B$ or the corresponding baryonic "chemical" potential $\mu_B$. So far, quantitative predictions from QCD exist only for $\mu_B = 0$. Extending these results to nonzero baryon densities is obviously one of the most urgent problems to be addressed by statistical QCD. The partition function (Equation 10) then becomes more generally

$$Z(T, \mu_B, V) = \text{Tr} \{\exp[-\beta(H - \mu N_B)]\}, \qquad\qquad 48.$$

with $N_B$ denoting the operator for the overall baryon number. Only the first steps toward a viable lattice evaluation have yet been taken (62–65).

In Figure 10, we show a possible schematic phase diagram for strongly interacting matter, which at $\mu_B = 0$ agrees with the results of the previous section. The coincidence of deconfinement and chiral symmetry restoration in that case does not imply similar behavior for $T = 0$ as a function of $\mu_B$. When, in a Mott transition, Debye screening dissolves the local binding between charges, this does not necessarily mean that a state of completely unbound constituents is the energetically most favorable one. Even beyond the transition point, collective binding mechanisms are still possible. The Cooper pairs in superconducting materials provide an example of a such bound state; they can, however, exist only at very low temperatures because thermal motion quickly overcomes the binding force. Something similar could, at least in principle, also occur in strongly interacting matter at low temperature: beyond the deconfinement point, chiral symmetry could still be broken, thus providing us with a system of massive colored quarks as constituents. Increasing either the temperature or the baryonic chemical potential would convert this stage into the true chromoplasma of massless colored quarks and gluons. Statistical QCD will eventually tell us if such an intermediate phase—something like a color superconductor—really exists. For the time being, the question is completely open.

## Deconfinement and Nuclear Collisions

To conclude our survey, we take a short look at the possibility of creating dense strongly interacting matter in the laboratory. For this, the collision of heavy nuclei seems to be our only, certainly not perfect, tool.

When two energetic hadrons undergo a central collision, they essentially pass through each other, leaving behind a streak of energy deposited in a certain space-time region. This energy subsequently decays into the observed hadronic secondaries. In a typical proton-proton collision, the energy deposited is about 0.3 GeV fm$^{-3}$. For the formation of the quark-gluon plasma, we need at least 2.5 GeV fm$^{-3}$, since $\varepsilon/T^4 \approx 12$ and $T_c \approx 200$

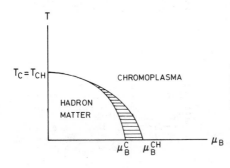

*Figure 10*  Schematic view of a possible phase diagram in QCD, as function of temperature $T$ and baryonic chemical potential $\mu_B$.

MeV. The typical energy deposit achieved in a collision of two identical nuclei of mass number $A$ is expected (24) to be about $(1-3) A^{1/3}$ GeV fm$^{-3}$; for heavy nuclei, this is well above the critical value. This estimate assumes a constant deposit per unit rapidity and hence becomes independent of the collision energy.

A sufficient local energy deposit does not, of course, guarantee the applicability of equilibrium thermodynamics to the process. The volume of space-time in question here is far from infinite. It is therefore crucial that the mean free path of quarks and gluons in the resulting medium be sufficiently small in comparison to the size of the system. Only then can we expect a description in terms of an expanding plasma with a subsequent transition to hadronic matter to be meaningful.

Assuming that in energetic nuclear collisions there is indeed plasma formation, how can we verify this experimentally? The discussion of signatures has so far not led to any unambiguous test, but rather to a number of supporting predictions (66). Since photons, because of the comparatively weak electromagnetic interaction, can escape unmodified if produced in the primordial plasma, they should carry information about this state (67). The transverse momenta of hadronic secondaries act as a measure of the temperatures involved, and studying their dependence on initial energy density should produce evidence for the latent heat of deconfinement (68) depicted in Figures 2 and 6. Finally, if the initial plasma achieves an equilibrium even between the different quark flavors ("chemical equilibrium"), then the resulting flavor ratios should lead to observable consequences for the measured hadron ratios (69).

These and a variety of other aspects of plasma formation in nuclear collisions have been and are at present the subject of intensive studies. The interest they trigger is perhaps best explained by recalling the ultimate aim of this field of research: an understanding of the states of matter in strong interaction physics.

ACKNOWLEDGMENT

It is a pleasure to thank R. Baier, J. Engels, and B. Petersson for a critical reading of the manuscript.

*Literature Cited*

1. Kajantie, K., ed. *Quark Matter 1984*; Proc. 4th Int. Conf. on Ultra-Relativistic Nucleus-Nucleus Collisions. Berlin/Heidelberg/New York: Springer-Verlag (1985)
2. *RHIC and Quark Matter*; Proposal for a Relativistic Heavy Ion Collider at BNL, *Rep. BNL 51801*. Upton, NY: Brookhaven Natl. Lab. (1984)
3. Pomeranchuk, I. *Dokl. Akad. Nauk. SSSR* 78:889 (1951); Hagedorn, R. *Nuovo Cimento Suppl.* 3:147 (1965)

4. Morley, P. D., Kislinger, M. B. *Phys. Rep.* 51:63 (1979); Shuryak, E. V. *Phys. Rep.* 61:71 (1980)
5. Kapusta, J. *Nucl. Phys. B* 148:461 (1979)
6. Khalashnikov, O. K., Klimov, V. V. *Phys. Lett.* 88B:328 (1979)
7. Wilson, K. *Phys. Rev. D* 10:2445 (1974); Wilson, K. In *New Phenomena in Subnuclear Physics*, ed. A. Zichichi. New York: Plenum (1977)
8. Borgs, C., Seiler, E. *Nucl. Phys. B* 215:125 (1983)
9a. Polyakov, A. M. *Phys. Lett.* 72B:477 (1978)
9b. Susskind, L. *Phys. Rev. D* 20:2610 (1979)
10. Creutz, M. *Phys. Rev. Lett.* 43:553 (1979); *Phys. Rev. D* 21:2308 (1980)
11. McLerran, L. D., Svetitsky, B. *Phys. Lett.* 98B:195 (1981); *Phys. Rev. D* 24:450 (1981)
12. Kuti, J., Polónyi, J., Szlachányi, K. *Phys. Lett.* 98B:199 (1981)
13. Engels, J., Karsch, F., Montvay, I., Satz, H. *Phys. Lett.* 101B:89 (1981); *Nucl. Phys. B* 205[FS5]:545 (1982)
14. Politzer, H. D. *Phys. Rep.* 14:129 (1974); Altarelli, G. *Phys. Rep.* 81:1 (1982)
15. Hasenfratz, A., Hasenfratz, P. *Ann. Rev. Nucl. Part. Sci.* 35:559 (1985)
16. Satz, H. *Nucl. Phys. A* 418:447c (1984)
17. Mott, N. F. *Rev. Mod. Phys.* 40:677 (1968) and references given there
18. Marciano, W., Pagels, H. *Phys. Rep.* 36:137 (1978); Wilczek, F. *Ann. Rev. Nucl. Part. Sci.* 32:177–209 (1982)
19. Çelik, T., Engels, J., Satz, H. *Nucl. Phys. B* 256:670 (1985)
20. Bernard, C. *Phys. Rev. D* 9:3312 (1974)
21. Wegner, F. J. *J. Math. Phys.* 10:2259 (1971)
22. Hasenfratz, A., Hasenfratz, P. *Nucl. Phys. B* 193:210 (1981)
23. Kawamoto, N. *Nucl. Phys. B* 190 [FS3]:617 (1981)
24. Satz, H. See Ref. 1
25. Villain, J. *J. Phys.* 36:581 (1975); Drouffe, M. *Phys. Rev. D* 18:1174 (1978)
26. Manton, N. S. *Phys. Lett.* 96B:328 (1980)
27. Bhanot, G., Creutz, M. *Phys. Rev. D* 24:3212 (1981)
28. Christ, N. H., Friedberg, R., Lee, T. D. *Nucl. Phys. B* 10[FS6]:310 (1982)
29. Szymanzik, K. *Nucl. Phys. B* 226:187 (1983)
30. Kogut, J., Susskind, L. *Phys. Rev. D* 11:395 (1975); Susskind, L. *Phys. Rev. D* 16:3031 (1977)
31. Drell, S. D., Weinstein, M., Yankielowicz, S. *Phys. Rev. D* 14:487 (1976)
32. Gavai, R. V., Karsch, F., Satz, H. *Nucl. Phys. B* 220[FS8]:223 (1983)
33. Nielsen, H. B., Ninomiya, M. *Nucl. Phys. B* 185:20 (1981)
34. Binder, K., ed. *Monte Carlo Methods in Statistical Physics*. Berlin/Heidelberg/ New York: Springer-Verlag (1979)
35. Karsch, F. *Nucl. Phys. B* 205[FS5]:285 (1982)
36. Engels, J., Karsch, F., Satz, H. *Phys. Lett.* 102B:332 (1981)
37. Çelik, T., Engels, J., Satz, H. *Phys. Lett.* 129B:323 (1983)
38. Bhanot, G., Rebbi, C. *Nucl. Phys. B* 180[FS2]:469 (1981); Gutbrod, F., Hasenfratz, P., Kunszt, Z., Montvay, I. *Phys. Lett.* 128B:415 (1983)
39. Kogut, J., et al. *Phys. Rev. Lett.* 50:393 (1983)
40. Çelik, T., Engels, J., Satz, H. *Z. Phys. C* 22:301 (1984)
41. Çelik, T., Engels, J., Satz, H. *Phys. Lett.* 125B:411 (1983)
42. Polónyi, J., Szlachányi, K. *Phys. Lett.* 110B:395 (1982)
43. Svetitsky, B., Yaffe, L. G. *Nucl. Phys. B* 210[FS6]:423 (1982)
44. Blöte, H. W. J., Swendsen, R. H. *Phys. Rev. Lett.* 43:799 (1979); Knak Jensen, S. J., Mouritsen, O. G. *Phys. Rev. Lett.* 43:1736 (1979)
45. Gavai, R. V., Satz, H. *Phys. Lett.* 145B:248 (1984)
46. Engels, J., Karsch, F., Satz, H. *Nucl. Phys. B* 205[FS5]:239 (1982)
47. Kennedy, A. D., Kuti, J., Meyer, S., Pendleton, B. *Phys. Rev. Lett.* 54:87 (1985)
48. Çelik, T., Engels, J., Satz, H. *Phys. Lett.* 133B:427 (1983)
49. Banks, T., Ukawa, A. *Nucl. Phys. B* 225[FS9]:145 (1983)
50. Shuryak, E. V. *Phys. Lett.* 107B:103 (1981); *Nucl. Phys. B* 203:140 (1982)
51. Pisarski, R. D. *Phys. Lett.* 110B:155 (1982)
52. Alessandrini, V. *Orsay Rep. LPTHE* 84/14 (1984)
53. Mathews, T., Salam, A. *Nuovo Cimento* 12:563 (1954)
54. Lang, C. B., Nicolai, H. *Nucl. Phys. B* 200[FS4]:135 (1982)
55. Hasenfratz, A., Hasenfratz, P., Kunszt, Z., Lang, C. B. *Phys. Lett.* 110B:289 (1982)
56. Fucito, F., Marinari, E., Parisi, G., Rebbi, C. *Nucl. Phys. B* 180[FS2]:369 (1981)
57. Polónyi, J., Wyld, H. W. *Phys. Rev. Lett.* 51:2257 (1983)
58. Fucito, F., Solomon, S. *Cal. Tech. Rep.* CALT-68-1084 (1984); Fucito, F., Rebbi, C., Solomon, S. *Cal. Tech. Rep.* CALT-68-1127 (1984)
59. Gavai, R. V., Lev, M., Petersson, B. *Phys. Lett.* 140B:397 (1984); 149B:492 (1984)

60. Polónyi, J., Wyld, H. W., Kogut, J. B., Shigemitsu, J., Sinclair, D. K. *Phys. Rev. Lett.* 53:644 (1984)
61. Hasenfratz, P., Karsch, F., Stamatescu, I. O. *Phys. Lett.* 133B:211 (1983)
62. Hasenfratz, P., Karsch, F. *Phys. Lett.* 125B:308 (1983)
63. Kogut, J., et al. *Nucl. Phys. B* 225:93 (1983)
64. Gavai, R. V., Ostendorf, A. *Phys. Lett.* 132B:137 (1983)
65. Bilić, N., Gavai, R. V. *Z. Phys. C* 23:77

(1984)
66. Gyulassy, M. *Nucl. Phys. A* 418:59c (1984); Cleymans, J., Gavai, R. V., Suhonen, E. *Phys. Rep.* In press (1985)
67. Shuryak, E. V. *Sov. J. Nucl. Phys.* 28:408 (1978); Kajantie, K., Miettinen, H. I. *Z. Phys. C* 9:341 (1981); 14:357 (1982)
68. Van Hove, L. *Phys. Lett.* 118B:138 (1982)
69. Rafelski, J. *Nucl. Phys. A* 418:215c (1984)

*Ann. Rev. Nucl. Part. Sci. 1985. 35 : 271–94*

# D-STATE ADMIXTURE AND TENSOR FORCES IN LIGHT NUCLEI

*T. E. O. Ericson*

CERN, CH-1211 Geneva 23, Switzerland


*M. Rosa-Clot*

Department of Physics, University of Pisa, and INFN, Sezione di Pisa, I-56100 Pisa, Italy

CONTENTS

0163–8998/85/1201–0271$02.00

# 1.  INTRODUCTION

## 1.1   Early Evidence for Tensor Interaction

As early as 1939 Schwinger (1) realized that the quadrupole moment of the deuteron (2) provided decisive evidence for a tensor component in the nuclear force : the bound state of the n-p system is not spherically symmetric as required by a central potential interaction, but has a D-state admixture in the wave function so that the deuteron is prolate with respect to its spin direction. Even in those early days of nuclear physics, the phenomenological analysis of the low-energy n-p scattering and the deuteron properties revealed that the tensor force was crucial, accounting for 70% of the binding interaction in spite of the small D-state admixture (3, 4).

The close relation between the deuteron quadrupole moment and the long-range tensor potential generated by vector or pseudoscalar meson exchange was realized by Bethe (3) years before the discovery of the $\pi$ meson (5). Pauli (6) used this fact as evidence for the exchange of a light pseudoscalar isovector meson.

Stimulated by these results, Gerjuoy & Schwinger (7) explored the consequences of the phenomenological tensor force for the 3- and 4-nucleon systems using variational techniques. The conclusion was that the tensor force was also crucial to the binding of these systems, accounting for 50% and 40% of the binding interaction respectively. The wave functions were found to contain D-state components generated by the tensor force with probability similar to that in the deuteron. Unlike the deuteron, however, the $^3$H, $^3$He, and $^4$He systems have spins 1/2 and 0, so that there is no longer any qualitative signature of the D-state components in the ground-state observables (e.g. quadrupole moment).

## 1.2   Qualitative Importance of Pion Tensor Interaction

The next major advance was the systematic introduction in the mid 1950s (8) of the static one-pion exchange (OPE) potential in the description of the long-range N-N potential : the pion exchange gives a well-defined and natural dynamical origin to the strong tensor force in the long- and intermediate-range region. Although the uncertainties in the description could not be well assessed, it became clear in numerous descriptions using N-N potentials with OPE that the deuteron quadrupole moment is given to a precision of 10% (9). These results were supported by dispersion relations (10). The importance of OPE for the deuteron is also clear from the fact that such OPE models consistently give an effective range close to the experimental value. An additional quantity that could give further information was the asymptotic D/S ratio $\eta$ in the deuteron, but lack of data made comparison with theory impossible (11, 12).

A major experimental advance occurred with the introduction of tensor polarized deuteron beams. This technique has permitted precise measurements of the deuteron D/S ratio by a variety of methods (13–15). In addition, direct measurements of the asymptotic D-state components in both 3- and 4-nucleon systems have become available (16–19), leaving it beyond doubt that these components exist and have qualitatively the expected magnitudes.

For the deuteron these experimental advances were matched and even surpassed on the theoretical level by the recent development of an approach in which the theoretical uncertainty can be explicitly assessed and in which the values of $Q$ and $\eta$ can be directly linked to details of the N-N interaction, notably OPE (20, 21).

## 2.   THE ONE-PION EXCHANGE TENSOR POTENTIAL

The tensor interaction, responsible for D-state phenomena in light nuclei, originates mainly in the OPE interaction. Consider the limit of two nucleons, 1 and 2, with large mass $M$ at a relative distance $\mathbf{r} = r\hat{\mathbf{r}} = \mathbf{r}_1 - \mathbf{r}_2$. The pion exchange gives rise to a potential that has the characteristic form of a dipole-dipole interaction:

$$V_\pi(\mathbf{r}) = f^2(\boldsymbol{\tau}_1 \cdot \boldsymbol{\tau}_2)(\boldsymbol{\sigma}_1 \cdot \mathbf{V})(\boldsymbol{\sigma}_2 \cdot \mathbf{V}) \frac{e^{-m_\pi r}}{r}.$$    1.

Here $f^2 \approx 0.08$, $\tau_i$ and $\sigma_i$ denote the nucleon isospin and spin operators, and $m_\pi$ is the pion mass. The structure of this potential is identical to the interaction between two classical electric dipoles $\mathbf{d}_1$ and $\mathbf{d}_2$ with the replacements $\mathbf{d}_i \to \tau_i f \sigma_i$, where $i = 1, 2$ and $r^{-1} \to \exp(-m_\pi r)/r$.

The well-known tensor interaction resulting from two electric (or two magnetic) dipoles therefore has an analogy in the pionic tensor potential $(\boldsymbol{\tau}_1 \cdot \boldsymbol{\tau}_2) S_{12}(\hat{\mathbf{r}}) V_T(r)$ with the tensor operator $S_{12}(\hat{\mathbf{r}}) = 3(\boldsymbol{\sigma}_1 \hat{\mathbf{r}})(\boldsymbol{\sigma}_2 \cdot \hat{\mathbf{r}}) - (\boldsymbol{\sigma}_1 \cdot \boldsymbol{\sigma}_2)$ and

$$V_T(r) = f^2 \frac{1}{3}\left[ 1 + \frac{3}{(m_\pi r)} + \frac{3}{(m_\pi r)^2} \right] \frac{e^{-m_\pi r}}{r}.$$    2.

This form of the tensor potential results both from a pseudoscalar and a pseudovector $\pi NN$ coupling in the static limit with the well-known relation between the corresponding coupling constant $f = g m_\pi / 2M$ (22).

The basic property of the tensor interaction is that it mixes orbital angular momentum states. In its absence the deuteron would be in a pure S-state and it would have no quadrupole moment. The OPE tensor interaction is very strong and it is mainly responsible for the deuteron binding. In fact the central OPE potential in the deuteron channel ($S = 1$, $T = 0$) is three times too weak to bind the system in the absence of the tensor force.

## 3. DEUTERON EQUATIONS AND D-STATE OBSERVABLES

### 3.1  *Asymptotic Amplitudes and Effective Range*

The normalized deuteron wave function can be written, using the S- and D-state radial wave functions $u(r)$ and $w(r)$, as

$$\psi_{J,M} = \frac{1}{\sqrt{4\pi}} \left\{ \frac{u(r)}{r} + \frac{w(r)}{r} \frac{1}{\sqrt{8}} S_{12}(\hat{\mathbf{r}}) \right\} \chi_{1M}, \qquad\qquad 3.$$

where $\chi_{1M}$ is the normalized spin wave function.

Outside the range of the nuclear interaction, the radial wave functions are completely determined by the deuteron binding energy $\varepsilon$ and the asymptotic normalization constant $A_S$ and $A_D = \eta A_S$:

$$u(r) \equiv A_S \tilde{u}(r) \xrightarrow[r \to \infty]{} A_S e^{-\alpha r}, \qquad\qquad 4a.$$

$$w(r) \equiv A_D \tilde{w}(r) \xrightarrow[r \to \infty]{} \eta A_S [1 + 3/\alpha r + 3/(\alpha r)^2] e^{-\alpha r}, \qquad\qquad 4b.$$

where $\alpha^2 = M\varepsilon$ and $M_R = M$ is the reduced mass of the nucleons. Both $A_S$ and $A_D$ are external properties of the deuteron, like the quadrupole moment, and they can be determined experimentally. The asymptotic S-wave amplitude $A_S$ is deduced to high precision both from the deuteron radius (23) and from effective range theory (24):

$$(1 + \eta^2) A_S^2 = 2\alpha / [1 - \alpha \rho(-\varepsilon, -\varepsilon)], \qquad\qquad 5.$$

where $\rho(-\varepsilon, -\varepsilon)$ is the diagonal effective range parameter defined by

$$\rho(-\varepsilon, -\varepsilon) = 2 \int_0^\infty dr \, [e^{-2\alpha r} - \tilde{u}(r)^2 - \tilde{w}(r)^2]. \qquad\qquad 6.$$

The D/S asymptotic ratio $\eta$ has only recently been measured to high precision. The experimental values are given in Table 1 (Section 6).

### 3.2  *The Coupled Channel Equations*

Consider now the deuteron in a potential description. The S- and D-wave functions then satisfy the coupled equations:

$$u''(r) = [\alpha^2 + U_{00}(r)] u(r) + U_{02}(r) w(r), \qquad\qquad 7a.$$

$$w''(r) = [\alpha^2 + 6/r^2 + U_{22}(r)] w(r) + U_{20}(r) u(r), \qquad\qquad 7b.$$

where the terms $U_{ij} = MV_{ij}$ are linear combinations of the central, tensor, spin-orbit, and quadratic spin-orbit potentials $V_C$, $V_T$, $V_{LS}$, and $V_{LL}$: $V_{00} = V_C$; $V_{02} = V_{20} = \sqrt{8}V_T$; $V_{22} = V_C - 2V_T - 3V_{LS} - 3V_{LL}$.

One notes from Equations 7a,b that the S-D coupling is produced only by the nondiagonal tensor interaction. The D-state admixture in the deuteron is therefore generated from the S-state by the tensor term $U_{20}(r)u(r)$, which acts as a D-wave source in Equation 7b.

The interaction in the D-state itself is otherwise repulsive, because of both the centrifugal barrier and the interaction potential $V_{22}$. This repulsion reduces the contributions to the asymptotic D-wave parameter $\eta$ from the short-range interaction region.

### 3.3 The Quadrupole Moment

In a potential description, the quadrupole moment is usually broken into two terms $Q = Q_1 + Q_2$:

$$Q_1 = 1/\sqrt{50} \int_0^\infty dr \, r^2 u(r)w(r), \qquad \text{8a.}$$

$$Q_2 = -1/20 \int_0^\infty dr \, r^2 w^2(r). \qquad \text{8b.}$$

The main term $Q_1$ is linear in the D-wave function $w(r)$, while the term $Q_2$, which is quadratic in $w(r)$, gives only a $-6\%$ correction. Since the integrand of $Q_1$ is weighted by $r^2$, its main contributions are from the asymptotic region of the D-wave function. Therefore, the quadrupole moment is closely related to the asymptotic D/S parameter $\eta$. In addition, meson exchange currents give small corrections of a few percent originating at rather short range (25–28).

## 4. THE DEUTERON WAVE FUNCTION AND OPE

### 4.1 The Pure OPE Wave Functions, Q and $\eta$

All modern N-N potentials (29–33) contain the OPE as the long-range interaction. They all agree that the tensor component is responsible for about 70% of the deuteron binding interaction, the bulk of it coming from OPE. On the other hand, the binding energy $\varepsilon$ is quite small, and results from a delicate cancellation between potential attraction and kinetic energy repulsion. As a consequence, the precise value of $\varepsilon$ is very sensitive to details of the short-range interaction, since these provide the fine-tuning of $\varepsilon$ to the observed value.

Since the binding energy also sets the scale of the deuteron size, it is extremely important to discuss other observables in the deuteron within a description that reproduces it correctly. This was done by Glendenning & Kramer (9) for a variety of models with the correct long-range OPE potential. The remarkable conclusion was that observables such as $Q$, $A_S$, and $\eta$ are insensitive to the short-range features and in good agreement with experiments. Even a pure OPE model with a short-range cut-off radius $r_0 = 0.48$ fm so as to reproduce the correct binding energy gives values for these quantities similar to more sophisticated approaches. The main differences occur in the effective range parameter and in the D-state probability. The stability of the results are such that we can conclude with the authors "it just happens to be one of the quirks of nature that the force that binds the deuteron does so in such a way as not to reveal itself too intimately" (9).

This conclusion can be better understood and improved upon by studying the solutions to the deuteron equations 7a,b with a pure OPE interaction and using the correct binding energy $\varepsilon$ (1, 34, 54). The coupled Schrödinger equation can then be integrated inward from large $r$ for any given value of the D/S ratio $\eta$ (see Figure 1). The solution will, of course, generally be irregular near the origin both for $\tilde{u}(r)$ and $\tilde{w}(r)$, since OPE does not reproduce $\varepsilon$. It is remarkable, however, that either the S-wave function or the D-wave function can be made regular for values of $\eta$ differing by only 1%, with $\eta_{\mathrm{OPE}} = 0.0274$ giving the regular D-wave function. A second remarkable feature is that the S-wave function $\tilde{u}(r)$ is nearly identical to sophisticated modern wave functions apart from the region inside 0.6 fm.

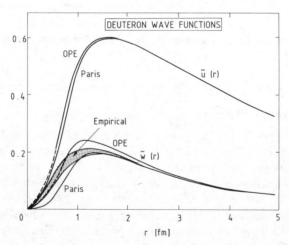

*Figure 1*    Deuteron S- and D-wave functions for a modern N-N potential (29) compared to the OPE deuteron "wave functions" (from 34). The empirical wave functions are deduced from (34b).

This is nearly entirely a consequence of the long-range part of the OPE interaction. On the other hand, the same comparison for the D-wave function reveals clear differences in the inner region in spite of the excellent asymptotic behavior.

These OPE S- and D-"wave functions" can be normalized. Although they do not correspond to eigensolutions of the Schrödinger equation, they can then be used to calculate a OPE quadrupole moment using Equations 8a,b; the result is $Q_{OPE} = 0.284$ fm$^2$ for $f^2 = 0.078$. The corresponding OPE diagonal effective range parameter from Equation 6 gives $\rho_{OPE}(-\varepsilon, -\varepsilon) \approx 1.74$ fm.

These OPE results are close to the experimental values $\eta = 0.0271$, $Q = 0.2859$ fm$^2$, and $\rho(-\varepsilon, -\varepsilon) = 1.767$ fm (21), which emphasizes that all these quantities primarily follow from the OPE interaction. The main deviation occurs in the effective range parameter $\rho(-\varepsilon, -\varepsilon)$, which experimentally is 2% larger than the OPE value.

## 4.2  OPE Effective Range and D-State Probability

Since the S-wave function $\tilde{u}(r)$ is very stable, we can assume it to be known with sufficient accuracy for an approximate determination of the D-state probability. Define the effective range discrepancy $\delta\rho$ as

$$\delta\rho = \rho_{exp}(-\varepsilon, -\varepsilon) - 2 \int_0^\infty dr\, [e^{-2\alpha r} - \tilde{u}(r)^2].$$ 9.

The D-state probability $P_D$ is then (34)

$$P_D = \tfrac{1}{2} A_S^2 \delta\rho.$$ 10.

Using the Paris wave function (29) and the OPE wave function as two limits, one obtains from the experimental effective range an 8% discrepancy $\delta\rho = 0.0147 \pm 0.0020$, where the uncertainty is mainly due to the model dependence of $\tilde{u}(r)$ in the region 1 to 1.5 fm. From this value, the D-state probability is obtained very directly (but not approximately) as $P_D = (6 \pm 1)\%$, while $P_D^{OPE} = 7.85\%$ using the OPE D-"wave function."

This result should be compared to the historical extraction of $P_D$ from the orbital contribution to the small $(-2.6\%)$ discrepancy between the deuteron magnetic moment and the sum of the proton and neutron ones, which is very model dependent with many incontrollable short-range corrections (23).

The D-state probability is not an observable, as was emphasized by Amado (36) and Friar (37), but depends on the theoretical description. In fact, unitary transformations can be applied to the wave functions, thereby preserving phase shifts and observables while changing the D-state probability. Such results are also obtained from equivalent descriptions with energy-dependent potentials. These limitations do not negate the usefulness of the concept of $P_D$ for practical purposes, but indicate that accurate theoretical discussions should concentrate on direct observables like $Q$ and $\eta$.

### 4.3    The Empirical Deuteron Wave Function

The experimental invariant structure functions $A(q^2)$ and $B(q^2)$ in elastic electron-deuteron scattering are consistent with these results. Both $A(q^2)$ and $B(q^2)$ are dominated by D-state contributions for squared momentum transfers $q^2 \gtrsim 15\,\text{fm}^2$, with significant additional contributions from meson exchange currents in $B(q^2)$ (34a). Locher and Švarc (34b) have parametrized the corresponding deuteron S and D vertex functions and deduced their empirical form using the static deuteron observables $[Q, \eta, \rho(-\varepsilon, -\varepsilon), \varepsilon]$ as additional constraints, as well as the condition of deuteron wave functions that are regular at the origin. The corresponding empirical D-wave function (see Figure 1) has a D-state probability $P_D = (6.3 \pm 0.5)\%$, very similar to the previous value. It is, however, difficult at present to evaluate the systematic uncertainty in this empirical wave function.

## 5.    MODERN THEORY OF THE DEUTERON D-STATE OBSERVABLES

The previous discussion emphasizes the intimate relation between the D-state observables and the OPE tensor interaction. The standard evaluations from the N-N interaction solving the Schrödinger equation do not permit a clear estimate of the systematic uncertainties in the theory nor do they clearly exhibit the nature and reliability of the correction terms. It is clear, however, that the agreement between theory and experiment is excellent.

Recent developments have remedied these defects, so that the precisely known D-state observables become powerful tests of models of the N-N force in the intermediate- and long-range region. In particular, Ericson & Rosa-Clot (20, 21, 38) have developed a rigorous constructive approach to the D/S ratio and quadrupole moment, both of which are expressed directly in terms of the underlying potential. This permits one to visualize the finer details in the N-N force directly, such as the intermediate-range region contributions and the strong suppression of the short-range contribution.

### 5.1    The Integral Formulation for $\eta$

The D-state wave function of Equation 7b can be formally obtained using a Green function technique with the tensor coupling $U_{20}(r)\tilde{u}(r)$ to the S-state viewed as a source term (39). Ericson & Rosa-Clot (20, 21) realized that the asymptotic D/S ratio $\eta$ is given by the following simple and exact expression:

$$\eta = \int_0^\infty dr\ \eta(r) = \sqrt{8M} \int_0^\infty dr\ r \mathscr{J}_2(r) V_T(r) \tilde{u}(r).$$    11.

Here $\mathscr{J}_2(r)$ is the regular solution of the homogeneous D-wave equation normalized to the corresponding spherical Bessel function at large $r$:

$$\mathscr{J}_2(r) \xrightarrow[r\to\infty]{} j_2(i\alpha r).$$

Two features of this formula are important for further discussion and make it particularly convenient. First, as already mentioned, the S-wave function $\tilde{u}(r)$ in the integrand is only very weakly dependent on the detailed description of the N-N interaction. Second, the diagonal D-state potential $V_{22}$ is strongly repulsive as a result of the OPE tensor potential. Consequently $\mathscr{J}_2(r)$ is suppressed for small $r$, since it represents an exponentially decreasing tunneling solution toward the origin produced by the repulsive barrier of the centrifugal and N-N forces. The contribution from the short-range region to the integrand in Equation 11 is controllably suppressed by this mechanism.

## 5.2    Discussion of Corrections and Uncertainties

The characteristic shape of the D/S density curve $\eta(r)$ is shown in Figure 2 for the OPE interaction and for a typical modern N-N potential (29) using the same $\pi$NN coupling constant $f^2 = 0.078$ and the same S-wave function $\tilde{u}_{\text{Paris}}(r)$ in both cases. The curves are strikingly similar, differing only by

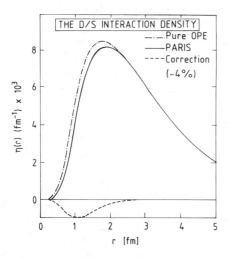

*Figure 2*   Comparison of $\eta(r)$ for OPE and Paris potential, as well as the difference between them. The total area of the difference is $-4.6\%$ (from 21).

4.6% with $\eta_{OPE} = 0.02762$ and $\eta_{Paris} = 0.02633$. Only a few percent of the contribution to $\eta$ originate inside of 1 fm and less than 12% inside of 1.5 fm. The entire sophistication of the modern N-N potential, including theoretical $2\pi$ exchange, amounts therefore to only a 4–5% correction arising nearly entirely from the intermediate-range region. We now discuss how to control the leading corrections and uncertainties without appealing to a particular model (21).

5.2.1   THE S-WAVE FUNCTION $\tilde{u}(r)$   Different wave functions have little effect on $\eta$ with a variation that is approximately proportional to the variation in the D-state probability: $\delta\eta/\eta \approx 1/3\delta P_S$. Since a variation of about 1% is the largest that can be envisioned in $P_D = (1 - P_S)$, the uncertainties in $\eta$ of this origin are well below 1%.

We note at this point that the small difference between $\eta_{OPE} = 0.0275$ of the previous section and $\eta = 0.02762$ from the integral relation in Equation 11 is due to the different S-wave functions.

5.2.2   THE $\pi$NN COUPLING CONSTANT   In the OPE potential it is important to account consistently for the masses and coupling constants of neutral ($\pi^0$) and charged ($\pi^c$) pions. To a precision of a few parts per thousand this is achieved simply by using the averaged pion mass $m_\pi = 1/3(m_{\pi^0} + 2m_{\pi^c})$ = 138.032 MeV and $f^2 = 1/3(f_0^2 + 2f_c^2)$ (21). The most recent compilation of $\pi$NN coupling constant (40) gives an experimental value $f^2 = 0.0776\,(9)$. This number includes two new determinations of $f_0^2$ from p-p scattering (40, 41) that substantially improve the precision in this quantity. The D/S parameter $\eta$ is, to leading order, proportional to $f^2$, which therefore contributes $\pm 1.2\%$ to its theoretical uncertainty.

5.2.3   THE $2\pi$ EXCHANGE POTENTIAL   The principal correction to the OPE result arises from the $2\pi$ exchange potential (30–33, 42), which is a theoretical ingredient in modern N-N potentials. The two pions have a continuous mass distribution. The weight factor for any given mass is closely linked to the $\pi$-N scattering amplitude, as well as to the $\pi$-$\pi$ scattering amplitude in S- and P-states. The mass distribution can be quantitatively evaluated using a dispersion approach for masses up to about $7m_\pi$ using the experimental information on $\pi$-N and $\pi$-$\pi$ scattering. Its shape is quite well known, but it has a 15% overall normalization uncertainty (B. Loiseau, private communication). The dominant correction to $\eta$ comes from the $2\pi$ isovector tensor potential, which is strongly dominated by P-wave $\pi$-$\pi$ exchange (the "$\rho$ meson" channel).

These contributions to $\eta$ come mainly from distances larger than 1 fm. The $2\pi$ exchange terms give a theoretically well-founded $(-4.3\pm0.6\%)$ overall correction (21).

5.2.4    THE πNN FORM FACTOR    In the description of the OPE potential, the nucleons can be considered to be point-like only in the idealized, long-wavelength approximation. In reality the πNN vertex has a form factor, which can be parametrized as

$$K(t) = \frac{\Lambda^2 - m_\pi^2}{\Lambda^2 + t} \qquad\qquad 12.$$

corresponding to an equivalent uniform radius $R = \sqrt{10}/\Lambda$. The πNN form factor is associated with at least $3\pi$ exchange because of quantum numbers, so that the previous investigation of $2\pi$ contributions gives no indication of its importance.

Empirically, the consequence of form factors with varying cut-off parameters $\Lambda$ for the tensor potential and for $\eta(r)$ is displayed in Figures 3 and 4. In comparison with the corresponding contributions from a modern N-N potential (21), one concludes that its shape is simulated by $\Lambda \approx 750$ MeV/$c$. This should not lead one to believe, however, that this provides evidence for a πNN form factor with this $\Lambda$, since both $\eta(r)$ and $V_T(r)$ have nearly this shape using only the theoretical $2\pi$ contributions with a point-like πNN vertex.

Direct experimental evidence of the πNN form factor is poor. Dominguez & VerWest (44) deduce from nearly forward n-p and p-p charge exchange reactions at 8 GeV/$c$ a value $\Lambda \approx 900$ MeV/$c$ (i.e. $R = 0.6$ fm). Holinde (30) points out that a low value for $\Lambda$ would lead to dramatic effects in the lower partial waves in N-N scattering. In particular the $^3D_1$ phase shift would be profoundly modified, even changing its sign for $\Lambda < 1000$ MeV/$c$, which indicates a value in the range $1200 < \Lambda < 1400$ MeV/$c$.

The strongest empirical argument for a low value of $\Lambda$ is the fact that the

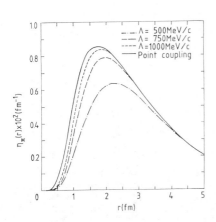

*Figure 3* Dependence of $\eta(r)$ on monopole form factors for the πNN vertex (Equation 12). The integrand decrease by 3.1, 8.5, and 19% respectively for the three cases. The corresponding uniform source radii are $R = 0.63$, 0.84, and 1.26 fm (from 21).

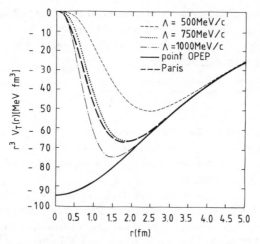

*Figure 4* The OPE tensor potential for the various monopole form factors compared with the Paris tensor potential. Same radii as for Figure 3 (from 21).

axial form factor for the nucleon has $\Lambda \approx 700$ MeV/$c$ (45) and that pionic and axial phenomena are closely related because of their similarity in quantum numbers. Ericson & Rosa-Clot (21) point out that such a low value for $\Lambda$ would lead to a large ($-10\%$) correction to $\eta$ and an even larger ($-16\%$) correction to $Q$, which would be very difficult to accommodate.

Theoretically there are two different approaches to the form factor leading to very different predictions. The Quark Bag models, like the MIT Bag model (46) or the Cloudy Bag (47) model, give a characteristic form factor, with $\Lambda < \Lambda_{\text{axial}}$, that is a quite soft $\pi NN$ vertex with $R > 0.88$ fm. Such models lead therefore to an important modification in $\eta$ and $Q$. Dispersion and field theory approaches (48, 49) typically give rather point-like form factors with $\Lambda > 1000$ MeV/$c$ or even larger, with small corrections to $\eta$ and $Q$; this is also typical of the Little Bag quark model (50).

The experimental values for both $\eta$ and $Q$ (see Table 1) make it difficult, even impossible, inside the present description to accept a pion source distribution for the nucleon as soft as the axial one without compensating dynamical effects (51). There is therefore a conflict between the experimental values of $\eta$ and $Q$ and the quark bag form factor, but there is no consensus on how this conflict should be resolved.

5.2.5   MINOR EFFECTS   The magnetic moment tensor interaction between the neutron and proton gives a well-defined $-0.34\%$ correction to $\eta$. There are small differences of $0.1\%$ between the results from pseudoscalar and pseudovector $\pi NN$ coupling using the charge-averaged mass and coupling constant (21).

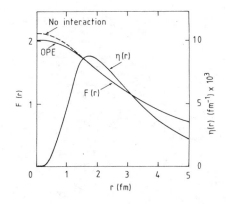

*Figure 5* Model dependence of the quadrupole weight function $F(r)$ with no interaction and with the OPE potential. A realistic wave function $\tilde{u}(r)$ has been used. The shape of $\eta(r)$ is given for comparison (adapted from 21).

## 5.3    The Integral Formulation for $Q$

A similar analysis can be made of the quadrupole moment $Q = Q_1 + Q_2$ (Equations 8a,b). The correction term $Q_2 = -0.0182$ fm$^2$ can easily be evaluated to the required accuracy. The leading term $Q_1$ can be identically rewritten as

$$Q_1 = \frac{1}{\sqrt{50}} \frac{A_S^2}{\alpha^3} \int_0^\infty dr \, F(r)\eta(r). \qquad 13.$$

Its integrand consists of the previous integrand $\eta(r)$ multiplied by the smooth weight function $F(r)$, which is nearly model independent apart from the unimportant region at very short range (see Figure 5). The corrections to $Q$ are very similar to those to $\eta$, with the $2\pi$ exchange correction ($-6.3 \pm 0.9$)% the most important one. Like $\eta$, the quadrupole moment is very sensitive to the $\pi NN$ form factor.

## 5.4    Universal Relation Between $Q$ and $\eta$

The integrands for $Q$ and $\eta$ have characteristically a long- and intermediate-range part dominated by OPE. In the intermediate- and short-range regions a modification in one of the integrands leads to a corresponding modification in the other one. There should therefore be a general linear relation between $\eta$ and $Q$, once the external proportionality factors $A_S^2$ and $f^2$ have been removed (21). Figure 6 shows that this indeed is the case for a series of standard major models where $Q/A_S^2$ is plotted against $\eta$ in reduced units. For models with the same effective range, i.e. the same $A_S$, the linear relation is $\delta Q/Q \approx 1.7\delta\eta/\eta$.

The $Q$-$\eta$ values using a $\pi NN$ form factor with $\Lambda = \Lambda_{\text{axial}}$ also lie on this universal curve, but the corresponding point falls outside Figure 6. This is a very strong indication that quark models using a $\pi NN$ form factor with

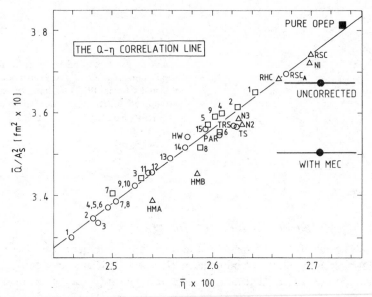

*Figure 6* Variations of $Q/A_S^2$ versus $\eta$ in various potential models. Both $Q$ and $\eta$ have been corrected for the proportionality to the $\pi NN$ coupling constant with $\bar{Q} = (0.078/f^2)Q$ and $\bar{\eta} = (0.078/f^2)\eta$ (adapted from 21 and 54). The solid and open squares, open circles, and triangles indicate results for different theoretical models shown by letters and numbers (54). The solid circles refer to the experimental results uncorrected or with meson exchange corrections (MEC) (25).

inert nucleons are not a satisfactory approach to the N-N interaction. Guichon & Miller (51) suggested a possible resolution of this problem: the exchange terms from $\pi qq$ couplings can provide the compensating mechanism leading to an apparent nearly point-like N-N interaction. This idea has not been systematically explored.

There will be an additional small contribution to the quadrupole moment from meson exchange currents (25–28). The theoretical value of this correction is $+0.0083$ (30) fm$^2$ (i.e. $+3 \pm 1\%$) according to Kohno (25); this is smaller than the correction from the $\pi NN$ form factor.

The very exact universal relation between $Q$ and $\eta$ in Figure 6 determines $\eta$ more accurately at present than direct measurements of $\eta$ and suggests the value $\eta = 0.0264(3)$. This is two standard deviations below the most accurately quoted experimental value for this quantity (see Table 1) and indicates a possible small inconsistency.

## 5.5    *Mathematical Constraints on* $\eta$

The previous discussion was based on a constructive approach to the quantities $\eta$ and $Q$. Klarsfeld, Martorell & Sprung (52–54a) explored instead the consequences of OPE when the experimental values of $Q$ and the deuteron RMS radius $r_d$ are imposed as constraints. It is then possible to establish rigorous Schwarz inequalities for the contributions of $Q$ and $r_d$ from the region inside a radius $R$, assuming a potential description outside. The input for the inequalities is obtained by integrating the deuteron equations from infinity inward using the correct binding energy and assuming that the long-range interaction (mainly OPE) is known. The inequalities determine $\eta$ to within narrow limits independent of direct experiments with $0.0261 < \eta < 0.0275$. This approach also gives a good description of the S-wave function outside 1 fm and determines the D-wave function as well, although with much larger uncertainties. These results confirm the conclusion in Section 4 that $\eta$ and $u(r)$ are very stable with respect to different approaches. The special merit of this approach is that it makes very weak assumptions concerning the short-range region. It is no accident that its limits on $\eta$ are in excellent agreement with the ones deduced from the universal $Q$-$\eta$ relation.

# 6.  PRECISE DETERMINATIONS OF THE D-STATE OBSERVABLES

## 6.1  *The Deuteron Quadrupole Moment*

The quadrupole moment is deduced from the hyperfine splitting in the HD and $D_2$ molecules. The experimental precision is mainly given by the $D_2$ ($J = 1$) state for which the splitting is measured to an accuracy of 0.01% (55). The theoretical analysis uses variational electron wave functions. The systematic uncertainty in this procedure is the present limit to the accuracy of $Q$. The earlier result by Reid & Vaida (56, 57), with $Q = 0.2860(15)$ fm$^2$, has recently been improved by Bishop & Cheung (58) using more sophisticated wave functions; their result is $Q = 0.2859(3)$ fm$^2$. This result should be quite reliable because the quadrupole interaction producing the splitting contributes from the region of large electron probability. The value includes a small correction for the nonadiabatic electron motion. The corrections due to finite deuteron size, polarizability, and relativistic effects are very small (21).

## 6.2  *The D/S Ratio and Tensor Polarization Experiments*

The quantity $\eta$ has been notoriously difficult to determine experimentally. In the past few years three different precise methods have become available. They are all based on the orientation of the deuteron using tensor polarization. This singles out the D-state by orienting the deuteron along the major axis of its prolate matter distribution. The experimentally determined values are summarized in Table 1.

**Table 1**   Experimental values of the D-state observables

| Quantity | $^2$H | $^3$H | $^3$He | $^4$He |
|---|---|---|---|---|
| $\eta$ | 0.0272(4)[a] | −0.051(5)[a] | | −0.5 < $\eta$ < −0.4[e] |
| | 0.0271(8)[b] | | | |
| | 0.0263(9)[c] | | | |
| $\bar\eta$ | 0.0271(4)[g] | | | |
| $D_2$ (fm$^2$) | 0.5055(75) | −0.279(12)[d] | −0.344[d] | −0.3[f] |
| $Q$ (fm$^2$) | 0.2859(3) | − | − | − |
| $\alpha$ (fm$^{-1}$) | 0.23154 | 0.4186 | 0.4467 | 1.072 |

[a] Pole extrapolation method $\dot d d \rightarrow p^3H$ (68); statistical uncertainties only.
[b] Sub-Coulomb stripping (21, 72).
[c] Pole extrapolation: $\dot d p$ elastic scattering (21, 65).
[d] Sub-Coulomb pick-up (94).
[e] Pick-up on $^{89}$Y (17).
[f] Radiative capture (18).
[g] Average value (21).

6.2.1   POLE EXTRAPOLATION METHODS   Amado et al (15, 59) pointed out that an analytical extrapolation in the angular variable cos $\theta$ to the nucleon exchange pole permits a direct determination of $\eta$ from tensor polarized angular distributions. At the pole position, the process has infinite range so that it directly measures the asymptotic amplitudes (60, 61).

6.2.1.1   *Elastic $\bar p d$ and $\dot d p$ scattering*   The pole occurs in the exchange graph at a position cos $\theta = -1.25 - 2.25\ \varepsilon/E$, where $E$ is the deuteron kinetic energy. The process measures the quantity $A_S^4\eta$. The main problems in deducing $\eta$ are

1. Coulomb effects, which according to Santos & Colby (62, 63) and Londergan, Price & Stephenson (64) can be correctly included on a level of few percent.
2. Uniqueness and stability of the extrapolation procedure. This is the main source of uncertainty and debate (65–67). In addition it is important to use a sufficiently accurate value for $A_S$ in the analysis. The value of $\eta$ obtained in this way is 0.0263(9) (21).

6.2.1.2   *Stripping reactions*   The analytical extrapolation in cos $\theta$ to the deuteron pole determines the quantity $\eta$. A particularly favorable case is the process $\dot d d \rightarrow p^3H$; Coulomb corrections are in this case less important than for the elastic case, and the extrapolation procedure is more stable (68). Borbely et al (69) deduce the value $\eta = 0.0272(4)$ by this method.

6.2.2   SUB-COULOMB STRIPPING   Stripping reactions below the Coulomb barrier provide $\eta$ very accurately. The repulsive Coulomb field keeps the

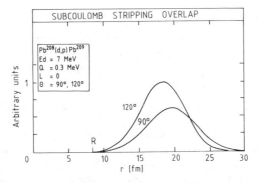

*Figure 7*    Contribution to the sub-Coulomb stripping amplitude according to Goldfarb (70).

deuteron far from the nuclear interaction region (see Figure 7) (70). In a
$(\vec{d}, p)$ reaction the neutron must tunnel to the nucleus, which is facilitated by
orienting the deuteron with the neutron as close to the nucleus as possible.
Processes with nearly zero $Q$ value are particularly interesting since the
range becomes particularly large (resonance condition). Only a small
number of suitable cases exist.

Such processes determine the quantity

$$D_2 = 1/15 \int_0^\infty dr \, r^4 w(r) \bigg/ \int_0^\infty dr \, r^2 u(r), \qquad\qquad 14.$$

which coincides to 0.1% with the corresponding quantity $\tilde{D}_2 = \eta/\alpha^2$ using
asymptotic deuteron wave functions (13, 71, 72). Goddard et al (73) studied
the influence of the deuteron structure and the Coulomb field: the dipole
polarizability and the electric quadrupole interaction give small, well-
defined corrections (74–76). The main uncertainty in the analysis is due to
the deuteron nuclear tensor interaction.

The most accurate determination of $\eta$ by this method (73) gives a value of
$\eta = 0.0271(8)$, corresponding to $D_2 = 0.506(15)$ from the $\vec{d}^{208}\text{Pb} \rightarrow p^{209}\text{Pb}$
reaction (77) with a $Q$ value of 0.3 MeV.

# 7.   THEORY OF THE D-STATE ADMIXTURE
# FOR $A = 3$ AND $A = 4$

As for the deuteron, the N-N tensor force also generates D-state com-
ponents in the wave functions of $^3\text{H}$, $^3\text{He}$, and $^4\text{He}$. These nuclei are
predominantly in a pure S-state. The $^{2S+1}\text{D}_J$-state generated by the tensor
force from this S-state is, to leading order, $^4\text{D}_{1/2}$ for $^3\text{H}$ and $^3\text{He}$, while $^4\text{He}$
acquires a $^5\text{D}_0$ component in this approximation. In addition, there are also
other weaker D-state amplitudes (78, 79).

## 7.1    *Magnetic Moments*

For $A = 3$ a rough indication of the D-state probability in the wave function is given by its orbital contribution to the isoscalar magnetic moment $\mu_s$ as compared to the isoscalar nucleon value $\mu_s^N = 1/2(\mu_p + \mu_n)$ (80):

$$\mu_s - \mu_s^N = -2P_D(\mu_s - 1/2) + \delta\mu_s^{MEC} = 0.851 - 0.880 = -0.029 \text{ nm}, \qquad 15.$$

where $\delta\mu_s^{MEC}$ represents the isoscalar meson exchange current, which is small and positive. From this 3% correction the impulse approximation ($\delta\mu_s^{MEC} = 0$) gives $P_D = 4\%$, while realistic values for $\delta\mu_s^{MEC}$ indicate $P_D \approx 8\%$ (81–83). This deduced D-state probability is, however, based on a small correction term. As for the deuteron, uncontrollable additional short-range terms can easily change the result.

## 7.2    *Asymptotic D-Wave Function*

In contrast, the asymptotic amplitudes for the virtual processes $^3\text{He} \to \text{pd}$, $^3\text{H} \to \text{nd}$, $^4\text{He} \to \text{dd}$ are model independent. Each of these can be separated into the S- and D-wave amplitudes $A_S$ and $A_D$ given by the asymptotic behavior of the normalized wave function:

$$u(r) \to A_S \exp(-\alpha r), \quad w(r) \to \eta A_S[1 + 3/\alpha r + 3/(\alpha r)^2] \exp(-\alpha r). \qquad 16.$$

Here $r$ is the relative distance between the center of mass of the two components and $\alpha^2 = 2M_R\varepsilon$ in terms of their reduced mass $M_R$ and binding energy $\varepsilon$ in the channel. In a more accurate treatment the asymptotic Coulomb wave functions should be used (84, 85). The D-state amplitudes in the $A = 3$ and $A = 4$ systems are less transparently connected to the underlying tensor force than in the deuteron case owing to the many-body nature of these systems. The qualitative features appear, however, using the following simple argument formulated by Santos (86, 87).

Consider the average tensor potential $\bar{V}_T(r)$ acting between the asymptotic constituents in terms of their relative distance $r$. If the folding correction from the extended deuteron is neglected, the tensor potential for the coupled S-D channel equation becomes $\bar{V}_T(r) = \sqrt{8}S(A)V_T(r)$, where $V_T(r)$ is the N-N tensor potential in the $I = 0$ channel. The kinematical factor $S(A)$ accounts for the spin summation and takes the values $1, -1, -2$ for $A = 2, 3$, and 4 respectively.

In the Born approximation, the D-state is then generated from the initially symmetric S-wave function, as in Equation 11, by the expression

$$\eta = \sqrt{8}\, M\, S(A) \int_0^\infty dr\, rj_2(i\alpha r)V_T(r)\tilde{u}(r). \qquad 17.$$

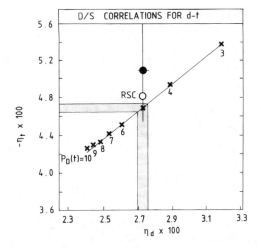

*Figure 8*   Plot of the tritium and deuteron D/S ratios $\eta_t$ versus $\eta_d$ in a separable potential model. The crosses correspond to different percentages of the D-state probability (85), while the open dot results from using a Reid soft-core interaction (89). The experimental value is also given (solid dot).

This expression is typically accurate to about 30%. It emphasizes that the different D/S ratios in the light nuclei are related.

For the three-body system it is possible to generate the asymptotic wave functions exactly from a given interaction using Faddeev equations. A systematic but simplified exploration of the asymptotic $^3$H $\rightarrow$ nd amplitude has been performed (84, 85) using a separable interaction with the deuteron binding energy and quadrupole moment constrained to the observed values. From the results of this oversimplified interaction one can deduce a strong linear correlation between the deuteron D/S ratio $\eta_d$ and the triton one $\eta_t$ as the interaction parameters vary (see Figure 8). The likelihood of such a correlation is strongly increased by the fact that a Faddeev calculation with the Reid soft-core potential (89) gave parameters falling nearly on the correlation line. This is consistent with experimental findings (68). This result suggests that the relation also holds for realistic forces: the $A = 2$ and $A = 3$ D/S ratios give analogous information on the tensor force.

## 7.3   D-state Probability in Recent Calculations

Modern realistic calculations are in substantial agreement with the result of Gerjuoy & Schwinger (7) and predict a contribution to the binding of the order of 50% by the tensor potential. There is, however, a large variation in the predicted D-state probability, which ranges from 6 to 8% in the three-body nuclei (83) and from 4 to 12% in $^4$He (90–92). Unfortunately, these calculations with realistic forces do not quote the asymptotic value $\eta$. This deficiency in the description can and should be remedied.

## 8. EXPERIMENTS ON THE D/S RATIO FOR $A = 3$ AND $A = 4$

### 8.1 *Pick-up Reactions*

Direct measurements of D-wave amplitudes were first obtained using $(\vec{d}, {}^3\text{He})$ and $(\vec{d}, {}^3\text{H})$ pick-up reactions on nuclei (16, 93, 94). The early results were obtained using rather high incident energies, which introduces large uncertainties in the theoretical analysis. More recent systematic experiments have therefore concentrated on the sub-Coulomb energy region, which permits $D_2$ values (Equation 14) to be deduced to a precision of 5–10% (95). The technique was reviewed by Knutson & Haeberli (14). Typical values for $D_2$ for ${}^3\text{H}$ and ${}^3\text{He}$ are $D_2 \approx -0.27\,\text{fm}^2$, with the uncertainty due mainly to systematic effects in the analysis.

Santos et al (86) and Tostevin (95) recently obtained the first clear information on the D-state in ${}^4\text{He}$ from the $(\vec{d}, {}^4\text{He})$ pick-up reaction on ${}^{32}\text{S}$ and ${}^{36,38}\text{Ar}$ at an energy well above the Coulomb barrier (96). The pick-up reaction on ${}^{89}\text{Y}$ just above the Coulomb barrier, which is more easily interpretable, was also recently analyzed, with the result $D_2({}^4\text{He}) \approx -0.3(1)\,\text{fm}^2$ (19, 97).

This pick-up analysis does not permit a direct determination of the asymptotic amplitude $\eta$, but $D_2$ differs from $\tilde{D}_2 = \eta/\alpha^2$ by less than 10%. The D/S ratio can be obtained by the pole extrapolation method, but this method has been used much less frequently than for the deuteron. Borbély et al (68, 69, 88) measured $\eta$ for the ${}^3\text{H}$ system from the $\vec{d}d \rightarrow p{}^3\text{H}$ reaction. They found $\eta = -0.051(5)$, corresponding to $D_2 = -0.21(4)\,\text{fm}^2$, which is consistent with the $D_2$ values from the pick-up reactions on heavy nuclei.

### 8.2 *Radiative Capture*

A very promising method for exploring the asymptotic D-state in ${}^4\text{He}$ and also in ${}^3\text{He}$ is the determination of the E2 component in the radiative capture reactions $\vec{d}d \rightarrow \gamma{}^4\text{He}$ and $\vec{d}p \rightarrow \gamma{}^3\text{He}$. In the case of $\vec{d}d \rightarrow \gamma{}^4\text{He}$, the two incident deuterons are identical bosons. This fact and spin-isospin selection rules lead to a very strong dominance of the E2 radiation with the principal contributions from the transitions ${}^1D_2 \rightarrow {}^1S_0$ and ${}^5S_2 \rightarrow {}^5D_0$. The quadrupole matrix elements depend mainly on the asymptotic wave functions, i.e. on $\eta$, so that the angular distribution becomes directly sensitive to the D-state asymptotic amplitude. With this technique Weller et al (18) recently obtained extremely clear evidence for the asymptotic D-state amplitude in ${}^4\text{He}$ (see Figure 9); an analysis of the data by Santos et al

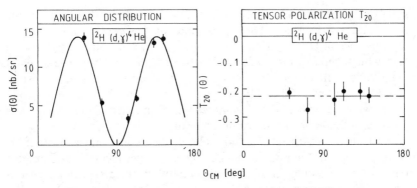

*Figure 9*  Differential cross section $\sigma(\theta)$ and tensor analyzing power for the reaction $\overset{\leftrightarrow}{d}d \to \gamma^4He$. The $T_{20}(\theta)$ values are manifestly nonvanishing and indicate the $^4He$ D-state.

(98) gives $-0.5 < \eta_\alpha < -0.4$. The analogous method applied to pd $\to \gamma^3He$ (99, 100) leads to a much more complicated analysis, since M1 transitions give important contributions. The results are consistent with other approaches. The radiative capture technique appears the best and most direct approach for determining $\eta_\alpha$ (101).

## 9.  OUTLOOK AND OUTSTANDING PROBLEMS

The independent development of highly precise theoretical and experimental methods for studying the asymptotic D-state amplitudes in the lightest elements has had a powerful impact on our perception of the structure of these systems. The use of integral methods in the deuteron forcefully emphasizes the crucial role of the tensor interaction and in particular of the tensor OPE interaction in this system (102). The fact that $Q$ and $\eta$ can be understood directly in these terms to an accuracy of a few percent without free parameters makes them the best understood dynamical quantities in nuclear physics. In addition, the agreement with the accurate experimental data implies very strong constraints on models for exotic components in the N-N interaction, and in particular on quark bag descriptions. Although the experimental precision on $\eta$ is excellent, the theoretical understanding has reached a level in which improving this quantity to the 1% level becomes desirable. In particular, this would permit a detailed test of the universal $Q$-$\eta$ correlation.

As a consequence of the highly successful description of the asymptotic deuteron quantities, it becomes natural to view the entire deuteron wave function as dominated by OPE, with the exact binding energy fine-tuned by short-range interactions with a D-state probability $P_D \approx 6\%$. A similar situation, although on a more primitive level, has also developed in the $A = 3$ and $A = 4$ systems. The experimental data establish unambiguously the existence of D-state components not only in $^3$He and $^3$H, but also in $^4$He, which therefore must be considered a deformed nucleus. The sign and magnitude of these components are consistent with the deuteron D-state amplitude and with OPE dominance. These observations provide strong direct evidence for the OPE-dominated tensor interaction in these systems in which it theoretically contributes half the binding energy.

It now becomes urgent to obtain reliable theoretical values for these D-wave asymptotic amplitudes, matching the experimental data. In this respect an interesting linear correlation between the deuteron and triton D/S ratios has been found to hold to high precision in a theoretical model. It should be established whether this relation is model independent or not, as well as how accurately the relation relates to experiments.

Present investigations in light elements have larger implications. It is well known that the nuclear tensor force should contribute nearly three fourths of the binding in nuclear matter, but it is difficult to isolate its effects transparently and unambiguously (103). In this perspective, the D-state components in the lightest nuclei vividly and directly demonstrate the importance of the tensor interaction to nuclear physics in general.

*Literature Cited*

1. Schwinger, J. *Phys. Rev.* 55:235 (1939)
2. Kellog, J. M. B., Rabi, I. I., Ramsey, N. F., Zacharias, J. R. *Phys. Rev.* 56:728–30 (1938)
3. Bethe, H. A. *Phys. Rev.* 57:390–413 (1940)
4. Rarita, V., Schwinger, J. *Phys. Rev.* 59:436–52 (1941)
5. Lattes, C. M. G., Occhialini, G. P. S., Powell, C. P. *Nature* 160:453–86 (1947)
6. Pauli, W. *Meson Theory of Nuclear Forces*. New York: Interscience (1946)
7. Gerjuoy, E., Schwinger, J. *Phys. Rev.* 61:138–46 (1941)
8. Taketani, M., ed. Meson Theory III. Nuclear Forces, *Suppl. Prog. Theor. Phys.* 3:1–171 (1956)
9. Glendenning, N. K., Kramer, G. *Phys. Rev.* 126:2159–68 (1962)
10. Wong, D. Y. *Phys. Rev. Lett.* 2:406–10 (1959)
11. Iwadare, J., Otsuki, S., Tanagaki, R., Watari, W. *Prog. Theor. Phys.* 15:86–88 (1955)
12. Iwadare, J., Otsuki, S., Tanagaki, R., Watari, W. *Prog. Theor. Phys.* 16:455–71 (1956)
13. Knutson, L. D. *Ann. Phys.* 106:1–25 (1977)
14. Knutson, L. D., Haeberli, W. *Prog. Part. Nucl. Phys.* 3:127–61 (1982)
15. Amado, R. D., Locher, M. P., Simonius, M. *Phys. Rev. C* 17:403–10 (1978)
16. Knutson, L. D., Hichwa, B. P., Barroso, A., Eiro, A. M., Santos, F. D., et al. *Phys. Rev. Lett.* 40:1570–74 (1975)
17. Roman, S., Basak, A. K., England, J. B. A., Nelson, J. M., Sanderson, N. E., et al. *Nucl. Phys. A* 287:269–74 (1977)
18. Weller, H. R., Colby, P., Roberson, N. R., Tilley, D. R. *Phys. Rev. Lett.* 53:1325–28 (1984)
19. Karp, B. C., Ludwig, E. J., Thomson, W. J., Santos, F. D. *Phys. Rev. Lett.* 53:1619–23 (1984)
20. Ericson, T. E. O., Rosa-Clot, M. *Phys. Lett.* 110B:193–98 (1982)
21. Ericson, T. E. O., Rosa-Clot, M. *Nucl. Phys. A* 405:497–533 (1983)
22. Dyson, F. J. *Phys. Rev.* 73:929–30 (1948)
23. Ericson, T. E. O. *Nucl. Phys. A* 416:281c–96c (1984)
24. Noyes, H. P. *Ann. Rev. Nucl. Sci.* 22:465–83 (1972)

25. Kohno, M. *J. Phys. G* 9:85–89 (1983)
26. Hadjimichael, E. *Nucl. Phys. A* 312: 341–60 (1978)
27. Friar, J. L. *Phys. Rev. C* 20:325–30 (1979)
28. Vassanji, M. G., Khanna, F. C., Towner, I. S. *J. Phys. G.* 7:1029–43 (1981)
29. Lacombe, M., Loiseau, B., Richard, J. M., Vinh Mau, R., Côté, J. *Phys. Rev. C* 21:861–73 (1983)
30. Holinde, K. *Phys. Rep. C* 68:121–57 (1981)
31. Nagel, M. M., Rijken, T. A., deSwart, J. *J. Phys. Rev. D* 20:1633–45 (1979)
32. Partovi, M. H., Lomon, E. L. *Phys. Rev D* 2:1999–2032 (1969)
33. Lagaris, I. E., Pandharipande, V. R. *Nucl. Phys. A* 359:331–48 (1977)
34. Ericson, T. E. O., Rosa-Clot, M. Unpublished
34a. Frois, B. *Nucl. Phys. A* 416:583C–604C (1984)
34b. Locher, M. P., Švarc, A. *Z. Phys. A* 316:55–60 (1984)
35. Ericson, T. E. O. *Prog. Part. Nucl. Phys.* 11:245–76 (1984)
36. Amado, R. D. *Phys. Rev. C* 19:1473–81 (1978)
37. Friar, J. L. *Phys. Rev. C* 22:796–812 (1980)
38. Ericson, T. E. O., Rosa-Clot, M. *J. Phys. G.* 10:L201–3 (1984)
39. Jackson, A. D., Riska, D. O. *Phys. Lett. B* 50:207–10 (1974)
40. Dumbrajs, O., Koch, R., Pilkuhn, H., Oades, G. C., Behrends, H., et al. *Nucl. Phys. B* 216:277–300 (1983)
41. Kroll, P. *Phys. Data 22-1, Fachinform. Zentrum,* ed. H. Behrens, G. Ebel. Karlsruhe (1981)
42. Lacombe, M., Loiseau, B., Richard, J. M., Vinh Mau, R., Pirès, P., et al. *Phys. Rev. D* 8:800–19 (1973)
43. Deleted in proof
44. Dominguez, C. A., VerWest, B. J. *Phys. Lett.* 89B:333–35 (1980)
45. Amaldi, E., Fubini, S., Furlan, G. *Pion electroproduction Springer Tracts in Mod. Phys.* 83:1–162 (1979)
46. Chodos, A., Thorn, B. *Phys. Rev. D* 12:2733–43 (1975)
47. Thomas, A. W. *Adv. Nucl. Phys.* 13:1–75 (1983)
48. Durso, J. W., Brown, G. E. *Nucl. Phys. A* 430:53–669 (1984)
49. Durso, J. W., Brown, G. E., VerWest, B. J. *Nucl. Phys. A* 282:404–24 (1977)
50. Brown, G. E., Rho, M. *Phys. Lett.* 82B:177–80 (1979)
51. Guichon, P. A. M., Miller, G. A. *Phys. Lett.* 134B:15–28 (1984)

52. Klarsfeld, S., Martorell, J., Sprung, D. W. L. *Nucl. Phys. A* 352:113–24 (1981)
53. Klarsfeld, S., Martorell, J., Sprung, D. W. L. *J. Phys. G* 10:L165–79 (1984)
54. Klarsfeld, S., Martorell, J., Sprung, D. W. L. *J. Phys. G* 10:2205–7 (1984)
54a. Klarsfeld, S., Martorell, J., Sprung, D. W. J. *J. Phys. G* 10:L205–7 (1984)
55. Code, R. F., Ramsey, N. F. *Phys. Rev. A* 4:1945–50 (1971)
56. Reid, R. V., Vaida, M. L. *Phys. Rev. A* 7:1841–48 (1973)
57. Reid, R. V., Vaida, M. L. *Phys. Rev. Lett.* 34:1064 (1975)
58. Bishop, D. M., Cheung, L. M. *Phys. Rev. C* 20:381–95 (1979) and cited by Ericson, T. E. O., Rosa-Clot, M. *Nucl. Phys. A* 405:497–533 (1983)
59. Amado, R. D., Locher, M. P., Martorell, J., Koenig, V., White, R. E., et al. *Phys. Lett.* 79B:368–70 (1981)
60. Ericson, T. E. O., Locher, M. P. *Nucl. Phys. A* 148:1–61 (1970)
61. Locher, M. P., Mizutani, T. *Phys. Rep.* 46:287–350 (1978)
62. Santos, F. D., Colby, P. C. *Nucl. Phys. A* 367:197–205 (1981)
63. Santos, F. D., Colby, P. C. *Phys. Lett.* 101B:291–93 (1981)
64. Londergan, J. P., Price, L. E., Stephenson, E. J. *Phys. Lett.* 120B:270–73 (1983)
65. Conzett, H. E., Hinterberger, F., von Rossen, P., Seiler, F., Stephenson, E. J. *Phys. Rev. Lett.* 43:572–76 (1979)
66. Gruebler, W., Koenig, V., Schmelzbach, P. A., Jenny, B., Sperisen, F. *Phys. Lett.* 92B:279–82 (1980)
67. Borbély, I. *J. Phys. G* 5:937–59 (1979)
68. Borbély, I., Gruebler, W., Koenig, V., Schmelzbach, P. A., Jenny, B. *Nucl. Phys. A* 351:107–11 (1981)
69. Borbély, I., Koenig, V., Gruebler, W., Jenny, B., Schmelzbach, P. A. *Phys. Lett. B* 109:262–64 (1982)
70. Goldfarb, L. J. D. *Lect. Theor. Phys. Boulder, Colo.* 8C:445–517 (1965)
71. Johnson, R. C. *Nucl. Phys. A* 90:289–310 (1967)
72. Johnson, R. C., Santos, F. D. *Part. Nucl.* 2:285–325 (1981)
73. Goddard, R. P., Knutson, L. D., Tostevin, J. A. *Phys. Lett.* 118B:241–44 (1982)
74. Tostevin, J. A., Johnson, R. C. *Phys. Lett.* 85B:14–16 (1979)
75. Tostevin, J. A., Johnson, R. C. *Phys. Lett.* 124B:135–38 (1983)
76. Lopes, M. H., Tostevin, J. A., Johnson, R. C. *Phys. Rev. C* 28:1779–82 (1983)
77. Stephenson, K., Haeberli, W. *Phys. Lett.* 45:520–23 (1980)

78. Beam, E. J. *Phys. Rev.* 158:907–16 (1967)
79. Irving, J. *Proc. Phys. Soc.* 66:17–27 (1952)
80. Friar, J. L., Gibson, B. F., Payne, G. L. *Ann. Rev. Nucl. Part. Sci.* 34:403–33 (1984)
81. Towner, I. S., Khanna, F. C. *Nucl. Phys. A* 399:344–54 (1983)
82. Jaus, W. *Nucl. Phys. A* 314:287–316 (1979)
83. Akaishi, Y., Sakai, M., Hiura, J., Tanaka, H. *Suppl. Prog. Theor. Phys.* 56:6–61 (1974)
84. Friar, J. L., Gibson, B. F., Lehman, D. R., Payne, G. L. *Phys. Rev. C* 25:1616–31 (1982)
85. Gibson, B. F., Lehman, D. R. *Phys. Rev. C* 29:1017–32 (1984)
86. Santos, F. D., Tonsfeldt, S. A., Clegg, T. B., Ludwig, E. J., Tagishi, Y., et al. *Phys. Rev. C* 25:3243–46 (1982)
87. Santos, F. D., Eiró, A. M. *Portugaliae Phys.* 15:65–88 (1984)
88. Borbély, I., Doleschall, P. *Phys. Lett.* 113B:443–46 (1982)
89. Kim, Y. E., Muslim. *Phys. Rev. Lett.* 42:1328–31 (1979)
90. Ballot, J. L. *Phys. Lett.* 127B:399–402 (1983)
91. Sakai, M., Shimodaya, I., Akaishi, Y.,

Hiura, J., Tanaka, H. *Suppl. Prog. Theor. Phys.* 56:32–53 (1974)
92. Goldhammer, P. *Phys. Rev. C* 29:1444–49 (1984)
93. Sen, S., Knutson, L. D. *Phys. Rev. C* 26:257–59 (1982)
94. Knutson, L. D., Colby, P. C., Hichwa, B. P. *Phys. Rev. C* 24:411–19 (1981)
95. Tostevin, J. A. *Phys. Rev. C* 28:961–64 (1983)
96. Tonsfeldt, S. A., Clegg, T. B., Ludwig, E. J., Tagishi, Y., Wilkerson, J. F. *Phys. Rev. Lett.* 45:2008–11 (1980)
97. Silvermann, B. L., Boudard, A., Briscoe, W. J., Bruge, G., Couvert, P., et al. *Phys. Rev. C* 29:35–41 (1984)
98. Santos, F. D., Arriaga, A., Eiró, A. M., Tostevin, J. A. *Phys. Rev. C* 31:707–9 (1985)
99. Arriaga, A., Santos, F. D. *Phys. Rev. C* 29:1945–47 (1984)
100. King, S., Roberson, N. R., Weller, H. R., Tilley, D. R., Engelbert, H. P., et al. *Phys. Rev. C* 30:1335–38 (1984)
101. Tostevin, J. A., Nelson, J. M., Karban, O., Basak, A. K., Roman, S. *Phys. Lett.* 149B:9–12 (1984)
102. Ericson, T. E. O. *Comments Nucl. Part. Phys.* 13:157–70 (1984)
103. Pandharipande, V. R., Wiringa, R. B. *Rev. Mod. Phys.* 51:821–59 (1979)

*Ann. Rev. Nucl. Part. Sci. 1985. 35 : 295–320*

# NUCLEON-NUCLEON PHYSICS UP TO 1 GeV

*D. V. Bugg*

Queen Mary College, London E1 4NS, United Kingdom

CONTENTS

## 1.  INTRODUCTION

The old fashioned way of describing nucleon-nucleon (N-N) forces is in terms of meson exchanges (Figure 1a). All the known mesons can in principle contribute. The interaction depends distinctively on the quantum numbers $J^{PG}$ of each meson. Thus, to unravel these exchanges, a quantitative determination of the spin dependence of amplitudes is crucial in both elastic and inelastic channels. The progress made experimentally in the last decade has depended on technology providing polarized targets, polarimeters, and intense, clean polarized beams of protons and monoenergetic

295

0163–8998/85/1201–0295$02.00

neutrons. The spin dependence of the N-N interaction turns out to be rich and complicated, because many mesons contribute strongly, but with delicate cancellations in some amplitudes.

In the 1960s, many small accelerators contributed to determinations of p-p elastic phase shifts up to 425 MeV and n-p phase shifts around 150 and 210 MeV (1). The pioneering work of Scotti & Wong (2) showed that these phase shifts could be fitted remarkably well by exchange of the lightest mesons, $\pi, \eta, \rho, \omega,$ and $\sigma$, with coupling constants broadly in line with other determinations (except the $\eta$). Today, the picture has expanded and changed quantitatively. Since 1974, the meson factories have made major progress. We now know p-p phase shifts uniquely and accurately to 1 GeV and n-p phase shifts likewise to 500 MeV and tentatively at 800 MeV. Table 1 gives a subjective impression of the completeness of the data, graded on a scale up to 10, with a passmark of 5. The resulting amplitudes are firm input to nucleon-nucleon calculations up to 500 MeV.

Progress since 1980 has been mainly in understanding inelasticity. Experiments at LAMPF and SIN have determined elastic amplitudes well above the inelastic threshold and have also provided extensive data on pp $\rightarrow d\pi^+$. With a little guidance from theory, the six amplitudes of the latter process can now be determined with confidence to 800 MeV. A start has been made on unravelling the spin dependence of the dominant inelastic channel, pp $\rightarrow$ pn$\pi^+$.

What emerges is not qualitatively surprising. The strong inelastic threshold at about 600 MeV due to pp $\rightarrow$ N$\Delta$ exerts a decisive influence on elastic amplitudes, complicating the determination of meson exchanges. It has become clear, for example, that excitation of an intermediate isobar (Figure 1c) is largely responsible for the weak central attraction that binds nuclei; there is a close analogy with Van der Waals forces in atomic physics.

The long-range part of the N-N interaction, due to $\pi$ and $2\pi$ exchange, is now calculated in terms of just two parameters, the $\pi$NN coupling constant $g^2$ and the $\pi$-$\pi$ isospin 0 s-wave scattering length $a_s^0$. Both are narrowly restricted by other data. All interpretations of data now take these long-range components as known input. The essence of the calculation of $2\pi$ exchange is that the top and bottom halves of Figures 1b and 1c are both $\pi$-

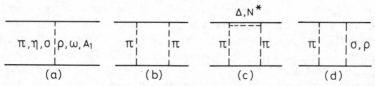

Figure 1 Processes contributing to N-N elastic scattering. (a) Single-particle exchanges. (b,c,d) Two-particle exchanges.

N scattering amplitudes known from experiment. Details of the rather tortuous calculation are to be found in reviews by Vinh Mau (3) and Bugg (4). On this basis, the Paris group has developed a potential (5) fitting N-N elastic data nearly as well as phase shift analysis up to 425 MeV. They take the interaction for $r > 0.8$ fm from $\pi$, $2\pi$, $\omega$, and $A_1$ exchange (four adjustable parameters); for $r \leqslant 0.8$ fm, they use phenomenological potentials, independent of $r$ and linearly dependent on energy, one for each of the ten amplitudes (5 spin × 2 isospin). The objective of this work is to derive a potential for nuclear physics with the right theoretical constraints built into long-range forces.

For $r < 0.8$ fm, there is a dilemma in interpreting the physics. Quark bag models use a radius close to 0.8 fm for nucleons, so there are obvious reservations about a fit in terms of meson exchanges. Nevertheless, two groups have obtained impressive fits in terms of $3\pi$, $\rho\pi$, $\sigma\pi$, and $\omega\pi$ exchanges. Grein & Kroll (6) used forward dispersion relations to show that these short-range contributions are definitely required, and were able to account quantitatively for six forward amplitudes (3 spin × 2 isospin). Holinde & Machleidt (7) also obtained an impressive fit to all partial waves below 300 MeV, as well as the properties of the deuteron; they included

**Table 1**  A survey of N-N data

| Channel | Lab energy (MeV) | Source | Date | Completeness[a] |
|---------|------------------|--------|------|-----------------|
| pp → pp | 25, 50 | Rutherford PLA | 1960–1965 | 8 |
|  | 140 | Harvard, Harwell, Orsay | 1954–1968 | 8 |
|  | 210 | Rochester | 1957–1966 | 8 |
|  | 325 | Berkeley | 1954–1957 | 6 |
|  | 425 | Chicago, Carnegie | 1954–1968 | 7 |
|  | 445–580 | SIN | 1976–1984 | 9 |
|  | 650–800 | LAMPF | 1976–1984 | 8 |
|  | 970 | Gatchina | 1980–1982 | 7 |
| np → np | 50 | UC Davis | 1975–1978 | 4 |
|  | 140 | Harvard, Harwell, Orsay | 1952–1967 | 7 |
|  | 210 | Rochester | 1952–1968 | 6 |
|  | 220–500 | TRIUMF | 1976–1980 | 8 |
|  | 630 | Dubna | 1967–1976 | 3 |
|  | 800 | LAMPF | 1976–1984 | 5 |
| pp → d$\pi^+$ | 300–500 | TRIUMF | 1976–1984 | 5 |
|  | 445–580 | SIN | 1979–1984 | 9 |
|  | 500–800 | LAMPF, SIN | 1979–1984 | 9 |
| pp → pn$\pi^+$ | 420–510 | TRIUMF | 1980–1983 | 6 |
|  | 650, 800 | LAMPF | 1978–1984 | 3 |

[a] The column labelled completeness gives a subjective impression on a scale of 10, with a passmark of 5.

reasonable form factors cutting off the divergent meson exchanges at small $r$. Despite the success of these three theoretical fits of N-N data, there are points requiring further work, as discussed in Section 3.

Quark models do not yet make quantitative predictions. This is hardly surprising. The long-range N-N interaction is dominated by $\pi$ and $2\pi$ exchanges, but QCD does not lead directly to an interaction with quantum numbers $J^P = 0^-$ or $0^+$. An interesting approach (8) is to use the quark model for $r < 0.8$ fm and a meson exchange tail for $r > 0.8$ fm. The well-known short-range repulsion in the N-N interaction arises naturally in a quark model. If six quarks overlap, two of them are prevented by the Pauli principle from going into s-wave orbitals; they have to be excited into p-states or into color combinations different from those in the free nucleon, and this excitation leads to an effective repulsion. A very interesting paper by Maltman & Isgur (9) attempts a calculation of the 1S0 and 3S1 ($^1S_0$ and $^3S_1$ in an older notation) potentials in a six-quark model, direct from a QCD-inspired Hamiltonian; results are in semi-quantitative agreement with experiment and are qualitatively very revealing.

In the naive bag model, an attraction arises between two colored octets, each constructed from three quarks. This has stimulated interest in the possibility that the N-N system might couple to such hidden-color systems, which would lead to exotic dibaryon resonances. Indeed, an embarrassing proliferation of such resonances has been predicted (10, 11). On the experimental side, interest was aroused when the Argonne group of Yokosawa and collaborators (12) discovered striking peaks in spin-dependent total cross sections. However, over the last four years it has gradually become clear that these peaks are closely related to the strong inelastic threshold NN $\rightarrow$ N$\Delta$, and not immediately related to hidden color. What has emerged is that the long-range N-$\Delta$ interaction is strongly attractive, possibly sufficiently so to generate a quasi-deuteron N-$\Delta$ resonance or a virtual state close to the threshold. This topic is reviewed in Section 4.

This review attempts two tasks: (a) a general survey for the amateur of our present understanding of the N-N and N-$\Delta$ interactions and the status of the data; and (b) a few details for the professional, attempting to expose weaknesses and controversies and giving an account of progress since 1980. At that date the subject was reviewed by Bugg (4), with particular attention to experiments and techniques, by Kroll (13), who gave extensive detail of theoretical formalism, and by Signell (14), who gave a valuable account of the status of potential models. The reader is referred to these three reviews for those details that remain unchanged since 1980. For brevity, references are limited to the most recent work on each topic; entry points to earlier, more extensive literature are References (3, 7, 9, 15, 17, 21, 29, and 31). This review was completed in January 1985.

## 2.  INELASTICITY

Inelasticity is the area that has progressed the most since 1980. It is interesting in its own right, also for what it says about dibaryon resonances (highly inelastic if they exist), and thirdly for its quantitative feedback to elastic scattering. In extracting meson exchanges from elastic amplitudes $f(s)$, one has to evaluate dispersion integrals of the form

$$I(s) = \frac{1}{\pi} \int_{4M^2}^{\infty} \frac{\text{Im } f(s') \, ds'}{s' - s},$$    1.

which have large inelastic contributions.

The NN → NΔ threshold at 600 MeV dominates; inelasticity in the channel with isospin $I = 0$ is negligible up to 800 MeV because NN → ΔΔ and NN → NN*(1420) have higher thresholds. For $I = 1$, one expects NΔ to appear initially with $L_{N\Delta} = 0$ in the N-N 1D2 state, then at higher energy (about 750 MeV) with $L_{N\Delta} = 1$ in 3F3, 3F2, 3P2, 3P1, and 3P0. At still higher energies, $L_{N\Delta} = 2$ will feed 1G4 and 1S0. This is precisely what one finds from N-N elastic data, pp → dπ$^+$ and pp → pnπ$^+$, but with two minor additions. Close to the inelastic threshold (from 300 to 500 MeV), there is significant competition from 3P1 NN → NZ, where Z is my shorthand for a repulsive π-N S-wave amplitude. Secondly, in the NNπ final state, there is a limited part of phase space where the N-N S-wave final-state interaction plays a strong role.

In analyzing the data, one mimimizes unknowns by constraining high partial waves to values calculated from pp → NΔ and pp → NZ mediated by π and ρ exchange. Many authors have done such calculations and all agree that π exchange dominates completely for $L_{N\Delta} \geqslant 2$ up to 800 MeV. Kloet & Tjon (15) review these calculations and show that quantitative differences (up to 40% in amplitude) arise from different vertex form factors and off-shell variation of π-N amplitudes. The most reliable results are probably those of Gruben & VerWest (16), because they adjusted their form factors to fit the limited inelastic data from 650 to 800 MeV, hence ensuring correct normalization. A touchstone in comparing theory with experiment is the 1G4 amplitude. It is free of spin complications, and is large enough to be clearly visible in both inelastic data and the N-N elastic diffraction peak. In my elastic phase shift analysis, inelasticity in this wave has 8% errors at both 800 and 970 MeV, and agrees within this error with Gruben & VerWest. Other theoretical predictions fail this test. Figure 2 shows

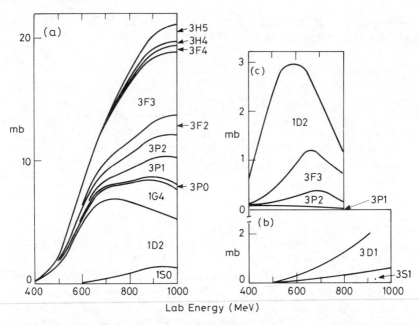

*Figure 2* Contributions of different partial waves to inelastic cross sections: (*a*) and (*b*) NN → NNπ calculated by Gruben & VerWest (16), (*c*) pp → dπ⁺ from the amplitudes fitted by Bugg (17).

inelastic cross sections predicted for the NNπ channel by Gruben & VerWest and also those for the pp → dπ⁺ channel from my analysis described in the next subsection. Quantitatively, predictions in Figure 2*a* are good for F-waves, but may be unreliable for S-, P-, and D-waves, where short-range effects enter.

## 2.1    $pp \to d\pi^+$

Below 450 MeV, the only data are the polarizations, $P$, the differential cross sections, $d\sigma/d\Omega$, total cross sections, and a handful of spin-spin asymmetry ($A_{LL}$) points. (For this reason, the data set is given a bare passmark for completeness in Table 1.) From 450 to 800 MeV, there are extensive and accurate data from SIN and LAMPF on $d\sigma/d\Omega$, $P$, and spin correlations $A_{NN}$, $A_{LL}$, $A_{SS}$ and $A_{SL}$, and from SIN on $iT_{11}$, the vector polarization of the deuteron. Of these, only $A_{SL}$ and $P$ are independent in forward and backward hemispheres, so one might conclude that the nine data sets cannot determine the six amplitudes. Without any help from theory, this is so. However, calculations by many authors concur that amplitudes for $L_{\pi d} \geq 3$ are either negligibly small or, in the case of 1G4, accurately predictable (and in agreement within 10% errors with experiment at 800 MeV). With this theoretical input, an amplitude analysis is possible up to 800 MeV (17); some minor energy smoothing is necessary for (a) the phase of the small 3F2 amplitude (the most poorly defined feature at present) and (b) the threshold dependence of amplitudes below 450 MeV. A second analysis by Watari and collaborators (18) arrives at similar results for the three largest amplitudes, but, because it frees three more partial waves, finds more erratic behavior in small amplitudes. Results of these two analyses are shown in Figure 3.

Most features of these results are close to expectation. The 1D2, 3F3, and 3P2 amplitudes are largest, and have the expected phase variation. These phases arise from (a) the broadened $\Delta$ in the $\pi$-d channel, and (b) the elastic N-N phase shift $\delta_{NN}$, via initial-state interaction. On Figure 3, a small overall phase difference is apparent between the two solutions; since the overall phase is determined only relative to theoretical high partial waves, it could be adjusted to take up this difference with very small change in $\chi^2$. The magnitude of the 3P2 amplitude is larger by a factor of two than any of the calculations; this feature is also found in the total inelasticity in N-N elastic phase shift analysis, and is the major outstanding point requiring clarification.

My 3P1 amplitudes (both $L_{\pi d} = 0$ and 2) show a distinctive pattern of destructive interference between a threshold NN → NZ amplitude and a loop due to NN → N$\Delta$. This is analogous to the $\pi$d → $\pi$d S-wave amplitude calculated by the Lyon group (C. Fayard, private communication) and shown in Figure 4; the repulsive character of the S-wave $\pi$-d amplitude is well known (19). Close to threshold, each NN → $\pi$d amplitude should have a phase $\Phi = \delta_{NN} + \delta_{\pi d}$, where $\delta$ are elastic phase shifts (Watson's theorem). This relation breaks down at energies where the three-body final state has significant K-matrix elements. My analysis respects Watson's theorem for all amplitudes below 450 MeV, but the analysis of Hiroshige et al (18) ignores this requirement, notably for the 3P1 amplitudes, which should both approach a phase $\delta_{NN} \approx -30°$ near threshold.

In my analysis, 3F2 and 1S0 amplitudes are weak, though nonzero. The latter is compatible with the energy dependence and phase expected for production via an intermediate $\Delta$ with $L_{N\Delta} = 2$.

## 2.2   $pp \rightarrow pn\pi^+$

There are extensive data (20) from TRIUMF at 420, 465, and 510 MeV, sampling a good part of phase space for observables $A_{N0}$ (polarization using polarized beam), $A_{0N}$ (polarization using polarized target), $A_{NN}$, $A_{LL}$, $A_{SS}$, and $A_{SL}$. A selection of the data is shown on Figure 5. There are five striking features.

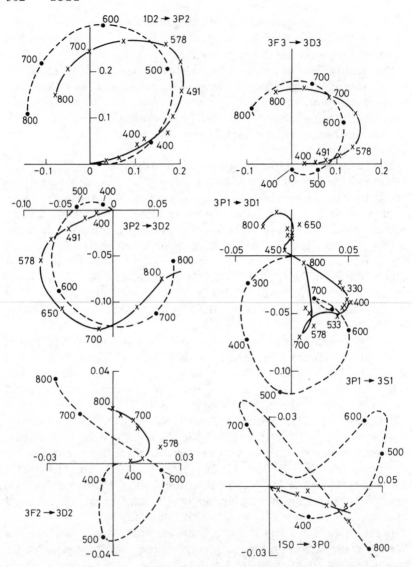

*Figure 3* Argand diagrams for pp → dπ⁺ amplitudes: solid curves are from Bugg (17); dashed curves are from Hiroshige et al (18). Numbers indicate lab energies (MeV).

1. Over most of phase space, $A_{NN}$, $A_{LL}$, and $A_{SS}$ are large and negative. This is a consequence of 1D2 dominance: for this state in isolation $A_{NN} = A_{LL} = A_{SS} = -1$.

2. For low $\pi^+ p$ masses, there is a significant tendency for $A_{LL}$ to be much more positive, and to a lesser extent this is true for $A_{SS}$ and $A_{NN}$. This is just

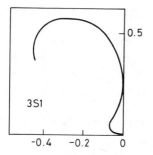

*Figure 4*  The Argand diagram for the S-wave $\pi d \rightarrow \pi d$ amplitude, from the calculation of the Lyon group (C. Fayard, private communication).

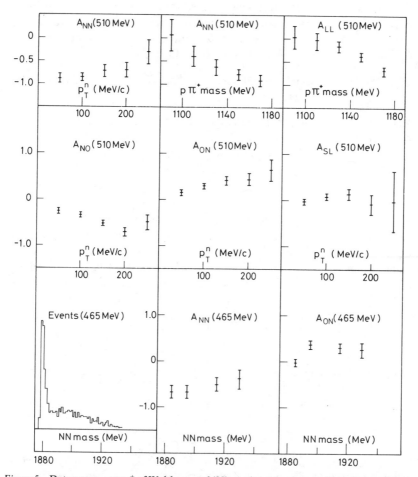

*Figure 5*  Data on $pp \rightarrow pn\pi^+$ of Waltham et al (20). $A_{ij}$ denotes spin-correlation parameters, where $i$ and $j$ are beam and target polarization directions; $N$ is normal to the plane of the $pp \rightarrow n\Delta^{++}$ reaction, $S$ is sideways and $L$ longitudinal. $p_T^n$ is the transverse momentum of the neutron. In each figure, results are summed over remaining kinematic variables.

what one expects from a threshold 3P1 NN → NZ amplitude, for which in isolation $A_{LL} = +1$, $A_{SS} = A_{NN} = 0$.

3. Both $A_{N0}$ and $A_{0N}$ increase roughly linearly with transverse momentum of the neutron, $p_T^n$, and become quite large ($>0.5$ in magnitude). This indicates large interfering amplitudes. The fact that $A_{N0} \approx -A_{0N}$ indicates interference between states of opposite parity, i.e. between 1D2 and any or all of 3P1, 3P2, and 3F3.

4. The most interesting and unexpected observation is that $A_{SL} \approx 0$ over most of phase space. This is quite different from pp → d$\pi^+$, in which large negative $A_{SL}$ is observed. It transpires that $\sigma A_{SL}$ receives contributions from the real parts of certain interference terms, while $\sigma A_{0N} = -\sigma A_{N0}$ is given by the imaginary part of precisely the same interference terms. Together, these data give an important phase determination. In an isobar analysis, pp → n$\Delta^{++}$ partial wave amplitudes have the form

$$f = \frac{A(s) \exp i[\delta_{NN}(s) + \delta_{N\Delta}(s)]}{w_0 - w - iM\Gamma(w)}, \qquad\qquad 2.$$

where the denominator expresses a Breit-Wigner dependence on $w = $ mass$^2$ of the $\Delta$ (resonant at $\sqrt{w_0}$ with width $\Gamma$); $M$ is the nucleon mass. For weak K-matrix elements, $\delta_{N\Delta}$ is the phase shift for elastic N-$\Delta$ scattering; for strong scattering, that relation is only approximate and $\delta_{N\Delta}$ may be regarded as a parameterization of the phase of $f$. Figure 6a shows the magnitudes of amplitudes fitting the data, but with phases given purely by $\delta_{NN}$, i.e. the initial-state interaction. With these phases, the amplitudes would give large $A_{SL}$ and slightly smaller $A_{N0}$, in contradiction with the data. In order to fit small $A_{SL}$, a large $\delta_{N\Delta}$ must be introduced somewhere. At such low energies, this must be in the N-$\Delta$ S-wave, i.e. N-N 1D2. If one sets $\delta_{N\Delta} = 0$ in all other waves (Occam's razor), the result is $\delta_{N\Delta} = 45 \pm 10°$ for

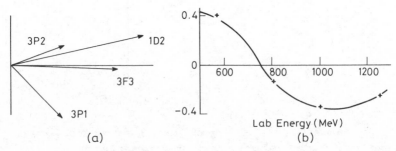

Figure 6    (a) Argand diagram for pp → n$\Delta^{++}$ partial wave amplitudes at 510 MeV, assuming phases are given only by the p-p initial-state reaction, (b) $A_{N0}$ observed by Calkin (22) in the pp → n$\Delta^{++}$ reaction for events within 30 MeV of the $\Delta$ peak and a Gottfried-Jackson angle $\theta = 90°$ for $\Delta^{++}$ decay.

the S-wave at each of the three energies 420, 465, and 510 MeV, i.e. a strongly attractive N-Δ interaction near threshold.

5. A strong final-state interaction is observed for n-p masses within 5 MeV of $2M_N$. This in itself is not surprising. The point of interest is that $A_{NN}$ has within errors at all three energies the same large negative value in this peak as in neighboring mass bins, while $A_{N0}$ and $A_{0N}$ drop precipitously in the final-state interaction peak. This implies that the peak is reached dominantly from the N-N 1D2 state, and implies that the strong amplitude in this partial wave is driven coherently by the N-N final-state interaction as well as by Δ formation in both π-N states.

From LAMPF there are isolated sets of data on $d\sigma/d\Omega$, $A_{N0}$, $A_{NN}$, and $A_{LL}$. At present the best one can do is to compare with calculations (21). Although these calculations are dependable for high partial waves present data do not dictate unambiguously the modifications required for low partial waves.

High statistics data on $d\sigma/d\Omega$ and $A_{N0}$ are available from an Argonne experiment (22) and show an intriguing result, displayed on Figure 6b. The polarization in the Δ peak shows a strong energy dependence and a sign change at 750 MeV. Remember that polarization is due to interference and is strongly phase sensitive. Unfortunately, one cannot say from $A_{N0}$ data alone which amplitude varies rapidly in phase, nor whether the movement is clockwise or anticlockwise around the Argand diagram; $A_{SL}$ data would resolve the latter ambiguity. This question is discussed further in Section 4 in connection with the possibility of a dibaryon resonance.

## 2.3  Summary

Inelasticity is predominantly in 1D2 (peaking at $\sim 600$ MeV) and at higher energies in 3F3 and 3P2 (which is larger than theory). Near threshold, there is significant 3P1 inelasticity due to NN → NZ, interfering destructively with the NN → NΔ amplitude. In pp → dπ$^+$, there is a modest 3F2 amplitude and a very small 1S0 inelasticity, both invisible in cross section on the scale of Figure 2c. TRIUMF pnπ$^+$ data also require little or no 3P0 inelasticity up to 510 MeV. These results are a strong guide as to what inelasticities to expect in elastic scattering. There is evidence for an attractive N-Δ S-wave interaction near threshold ($\delta_{NA} \approx 45°$), and a rapid phase variation at higher energy in this wave or another interfering with it.

## 3.  ELASTIC SCATTERING

### 3.1  Data

There are now over 10,000 data points below 1 GeV, most of them of high accuracy. Even allowing for the recent tendency to report points at angular intervals much smaller than significant variations in the data, this is by far the most complete data set for any reaction in nuclear or particle physics.

With a polarized beam and target, one can measure $P$ and four spin correlation parameters $A_{NN}$, $A_{LL}$, $A_{SS}$, and $A_{SL}$. For p-p scattering, these and the differential cross section $\sigma$ are all symmetric or antisymmetric about $90°$. With a polarized beam and a polarimeter, one can measure four independent Wolfenstein parameters (spin transfer from initial to final state); these are different in forward and backward hemispheres. This makes a total of 14 measurements. Rodebaugh & Bonner (23) demonstrate that nine of them in principle determine the five amplitudes completely. Experience shows that, with allowance for errors and ambiguities, eleven are usually desirable. Present data have statistical errors of typically $\pm 0.5\%$ for $P$, $\pm 1\%$ for correlation parameters, and $\pm 1.5\%$ for Wolfenstein parameters. Cross calibrations between TRIUMF, SIN, and LAMPF agree closely (24) and indicate normalization errors of about $\pm 2\%$. A detail is that accurately normalized $\sigma$ data are missing, except for those of Chatelain et al (25) from 515 to 580 MeV. This measurement would be valuable for two reasons. Firstly, formulas for all spin-dependent observables $X$ take the form

$\sigma X$ = bilinear combination of amplitudes,

with the result that errors in $\sigma$ feed through in haphazard fashion into the determination of spin-dependent amplitudes. Secondly, inelasticities are determined largely by the shape of the diffraction peak through $\sigma A_{NN}$, $\sigma A_{LL}$, $\sigma A_{SS}$, and $\sigma$ itself.

For n-p scattering, inelasticity is negligible up to 800 MeV, so one needs in principle only five experiments in one hemisphere. In practice, it is helpful to make use of interferences between isospin 0 and 1 amplitudes; these change sign at $90°$. At TRIUMF (200–500 MeV), $\sigma$, $P$, and three spin transfer parameters have been measured in both hemispheres with errors of $1\%$ ($\sigma$ and $P$) to $4\%$. These data overdetermine $I = 0$ amplitudes and even improve $I = 1$ waves to some extent (because of missing p-p $\sigma$ data). Above 500 MeV, n-p data are limited. At 800 MeV (LAMPF), most spin transfer data are confined to extreme forward and backward angles. In consequence, $I = 0$ phase shifts are very shaky at 800 MeV. Measurements of $A_{SS}$ and $A_{LL}$ over a large angular range are in progress from 500 to 800 MeV (H. Spinka, private communication) and will radically improve the situation.

Experience shows that Wolfenstein parameters (25a) $R$ and $R_t$ fix singlet amplitudes; $P$ fixes the spin-orbit combination of triplet amplitudes; $A$, $A'$, $A_t$, and $A'_t$ fix the tensor combination; and $\sigma$ fixes the spin-averaged (central) combination. Mixing parameters $\bar{\varepsilon}$ (connecting initial and final

$L = J \pm 1$ states) are sensitive to $D$ and $D_t$ and to $A_{NN}$, which also helps to determine several other waves; $A_{LL}$ is sensitive to triplet waves with $L = J$.

## 3.2   Phase Shift Analysis

Three groups are active in phase shift analysis up to 1 GeV: Arndt and collaborators, the Saclay-Geneva group, and myself. The analyses differ in the freedom given to (a) high partial waves, (b) inelasticity, and (c) energy dependence. Care is required in selecting the appropriate number of free parameters. If too many are free, noise and systematic errors in data appear in phase shifts and the benefit of the analysis as a filter can be lost. A warning flag is the appearance of correlation coefficients $>0.3$ in the error matrix. Conversely, too little freedom can miss or bias the physics.

The Saclay-Geneva group (26) allows the greatest flexibility in both the number of parameters and their energy dependence. For example, near 500 MeV, they free eight more $I = 0$ waves and four more $I = 1$ waves than does my analysis. Arndt et al (27) steer a middle course. My analysis (unpublished, available on request) is the most constrained, the motive being to inject theoretical input where it appears superior or equal to experiment. Several high partial waves are constrained by $\pi + 2\pi$ exchange, and the threshold behavior of inelasticities is constrained to behavior suggested by centrifugal barriers. This is done by including in the $\chi^2$ minimization penalty terms of the form

$$\chi_{th}^2 = \left(\frac{\text{theoretical value} - \text{fitted value}}{\text{theoretical error}}\right)^2$$

with a generous denominator to allow maximum leverage to experiment. Details are to be found in the work of Dubois et al (28). Typically, $\chi_{th}^2$ contributes 10 at one energy, comparable with a bad data point. Figure 7 illustrates for some waves the measure of agreement between the three analyses.

From 25 to 100 MeV, n-p data are not quite sufficient to fit $I = 0$ phase shifts. Results exhibit a poorly defined $\chi^2$ minimum and a nasty correlation between $\bar{\varepsilon}1$ and 1P1. Recently, Mathelitsch & VerWest (29) looked afresh at this energy range. Using recent accurate determinations of $\eta$, the asymptotic ratio of D- and S-wave functions of the deuteron, they constrained $\bar{\varepsilon}1$ to a reasonable energy dependence and hence lifted the ambiguity in phase shifts in this energy range.

INELASTICITY   From pp $\rightarrow$ d$\pi$ data, we know inelasticity appears first in 3P1 then 1D2. Up to 425 MeV, these are the only inelasticities determined by elastic data. At higher energies, N-$\Delta$ production with $L = 1$ can appear in 3F3, 3F2, 3P2, 3P1, and 3P0. From 515 to 970 MeV, the data determine the large inelasticities in 3F3, 3P2, and 1D2 well. At 580 MeV, one gets weak determinations of the small 3P1 and 3P0 inelasticities, and at 800 MeV reasonably firm values for 3F2, 3P1, and 3P0. At other energies, inelasticities for 3F2, 3P1, and 3P0 need to be constrained to energy dependence given by centrifugal barrier factors. The small inelasticity in

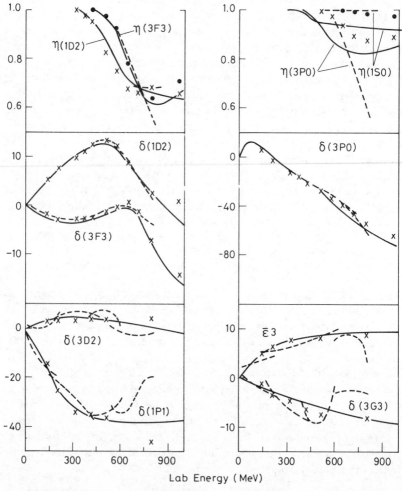

*Figure 7*   N-N phase shifts $\delta$ (degrees), elasticity parameters $\eta$, and mixing parameters $\bar{\varepsilon}$ (degrees): solid curves are from Arndt et al (27); dashed curves are from Bystricky et al (26); crosses and circles are from Bugg (unpublished). In the top diagrams, circles refer to $\eta$(3F3) and $\eta$(1S0), crosses to $\eta$(1D2) and $\eta$(3P0).

1S0 is poorly determined at all energies, because of the low multiplicity of this wave.

RESULTS    The outcome is that the three groups agree reasonably closely on phase shifts and rather less well on inelasticities. Reconstructed amplitudes agree better than phase shifts, which shows the existence of strong correlations in the error matrix in some analyses.

There is, however, one point of qualitative disagreement, namely in inelasticities for 3P0 and 1S0. As Figure 7 shows, Arndt et al find significant inelasticity in these waves. My fit and that of the Geneva-Scalay group give low inelasticity up to 650 MeV in 3P0 and none in 1S0. Arndt's $\eta$ parameters reach half-height at 520 MeV for 1D2 and 3P0, and at 420 MeV for 1S0. This is hard to reconcile with plausible physics. Inelasticity due to N-$\Delta$ with $L = 1$ (3P0) should reach half-height at about 640 MeV, like 3F3, and at higher energy for $L = 2$ (1S0). The d-$\pi$ channel does not couple to 3P0 because of parity and $J$ conservation. A possible source of inelasticity in 3P0 would be coupling to a final 1S0 p-p pair with an $L = 0$ pion; but this should lead to a strong pp$\pi^0$ channel at threshold, in contradiction with experiment. There is no plausible source of 1S0 inelasticity, in view of the very weak pp $\rightarrow$ d$\pi$ amplitude. Both $\eta$(1S0) and $\eta$(3P0) are sensitive to the normalization adopted for $\Delta\sigma_L$ and $\Delta\sigma_T$ data; this is discussed in Section 4. Even Arndt's inelasticity in 3P0 contributes only 0.55 mb to $\sigma_{tot}$ at 515 MeV.

ANALYTICITY    A virtue of Arndt's parameterization is that he uses analytic functions with branch cuts positioned to account qualitatively for inelastic thresholds. Although analyticity is not imposed in my analysis, amplitudes are checked to make sure there is no obvious discrepancy. The 3F3 amplitude in Figure 7 illustrates vividly the considerable influence of the inelastic threshold on the real part of the amplitude as low as 200 MeV. This must be taken into account explicitly in the determination of meson exchanges.

## 3.3   Analogy with Van der Waals Forces

As for 3F3, inelasticity causes attraction in all partial waves below the inelastic threshold. This is responsible for much of the weak central (spin-averaged) attraction that binds nuclei. Tarrach & Ericson (30) have, in fact, developed a precise analogy between 2$\pi$ exchange (Figures 1b and 1c) and Van der Waals forces. In Figure 1c, one nucleon excites (polarizes) the other, and the polarized nucleon reacts on the first via a second $\pi$ exchange. This is analogous to the electric field of an atom polarizing a second by mixing into the wave function some excited state. The pion field is strictly analogous to the atomic electric field, and the atomic polarizability is analogous to the combination of $\pi$-N p-wave scattering lengths corresponding to $I = 0$, $J^P = 0^+$ exchange (6). The saturation of nuclear attraction at high density arises because of relativistic effects (31).

## 3.4  Potentials

Vinh Mau and collaborators (3, 5) constructed a potential, known as the Paris potential, by fitting directly to elastic data up to 425 MeV. They take the long-range part of the potential ($r > 0.8$ fm) from $\pi$, $2\pi$, $\omega$, and $A_1$ exchange. Coupling constants of the $\omega$ and $A_1$ are fitted, and $2\pi$ exchange is calculated as explained in the Introduction. For $r < 0.8$ fm, each of the ten N-N potentials (5 spin combinations $\times$ 2 isospins) is fitted to $V = V_0 + V_1 E/M$, where $E$ is center-of-mass kinetic energy; where possible, $V_1$ is set to zero.

The Paris potential gives a fit to the data with $\chi^2 = 1.99$ for p-p scattering and 2.17 for n-p, compared with $\sim 1.5$ for phase shift analysis. It is a large improvement in terms of $\chi^2$ on earlier potentials (e.g. Hamada-Johnson and Reid soft-core). The reader is referred to the reviews of Signell (14) and Bugg (4) for a detailed graphical comparison of these potentials. As one might expect, the Paris p-p potentials are not too different from those of the 1960s, but the $I = 0$ potentials differ in major respects, because of the qualitative progress in the 1970s in determining n-p phase shifts.

For nuclear physics calculations, one should be aware of two points about the Paris potential. First, the energy dependence is awkward to handle. Second, any potential model calculation implicitly contains high momentum components, and the Paris potential is not a good fit to some partial waves above 425 MeV (32). The Paris group has calculated N-N inelasticities (33) with the same ingredients as those going into the $2\pi$ exchange part of their potential. For 1D2 and 3F3, agreement with experiment is good. However, for high partial waves, inelasticities are high (e.g. 1G4) and for 3P2 badly wrong (like all other calculations); this implies corresponding errors in the high energy behavior of phase shifts in these waves.

Love & Franey (34) provided a direct parameterization of N-N t-matrices for the purpose of nuclear physics calculations. In particular, n-p charge exchange amplitudes, well known from 140 to 500 MeV, are almost purely real and have a distinctive spin dependence that is useful for studying collective excitation of nuclei in (p,n) and (d,2p) reactions.

## 3.5  Forward Dispersion Relations

Dispersion relations are an unambiguous prescription for finding meson exchanges, provided one can evaluate dispersion integrals accurately. The best modern calculation along these lines is that of Grein & Kroll (6). They take $2\pi$ exchanges from $\pi$-N data [as do Vinh Mau et al (3, 5)]. After accounting for $\pi$ and $2\pi$ exchange and dispersion integrals, the remaining amplitudes unambiguously require further contributions, from $3\pi$, $\omega$, and $A_1$ exchange. Grein & Kroll are in fact able to obtain an impressive fit to six amplitudes (3 spin combinations × 2 isospins) with just two adjustable coupling constants for the $\omega$ and $A_1$. The $3\pi$ exchange terms, they claim to be able to evaluate from $\pi$-$\sigma$ and $\pi$-$\rho$ exchange, with results in remarkable agreement with experiment. The reader is referred to their paper and to the review by Bugg (4) for figures comparing calculations and experiment.

Unfortunately a later set of calculations by Grein & Kroll (35) obtained n-p spin-dependent amplitudes in serious disagreement with phase shift analysis. The latter will not accommodate to these amplitudes without very large increases in $\chi^2$ (36). These amplitudes are the ones quoted in Kroll's review article (13). Independent evidence of trouble is that these amplitudes predict a value of $K_{LL} = -0.08$ for n-p scattering at $0°$ and 800 MeV, with a small error. The polarized neutron beam at LAMPF is made via n-p charge exchange at $0°$ through precisely this $K_{LL}$ parameter and requires $K_{LL} = -0.65 \pm 0.05$. Clearly, a fresh look is needed at forward dispersion relations. Ideally this should use all five amplitudes and should cover $t \neq 0$ as well.

## 3.6  Fits of Holinde and Machleidt

Holinde & Machleidt fitted all N-N partial waves up to 300 MeV and also the properties of the deuteron using $\pi$, $2\pi$, $\omega$, $\sigma\pi$, $\rho\pi$, and $\delta$ exchange and diagrams containing intermediate $\Delta$s (7). In order to cut off divergent exchanges at small $r$, they included form factors $(\Lambda^2 - m^2)^2/(\Lambda^2 + t)^2$ for $\pi$, $\rho$, and $\omega$ exchanges with fitted values of $\Lambda$ ($\sim 1.3$ to 1.5 GeV) in reasonable agreement with other sources. In the form factor, $m$ stands for meson mass. An interesting feature of the calculations is that $\rho\pi$ exchange provides general repulsion, but also important spin dependence (e.g. in 3P0 and 3P1), which cannot be given by $\omega$ exchange alone. In addition to the well-known strong cancellation between attractive $\sigma$ exchange and repulsive $\omega$ exchange, they find a cancellation between $\sigma$-$\pi$ and $\omega$-$\pi$ exchange (except in $\bar{\varepsilon}1$ where both are needed to fit experiment). One gets an impression of cancellation in short-range exchanges giving rise to something like a convergent series. Impressive though the fit is (Figure 8), some further clarification is desirable. The calculations use a narrow $\Delta$ approximation. This prevents comparison with experiment above 300 MeV, and with N-N inelasticities. It would be very interesting to see the effects of the width of the $\Delta$. Secondly, because of their form factors, Holinde & Machleidt arrive at much larger coupling constants than do Grein & Kroll. It would be instructive to clarify the relation between the amplitudes emerging from these two apparently different sets of coupling constants.

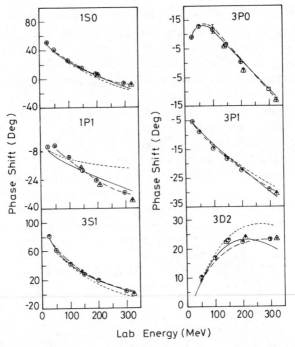

*Figure 8*   N-N phase shifts for low partial waves: solid curves represent fits of Holinde & Machleidt (7); dotted curves are typical one-boson exchange fits; and dashed curves are from the energy-dependent phase shift analysis of Arndt et al (27); circles are from Arndt et al. Energy-independent analysis; triangles are from Dubois et al (28). Reproduced, with permission, from Reference 7.

## 3.7   *The Quark Model*

There have been many attempts to describe the short-range N-N inter-actions in terms of a quark model. One paper, by Maltman & Isgur (9), is particularly illuminating (and readable); references to earlier work are found there.

These authors start with precisely the same nonrelativistic quark-quark Hamiltonian as is used by Isgur and Karl in fitting the spectrum of baryon and meson resonances. They consider a six-quark system with the overall charge and flavor of the deuteron. The six quarks are allocated five arbitrary relative coordinates, chosen as the internal coordinates of two clusters of three quarks and an intercluster coordinate. Although cluster coordinates are used, clustering is not imposed on the dynamics; all quark-quark interactions are taken into account. The wave function is now expanded in terms of 115 basis states, properly antisymmetrized, having all possible color, spin, flavor, and space combinations with up to two units of orbital excitation within either cluster or between clusters (e.g. the N-N D-wave). The objective is to allow very wide freedom to the wave function, and allow all possible quark configurations (2q, 3q, colored, etc) into the wave function. The ground state is now determined by a variational calculation.

What emerges is very striking. The 3u-3d system clusters strongly into a neutron-proton configuration. Other six-quark systems do not exhibit such clustering. The n-p component nearly saturates the full wave function; i.e. mixing with hidden-color states, $\Delta$-$\Delta$ and N-N* is small. An effective potential is derived, which, when inserted into the Schrödinger equation, gives the n-p component of the wave function. This effective potential shows three features in semiquantitative agreement with experiment: (a) a strong repulsion for $r < 1$ fm rising to 1–2 GeV at $r = 0$; (b) intermediate-range weak attraction ($\sim 5$ MeV) for $r > 1$ fm; and (c) slightly more attraction for 3S1 than for 1S0 for all $r$.

In the final phase of this work, a pion exchange tail is added for $r > 2$ fm, to account for q-$\bar{q}$ clustering into mesons. With this addition, the properties of the deuteron are well reproduced; however, it is not clear how much of this success is to be attributed to the pion tail.

## 4. DIBARYONS

There has been considerable speculation on the possible existence of dibaryon resonances, of either a conventional variety or in an exotic hidden-color configuration. The latter would have a wave function that is a linear combination of N-N, N-$\Delta$, and $C_1$-$C_2$ where C are colored octets combining to net zero color. The present situation is that there is no firm evidence for such exotic states, but the possibility remains that conventional resonances might exist in association with inelastic thresholds.

The story began when Auer et al (12) at the Argonne National Laboratory (ANL) observed striking peaks in $\Delta\sigma_L = \sigma(\rightleftarrows) - \sigma(\rightrightarrows)$, the difference in total cross sections measured with longitudinally polarized beam and target with spins parallel or opposed. A similar peak has since been found in $\Delta\sigma_T = \sigma(\uparrow\downarrow) - \sigma(\uparrow\uparrow)$, the corresponding measurement with transverse spin orientations. Latest data are shown on Figure 9, with scaling factors described in Section 4.1. Hidaka et al (37) proposed that the peak at 550 MeV is due to a 1D2 resonance and the inverted peak in $\Delta\sigma_L$ at 800 MeV to a 3F3 resonance. Since then, phase shift analysis of elastic data has revealed resonance-like half-loops in the Argand diagrams of 1D2, 3F3, and 3P2. In the pp → d$\pi$ reaction, nearly complete loops are also observed in the same partial waves (Figure 3).

There are two interpretations. One view is that these peaks and Argand diagrams look like resonances, can be parameterized as resonances, and therefore are resonances (38, 39). The conservative viewpoint, however, is that one should first make sure that conventional nonresonant physics is not capable of explaining the data. In fact, the process NN → NΔ can be shown to give at least a semiquantitative account of all the results.

Consider first the loops in pp → dπ. In the π-d total cross section there is an enormous peak due to πN → Δ at a mass corresponding to 650 MeV in the N-N channel. With the inclusion of Fermi motion, double scattering, and spreading width due to the extra open channel ΔN → NN, Glauber theory fits the total cross section and the elastic diffraction peak to high accuracy (1–2%). This leaves little room for extra coupling to a dibaryon resonance. As early as 1974, Hoenig & Rinat (40) did a partial wave decomposition of the Glauber amplitude, and demonstrated the existence of loops in many π-d partial waves. Each loop has a phase variation caused by the Δ, but the partial waves are not themselves resonant. They contain logarithmic singularities, not poles. Hoenig & Rinat called them pseudo-resonances. Essentially what happens is this: because of the size of the deuteron and the fact that the Δ is not at rest in the π-d center of mass, the Δ amplitude gets distributed coherently among many π-d partial waves. There is a 5% branching ratio from the π-d channel to N-N, and similar loops appear in the pp → dπ amplitudes.

In extracting physics from these amplitudes, one must first account quantitatively for this foreground looping behavior due to the Δ. The snag

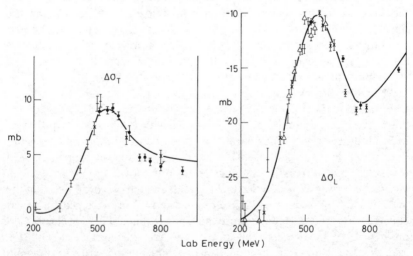

*Figure 9*   A compilation of p-p $\Delta\sigma_L$ and $\Delta\sigma_T$ data: ● Saclay (41), △ Geneva (43) scaled by 1.04, + TRIUMF (42) scaled by 0.90, × LAMPF (44, 45) scaled by 1.10. Curves are from my phase shift analysis.

is that one does not know, except from a model, the precise N-N couplings. A dibaryon could be formed between the $\Delta$ and the spectator nucleon in the deuteron via a strong final-state N-$\Delta$ interaction. This would superpose a further phase (and amplitude) variation on pp $\rightarrow$ d$\pi$ amplitudes. Clearly this is hard to identify unless the dibaryon couples strongly to both channels; present evidence is that such coupling is $\leqslant 20\%$ to N-N and $\leqslant 10\%$ to $\pi$-d.

The process NN $\rightarrow$ N$\Delta$ gives rise to half-loops in N-N elastic scattering closely resembling the data, and again making it hard to tell whether there is an additional resonance superposed. It is a difficult quantitative question whether or not these two conventional pieces of physics account for all of the structure in $\Delta\sigma_L$ and $\Delta\sigma_T$. What is clear is that if dibaryons exist, they have branching ratios $\geqslant 70\%$ to NN$\pi$, which is therefore the best channel to study.

## 4.1   Peaks in $\Delta\sigma_L$ and $\Delta\sigma_T$

In the early days, there was some controversy over the data themselves. The three lowest energy points of the original data set of Auer et al lay at or above the limit to which phase shift analysis could be stretched by juggling inelasticities, and these inelasticities looked odd. Remeasurements of $\Delta\sigma_L$ and new measurements of $\Delta\sigma_T$ were made at TRIUMF, SIN, LAMPF, and Saclay. They confirmed the existence of peaks in both $\Delta\sigma_L$ and $\Delta\sigma_T$ at 570 MeV, but there were obvious discrepancies of normalization between the experiments. Today, with the aid of the very complete set of p-p elastic data from 445 to 800 MeV and phase shift analysis, the situation seems to be as follows.

1. Saclay $\Delta\sigma_L$ (41) have the correct normalization.
2. TRIUMF $\Delta\sigma_L$ and $\Delta\sigma_T$ (42) are 11% high in normalization, compared with quoted errors of $\pm 6.5\%$; this probably arises from an error in target density.
3. SIN values (43) are 3.7% low (quoted error $\pm 6.2\%$).
4. LAMPF $\Delta\sigma_L$ and $\Delta\sigma_T$ (44, 45) are 10% low (quoted error $\pm 4.1\%$).
5. The original ANL $\Delta\sigma_L$ value at 561 MeV has been withdrawn (because of a depolarization resonance in the accelerator) and the points at 433 and 508 MeV are low by 2 and 15% respectively.

Now let us turn to interpretation. The N-N 1D2 wave contributes positively to both $\Delta\sigma_L$ and $\Delta\sigma_T$ with weight $(2J+1)$; 3F3 contributes negatively to $\Delta\sigma_L$ alone with weight $(2J+1)$; and 3P2 contributes negatively to both $\Delta\sigma_L$ and $\Delta\sigma_T$ with weights 1 and $J$ respectively. The shapes of $\Delta\sigma_L$ and $\Delta\sigma_T$ are then explained qualitatively by the inelastic threshold in 1D2 (N-$\Delta$ with $L=0$) appearing at lower energy than in 3F3 and

3P2 ($L_{N\Delta} = 1$). At present, this argument cannot be made completely quantitative because of lack of data on the $NN\pi$ channels.

## 4.2   Half-loops in $NN \to NN$ Argand Diagrams

Amplitudes from current phase shift analysis are shown in Figure 10. They resemble qualitatively many $\pi$-N resonances. Bhandari et al (46) and Edwards (47) claim they require the existence of poles in 1D2 and 3F3 with masses close to the N-$\Delta$ branch point 2.15–0.05$i$ GeV. On the other hand, Kloet, Tjon & Silbar (48) demonstrate in a model calculation that the strong inelastic thresholds due to NN $\to$ N$\Delta$ produce half-loops shown by the dashed curves on Figure 10, rather close to experiment. Clearly the distinction between resonant and nonresonant interpretations is a fine one when made on the basis of N-N data alone. Lee (49) also achieves an impressively good account of N-N phases and inelasticities with a model not containing resonances. The essential point is that the dispersion integral of Equation 1 makes positive contributions to the real part of partial wave amplitudes below either a resonance or an inelastic threshold; only above resonance (or threshold) does the distinction appear between a complete loop (resonance) or half-loop (threshold). It is, of course, possible that the strong threshold may help generate a resonance superimposed on the threshold.

## 4.3   $pp \to pn\pi^+$ Data

The TRIUMF data of Section 2.2 indicate an attractive N-$\Delta$ interaction from 420 and 510 MeV. This might correspond to ($a$) a quasi-deuteron at the N-$\Delta$ threshold, ($b$) a virtual state like the N-N 1S0 interaction, or ($c$) a dibaryon above the energy range yet explored. In view of the energies of the poles found by Bhandari et al and Edwards, the third possibility looks less likely. TRIUMF values of $\delta_{N\Delta} = 45 \pm 10°$ at 420, 465, and 510 MeV resemble possibility ($b$). Either of the first two possibilities would accord

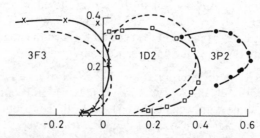

*Figure 10*   Argand diagrams for N-N elastic scattering. Solid curves guide the eye through data points of Bugg; dashed curves represent the calculations of Kloet et al (48).

simply with quark models, which, as for the N-N channel, predict long-range attraction and short-range repulsion. In $\Lambda N \to \Lambda N$ and $\Sigma N \to \Sigma N$ data (50) there are strong rises in low-energy cross sections, and in $K^-d \to \pi^-\Lambda p$ a strong $\Lambda p$ enhancement (51) at the $\Sigma$-N threshold. It seems plausible that long-range attraction giving rise to quasi-deuterons or virtual states may be a feature common to N-N, Y-N and N-$\Delta$ interactions at threshold.

The Argonne $A_{N0}$ data of Figure 6 are intriguing. They undoubtedly indicate a rapid variation of either the states that interfere or the phase of at least one amplitude. Possibilities (a) and (b) above would give rise to large $\delta_{N\Delta}$ near threshold, falling with increasing energy; it would also be likely that $L = 1$ N-$\Delta$ amplitudes would develop positive phase shifts as the effect of the centrifugal barrier drops with increasing energy. One would in this case expect the interfering amplitudes to converge in phase; if $A_{N0}$ were due to interference between 1D2 and either or both of 3F3 and 3P2, this could lead to the observed drop in $A_{N0}$ and would lead to large negative $A_{SL}$ (the sign arising, as in pp $\to$ d$\pi^+$, from Clebsch-Gordan coefficients). Possibility (c) would conversely lead to an anticlockwise movement of the resonant 1D2 amplitude around the Argand diagram, and hence large $A_{SL}$ of the opposite sign. A measurement of $A_{SL}$ up to 800 MeV will be made at LAMPF during 1985.

## 4.4   Narrow Dibaryons

Three searches at LAMPF for narrow dibaryons with a mass resolution $<2$ MeV have yielded negative results. Lisowsky et al (52) measured n-p total cross sections with $<1\%$ statistics using a continuous neutron spectrum and time of flight. Seth and collaborators measured $A_{N0}$ at $90°$ in pp $\to$ d$\pi^+$ over a wide energy range and also $A_{N0}$ for p' in the d(p,p')pn reaction at 700 MeV, with high statistics but negative results (53).

There is a claim that the tensor polarization $t_{20}$ has narrow structure in backward $\pi$-d elastic scattering at 134 MeV (54); but in view of discrepancies between SIN and LAMPF data (55) at neighboring energies, this result is not currently regarded as definitive. Also, it cannot be explained by a resonance in a single partial wave; in order to fit other data, narrow cooperative perturbations would be required in several partial waves. Elastic N-N data at the corresponding energy (560 MeV) show no unexpected effects (56).

## 5.   SUMMARY AND OUTLOOK

Nucleon-nucleon physics is like a complicated multisheeted jigsaw puzzle. An intimate blend of theory and experiment is needed to solve the puzzle.

Spin-dependent data from the meson factories have filled in much of the picture up to 1 GeV. Inelasticity is qualitatively as expected, but it is not completely understood quantitatively, for example, the large inelasticity in 3P2. Further progress depends on coupled-channel calculations, containing all the known ingredients: production of $\Delta$ and other $\pi$-N partial waves, $\pi$ and $\rho$ exchange, $\Delta$-broadening, N-N initial- and final-state interactions, and the deuteron D-state in the $\pi$-d channel. Only then can we expect to see if there are missing pieces. Understanding of inelasticity is essential for quantitative understanding of meson exchanges.

The long-range part of the NN $\rightarrow$ NN interaction due to $\pi$ and $2\pi$ exchange is settled. At shorter range, $\omega$, $\rho$-$\pi$, and $\sigma$-$\pi$ exchanges are certainly required, but there is not yet universal agreement on the strengths of all the pieces. This requires a fit to the whole $t$ range up to 800 MeV.

The work of Maltman & Isgur shows that the quark model with their Hamiltonian has the right properties to explain simultaneously the spectrum of baryon and meson resonances and the features of the N-N S-wave interaction; hidden-color components of the wave function seem to be small. It would be of great interest to extend this type of calculation to N-N P-waves and to N-$\Delta$.

A recent development is the Skyrmion model of the nucleon. This model uses a nonlinear chiral-invariant Lagrangian (QCD inspired), in which the nucleon appears as a soliton. Excited states, such as the $\Delta$ and N*(1420) appear naturally. The N-N and N-$\Delta$ interactions may be derived (57, 58) by including in the Lagrangian terms corresponding to the $\pi$, $\sigma$, $\rho$, and $\omega$ fields. This approach, though still in its infancy, shows real promise of accounting quantitatively and coherently for low-energy N-N, N-$\Delta$, and $\pi$-$\pi$ interactions.

In the near future, one foresees elastic data from LAMPF settling n-p phase shifts up to 800 MeV and inelastic data pinning down NN$\pi$ amplitudes in both magnitude and phase. A comprehensive study of p-p and n-p elastic amplitudes at higher energies has begun at Saclay. At the Argonne National Laboratory, a heroic attempt was made to determine p-p elastic amplitudes at 6 and 12 GeV out to $t = 1$ (GeV/$c$)$^2$. Analysis by Moravcsik et al (59) arrives at four possible solutions at 6 GeV, all very different from predictions of the latest Regge models. This program is tantalizingly close to completion and deserves to be finished, for example at Brookhaven National Laboratory, where a polarized beam is now available. At large $t$, near 90°, there are dramatically large effects in $A_{NO}$ and $A_{NN}$ for p-p scattering (60, 61) and in $A_{NO}$ for n-p, much larger than Regge theory or naive parton models predict. At present there is little or no understanding of the region above 1 GeV, which therefore deserves further study.

*Literature Cited*

1. MacGregor, M. H., Arndt, R. A., Wright, R. *Phys. Rev.* 182:1714–28 (1969)
2. Scotti, A., Wong, D. Y. *Phys. Rev.* 138:B145–62 (1965)
3. Vinh Mau, R. In *Mesons in Nuclei*, ed. M. Rho, D. Wilkinson, 1:151–96. Amsterdam: North-Holland (1979)
4. Bugg, D. V. In *Progress in Particle and Nuclear Physics*, ed. D. H. Wilkinson, 7:47–112. Oxford: Pergamon (1981)
5. Lacombe, M., et al. *Phys. Rev. C* 21:861–73 (1980)
6. Grein, W., Kroll, P. *Nucl. Phys. A* 338:332–48 (1980)
7. Machleidt, R. In *Lecture Notes in Physics*, ed. H. Araki, et al, 197:352–89. Berlin: Springer-Verlag (1984)
8. Lomon, E. In *The Interaction Between Medium Energy Nucleons in Nuclei*, ed. H. O. Meyer, *AIP Conf. Proc.* 97:1–19. New York: Am. Inst. Phys. 1983)
9. Maltman, K., Isgur, N. *Phys. Rev. D* 29:952–77 (1984)
10. Mulders, P. J., Aerts, A. T., de Swart, J. J. *Phys. Rev. D* 21:2653–71 (1980)
11. Ueda, T. *Phys. Lett. B* 79:487–91 (1978)
12. Auer, I. P., et al, *Phys. Rev. Lett.* 41:354–56 (1978)
13. Kroll, P. *Phys. Data* 22:1–274 (1981)
14. Signell, P. In *Proc. Telluride Conf. on the (p,n) Reaction and the Nucleon-Nucleon Force*, ed. C. D. Goodman, et al, pp. 1–21. New York: Plenum (1980)
15. Kloet, W. M., Tjon, J. A. *Phys. Rev. C* 30:1653–61 (1984)
16. Gruben, J. H., VerWest, B. J. *Phys. Rev. C* 28:836–47 (1983)
17. Bugg, D. V. *J. Phys. G* 10:717–26 (1984); *Nucl Phys. A* 437:534–40 (1985)
18. Hiroshige, N., Watari, W., Yonezawa, M. Partial-wave amplitudes of pp → $\pi^+$d process from its threshold to $T_p^L = 810$ MeV. *Univ. Hiroshima preprint* (1984)
19. Krell, M., Ericson, T. E. O. *Nucl. Phys. B* 11:521–50 (1969)
20. Waltham, C. E., et al. *Nucl. Phys. A* 433:649–70 (1985)
21. Silbar, R. R. *Comments Nucl. Part. Phys.* 12:177–89 (1984)
22. Calkin, M. M. PhD thesis, Rice Univ. (1984)
23. Rodebaugh, R. F., Bonner, B. E. In *Few Body Problems in Physics, II*, ed. B. Zeitnitz, pp. 43–44. Amsterdam: North Holland (1984)
24. Aprile-Giboni, E., et al. *Nucl. Instrum. Methods* 215:147–57 (1984)
25. Chatelain, P., et al. *J. Phys. G* 8:643–48 (1982)
25a. Wolfenstein, L. *Phys. Rev.* 98:1870–75 (1955)

26. Bystricky, J., Lechanoine-Leluc, C., Lehar, F. *Saclay preprint D.Ph.P.E. 82-12* (1984)
27. Arndt, R. A., et al. *Phys. Rev. D* 28:97–122 (1983)
28. Dubois, R., et al. *Nucl. Phys. A* 377:554–84 (1982)
29. Mathelitsch, L., VerWest, B. J. *Phys. Rev. C* 29:739–46 (1984)
30. Tarrach, R., Ericson, M. *Nucl. Phys. A* 294:417–34 (1978)
31. Brockmann, R., Machleidt, R. *Phys. Lett. B* 149:283–87 (1984)
32. Matin, M. A., Gandhi, D. C. *Can. J. Phys.* 61:1003–12 (1983)
33. Côté, J., et al. *Nucl. Phys. A* 379:349–68 (1982)
34. Love, W. G., Franey, M. A. *Phys. Rev. C* 24:1073–94 (1981)
35. Grein, W., Kroll, P. *Nucl. Phys. A* 377:505–17 (1982)
36. Bugg, D. V. See Ref. 8, pp. 1–19
37. Hidaka, K., et al. *Phys. Lett. B* 70:479–81 (1977)
38. Kravtsov, A. V., Ryskin, M. G., Strakovsky, I. I. *J. Phys. G.* /:L187–90 (1983)
39. Hiroshige, N., et al. A Three-Channel (pp, N$\Delta$, $\pi$d) K-Matrix Analysis of the Dibaryon Resonances with $J^P = 2^+$ and $3^-$. *Univ. Hiroshima preprint* (1984)
40. Hoenig, M. M., Rinat, A. S. *Phys. Rev. C* 10:2102–5 (1974)
41. Bystricky, J., et al. *Phys. Lett.* 142:130–34 (1984)
42. Stanley, J. P., et al. *Nucl. Phys. A* 403:525–52 (1983)
43. Aprile-Giboni, E., et al. *Nucl. Phys. A* 431:637–68 (1984)
44. Auer, I. P., et al. *Phys. Rev. D* 29:2435–68 (1984)
45. Ditzler, W. R., et al. *Phys. Rev. D* 27:680–83 (1983)
46. Bhandari, R., et al. *Phys. Rev. Lett.* 46:1111–14 (1981)
47. Edwards, B. J. *Phys. Rev. D* 23:1978–86 (1981)
48. Kloet, W. M., Tjon, J. A., Silbar, R. R. *Phys. Lett. B* 99:80–84 (1981)
49. Lee, T. S. H. *Phys. Rev. C* 29:195–203 (1984)
50. Nagels, M. M., Rijken, T. A., de Swart, J. J. *Phys. Rev. D* 20:1633–45 (1979)
51. Pigot, C., et al. *Saclay preprint D.Ph.P.E. 84-03* (1984)
52. Lisowsky, P. W., et al. *Phys. Rev. Lett.* 49:255–59 (1982)
53. Seth, K. K. Searching for Dibaryons with Hadronic Probes: The Inelastic Reaction. *Northwestern Univ. preprint* (1983)

54. Grüebler, W., et al. *Phys. Rev. Lett.* 49: 444–47 (1984)
55. Holt, R. J., et al. *Phys. Rev. Lett.* 47: 472–75 (1981)
56. Aprile, E., et al. See Ref. 23, pp. 163–64
57. Jackson, A., Jackson, A. D., Pasquier, V. *Nucl. Phys. A* 432: 567–609 (1985)
58. Vinh Mau, R., et al. The Static Baryon-Baryon Potential in the Skyrme Model.

*Univ. Paris Orsay preprint* (1984)
59. Moravcsik, M. J., Ghahramany, N., Goldstein, G. G. *Phys. Rev. D* 30: 1899–1903 (1984)
60. Peaslee, D. C., et al. *Phys. Rev. Lett.* 51: 2359–61 (1983)
61. Crosbie, E. A., et al. *Phys. Rev. D* 23: 600–3 (1981)

*Ann. Rev. Nucl. Part. Sci. 1985. 35:321–49*

# OBSERVATORY FOR ULTRA HIGH-ENERGY PROCESSES: THE FLY'S EYE

*George L. Cassiday*

Department of Physics, University of Utah, Salt Lake City, Utah 84112

CONTENTS

## 1. INTRODUCTION

The University of Utah has constructed a large high-energy physics/ astrophysics observatory on top of Little Granite Mountain, Dugway, Utah, approximately 160 km southwest of Salt Lake City, Utah. The observatory was designed to detect the passage of extensive air showers (EAS) through the atmosphere by means of the nitrogen fluorescence light given off after excitation by the relativistic charged particles in the shower. Experiments being carried out with the detector include (*a*) a direct measurement of the proton-air cross section at $s^{1/2} = 30$ TeV; (*b*) an

321

0163–8998/85/1201–0321$02.00

analysis of the primary cosmic-ray spectrum in the energy range 0.01–100 EeV (1 EeV = $10^{18}$ eV), with emphasis on the 3 K blackbody cutoff region near 60 EeV; (c) an extraction of the composition of high-energy cosmic-ray primaries; (d) a search for anisotropies in arrival directions; (e) a search for deeply penetrating showers indicative of primary neutrinos, heavy-lepton production, or quark matter in the primary flux; and (f) a search for sources of $\gamma$ rays near 1 PeV ($10^{15}$ eV).

Previous experiments with EAS (1, 2) involved ground-based particle-detector arrays in which the properties of the showers were inferred from a sample of the secondaries taken at a few locations along the shower front at a single atmospheric depth. Other experiments involved Cherenkov detector arrays, which employ fast timing measurements of samples of the shower-generated Cherenkov wavefront to obtain an integral history of the shower. In each of these cases, observations are limited to small areas and low event rates for high-energy showers as well as poor, if any, resolution of shower longitudinal development. The Fly's Eye detector was designed to overcome these two experimental deficiencies. The experimental goals to be carried out by the Fly's Eye experiment depend crucially upon its ability not only to "see" the highest energy cosmic rays at distances on the order of 20 km from the detector, but also to see each event with enough clarity to ascertain in a fairly model-independent way the energy of the primary cosmic ray and longitudinal structure of the resultant EAS. In the following sections of this paper, we present design and operational details of the Fly's Eye detector along with those experimental results that illustrate how far this novel experimental technique has evolved in pursuit of the above-mentioned goals.

## 1.1. Brief Description of the Fly's Eye

The Fly's Eye observatory (Figure 1) consists of two experimental stations (Fly's Eye I and Fly's Eye II) separated by 3.3 km. Fly's Eye I consists of 67 62-inch front aluminized spherical section mirrors, associated Winston light collectors, photomultiplier tubes (PMTs), and data acquisition electronics. The Winston light collectors and PMTs are hexagonally packed in groups of either 12 or 14 light-sensing "eyes" mounted in the focal plane of each mirror. A motorized shutter system keeps the PMTs both light tight and weatherproof during the day and permits exposure to the sky at night. Each mirror unit and associated light-sensing cluster is housed in a single, motorized corrugated steel pipe about 2.13 m long and 2.44 m in diameter. The units are turned down, with mirror and open end facing the ground, during the day (to protect cluster and mirror from light and weather) and turned up at night to a predetermined position so that each "eye" observes a designated angular region of the sky. In all, there are 880 "eyes" at Fly's Eye I arranged to image completely the entire night sky. The projection of each hexagonal "eye" onto the celestial sphere resembles the compound eye of an insect; hence the name Fly's Eye. Shown in Figure 2 is a picture of a single mirror housing unit and associated optical cluster. Fly's Eye II is a smaller array of identical units, 8 in all, with 120 total "eyes." Fly's Eye II observes roughly one azimuthal quadrant of the night sky with elevation angles ranging between 2° and 38° above the horizon.

On clear, moonless nights an operator at each site activates the detector, which records the pulse integral and arrival times generated by each visible EAS as it progresses across the celestial sphere. Whenever Fly's Eye I triggers, it sends an infrared flash of light toward Fly's Eye II, which, if also triggered by the "simultaneous" observation of something in the night sky, records its own pulse integral and arrival times. EAS track geometry may then be reconstructed either from hit patterns and timing by a single Fly's Eye detector or by stereoscopic viewing and relative timing by both Fly's Eyes. Once the geometry of an EAS track is determined, shower longitudinal development profiles and total shower energy can be obtained on an event-by-event basis from measured pulse integrals after suitable correction for light attenuation and Cherenkov light contamination.

*Figure 1*  Aerial view of the Fly's Eye I detector. The detector is located on top of Little Granite Mountain, elevation 860 g cm$^{-2}$, Dugway, Utah.

*Figure 2* Single mirror unit. A 67-inch mirror is housed in a large 2.13 × 2.44 m motorized, corrugated steel pipe. A cluster of 14 Winston cone light collectors, PMTs, and associated preamps is mounted in the mirror's focal plane.

## 2.   PRODUCTION OF LIGHT BY EAS

The fundamental problem of detecting an EAS via air fluorescence can best be imagined as follows: Consider an apparently blue (actually near UV) 5-watt light bulb streaking through the sky at the speed of light against a continuous backdrop of starlight, atmospheric airglow, and man-made light pollution. Add to this sporadic sources of light such as lightning, auroras, and airplane and smokestack strobe lights (visible for hundreds of miles) creating a certain visual havoc. The task of the Fly's Eye is to pick out the faint, fast signal from the background noise.

Four basic mechanisms contribute to the generation of the light signal seen by the Fly's Eye detector: fluorescence, direct Cherenkov light, Rayleigh-scattered Cherenkov light, and Mie-scattered Cherenkov light. Of these, fluorescence is the one relating most directly to the numbers of charged particles in an EAS seen by any particular PMT.

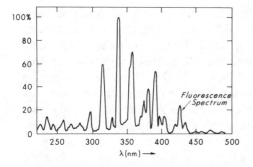

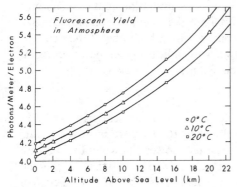

*Figure 3* (*a*) Atmospheric fluorescence spectrum emanating mostly from the 2P band of molecular nitrogen and the 1N band of the $N_2^+$ molecular ion. (*b*) Fluorescent yield (equivalent 360 nm photons/electron/meter) as a function of altitude and temperature.

## 2.1   Atmospheric Fluorescence

Most of the energy of a primary cosmic ray is dissipated in the atmosphere by the ionization and excitation of air molecules. Nearly all of the optical fluorescence comes from the 2P band system of molecular nitrogen and the 1N band system of the $N_2^+$ molecular ion (3–5). The measured fluorescence spectrum is shown in Figure 3*a*. Most of the light is emitted in the spectral region 310 to 440 nm. The resultant fluorescent yield as a function of altitude integrated over our spectral response is shown in Figure 3*b*. It is mildly altitude and temperature dependent. The resultant light yield corresponds to a scintillation efficiency of only 0.5%. However, the poor efficiency is compensated for by the overwhelming amount of energy being dissipated by a 100-EeV EAS—more than 1 joule in 30 $\mu$s!

## 2.2   Cherenkov Light

Electrons in an EAS generate a prodigious amount of Cherenkov light that is primarily beamed in the forward direction (6). The amount of Cherenkov light at any point along the shower front depends upon the previous history of the shower, and thus is not strictly proportional to local shower size as is the case for scintillation light. Directly beamed Cherenkov light dominates the light seen by the Fly's Eye detector at emission angles of up to 25° relative to the EAS axis; this makes the inference of size difficult for early stages of shower development. Moreover, as the Cherenkov component builds up with the propagating shower front, the resultant intense beam can generate enough scattered light at low altitudes to compete with the locally produced scintillation light from the rapidly dying shower. Even so, scattered Cherenkov light usually constitutes in worst-case situations no more than about 30% of the total light seen, whereas the directly beamed Cherenkov light at small angles may swamp scintillation light by a factor of $10^2$. These considerations severely limit the accuracy of shower size measurement for those EAS that strike within 1 km or so of the detector since much of the developing shower can only be observed at emission angles less than 25° under such circumstances.

## 2.3    *Resultant Signal*

A detailed treatment of the complete production and scattering of EAS-generated light has been presented elsewhere (6). Here we show the results of numerical calculations made for a sample shower. Shown in Figure 4 are relative photoelectron yields generated by a shower of size $N_e$ via the various light-generating mechanisms mentioned above. The effects of atmospheric attenuation are also included. The curves labelled Sc, $C^v$, R, and M refer to scintillation, direct Cherenkov, Rayleigh-, and Mie-scattered Cherenkov light. In general, fluorescence, or scintillation, light dominates during most of the observed trajectory. Scattered light is always inconsequential during the early stages of the shower. Hence, the initial size of the shower at the first observation direction can be deduced by assuming that the received light is all direct Cherenkov and scintillation light. This permits an accurate estimate of the Cherenkov beam buildup, which is necessary for calculating those scattering corrections that must be applied during the latter stages of shower development.

## 2.4    *Noise*

Essential noise mechanisms limiting detector sensitivity include scattered starlight, diffuse radiation of the galaxy, light from subthreshold stars, light from other galaxies and intergalactic matter, sunlight scattered by interplanetary matter, and light produced by photochemical processes in the earth's ionosphere (7, 8). This background light induces a dc signal in any phototube aimed at the night sky and fluctuations in this dc signal constitute the noise.

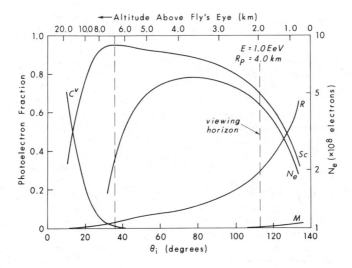

*Figure 4* Relative fractions of light received at the Fly's Eye detector via all light-generating mechanisms as a function of altitude above the Fly's Eye (*upper scale*) and beam emission angle $\theta$ (*lower scale*). $S_c$ is the scintillation light, C for direct Cherenkov, R for Rayleigh-scattered and M for Mie-scattered Cherenkov light. The curve labelled $N_e$ is the size of the shower that generated the light.

In the relevant spectral region of our sensitivity, the night-sky background averages about a fourth magnitude per square degree (9, 10). This corresponds to a brightness of about $5 \times 10^5$ photons $m^{-2} sr^{-1} \mu s^{-1}$. This value should be roughly doubled to account for long-term atmospheric airglow. Moonlight at quarter phase doubles the amount of background light, while at full phase it makes the night sky more than 10–20 times as bright, completely eliminating the possibility of detector operation during such times.

The average night-sky brightness fluctuates during the course of a night but typically by no more than a factor of two from detector turn-on-to turn-off. More troublesome are local wandering "hot spots" caused by certain stars, planets, and the Milky Way disk. The average brightness seen by a Fly's Eye PMT corresponds to about a first-magnitude star. Hence, when a star or planet brighter than this (about 15 or so in the Northern Hemisphere) enters a PMT field of view, the dc background more than doubles (in the case of Venus, it climbs ten-fold!). This situation has been handled by programming all discriminator input thresholds to keep count rates constant.

**Table 1**  Fly's eye detector parameters

| Parameter | Fly's Eye I (II) |
|---|---|
| 1. Number of mirrors | 67 (8) |
| 2. Diameter of mirror | 1.575 m |
| 3. Focal length | 1.50 m |
| 4. Number of PMTs (and Winston Cones) | 880 (112) |
| 5. Mirror obscuration | 13% |
| 6. Mirror-cone efficiency product | ~70% |
| 7. PMT | EMI 9861B |
|  | 90-mm Super S-11 |
| 8. Peak quantum $\varepsilon$ at 360 nm | 0.21 |
| 9. Angular aperture per PMT | 91.5 mr |
| 10. Solid angle per PMT | 6.57 msr |
| 11. Number of electronics channels | 2640 (336) |
| 12. Charge dynamic range | $10^5$ linear |
| 13. Time resolution | 25 ns |

## 3.   DETECTOR PARAMETERS

Shown in Table 1 are the Fly's Eye detector parameters chosen on the basis of optimizing signal-to-noise ratio and resolution. We note that signal-to-noise ratio and resolution scale as $(\varepsilon D^3/d)^{1/2}$ and $d/D$, respectively, where $D$ and $d$ are the diameters of the mirror and phototube apertures and $\varepsilon$ is the overall light collection and photoelectron conversion efficiency. Clearly one maximizes $\varepsilon$ and $D$ and minimizes $d$ within budgetary constraints. Final choices of the parameters listed in Table 1 were essentially dictated by cost and estimated construction time as well as by diminishing returns on resolution.

### 3.1   *Kinematics*

Since each Fly's Eye sensing element subtends a specific solid angle in the night sky, an EAS trajectory appears as a sequential track propagating along a great circle projected upon the celestial sphere. The PMT "hit pattern" determines a plane in space within which the EAS trajectory lies (Figure 5). The orientation and "distance" of the EAS trajectory relative to the Fly's Eye in the shower-detector plane can be determined from the timing sequence of the light-pulse arrival times. The expected timing sequence stems from the fact that an EAS propagates in a straight line at the speed of light. The relationship between observation directions $\chi_i$ for each PMT as a function of time is given (6) by

$$\chi_i(t_i) = \chi_0 - 2 \tan^{-1} [c(t_i - t_0)/R_p]. \qquad\qquad 1.$$

A best fit of the observed $\chi_i(t_i)$ to this function yields the parameters $R_p$ and $\chi_0$, which (given the orientation of the shower-detector plane) completes the specification of the EAS trajectory.

If a shower is seen simultaneously by Fly's Eyes I and II (FEI and FEII), a

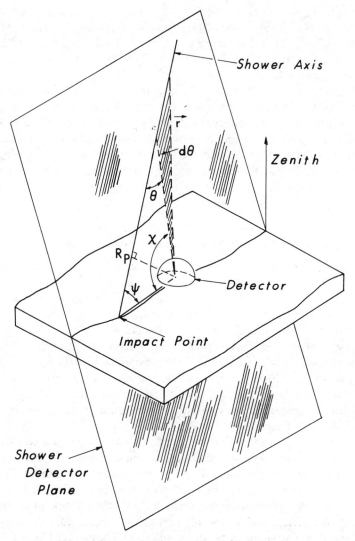

*Figure 5*   Geometry of an EAS trajectory as seen by the Fly's Eye. The shower-detector plane contains the EAS and the center of the Fly's Eye detector. It is specified by fits to the spatial pattern of "hit" PMTs, which must lie along a great circle on the celestial sphere. The angle $\psi$ and impact parameter $R_p$ are obtained by fits to observation angles $\chi_i$ vs time of observation.

best shower-detector plane for each detector can be determined and the intersection of those planes defines the shower trajectory. In addition, for each event observed by the two detectors in coincidence, a timing marker is sent via an infrared optical flash from FEI to FEII; this permits clock synchronization to within $\pm 100$ ns. Thus, for each such event, including the degenerate coplanar case in which direct stereoscopic reconstruction fails, a technique relying simultaneously upon the extended geometry of both detectors and absolute time recordings of all hit PMTs regardless of position (either at FEI or FEII) can be used to obtain the final geometry of the observed track.

The techniques of geometrical reconstruction have been checked by applying them to fiducial marker tracks of known geometry generated by pulsed xenon flashers permanently deployed around the Fly's Eye and by a mobile nitrogen laser. Comparisons have also been made between the geometries obtained stereoscopically and from single-eye timing reconstruction. In general, no systematic problems exist with geometrical reconstruction within quoted error limits (6). Depending primarily upon measured track length, angular resolution ranges from on the order of $\pm 10°$ down to about $\pm 1°$.

## 3.2   Size and Total Energy Measurements

Clearly, the chief virtue of the Fly's Eye detector is its advertised ability to make total energy measurements calorimetrically in a model-independent way. The energy measurement involves basically a three-step procedure. First, trajectory parameters are reconstructed using the "hit" phototube direction vectors and pulse arrival times. An example of applying both single-eye as well as stereoscopic reconstruction to an event observed simultaneously by Fly's Eye I and II is shown in Figures 6 and 7. Pulse integrals for each PMT are corrected for pedestals and relative efficiencies and then converted into photoelectron yields. The accuracy of the photoelectron measurement is discussed in Baltrusaitis et al (6) and is on the order of $\pm 15\%$ for each PMT.

Secondly, the EAS longitudinal size distribution is calculated using an iterative process to remove the photoelectron contributions due to direct and scattered Cherenkov light. Finally, each resultant longitudinal profile is fitted with two functions in order to obtain best estimates of (a) the shower size at maximum development $N_{max}$, (b) the location of maximum development $X_{max}$, and (c) the integral of the longitudinal development profile $\int N_e(x)\, dx$. There are two functions used. The first is a Gaisser-Hillas shower development function of the form (11):

$$N_e(x) = N_{max}\{(x - x_0)/(X_{max} - x_0)\}^{(X_{max} - x_0)/\lambda} \exp (X_{max} - x)/\lambda, \qquad 2.$$

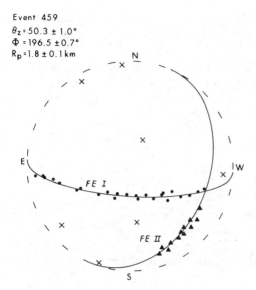

Event 459
$\theta_z = 50.3 \pm 1.0°$
$\Phi = 196.5 \pm 0.7°$
$R_p = 1.8 \pm 0.1$ km

*Figure 6* Stereoscopic reconstruction of an event seen simultaneously by Fly's Eye I and II. The lines drawn through the dots and triangles (indicating "hit" PMTs at each eye) represent the best fit for each of the two shower-detector planes. The intersection of the two planes gives the EAS trajectory whose parameters are indicated above the figure.

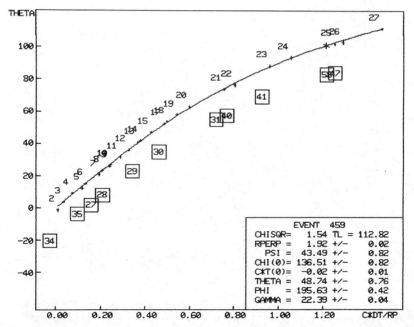

*Figure 7*    Timing fit for Fly's Eye I data only for the same event reconstructed stereoscopically in Figure 6.

where $x_0$ is the depth of first interaction and $\lambda = 70$ g cm$^{-2}$. The second function used is a Gaussian. An example of fitting the size data is shown in Figure 8.

The electromagnetic energy deposited by the primary article is estimated from the integral of either the Gaussian or Gaisser-Hillas fit (whichever is statistically favored) to the shower development data points

$$E_{em} = \frac{\varepsilon_0}{x_0} \int N_e(x) \, dx, \qquad\qquad 3.$$

where $\varepsilon_0/x_0$ is the ratio of the critical energy of an electron to the radiation length in air (12). This ratio converts track length integrals into total energy loss by excitation and ionization (13). We take $\varepsilon_0 = 81$ MeV and $x_0 = 37.1$ g cm$^{-2}$; this gives an energy loss rate of 2.18 MeV/electron g cm$^{-2}$ (12). We apply roughly a 10% correction (6) to account for undetected energy lost via (*a*) undetected neutrals that fail to decay quickly into detectable charged particles before striking the ground, (*b*) a significant number of high-energy

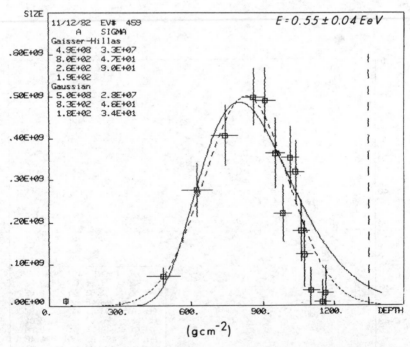

*Figure 8*  Shower development curve with a Gaisser-Hillas (*solid line*) and a Gaussian (*short-dashed line*) functional fit applied. Parameters and associated errors listed for each fit in the insert include (*a*) size at maximum, (*b*) depth of maximum $X_{max}$, and (*c*) shower width $\sigma$. (Parameters *c* and *d* for the Gaisser-Hillas fit are the shower starting point $x_0$ and width $\sigma$.)

muons, or (c) nuclear excitation by the hadrons in the shower. This method of assigning energy to the incoming primary cosmic ray is virtually independent of any high-energy physics models, a situation that holds true for no other EAS detector of which we are aware.

The energy of the shower depicted in Figures 13–15 is $E = 0.55 \pm 0.04$ EeV, while its size at maximum is $N_{max} = 5.0 \pm 0.28 \times 10^8$ electrons. The ratio $E/N_{max} = 1.1$ GeV/electron is close to a value based upon Monte Carlo simulations of EAS in this energy region using high-energy cross sections measured at accelerator energies along with radial scaling (14, 15).

## 3.3    Real vs Monte Carlo Response

A Monte Carlo simulation of the Fly's Eye detector has been performed in order to calculate the energy-dependent detector aperture as well as to check the validity of the overall analysis procedure. An isotropic cosmic-ray flux incident upon a model atmosphere (17) is generated by selection from quasi-random trajectories and depths of first interactions. The resultant EAS is then developed according to a Gaisser-Hillas shower development function. All light-production mechanisms are invoked to calculate numbers of photons striking each viewing PMT. Those PMTs that generate pulse voltages above triggering threshold are registered as "hits." Pulse integrals and arrival times are stored for the "hit" PMTs. The resultant simulated "data" are written onto a data file that can then be subjected to the same analysis procedure as that of the real data. Shown in Figures 9 and 10 are response functions for shower impact parameter $R_p$ and energy $E$. The agreement between Monte Carlo input and analyzed output is quite good. No systematic errors in the estimation of the primary particle parameters result from the analysis procedure.

The accuracy with which the Monte Carlo simulation describes a correct model of the Fly's Eye detector can best be seen by comparing real and Monte Carlo–generated data distributions. A sample of such distributions is shown in Figures 11–13. Shower energies were selected from an $E^{-3}$ differential spectrum. The agreement obtained is quite good, though not exact. At lower energies, which can be seen only at nearby distances, the finite spatial extent of the shower results in a dispersed light source that degrades the signal-to-noise ratio from that calculated by the Monte Carlo. Furthermore, the rapidly rising acceptance in that energy regime is quite dependent on precise triggering details that have not yet been accurately modelled. At higher energies ($E > 10$ EeV) where events are seen mostly at large distances, the electronic response of the detector is well understood and triggering uncertainties introduce no inaccuracies into acceptance calculations. The acceptance, thus, depends solely on applied data cuts and their effect on geometry. For example, relaxing the track length cut from 50°

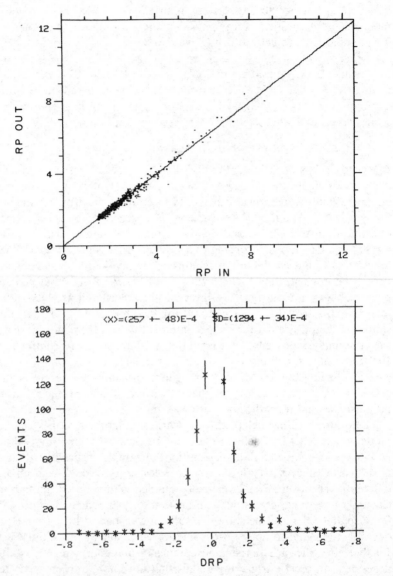

*Figure 9*   Monte Carlo response function for shower impact parameter $R_p$, i.e. analyzed output value vs known Monte Carlo input value. Lower figure shows how the distribution differs between output and input.

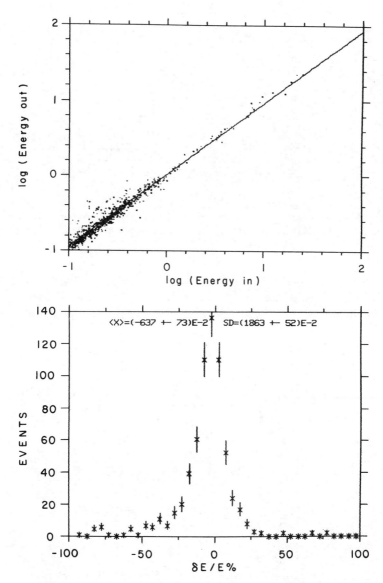

*Figure 10* Monte Carlo response function for shower energy *E*.

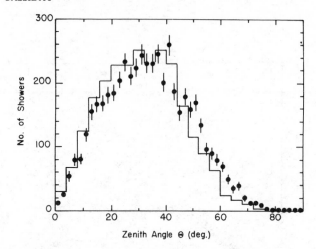

*Figure 11*    Distribution of shower zenith angles $\theta_z$. Real data are shown as a histogram without error bars. Monte Carlo data are shown as points with error bars.

to 35° opens up the high-energy acceptance where events have been detected out to distances of $R_p = 22$ km. The resultant detector acceptance is shown in Figure 14. At energies above 100 EeV, the acceptance is on the order of 1000 km² sr.

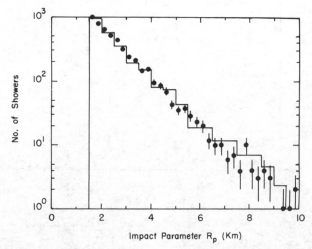

*Figure 12*    Distribution of shower impact parameters $R_p$ for real (*histogram*) and Monte Carlo (*points*) data.

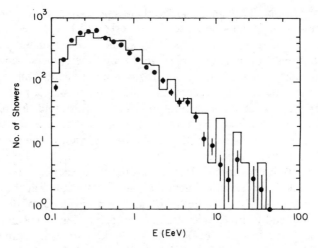

*Figure 13*  Distribution of shower energies for real (*histogram*) and Monte Carlo (*points*) data.

# 4.  EXPERIMENTAL RESULTS

## 4.1  *The UHCR Spectrum*

Roll & Wilkinson (18) reported the first confirmation that the intense isotropic radiation observed earlier by Penzias & Wilson (19) was thermal in nature and characteristic of a universe cooled to a temperature of about 3 K. Given the density ($\sim 500$ photons/cm³) and pervasiveness of this

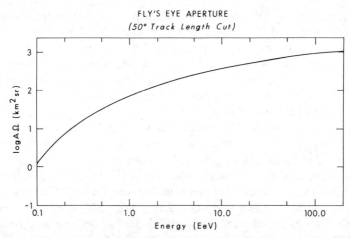

*Figure 14*  Monte Carlo–calculated Fly's Eye I aperture $A\Omega$ in km² sr for a 50° track length cut.

radiation, Greisen (20) and, independently, Zatsepin & Kuzmin (21) realized that if ultra high-energy cosmic rays (UHCR) were predominately of extragalactic origin then their spectrum should steepen abruptly because of the rapid onset of opacity to passage through the universe, an opacity caused by energy degradation from photopion production off the 2.7 K blackbody radiation. This steepening should commence at or near 60 EeV, which corresponds roughly to photoproduction threshold. More recent calculations by Hill and Schramm (22) indicate the development of a spectral bump, due to pile up of the recoil protons, prior to the onset of the "Greisen-Zatsepin" cutoff.

The current observational picture regarding the existence of the cutoff is confusing. The Yakutsk experiment (16f) supports a cutoff at the surprisingly low energy of about 40 EeV, with almost no evidence of the "pileup bump" predicted by Hill & Schramm. The Volcano Ranch, Haverah Park, and Sydney experiments (16) indicate an "ankle" or a flattening of the UHCR spectrum from a differential power law index of about 3.0 to about 2.3 at 10 EeV and continuing on beyond the Greisen-Zàtsepin cutoff to energies in excess of 100 EeV. In addition, the Haverah Park experiment indicates that the onset of this surprising spectral feature coincides with the onset of a cosmic-ray anisotropy (23) directed toward Northern galactic latitudes. Indeed, the Haverah Park group notes that the Virgo Cluster of galaxies is only 20 Mpc away and if sources in this cluster were responsible for the locally observed UHCR in the Northern Hemisphere then the problem of the Greisen-Zatsepin cutoff could be nicely circumvented. However, the energy poured into the UHCR flux at $E > 100$ EeV would be roughly equivalent to the entire radio output of M87 and Centaurus A combined, and a solution to this difficulty would severely tax the imagination even of astrophysicists. On the other hand, the Sydney group reports no such anisotropy (24). If anything, their results indicate a deficit of UHCR events relative to expectations from Northern galactic latitudes.

The Fly's Eye experiment was designed and built primarily to address the situation described above. Shown in Figures 15$a$ and $b$ are the resultant differential and integral spectra $j(E)$ and $I(>E)$ derived from Fly's Eye data. Figure 15$b$ also contains the Yakutsk and Haverah Park data. The Fly's Eye data was accumulated during 0.145 years live time, 33 months elapsed time for a duty cycle of about 6.3%. The differential spectrum [plotted as $E^3j(E)$] is essentially flat, with the appearance of a bump roughly between 10 and 50 EeV. The number of events in that interval is 62. If the spectrum continued with the same slope ($\gamma = 2.94 \pm 0.02$) as obtained at energies < 10 EeV, the expected number of events would be 46. The uncertainty in the expected number is small, hence the significance of the bump is $16/\sqrt{46} \approx 2.4\sigma$. The spectrum between 10 and 50 EeV exhibits a slope of

$\gamma = 2.42 \pm 0.27$, about $2\sigma$ flatter than the lower energy value. Furthermore, we have obtained only one event in excess of 50 EeV, while based on the Haverah Park data we should have seen $7 \pm 2$ events with two of them at $E > 100$ EeV. Thus, the discrepancy between Fly's Eye and Haverah Park event rates at $E > 50$ EeV is about $3\sigma$. We infer that the "ankle" in the UHCR spectrum at 10 EeV is most probably a bump, as predicted by Hill & Schramm (22), followed by a cutoff at 70 EeV as predicted by Greisen (20).

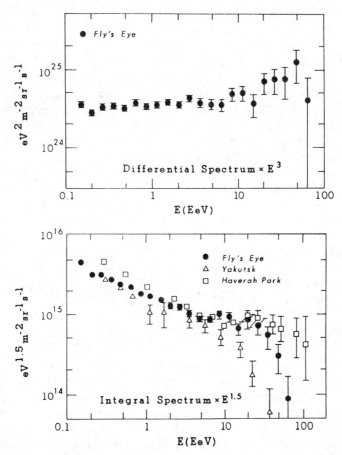

*Figure 15*   (*a*) Differential spectrum $j(E)$ plotted as $E^3 j(E)$. A power law best fit of the form $j(E) = aE^{-\gamma}$ yields: $a = 109.6 \pm 2.2$ EeV$^{-1}$ km$^{-2}$ sr$^{-1}$ yr$^{-1}$ and $\gamma = 2.94 \pm 0.02$ for $E < 10$ EeV. Between 10 and 50 EeV, $a = 34 \pm 17$ EeV$^{-1}$ km$^{-2}$ sr$^{-1}$ yr$^{-1}$ and $\gamma = 2.42 \pm 0.27$. The lack of events at $E > 50$ EeV indicates that the flattened spectrum does not continue.

(*b*) Integral spectrum $I(>E)$ plotted as $E^{1.5} I(>E)$. Data from Haverah Park and Yakutsk (18) are shown for comparison.

These spectral features imply that UHCR have travelled distances on the order of 70–150 Mpc, or roughly the scale of intercluster spacing.

## 4.2   Total Proton-Proton Cross Section

Fly's Eye data can be used to determine the proton-air inelastic cross section in a more or less model-independent fashion. Using Glauber theory (25) and a Gaussian profile distribution for the nucleus (26) and by assuming that the elastic-scattering slope parameter $b$ is proportional to $\sigma_{pp}^{tot}$, as is the case for geometric scaling (27), one can derive a value for $\sigma_{pp}^{tot}$ from $\sigma_{p\text{-air}}^{inel}$. The technique used to obtain $\sigma_{p\text{-air}}^{inel}$ is based on the notion that cosmic protons interact in the atmosphere at rates that decrease exponentially with increasing slant depth. Thus, a direct observation of the distribution of first interactions would yield the nucleon-air interaction length $\lambda_n$. The point of first interaction cannot be observed. However, the depth maximum can and its distribution also has an exponential tail whose slope is $\lambda = 1.6\lambda_n$ (28). Consequently, a measurement of $\lambda$ permits the direct derivation of $\sigma_{p\text{-air}}^{inel}$.

Events were selected such that $X_{max} > 830$ g cm$^{-2}$, where changes of slope due to resolution effects are minimal. In addition, the selection of such penetrating events insures that their parents are most likely protons (16, 28). Based on a number of different resolution cuts applied to the data, our current best estimate for $\lambda$ is $72 \pm 9$ g cm$^{-2}$. This result implies that $\lambda_n = 45 \pm 5$ g cm$^{-2}$ or $\sigma_{p\text{-air}}^{inel} = 530 \pm 66$ mb. The derived value for the total pp cross section using the above-mentioned assumptions is $\sigma_{pp}^{tot} = 120 \pm 15$ mb at an energy of $\bar{E} = 0.5$ EeV or $S^{1/2} = 30$ TeV. The result is pictured in Figure 16 along with various model predictions and extrapolations from lower energy regions.

## 4.3   Penetrating Particles

We have also searched for deeply penetrating particles in the UHCR flux as part of normal detector operation. None have been found in $6 \times 10^6$ s of on-time, which allows us to place limits on the occurrence of certain types of events such as (a) the presence of metastable quark matter in the primary cosmic ray flux, (b) the production of taus and other long-lived particles at $E > 0.1$ EeV, and (c) the flux of weakly interacting particles of astrophysical origin, such as neutrinos and photinos. These limits have been reported elsewhere (29). Here we outline the details relevent to item (c), which is intimately tied to measurements of the UHCR spectrum previously discussed.

Deeply penetrating (weakly interacting) particles make their presence known by generating EAS that occur within the Fly's Eye detectable fiducial volume but only after traversing many interaction lengths of

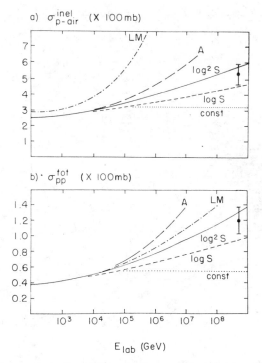

*Figure 16*  (*a*) The $\sigma_{p\text{-air}}$ inelastic and (*b*) $\sigma_{pp}$ total cross sections. Log *s* and log² *s* curves are extrapolations of fits up to CERN ISR energies. The curve labelled A is an extrapolation of the estimate of Afek et al (53) and LM stands for Leader & Maor (54).

material. This signature effectively rules out hadrons, electrons, or gamma rays as the parents of such EAS. The detectable fiducial volume is energy dependent, as can be inferred from Figure 17 in which the distribution of EAS energy vs $R_p$ is shown. At $E = 10$ EeV, showers with $R_p = 20$ km are detectable, whereas at 1 EeV, the maximum $R_p$ is 5 km. The approximate fiducial volume throughout which EAS are detected with good efficiency is roughly a vertical cylinder of height 15 km and radius $R = R_p^{\max}(E)$, centered on the Fly's Eye.

We search for deeply penetrating events by demanding that the observed EAS shower have a zenith angle $\theta_z > 80°$, which implies that the parent of the EAS must have traversed at least $x > 3000$ g cm$^{-2}$ before interacting within the fiducial volume. Since the interaction length for protons is $\sim 45$ g cm$^{-2}$, no such events are expected (see Figure 11). Events occurring within the zenith angle range 80°–90° could signify quark matter, tau or tau-like production, or the neutrino flux if the neutrino has an anomalously large cross section ($\sigma_v \gtrsim 10^{-31}$ cm²). Events could also occur at zenith angles $\theta_z$

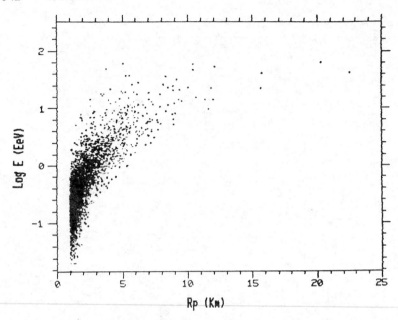

*Figure 17*    Distribution of EAS energy vs impact parameter $R_p$.

> 90° if the EAS parent was a neutrino that interacted somewhere in the Earth below the detector.

The UHE neutrino flux has been calculated by several authors (22, 30) and the values range from $10^{-17}$ to $10^{-18}$ $v$ cm$^{-2}$ s$^{-1}$ sr$^{-1}$ at $E_v = 10$ EeV. The most copious source of such neutrinos results from interactions of the primary cosmic-ray flux with the 2.7-K blackbody radiation; in other words, the very same interaction that leads to a cut off in the cosmic-ray spectrum serves as the source of an intense, energetic, and isotropic neutrino flux of the same order of magnitude as the primary cosmic-ray flux at 10 EeV. The $v_\mu$'s are not detectable, but the expected signature of $v_e$ would be an EAS of energy $E > 0.1$ EeV produced by an electron in the fiducial region $80° < \theta_z < 90°$. Table 2 gives the flux limits based on no observed events.

Upward-going events ($\theta_z \gtrsim 90°$) can be uniquely interpreted as neutrino-induced since there is no competition from any other known or predicted process. The sensitivity to the neutrino flux depends primarily on two factors: the attenuation of neutrinos by the Earth as a function of $\theta_z$, and the depth at which a neutrino interaction can occur underground and still produce an atmospheric EAS sufficiently energetic to trigger the detector. The first factor has been calculated assuming that $\sigma_v \approx 10^{-33}$ cm$^2$ as

**Table 2**  Limits on neutrino flux based on downward events $(\nu/cm^2 \cdot s \cdot ster)$

| $\sigma_\nu$ (cm$^2$) | $E_\nu$ (EeV) | | | |
|---|---|---|---|---|
| | 0.1 | 1 | 10 | 100 |
| $1 \times 10^{-31}$ | $1.0 \times 10^{-14}$ | $3.8 \times 10^{-15}$ | $1.0 \times 10^{-16}$ | $3.8 \times 10^{-16}$ |
| $1 \times 10^{-30}$ | $1.0 \times 10^{-15}$ | $3.8 \times 10^{-16}$ | $1.0 \times 10^{-17}$ | $3.8 \times 10^{-17}$ |
| $1 \times 10^{-29}$ | $1.0 \times 10^{-16}$ | $3.8 \times 10^{-17}$ | $1.0 \times 10^{-18}$ | $3.8 \times 10^{-18}$ |

predicted by the standard model. As a result of the Earth's nonnegligible neutrino opacity, the resultant angular distribution of events will be peaked toward the horizontal, essentially contained within $90° < \theta_z < 105°$. The second factor, the interaction depth, is surprisingly large as a result of the Landau-Pomeranchuk-Migdal effect (31), which predicts a suppression of the high-energy pair production and bremsstrahlung cross section relative to Bethe-Heitler in dense materials. The effect is much less pronounced in the atmosphere (32) than in the dense crust below. The net result is a tremendous elongation of shower development while the shower is in the Earth's crust, followed by rapid development upon entering the atmosphere. Showers are detected with good efficiency if the neutrino interacts at depths underground of 40 m for $E_\nu = 1$ EeV, 100 m for $E_\nu = 10$ EeV, 300 m for $E_\nu = 100$ EeV, and 1200 m for $E_\nu = 1000$ EeV. Shown in Table 3 are the resultant neutrino flux limits based on the nonobservation of such upward-going events. These limits, though more stringent than those in Table 2, are still an order of magnitude below those calculated by Hill & Schramm (22).

Although the limits are not yet very restrictive, they are the first limits for such processes at these energies. With improvements in Fly's Eye sensitivity and increased running time, the detection of UHE astrophysical neutrinos becomes a possibility that might offer additional confirmation of the

**Table 3**  Limits on neutrino flux based on upward events $(\nu/cm^2 \cdot s \cdot ster)$

| $\sigma_\nu$ (cm$^2$) | $E_\nu$ (EeV) | | | |
|---|---|---|---|---|
| | 1 | 10 | 100 | 1000 |
| $1 \times 10^{-33}$ | $7.2 \times 10^{-14}$ | $9 \times 10^{-15}$ | $3.8 \times 10^{-16}$ | $5.0 \times 10^{-17}$ |
| $3 \times 10^{-33}$ | $1.4 \times 10^{-13}$ | $7.8 \times 10^{-15}$ | $7.2 \times 10^{-16}$ | $1.1 \times 10^{-16}$ |
| $5 \times 10^{-33}$ | $4.1 \times 10^{-13}$ | $2.3 \times 10^{-14}$ | $2.2 \times 10^{-15}$ | $3.3 \times 10^{-16}$ |
| $1 \times 10^{-32}$ | $3.7 \times 10^{-12}$ | $2.1 \times 10^{-13}$ | $2.0 \times 10^{-14}$ | $3.0 \times 10^{-15}$ |

universality of the 2.7-K blackbody radiation and the UHE cosmic-ray flux.

## 4.4   Gamma-Ray Astronomy

Little is known about the sources of $E > 1$ PeV $\gamma$ rays. A variety of models has been proposed for ultra high-energy particle acceleration and $\gamma$-ray production by specific sources such as Cygnus X-3. It is probable that, at these energies, $\gamma$ rays are produced by energetic protons or nuclei, not electrons (33). Thus, any identified $\gamma$-ray sources are likely also to be the first identified sources of galactic PeV cosmic rays. So far, only a few PeV $\gamma$-ray sources have been observed or suggested by observations (34–42a).

The Fly's Eye detector was built to detect scintillation light emitted by extremely energetic and remote showers. It also detects Cherenkov light from nearby $(R_p \lesssim 1$ km) showers of much lower energy $(E \approx 1$ PeV). Such showers essentially strike the detector head-on and their directionality can be inferred typically to within $\pm 3.5°$ from the geometrical distribution of struck photomultiplier tubes. The entire night sky can be scanned for the existence of astronomical point sources producing a directional flux in excess of the isotropic cosmic-ray background. The observation of a directional flux would imply that a neutral particle, most likely a high-energy $\gamma$ ray with $E > 1$ PeV, would be the parent of the observed directional EAS excess.

The energy of the primary gamma ray can be determined in two independent ways: (a) comparing calculated Cherenkov photon densities (43) for showers of a given energy to measured ones, and (b) comparing the measured rate of background triggers to the rate anticipated from the known cosmic-ray energy-dependent flux. The acceptance of the detector, about $10^5$ m$^2$ sr, and the measured triggering rate of about 1 Hz imply a threshold triggering energy on the order of 1 PeV, which agrees closely with the estimate based on photon densities.

Simultaneous drift scans by the 67 mirror units of the Fly's Eye detector have so far given nearly full coverage of the entire Northern Hemisphere. Two sources, Cygnus X-3 and the Crab Nebula have been detected at marginal signal-to-background levels ($3.5\sigma$ and $3.1\sigma$ respectively) while a third, Hercules X-1, has been seen at a confidence level of 99.98% with the same period of 1.24 s as observed in the x-ray region (44). Shown in Figure 18 are the upper limits for the steady flux of PeV $\gamma$ rays produced from Northern Hemisphere point sources. The upper flux limits are quoted at the 95% confidence level and range from roughly $10^{-12}$ to $10^{-13}$ cm$^{-2}$ s$^{-1}$.

Shown in Figure 19 is a plot of count rate vs right ascension centered about the Crab Nebula. The data showing the $3.1\sigma$ signal were obtained on December 9, 1980, while the data exhibiting no signal were obtained on

February 1, 6, and 7, 1981. The result implies that the Crab Nebula is quite likely (97% confidence) a sporadic high-energy $\gamma$ source. Maximum signal for Cygnus X-3 was obtained in the phase interval 0.2–0.3 within its 4.8-h period, in agreement with the results of Lloyd-Evans et al (36) and Samorski & Stamm (35) if the same ephemeris (45) is used. The excess signal implies to a confidence level of 99.6% that Cygnus was observed and that it, too, is quite likely a variable source of PeV $\gamma$ rays.

Finally, shown in Figure 20a is the count rate obtained in the direction of Hercules X-1 as a function of the phase within its known 1.24-s period. The phase of maximum rate was obtained by comparing the data to a constant background prediction and maximizing $\chi^2$. Also, as can be seen in Figures 20b and c, the data are quite specific in preferring a period within about 10 $\mu$s of that observed by Nagase et al (44) in the x-ray region. The confidence level of the resulting observation is 99.98%.

The approximate average $\gamma$-ray flux within the 1.24-s period is $3.3 \pm 1.1 \times 10^{-12}$ cm$^{-2}$ s$^{-1}$. The mean energy is 500 TeV. This implies that the luminosity for Hercules X-1 at $E > 500$ TeV is about $10^{37}$ ergs s$^{-1}$, assuming that the $\gamma$ rays are emitted isotropically. This is close to the total estimated luminosity for Hercules X-1 (46). Given that the source is probably sporadic and beamed, the high-energy $\gamma$ luminosity may be several orders of magnitude lower. On the other hand, peak $\gamma$-ray activity may be higher at times other than during our observation. In other words, PeV charged-particle production apparently makes up a large fraction of the total luminosity of the entire system!

Certain models of charged-particle acceleration can be eliminated. For example, the models of Goldreich & Julian (47), Cheng & Ruderman (48),

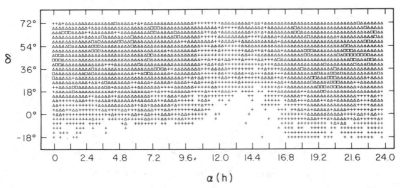

*Figure 18*   Flux upper limits at the 95% confidence level: *squares* = $F < 3 \times 10^{-13}$ cm$^{-2}$ s$^{-1}$; *triangles* = $3 \times 10^{-13} < F < 10^{-12}$ cm$^{-2}$ s$^{-1}$; + *signs* = $10^{-12} < F < 3 \times 10^{-12}$ cm$^{-2}$ s$^{-1}$. Overlapping bins have declination ($\delta$) intervals of 7.2° and right ascension ($\alpha$) intervals of 0.48 cos $\delta$ hours. Bin centers are separated by half those intervals.

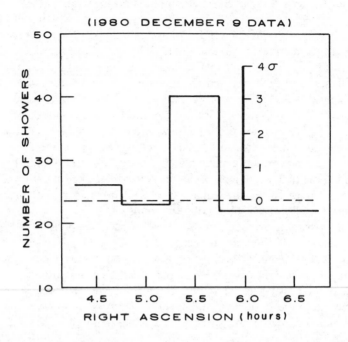

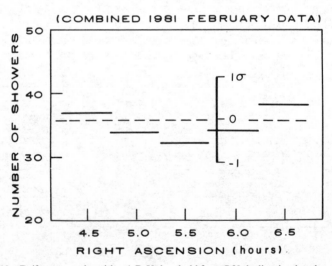

*Figure 19* Drift scan results with a 1-PeV threshold for a 7.2° declination band centered on the Crab Nebula.

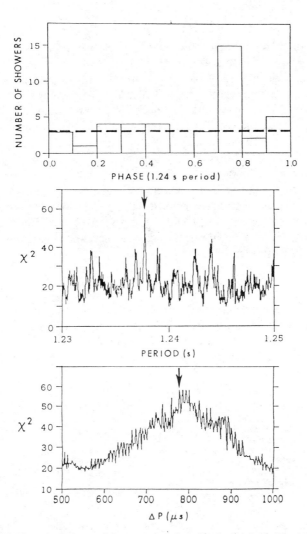

*Figure 20*    (*a*) Phase dependence of shower arrival times for the first half of the data received on 1983 July 11 from the direction of Hercules X-1. The dashed line is the expected number of events in each bin.

(*b*) The $\chi^2$ dependence on the period. Maximum $\chi^2$ is obtained at greatest deviation of observed rate from background expectation.

(*c*) The period obtained is $1.237s + \Delta P$ $\mu$s. The arrow marks the period obtained from the x-ray observations of Nagase et al (44).

and Gunn & Ostriker (49) each predict a maximum charged-particle energy of about 10 TeV for Hercules X-1; this is too low. The models of Kundt (50) and Chanmugan & Brecher (51), however, predict sufficient energies in the PeV region comparable to observed threshold values.

Finally, we note that our time of observation of $\gamma$-ray production coincided with a time of low x-ray production (5% of peak activity). However, optical observations indicate a time of normal x-ray activity at the source (52). Moreover, at the time of observation, the companion star was not in the line of sight of the pulsar. The inference is that the surrounding accretion disk may have thickened to on the order of 10–100 g cm$^{-2}$ sufficient to absorb keV x rays but necessary as a target for the production of PeV $\gamma$ rays.

We conclude that the future is quite exciting for $\gamma$-ray observations in the PeV region. Future observations with a higher-resolution, more sensitive Fly's Eye detector should provide an answer to the question of UHCR origin. We may be beginning to see the very first direct observational evidence of UHCR sources.

## ACKNOWLEDGMENTS

I gratefully acknowledge the National Science Foundation for support of this experiment. I am especially indebted to my colleagues: E. C. Loh, J. Elbert, P. Sokolsky, R. Baltrusaitis, P. Gerhardy, Y. Mitsumoto, D. Steck, S. Ko, and R. Cooper, without whose great effort this review would not have been possible.

*Literature Cited*

1. Rochester, G. D., Turver, K. E. *Contemp. Phys.* 22:425 (1981)
2. Hammond, R. T., et al. *Nuovo Cimento* 1C:315 (1978)
3. Bunner, A. N. *The Atmosphere As a Cosmic Ray Scintillator*, PhD thesis, Cornell Univ., Ithaca, NY (1964)
4. Nicholls, R. W., Reeves, E. M., Bromley, D. A. *Proc. Phys. Soc.* 74:87 (1959)
5. Hughes, R. H., Philpot, J. L., Fan, C. Y. *Phys. Rev.* 123:2084 (1961)
6. Baltrusaitis, R. M., et al. *Nucl. Instrum. Methods.* In press (1985)
7. Baum, N. A. *Stars and Stellar Systems, Astronomical Techniques, The Detection and Measurement of Faint Astronomical Sources*, Vol. 2. Chicago and London: Univ. Chicago Press (1962)
8. Chamberlain, J. W. *Physics of the Aurora and Air Glow.* New York: Academic (1961)
9. Allen, C. W. *Astrophysical Quantities.* Athlone Press, Univ. London (1976)
10. Roach, F. E. *Adv. Electron. Electron Phys.* 18:1 (1963)
11. Gaisser, T. K., Hillas, A. M. *Proc. 15th Int. Conf. Cosmic Rays*, Plovdiv, Bulgaria, 8:353. Sofia: Bulg. Acad. Sci. (1977)
12. Dozkhenko, O. I., Pomanskii, A. A. *Sov. Phys. JETP* 18:187 (1964)
13. Rossi, B. *High Energy Particles*, Ch. 5. Englewood Cliffs, NJ: Prentice-Hall (1952)
14. Hillas, A. M. *Proc. Cosmic Ray Workshop, Univ. Utah*, ed. T. K. Gaisser, Bartol Res. Found. Franklin Inst., Univ. Delaware, Newark, p. 16 (1983)
15. Hillas, A. M. *Proc. 16th Int. Conf. Cosmic Rays*, Kyoto, Japan, 6:13. Sofia: Bulg. Acad. Sci. (1979)
16. *For a summation see:*
    (a) Linsley, J. Spectra, Anisotropies and Composition of Cosmic Rays Above 1000 GeV, *Proc. 18th Int. Conf. Cosmic Rays*, Bangalore, India, 12:135. Bom-

bay: Tata Inst. Fundamental Res. (1983)
*For specific experiments see:*
(b) Cunningham, G. et al. *Astrophys. J.*
236:L71 (1980)
(c) Bower, A. J., et al. *J. Phys. G* 9:L53
(1983)
(d) La Pointe, M., et al. *Can. J. Phys.*
46:S68 (1968)
(e) Horton, L., et al. See Ref. 16a, 2:128
(f) Diminstein, O. S., et al. *Nauchno-Tekh. Inf. Byull. Yakutsk, USSR* 9:537
(1982)
(g) Hara, T., et al. See Ref. 16a, 9:198
17. *Handbook of Chemistry and Physics,*
Sect. F, p. 190. Cleveland: CRC Press.
54th ed. (1973–74)
18. Roll, P. G., Wilkinson, D. R. *Phys. Rev.
Lett.* 16:405 (1966)
19. Penzias, A. A., Wilson, R. W. *Astrophys.
J.* 142:419 (1965)
20. Greisen, K. *Phys. Rev. Lett.* 16:748
(1966)
21. Zatsepin, G. T., Kuzmin, V. A. *Sov. Phys.
JETP Lett.* 4:78 (1966)
22. Hill, C. T., Schramm, D. N. *Phys. Rev. D*
31:564 (1985); *Phys. Lett.* 131B:247
(1983)
23. Cunningham, G., Lloyd-Evans, J., Reid,
R. J. O., Watson, A. A. See Ref. 16a,
33:157
24. Horton, L., McCusker, C. B. A., Peak, L.
S., Ulrichs, J., Winn, M. M. See Ref. 16a,
32:153
25. Glauber, R. J., Matthias, G. *Nucl. Phys.
B* 21:135 (1970)
26. Barger, V., et al. *Phys. Rev. Lett.* 33:1051
(1974)
27. Dias de Deus, J. D. *Nucl. Phys. B* 59:231
(1974); Buras, A. J., Dias de Deus, J. D.
*Nucl. Phys. B* 71:481 (1974)
28. Ellsworth, R. W., et al. *Phys. Rev. D*
27:1183 (1983)
29. Baltrusaitis, R. M., et al. *Phys. Rev. D.* In
press (1985)
30. Stecker, F. W. *Phys. Rev. Lett.* 21:101
(1968); Berezinsky, V. S., Zatsepin, G. T.
*Sov. J. Nucl. Phys.* 11:111 (1970);
Margolis, S. H., Schramm, D. N.,
Silverberg, R. *Astrophys J.* 221:990
(1978); Stecker, F. W. NASA Tech. Mem.
79609. Unpublished (1979)
31. Landau, L. D., Pomeranchuk, I. Y. *Dokl.
Akad. Nauk CCCR* 92:535 (1953);

Migdal, A. B., *Sov. Phys. JETP* 5:527
(1957)
32. Konishi, E., Misaki, A., Fujimaki, N.
*Nuovo Cimento* 44A:509 (1978)
33. Vestrand, W. T., Eichler, D. *Astrophys. J.*
261:251 (1982)
34. Stepanian, A. A. *Proc. 17th Int. Cosmic
Ray Conf.* 1:50. Paris: Commissariat
Energ. Atom. (1981)
35. Samorski, M., Stamm, W. *Astrophys. J.
Lett.* 268:L17 (1983)
36. Lloyd-Evans, J., et al. *Nature* 305:784
(1983)
37. Morello, C., Navarra, G., Vernett, S. See
Ref. 16a, 1:127
38. Protheroe, R. J., Clay, R. W., Gerhardy,
P. R. *Astrophys. J. Lett.* 280:L47 (1984)
39. Protheroe, R. J., Clay, R. W. *Nature.*
Submitted for publication (1985)
40. Dzikowski, T., et al. See Ref. 16a, 2:132
41. Boone, J., et al. *Astrophys. J.* 285:264
(1984)
42. Lambert, A., Lloyd-Evans, J., Watson, A.
A. See Ref. 16a, 9:219
42a. Baltrusaitis, R. M., et al. *Astrophys. J.
Lett.* In press (1985)
43. Smith, G. J., Turver, K. E. *Proc. 13th Int.
Cosmic Ray Conf.,* Denver, Colo.,
4:2369. Univ. Denver
44. Nagase, F., et al. *Proc. Workshop on High
Energy Transients,* Santa Cruz, p. 131.
Am. Inst. Phys. New York (1984)
45. Van der Klis, M., Bonnet-Bidaud, J. M.
*Astron. Astrophys. Lett.* 95:L5 (1981)
46. Bradt, H., Doxset, R., Jernigan, J.
*COSPAR X-Ray Astronomy.* Oxford:
Pergamon (1979)
47. Goldreich, P., Julian, W. H. *Astrophys. J.*
157:869 (1969)
48. Cheng, A. F., Ruderman, M. A. *Astrophys. J.* 216:865 (1977)
49. Gunn, J. E., Ostriker, J. P. *Phys. Rev.
Lett.* 22:728 (1969)
50. Kundt, W. *Astrophys. Space Sci.* 90:59
(1983)
51. Chanmugan, G., Brecher, K. Preprint
(1984)
52. Delgada, A. J., et al. *Astron. Astrophys.*
127:L15 (1983); Parmer, A. N., et al.
*Nature* 313:119 (1985)
53. Afek, Y., et al. *Phys. Rev. Lett.* 45:85 (1980)
54. Leader, E., Maor, V. *Phys. Lett.* 43B:505
(1973)

*Ann. Rev. Nucl. Part. Sci. 1985. 35:351–95*

# ELASTIC SCATTERING AND TOTAL CROSS SECTION AT VERY HIGH ENERGIES

*Rino Castaldi and Giulio Sanguinetti*[1]

INFN, Sezione di Pisa, Italy

CONTENTS

## 1. INTRODUCTION

The unexpected rise of the proton-proton total cross section (1, 2) discovered many years ago, when the CERN Intersecting Storage Rings (ISR) began operation, excited renewed interest in the asymptotic behavior of hadron interactions. An old common prejudice expects that at very high energies the mechanisms that control hadron scattering become simpler and can be interpreted in terms of a few basic principles, more general than any specific model. Indeed, at low energy, hadron cross sections show a

[1] Visitors at CERN, Geneva, Switzerland.

351

0163–8998/85/1201–0351$02.00

pattern rich in bumps and structures, but at higher energies their behavior appears to be smooth, which suggests that the laws of strong interactions become simpler and merely follow from general assumptions independent of the details of particle dynamics. Such a high-energy limit appears, however, to be approached very slowly, and a large span in energy is certainly required in order to speculate upon this subject.

The successful cooling technique (3) of antiproton beams at CERN has recently allowed the acceleration of proton and antiproton bunches simultaneously circulating in opposite directions in the Super Proton Synchrotron (SPS). Hadron-hadron collisions could so be produced at a center-of-mass (c.m.) energy one order of magnitude higher than previously available; this opened a new wide range of energies to experimentation. This technique also made it possible to replace one of the two proton beams in the ISR by a beam of antiprotons, which allowed a direct precise comparison, by the same detectors, of p-p and p̄-p processes at the same energies.

The aim of this review is to summarize the recent progress in the field of elastic scattering and total cross section in this new energy domain. In Section 2 a survey of the experimental situation is outlined. The most significant data are presented, with emphasis on the interpretation, not the specific details or technicalities. This section is therefore intended to give a self-contained look at the field, especially for the nonspecialist. In Section 3, hadron scattering at high energy is described in an impact parameter picture, which provides a model-independent intuitive geometrical representation. The diffractive character of elastic scattering, seen as the shadow of inelastic absorption, is presented as a consequence of unitarity in the s-channel. Spins are neglected throughout this review, inasmuch as the asymptotic behavior in the very high-energy limit is the main concern here. In Section 4 some relevant theorems are recalled on the limiting behavior of hadron-scattering amplitudes at infinite energy. There is also a brief discussion on how asymptotically rising total cross sections imply scaling properties in the elastic differential cross sections. A quick survey of eikonal models is presented and their predictions are compared with ISR and SPS Collider data.

We apologize that space does not allow us to cover comprehensively this wide subject. In particular, we have preferred to discuss the general aspects of this field rather than to enter into the details of the many current phenomenological models. Furthermore, we have not tried to give exhaustive references and we apologize to those authors whose contributions have not been mentioned. Many excellent reviews on hadron scattering at high energies can be found in the literature. A sample of the most recent ones is listed in Reference (4a–j) for the convenience of the interested reader.

# 2.  REVIEW OF THE EXPERIMENTAL DATA: A GENERAL LOOK AT THE FIELD

## 2.1  *Elastic Scattering*

At very high energy, hadron elastic scattering is believed to be well described by a single scalar, mainly imaginary, amplitude and one usually assumes that, as energy increases, spin tends to play a negligible role (see Section 2.2). The interaction becomes mostly absorptive, dominated by the many open inelastic channels. Elastic scattering is then essentially the shadow of the inelastic cross section. Therefore the elastic amplitude has a mainly diffractive character, in close analogy with the diffraction of a plane wave in classical optics, and its dependence on the momentum transfer $t$ is the Fourier transform of the spatial distribution of the hadronic matter inside the interacting particle. In the simple case of an absorbing disc of radius $R$ and uniform greyness $\xi \leq 1$, to the inelastic cross section $\sigma_{inel} = \pi R^2 \xi$ corresponds an elastic hadronic diffraction pattern that is a steep function of $tR^2$. Its integral $\sigma_{el}$ is bigger for larger $\xi$ up to a maximum value $\sigma_{el} = \sigma_{inel}$ in the limit of a completely black disc ($\xi = 1$).

The elastic differential cross section $d\sigma/dt$ measured for p-p scattering at the ISR (5) is shown in Figure 1 as a function of the four-momentum transfer $t$, at five values of the total c.m. energy $\sqrt{s}$. A salient feature of these distributions is the presence of a pronounced narrow peak around the forward direction, which decreases almost exponentially in $t$ by more than six orders of magnitude, down to $-t \approx 1.4\ \text{GeV}^2$. In a geometrical picture, as discussed in Section 3.4, such a smooth diffraction pattern over so wide a $t$ interval indicates that the transverse distribution of matter inside the proton is nearly Gaussian. At larger values of $|t|$, a prominent structure is observed, consisting of a sharp minimum followed by a secondary maximum and by a subsequent exponential fall-off by three more orders of magnitude with a much lower slope. As energy increases, the slope of the forward peak becomes steeper and the dip moves toward smaller values of $|t|$. Following the optical analogy, this shrinkage of the diffraction peak indicates an expansion of the interaction radius. A closer look at the diffraction peak reveals a slight variation of slope presumably localized around $-t \approx 0.13\ \text{GeV}^2$.

In the very forward direction ($-t < 10^{-3}\ \text{GeV}^2$) the elastic differential cross section is dominated by the almost real Coulomb amplitude, which is well understood theoretically and can easily be calculated. In the $|t|$ region between $10^{-3}$ and $10^{-2}$ GeV², the Coulomb and the nuclear amplitudes are of the same order of magnitude and may give rise to a nonnegligible interference effect, if the hadronic amplitude is not purely imaginary. The measurement of the interference term allows the determination of the phase of the nuclear amplitude $F$ in the forward direction, which is usually expressed as the ratio of the real to the imaginary part of the amplitude at $t = 0$: $\rho(s) = [\text{Re } F(s,t)/\text{Im } F(s,t)]_{t=0}$.

2.1.1 THE REGION OF COULOMB-NUCLEAR INTERFERENCE (0.001 < $-t$ < 0.01 GeV²) In this $t$ region, the elastic differential cross section is determined by both the nuclear and the Coulomb amplitudes: $d\sigma/dt = |f_N \pm f_C \exp[i\phi(t)]|^2$, where the upper sign is used for p-p and the lower sign for p̄-p scattering. The nuclear amplitude $f_N$ can be parametrized in terms

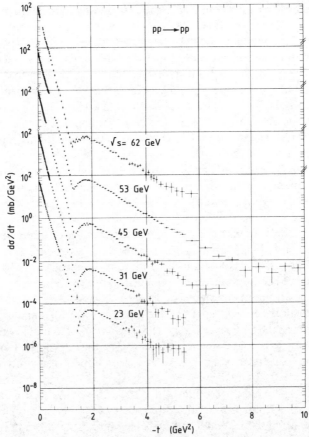

*Figure 1*   Differential cross section for p-p elastic scattering at the five ISR energies (from 5).

of the total cross section $\sigma_{tot}$ and of the slope $B$ of the forward elastic peak, by means of the optical theorem (see Section 3.2), as $f_N = (1/4\sqrt{\pi})$ $(i + \rho)\sigma_{tot} \exp(Bt/2)$, disregarding spin effects and assuming that the real and imaginary parts have the same exponential $t$ dependence in this region. The Coulomb amplitude $f_C$ represents the well-known Rutherford scattering and is expressed by $f_C = -2\sqrt{\pi}\alpha G^2(t)/|t|$, where $\alpha$ is the fine structure constant and $G(t)$ is the proton electromagnetic form factor. The small phase factor $\phi(t)$ arises from the simultaneous presence of both hadronic and electromagnetic exchanges in the same diagram (6) and has opposite

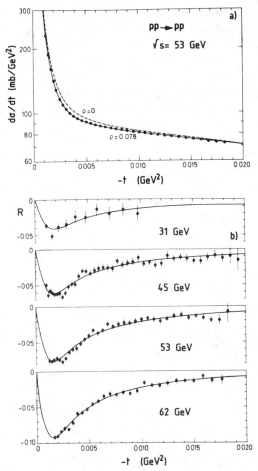

*Figure 2*   (*a*) Differential cross section for p-p elastic scattering in the Coulomb region at $\sqrt{s} = 53$ GeV (the $\rho = 0$ curve is shown for comparison); (*b*) destructive interference observed at four ISR energies; $R$ is defined as $(d\sigma/dt$ measured$)/(d\sigma/dt$ for $\rho = 0) - 1$ (from 7).

sign in the p-p and p̄-p channels. The relative importance of the interference term is maximum when the nuclear and the Coulomb amplitudes are comparable (at $-t \approx 8\pi\alpha/\sigma_{tot}$). Its contribution to the cross section in a first approximation is $\approx -(\pm)(\rho+\phi)\alpha\sigma_{tot}/|t|$ and can easily be distinguished from the steeper $1/t^2$ dependence of the Coulomb term and from the relatively flat nuclear contribution. If the factor $(\rho+\phi)$ is positive, the interference is destructive for p-p and constructive for p̄-p scattering.

A sizeable destructive interference, which becomes more and more pronounced as energy increases, is indeed observed in the ISR p-p data (7) shown in Figure 2. This is the experimental evidence that $\rho$, which was observed to go through zero at Fermilab energies (8), increases toward higher and higher positive values over the whole ISR energy range. Data on the p̄-p elastic differential cross section in this $t$ region have also become available recently at the ISR (9). In this case a constructive interference is observed, which indicates a rising positive value of $\rho$ in this energy range for p̄-p scattering too. A summary of ISR data on $\rho$ for both p-p and p̄-p scattering is shown in Figure 3, together with lower energy data.

The real part of the elastic amplitude is related to the imaginary part via dispersion relations. On the other hand, the imaginary part at $t = 0$ is related to the total cross section by the optical theorem. As a consequence, it is possible to write the parameter $\rho$ at a given energy as an integral of the total cross section over energy. Such an integral relation can be approximated by a local expression relating $\rho$ to the derivative of $\sigma_{tot}$ with respect to energy (10, 11):

$$\rho(s) \approx (\pi/2\sigma_{tot})\, d\sigma_{tot}/d \ln s. \qquad\qquad 1.$$

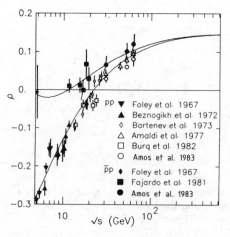

*Figure 3* Summary of high-energy data on the real part $\rho$ for p-p and p̄-p forward elastic scattering (from 9); the curve represents the dispersion relation fit of Amaldi et al (7).

Equation 1 is not really adequate to compute the values of $\rho$ as a function of energy in a quantitative way but has the merit of describing in a transparent way the qualitative connection between $\rho$ and $\sigma_{tot}$ at asymptotic energies. In particular, it allows one to understand easily the result, rigorously proved in (12), that a rising total cross section implies a positive value of $\rho$. For instance, in the case of an asymptotic behavior that saturates the Froissart bound ($\sigma_{tot} \sim \ln^2 s$), $\rho$ goes to zero from positive values as $\pi/\ln s$.

2.1.2  THE DIFFRACTION PEAK REGION ($0.01 < -t < 0.5$ GeV$^2$)    Recent data on p̄-p elastic scattering at the ISR are compared in Figure 4 with p-p data taken at the same energies by the same experiment (13a). It is remarkable how the shape of the p̄-p distribution, which is notably steeper at lower energies, becomes similar to that of p-p distribution in the ISR energy range. The

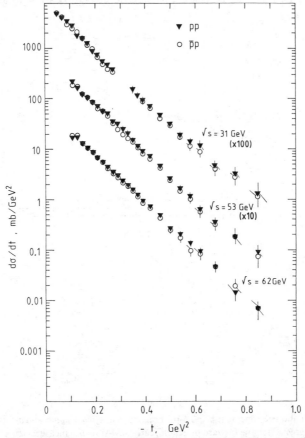

*Figure 4*   Comparison of p-p and p̄-p diffraction peaks at three ISR energies (from 13a).

energy dependence of the slope of the elastic differential cross section for p̄-p and p-p scattering (Figure 5) indicates a shrinkage of the diffraction peak at a rate of at least ln $s$ (14a–c). It appears evident from there that, as energy increases, the antiproton and the proton tend to behave in exactly the same way. Such behavior is indeed expected at asymptotic energies, since the Cornille-Martin theorem (see Section 4.1.3) states that the elastic differential cross sections of particle and antiparticle in the region of the diffraction peak tend to be the same for $s \to \infty$.

In Figure 5, one can also see that the value of the slope $B$ is different at different values of $t$. As a matter of fact, the forward elastic peak deviates from a pure exponential in $t$, and rather than a constant slope $B$ one should introduce a local slope $B(t) = (d/dt) \ln (d\sigma/dt)$. A slope that continuously decreases with increasing $|t|$ is observed up to Fermilab energies, where the elastic diffraction peak is reasonably well described by an exponential with

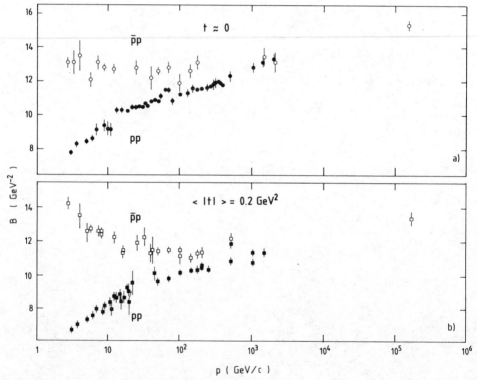

*Figure 5*   A compilation of slopes of p-p and p̄-p diffraction peaks (*black points* and *open points*, respectively) as a function of energy; (*a*) in the forward direction [early, low-statistics measurements at the SPS Collider (17, 18) are not included]; (*b*) around $-t \approx 0.2$ GeV$^2$ (from 13a, 14a).

the addition of a small quadratic term (15), of the kind exp $(Bt + Ct^2)$. On the other hand, ISR data exhibit a rather sudden break localized around $-t \approx 0.13$ GeV$^2$. A similar feature is observed in the $\bar{p}$-p elastic differential cross section measured at the SPS Collider at $\sqrt{s} = 546$ GeV. As illustrated in Figure 6, in the region $0.03 < -t < 0.15$ GeV$^2$ the data show no hint of curvature and are well fitted by a single exponential (16) of slope $B = 15.2 \pm 0.2$ GeV$^{-2}$ with no need for a quadratic term. A significant quadratic dependence can only be obtained from fits over a $t$ region that extends up to at least 0.3 GeV$^2$. However, such a parametrization appears to be

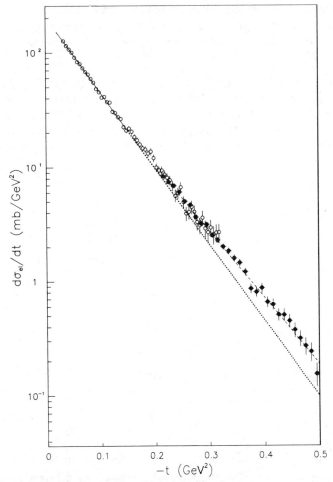

*Figure 6*   Elastic differential cross section for $\bar{p}$-p at $\sqrt{s} = 546$ GeV; the lines are single-exponential fits below and above $-t \approx 0.15$ GeV$^2$ (from 16).

inadequate to represent the data in the $t$ range from 0.03 to 0.5 GeV². On the other hand, a single exponential also fits the data in the region $0.2 < -t < 0.7$ GeV² : it is remarkable that in such a wide $t$ interval the slope appears to remain constant (16–19), with a value of $13.4 \pm 0.3$ GeV$^{-2}$. The phenomenon of a fast slope variation by about two units in a small $t$ interval between 0.1 and 0.2 GeV² is represented in Figure 7 (from 20), where the local slope of the diffraction peak is shown in various $t$ intervals, at $\sqrt{s} = 53$ and 546 GeV. Independently of this double-slope structure, the antishrinkage observed in p̄-p scattering at lower energy turns now into a shrinkage similar to that of the p-p diffraction peak. As discussed in Section 3.4, in a geometrical picture the slope of the diffraction peak is proportional to the square of the mean interaction radius: the shrinkage of the diffraction peak therefore reflects an increase of the dimension of the nucleon.

2.1.3    THE LARGE-$t$ REGION $(-t > 0.5\,\text{GeV}^2)$    The main feature of the p-p elastic differential cross section in this $t$ range is the progressive development, up to ISR energies, of a sharp dip around $-t \approx 1.4$ GeV², as illustrated in Figure 8a. In the ISR energy range, the position of the dip is observed to move toward smaller $|t|$ values, while the height of the secondary maximum

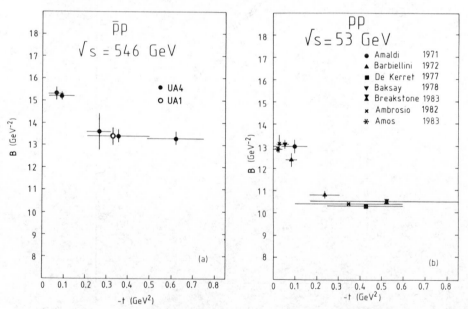

Figure 7    Local slope of the elastic diffraction peak as a function of $t$ (from 20): (a) for p̄-p at $\sqrt{s} = 546$ GeV; (b) for p-p at $\sqrt{s} = 53$ GeV. The horizontal bars indicate the $t$ interval in which the exponential fit was performed.

grows, as already shown in Figure 1. It is natural to relate this behavior to the rising total cross section observed in the same energy interval. Indeed the diffraction pattern has to shrink, if the increase of $\sigma_{tot}$ is at least partially due to an expansion of the proton radius. The geometrical scaling hypothesis actually assumes that the increase of $\sigma_{tot}$ is only due to the increase of the interaction radius $R$, while the nucleon opacity remains constant. This implies a scaling property in the elastic differential cross

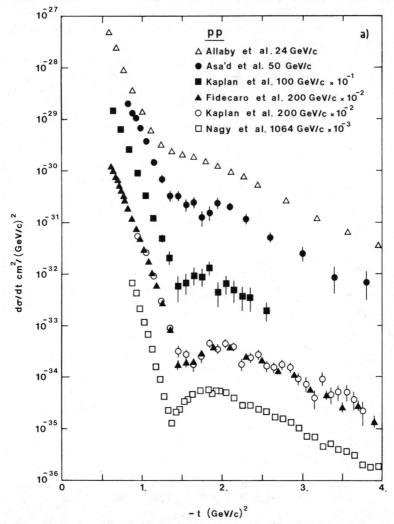

*Figure 8a*  Progressive development of the dip-bump structure in p-p elastic differential cross section from 24 to 1064 GeV/$c$ (from 22).

section as a function of the variable $\tau = t\sigma_{tot}$, which indeed seems to be reasonably verified by the ISR data (21) as shown in Figure 8b. However, it can be seen in Figure 1 that the dip appears to get more pronounced at the lower ISR energies, and gradually fills in with increasing energy. As a consequence, a scaling violation in a limited region near the dip can be seen in Figure 8b. In a simple diffractive picture the imaginary part is assumed to be the dominant scaling component of the elastic amplitude and is supposed to vanish at the position of the dip, where only the small real part contributes to the cross section. This interpretation is supported by the fact that, as shown in Figure 3, $\rho$ goes through zero at about the energy at which the dip is sharpest (see Section 4.2).

The p̄-p elastic differential cross section already at an incident beam momentum $p_{lab}$ as low as 30 GeV/c (22) starts to exhibit a pronounced dip around $-t \approx 1.7\,\text{GeV}^2$. At these low energies p̄-p and p-p elastic scattering

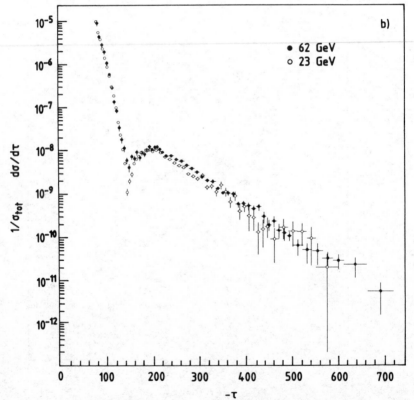

*Figure 8b*  Geometrical scaling property of p-p elastic differential cross section in the ISR energy range (from 21). The scaling variable $\tau$ is expressed in absolute units.

therefore look quite different. A direct comparison is shown in Figure 9a at $p_{lab} = 50$ GeV/$c$, where the dip structure in p-p scattering is not yet developed. These differences tend to disappear as energy increases. In particular, at $p_{lab} = 200$ GeV/$c$ there is a well-developed p-p dip structure around $-t \approx 1.5$ GeV$^2$, similar to that observed in p̄-p scattering at the same energy (23). A direct comparison of p̄-p (13b) and p-p (21) elastic differential cross sections at an ISR energy equivalent to $p_{lab} = 1500$ GeV/$c$ is shown in Figure 9b. Within the limited statistics of the p̄-p data, the similarity of the diffraction peaks also extends to the region of the dip

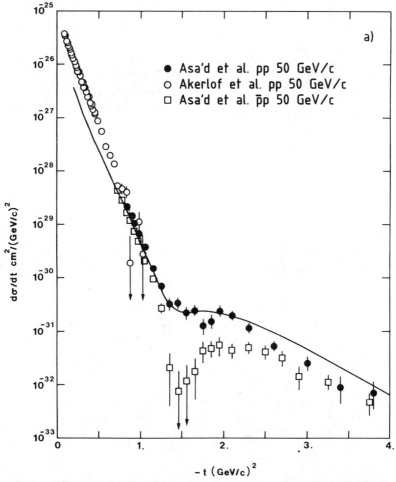

*Figure 9a* Comparison of the dip-bump structure in p-p and p̄-p elastic scattering at 50 GeV/$c$ (from 22).

(13b,c), if one considers that the higher value of the real part of the p̄-p elastic amplitude at this energy may account for a larger nondiffractive term filling the dip structure more in p̄-p than in p-p. The possible connection between the depth of the dip and the value of $\rho$ is discussed in detail in Section 4.2.

In conclusion, the differences in the p̄-p and p-p differential cross sections appear to diminish as energy increases, and it is reasonable to assume that the Cornille-Martin prediction (of the two diffraction peaks becoming equal) is essentially verified beyond ISR energies. Data on p̄-p elastic

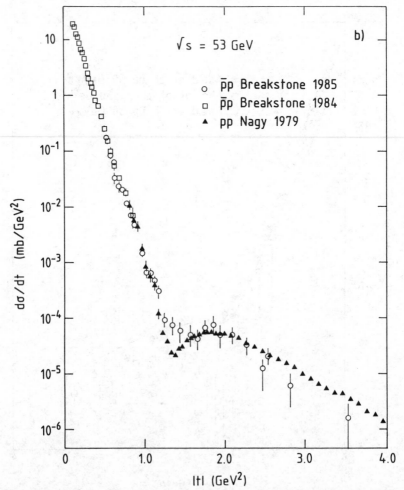

*Figure 9b* Comparison of the dip-bump structure in p-p and p̄-p elastic scattering at 1500 GeV/c (from 13b).

scattering from the SPS Collider at $\sqrt{s} = 546$ GeV are shown in Figure 10a, compared with ISR p-p data at $\sqrt{s} = 53$ GeV. The conspicuous shrinkage of the diffraction peak is accompanied by a forward movement of the dip structure that now only appears as a simple break of the exponential fall-off at $-t \approx 0.9$ GeV$^2$. Furthermore, the shoulder observed in $\bar{p}$-p at the Collider is more than one order of magnitude higher than the secondary maximum measured at the ISR.

There are models that attribute the disappearance of the dip in $\bar{p}$-p elastic scattering to an intrinsic difference between the $\bar{p}$-p and the p-p channels rather than to the evolution of the diffraction peak as energy increases. Indeed, QCD diagrams with three-gluon exchanges have been considered (24, 25a). In principle they may give rise to a crossing-odd component in the diffractive peak that can account for a shoulder in $\bar{p}$-p and for a dip in p-p at the same time. Good fits can be obtained to ISR p-p data, although one might question whether it is plausible to isolate a few QCD diagrams in this

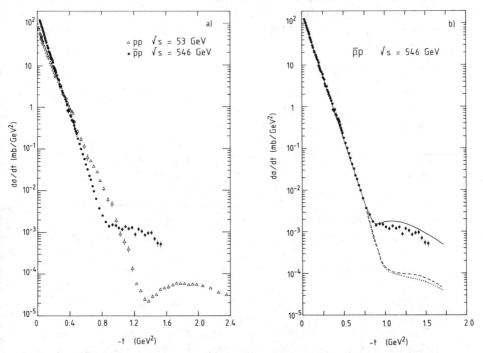

*Figure 10*   (a) Differential cross section for $\bar{p}$-p elastic scattering at the CERN SPS Collider at $\sqrt{s} = 546$ GeV (from 19). Data on p-p elastic scattering at $\sqrt{s} = 53$ GeV are also shown. (b) Comparison of different models for $\bar{p}$-p elastic scattering at $\sqrt{s} = 546$ GeV. Solid line from (28b); dashed line from (24); dotted line from (27). Data from (19).

$t$ region. However, the height of the shoulder predicted by this model for p̄-p at $\sqrt{s} = 546$ GeV is about one order of magnitude lower than the SPS Collider data (see dashed line in Figure 10$b$).

The correct prediction of the height of the shoulder is a serious challenge for all models. Good fits to the large-$t$ structures can be obtained in the "nucleon core" model (26), but are restricted to this $t$ region. In the geometrical scaling picture (see Section 4.2), the disappearance of the dip can easily be accounted for by the growing real part of the elastic amplitude (27). In this model, however, the value of the scaled d$\sigma$/d$t$ grows proportionally to $\sigma_{tot}^2$ and is inadequate by one order of magnitude to reproduce the fast growth of the shoulder from ISR to Collider energy (see dotted line in Figure 10$b$).

A stronger energy dependence of the height of the shoulder is predicted by the eikonal-type models, which can reasonably reproduce ISR and Collider data in the large-$t$ region also (28a–d) (see solid line in Figure 10$b$). Indeed the prediction of a shoulder at this energy, with about the correct height, ten years before the measurement at the Collider, has to be considered as a remarkable success of factorizing eikonal models, which, as discussed in Section 4.3, are able to reproduce many qualitative features of elastic scattering.

## 2.2   *Total Cross Section*

The total cross section is one of the basic parameters of hadron-scattering processes. Although this quantity is directly the sum of the many cross sections of all accessible final states, it is simply related via the optical theorem to the imaginary part of the amplitude of forward elastic scattering. To such a simple process, axiomatic approaches are possible, exploiting the general principles of scattering theory such as unitarity, analyticity, and crossing symmetry. Therefore a number of theorems and bounds can be derived, which constrain the asymptotic behavior of the total cross section in the limit $s \to \infty$. Well-known examples (see Section 4.1) are the Froissart bound, which limits the asymptotic energy dependence of $\sigma_{tot}$ to a $\ln^2 s$ growth at maximum, and the Pomeranchuk theorem, which states that the ratio of the total cross sections of particle and antiparticle should tend to unity. Experimental measurements of $\sigma_{tot}$ at the highest available energies therefore play a fundamental role in probing the onset of an eventual asymptotic regime.

Total cross sections cannot be measured at colliding-beam machines by traditional transmission techniques. Three different methods have been used instead, by measuring simultaneously two of the three following quantities: total interaction rate, forward elastic rate, and machine luminosity. The most direct method (2) determines $\sigma_{tot}$ by the ratio between the total interaction rate, measured in a detector with full solid-angle coverage, and the luminosity. A second method (1) exploits the optical theorem, extrapolating to $t = 0$ the measured rate of small-angle elastic scattering. In both cases an absolute calibration of the machine luminosity is required, usually performed either by the van der Meer method (29, 30) or by direct measurements of beam profiles (31, 32), or even by the measurement of a known electromagnetic process such as Coulomb scattering (33). When a high-precision luminosity calibration is not available, the simultaneous measurement of low-$t$ elastic scattering and of the total rate allows the determination of $\sigma_{tot}$ without the need for an independent luminosity measurement (34–36). The use of the optical theorem in the second and third methods implicitly assumes that spin effects at small $t$ can be disregarded. The plausibility of this assumption was firmly supported by the precise measurements performed at the ISR, where all three methods were exploited simultaneously (34) to obtain consistent results with an accuracy of better than 1% [see a discussion in (25b)].

Until the coming into operation of the ISR, total cross sections were believed to approach a finite limit with increasing energy. Actually some hints of growth were present in cosmic-ray data at very high energy (37) in addition to the well-known rising trend of $\sigma_{tot}(K^+p)$ that was considered as a transient feature. Moreover, the possibility of indefinitely rising hadron cross sections had also been considered theoretically (38, 39). Nonetheless, the generally accepted prejudice was that total cross sections should tend to constant values, and the discovery of the p-p rising cross section at the ISR (1, 2) undeniably came as a surprise. Afterward, this trend was found to be a general feature of all hadronic total cross sections. As discussed in Section 2.1.1, the value of $\rho$ is sensitive to the variation of $\sigma_{tot}$ with energy, via dispersion relations. The simultaneous study of $\rho$ and $\sigma_{tot}$ therefore provides a better understanding of the energy dependence of $\sigma_{tot}$ and allows a sensible extrapolation in a domain that extends well beyond the accessible energy range. This kind of analysis, pioneered by Amaldi et al (7) on $\rho$ and $\sigma_{tot}$ data then available up to ISR energies, was indeed able to predict that the p-p and p̄-p total cross sections should keep rising at least up to $\sqrt{s} \approx 300$ GeV at a rate very close to $\ln^2 s$.

Recently, the p̄-p total cross section was measured at the SPS Collider (36) at an energy as high as $\sqrt{s} = 546$ GeV. The observed value of $61.9 \pm 1.5$ mb is $\sim 50\%$ higher than at the ISR and indicates that a $\sigma_{tot}$ rise at a rate compatible with $\ln^2 s$ persists also in this new energy domain. This Collider measurement is shown in Figure 11 together with a compilation of lower energy p-p and p̄-p total cross-section data that includes the recent p̄-p results from the ISR (9, 40a,b). The dispersion relation fit (7) mentioned above is also shown in Figure 11 for comparison (the simultaneous fit to the real part is the curve shown in Figure 3). One can see that the recent p̄-p data at the highest energies agree well with the prediction of this analysis and their inclusion in fits of this kind confirms the $\ln^2 s$ behavior of $\sigma_{tot}$ (4j, 14c). It is striking that data at present energies indicate a rate of growth of $\sigma_{tot}$ that is just the fastest function allowed at asymptotic energies by the Froissart bound. A suggestive hypothesis is that the observed qualitative saturation of the Froissart bound could be the manifestation of an asymptotic regime, already appearing at present energies, which leads hadron total cross sections to increase indefinitely at this rate.

At the energy of the Collider the p-p and p̄-p total cross sections are expected to have practically the same value. The total cross-section difference $\Delta\sigma = \sigma_{tot}(\bar{p}p) - \sigma_{tot}(pp)$ is shown in Figure 12. Its energy dependence exhibits a Regge behavior of the kind $s^{-\alpha}$, with $\alpha \approx 1/2$, which makes it tend to zero rather rapidly. Actually, if total cross sections rise indefinitely, their difference does not necessarily have to vanish and might even increase logarithmically with energy, still preserving the limit

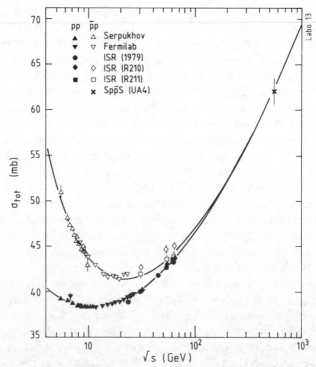

*Figure 11* Total cross-section data for p-p and p̄-p scattering [early, low-statistics measurements at the SPS Collider (18, 35) are not shown]. The curve represents the dispersion relation fit of Amaldi et al (7).

$\sigma_{tot}(\bar{p}p)/\sigma_{tot}(pp) \to 1$ as required by the Pomeranchuk theorem at infinite energy. The literature (41) has indeed considered nonvanishing asymptotic contributions (odderons) to the part that is odd under crossing of the forward elastic amplitude, which determines the behavior of $\Delta\sigma$. The operation of the ISR both with proton and antiproton beams allowed the study of the convergence of p-p and $\bar{p}$-p total cross sections in an energy interval where the difference is expected to become very small. While $\sigma_{tot}(\bar{p}p)$ starts to rise at these energies, the difference $\Delta\sigma$ keeps decreasing following the same inverse power law as observed at lower energy. The ISR data from (40a,b) (black points in Figure 12) could actually suggest a small, systematic deviation above the Regge fit. However, this measurement could include a small but significant electromagnetic inelastic contribution to $\Delta\sigma$—up to 0.4 mb according to the authors (40b)—from which the data of (9) (open circles) are free. As a conclusion, the data are well compatible with the behavior expected in the framework of a standard Regge exchange picture, with no need for odderons, though a sufficiently small contribution of this kind certainly cannot be excluded (see Section 4.1.2).

Data on p-p and $\bar{p}$-p total elastic cross section $\sigma_{el}$ measured at the ISR and at the SPS Collider, and the corresponding ratio $\sigma_{el}/\sigma_{tot}$, are shown in Figures 13a and 13b, respectively, together with lower energy data. Similar to $\sigma_{tot}$, $\sigma_{el}$ also starts rising at ISR energies, reaching a value of $13.3 \pm 0.6$ mb

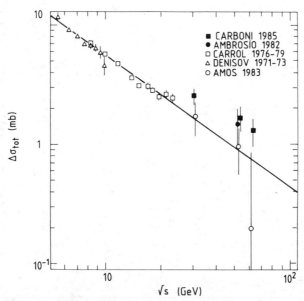

*Figure 12*   Total cross-section difference $\Delta\sigma = \sigma_{tot}(\bar{p}p) - \sigma_{tot}(pp)$ as a function of energy; the line is a Regge-like fit (from 40b).

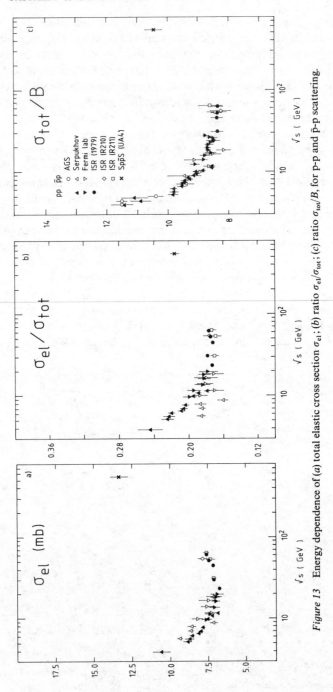

*Figure 13*  Energy dependence of (*a*) total elastic cross section $\sigma_{el}$; (*b*) ratio $\sigma_{el}/\sigma_{tot}$; (*c*) ratio $\sigma_{tot}/B$, for p-p and p̄-p scattering.

at the SPS $\bar{p}$-p Collider (36). More than $\sigma_{el}$ itself, the ratio $\sigma_{el}/\sigma_{tot}$ has a crucial role in the investigation of the high-energy regime of hadron scattering, since this parameter is directly sensitive to the hadron opacity. Great interest was aroused a few years ago, when this ratio was found to reach a constant value of about 0.175 over the whole energy range of the ISR (42). This value is below the saturation limit of 0.5, which is typical of a fully absorbing black disc. This was interpreted as the attainment of a "geometrical scaling" regime in which the proton blackness stabilizes at a rather "grey" value and the observed $\ln^2 s$ rise of $\sigma_{tot}$ only reflects a steady expansion of the proton radius like $\ln s$ (see a more detailed discussion in Sections 3.4 and 3.5). However, the validity of this attractive simple picture has been denied by the recent result from the SPS $\bar{p}$-p Collider. The measured value $\sigma_{el}/\sigma_{tot} = 0.215 \pm 0.005$ at $\sqrt{s} = 546$ GeV (36) definitely indicates an increase of the opacity with increasing energy.

A summary of the most relevant parameters of p-p and $\bar{p}$-p scattering, at ISR and SPS Collider energies, can be found in Tables 1 to 4. They also include, when available, results on the direct comparison of p-p and $\bar{p}$-p data measured by the same experiment, in particular on the following quantities: $\Delta\sigma_{tot} = \sigma_{tot}(\bar{p}p) - \sigma_{tot}(pp)$, $\Delta\rho = \rho(\bar{p}p) - \rho(pp)$, and $\Delta B = B(\bar{p}p) - B(pp)$.

It is worth stressing that the energy dependence of the parameter $\sigma_{el}/\sigma_{tot}$ is crucial in order to discriminate between the various theoretical models that claim to describe the asymptotic behavior of hadron scattering. These can be classified coarsely into three groups (43): in addition to the "geometrical scaling" approach mentioned above, in which the ratio $\sigma_{el}/\sigma_{tot}$ is expected to be energy independent, other models have been developed in which this parameter is predicted either to decrease or to increase with energy. In the Reggeon Field Theory with "critical Pomeron" (44) the opacity, and thus $\sigma_{el}/\sigma_{tot}$, decreases with increasing energy, while the expansion of the radius prevents $\sigma_{tot}$ from decreasing. This approach seems

**Table 1**   p-p scattering

| $\sqrt{s}$ (GeV) | $\sigma_{tot}$ (mb) | $\sigma_{el}$ (mb) | $\sigma_{el}/\sigma_{tot}$ | $\sigma_{tot}/B$ |
|---|---|---|---|---|
| 23 | $38.94 \pm 0.17$ | $6.73 \pm 0.08$ | $0.1728 \pm 0.0016$ | $8.47 \pm 0.15$ |
| 31 | $40.14 \pm 0.17$ | $7.16 \pm 0.09$ | $0.1784 \pm 0.0017$ | $8.45 \pm 0.14$ |
| 45 | $41.79 \pm 0.16$ | $7.17 \pm 0.09$ | $0.1716 \pm 0.0018$ | $8.38 \pm 0.13$ |
| 53 | $42.67 \pm 0.19$ | $7.45 \pm 0.09$ | $0.1746 \pm 0.0016$ | $8.37 \pm 0.13$ |
| 62 | $43.32 \pm 0.23$ | $7.66 \pm 0.11$ | $0.1768 \pm 0.0021$ | $8.36 \pm 0.13$ |
| Ref. | (42)[a] | (42)[b] | (42) | (34, 42) |

[a] Scale error $\pm 0.25$ mb.
[b] Scale error $\pm 0.09$ mb.

**Table 2**   p-p elastic scattering

| $\sqrt{s}$ (GeV) | $\rho$ | $B$ (GeV$^{-2}$) | $B'(-t > 0.15)$ (GeV$^{-2}$) |
|---|---|---|---|
| 23 | 0.02 ± 0.05 | 11.8 ± 0.2 | 10.3 ± 0.2 |
| 31 | 0.042 ± 0.011 | 12.2 ± 0.2 | 10.9 ± 0.2 |
| 45 | 0.062 ± 0.011 | 12.8 ± 0.2 | 11.0 ± 0.2 |
| 53 | 0.078 ± 0.010 | 13.1 ± 0.2 | 10.7 ± 0.2 |
| 62 | 0.095 ± 0.011 | 13.3 ± 0.2 | 10.4 ± 0.2 |
| Ref. | (42)[a] | (34) | (34) |

[a] Scale error ± 0.015.

now to be excluded by the Collider measurements. In the "factorizing eikonal" models, as discussed in Section 4.3, the asymptotic regime is still far away in energy, although $\sigma_{tot}$ already increases with a dependence close to $\ln^2 s$: the opacity is also increasing slowly with energy, so that hadrons tend gradually to become completely black discs. The ratio $\sigma_{el}/\sigma_{tot}$ is therefore predicted to reach the value 1/2 at infinite energy [the "super-critical string" model (45) also exhibits similar features]. The recent findings at the SPS $\bar{p}$-p Collider clearly favor the latter class of models.

Another parameter often used to probe the asymptotic behavior of hadron scattering, which, however, is not independent of $\dot{\sigma}_{el}/\sigma_{tot}$ (see Section 4.1.4), is the ratio $\sigma_{tot}/B$. As with the former, the latter is also sensitive to the hadron opacity. The trend of this quantity is illustrated in Figure 13c up to Collider energy and appears to be similar to that of $\sigma_{el}/\sigma_{tot}$. After reaching a constant plateau in the ISR energy range at about 8.4, it rises again up to $\sigma_{tot}/B = 10.5 \pm 0.3$ at $\sqrt{s} = 546$ GeV.

As a conclusion, the measurements at the highest currently available energies consistently indicate that the hadron opacity, which was believed to have definitively reached a constant value already at ISR energies, is

**Table 3**   $\bar{p}$-p scattering

| $\sqrt{s}$ (GeV) | $\sigma_{tot}$ (mb) | $\sigma_{el}$ (mb) | $\sigma_{el}/\sigma_{tot}$ | $\sigma_{tot}/B$ | $\Delta\sigma_{tot}$ (mb) | Ref. |
|---|---|---|---|---|---|---|
| 31 | 42.0 ± 0.5 | 7.14 ± 0.17 | 0.170 ± 0.005 | — | 1.7 ± 0.5 | (9) |
| 31 | 42.8 ± 0.35 | — | — | — | 2.58 ± 0.41 | (40b) |
| 53 | 43.65 ± 0.41 | 7.36 ± 0.30 | 0.169 ± 0.007 | 8.39 ± 0.34 | 0.98 ± 0.36 | (9) |
| 53 | 44.71 ± 0.46 | 7.89 ± 0.28 | 0.176 ± 0.007 | 8.25 ± 0.36 | 1.70 ± 0.53 | (40a,b) |
| 62 | 43.9 ± 0.6 | 7.62 ± 0.19 | 0.174 ± 0.005 | 8.61 ± 0.41 | 0.2 ± 0.6 | (9) |
| 62 | 45.14 ± 0.38 | — | — | — | 1.32 ± 0.48 | (40b) |
| 546 | 61.9 ± 1.5 | 13.3 ± 0.6 | 0.215 ± 0.005 | 10.5 ± 0.3 | — | (36) |

**Table 4** $\bar{p}$-p elastic scattering

| $\sqrt{s}$ (GeV) | $\rho$ | $B$ (GeV$^{-2}$) | $B'(-t > 0.15)$ (GeV$^{-2}$) | $\Delta\rho$ | $\Delta B$ (GeV$^{-2}$) | Ref. |
|---|---|---|---|---|---|---|
| 31 | $0.065 \pm 0.025$ | — | — | $0.036 \pm 0.027$ | — | (9) |
| 31 | — | — | $11.16 \pm 0.20$ | — | — | (13a) |
| 53 | $0.101 \pm 0.018$ | $13.36 \pm 0.53$ | — | $0.042 \pm 0.020$ | $0.51 \pm 0.54$ | (9) |
| 53 | — | $13.92 \pm 0.59$ | $10.68 \pm 0.26$ | — | $0.83 \pm 0.83$ | (40a) |
| 53 | — | — | $11.50 \pm 0.15$ | — | — | (13a) |
| 62 | $0.12 \pm 0.03$ | $13.1 \pm 0.6$ | — | $0.04 \pm 0.03$ | $0.1 \pm 0.6$ | (9) |
| 62 | — | — | $11.12 \pm 0.15$ | — | — | (13a) |
| 546 | — | $15.2 \pm 0.2$ | $13.4 \pm 0.3$ | — | — | (16) |

actually increasing: hadrons become not only larger but also darker as energy increases. Despite the observed qualitative saturation of the Froissart bound, there is still a long way to go in approaching the elusive asymptotia.

# 3.   INTERPRETATION OF HADRON SCATTERING AT HIGH ENERGY

## 3.1   *Notations and Kinematics of Scattering Processes*

In the *S*-matrix formalism, the most general operator leading from the initial state $|i\rangle$ to a final state $|f\rangle$ can be written as

$$\langle f|1 + iT|i\rangle = \delta_{fi} + i(2\pi)^4 \delta^4(P_f - P_i) \frac{\langle f|F|i\rangle}{\sqrt{\Pi_n^{(f)}(2E_n)\Pi_m^{(i)}(2E_m)}}. \qquad 2.$$

In Equation 2, the running products $\Pi_j$ of the initial and final energies ($E_m$, $E_n$, respectively) separate the Lorentz-invariant part $F$ of the transition matrix $T$, and the energy and momentum conservation is explicitly required by the $\delta$ function of the initial and final total four-momenta $P_i$ and $P_f$ (spins are disregarded, as discussed previously). The cross section for a process where two particles with masses $m_1$, $m_2$ give rise to a final state with an arbitrary number of particles $(1 + 2 \rightarrow 3 + 4 + \cdots + N)$ takes the following form:

$$\sigma_{fi} = \frac{1}{4\sqrt{(p_1 \cdot p_2)^2 - (m_1 m_2)^2}} \int (2\pi)^4 \delta^4(P_f - P_i)|F_{fi}|^2 \prod_{n=3,N} \frac{d^3 \mathbf{p}_n}{(2\pi)^3 2E_n}, \qquad 3.$$

where $p_j$ is the four-momentum of the $j$th particle.

When one considers the particle case of a two-body reaction $(1 + 2 \rightarrow$

3+4), Equation 3 becomes

$$\sigma_{\mathrm{fi}} = (1/k_{\mathrm{i}}) \int (1/8\pi W)^2 |F_{\mathrm{fi}}|^2 k_{\mathrm{f}} \, d\Omega_{\mathrm{cm}}. \qquad 4.$$

Here $k_{\mathrm{i}}$ and $k_{\mathrm{f}}$ are the c.m. momenta of the particles in the initial and final states, respectively, and $W$ is the total c.m. energy. For elastic scattering, one has $k_{\mathrm{f}} = k_{\mathrm{i}} = k$ and therefore the differential cross section per unit solid angle $d\Omega_{\mathrm{cm}}$ around the scattering angle $\theta$ can be expressed as

$$d\sigma/d\Omega_{\mathrm{cm}} = (1/8\pi W)^2 |F(W,\theta)|^2. \qquad 5.$$

It is convenient to write all physical quantities in terms of the Lorentz-invariant Mandelstam variables $s, t, u$:

$$s = (p_1 + p_2)^2 = W^2$$

$$t = (p_1 - p_3)^2 = -2k^2(1 - \cos\theta)$$

$$u = (p_1 - p_4)^2 = -2k^2(1 + \cos\theta) + (m_1^2 - m_2^2)^2/W^2$$

$$= m_1^2 + m_2^2 + m_3^2 + m_4^2 - s - t.$$

Since we are interested in the asymptotic behavior at high energies, from now on we assume $k^2 \gg m_{\mathrm{i}}^2$ and thus $s \approx 4k^2$ and $dt \approx (s/4\pi)\, d\Omega_{\mathrm{cm}}$. In this limit, Equation 5 reads

$$d\sigma/dt = (1/16\pi s^2)|F(s,t)|^2. \qquad 6.$$

## 3.2   Unitarity and the Optical Theorem

Probability conservation implies the unitarity of the $S$-matrix: $S^{\dagger}S = SS^{\dagger} = 1$. Since $S = 1 + iT$, one has $i(T^{\dagger} - T) = TT^{\dagger}$, and therefore

$$2\,\mathrm{Im}\,\langle f|T|i\rangle = \sum_l \langle f|T|l\rangle\langle l|T^{\dagger}|i\rangle$$

$$= \sum_m^{(\mathrm{inel})} \langle f|T|m\rangle\langle m|T^{\dagger}|i\rangle + \sum_n^{(\mathrm{el})} \langle f|T|n\rangle\langle n|T^{\dagger}|i\rangle, \qquad 7.$$

where the sum over all possible intermediate states $|l\rangle$ has been split into separate sums over the inelastic and the elastic channels. Equation 7 acquires a special meaning when the initial and final states are the same ($|i\rangle \equiv |f\rangle$), as for elastic scattering at $\theta = 0$):

$$2\,\mathrm{Im}\,F_{\mathrm{ii}} = \sum_m^{(\mathrm{inel})} \int (2\pi)^4 \delta^4(P_{\mathrm{i}} - P_{\mathrm{m}})|F_{\mathrm{im}}|^2 \prod_j \frac{d^3\mathbf{p}_j}{(2\pi)^3 2E_j}$$

$$+ \frac{4W^2 k^2}{E_1 E_2} \int \frac{|F_{\mathrm{in}}(\theta)|^2}{(8\pi W)^2} \, d\Omega. \qquad 8.$$

The two terms in Equation 8 can be recognized as the quantities defined in Equations 3 and 5, i.e. the inelastic and elastic cross sections respectively. Equation 8 is therefore the well-known optical theorem, which is more simply written as

$$\text{Im } F(\theta = 0) = 2Wk(\sigma_{\text{inel}} + \sigma_{\text{el}}) \approx s\sigma_{\text{tot}}. \qquad 9.$$

In general, as a consequence of unitarity, for elastic scattering at $t \neq 0$ (i.e. $|i\rangle \neq |f\rangle$) one can write

$$\text{Im } F(t) = s[G_{\text{inel}}(t) + G_{\text{el}}(t)], \qquad 10.$$

where $G_{\text{inel}}(t)$ and $G_{\text{el}}(t)$ are the inelastic and elastic overlap functions introduced by Van Hove (46). These functions are so normalized that $G_{\text{inel}}(0) = \sigma_{\text{inel}}$ and $G_{\text{el}}(0) = \sigma_{\text{el}}$.

## 3.3  Impact Parameter Representation

Let us consider the standard partial-wave expansion of the scattering amplitude $F$:

$$F(s, \cos \theta) = (8\pi W/k) \sum_l (2l + 1) f_l(s) P_l (\cos \theta), \qquad 11.$$

where $f_l(s)$ is the $l$th partial-wave amplitude and $P_l (\cos \theta)$ is the Legendre polynomial of order $l$. Because at very high energy a great number of partial waves contribute to $F$, it is legitimate to replace the sum over $l$ by an integral over the bidimensional impact parameter $\mathbf{b}$ [with $|\mathbf{b}| = (l + 1/2)/k$]. Equation 11 thus becomes

$$\frac{F(s, q^2)}{8\pi s} = \frac{1}{2\pi} \int f(s, b) \exp(i\mathbf{q} \cdot \mathbf{b}) \, d^2\mathbf{b} = \int f(s, b) J_0(qb) b \, db, \qquad 12.$$

where $q = k_1 - k_3$ is the momentum transfer ($q^2 = -t$), and the Bessel function $J_0$ may be recognized as an asymptotic form of the Legendre polynomials, valid near the forward direction. The amplitude $f(s, b)$ can then be obtained from the inversion of Equation 12, by the known properties of Fourier transforms:

$$f(s, b) = \frac{1}{2\pi} \int \frac{F(s, q^2)}{8\pi s} \exp(-i\mathbf{q} \cdot \mathbf{b}) \, d^2\mathbf{q} = \int \frac{F(s, q^2)}{8\pi s} J_0(qb) q \, dq. \qquad 13.$$

The unitarity condition (Equation 10) in the impact parameter space then reads

$$\text{Im } f(s, b) = |f(s, b)|^2 + \tfrac{1}{4} G_{\text{inel}}(s, b). \qquad 14.$$

Im $f(s, b)$ is usually referred to as the profile function, and represents the hadronic opacity as a function of the impact parameter $b$. The meaning of unitarity relation shown as Equation 14 is particularly simple when integrated over $b$. It reduces to $\sigma_{tot} = \sigma_{el} + \sigma_{inel}$, since from Equations 9 and 12 one has

$$\sigma_{tot} = 8\pi \int \text{Im } f(s,b)b \, db, \qquad\qquad 15.$$

$$\sigma_{el} = 8\pi \int |f(s,b)|^2 b \, db, \qquad\qquad 16.$$

$$\sigma_{inel} = 8\pi \int \tfrac{1}{4}G_{inel}(s,b)b \, db. \qquad\qquad 17.$$

In its differential form (Equation 14), unitarity relates the elastic and inelastic cross sections at the same value of $b$: $d^2\sigma_{tot}/db^2 = d^2\sigma_{el}/db^2 + d^2\sigma_{inel}/db^2$. This is often referred to by saying that "unitarity is diagonal in impact parameter space." Equation 14 puts a limit on the inelastic overlap function: $G_{inel} \leq 1$. In addition, since $G_{inel}$ is positive, the unitarity condition 14 implies:

$$0 \leq |f(s,b)|^2 \leq \text{Im } f(s,b) \leq 1. \qquad\qquad 18.$$

A relevant physical consequence of Equation 14 is that no scattering process can be uniquely inelastic: a nonvanishing elastic scattering amplitude must always be present as a "shadow" of all the inelastic channels open at that energy. The forward elastic peak can thus be seen as analogous to the diffraction pattern arising in classical optics when a plane wave encounters an absorbing disc. As energy increases, elastic scattering is indeed expected to become purely diffractive, as small as allowed by Equation 14 for a given $G_{inel}$. The elastic amplitude then tends to be purely imaginary, and the smaller of the two solutions of Equation 14 is assumed to correspond to the physical situation. In this case, the upper bound of Equation 18 is lowered to Im $f(s,b) \leq 1/2$.

In order to satisfy the unitarity relation (Equation 14) it is convenient to express the amplitude $f(s,b)$ in terms of the complex eikonal function $\chi(s,b)$:

$$f(s,b) = i\{1 - \exp[i\chi(s,b)]\}/2, \qquad\qquad 19.$$

where Im $\chi \geq 0$ and the inelastic overlap function is recognized to be $G_{inel} = 1 - \exp(-2 \text{ Im } \chi)$. For Im $\chi = 0$, $G_{inel}$ vanishes: this is the case when the reaction is below the threshold of all inelastic channels. At high energy we are dealing with the opposite case: elastic scattering is essentially diffractive, and Re $\chi$ is small. If Re $\chi = 0$, the amplitude $f$ is purely imaginary and determined by the opaqueness $\Omega(s,b) = \text{Im } \chi$. In the limit $\Omega \to \infty$ the inelastic absorption is maximum and Im $f$ approaches the reduced unitarity bound of 1/2. This is the case for a completely black disc; the elastic cross section is then also maximum and reaches the value (see Equations 15 to 17):

$$\sigma_{el} = \sigma_{inel} = \tfrac{1}{2}\sigma_{tot}. \qquad\qquad 20.$$

## 3.4   Overlap Function Analysis of p-p Scattering at the ISR

The geometrical representation outlined in the previous section provides an easy, intuitive framework that is independent of the underlying dynamics. Hadron elastic differential cross sections do exhibit diffraction patterns quite similar to those observed in optics. As shown in Figure 14c, this phenomenon becomes surprisingly evident when the scatterer is a complex nucleus, which has rather well-defined dimensions. The profile function $f(b)$ of a heavy nucleus closely resembles that of a uniformly opaque disc, whose Fourier transform (Equation 12) is a Bessel function $J_1(qR)$, $R$ being the disc radius. Such an amplitude indeed shows a series of zeros with the increase of the momentum transfer $q$, as illustrated in Figure 14a.

A qualitatively different picture is observed when the scatterer is a single proton. In this case the elastic differential cross section exhibits a monotonous, exponential decrease in $t$, down to $|t| \approx 1.4\,\text{GeV}^2$, over about six orders of magnitude (see Figure 14d). This apparently inconsistent behavior is, however, easily reconciled when one considers that an exponential in $t$ is precisely the Fourier transform of a Gaussian profile function, as shown in Figure 14b. From this point of view, the difference between a massive nucleus and a single proton lies just in the different sharpness of the edges! In a simplified calculation, if one neglects the real part of the elastic scattering amplitude and moreover assumes that the differential cross section can be parametrized by a single exponential with slope $B$, using the optical theorem (Equation 9) one can write the elastic amplitude in the form $F(s, t) = is\sigma_{tot}\exp(-B|t|/2)$. The profile function, derived by Equation 13, is then

$$f(b) = (i/8\pi)(\sigma_{tot}/B)\exp(-b^2/2B). \qquad\qquad 21.$$

Let us go further, and consider a quantitative analysis of p-p elastic scattering data (Figure 1). In order to compute the profile function Im $f(s, b)$ from Equation 13,

$$\text{Im } f(s,b) = (1/4\sqrt{\pi})\int dt\, J_0(b\sqrt{-t})\{d\sigma/dt - (1/16\pi s^2)[\text{Re }F(s,t)]^2\}^{1/2},$$

$$22.$$

an estimate of the real part of the elastic scattering amplitude at $t \neq 0$ is needed. It should, however, be noticed that the real part has little effect, because at any given value of $b$ the integral in Equation 22 is dominated by the contribution at low $t$, where the real part is known to be small. Therefore the determination of the profile function does not depend critically on any sensible choice.

A reasonable parametrization of the real part of the elastic scattering amplitude at ISR energies [see, for instance, (42)] takes into account the measured values of $\rho$, vanishes at $t \approx 0.2 \, \text{GeV}^2$ to satisfy dispersion relation analysis (47), and correctly fills the dip at $t \approx 1.4 \, \text{GeV}^2$, where the imaginary part is supposed to be vanishing. At all values of $t$, the real part turns out to

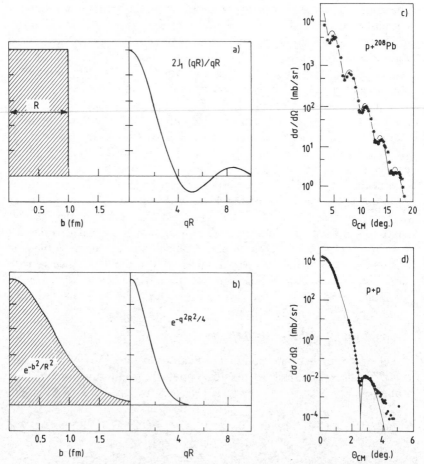

*Figure 14* (a), (b) Examples of profile functions and corresponding scattering amplitudes; (c), (d) diffraction patterns observed in hadron scattering (from 4g).

be much smaller than the imaginary part, except of course in the dip region. With a reasonable assumption for the real part, by Equation 22 the profile function Im $f(s, b)$ can be computed numerically from the $d\sigma/dt$ data: the inelastic overlap function $G_{\text{inel}}(s, b)$ is then obtained by Equation 14.

The results of such an analysis (48–52) performed on the data at $\sqrt{s} = 53$ GeV are displayed in Figure 15. The profile function exhibits a shape very close to a Gaussian in $b$, as expected from the approximately exponential behavior of $d\sigma/dt$. Its r.m.s. width, corresponding to the effective proton radius, is found to be slightly less than 1 fm at this energy. One can notice that Im $f(b)$ has a maximum value of 0.36, which is far from the unitarity limit of 0.5. This means that even in a central collision at $b = 0$ a target proton is not completely dark, but the projectile has a nonvanishing probability of going straight through without interacting. The presence of the dip structure in $d\sigma/dt$ at $t \approx 1.4 \, \text{GeV}^2$ produces a slight flattening of the profile function with respect to a Gaussian (too small to be visible in the figure) at very small values of $b$. One should realize that a sharp black disc with a radius of 1 fm would produce a first dip in $d\sigma/dt$ at $t \approx 0.6 \, \text{GeV}^2$ (see Figure 14a). In the interval from this value of $t$ to the measured position of the dip, the elastic amplitude decreases by two other orders of magnitude. It should not be surprising then that the striking structure in $d\sigma/dt$ can hardly

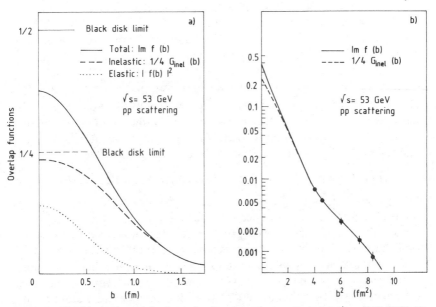

*Figure 15*   Total, inelastic, and elastic overlap functions of p-p scattering at $\sqrt{s} = 53$ GeV (50, 52). The deviation of the tail from a Gaussian is evidenced by the logarithmic scale.

be noticed in the profile function: the dip at 1.4 GeV² is simply too far to modify significantly the basic Gaussian shape of Im $f(b)$. On the other hand, the slight variation of the slope of $d\sigma/dt$ at $-t \approx 0.13$ GeV², though less striking, brings a distortion to the tail of Im $f(b)$ above 2 fm that is almost one order of magnitude higher than the Gaussian (see Figure 15b).

The same qualitative features are exhibited by the inelastic overlap function $G_{inel}(b)$, which is related to the profile function by the unitarity relation (Equation 14). According to this equation, the tails of 1/4 $G_{inel}$ and of IM $f$ are essentially identical at large values of $b$, where the term $|f|^2$ is negligible, while near $b = 0$ the inelastic overlap function flattens with respect to the profile function.

Detailed analyses of the energy dependence of $G_{inel}$ over the ISR range have been performed (42, 53). In particular, a systematic study based on a critical review of available ISR data has been carried out by Amaldi & Schubert (42). They selected elastic $d\sigma/dt$ data from several ISR experiments, readjusted the absolute normalizations taking into account the general trend of the differential cross section, and interpolated between neighboring data sets in those $t$ ranges in which no measurement was available. The experimental data they used at the five ISR energies were shown in Figure 1. Their results indicate that the observed $\ln^2 s$ rise of $\sigma_{tot}$ over the ISR energy range only comes from an increase of the interaction radius, which is linear in $\ln s$, while the central opacity remains practically constant. As a matter of fact, the variation $\Delta G_{inel}(b)$, which from zero at $b = 0$ reaches a maximum around $b \approx 1$ fm (see Figure 16a), essentially disappears when the impact parameter is rescaled according to the mean interaction radius or, equivalently, to the square root of the total cross section. Indeed, if $G_{inel}$ is expressed in terms of the reduced impact parameter $b' = b/\sqrt{\sigma_{tot}}$, one obtains an energy-independent pattern. This is illustrated in Figure 16b, where the energy variation $\Delta G_{inel}(b')$ of the inelastic overlap function over the ISR energy range, as a function of the reduced impact parameter, appears to be consistent with zero.

## 3.5 The Geometrical Scaling Hypothesis Confronted with SPS Collider Data

As discussed in the previous section, the ISR p-p elastic scattering data suggest that, as energy increases, the central blackness remains constant, while the edge of the overlap function moves toward larger values of $b$. It is therefore natural to consider a simple model in which the only effect produced on the profile function $f(s, b)$ by varying the energy is a change in the mean interaction radius $R$. This so-called geometrical scaling hypothesis can be expressed as follows:

$$f(s, b) = f'(b/R). \tag{23}$$

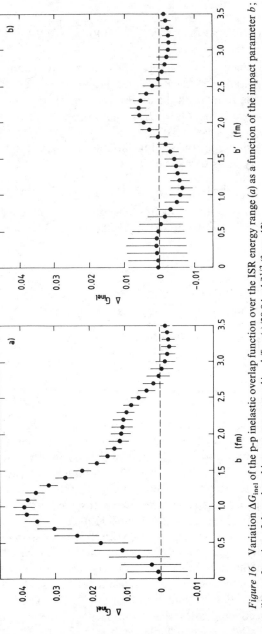

*Figure 16* Variation $\Delta G_{inel}$ of the p-p inelastic overlap function over the ISR energy range (*a*) as a function of the impact parameter *b*; (*b*) as a function of the reduced impact parameter $b' = b/[\sigma_{tot}(s)/38.94 \text{ mb}]^{1/2}$ (from 42).

After defining the reduced impact parameter $b' = b/R$, Equation 12 becomes

$$F(s,t)/8\pi s = R^2 \int_0^\infty f'(b')J_0(Rb'\sqrt{-t})b'\ db' = R^2 F'(R^2 t),\qquad 24.$$

where $F'$ no longer depends separately on $s$ and $t$, but only on the adimensional variable $tR^2$, with $R$ depending on energy. A direct consequence of Equation 24, via the optical theorem (Equation 9), is that the energy dependence of the total cross section comes only from the variation of $R$ with energy: $\sigma_{\text{tot}}(s) \propto R^2(s)$.

If $\tau = t\sigma_{\text{tot}}$ is chosen as the scaling variable, it follows from Equation 24 that the quantity (see Equation 6)

$$(1/\sigma_{\text{tot}})\ d\sigma_{\text{el}}/d\tau = \phi(\tau)\qquad 25.$$

has a universal shape independent of energy. As a consequence, the local slope parameter $B = (d/dt) \ln (d\sigma/dt)$ can be written as $\sigma_{\text{tot}}(d/d\tau) \ln [\phi(\tau)]$ and, therefore,

$$\sigma_{\text{tot}}/B = \text{const.}\qquad 26.$$

In particular, it also follows that the position of the dip and the height of the secondary maximum, respectively, go as

$$|t_{\text{dip}}| \propto 1/\sigma_{\text{tot}} \quad \text{and} \quad d\sigma/dt|_{t=t_2} \propto \sigma_{\text{tot}}^2.$$

In addition

$$\sigma_{\text{el}}/\sigma_{\text{tot}} = \int \phi(\tau)\ d\tau = \text{const.}\qquad 27.$$

Indeed elastic scattering data at the ISR seem to support this hypothesis. Figure 8b shows how elastic differential cross sections $(1/\sigma_{\text{tot}})\ d\sigma/d\tau$ measured at the two extreme ISR energies fall one on top of the other. Also the ratio $\sigma_{\text{el}}/\sigma_{\text{tot}}$, shown in Figure 13b, although decreasing at lower energies, seems to become constant in the ISR energy range. The same tendency of approaching a constant value is observed in the behavior of the ratio $\sigma_{\text{tot}}/B$ up to ISR energies (see Figure 13c).

This nice picture unfortunately fails when the energy span is extended far beyond the ISR range. As a matter of fact, the recent measurement of the ratio $\sigma_{\text{el}}/\sigma_{\text{tot}}$ at the p̄-p Collider is definitely larger than at the ISR and incompatible with the assumption of constancy (expressed by Equation 27) as discussed in Section 2.2 and shown in Figure 13b. The increase of this parameter directly indicates that the opacity of the p̄-p interaction is actually rising, after the seeming plateau observed in the ISR energy range.

The same conclusion is, of course, drawn from the behavior of the parameter $\sigma_{tot}/B$, which is also proportional to the nucleon opacity (see the simplified example of Equation 21) and is found to increase sensibly in the same energy interval, as can be seen in Figure 13c.

This violation of geometrical scaling appears quite evident in a detailed analysis of the overlap function $G_{inel}(b)$ (54). Here, in addition to a clear expansion of the ineraction radius, the central opacity is also found to be definitely larger than at ISR energies, closely approaching the unitarity limit of one. Even rescaled in the reduced impact parameter $b'$ [as was done for the ISR data in the analysis (42) quoted in the previous section], $G_{inel}(b')$ exhibits a clear energy dependence when going from the ISR to the SPS Collider. This behavior is shown in Figure 17, in which the energy variation $\Delta G_{inel}(b')$ from $\sqrt{s} = 53$ GeV to $\sqrt{s} = 546$ GeV is found to be incompatible with a zero level. This picture has to be compared with Figure 16b, where the same quantity, evaluated over the ISR energy span, appears to be consistent with zero. One can conclude that there is clear evidence that the nucleon, which at the ISR just seemed to become larger, is also getting blacker as energy increases. The observed rise of $\sigma_{tot}$ is therefore not only due to the expansion of the interaction radius but also to an increase of the nucleon darkness.

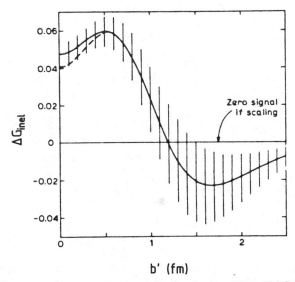

*Figure 17*   Variation $\Delta G_{inel}$ of the inelastic overlap function from ISR to SPS Collider energy, as a function of the reduced impact parameter $b' = b/[\sigma_{tot}(s)/61.9 \text{ mb}]^{1/2}$ (from 54).

## 4.  THE VERY HIGH-ENERGY LIMIT

It is reasonable to assume that at sufficiently high energy the features of hadron scattering can be interpreted in a simple way in terms of a few general theoretical principles, independently of specific models. Asymptotic properties of the scattering amplitudes can indeed be derived from the requirements of unitarity, analyticity, and crossing symmetry. This kind of approach, pioneered by Pomeranchuk (55) and by Froissart (56) has led to several theorems that impose nontrivial constraints on hadron behavior at asymptotic energies [the interested reader can find a complete review in (4d) and (4i)].

### 4.1   *Asymptotic Theorems*

4.1.1   FROISSART-MARTIN BOUND   The well-known Froissart-Martin upper bound (56) on total cross section has been derived from axiomatic field theory (57, 58a) and states that

$$\sigma_{\text{tot}} \leq (\pi/m_\pi^2) \ln^2 s. \qquad\qquad 28.$$

An intuitive idea of the physical meaning of this theorem can be obtained in impact parameter representation. In order to satisfy analyticity, the profile function has been shown to be limited by an exponential decrease with increasing values of the impact parameter, and by a rather slow increase with energy, of the kind Im $f(s, b) \leq$ const $s^\beta \exp(-2m_\pi b)$ with $\beta \leq 1$. In the case of diffraction scattering, however, the reduced unitarity limit Im $f(s, b) \leq 1/2$ cannot be exceeded (see Section 3.3). The upper bound shown in Figure 18 is thus obtained:

$$\text{Im } f(s, b) = 1/2 \qquad\qquad \text{for } b \leq b_0$$

$$\text{Im } f(s, b) = \exp[-2m_\pi(b-b_0)] \quad \text{for } b > b_0,$$

where $b_0 = (1/2m_\pi) \ln (s/s_0)$. Using Equation 15, one then obtains the upper limit in Equation 28, which corresponds to a full absorption of all partial waves up to $b_0$: the surface of this black disc increases as $\ln^2 s$. The grey fringe for $b > b_0$ imposed by analyticity has a constant width, so that its surface increases only as $\ln s$ and has a negligible role in the upper bound (Equation 28) at infinite energy.

It is worth noticing that Equation 28 is indeed a minimum upper bound that can by no means be further restricted. For some time, before rising cross sections were discovered at the ISR, when people believed that hadrons had an asymptotically constant size, theoreticians tried to lower the bound and to prove that total cross sections were asymptotically limited by a constant. This effort was revealed to be hopeless when a crossing

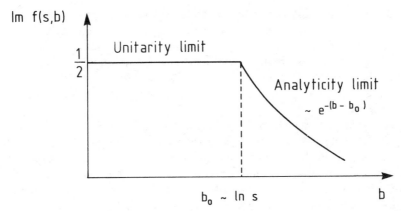

*Figure 18*   Upper bound to the profile function.

symmetric amplitude (59) was constructed that satisfied unitarity and analyticity and behaved like $s \ln^2 s$ in the forward direction.

ISR and Collider $\sigma_{\text{tot}}$ data are indeed compatible with a $\ln^2 s$ rise (see Figure 11), although with a coefficient much smaller than in Equation 28. This behavior may lead to the belief that an asymptotic regime is already established at presently available energies and that the Froissart bound is indeed saturated by nature qualitatively, if not quantitatively. This might suggest that, for some reason, the profile function of the nucleon is bound not to exceed the "greyness" we observe today and the rise of the total cross section simply reflects the expansion of the proton radius. This hypothesis was indeed supported by the geometrical scaling behavior found in the ISR data, as discussed in Section 3.4. The recent measurement at the SPS Collider indicates, however, that the ratio $\sigma_{\text{el}}/\sigma_{\text{tot}}$ does not remain constant beyond ISR energies and that the hadronic opacity is actually slowly increasing with $s$. One must remember though that in the limit of a completely saturated black disc the ratio $\sigma_{\text{el}}/\sigma_{\text{tot}}$ must approach $1/2$, which at the Collider is also far from being the case. Even at $b = 0$ the profile function is still sensibly lower than the reduced diffractive limit of $1/2$ imposed by unitarity. Therefore, if the nucleon becomes blacker and blacker as energy increases, since it is still far from the quantitative saturation of the Froissart bound, there might be a very long way to an asymptotic regime and the present $\sim \ln^2 s$ behavior could just be a transient accident with no deep meaning.

4.1.2   POMERANCHUK THEOREM   On a phenomenological basis, one observes that the total cross section $\sigma_{\text{tot}}(ab)$ of hadron "a" hitting target "b" and the total cross section $\sigma_{tot}(\bar{a}b)$ of the corresponding antiparticle hitting the same target approach each other more and more as energy increases (see, for instance, Figures 11 and 12). In its original formulation, based on dispersion relations for the forward-scattering amplitude, the Pomeranchuk theorem (55) stated that if the total cross sections of particle and antiparticle tend to a constant with increasing energy, their difference tends to zero.

At the time in which the total cross sections were indeed believed to approach a constant limit, everybody was happy with this formulation of the theorem, which seemed to be confirmed by experiment. However, this formulation is of no use if the rising total cross sections discovered at the ISR are interpreted as the onset of an asymptotic behavior of everlasting growth. Therefore many authors (60a–c) have worked at a reformulation of the theorem in order to preserve the final statement, which still appeared to be supported by experimental evidence, by trying to exploit analyticity and unitarity without adopting additional restrictive assumptions. This method finally led to proving that if at least one cross section rises indefinitely [either $\sigma_{tot}(\bar{a}b) \to \infty$ or $\sigma_{tot}(ab) \to \infty$], then

$$\sigma_{tot}(\bar{a}b)/\sigma_{tot}(ab) \to 1 \qquad\qquad 29.$$

for $s \to \infty$. It should be noted that this formulation does not imply that the total cross-section difference $\Delta\sigma = \sigma_{tot}(\bar{a}b) - \sigma_{tot}(ab)$ goes to zero. Equation 29 in fact cannot even exclude that $\Delta\sigma$ goes to infinity as energy increases. Some authors have indeed considered the possibility that the imaginary part of the odd-signature amplitude $F^- = F^{\bar{a}b} - F^{ab}$ at $t = 0$, to which $\Delta\sigma$ is proportional, grows as fast as $\ln s$ (41). The recent data on $\sigma_{tot}(pp)$ and $\sigma_{tot}(\bar{p}p)$ from the ISR, however, disfavor this hypothesis (see Figure 12). Another possibility is that the asymptotic odd-signature contribution to $\Delta\sigma$ remains constant with increasing energy [see a discussion in (61)]. An even more subtle hypothesis assumes that the amplitude $F^-$ has an asymptotically vanishing imaginary part and thus does not contribute to $\Delta\sigma$, but only affects the behavior of the real part $\rho$ of the elastic scattering amplitude in the forward direction (25a, 62). Recent analyses (61, 63) of ISR data on $\sigma_{tot}$ and $\rho$, however, leave little room for exotic contributions by such "odderon" terms. Actually it can be proved (64) that $\Delta\sigma$ must go to zero, on the hypothesis that asymptotically Re $F^-$ and Im $F^-$ have the same sign (Fischer theorem). Measurements of $\rho$ indicate that this hypothesis is probably satisfied, despite the large errors, at ISR energies. Although one cannot exclude in principle that Re $F^-/$Im $F^-$ changes sign at higher energy, it is reasonable to assume that the cross-section difference really goes to zero.

In conclusion, the data seem to support a stronger picture than the one conjectured by the Pomeranchuk theorem, in the sense that not only the cross-section ratio tends to unity, but also the difference appears to tend to zero.

4.1.3 CORNILLE-MARTIN THEOREM     Some effort was made to extend the validity of the Pomeranchuk theorem to elastic scattering. It was first recognized (60b) that analyticity and unitarity imply that the ratio of the forward elastic differential cross sections of particle and antiparticle should tend to unity for $s \to \infty$. (If $\rho$ tends to zero, this statement is equivalent to the Pomeranchuk theorem.) Furthermore, based on the positivity of the absorptive parts of the elastic amplitude (Equation 11), Cornille & Martin (65) have shown that the above property is valid for the whole diffraction peak, namely

$$\frac{d\sigma^{\bar{a}b}}{dt}\left[s, t(s)\right] \bigg/ \frac{d\sigma^{ab}}{dt}\left[s, t(s)\right] \to 1 \qquad\qquad 30.$$

for any smooth $t(s)$ such that the ratio $\{d\sigma[s, 0]/dt\}/\{d\sigma[s, t(s)]/dt\}$ stays finite as $s \to \infty$. As a consequence, the ratio of the slopes of the diffraction peak in p-p and p̄-p elastic scattering must approach unity as energy increases: a persistent shrinking in p-p elastic scattering must be accompanied by a similar shrinking in p̄-p. This picture is indeed supported by experimental data, as can be seen in Figures 4 and 5. In particular, the slope measurement at the SPS Collider indicates that also for p̄-p scattering a shrinkage of the elastic diffraction peak sets in at a rate compatible with ln $s$.

The theorem extends naturally to the asymptotic equality of the total elastic cross sections of particle and antiparticle: $\sigma_{el}(\bar{a}b)/\sigma_{el}(ab) \to 1$ for $s \to \infty$. This feature is in good agreement with the experimental data, as shown in Figure 13a.

Another important consequence of the Cornille-Martin theorem is that, if the dip of the p-p elastic scattering around $-t \approx 1.4\,GeV^2$ at ISR energies belongs to the diffraction peak, then a similar dip should also be present in p̄-p scattering (provided that ISR energies can really be considered as asymptotic). Models (24) that predict a different dip in p-p and p̄-p elastic scattering have to apply to a nonasymptotic situation: when diffraction dominates, the p-p and p̄-p dips must behave in the same way for $s \to \infty$ (43, 61). The data shown in Figure 9b point to a significant difference in the large-$t$ region, where a shoulder seems to be favored rather than a dip in p̄-p scattering. However, the limited statistics does not allow one to draw a definite conclusion, particularly when one realizes, as discussed in Section 2.1.3, that at these energies the p̄-p dip may be partially filled by a higher value of $\rho$.

4.1.4 MacDOWELL-MARTIN BOUND   MacDowell & Martin (66) showed that the shape of the forward elastic peak must satisfy the following bound, based on unitarity (disregarding terms of order $1/s$):

$$(d/dt) \ln [\text{Im } F(t)]|_{t=0} \geq (1/36\pi)\sigma_{tot}^2/\sigma_{el}.\qquad 31.$$

If the real part of the elastic amplitude can be neglected, this inequality gives a bound on the forward slope $B$ of the elastic peak:

$$B \geq (1/18\pi)\sigma_{tot}^2/\sigma_{el}.\qquad 32.$$

Experimental data show that this bound is almost saturated (within $\sim 15\%$). This should not be surprising since the equality $B = (1/16\pi)\sigma_{tot}^2/\sigma_{el}$ holds for a purely imaginary exponential amplitude.

An interesting consequence of this bound on $B$, together with the requirement of analyticity of the scattering amplitude and the assumption of a qualitatively saturated Froissart bound, is the prediction of a nonuniform behavior of the slope of the diffraction peak near the forward direction (61, 67). One should notice that, if $\sigma_{tot} \sim \ln^2 s$, then $B$ must increase at least as $\ln^2 s$: since analyticity constrains $B$ not to exceed a $\ln^2 s$ behavior, one can conclude that the forward slope $B$ must go just like $\ln^2 s$. Experimental data up to present energies do not give a clear indication of a $\ln^2 s$ behavior of the forward slope, which still may be consistent with an increase linear in $\ln s$ (see Figure 5a). It must be clear, however, that if $\sigma_{tot}$ continues to rise as $\ln^2 s$, at some finite energy the forward slope must start to increase faster than $\ln s$ in order to prevent a violation of inequality 32 above.

4.1.5 AUBERSON-KINOSHITA-MARTIN THEOREM   From the hypothesis of a qualitatively saturated Froissart bound follows a remarkable property of the scattering amplitude. In fact, if $\sigma_{tot}$ grows asymptotically like $\ln^2 s$, then we also find $\sigma_{el} \sim \ln^2 s$ (58a,b). By the optical theorem, $d\sigma/dt|_{t=0}$ increases like $\ln^4 s$, so that the width of the diffraction peak must shrink like $1/\ln^2 s$ (note that a weaker shrinkage is excluded since it would produce $\sigma_{el}$ larger than $\sigma_{tot}$ at sufficiently high energy). As a matter of fact, it has been shown (68) that with this limiting behavior not only the width but also the whole shape of the diffraction peak is constrained to exhibit a scaling property. More rigorously, neglecting odd-signature contributions, if $\sigma_{tot}/\ln^2 s \rightarrow \text{const for } s \rightarrow \infty$, then we obtain

$$\text{Im } F(s,t)/\text{Im } F(s,0) \rightarrow \Phi(\tau), \quad \text{Re } F(s,t)/\text{Re } F(s,0) \rightarrow (d/d\tau)[\tau\Phi(\tau)], \quad 33.$$

where $\tau = t \ln^2 s$, and $\Phi(\tau)$ is an entire function of order $1/2$, i.e. analytic in $\tau$ and bounded by $\exp(c\sqrt{|\tau|})$. In other words, if the Froissart bound is saturated, the diffraction amplitude $F(s,t)$ asymptotically reduces to a function of a single scaling variable, proportional to $t\sigma_{tot}$. This is a useful result for applications in phenomenology; it leads to predictions on the asymptotic behavior of diffraction scattering that can be compared with experimental data and thereby allow one to test whether an asymptotic regime has already set in at present energies.

## 4.2   Scaling as an Asymptotic Property: Comparison with High-Energy Data

If one believes that a qualitative saturation of the Froissart bound is already in effect at present energies, then Equations 33 allow one to write the elastic scattering amplitude in terms of two input parameters, $\sigma_{tot}$ and $\rho$, that can be determined by experiment at each energy, and of a universal function $\Phi(\tau)$ (69):

$$d\sigma/dt = (1/16\pi)\sigma_{tot}^2\{\Phi^2(\tau)+\rho^2[(d/d\tau)\tau\Phi(\tau)]^2\}. \qquad 34.$$

This differential equation can be solved numerically for $\Phi(\tau)$ if all other quantities ($d\sigma/dt$, $\sigma_{tot}$, $\rho$) are accurately known at a given energy. The elastic differential cross section $d\sigma/dt$ can then be computed at other energies by inserting the measured values of $\sigma_{tot}$ and $\rho$. Note that Equation 34 exhibits the same scaling property as Equation 25, apart from the presence of the quantity $\rho$ that depends on $s$ alone. If one considers, however, that the real part of the amplitude is much smaller than the imaginary part (at all values of $t$ except of course in the region of the dip) the geometrical scaling hypothesis formulated on the basis of the ISR data (see Section 3.5) appears to be more than a phenomenological description of diffraction scattering at high energy: it comes out naturally as an asymptotic property, provided that the Froissart bound is saturated. Should this be the case, Equations 25 to 27 would hold asymptotically in a model-independent way.

The scaling function $\Phi(\tau)$ was computed (70) by solving numerically Equation 34 at $\sqrt{s} = 53$ GeV, where the most accurate data exist. The solution obtained for $\Phi(\tau)$ is shown in Figure 19a. It goes through zero at the position of the dip, which is partially filled by the real part of the amplitude. Owing to this feature, the behavior of $d\sigma/dt$ at the dip is very sensitive to the real part $\rho$ in the forward direction, since Equation 34 states that the value of the cross section becomes proportional to $\rho^2$ where the imaginary part vanishes. This nontrivial connection between $\rho$ and the dip is illustrated in Figure 19b.

The ISR data at the other energies are well reproduced by Equation 34 also in the region of the dip, which is predicted to be deepest where $\rho$ is zero and which gradually fills in with increasing energy as the measured value of $\rho$ increases. The scaling violation at the dip, which appeared in the original geometrical hypothesis of Section 3.5, is therefore taken into account by the

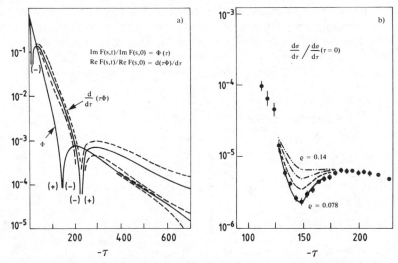

*Figure 19* (*a*) The scaling function $\Phi$ and its derivative computed by Equation 34 for p-p elastic scattering at $\sqrt{s} = 53$ GeV (the dashed lines represent the uncertainty of the calculation); (*b*) sensitivity of the dip-bump structure to the value of the real part $\rho$ (the four curves correspond to $\rho = 0.078, 0.10, 0.12,$ and 0.14, respectively) (from 70).

nonscaling term in Equation 34 due to the energy dependence of $\rho$. At the energy of the SPS Collider, $\rho$ should attain a value high enough ($\rho \geq 0.15$) to fill in completely the dip, which is expected to turn into a shoulder. Indeed, recent data on $\bar{p}$-p elastic scattering at $\sqrt{s} = 546$ GeV (19) exhibit a shoulder at the predicted position. However, the measured value of the cross section in this $t$ region is almost one order of magnitude higher than predicted (27), as shown in Figure 10*b*. This serious discrepancy is hard to reconcile. Furthermore, the rise of the ratios $\sigma_{el}/\sigma_{tot}$ and $\sigma_{tot}/B$ found at the Collider indicates that the nucleon opacity is increasing and provides clear evidence for violation of geometrical scaling. This growing behavior is detailed in Figure 17, which illustrates the energy dependence of the inelastic overlap function from the ISR to the SPS Collider, obtained in the analysis (54) discussed in Section 3.5.

In conclusion, the onset of an asymptotic geometrical scaling regime, which was believed to already be in effect at the ISR, has to be pushed far beyond Collider energy, in spite of the observed $\sim \ln^2 s$ behavior of $\sigma_{tot}$ that had suggested a precocious saturation of the Froissart bound. This conflict with Equation 33 may raise a doubt as to whether the hypothesis of the Auberson-Kinoshita-Martin theorem is really verified; the present $\ln^2 s$ rise of $\sigma_{tot}$ might be only a transient trend, different from the asymptotic behavior.

## 4.3   *Factorizing Eikonal Models*

Another interesting approach stems from the original idea that was at the basis of the Chou-Yang model. In this geometrical picture, elastic scattering is considered as the result of the attenuation of the incoming wave through the extended distribution of matter inside the hadron (in close analogy with Glauber's optical model of hadron scattering on nuclei). In the pioneering works listed in (71), the eikonal representation in impact parameter space (Equation 19) was adopted with the assumption that the matter distribution $G(b)$ is proportional to the charge distribution, i.e. the Fourier transform of the form factor. In subsequent works (72) after the discovery of rising cross sections at the ISR, the energy dependence of $\sigma_{tot}$ was taken into account by introducing a nonconstant factor $K(s)$ in the opaqueness: $\Omega(s, b) = K(s)G(b)$. At a given energy, the factor $K$ is empirically determined by the value of $\sigma_{tot}$ via Equation 15. Such a factorized form of the eikonal function allowed the generalization of the model, still preserving its original physical concept.

Several years before the ISR results, on the other hand, Cheng & Wu (39) had deduced in the framework of massive QED that the asymptotic behavior of the elastic amplitude should be of the eikonal kind with $K(s) \propto s^c$ ($c$ is a positive contant) and $G(b)$ decreasing exponentially at large $b$. It should be noticed that this result has a remarkable predictive power because it implies that the edge of the profile function (Equation 19) moves toward larger values of $b$ proportionally to $\ln K(s)$, while its central part approaches the unitarity limit of $1/2$ as the energy increases. The interacting hadron tends to appear as a black core whose radius grows as $\ln s$ surrounded by a grey fringe of constant width, and to behave just as in the limiting case discussed in Section 4.1.1, completely saturating the Froissart bound at infinite energy. This model therefore predicts that for all hadrons at extremely high energies the total cross section grows as $\ln^2 s$, the elastic diffraction peak shrinks indefinitely, and the ratio $\sigma_{el}/\sigma_{tot}$ rises with energy toward the limit $1/2$.

All these qualitative predictions seem indeed to be supported by experimental data. In order to improve the quantitative agreement with the data, the model was subsequently implemented (28a–d) with various phenomenological parametrizations of the hadronic matter distribution $G(b)$. Sometimes an expression of the kind $\exp[-\lambda(b^2 + b_0^2)^{1/2}]$ has been adopted (28a), which essentially is a Gaussian at small $b$ with an exponential tail at $b \gg b_0$, in close analogy with the behavior of the Fourier transform of a dipole form factor $1/(1 - t/\mu^2)^2$. Other authors (28b) have tried to multiply a dipole form factor by an ad hoc slowly varying function of $t$. In all cases, good fits to the data are obtained, and the shoulder at large

$t$ in p̄-p elastic scattering at the SPS Collider is predicted with about the correct height (see Figure 10$b$). Regardless of a quantitative comparison of these specific parametrizations with experimental data, it is a fact that the qualitative predictions of the bare eikonal formulation of the original models have had remarkable confirmation, first by the $\sim\ln^2 s$ rise of the total cross section found at the ISR, and now at the Collider by the observation that the ratio $\sigma_{el}/\sigma_{tot}$ also grows with energy.

## 5. CONCLUSIONS

The operation of the CERN ISR with protons and antiprotons has allowed the study of the expected similarity of the dominant features of particle and antiparticle in a new higher energy range. The observed convergence of the properties of p-p and p̄-p scattering reinforces the feeling that the first signs of an asymptotic regime have already appeared at ISR energies.

The CERN SPS Collider has made accessible to experimentation an energy domain that is still one order of magnitude higher, making a long leap toward the land where the general principles prevail. In order to investigate the asymptotic features of hadron scattering, it is sensible to compare ISR p-p data directly with Collider p̄-p data, insofar as the two reactions should at such energies exhibit practically the same behavior.

In spite of the $\sim\ln^2 s$ growth of the total cross section, the predicted asymptotic property of geometrical scaling is not observed in Collider p̄-p data: the geometrical scaling behavior found at the ISR should therefore be considered a merely transient feature. The apparent contradiction with the Auberson-Kinoshita-Martin theorem may lead one to question whether the $\sigma_{tot}$ rise at this rate will really persist.

New light will be shed by the measurement at the SPS Collider of the real part $\rho$, which is sensitive to the derivative of $\sigma_{tot}$. This will indicate the trend of the total cross section beyond presently available energies, waiting for direct measurements at the forthcoming colliders. Should the $\sim\ln^2 s$ rise persist, the picture presented by the eikonal models could really be true: a steadily increasing opaqueness would lead all hadrons to reach the black disc limit at infinite energy. The Froissart bound would then tend to be quantitatively saturated very slowly with increasing energy. In this case the $\sim\ln^2 s$ dependence of $\sigma_{tot}$ observed already at energies as low as those at the ISR should be interpreted as a precocious qualitative saturation of the Froissart bound, and should persist more and more quantitatively at higher and higher energy. The hypothesis of the Auberson-Kinoshita-Martin theorem would actually be true asymptotically, and geometrical scaling should indeed be regarded as only an asymptotic property.

Measurements at future big hadron colliders will be exciting and may eventually settle these tantalizing questions, making possible another step toward comprehending the underlying dynamics.

## ACKNOWLEDGMENTS

It is a pleasure to thank all our colleagues of the UA4 Collaboration, and Professors G. Bellettini and A. Martin for useful discussions.

*Literature Cited*

1. Amaldi, U., et al. *Phys. Lett.* 44B: 112–18 (1973)
2. Amendolia, S. R., et al. *Phys. Lett.* 44B: 119–24 (1973); *Nuovo Cimento* 17A: 735–55 (1973)
3. Möhl, D., et al. *Phys. Rep.* 58: 73–119 (1980) and references therein
4a. Giffon, M., Predazzi, E. *Riv. Nuovo Cimento* 7(5): 1–62 (1984)
4b. Kamran, M. *Phys. Rep.* 108: 275–399 (1984)
4c. Alberi, G., Goggi, G. *Phys. Rep.* 74: 1–207 (1981)
4d. Fischer, J. *Phys. Rep.* 76: 157–214 (1981)
4e. Giacomelli, G., Jacob, M. *Phys. Rep.* 55: 1–132 (1979)
4f. Predazzi, E. *Riv. Nuovo Cimento* 2(11): 1–43 (1979); 6: 217–93 (1976)
4g. Amaldi, U., Jacob, M., Matthiae, G. *Ann. Rev. Nucl. Sci.* 26: 385–456 (1976)
4h. Giacomelli, G. *Phys. Rep.* 23: 123–235 (1976)
4i. Roy, S. M. *Phys. Rep.* 5C: 125–96 (1972)
4j. Block, M. M., Cahn, R. N. *Preprint LBL-17522* (1984); *Rev. Mod. Phys.* Submitted (1985)
5. Schubert, K. R. Tables on pp elastic scattering. In *Landolt-Börnstein, New Series*. Vol. I/9a, pp. 216–86. Berlin: Springer-Verlag (1979), and references therein. See also Ref. 42
6. West, G. B., Yennie, D. R. *Phys. Rev.* 172: 1413–22 (1968); Locher, M. P. *Nucl. Phys. B* 2: 525–31 (1967); Cahn, R. N. *Z. Phys. C* 15: 253 (1982)
7. Amaldi, U., et al. *Phys. Lett.* 66B: 390–94 (1977)
8. Bartenev, V., et al. *Phys. Rev. Lett.* 31: 1367–70 (1973)
9. Amos, N., et al. *Phys. Lett.* 120B: 460–64 (1983); 128B: 343–48 (1983)
10. Bronzan, J. B., Kane, G. L., Sukhatme, U. P. *Phys. Lett.* 49B: 272–76 (1974)
11. Sidhu, D. R., Sukhatme, U. P. *Phys. Rev. D* 11: 1351–53 (1975)
12. Khuri, N. N., Kinoshita, T. *Phys. Rev.* 137: B720–29 (1965)

13a. Breakstone, A., et al. *Nucl. Phys. B* 248: 253–60 (1984)
13b. Breakstone, A., et al. *Phys. Rev. Lett.* 54: 2180–83 (1985)
13c. Erhan, S., et al. *Phys. Lett.* 152B: 131–34 (1985)
14a. Burq, J. P., et al. *Phys. Lett.* 109B: 124–28 (1982)
14b. Block, M. M., Cahn, R. N. *Phys. Lett.* 120B: 229–32 (1983)
14c. Gauron, P., Nicolescu, B. *Phys. Lett.* 143B: 253–58 (1984)
15. Schiz, A., et al. *Phys. Rev. D* 24: 26–45 (1981); Ayres, D. S., et al. *Phys. Rev. D* 15: 3105–38 (1977); Akerlof, C. W., et al. *Phys. Rev. D* 14: 2864–77 (1976)
16. Bozzo, M., et al. *Phys. Lett.* 147B: 385–91 (1984)
17. Battiston, R., et al. *Phys. Lett.* 127B: 472–75 (1983); 115B: 333–37 (1982)
18. Arnison, G., et al. *Phys. Lett.* 128B: 336–42 (1983)
19. Bozzo, M., et al. *Int. Conf. on High Energy Physics, Brighton, 1983*, Pap. 116; *Phys. Lett. B* 155B: 197–202 (1985)
20. Matthiae, G. *15th Symp. on Multiparticle Dynamics, Lund, 1984*
21. Nagy, E., et al. *Nucl. Phys. B* 150: 221–67 (1979)
22. Asa'd, Z., et al. *Phys. Lett.* 108B: 51–54 (1982); 128B: 124–28 (1983); 130B: 335–39 (1983); Preprint CERN-EP/84-144 (1984); *Nucl. Phys. B*. Submitted (1985)
23. Rubinstein, R., et al. *Phys. Rev. D* 30: 1413–31 (1984)
24. Donnachie, A., Landshoff, P. *Nucl. Phys. B* 231: 189–204 (1984); *Phys. Lett.* 123B: 345–48 (1983)
25a. Fukugita, M., Kwiecinsky, J. *Phys. Lett.* 83B: 119–22 (1979)
25b. Martin, A. *CERN-TH 4082/84* (1984)
26. Islam, M. M., Fearnley, T., Guillaud, J. P. *Nuovo Cimento* 81A: 737 (1984)
27. Dias de Deus, J., Kroll, P. *J. Phys. G* 9: L81–84 (1983)
28a. Cheng, H., Walker, J. K., Wu, T. T. *Phys. Lett.* 44B: 97–101 (1973)
28b. Bourrely, C., Soffer, J., Wu, T. T. *Phys.*

*Rev. D* 19 : 3249–60 (1979): *Nucl. Phys. B* 247 : 15–28 (1984); Marseille Preprint *CPT 84/PE.1674* (1984)

28c. Chiu, C. *Phys. Lett.* 142B : 309–14 (1984)

28d. Glauber, R. J., Velasco, J. *Phys. Lett.* 147B : 380–84 (1984)

29. van der Meer, S. CERN Intern. Rep. ISR-PO/68-31 (1968)

30. Rubbia, C. CERN p̄p NOTE 38 (1977)

31. Amendolia, S. R., et al. *Proc. Int. Conf. on Instrumentation for High Energy Physics, Frascati, 1973*, pp. 397–401. Frascati : Lab. Nazionali Comitato Nazionale Energia Nucl. (1973)

32. Bosser, J., et al. *Nucl. Instrum. Methods A* 235 : 475–80 (1985)

33. Amaldi, U., et al. *Phys. Lett.* 43B : 231–36 (1973)

34. Amaldi, U., et al. *Phys. Lett.* 62B : 460–66 (1976); *Nucl. Phys. B* 145 : 367–401 (1978)

35. Battiston, R., et al. *Phys. Lett.* 117B : 126–30 (1982)

36. Bozzo, M., et al. *Phys. Lett.* 147B : 392–98 (1984)

37. Yodh, G. B., Pal, T., Trefil, S. J. *Phys. Rev. Lett.* 28 : 1005–8 (1972)

38. Heisenberg, W. *Kosmische Strahlung*, pp. 148–64. Berlin : Springer-Verlag (1953)

39. Cheng, H., Wu, T. T. *Phys. Rev. Lett.* 24 : 1456–60 (1970)

40a. Ambrosio, M., et al. *Phys. Lett.* 115B : 495–502 (1982)

40b. Carboni, G., et al. Preprint CERN-EP/84-163 (1984); *Nucl. Phys. B.* Submitted (1985)

41. Lukaszuk, L., Nicolescu, B. *Lett. Nuovo Cimento* 8 : 405–13 (1973); Kang, K., Nicolescu, B. *Phys. Rev. D* 11 : 2461–65 (1975); Gauron, P., Nicolescu, B. *Phys. Lett.* 124B : 429–34 (1983)

42. Amaldi, U., Schubert, K. R. *Nucl. Phys.* B166 : 301–20 (1980)

43. Martin, A. *Proc. 4th Topical Workshop on Proton-Antiproton Physics, Berne, 1984.* CERN 84-09, pp. 308–13. Geneva : CERN (1984)

44. Baumel, J., Feingold, M., Moshe, M. *Nucl. Phys. B* 198 : 13–25 (1982); White, A. R., Fermilab Preprint CONF 82/16-THY (1982); Baig, M., Bartels, J., Dash, J. W. *Nucl. Phys. B* 237 : 502–24 (1984)

45. Ter-Martirosyan, K. A. *22nd Int. Conf. on High Energy Physics, Leipzig, 1984*, pap. 922 and references therein

46. Van Hove, L. *Nuovo Cimento* 28 : 798–817 (1963); *Rev. Mod. Phys.* 36 : 655–65 (1964)

47. Grein, W., Guigas, R., Kroll, P. *Nucl. Phys. B* 89 : 93–108 (1975); Kroll, P. *Fortschr. Phys.* 24 : 565 (1976)

48. Amaldi, U. *Proc. 2nd Int. Conf. on Elementary Particles Aix-en-Provence,* 1973, *J. Phys. (France) Suppl.* 10(C1): 241–59 (1973)

49. de Groot, E. H., Miettinen, H. I. *Proc. 8th Rencontre de Moriond, Méribel-les-Allues, 1973*, pp. 193–234. Orsay : Lab. Phys. Théor. Part. Elément. (1973)

50. Henyey, F. S., Hong Tuan, R., Kane, G. L. *Nucl. Phys. B* 70 : 445–60 (1974)

51. Henzi, R., Valin, P. *Phys. Lett.* 48B : 119–24 (1974)

52. Miettinen, H. I. *Proc. 9th Rencontre de Moriond, Méribel-les-Allues, 1974*, pp. 363–402. Orsay : Lab. Phys. Théor. Part. Elément. (1974)

53. Henzi, R., Valin, P. *Nucl. Phys. B* 148 : 513–37 (1979)

54. Henzi, R., Valin, P. *Phys. Lett.* 132B : 443–48 (1983); 149B : 239–44 (1984); *22nd Int. Conf. on High Energy Physics, Leipzig, 1984*; Henzi, R. *Proc. 4th Topical Workshop on Proton-Antiproton Physics, Berne, 1984*, pp. 314–21 (1984); Valin P. *Z. Phys. C* 25 : 259–67 (1984)

55. Pomeranchuk, I. Ia. *Zh. Eksp. Teor. Fiz.* 34 : 725–28 (1958); *Sov. Phys. JETP* 7 : 499–501 (1958)

56. Froissart, M. *Phys. Rev.* 123 : 1053–57 (1961)

57. Martin, A. *Phys. Rev.* 129 : 1432–36 (1963); *Nuovo Cimento* 42 : 930–54 (1966)

58a. Lukaszuk, L., Martin, A. *Nuovo Cimento* 52A : 122–45 (1967)

58b. Martin, A. In *A Discussion on pp Scattering at Very High Energies, London, 1973. Proc. R. Soc. London Ser. A* 533 : 503–7 (1973)

59. Kupsch, J. *Nuovo Cimento* 71A : 85–103 (1982)

60a. Eden, R. J. *Phys. Rev. Lett.* 16 : 39–41 (1966)

60b. Kinoshita, T. *Phys. Rev. D* 2 : 2346–48 (1970); in *Perspectives in Modern Physics*, ed. R. E. Marshak, pp. 211–13. New York : Wiley (1966)

60c. Volkov, G. G., et al. *Teor. Mat. Fiz.* 4 : 196–201 (1970); Truong, T. N., Lam, W. S. *Phys. Rev. D* 6 : 2875–83 (1972); Grunberg, G., Truong, T. N. *Phys. Rev. Lett.* 31 : 63–66 (1973); *Phys. Rev. D* 9 : 2874–93 (1974)

61. Martin, A. *Z. Phys. C* 15 : 185–91 (1982)

62. Joynson, E., et al. *Nuovo Cimento* 30A : 345–84 (1975)

63. Block, M. M., Cahn, R. N. *Phys. Lett.* 120B : 224–28 (1983)

64. Fischer, J., Saly, R., Vrkoc, I. *Phys. Rev. D* 18 : 4271–81 (1978)

65. Cornille, H., Martin, A. *Phys. Lett.* 40B : 671–74 (1972); *Nucl. Phys. B* 48 : 104–16 (1972); 49 : 413–40 (1972); 77 : 141–62 (1974)

66. MacDowell, S. W., Martin, A. *Phys. Rev.* 135: B960–62 (1964)
67. Martin, A. *Proc. 3rd Topical Workshop on Proton-Antiproton Physics, Rome, 1983.* CERN 83-04, pp. 351–71. Geneva: CERN (1983)
68. Auberson, G., Kinoshita, T., Martin, A. *Phys. Rev. D* 3: 3185–94 (1971)
69. Martin, A. *Lett. Nuovo Cimento* 7: 811–12 (1973)
70. Dias de Deus, J., Kroll, P. *Acta Phys. Pol. B* 9: 157–65 (1978)
71. Wu, T. T., Yang, C. N. *Phys. Rev.* 137: B708–16 (1965); Byers, N., Yang, C. N. *Phys. Rev.* 142: 976–81 (1966); Chou, T. T., Yang, C. N. *Proc. 2nd Int. Conf. on High Energy Physics and Nuclear Structure, Rehovoth, 1967,* ed. G. Alexander, pp. 348–59. Amsterdam: North Holland (1967); Chou, T. T., Yang, C. N. *Phys. Rev.* 170: 1591–96 (1968); Durand, L., Lipes, R. *Phys. Rev. Lett.* 20: 637–40 (1968)
72. Hayot, F., Sukhatme, U. P. *Phys. Rev. D* 10: 2183–85 (1974); Chou, T. T., Yang, C. N. *Phys. Rev. D* 19: 3268–73 (1979)

*Ann. Rev. Nucl. Part. Sci. 1985. 35 : 397–454*

# HIGH-ENERGY PHOTOPRODUCTION OF CHARMED STATES

*Stephen D. Holmes*

Fermi National Accelerator Laboratory, Batavia, Illinois 60510

*Wonyong Lee*

Department of Physics and Nevis Laboratory, Columbia University, New York, New York 10027

*James E. Wiss*

Department of Physics, University of Illinois, Urbana, Illinois 61801

CONTENTS

397

0163–8998/85/1201–0397$02.00

## 1. INTRODUCTION

Following the first observation of the $J/\psi(3100)$ in pp interactions at BNL and in electron-positron collisions at SPEAR (13, 17), speculation centered on whether this new particle might be identified as the lowest mass vector meson associated with a new species of heavy quark. The subsequent observation of the $\psi'(3700)$ (5) lent credence to this interpretation and led to a tentative assignment of the $\psi$ as the lowest state and $\psi'$ as the first radially excited state of a $^3S_1$ heavy quarkonium system. If the vector meson interpretation for the $\psi$ family were indeed correct, it was expected that these particles would be photoproduced at levels comparable to previously known vector mesons. This expectation was confirmed in a relatively short time (28, 43, 44, 51, 53, 63) by experiments showing that the $\psi$ interacted with nucleons at a level characteristic of a hadron. Subsequent measurements at SPEAR (26, 58) soon established $JPC$ quantum numbers of the $\psi$ and $\psi'$ as $1^{--}$. Shortly thereafter states exhibiting the charm quantum number explicitly were identified in $e^+e^-$, $v$, and $\gamma$-nucleon interactions (29, 46, 52, 62).

The predominant role played by $e^+e^-$ annihilation experiments, where a time-like photon materializes as a charmed quark-antiquark pair, and by real photoproduction experiments in the development of our current understanding of charmed particle properties and production dynamics is no accident. Since photons couple to charm quarks through the electric charge, charm production rates are proportional to $q_c^2/E_{cm}^2$ in $e^+e^-$ interactions and $q_c^2/m_c^2$ in $\gamma$-N interactions. Thus, well above the charm production threshold approximately 40% of the $e^+e^-$ and 1% of the photoproduction total cross sections contain charmed particles in the final state. In contrast, the relative level of charm production in hadronic interactions is an order of magnitude lower. While background levels to charm production are much lower in $e^+e^-$ interactions, the use of real photons to produce either an open or a closed charmed state affords several advantages to the experimenter. Absolute production rates are orders of magnitude higher in photon beams, and one gains the ability to observe simultaneously all portions of the charmed quark invariant mass spectrum.

Photoproduction experiments studying the formation of bound and open charm have been carried out over the last ten years using both real and virtual (space-like) photon beams. Figure 1 (*left*) illustrates the kinematic variables used to describe the real photoproduction of charmed quark-antiquark pairs. The variable $E_\gamma$ is the energy of the photon as measured in the rest frame of the target nucleon. The total energy in the $\gamma$-N center of mass is called $\sqrt{s}$. The variable $t$, called the momentum transfer, is most relevant for events in which the target nucleon elastically scatters without breaking up into fragments. In the high-energy limit, $t$ approaches $-P_\perp^2$, where $P_\perp$ is the momentum component of the scattered nucleon transverse to the direction of the incident photon.

Figure 1 (*right*) illustrates several variables used in the description of the virtual photoproduction of a $c\bar{c}$ pair. Virtual photoproduction refers to a certain kinematic limit of charm production in lepton-nucleon scattering. In this process, one incident lepton (usually a muon) scatters to produce a virtual photon of 4-momentum $q$. The negative squared mass of the virtual photon is $Q^2 = -q \cdot q$, and its lab energy (or energy in the rest frame of the target nucleon) is called $\nu$. Both $Q^2$ and $\nu$ can be determined by measuring the momentum of the incident and scattered muon. The virtual photoproduction limit, which is described further in Section 3, is low $Q^2$ at high $\nu$.

The early measurements of charm photoproduction were usually interpreted using the vector meson dominance model (VDM). Use of the VDM allowed extraction of the $\psi$-N total cross section from the $\psi$ photoproduction measurements and provided a qualitative understanding

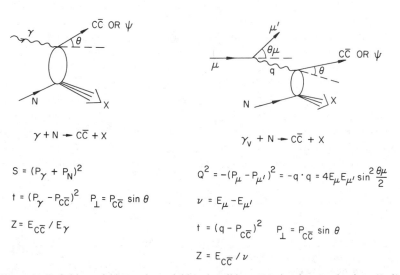

$$\gamma + N \rightarrow c\bar{c} + X$$

$$S = (P_\gamma + P_N)^2$$

$$t = (P_\gamma - P_{c\bar{c}})^2 \qquad P_\perp = P_{c\bar{c}} \sin\theta$$

$$Z = E_{c\bar{c}} / E_\gamma$$

$$\gamma_v + N \rightarrow c\bar{c} + X$$

$$Q^2 = -(P_\mu - P_{\mu'})^2 = -q \cdot q = 4E_\mu E_{\mu'} \sin^2\frac{\theta_\mu}{2}$$

$$\nu = E_\mu - E_{\mu'}$$

$$t = (q - P_{c\bar{c}})^2 \qquad P_\perp = P_{c\bar{c}} \sin\theta$$

$$Z = E_{c\bar{c}} / \nu$$

*Figure 1* Definition of kinematic variables describing $c\bar{c}$ pair photoproduction. (*Left*) Variables used in real photoproduction. $P_\gamma$ and $P_{c\bar{c}}$ are the 4-vectors of the incident photon and the produced $c\bar{c}$ pair. The momentum transfer from the photon to the $c\bar{c}$ pair is called $t$, while $Z$ is the elasticity or fraction of the incident energy transferred to the $c\bar{c}$ pair. (*Right*) Variables used in virtual photoproduction. The 4-momentum of the virtual photon, $q$, equals the difference between the 4-momenta of the incident muon ($P_\mu$) and scattered muon ($P_{\mu'}$). The lab energy of the virtual photon is called $\nu$. The elasticity and momentum transfer are defined analogously to the real photoproduction case.

of the diffractive nature of the production. VDM attributes both open and closed charm production to direct coupling between the photon and vector mesons containing $c\bar{c}$ pairs ($\psi, \psi', \psi'' \ldots$) followed by subsequent interaction of vector mesons with the nucleon (Figure 2).

Because the VDM process is essentially $\psi$ diffractive dissociation, one expects the charm cross section to become energy independent at sufficiently large $s$. It is not clear beyond what $s$ this limiting behavior will occur, but the frequently applied criteria in hadronic diffractive dissociation is $M_{th}^2/s < 0.1$ where $M_{th} \approx 2M_D \approx 4\,\text{GeV}$. This argument suggests that the limiting behavior might not develop until $E_\gamma \approx 100$ GeV. VDM makes a definite prediction about the $Q^2$ dependence of virtual photoproduction of charm. In the VDM model all virtual photoproduction $Q^2$ dependence should come from the vector meson propagator and thus be of the form $\sigma \propto (1 + Q^2/\Lambda^2)^{-2}$ with $\Lambda = M_\psi$ for all $v$. This $Q^2$ dependence is predicted for both open and bound charm photoproduction.

More recently the efforts of experimenters have focused on precise measurements of the energy dependence, elasticity, and, through the use of "virtual photon beams," the $Q^2$ dependence of both bound and open charm production. These data are now interpreted through a perturbative model based on quantum chromodynamics (QCD); the model is known as photon-gluon fusion (PGF), and the data are used to extract information on the gluon distribution function within the nucleon in addition to providing tests of the predictive power of QCD. PGF calculates charm production in terms of simple diagrams (Figure 3) in which the photon fuses with a gluon to form a $c\bar{c}$ pair. Inputs to this model include a value for the charmed quark mass and a parameterization for the gluon momentum distribution functions. In PGF, the energy dependence of charm photoproduction is directly related to the momentum distribution of gluons within the nucleon. PGF makes semi-quantitative predictions about the

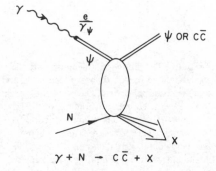

*Figure 2* The vector dominance model (VDM) for photoproduced charm. In this model, a photon couples to a $\psi$ with a coupling constant of $e/\gamma_\psi$. The $\psi$ propagates with a propagator factor of $(1 + Q^2/M_\psi^2)$ before interacting with the target nucleon.

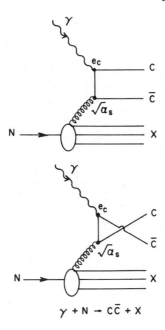

*Figure 3*  The two lowest-order diagrams for charmed quark production in the photon-gluon fusion (PGF) model. The gluon (assumed to be massless) carries a fraction $X$ of the target nucleon 4-vector. Because the gluon transforms as a color octet, the created charm-anticharm quark pair presumably must radiate soft gluons (not shown) before hadronizing into color singlet final states of either bound or open charm. (*Top*) This diagram has a $t$-channel pole where $t$ is the momentum transfer from the photon to the charmed quark. (*Bottom*) This diagram has a $u$-channel pole where $u$ is the momentum transfer from the photon to the anti-charmed quark.

ratio of inelastic to elastic charm production and fairly definite predictions about the $Q^2$ dependence. This model has difficulty predicting the absolute level of bound and open charm production. As we shall see, the present data do not clearly discriminate between PGF and VDM.

In this paper we review the photoproduction of both bound and open charm, by both real and virtual photon beams, and compare the experimental data with the current theoretical models. We have chosen to include only those data appearing published in either journals or conference proceedings prior to September 1984.

## 2.  MODELS OF CHARM PHOTOPRODUCTION

Charm photoproduction results are most often interpreted within the context of either the vector dominance model (VDM) or the photon-gluon fusion model (PGF).

VDM is a phenomenology that allows one to relate measurements of $e^+e^- \to \psi$, $\psi$ elastic photoproduction, and $\psi$-N scattering. Within the framework of the VDM (22, 40) the photoproduction of vector mesons proceeds through the mechanism of the photon coupling directly to the vector meson, V, which subsequently scatters off the nucleon.

In the "diagonal" approximation it is assumed that terms of the form $V'N \rightarrow V+N$ do not contribute and so the vector meson photoproduction cross section can be directly related to the vector meson–nucleon scattering cross section once the coupling to the photon is known. Fortunately VDM provides a means of independently measuring the coupling using $e^+e^- \rightarrow V$. The $e^+e^- \rightarrow \psi$ production cross section is given by

$$\int \sigma(e^+e^- \rightarrow \psi)\,dE = \frac{2\pi(2J+1)}{M_\psi^2}\,\Gamma_{ee}, \qquad\qquad 1.$$

where the left-hand side is the measured integrated resonant cross section. Knowledge of the spin, $J$, of the $\psi$ allows calculation of the dielectron partial width. This way the width, $\Gamma_{ee}$, is measured to be $4.6 \pm 0.8$ keV (61). The width is related to the $\psi$-photon coupling through

$$\Gamma_{ee} = \frac{\alpha^2}{3}\left(\frac{4\pi}{\gamma_\psi^2}\right) M_\psi. \qquad\qquad 2.$$

Solving for the coupling yields

$$\frac{4\pi}{\gamma_\psi^2} = 0.085 \pm 0.014. \qquad\qquad 3.$$

With this knowledge, the $\psi$-N elastic scattering cross section can be deduced from the $\psi$ photoproduction cross section through the relation,

$$\sigma(\gamma N \rightarrow \psi N) = \left(\frac{4\pi\alpha}{\gamma_\psi^2}\right) \sigma(\psi N \rightarrow \psi N). \qquad\qquad 4.$$

Additionally, one can use the optical theorem,

$$\left.\frac{d\sigma}{dt}\right|_{t=0} (\gamma N \rightarrow \psi N) = \left(\frac{4\pi\alpha}{\gamma_\psi^2}\right)\frac{\sigma_T^2(\psi N)}{16\pi} \qquad\qquad 5.$$

to calculate the total $\psi$-N cross section from the measured value of the forward $\psi$ photoproduction cross section under the assumption that the scattering amplitude at $t=0$ is purely imaginary.

Within the framework of the description given in Figure 2, VDM also makes the very definite prediction that the $Q^2$ dependence of $\psi$ production by virtual photons should be given by the $\psi$ propagator,

$$\frac{\sigma(Q^2)}{\sigma(0)} = \left(1 + \frac{Q^2}{M_\psi^2}\right)^{-2}.$$

In the photon-gluon fusion model, charm production proceeds via the fusion of the photon with a gluon from one target nucleon to form a $c\bar{c}$ pair (20, 21, 35, 42, 45, 48, 50, 54–56, 69).

The fundamental photon-gluon cross section (for transverse photons and vector gluons) is given by

$$\sigma_{\gamma g \to c\bar{c}} = \frac{2\pi\alpha e_c^2 \alpha_s}{(\hat{s}+Q^2)^3} \left\{ [\hat{s}^2 + Q^4 + 4m_c^2(\hat{s} - 2m_c^2)] \ln\left(\frac{1+\beta}{1-\beta}\right) \right.$$

$$\left. - [(\hat{s} - Q^2)^2 + 4\hat{s}m_c^2]\beta \right\}, \qquad\qquad 6.$$

where $\alpha_s$ is the strong coupling constant at the mass scale of the charmed quark, $\sqrt{\hat{s}}$ is total energy in the photon-gluon center of mass, $e_c$ is the electric charge of the charmed quark, $m_c$ is the charmed quark mass, $\beta = (1 - 4m_c^2/\hat{s})^{1/2}$, and $Q^2$ is the virtual photon mass.

This elementary cross section can be convoluted with the gluon momentum fraction distribution of the nucleon $[G(x)]$ to predict the energy dependence of the charm photoproduction cross section. Of course it is necessary to partition this cross section between closed and open charm photoproduction. In doing so one must make some sort of assumption about how the $c\bar{c}$ pair produced in the final state dresses to form either charmed particles or charmonia states. This is done in a slightly ad hoc way since, as shown in Figure 3, the final-state $c\bar{c}$ pair do not form a color singlet and so there are necessarily final-state interactions that are not represented by the diagram. In the most common approach, such final-state effects are ignored as not contributing to the dynamics of the $c\bar{c}$ formation and one simply restricts the contributing region of phase space to that in which the mass of the $c\bar{c}$ pair is appropriately situated with respect to the mass of the lowest-lying open charm state, the D(1865). With this assumption, the $Q^2 \to 0$ limit of Equation 6 can be used to predict the level of the bound and open charm real photoproduction cross section:

$$\sigma_{\gamma N \to \psi x} = f \int_{4m_c^2}^{4M\hat{D}} \frac{d\hat{s}}{s} G\left(x = \frac{\hat{s}}{s}\right) \hat{\sigma}_{\gamma G \to c\bar{c}}(\hat{s}) \qquad\qquad 7.$$

$$\sigma_{\substack{\gamma N \to \text{open} \\ \text{charm}}} = \int_{4M_D^2}^{S} \frac{d\hat{s}}{s} G\left(x = \frac{\hat{s}}{s}\right) \hat{\sigma}_{\gamma G \to c\bar{c}}(\hat{s}). \qquad\qquad 8.$$

The factor $f$ included in Equation 7 gives the fraction of $c\bar{c}$ pairs produced below open charm threshold that will actually materialize as $\psi$'s rather than other bound charm states. The ansatz of "semi-local duality" is used to assert that the $c\bar{c}$ divide equally among the available bound states in this region, which leads to the prediction $f = 1/7$. As discussed below, this assumption cannot be exactly true since it fails to account for the factor-of-five difference between $\psi$ and $\psi'$ production.

The lowest-order PGF diagram makes no prediction about the $t$ distribution of elastic $\psi$ production since $t$ is equal to the mass squared of the gluon (assumed zero). Within a broader context the $P_\perp^2$ distribution of the $\psi$ can be related to the transverse monentum distribution of the gluon within the nucleon. If the gluons have enough intrinsic transverse momentum to impart a $P_\perp^2$ greater than 1 GeV to the $\psi$, then the PGF model also predicts a rather strong dependence of the production on $\phi_p$, the azimuthal angle between the muon scattering plane and the virtual photon-$\psi$ production plane (57). An asymmetry is predicted in the differential cross section,

$$\frac{d\sigma}{d\phi_p} \propto 1 + a_1 \cos \phi_p + a_2 \cos 2\phi_p. \qquad 9.$$

Values of $a_1 = -0.3$ are expected in the region $Q^2 > 1$ GeV$^2$, $P_\perp^2 > 1$ GeV. The term $a_1$ arises from an interference between transverse and longitudinal amplitudes and hence must go to zero as $Q^2 \to 0$. The $P_\perp$ cut arises from the fact that the angle $\phi_p$ is singular at $P_\perp^2 = 0$. PGF predicts a small (0.05) value of $a_2$.

More information can be obtained on $t$ and inelasticity distributions by considering higher-order diagrams (23, 36, 37, 68). We do not discuss these here but simply comment that such calculations typically predict ratios of inelastic to elastic production that are of order $\alpha_s$, i.e. in the range 0.2–0.4.

An interesting prediction of the PGF model applied to open charm photoproduction concerns the angular distribution of the $c\bar{c}$ quarks in the photon-gluon rest frame. If present ideas about hadronization are correct, the charmed particles produced as the c or $\bar{c}$ quark hadronize should appear in a direction very close to the quark direction. Hence the quark angular direction is in principle measurable in events with sufficient $\hat{s}$ to overcome any $P_\perp$ broadening effects that might appear in the hadronization process.

The angular distribution in the photon-gluon rest frame can be parameterized in terms of a variable $r = \frac{1}{2}(1 - \beta \cos \theta^*)$, where $\beta$ is the charmed quark velocity. In terms of this variable the elementary differential cross section is predicted to be

$$\frac{d\hat{\sigma}}{dr} = \frac{\pi\alpha\alpha_s}{\hat{s}} e_c^2 \left\{ -2 + \frac{1+\gamma}{r(1-r)} - \frac{\gamma^2}{4}\left[\frac{1}{r(1-r)}\right]^2 \right\}, \qquad 10.$$

where $\gamma = 4M_c^2/\hat{s}$ and $e_c$ is the quark charge. The singularities in $r$ and $1-r$ are t-channel and u-channel poles due to the presence of the charmed quark propagators in Figure 3. The effect of these poles is to create a highly forward-backward peaked quark angular distribution at large $\hat{s}$ as $\beta \to 1$. The prediction that the charm particle angular distribution in the charm-anticharm rest frame should become increasingly nonisotropic as $\hat{s}$

increases is in contrast to the VDM prediction that there will be at most a $\cos^2 \theta^*$ dependence because of the restricted spin of the initial state. Experimental evidence for this forward-backward peaking is presented in Section 5.5.

# 3. EXPERIMENTAL METHODS

Prior to the construction of the 400-GeV proton accelerators at Fermilab and CERN, both real and virtual photoproduction experiments were generally performed at electron accelerators. With the construction of high-energy proton accelerators it became possible to obtain photons with lab energies of several hundred GeV as opposed to the tens of GeV available at the previously constructed electron machines. All of the real photoproduction results reviewed here employed detectors appropriate to the so-called open geometry configuration in which final-state particles emerge from a thin target to be measured in a magnetic spectrometer, or in electromagnetic shower detectors. By contrast the virtual photoproduction results are obtained in $\mu$-N scattering experiments performed in the closed geometry configuration, where the interaction takes place in dense material that absorbs or ranges out all particles except final-state muons prior to momentum measurement. In this short section we give a brief overview of the beams and detectors used in the experiments reviewed in this article, and we discuss the virtual photoproduction limit of lepton-nucleon scattering.

## 3.1 Real Photoproduction

The first high-energy photoproduction experiments at electron accelerators employed bremsstrahlung beams. In the bremsstrahlung process, an incident electron scatters off the electromagnetic field of a target nucleus to produce a scattered electron and a final-state photon with an energy spectrum that roughly approximates $1/\omega$, where $\omega$ is the photon energy (except close to the endpoint). The photon and the scattered electron emerge in a direction nearly collinear with the direction of the incident electron and between them share nearly all of the incident beam energy since very little momentum is transferred to the target nucleus. As an example of a bremsstrahlung beam, we consider the beam used by Gittelman et al (43, 44) to study the photoproduction of $\psi$'s near threshold. An 11.8-GeV electron beam from the Wilson Electron Synchrotron at Cornell University impinges on a thin radiator. Immediately downstream of this radiator lies a sweeping magnet, which steers the scattered and uninteracted primary electrons into a lead dump and thereby allows the photons to pass through a central hole in the dump. The resulting photon beam is further defined through a set of upstream collimators before impinging on the experimenter's beryllium target. The photoproduced final state is analyzed by two lead glass hodoscopes that reconstruct the $\psi$ through its $e^+e^-$ decay mode.

One frequently chooses a beryllium target in real photoproduction experiments because beryllium is a dense material at room temperature with a large ratio of radiation length to interaction length (because of the low $Z$ of the beryllium nucleus). Having a low interaction length relative to the radiation length reduces the extent to which uninteresting electromagnetic processes such as Bethe-Heitler $e^+e^-$ pair production dominate over the hadronic processes, such as charm production, that these experiments are designed to study. The use of a low-$Z$ target thus allows the experiments to run with a higher incident photon flux before saturating their data acquisition or detector capabilities.

The photon beam used in the SLAC Hybrid Photon Facility Collaboration for their study of charm photoproduction near threshold (3) is an alternative to the conventional electron accelerator bremsstrahlung beam. In this beam, a neodymium-doped yttrium aluminum garnet laser produces a 4.68-eV photon beam that is directed against the SLAC 30-GeV electron beam at a crossing angle of 2 mrad. The photons that elastically backscatter from the electrons via the Compton effect appear in the laboratory with a maximum energy comparable to the incident electron energy, provided that the geometrical mean of the laser energy and incident electron energy becomes comparable to the electron mass. The energy spread of the backscattered photon beam is essentially dictated by the angular acceptance of the beam line and experimenter's target. After collimation, the photon beam impinges onto a bubble chamber located 170 meters downstream of the initial collision point with a 3-mm circular spot size. The advantage of the backscattered laser beam over the more conventional bremsstrahlung beam is that the photon spectrum can be made nearly monochromatic in a backscattered beam. The disadvantage of the backscattered laser beam is a relatively low flux. The SLAC Hybrid Photon Facility Collaboration ran with 20-GeV photons at a flux of 20 to 30 photons per accelerator pulse. This relatively low photon yield was well suited to the rate limitations implicit in the use of a bubble chamber.

The hydrogen-filled bubble chamber used in this experiment was run under special conditions that produced a small bubble size and high bubble density. These conditions provided sufficient resolution to observe charm particles by seeing the secondary vertices produced when charm particles decay with their short lifetime. After leaving the bubble chamber, charged particles passed through a magnetic spectrometer (in order to refine the momentum measurement made in the bubble chamber) and two multicell

Čerenkov counters with pion thresholds of 3.2 and 6.0 GeV. Čerenkov counters, which can discriminate between pions, protons, and kaons over large momentum ranges, are important components of open geometry charm photoproduction experiments because charm particles tend to decay into Cabibbo-favored final states that contain strange particles. Because charged kaons and protons are relatively rare in ordinary photon interactions, Čerenkov information tends to suppress combinatorial background. A highly segmented lead glass shower detector lying downstream of the Čerenkov counters completes the highly instrumented spectrometer of this collaboration by providing reconstruction and measurement of final-state photons and identification of final-state electrons.

We turn next to a discussion of the beams and detectors in experiments performed at proton accelerators. The ultimate source of photons produced at high-energy proton accelerators is the electromagnetic decay of high-energy secondaries that are produced when the pimary accelerator proton beam interacts with a target. The yield of photons is dominated by the two-photon decay of the $\pi^0$. The charged secondaries that accompany these photons can be essentially eliminated by having the beam pass through a series of collimated sweeping magnets. The resulting neutral beam consists primarily of long-lived neutral hadrons (i.e. $K_L^0$ and neutrons) and photons. Because the non-QED interaction rate for photons is less than 1% of the interaction rate for neutral hadrons, a dramatic reduction of the neutral hadron component is necessary before one can perform a reasonable photoproduction experiment with a neutral beam. Two methods have been successfully employed in order to reduce significantly the neutral hadron contamination in proton accelerator photon beams.

The Fermilab wideband photon beam used in experiments E87 and E401 illustrates one of these methods (51). The zero-degree neutral beam produced by 400-GeV primary protons impinging on a 30.5 cm long beryllium target, and purified of charged secondaries by magnets and collimators, passes through a liquid deuterium filter 34 meters long. The deuterium filter increases the photon-to-neutral-hadron yield by roughly a factor of 200 by virtue of its small ratio of interaction length to radiation length. The particles produced by neutral hadron interactions in the deuterium are removed by sweeping magnets and collimators located along the cryostat holding the liquid deuterium. The cryostat sweeping magnet system also tends to eliminate the $e^+e^-$ pairs created by photons converting in the deuterium before they can produce softer, secondary bremsstrahlung photons. The resultant beam suffers from an approximately 2% neutral hadron contamination. Except for a low-energy tail, the spectrum falls off approximately exponentially in photon energy.

For E87 this photon beam is collimated to form a 5 × 5 cm spot on the experimenter's target. The photoproduced final state is analyzed by a multiparticle magnetic spectrometer augmented with Čerenkov identification of charged tracks, muon identification, hadronic calorimetry, and a highly segmented, lead glass shower counter. The evolution of the E87 detector into this final form is described in more detail in Section 5.1.

An alternative approach to reducing hadronic contamination in a neutral secondary beam is illustrated by the Fermilab Tagged-Photon Beam employed by experiment E516 (38). In this beam, 400-GeV incident protons impinge on a 30 cm long Be target. Charged secondaries and noninteracting protons are swept away by a series of vertical dumping magnets. The photons produced by $\pi^0$ decay convert to $e^+e^-$ pairs in a 0.57 radiation length lead sheet. A 290 meter long beam line transports electrons of the desired energy (up to 200 GeV) with about a $\pm 2.5\%$ momentum spread to a direction 45 mrad from 0° where they strike a copper radiator that is 1% of a radiation length; this forms a bremsstrahlung photon beam. The electron yield per incident proton falls from $10^{-5}$ to $10^{-7}$ as the electron energy of the transport line is tuned from 50 to 200 GeV. The pion contamination is typically less than a few tenths of a percent of the electron yield. The muon halo striking the experimental apparatus is kept to tolerable levels by bending the beam line away from 0°.

The bremsstrahlung photon energy can be measured on an event-by-event basis by measuring the energy of the scattered electron and subtracting this energy from the energy of the transported electron beam. It is important to use a thin electron radiator in order to reduce photons produced in double bremsstrahlung and reduce the level of false tagging electrons from trident production. The scattered electrons are momentum analyzed by a set of magnets that deflect the electrons into a lead glass and scintillator hodoscope. The position of the scattered electron in the hodoscope serves as a measurement of the electron's momentum, while the total pulse height in the hodoscope serves as a measurement of the electron's energy. Agreement between the energy and momentum measurement is required in order to suppress the number of false tags due to beam contamination.

The Tagged-Photon Spectrometer used in E516 is a forward magnetic spectrometer consisting of 2 analysis magnets and 29 planes of drift chambers. Augmenting the forward spectrometer are two multicell Čerenkov counters, a highly granular shower counter, muon identification capability, and a hadronic calorimeter. In order to provide additional information, the E516 collaboration surrounded their liquid hydrogen target with a recoil detector consisting of cylindrical multiwire proportional chambers and scintillation counters for $dE/dx$ ionization measurement. One of the main E516 triggers demanded the presence of a single track, consistent with an elastically scattered proton in the recoil detector. The 4-momentum of the scattered proton can be measured by the recoil detector, and subtracted from the incident tagged-photon 4-momentum to form the invariant mass of the forward system. A trigger processor was used to require a forward mass exceeding 2.5 GeV for the particular trigger used to obtain the E516 D* sample. Information on the 4-vector of the state recoiling against the elastically scattered proton plays an important role in the analysis of the E516 charm photoproduction data.

## 3.2 Virtual Photoproduction

The dominant contribution to lepton-nucleon scattering can be viewed as the emission of a virtual photon of mass $Q^2$ and lab energy $v$ striking a target nucleon to create a hadronic state. The $Q^2$ and $v$ variables can be found by measuring the momenta of the initial- and final-state muons—no measurement of the final-state hadrons is necessary. As $Q^2$ diminishes, the virtual photon becomes closer in character to a real photon, and the lepton-nucleon scattering cross section can be related to the real photon-nucleon cross section through quantum electrodynamics. In this article we are interested in comparing real photoproduction results on charm to the results obtained in the virtual photoproduction limit of lepton-nucleon scattering. Just as the advent of high-energy proton accelerators changed the methodology used to obtain real photon beams, it shifted the beams used in high-energy lepton-nucleon scattering experiments away from electron beams and toward muon beams. The leptonic decays of charged pion and kaon secondaries produced by the incident proton beam provides the predominant source of muons at a proton accelerator. The properties of such muon beams were reviewed recently by Drees & Montgomery (33).

Unlike real photons, virtual photons can exist in longitudinal as well as transverse polarization states. Separate transverse and longitudinal virtual photon fluxes can be reliably computed as a function of $v$ and $Q^2$, which can (in principle) be used to extract $\sigma_T(v, Q^2)$ and $\sigma_L(v, Q^2)$—the transverse and longitudinal virtual photoproduction cross sections (49). One expects that as $Q^2 \to 0$ the longitudinal cross section will disappear while the transverse cross section will smoothly approach the real photoproduction cross section for a real photon of energy $E_\gamma = v$. In practice, the separate measurement of $\sigma_T(v, Q^2)$ and $\sigma_L(v, Q^2)$ is difficult and effectively their sum is measured in the low-$Q^2$ limit.

To lowest order, the differential lepton-nucleon scattering cross section can be written in the laboratory as

$$\frac{d^2\sigma}{dQ^2\,dy} = \frac{\alpha K}{4\pi Q^2} \frac{1}{E} \left(\frac{2}{1-\varepsilon}\right)(\sigma_T + \varepsilon\sigma_L),$$

11.

where $E$ is the incident lepton energy, $y = v/E$, $K$ is a convention-dependent flux factor usually taken to be $K = v - Q^2/2m_p$ with $m_p$ being the proton mass. The $\varepsilon$ parameter, which represents the ratio of the longitudinal to transverse flux depends on the lepton scattering angle as follows:

$$\varepsilon = \left[ 1 + 2 \tan^2 \frac{\theta}{2} (1 + v^2/Q^2) \right]^{-1}. \qquad 12.$$

In the limit $Q^2 \ll 2m_p v$ the virtual photoproduction cross section approaches the Weisäcker-Williams limit:

$$\frac{d\sigma}{dQ^2 \, dy} = \frac{\alpha}{2\pi} \frac{1}{Q^2 y} [1 + (1-y)^2] \sigma_T. \qquad 13.$$

Comparison of the above expression with the expression for the photon spectrum in a bremsstrahlung beam shows that the virtual photon flux spectrum is nearly identical to the real photon flux produced by a few-percent bremsstrahlung radiator.

Because high-energy muons are highly penetrating particles and the muon nucleus cross section is low, the two virtual photoproduction experiments reviewed here have employed an essentially closed geometry configuration, where the incident muon interacts in a large iron target that ultimately absorbs all final-state particles except the final-state muons. The final-state muons are momentum analyzed by a magnetic spectrometer. Both experiments instrument the iron target with scintillation counters, which allows the target to double as a calorimeter that measures the total hadronic final-state energy accompanying the final-state muons. The calorimetry information is used to trigger the detectors and plays an important role in the interpretation of the data.

## 4.  PHOTOPRODUCTION OF $\psi$ AND $\psi'$

Measurements of $\psi$ photoproduction span the range of photon energies from 9 GeV (threshold) to almost 300 GeV. The data come from six different real photoproduction experiments and from two virtual photoproduction (muon scattering) experiments. Data on $\psi'$ photoproduction are more limited, in part because of the weaker coupling to the photon and in part because of lower branching ratios into experimentally accessible states. The $\psi'$ data come from four real and one virtual photoproduction experiments. The experiments, the collaborating institutions, associated references, and beam types are given in Table 1.

### 4.1  $\psi$ Photoproduction

Psi photoproduction is generally described in terms of its dependence on the photon energy, the 4-momentum transfer squared ($t$), the elasticity of

**Table 1**   Measurements of $\psi/\psi'$ photoproduction

| Experiment | Institutions | Reference | Beam |
|---|---|---|---|
| E87 | Columbia/Hawaii/Cornell/Illinois/ | | |
| | Fermilab | (51, 53) | Wideband |
| SLAC | Wisconsin/SLAC | (28, 63) | Bremsstrahlung |
| Cornell | Cornell | (43, 44) | Bremsstrahlung |
| E25 | Fermilab/Lebedev/UCSB/Toronto | (59) | Tagged |
| E401 | Fermilab/Illinois | (24, 25) | Wideband |
| E516 | UCSB/Carleton/Colorado/Fermilab/ | | |
| | NRCC/Oklahoma/Toronto | (34, 60) | Tagged |
| BPF | Berkeley/Princeton/Fermilab | (31, 32, 67) | Virtual |
| EMC | European Muon Collaboration | (14, 15) | Virtual |

the interaction, and in the case of virtual photoproduction the invariant mass squared $(Q^2)$ of the photon. Table 2 lists the measured value of $d\sigma/dt$ at $t = 0$, the slope $b$ arising from a parameterization of the form $d\sigma/dt = d\sigma/dt$ $(t = 0)$ $e^{-bt}$, and the total cross section for the elastic scattering reaction $\gamma N \rightarrow \psi N$ as observed in eight different experiments. Included in the table are the photon energy (or range of energies) for which the production has been measured, the target used, the decay modes observed, the number of $\psi$ events upon which the measurements are based, the forward cross section, the slope, and the total cross section. Some experiments are better suited to quoting the total cross section while others are better suited to quoting the forward cross section. If the slope of the $t$ dependence is known, a direct translation between the two types of measurements can be made. We indicate by arrows in the table when we made such a translation. The arrows point from the measured to the inferred quantity. The absence of an arrow indicates that both the forward and total cross sections are measured. The forward cross section as a function of energy is displayed in Figure 4. The energy dependence of the total elastic cross section is given in Figure 5. The data give a remarkably consistent picture of the energy dependence of $\psi$ photoproduction. The cross section rises extremely rapidly in the region between threshold and 50 GeV. At higher energies, the rise with energy is much less pronounced, yet persists to the highest energies measured.

The VDM model makes no definite predictions as to the energy dependence of the $\psi$ photoproduction cross section except that it should become independent of energy considerably above threshold. The PGF model, on the other hand, makes explicit predictions about the cross section's dependence on energy in terms of the gluon momentum distri-

**Table 2**   Measured forward cross section for $\gamma + N \to \psi N$

| Exp. | $E$ (GeV) | Target | Modes | Events | $\frac{d\sigma}{dt}(t=0)$ (nb GeV$^{-2}$) | $b$ (GeV$^{-2}$) | $\sigma_T$ (nb) |
|---|---|---|---|---|---|---|---|
| E87 | 80–200 | Be | $\mu^+\mu^-$ | 60 | $45.1 \pm 14.0$ | 4.0 | $11.1 \pm 3.5$ |
| SLAC | 13.0 | $D_2$ | $e^+e^-$ | 170 | $7.5 \pm 1.6$ | $-2.9\to$ | $1.3 \pm 0.3$ |
| | | | $\mu^+\mu^-$ | 400 | | | |
| | 15.0 | $D_2$ | | | $9.4 \pm 1.6$ | | $2.0 \pm 0.4$ |
| | 16.0 | $D_2$ | | | $12.1 \pm 1.6$ | | $2.8 \pm 0.4$ |
| | 17.0 | $D_2$ | | | $15.1 \pm 1.4$ | | $3.7 \pm 0.4$ |
| | 19.0 | $D_2$ | | | $15.5 \pm 1.4$ | | $4.1 \pm 0.4$ |
| | 19.0 | $H_2$ | | | $13.9 \pm 1.4$ | | $3.7 \pm 0.4$ |
| | 21.0 | $D_2$ | | | $17.8 \pm 1.5$ | | $5.0 \pm 0.4$ |
| Cornell | 9.0–11.8 | Be | $e^+e^-$ | 470 | $1.7 \pm 0.4$ | $-1.3 + 0.2\to$ | $0.72 \pm 0.2$ |
| E25 | 30–80 | $D_2$ | $e^+e^-$ | 24 | $68.0 \pm 19.0$ | $\leftarrow 1.8 + 0.4$— | $37.5 \pm 8.2$ |
| E401 | 75 | $D_2$ | $e^+e^-$ | 700 | $53.0 \pm 8.0$ | $-5.6 + 1.2^a\to$ | $11.9 \pm 1.8$ |
| | | | $\mu^+\mu^-$ | 950 | | | |
| | 125 | $D_2$ | | | $64.0 \pm 8.0$ | | $14.4 \pm 1.8$ |
| | 175 | $D_2$ | | | $82.0 \pm 15.0$ | | $18.4 \pm 3.4$ |
| | 225 | $D_2$ | | | $105.0 \pm 15.0$ | | $23.6 \pm 3.4$ |
| E516 | 60–160 | $H_2$ | $e^+e^-$ | 63 | $71.0 \pm 21.0$ | $\leftarrow 5.0 + 2.0^b$— | $14.2 \pm 4.1$ |
| | | | $\mu^+\mu^-$ | 147 | | | |
| BPF | 40 | Fe | $\mu^+\mu^-$ | 6700 | $26.8 \pm 2.1$ | $\leftarrow 2.6^c$— | $10.3 \pm 0.8$ |
| | 58 | Fe | | | $37.2 \pm 2.3$ | | $14.3 \pm 0.9$ |
| | 80 | Fe | | | $45.5 \pm 2.3$ | | $17.5 \pm 0.9$ |
| | 108 | Fe | | | $53.8 \pm 3.1$ | | $20.7 \pm 1.2$ |
| | 140 | Fe | | | $61.9 \pm 4.2$ | | $23.8 \pm 1.6$ |
| | 173 | Fe | | | $62.4 \pm 13.0$ | | $24.0 \pm 5.0$ |
| EMC | 70 | Fe | $\mu^+\mu^-$ | 907 | $52.8 \pm 9.2$ | $\leftarrow 4.4^c$— | $12.0 \pm 2.1$ |
| | 90 | Fe | | | $56.4 \pm 8.8$ | | $12.8 \pm 2.0$ |
| | 110 | Fe | | | $50.6 \pm 8.8$ | | $11.5 \pm 2.0$ |
| | 130 | Fe | | | $79.5 \pm 13.5$ | | $18.0 \pm 3.0$ |
| | 150 | Fe | | | $57.5 \pm 11.0$ | | $13.0 \pm 2.5$ |
| | 170 | Fe | | | $94.8 \pm 19.5$ | | $21.5 \pm 4.5$ |
| | 190 | Fe | | | $107.8 \pm 19.5$ | | $24.5 \pm 4.5$ |

[a] The actual $t$ distribution is fit to the form $\exp(-bt + ct^2)$, with $b = 5.6 \pm 1.2$ GeV$^{-2}$ and $c = 2.9 \pm 1.3$ GeV$^{-4}$.

[b] This slope is actually derived from the quoted value of $\langle Pt \rangle = 0.39 \pm 0.11$ GeV/$c$.

[c] Effective $b$ value. They actually measure $\exp(4.3t) + 0.23 \exp(0.9t)$ (BPF) and $\exp(5.2t) + 0.03 \exp(0.66t)$ (EMC).

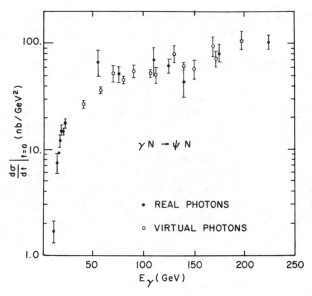

*Figure 4*   The forward cross section for $\psi$ photoproduction as a function of photon energy.

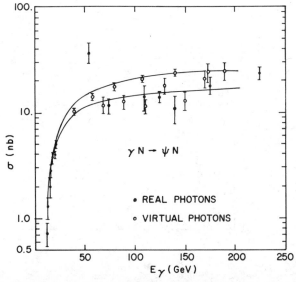

*Figure 5*   The total $\psi$ photoproduction cross section as a function of photon energy. The superimposed curves are predictions of the PGF model as described in the text.

bution $G(x)$, the bare charmed quark mass $m_c$, and the strong coupling constant $\alpha_s$. Unfortunately, many of these input parameters are poorly determined at present.

The two curves superimposed on Figure 5 are the PGF predictions with $m_c = 1.5$ GeV, $xG(x) = \frac{1}{2}(n+1)(1-x)^n$, $\alpha_s = 0.35$, and $f = 1/7$ for $n = 5.0$ (*upper*) and $n = 4.0$ (*lower*). We remark that $n = 5$ is the value preferred by counting rules and that we have normalized $G(x)$ such that gluons carry half the momentum of the nucleon. While the model does indeed give a very nice description of the data, we must point out that other combinations of $m_c$ and $n$ can reproduce the data equally well if we are willing to change the value of $f\alpha_s$. For example, the top upper curve can also be generated with $(m_c, n, f\alpha_s) = (1.2, 6.5, 0.011)$ or $(1.7, 4.4, 0.29)$, and the lower curve with $(1.2, 5.0, 0.009)$ or $(1.7, 3.4, 0.23)$. It is only a belief that either the value of $m_c$, $n$, or $f\alpha_s$ is known that allows the identification of the correct solution. Thus

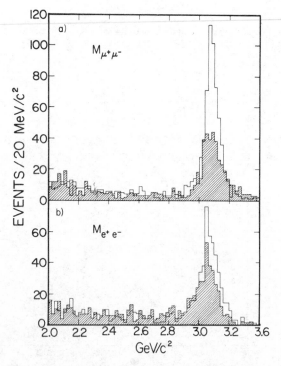

*Figure 6*   The dilepton mass distribution near the $\psi$ as measured in the real photoproduction experiment E401 for (*a*) dimuons and (*b*) dielectrons. The shaded distributions are for dileptons in which one lepton falls outside the aperture of the second of two analysis magnets.

within the context of a charmed quark of mass 1.5 GeV, the gluon distribution is measured to have $n = 4.5 \pm 0.5$.

While all the experiments listed in Table 2 observe the $\psi$ through its leptonic decay modes, the conditions under which the various data are taken and analyzed vary widely. For instance the first six experiments listed represent real photoproduction data, while the last two measure $\psi$ production by virtual photons and extrapolate to $Q^2 = 0$. Furthermore, all the experiments have different methods for subtracting the coherent scattering contributions, for separating elastic from inelastic interactions, and for presenting the measured cross sections.

Figure 6a,b shows the measured dimuon mass spectrum and dielectron mass spectrum from the Fermilab wideband beam (E401). Figure 7 shows the measured dimuon mass spectrum from the Berkeley-Princeton-Fermilab (BPF) muon scattering experiment. These are chosen as representative of real and virtual photoproduction data. As can be seen, the mass resolution in the open geometry of the real photoproduction experiments is typically an order of magnitude better than in the closed geometry of the muoproduction experiments (80 vs 600 MeV). As a consequence it is virtually impossible to resolve $\psi$ and $\psi'$ in the muon scattering experiments. However, as we discuss below the fraction of $\psi'$ events relative to $\psi$ events in the dimuon mass spectrum is less than 2%. The

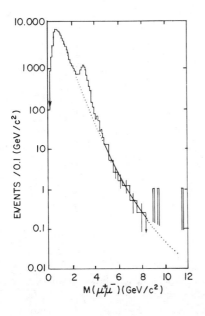

*Figure 7* The dimuon mass spectrum obtained in the virtual photoproduction experiment of the BPF collaboration.

contribution of $\psi'$ production to $\psi$ production through the decay $\psi' \to \psi\pi\pi$ produces a 10% correction to the inelastic production rates and is accounted for in all the experiments discussed.

The use of targets with $A > 1$ requires some sort of extrapolation to find the single-nucleon cross section. In general this involves more than just dividing the observed cross section per nucleus by the atomic number because coherent production from the entire nucleus must be accounted for first. The real photoproduction experiments can remove the coherent contribution directly since they have sufficient resolution in $t$ either to resolve the forward coherent peak (see Figure 8) or to eliminate it with a cut on $t$. The first tagged-photon experiment (E25), however, fails to make any coherent correction. This may explain why their total cross-section measurement is significantly ($\times 2$) higher than one might expect on the basis of the other data. In the low-energy photoproduction experiments (Figure 9), no coherent correction is made nor is any necessary since coherent production is heavily suppressed as a result of the relatively large $t_{min}$ for these experiments ($t_{min} \approx -M_\psi^4/4E_\gamma^2$). In the SLAC experiment, one data

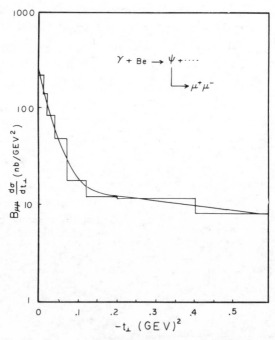

*Figure 8*    The $t$ distribution of $\psi$ photoproduction off a beryllium target as measured by the E87a group for photons with a mean energy of 140 GeV.

point is measured with $H_2$ as well as $D_2$ in order to verify this assumption and to establish the approximate equality of the scattering from protons and neutrons. The muoproduction experiments do not observe coherent production from iron directly because of their limited resolution in $t$. Instead BPF and the European Muon Collaboration (EMC) calculate a correction using Monte Carlo simulations that include coherent scattering in the form $A^2 e^{-at}$ where $a = 150$ and $135$ GeV$^{-2}$ respectively. The muoproduction experiments also correct for nuclear shadowing effects.

The experiments described above all use different criteria for defining and separating elastic and inelastic events. The Cornell experiment simply assumes that the production they are observing is completely elastic—not an unreasonable assumption considering that they are quite close to threshold and unable to distinguish directly anyway. They make the interesting observation, however, that within their data there appears to be no energy dependence to the cross section as measured over the range of energies 9.0–11.8 GeV. This is a rather surprising result considering the extremely rapid variation measured at SLAC. They themselves point out that the apparent contradiction could be explained if there were substantial inelastic production that would, because of the kinematics, contribute only to the lower energy range. The SLAC experiment effectively eliminates possible inelastic contributions from the data given in Table 2 by observing $\psi$'s with energies within 0.5 GeV of the bremsstrahlung endpoint. This limits the range of masses recoiling against the $\psi$ to be less than 1.25 GeV since $p/p_{max} > 0.95$. They do, however, also compare $\psi$ production at $E_\psi = 15$ GeV from 20-GeV and 16-GeV endpoint energies and conclude that the inelastic contribution is possibly 20–30%.

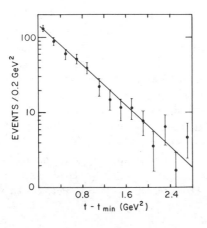

*Figure 9* The $t$ distribution of $\psi$ photoproduction off a beryllium target as measured by the Cornell group for photons with energies less than 12 GeV.

The initial tagged-photon experiment (E25) requires that the energy of the $\psi$ be at least 0.85 of the energy of the tagged photon. This requirement eliminates 1 of 25 events. The initial wideband photon experiment (E87) does not have the luxury of knowing the energy of the incident photon on an event-by-event basis. As such they can only eliminate inelastic events by asking if there are extra tracks accompanying the two leptons. They find that of 48 events in the region $t \leqslant 0.7 \text{ GeV}^2$, five are accompanied by extra tracks in the forward direction. Two of these are unambiguously identified as coming from $\psi'$ decays. They also observe a large number of events with large $t$ (12 of 60 events have $t > 0.7 \text{ GeV}^2$) and so attempt to minimize inelastic contributions by fitting the $t$ distribution only in the region $0.0 \rightarrow 0.5 \text{ GeV}^2$. The second wideband experiment (E401) attempts to extend the definition of inelasticity into the recoil region by adding a recoil detector and by requiring (a) no accompanying tracks in the forward direction and (b) a single recoil proton consistent with momentum conservation. They find approximately 30% of their events are inelastic in the recoil direction and 5% are inelastic in the forward direction. Finally, the more recent tagged-photon experiment (E516) combines knowledge of the incident photon energy with recoil detection to provide nearly complete knowledge of the event topology. They define elastic events as those with $E_\psi/E_\gamma > 0.9$ with no additional forward tracks. Approximately 31% of their events are found to be inelastic.

The two muoproduction experiments define elastic events in different ways. BPF, which employs a fully active target, cuts directly on the hadronic energy as measured through calorimetry. They require $E_{\text{had}} < 5 \text{ GeV}$. The EMC simply requires that the $\psi$ energy be at least 0.95 of the virtual photon energy. These two experiments observe 44% and 62% inelastic production respectively.

As discussed above, the ratio of inelastic to elastic $\psi$ production provides an important quantitative test of theoretical understanding. In Table 3 we list the existing data on the ratio of the inelastic to the total cross section. As seen, only four experiments have measured inelastic production carefully. Within the four experiments there seems to be a real discrepancy between the ratio as measured with real and virtual photons. Since, as was seen in Figure 6, the real and virtual photoproduction measurements agree for elastic production, this means that the virtual photoproduction experiments measure enhanced inelastic production relative to photoproduction by real photons. One might expect that this could be related to the finite values of $Q^2$ used in the muon scattering experiments, and the BPF measurement does indeed represent an average over all $Q^2$. However, the EMC result given has been explicitly extrapolated to $Q^2 = 0$. Perhaps the

**Table 3**  Ratio of inelastic to total $\psi$ photoproduction

| Experiment | Ratio | Inelastic definition |
|---|---|---|
| Cornell | — | Unresolved |
| SLAC | | "Possible 20–30% contribution" |
| E25 | 1/25 | $E_\psi/E_\gamma < 0.85$ |
| E87 | 15/60 | $>2$ forward tracks, or $t > 0.7$ GeV$^2$ |
| E401 | $0.30 \pm 0.04$ | $>2$ forward tracks, or inelastic recoil |
| E516 | $0.32 \pm 0.08$ | $>2$ forward tracks, or $E_\psi/E_\gamma < 0.90$ |
| BPF | $0.44 \pm 0.05$ | $E_{\text{had}} > 5$ GeV |
| EMC | $0.60 \pm 0.06$ | $E_\psi/E_\gamma < 0.95$ |

discrepancy is related to the manner in which coherent production is accounted for in the muon scattering experiments.

The final three experiments listed above also have measured the $Z = E_\psi/E_\gamma$ and $P_\perp^2$ distributions of inelastic scattering. Once again the results are not entirely consistent. The EMC finds inelastic scattering biased much more strongly toward high $Z$ than do BPF and E516. While the transverse momentum distributions as measured in elastic scattering are in reasonable agreement, there are again some discrepancies in the inelastic measurements. Both BPF and EMC find that at large $t$ values the data are not fit well by a single exponential in either the elastic or inelastic case. However, in comparing the inelastic and elastic distributions BPF and E516 find evidence for a much (factor of 2–4) broader $P_\perp^2$ distribution for inelastic scattering, while EMC finds little difference.

To summarize, there appears to be broad agreement over a large range of data as to the general features of $\psi$ elastic photoproduction. More recent measurements of inelastic production have not yet achieved this level of consistency. This is probably due both to the difficulty in making a "detector-independent" definition of inelasticity at high energies and to the fact that it is only very recently that experimenters have attempted to understand inelastic production, while more than a decade of experience has been accumulated doing the less difficult elastic measurements.

## 4.2   Extraction of the $\psi$-N Cross Section

As discussed in Section 2.1, the measured elastic cross section can be related to the $\psi$-N cross section within the context of VDM. Well above threshold (greater than 100 GeV) the total $\psi$-N cross section is found to be $\sim 1.3$ mb while the elastic $\psi$-N cross section is $\sim 25$ $\mu$b. In other words elastic $\psi$-N scattering represents only about 2% of the total $\psi$-N cross section.

One might ask what happens in the 98% of $\psi$-N interactions where the $\psi$ does not scatter elastically. As we have seen earlier, inelastic scattering can account for only a few percent of this. If we invoke an argument based on Zweig's rule, and assume that the $\psi$ is a bound state of a charmed $c\bar{c}$ pair, then we expect that nearly the entire (95%) $\psi$-N cross section is populated with charmed particles in the final state. Once again invoking VDM, we conclude that the rate of photoproduction of charm must be approximately 50 times the rate for elastic $\psi$ production:

$$\frac{\sigma(\gamma N \to \text{charm})}{\sigma(\gamma N \to \psi N)} = \frac{\sigma(\psi N \to \text{not } \psi)}{\sigma(\psi N \to \psi N)} \approx 50. \qquad 14.$$

Scaling the measured $\psi$ elastic scattering cross section leads us to an expected charm photoproduction cross section of 1 $\mu$b, or 1% of the total photon-nucleon cross section. This argument was of course presented after the observations made by the first four experiments listed in Table 1; it provided a primary impetus for the subsequent photoproduction experiments looking for bare charm (63, 66).

The traditional VDM analysis assumes that it is safe to use the photon-$\psi$ coupling constant measured at $Q^2 = -M_\psi^2$ for real photoproduction processes at $Q^2 = 0$. One way to check this is to measure the $\psi$-N total cross section by a different method and see how the results compare. An experiment at SLAC (8) attempts to do this by measuring the $A$ dependence of $\psi$ photoproduction. By comparing the yield of $\psi$'s (as measured by looking for single, large transverse momentum muons) from beryllium and tantalum targets, the SLAC group conclude that at 20 GeV the $\psi$-N total cross section is $3.5 \pm 0.8$ mb. They state that their measured value is "in general agreement with, but somewhat higher than, the value based on vector-dominance ideas."

## 4.3   $Q^2$ and Decay Angular Dependence

The use of virtual photon beams in the muon scattering experiments adds two degrees of freedom not found in the real photoproduction experiments. First, it allows one to explore the (space-like) region $Q^2 > 0$ where very definite predictions are made by both VDM and PGF. Second, since the virtual photons are polarized with all three helicity states accessible, one can gain information both by looking at angular decay distributions and by looking for contributions from longitudinally polarized photons.

The $Q^2$ dependence of $\psi$ photoproduction has been measured by the EMC and BPF experiments. The EMC divide their data into two energy bins, and into elastic and inelastic production. The BPF experiment presents data only on elastic production averaged over their entire energy range. Both experiments fit the energy dependence to the form suggested by VDM,

$$\frac{\sigma(Q^2)}{\sigma(0)} = \frac{1}{(1 + Q^2/\Lambda^2)^2}.$$     15.

The mass, $\Lambda$, is regarded as a free parameter in the fits. The fit values of $\Lambda$ are given in Table 4. The detailed method of extracting $\Lambda$ employed by both groups was not identical. The EMC simply fits the $Q^2$ dependence of the cross section averaged over all other variables, while BPF actually fits the $Q^2$ and angular decay distributions simultaneously. Since BPF allows a $Q^2$ dependence to the angular decay distribution (see below), variations in the angular acceptance with $Q^2$ can change the apparent $Q^2$ dependence. Indeed, BPF does find that ignoring such an effect raises their best-fit $\Lambda$ from $2.2 \pm 0.2$ GeV to $2.7 \pm 0.5$ GeV as indicated in the table. There is also some indication in the data of the EMC that the $Q^2$ dependence becomes less steep as the energy and elasticity increase. Unfortunately, the statistics are not sufficient to make a definitive statement in this regard.

In Figure 10 we display the $Q^2$ dependence of $\psi$ elastic production by virtual photons. Here we have normalized to the cross section at $Q^2 = 0$ and averaged over all energies. Three theoretical curves are superimposed on the data. First the $\psi$ propagator as suggested by VDM, and second the PGF prediction for two values of $m_c$. The VDM curve is seen to lie systematically above the data points while still giving a good qualitative description. The PGF curves (done for $E = 130$ GeV, $n = 5$) both give a somewhat nicer description of the data. While the data seem to favor the PGF description, the statistics really are not sufficient for discriminating between the alternate descriptions. It is worth noting that the PGF model does predict a flattening of the $Q^2$ distribution as the energy rises. A higher-

**Table 4**   $Q^2$ dependence of $\psi$ photoproduction

| Experiment | Energy (GeV) | $\Lambda$ (GeV) | Comments |
|:---:|:---:|:---:|:---:|
| BPF | 40–175 | $2.2 \pm 0.2$ | Decay distribution included, $E_{had} < 4.5$ GeV |
|  |  | $2.7 \pm 0.5$ | Decay distribution not included, $E_{had} < 4.5$ GeV |
| EMC | 60–120 | $2.9 \pm 0.3$ | $Z > 0.95$ |
|  | 120–200 | $3.6 \pm 0.4$ | $Z > 0.95$ |
|  | 60–120 | $2.3 \pm 0.2$ | $Z < 0.95$ |
|  | 120–200 | $2.7 \pm 0.3$ | $Z < 0.95$ |

statistics experiment measuring over an energy range of a few hundred GeV might be able to identify such a trend, already suggested in Table 4.

The angular decay distribution of the muoproduced $\psi$'s are studied in terms of the three angles $\phi_p$, $\phi$, and $\theta$: $\phi_p$ is the azimuthal angle between the planes containing the virtual photon/$\psi$ and the incident/scattered muon; $\phi$ is the azimuthal angle between the dimuons as viewed in the $\psi$ helicity frame and the muon scattering plane; and $\theta$ is the polar angle of the decay muons in the $\psi$ helicity frame.

A dependence of the $\psi$ production on $\phi_p$ would indicate interference between the two transverse spin states or between the transverse and longitudinal spin states of the virtual photon. The BPF experiment observes no $\phi_p$ dependence. The EMC observes a slight dependence if they restrict the data to elastic production with $P_\perp^2 > 0.5 \, \text{GeV}^2$ (to improve the resolution in $\phi_p$ since $\phi_p$ is undetermined at $P_\perp^2 = 0$). They fit to the form

$$\frac{d\sigma}{d\phi_p} = A(1 + B \cos \phi_p + C \cos 2\phi_p)$$

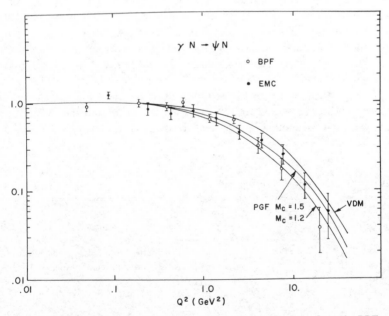

*Figure 10*   The $Q^2$ dependence of $\psi$ virtual photoproduction. Data are from the BPF and EMC groups. Plotted is the ratio of the total cross section to its value at $Q^2$. The superimposed curves are described in the text.

and find $B = -0.11 \pm 0.23$, $C = 0.43 \pm 0.19$. The statistical significance of this result is not overwhelming and, as seen from the data (Figure 11), the nonzero value for $C$ comes almost entirely from one low bin near $\phi_p = 90°$.

Both experiments fit the decay angular distribution to the s-channel helicity-conserving (SCHC) (65) form,

$$W(\theta, \phi) = \frac{3}{16\pi(1+\varepsilon R)}(1 + \cos^2 \theta + 2\varepsilon R \sin^2 \theta - \varepsilon \sin^2 \theta \cos 2\phi).$$

Here $\varepsilon$ is the polarization parameter (0.7–0.8 in these experiments) and $R$ is the ratio of the longitudinal to transverse cross sections. The $\phi$ distribution as measured by BPF shows a strong $\cos 2\phi$ dependence while the EMC data show no such effect. Neither distribution shows any strong $\cos \theta$ dependence. The BPF experiment is consistent with SCHC and an $R$ value of the form

$$R = 3.3^{+4.9}_{-3.0} \frac{Q^2}{M_\psi^2}.$$

As seen from the above expression, the nonzero value of $R$ is responsible for flattening the $\cos \theta$ dependence in their fits. The EMC data are at best marginally consistent with the SCHC form.

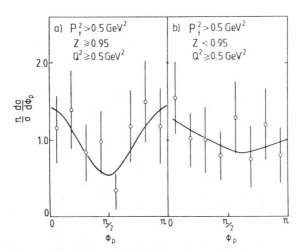

Figure 11   The dependence of $\psi$ virtual photoproduction on the angle $\phi_p$ (defined in the text) as measured by the EMC group: (a) is for elastic and (b) for inelastic scattering. The curves are described in the text.

### 4.4    *Photoproduction of ψ′*

Several of the experiments discussed above have observed the production of ψ′(3700). Since the ψ and ψ′ are both observed with the same apparatus in a given experiment, systematic errors are greatly reduced by quoting the ratio of ψ′ to ψ production rather than the absolute value of the ψ′ cross section. The ψ′ signals observed in all the real photoproduction experiments are almost completely background free (see for example Figure 12). In contrast, the muon scattering experiments do not really have sufficient resolution to observe the ψ′ (through its $\mu^+\mu^-$ decay) in the presence of the ψ. For this reason BPF does not attempt to measure ψ′ production. The EMC does, however, perform a fit to their dimuon mass distribution fixing the mass and width of the ψ′ relative to the ψ. They find an approximately two-standard-deviation shoulder by this method. This result is legitimized only through their observation of ψ′ production at an identical level, but with much lower statistics, in runs taken with hydrogen and deuterium targets. The data on ψ′ production are summarized in Table 5.

As seen, the data are quite consistent with ψ′ production being 15–20% of

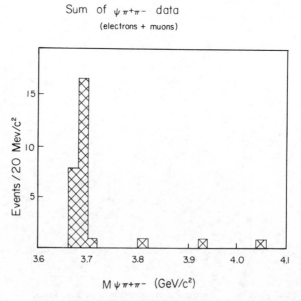

*Figure 12*    The mass spectrum of photoproduced $\psi\pi^+\pi^-$ from E401 showing a prominent ψ′ peak.

**Table 5**  $\psi'/\psi$ total cross sections

| Experiment | Mode | Number of $\psi'$ | $(\gamma + N \rightarrow \psi' + X)/(\gamma + N \rightarrow \psi + X)$ |
|---|---|---|---|
| E87 | $\psi\pi^+\pi^+$ | 2 | $0.1 \pm 0.14$ |
| SLAC | $\mu^+\mu^-, e^+e^-$ | 13 | $0.15 \pm 0.05$ |
| E401 | $\psi\pi^+\pi^+$ | 26 | $0.20 \pm 0.05$ |
| E516 | $\psi\pi^+\pi^+$ | 4 | "Consistent with E401 and EMC" |
| EMC | $\mu^+\mu^-$ | 25 | $0.22 \pm 0.10$ |

$\psi$ production. It is worth noting that the $\psi'$-to-$\psi$ ratio is consistent with predictions of VDM assuming $1/M_v^2$ scaling for $\sigma_{vN}$. It should also be noted that $\psi$ production followed by decay into $\psi\pi\pi$ represents a 10% background contribution to $\psi$ inelastic production. This effect is taken into account in the measurements of the $\psi$ inelastic cross sections made after 1975.

# 5. OPEN CHARM PHOTOPRODUCTION

We begin this section by discussing experimental results on the photo-production of open charm states by real (as opposed to virtual) photo-production beams with energies from 20 to 200 GeV. The data were collected by experiments using four photon beam facilities in the period from 1975 to 1982. Each of the photoproduction experiments we review runs in the so-called open geometry configuration where detailed measurements of the charm particle decay products is possible. This allows one to identify charm particles by reconstructing their invariant mass, or by measuring their lifetime. Besides investigating heavy flavor photoproduction mechanisms, these experiments have substantially contributed to information on the properties of charmed particles. We first review the charm signals observed by the various photoproduction groups, beginning with those experiments that observe charm particles through the invariant mass technique.

## 5.1  Charm Signals Obtained in the FNAL Wideband Beam

Charm signals were obtained in the two runs of Fermilab experiment E87a. The wideband photon beam suffers from a 1–2% neutral hadronic contamination due to neutrons and $K_L$ mesons. Because the multihadronic interaction cross section for neutrons and $K_L$'s is substantially larger than the multihadronic cross section for photons, even this low-level beam contamination produces the major background for the wideband beam charm signals.

The first run of E87a, occurring late in 1975, was designed to study multihadronic states produced by photons in the 50–200-GeV range impinging on a Be target. The E87a apparatus consisted of a magnetic spectrometer augmented with a hadron calorimeter, an electromagnetic shower counter, and a muon identification system. The spectrometer had negligible acceptance for most charm states produced with lab energies below 80 GeV, and hence was only sensitive to charm states carrying a substantial fraction of the incident photon energy. The first run of E87a did not use a Čerenkov counter, so strange particle identification was limited to decays involving neutral V's.

Evidence was obtained for the existence of a narrow, antibaryon state with a mass of 2.26 GeV decaying into the $\bar{\Lambda}\pi^-\pi^-\pi^+$ final state (52). Figure 13$a$ shows the $\bar{\Lambda}\pi^-\pi^-\pi^+$ invariant mass spectrum for events with fewer than nine observed tracks and a total charged track energy below 200 GeV. These two cuts serve to reduce combinatorial background as well as background from neutron interactions since the neutron spectrum is known to peak near 320 GeV. Unfortunately, no comparable signal for a 2.26-GeV baryon was observed in the $\Lambda\pi^+\pi^+\pi^-$ invariant mass distribution. The background level for baryon states was a factor of three higher than the level for antibaryons; hence a comparable-level baryon signal could plausibly be buried in neutron-induced background.

The observed narrow width of the mass peak, the proximity of the new state's mass to the mass of the then recently discovered charmed mesons, and the consistency of the observed decay mode with the $\Delta C = \Delta Q = \Delta S$ decay expected for a weakly decaying ground state charmed baryon all strongly identified the new state with the isosinglet anticharmed baryon— the $\bar{\Lambda}_c$. Two other arguments were proposed at the time of the antibaryon discovery that supported its charmed baryon interpretation. If the $\bar{\Lambda}(3\pi)^-$ state at 2.26 GeV were a conventional hadron, it would consist of $\bar{s}u u$ quarks and thus be a member of an antibaryon isotriplet of states. By isospin invariance, a positively charged antibaryon state near 2.26 GeV, consisting of $\bar{s}d d$ quarks, would exist and could decay via $\bar{\Lambda}(3\pi)^+$. No such isospin brother with a mass of around 2.26 GeV is seen in the $\bar{\Lambda}\pi^+\pi^+\pi^-$ invariant mass plot of Figure 13$b$, which thus argues against an excited $\bar{\Sigma}$ interpretation for the 2.26-GeV antibaryon. A final argument concerned the fact that the E87a $\bar{\Lambda}_c$ candidate was consistent with the interpretation of the previously published neutrino event (29), which appeared to violate the $\Delta S = \Delta Q$ selection rule of weak interactions unless interpreted as the charged current production of a $\Sigma_c^{++}$ with a subsequent decay via $\Lambda_c\pi^+$. One of the several possible $\Lambda_c$ masses in the Cazzoli event was consistent with the E87a $\bar{\Lambda}_c$ candidate. Evidence was obtained in E87a for the existence of a broader $\bar{\Sigma}_c^{--}$ state also consistent with the mass reported by Cazzoli et al (29). Since the observation of the 2.26-GeV antibaryon state by E87a, charmed baryon candidates have been observed at SPEAR and in the second run of E87a. The presently accepted mass value for this state is 2282 MeV, which is somewhat higher than the original E87a value,

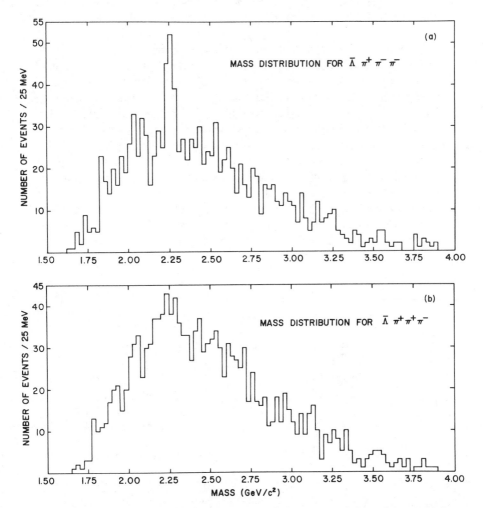

*Figure 13* Evidence for a photoproduced charmed baryon signal from E87a. (*a*) The $\bar{\Lambda}\pi^+\pi^-\pi^-$ invariant mass spectrum shows a narrow peak near 2.26 GeV. (*b*) The $\bar{\Lambda}\pi^+\pi^+\pi^-$ invariant mass spectrum shows no peak near 2.26 GeV.

but consistent given the systematic uncertainties of their spectrometer calibration.

The second run of E87a occurred from December 1977 to May 1978 with a considerably upgraded apparatus including two multicell Čerenkov counters and an additional analysis magnet designed to provide momentum analysis for low-momentum, wider angle tracks. The inclusion of the Čerenkov counters allowed one to expand the search for charm to states decaying into charged heavy particles as well as $V^0$'s. Three types of charm signals were obtained from the analysis of the data collected during the second run. Each type exploited a different strategy to suppress the considerable noncharm and beam contamination background.

The first charm signal observed in the second run of E87a was obtained by searching for D's decaying via K multipions within the subsample of events that contained both an identified $K^+$ and $K^-$ in the final state (12). The requirement of an additional oppositely charged kaon, presumably from the decay of another charmed particle, greatly suppresses the copious noncharm background to the K-multipion final state from $K_L$ beam contamination. Figure 14 (*left*) shows the $K\pi$ invariant mass distribution obtained in dikaon events with additional cuts on the observed number of tracks and total reconstructured event energy. A 4-sigma enhancement was found at the known mass of the D consisting of $99 \pm 24$ events above background. The signal-to-noise ratio is considerably improved by requiring that the total invariant mass of all observed tracks lie from 3 to 4 GeV; this results in a signal of $99 \pm 20$ events [see Figure 14 (*right*)].

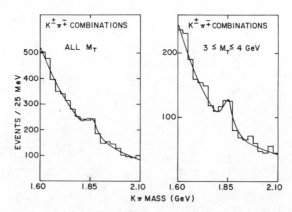

*Figure 14*   $D \to K\pi$ signals obtained in E87a in photoproduced events with an additional observed recoil kaon. (*Left*) has cuts on the number of tracks and total event energy. (*Right*) has these cuts and an additional requirement that the invariant mass of all observed tracks ($M_T$) lie between 3 and 4 GeV.

In addition to charmed mesons, charmed baryon signals were obtained in the data of the second E87a run (64). Figure 15 shows a signal for photoproduced charmed baryons decaying via $pK_S$ and $\bar{p}K_S$ where the proton is positively identified by the two Čerenkov counters as being unambiguous with either a kaon or pion. Cleanliness of the $pK_S$ decay mode reflects the fact that there is virtually no combinatorial background in this state owing to the small number of events with multiple protons or $K_S$'s.

The final charm signal observed in the second run of E87a was for the decay process $D^{*+} \rightarrow \pi^{+}D^{0}$, with the $D^{0}$ subsequently decaying via $K^{\mp}\pi^{\pm}$ or $K_S\pi^{+}\pi^{-}$ (18). In general one can measure the $D^{*+}$-$D^{0}$ mass difference in this process with great precision because the $Q$ value is so small (less than 5 MeV). Figure 16 illustrates this fact using data obtained in the second run of E87a by plotting $\Delta$ ($= M_{K\pi\pi} - M_{K\pi}$) for $K^{\mp}\pi^{\pm}$ combinations with invariant masses in a control region below, straddling, and above the known mass of the D. The shaded regions represent appropriately scaled beam contamination background. The background data were taken during special runs in which six radiation lengths of lead were inserted to remove the photons and allow neutral contaminants to interact in the target. The peak observed in Figure 16b, consisting of about 150 events, has an experimental width of about $\pm 1$ MeV and is in excellent agreement with the $D^{*+}$-$D^{0}$ mass difference of 145.3 MeV measured at SPEAR (39). Figure 16 contains nearly equal signal contributions from charmed and anticharmed mesons. A smaller signal consisting of $35 \pm 13$ events was observed for the $K_S\pi^{+}\pi^{-}$ decay of the $D^{0}$. Because the cascade pion from $D^{*+}$ decay has only about 7% of the $D^{*+}$ momentum, it generally has a low laboratory momentum. For this reason, it was only observable in the second run of E87a as a result of the enhanced acceptance provided by the installation of the second analyzing magnet.

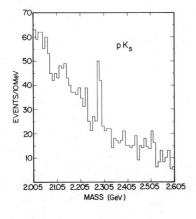

*Figure 15* $\Lambda_c^+(\bar{\Lambda}_c^-) \rightarrow pK_S(\bar{p}K_S)$ signals obtained in E87a.

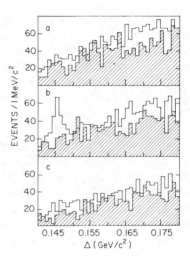

*Figure 16*  The D*-D mass difference plot for the decay modes $D^{*+} \to D^0\pi^+$ where $D^0 \to K\pi$. The histogrammed variable is $\Delta = M_{K\pi\pi} - M_{K\pi}$ for $K\pi$ mass combinations (*a*) below the known D mass; (*b*) straddling the known D mass; and (*c*) above the D mass. The shaded region shows background to the photoproduced signals due to neutral hadronic contamination in the photon beam.

## 5.2    Charm Signals Obtained in the CERN SPS Tagged-Photon Facility

Charm signals were obtained by the invariant mass technique from two experiments (NA 14 and WA-4) by a British, French, and German collaboration of physicists using the CERN Ω spectrometer (9). The Ω multiparticle spectrometer is a large-acceptance magnetic spectrometer augmented by a multicell Čerenkov counter and an electromagnetic shower counter. The WA-4 collaboration studied the interactions of photons from 40 to 70 GeV on a liquid hydrogen target. The spectrometer acceptance and photon spectrum were such that charm particles could be detected with reasonable acceptance in essentially the forward center-of-mass hemisphere ($X_f > 0$). By contrast, the FNAL wideband beam experiments only had acceptance in the $X_f$ region near 1/2. Evidence was obtained by this group for photoproduction of the $\bar{D}^0$, the $D^-$, and the $F^+$ and $F^-$.

The Ω spectrometer F signal (10, 11) appeared in the η-multipion and φ-multipion channels with masses consistent with the 2040-MeV mass reported by the DASP group studying $e^+e^-$ annihilations in the vicinity of 4 GeV (27) but inconsistent with the 1970-MeV mass reported recently by the CLEO group in their higher-energy study of $e^+e^-$ annihilation (30) and confirmed by other $e^+e^-$ experiments. It is our opinion that the recent evidence for the existence of the $F^+$ at a mass of 1970 MeV is inconsistent with the F results from the Ω spectrometer and outweighs the Ω results. For this reason we review only the D results from the Ω spectrometer.

The inclusive invariant mass spectra for the $K\pi$ and $K\pi\pi^0$ final states, where the kaon has been identified by the Čerenkov counter, showed enhancements with marginal statistical significance at the known D mass in both anticharm meson channels. No charmed meson signals were observed. If both the $K\pi$ and $K\pi\pi^0$ invariant mass spectra are added together, a 5-sigma enhancement was observed at the $D^0$ mass. Because only an anticharm meson signal is observed, the WA-4 collaboration argued that charmed meson photoproduction at their energies is dominated by an associated production process where an anticharmed meson is produced in association with a charmed baryon. Because charmed baryons are presumed to decay copiously into final states that include a proton, the WA-4 collaboration exploited their assumed production mechanism by searching for anticharmed mesons produced in association with a proton. Figure 17 shows several K-multipion invariant mass spectra subject to the requirement that a heavy, positively charged particle is observed in the same event as the plotted $\bar{D}$ candidate. The observed heavy, positive track generally lies in a momentum range where it is not possible to distinguish positive kaons from protons; however, only the proton hypothesis makes sense in the context of this analysis. The proton requirement enhances the signal-to-noise ratio in all three channels.

## 5.3    Charm Signals Obtained in the FNAL Tagged-Photon Beam

Signals for the process $D^{*+} \rightarrow D^0\pi^+$ process have been reported by Fermilab experiment E516 (60, 71). In E516, tagged photons with energies from 60 to 160 GeV impinge on a 1.5 meter long hydrogen target, creating photoproduced final states that are analyzed in a two-magnet multiparticle spectrometer augmented with Čerenkov counters, an electromagnetic shower detector, and a recoil proton detector. Figure 18 shows the D*-D mass difference plot for the $K^{\mp}\pi^{\pm}$ and $K^{\mp}\pi^{\pm}\pi^0$ D candidates within 50 MeV of the known D mass. D signals consist of $41 \pm 9$ events for the $K\pi\pi^0$ and $39 \pm 8$ events for the $K\pi$ mode. Both signals are consistent with the known D*-D mass difference of 145.3 MeV. A slight excess of $D^{*-}$ over $D^{*+}$ signal is observed in the E516 data.

## 5.4    Charm Signals by Finite Lifetime Methods

Two groups have reported on photoproduced charm signals by the finite lifetime technique. An early result, published by the WA58 experiment, shows an example of associated production where a $\bar{D}^0$ is produced in association with a $\Lambda_c$ (7, 41). The WA58 apparatus consisted of the $\Omega$ spectrometer upgraded with a rapidly recycling emulsion target.

The second experiment was performed in the SLAC backscattered laser beam from 1980 to 1982 by the SLAC Hybrid Facility Photon

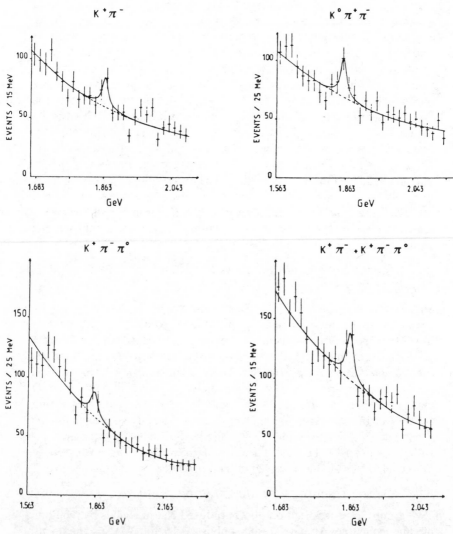

*Figure 17*    K multipion invariant mass spectra obtained by the WA-4 Collaboration. These spectra are subject to the requirement that there be a heavy, positive particle (presumably a proton) produced in association with the histogrammed D̄ candidate.

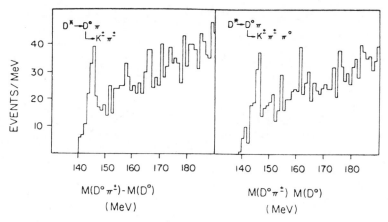

*Figure 18*  D*-D mass difference plot obtained by the E516 Collaboration.

Collaboration (1–4). In this experiment, the 20-GeV, nearly monochromatic photon beam interacted in a high-resolution hydrogen bubble chamber augmented with a magnetic spectrometer and shower counter. Evidence for a total of 62 charm events was obtained in the multiprong data on the basis of either a visible secondary decay vertex or the presence of a track that when extrapolated back to the primary vertex missed by an impact parameter outside of measurement error. A total of 49 events containing 51 charm decays remained after the imposition of cuts designed to ensure that events were detected with uniformly high efficiency. Background from three-body strange particle decay, multiple interactions, and other conspiratorial sources was estimated to be less than one event at the 90% confidence level. Figure 19 shows the effective decay length distributions for the 22 unambiguously neutral and 21 unambiguously charged secondary vertices. The curves are from Monte Carlo calculations using the group's best-fit lifetimes of $6.8 \times 10^{-13}$ s for the neutral decays and $7.4 \times 10^{-13}$ s for the charged decays. The group obtains a lifetime ratio of $\tau^+/\tau^0 = 1.1^{+0.6}_{-0.3}$, which is somewhat closer to unity than the previous world average of $\tau^+/\tau = 2.5$ and adds significantly to lifetime data. About one third of the total candidate sample was fully reconstructed. The eight neutral, fully reconstructed events were consistent in mass with the Cabibbo-allowed decays of the $D^0$, as one would expect. Eleven of 17 charged charm candidates, which in principle could be either $F^+$ mesons or $\Lambda_c$ baryons, had reconstructed masses consistent with the $D^+$ hypothesis. No discernable signal was present when the invariant mass was plotted under the $F^+$ hypothesis. The absence of reconstructed F candidates may

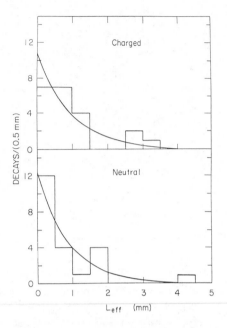

*Figure 19* The effective decay length distribution obtained by the SLAC Hybrid Photon Facility Collaboration for secondary vertices consistent with charmed particle decay. The curves are Monte Carlo calculations assuming lifetimes of $6.8 \times 10^{-13}$ s for neutral decays and $7.4 \times 10^{-13}$ s for charged decays.

reflect either a small F photoproduction cross section or a small F lifetime ($\tau < 2 \times 10^{-13}$ s). No clear $\Lambda_c$ signal was observed among the reconstructed charged decay sample, although there are indirect indications for a $35 \pm 20\%$ $\Lambda_c$ contribution to photoproduced charm (discussed in Section 5.5). Finally a search made for the $D^{*+} \rightarrow \pi^+ D^0$ process determined that the number of $D^{*\pm}$ per charm event was $17 \pm 11\%$, which implies that $D^*$ production near threshold is small.

## 5.5  *Comparing Open Charm, Real Photoproduction Data with Models*

In both the VMD and PGF models for charm photoproduction, a charm-anticharm quark pair is created by the incident photon in a symmetric fashion, with the valence quarks of the target nucleon playing the role of a passive spectator. In such models one expects an equal yield and identical momentum spectrum for charmed and anticharmed mesons, and charmed and anticharmed baryons. In addition to these symmetric models, one can imagine processes in which the valence quarks within the target play an active role in the charm photoproduction process. In such processes the quark-antiquark asymmetry within the target nucleon introduces asymmetries between the charmed and anticharmed particles that are produced. The associated production mechanism is an example of such an asymmetric production model. In this model, a forward-going anticharmed meson is produced against a backward-going charmed baryon in the overall center-of-mass frame. This is equivalent to the $t$-channel exchange of a D meson. The strangeness analogy of this process is known to play an important role in production of kaons and lambdas near low energies.

Because of acceptance limitations, most experiments are primarily sensitive to charm states produced in the forward hemisphere of the center of mass. For this reason, the ratio of anticharmed to charmed mesons and baryons gives the best indication as to whether symmetric or antisymmetric processes dominate charm production within the kinematic regime appropriate to the trigger and incident photon spectrum used in a given experiment. Each of the real photoproduction experiments can accommodate sizeable contributions from either class of model, but there is a considerable disagreement as to which process dominates. This disagreement began with the publication of results from the first two completed experiments—E87a and WA-4.

Each of the three charmed signals observed in the second run of E87a was consistent with equal charm-anticharm yields. These signals included production of the $\bar{\Lambda}_c$ decaying via $pK^0_S$ and $D^{*+}$ decaying via $D\pi^+$ with the D decaying via $K\pi$ and $K^0_S\pi^+\pi^-$. After acceptance corrections, $34 \pm 15\%$ of the D* were produced along with an additional kaon with a charge opposite that of the kaon from the observed D*. This large recoil kaon fraction suggests the photoproduced final state consists of a $c\bar{c}$ meson pair, rather than the anticharmed meson–charmed baryon final state expected in associated production. The cross section for diffractively produced $D^0$ mesons found in E87a at an average photon energy of 120 GeV was found to be $295 \pm 130$ nb per nucleon (19).

In contrast with the E87a results, the WA-4 experiment observes primarily associated production for D mesons. By comparing their inclusive yield of D mesons to the yield after requiring that there be a heavy positive particle accompanying the charm state, WA-4 finds that essentially their entire observed charmed meson yield is consistent $\bar{D}$'s produced in association with centrally produced charmed baryons that decay equally into proton and neutron final states. The cross section for associated production of $\bar{D}^0$ mesons at an average energy of 50 GeV obtained by the WA-4 experiment was $570 \pm 190$ nb per nucleon.

Although the characteristics of the photoproduced charmed final states observed in E87a and WA-4 differ substantially, there is no hard contradiction between the results owing to differences in triggering acceptance and beam energy. Although the E87a data placed an upper limit of $10\%$ in the fraction of $D^{*-}$ events with an additional accompanying proton, the acceptance for observing this proton under the assumptions of the WA-4 model is estimated to be about 25%. Under the assumption that protons are produced in 50% of charmed baryon decays in the associated production process, this upper limit is consistent with an associated production component equal to the observed symmetric, diffractive component. Alternatively, although the WA-4 data requires an associated production component, it does not exclude a cross section for symmetric, diffractive, charm photoproduction up to a limit of 410 nb per nucleon.

Because the D* signals obtained by the E516 collaboration were taken under a trigger that required an elastically recoiling proton detected in their recoil detector, their average photoproduced $D^{*+}$ cross section of $92 \pm 34$ nb per nucleon presumably represents symmetric, diffractive production rather than associated production. In order to investigate the level of associated production, E516 took additional data under a trigger that allowed for the production and decay of a charmed baryon in the recoil region. Data collected under this trigger was used to set a limit on the associated production process $\gamma p \to \Lambda_c^+ \bar{D}^* X$ of $<60$ nb per nucleon over the incident photon energy range from 40 to 160 GeV. Unfortunately, an E516 limit on $\bar{D}$ associated production that can be directly compared to the measured WA-4 cross section is currently unavailable.

We turn next to data on associated production obtained by finite lifetime techniques rather than spectrometer techniques. Lifetime techniques suffer from statistical limitations but compensate by having a much larger acceptance. The WA58 collaboration obtained one example of the associated production process in an emulsion exposure at the CERN $\Omega$ spectrometer in 1981. No fully reconstructed associated production charm events were observed by the SLAC Hybrid Facility Photon Experiment; however, there is indirect evidence for associated production. The excess of $\Lambda$'s observed in charm events over noncharm events implies that $40 \pm 20\%$ of the charmed events are associatedly produced. A similar estimate for associated production is obtained from the excess of $\bar{D}$ over D mesons observed in the data. On the other hand, a total of six events were observed with both a D and $\bar{D}$ candidate in the final state. Combining all this indirect evidence, the SLAC Hybrid Facility Photon Collaboration concludes that the level of associated production is $35 \pm 20\%$ of total charm production. Of course, such a level of associated production is hardly a surprise for production so close to threshold. Although there is evidence for the existence of a nondiffractive component to open charm photoproduction, the higher-energy experiments, E516 and E87, appear to be dominated by diffractive mechanisms.

Comparisons of the results of these experiments to the predictions of the VDM and PGF models are not completely unambiguous. Unfortunately, much of this data is too restrictive in statistics and kinematic coverage to pin down some of the important parameters of these models, and some of the data appear to be inconsistent with the models. Experimental data can be compared to the two diffractive models on the basis of the energy dependence of the charm photoproduction cross section, as well as through analysis of inclusive properties of the final state.

A common expectation in both the PGF and VDM models is that the momentum transfer between the incident photon and the final total $c\bar{c}$ state should fall exponentially in a manner reminiscent of hadronic, diffractive scattering. E516 obtains the $t$ distribution, shown in Figure 20, by measuring the elastically scattered target proton with their recoil detector for D* events. Their data are consistent with an exponential slope of 4.7 GeV$^{-2}$.

Both the PGF and VDM models predict that the $P_\perp$ of photoproduced charmed particles should be small (i.e. typically less than 1 GeV at the energies available in present experiments). Although VDM makes no specific prediction, one expects that the mass spectrum of the final state produced by a diffractively dissociated vector meson will peak sharply near threshold, which would thus limit the $P_\perp$ available to charmed particles.

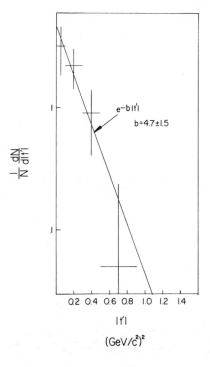

*Figure 20*  The $t$ distribution obtained by the E516 collaboration for events with an observed D*.

One also expects the $c\bar{c}$ final state produced by photon-gluon fusion to be peaked near threshold owing to the presumed softness of the gluon momentum distribution within the proton. Furthermore, the PGF model predicts that because of propagator effects the charmed quarks will be produced primarily along or against the incident photon direction; this would further limit available $P_\perp$ to the final-state charmed particles. The data from E87a, E516, and the SLAC Hybrid Facility Photon Collaboration show that the majority of photoproduced charmed particles have a $P_\perp$ of less than 1 GeV with respect to the beam axis.

As representative data we show the $P_\perp$ distribution for the D* produced in E87a in Figure 21 (70). The $D^{*+}$ $P_\perp$ data shown in Figure 21 are consistent with either the PGF or VDM model. Within the VDM context, the $P_\perp$ distribution is well fit to the isotropic decay of an intermediate state with a $1/M^N$ mass spectrum (where $N$ ranges from 5 to 10) decaying via $D^*\bar{D}^*$. Within the context of the PGF model, the $P_\perp$ distribution can be fit with a gluon momentum distribution of the form $xG(x) = (1-x)^n$, with $n$ ranging from 5 to 10.

The fact that the majority of $c\bar{c}$ pairs created by the PGF process have masses very close to threshold considerably complicates the analysis of the $P_\perp$ distribution. For example, the intrinsic $P_\perp$ of the gluon and the charmed quark dressing process may contribute significantly to the $P_\perp$ of the final-state D*'s. In addition, the usual assumptions about scaling of the dressing function may not apply near $c\bar{c}$ threshold and the distinction between scaling in energy or scaling in momentum becomes significant. For all these

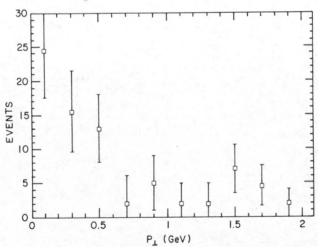

*Figure 21*   The $P_\perp$ distribution of D*'s obtained by the E87a collaboration. $P_\perp$ is taken with respect to the incident photon direction.

reasons it is felt that agreement between the $P_\perp$ distribution of D*'s and the PGF model is suggestive rather than conclusive (19).

E516 employed a hydrogen target, surrounded by a recoil detector, and a tagged incident photon beam. The recoil counter was capable of measuring the direction and energy of the elastically scattered recoil proton. In events where the target proton was elastically scattered into the final state, one could use the recoil counter information to determine the 4-vector of the $c\bar{c}$ system by subtracting the recoil proton 4-vector from the tagged 4-vector of the incident photon. This allowed for an event-by-event measurement of the $c\bar{c}$ system invariant mass, and the angle of the D* with respect to the incident photon in the $c\bar{c}$ center-of-mass system.

To enhance the fraction of charm data with useful recoil data, the bulk of the E516 data was taken under a trigger that required an elastically scattered proton in the final state and a deduced $c\bar{c}$ bar mass above charm threshold. This triggering requirement may have suppressed the contribution of charm production from the photon-gluon fusion process, which requires the interaction of a photon with a moderately high-momentum gluon within the proton. Once this gluon is removed from the proton to form a $c\bar{c}$ pair, the proton may have a small probability of elastically scattering into the final state and satisfying the E516 trigger. With this warning we proceed to compare the E516 data with the VDM and PGF models.

Figure 22 gives the mass distribution recoiling against the elastically

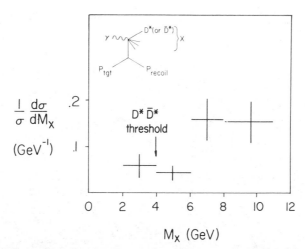

*Figure 22* The charmed system invariant mass distribution obtained by the E516 collaboration. The charmed system mass is the mass recoiling against elastically scattered target protons as measured by the recoil detector.

scattered proton in D* events. Although the statistics are limited, the mass distribution is surprisingly broad with a substantial number of events more than 2 GeV above threshold. Such large masses for the total charm system are unexpected in either the VDM or PGF model. Figure 23 gives the result of a Monte Carlo calculation for the mass of the $c\bar{c}$ system for photon-gluon fusion at a beam energy of 100 GeV (the average E516 beam energy). This calculation assumes a 1.6-GeV charmed quark bare mass, and $xG(x) = (1-x)^5$. The disagreement between this model calculation and the data appears to be too large to accommodate by a reasonable change of these parameters.

The large mass of the total charmed system and the limited $P_\perp$ of the D*'s with respect to the incident photon imply a very nonisotropic distribution for D*'s in the total charmed center-of-mass system. Figure 24 shows this distribution for the E516 data compared to distribution given by $d\sigma/d\Omega \propto \cos^4 \theta$, where $\theta$ is the angle between the incident photon and the D*. As previously discussed, the PGF model predicts a similar anisotropy for charmed mesons in the charmed center-of-mass system. However, our PGF Monte Carlo for 100-GeV incident photons gives a much flatter angular distribution than does $\cos^4 \theta$. The PGF-predicted distribution in this energy range is very close to $1 + \cos^2 \theta$.

The E516 mass and angular distributions appear to disagree significantly with the details of both the VDM and PGF models. Perhaps this is because the E516 trigger is biased against the PGF process because of the elastic

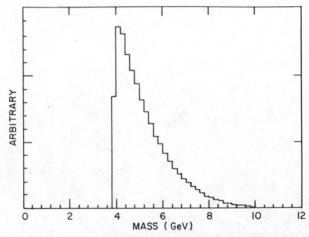

*Figure 23* The charmed system invariant mass distribution expected in the photon-gluon fusion model for 100-GeV incident photons. This calculation uses counting rule glue.

recoil requirement. Another possibility is that some fraction of the E516 D*
sample actually represents inelastic target breakup rather than elastically
scattered target protons. Inclusion of such data would only be possible if the
inelastic fragments escaped detection in the recoil detector. These inelastic
fragments would distort the recoil proton momentum enough to cause an
anomalously large apparent mass for the total charmed system. The E516
group believes such inelastic contamination to their D* sample is small.

Both the PGF and VDM models predict an open charm photoproduction cross section that slowly varies with incident photon energy once one is
sufficiently above threshold. Within the context of the PGF model, the rise
of the photoproduction cross section with increasing energy is strongly
dependent on the $x$ dependence of the gluon distribution function. Figure
25 (60) shows the world's data on the energy dependence of the open charm
photoproduction cross section. Data are combined from many experiments
in order to get a wide energy coverage. The figure also includes virtual open
charm photoproduction data. There is certainly danger in combining data
from different experiments with different final states, acceptances, and
trigger biases, but no single experiment on the real photoproduction of
open charm has a wide enough energy range in isolation. Superimposed on
this data is the PGF prediction using $xG(x) = (1 - x)^5$. The relative flatness
of data above 50-GeV incident photon energy is consistent with the
expectations of VDM, and the low-energy turn-on appears to be consistent
with the PGF and naive gluon distribution, except for the very low-energy
point from the SLAC Hybrid Photon Facility Collaboration. This
disagreement may indicate a broader gluon distribution than the naive
counting rule gluon distribution, or may indicate a significant degree of

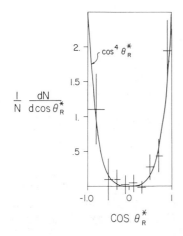

*Figure 24* The D* angular distribution
with respect to the incident photon direction measured in the charm system center
of mass obtained by the E516 collaboration. The solid curve is a $\cos^4 \theta$ distribution.

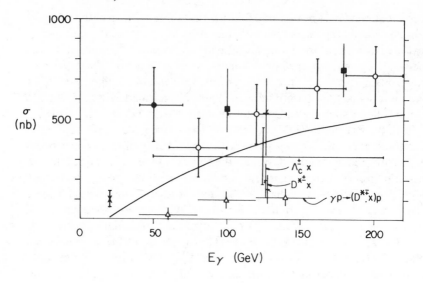

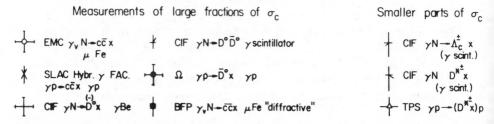

*Figure 25*  A compendium of the world's data on the real and virtual charm photoproduction cross section. This figure is reproduced from (60).

additional, non-PGF production. We recall this group's evidence for a considerable, additional, nondiffractive production mechanism.

## 6.  THE VIRTUAL PHOTOPRODUCTION OF OPEN CHARM STATES

Present experimental data on virtual photoproduction comes from experiments studying the interactions of high-energy muons on iron targets, which result in events with two or three muons in the final state. The semileptonic muon decays of charmed particles have been established as a significant source of multimuon events. Measurements of the virtual photon charm cross section have been published by both the Berkeley-Princeton-Fermilab collaboration (BPF) and the European Muon Collaboration (EMC).

The BPF data (47) are based on a high-statistics sample of $\mu^{\pm} Fe \rightarrow \mu^{\pm} \mu x$ collected at the multimuon spectrometer in the N1 beamline at Fermilab. In the BPF experiment, incident 209-GeV muons impinge on a 475-ton detector-target that is capable of identifying and measuring the momentum of final-state muons and measuring the energy in the final-state hadronic shower using calorimetric methods. This detector cannot identify or accurately measure the momenta of individual hadrons.

The EMC data are based on a somewhat smaller sample of $\mu^+ Fe \rightarrow \mu^+ \mu x$ and $\mu^+ Fe \rightarrow \mu^+ \mu^+ \mu^- x$ events collected in the CERN SPS muon beam using the EMC forward spectrometer. Analyses have been published on an earlier data run with 280-GeV incident muons, as well as on a later run with five times the luminosity using 250-GeV muons (16); we discuss here the data of the later run. The 250-GeV incident muons interacted in a 3.75 meter long iron scintillator target (STAC), which measured the hadronic shower energy accompanying the final-state muons. The final-state muons were momentum analyzed by a magnetic spectrometer. The essential information on multimuon events in both the BPF and EMC experiments consists of precise momentum determination of both the initial- and final-state muons and less precise information on the hadronic energy accompanying the final-state muons. Because neither experiment is able to reconstruct the invariant mass of charmed mesons and baryons by measuring the momenta of the decay products, the groups must eliminate the noncharmed sources of their dimuon and trimuon events. We begin with a discussion of backgrounds to the dimuon charm sample.

## 6.1  *Isolation of the Charm Sample*

Both groups of experimenters estimate that the most serious background to charm production in the $\mu Fe \rightarrow \mu \mu x$ sample after the imposition of all background-reducing kinematic cuts comes from the decay in flight of pions and kaons produced in $\mu Fe$ interactions. The level of this background after cuts are applied is calculated to be $\sim 20\%$ and is discussed in more detail below. Additional conventional background sources include the muon production of another $\mu^+ \mu^-$ pair through either the QED trident process or through the muon decay of a light vector meson. These sources can produce background to the charm signal when one member of the muon pair fails to be detected. After the kinematic cuts are employed, the residual contaminations from these sources are estimated to be less than 1%. Less conventional background sources include the QED production of

$\tau^+\tau^-$ pairs, and photoproduction of bottom mesons and baryons and their subsequent semileptonic decay. The level of this contamination is calculated to be less than 1%.

Both groups have imposed a set of cuts on their $\mu$Fe → $\mu\mu x$ data sample in order to reduce their background level and to measure charm events in kinematic regions free from rapid variations in acceptance. The cuts require a minimum momentum for both final-state muons, significant hadronic deposition in the calorimeter, and a minimum virtual photon energy. The photon energy cut and muon momentum cut suppress nonprompt dimuon background, while the hadronic energy cut suppresses QED tridents, which tend to transfer little energy to the target.

Because of differences in the EMC and BPF detectors, the EMC group must invoke additional cuts that suppress a large trident contamination to their dimuon sample. The EMC apparatus is made insensitive in the central region through which the beam passes. This central hole allows one to miss low-$Q^2$ tridents where the scattered muon is lost down the central hole and the produced dimuon mimics a $\mu^+$Fe → $\mu^+\mu^- x$ event. Low-$Q^2$ trident production accounts for a significant excess in the initial number of opposite-sign dimuons (49,791) compared to same-sign dimuons (9,971) before the imposition of clean-up cuts. These trident-suppressing cuts exclude events with hits in either a forward high-rate chamber or hodoscope (large amounts of missing energy) or a low amount of shower energy recorded in the STAC. After the imposition of these cuts there remain comparable numbers of opposite-sign and same-sign dimuon events.

The only remaining significant background to charm arises from nonprompt K, $\pi$ decays from inelastic $\mu$Fe interactions. The behavior and normalization of this background was determined from Monte Carlo calculations utilizing data from other muon production experiments and hadronic shower simulations. The contamination of $\pi$, K background in various kinematic regions is assessed in Figure 26 by the BPF group (47), and in Figure 27 by the EMC group (16).

Figure 26 shows the background-subtracted inclusive distributions for the ~20,000 dimuon events satisfying the BPF analysis cuts. The histograms plotted below the axis are the predicted $\pi$, K background inclusive distributions from their $\pi$, K Monte Carlo calculations. Overall the background level is about $19 \pm 10\%$ of the charm level. The prediction (*solid curve*) comes from the photon-gluon fusion model (see below). Figure 27 shows distributions of the 2,900 dimuon events satisfying the EMC analysis cuts. The variable $P_\perp^2$ is the square of the transverse momentum of the daughter muon relative to the direction of the virtual photon. The $\pi$, K– computed background are also shown in Figure 27. Overall the K, $\pi$ background level is estimated to be $14 \pm 7\%$ in the EMC dimuon sample.

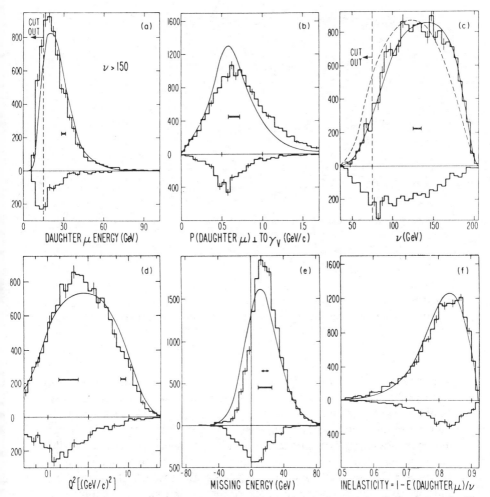

*Figure 26* Raw inclusive properties of the BPF dimuon virtual photoproduced charm sample. The histograms below the *x* axis are calculations of the charm sample background due to nonprompt muon sources (π and K decay). The solid curves are predictions of the PGF model.

## 6.2 *Extraction of the Charm Virtual Photoproduction Cross Section*

The multimuon experiments discussed in this section impose cuts on the minimum energy of daughter muons; neither experiment has acceptance for charm produced nearly at rest in the photon-nucleon center of mass. For this reason both the BPF and EMC experiments published cross sections for "diffractive" or "current fragmentation region" virtually photoproduced charm.

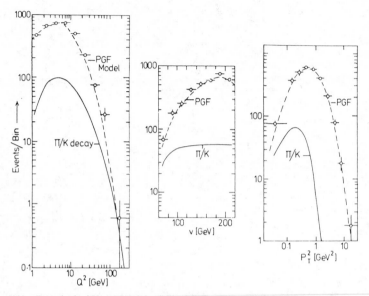

*Figure 27*   Raw inclusive properties of the EMC dimuon virtual photoproduced charm sample. The estimated nonprompt background is shown in the curves marked π/K. The data is well represented by the sum of this background and the predictions of the PGF model.

In order to extract the diffractive charm photoproduction cross section from multimuon data, one must make assumptions about the average semileptonic branching ratio of virtually photoproduced charm particles and the semileptonic decay mechanism. Both the EMC and BPF groups assume a branching ratio and lepton momentum spectrum consistent with measurements made in $e^+e^-$ annihilations near charm threshold. Although both the PGF and VDM models predict that the charm system invariant mass distribution will be peaked toward threshold, there is no compelling reason to expect that the virtual photoproduction $D^0$, $D^+$, $F^+$, and $\Lambda_c^+$ ratios are identical to the ratios produced in threshold $e^+e^-$ experiments. Furthermore, charm lifetime data and specific-state semileptonic branching data suggest unexpectedly large differences in the $D^0$, $D^+$, $F^+$, and $\Lambda_c^+$ semileptonic branching ratios.

This issue is addressed by the EMC group, who have isolated a charm trimuon sample consisting of 83 events after the removal of a large background from vector mesons and QED tridents. By comparing the yield of trimuon charm events (proportional to the square of the branching ratio) to the yield of dimuon charm events (linearly proportional to the branching ratio), the EMC experiment obtains an average semileptonic branching ratio of $7 \pm 3\%$, which is consistent with the assumptions of EMC and BPF.

Both the BPF and EMC experiments use the photon-gluon fusion model as a physics generator for a Monte Carlo simulation of diffractive charm virtual photoproduction. This simulation is used to compute separate acceptance correction factors for each $\Delta v$-$\Delta Q^2$ bin employed in the data. Necessary inputs to the photon-gluon fusion model include a prescription for partitioning the cross section between closed and open charm, the value of $\alpha_s$ and the bare charm quark mass, the gluon momentum distribution, and a prescription for dressing charm quarks into charmed hadrons. Acceptance calculations are most sensitive to the gluon momentum distribution function and the dressing procedure since they most seriously influence the lab momentum spectrum of the daughter muons.

Both groups use the naive counting rule gluon distribution $xG(x) = 3(1-x)^5$. This distribution will significantly influence the $v$ dependence of the charm cross section, which is well matched by the Monte Carlo models of both groups (see Figures 26c and 27b). The EMC and BPF Monte Carlo models differ substantially in their assumptions concerning charm quark dressing. However, both groups' Monte Carlos succeed in reproducing the daughter elasticity distributions ($Z_\mu = E_D/v$) of their respective data sets. This suggests that either dressing assumption gives an adequate fit to the daughter muon spectrum and thus should be reliable in estimating inefficiencies due to the minimum daughter energy cut. The experimenters ascribe a $\pm 20\%$ systematic uncertainty in their acceptance to dressing uncertainties.

Figure 28 illustrates a technique for extrapolating virtual photoproduction data to $Q^2 \to 0$ by showing the acceptance-corrected $Q^2$ dependence of the diffractive charm virtual photoproduction obtained by the BPF group in two large $v$ bins: $75 < v < 133$ GeV ($\langle v \rangle = 100$ GeV), and $v > 133$ GeV ($\langle v \rangle = 178$ GeV). The various curves show the results of several VDM-inspired fits to the measured $Q^2$ dependence of the form $\sigma(Q^2) = \sigma(0)/(1 + Q^2/\Lambda^2)^2$. The $\Lambda$ parameter is measured to be $2.9 \pm 0.2$ GeV at $\langle v \rangle = 100$ GeV and $3.3 \pm 0.2$ GeV at $\langle v \rangle = 178$ GeV.

The $v$ dependences of the charm photoproduction cross section obtained by the BPF and EMC experiments are included in Figure 25 and are found to be quite consistent with each other and the world's sample of real photoproduced cross sections.

Leptoproduction data do not appear to be strong enough at present to rule out a VDM picture for charm virtual photoproduction, although there are some negative indications. Two major VDM predictions are that the cross section becomes $v$ independent at large $v$ and that the $Q^2$ dependence is of the form $\sigma(Q^2) = \sigma(0)/(1 + Q^2/\Lambda^2)^2$, with $\Lambda = M_\psi$. The EMC and BPF groups observe a significant rise in the virtual photoproduction charm cross section from $v = 75$ to $200$ GeV but this $v$ range is sufficiently close to the threshold for diffractive dissociation limiting behavior to prevent the rising cross section from being used as a strong argument against VDM.

Figure 29 shows the dependence of $\Lambda$ on $v$ for both the BPF and EMC groups. Both data sets show a statistically marginal rise in $\Lambda$ with increasing $v$. The EMC data appear to be incompatible with $\Lambda = M_\psi$ but they also appear incompatible with the BPF data, which appear more consistent with $\Lambda = M_\psi$. The BPF group states that their fits suggest significant $v$ dependence in the $\Lambda$ parameter because their quoted errors include some systematic errors common to their two $v$ bins.

It is interesting to observe that $\Lambda$ would be expected to rise in the photon-gluon fusion model in which the dominant $Q^2$ dependence is of the form $(1 + Q^2/\hat{s})^{-2}$, where $\hat{s}$ is the effective photon-gluon fusion squared center-of-mass energy (which should grow slowly in $v$). More data will be required to discriminate between the $Q^2$ dependence expected in the VDM and PGF models.

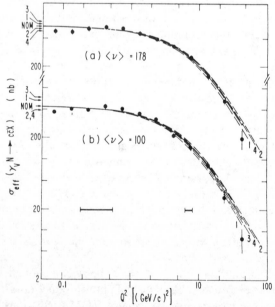

*Figure 28*  The BPF extrapolation of the virtual charm photoproduction cross section to $Q^2 \to 0$ for data in a bin centered at $\langle v \rangle = 100$ GeV and $\langle v \rangle = 178$ GeV. The various curves are fits with different assumed VDM-inspired propagator masses. The error bars at the bottom of the figure show typical errors in the two $Q^2$ regions.

Because of the success of the photon-gluon fusion model in fitting the charm virtual photoproduction of both the BPF and EMC group, one might hope to gain detailed information about some of the unspecified parameters of that model from leptoproduction data. Unfortunately, a large variation of model parameters can successfully fit the data. With some parameters this occurs because changes in several parameters can mutually compensate in such a way as to leave the prediction unchanged.

For example, measurement of the charm leptoproduction in cross section should in principle provide useful information on both $\alpha_s$ and $G(x, Q^2)$, the gluon momentum distribution within the nucleon. Because the yield of charmed particles is proportional to $\alpha_s$ in the PGF model, one would hope that the cross section might serve as a direct measure of this important quantity. Unfortunately, the predicted level of the cross-section function depends on $M_c$ (the bare charm quark mass), which strongly influences the elementary cross section for the process $\gamma g \to c\bar{c}$ at $\hat{s}$ near threshold where

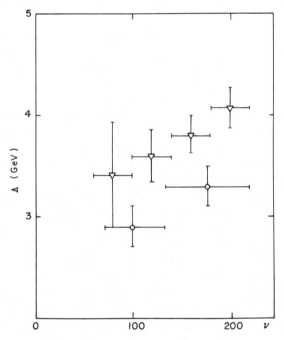

*Figure 29* Summary of fits to the $Q^2$ dependence of the virtual photoproduced charm cross section. We plot the value of $\Lambda$ where $\sigma(Q^2) = \sigma(0)/(1 + Q^2/\Lambda^2)^2$. The BPF points are circles; the EMC points are triangles. The values of $\Lambda$ are close to the $\psi$ mass but there is some suggestion of a rise of $\Lambda$ with increasing $\nu$.

the bulk of the cross section is predicted to be for the currently accepted gluon distributions. Both experimental groups claim that changes in $M_c$ primarily affect the normalization of the open charm cross section rather than its $v$ or $Q^2$ dependence. Uncertainties in both the value of $M_c$ and the prescription for determining the fraction of the charmed quark cross section appearing as open charm precludes a clean determination of $\alpha_s$ through measurements of the charm yield. Alternatively, one could attempt to measure $\alpha_s$ by measuring its effect on the $Q^2$ evolution of the charm leptoproduction cross section. The QCD mass scale determines $\alpha_s$ through the relation $\alpha_s = (12\pi/25) \ln [(Q^2 + \hat{s})/\bar{\Lambda}^2]$. Unfortunately, additional $Q^2$ dependence can arise from scale violations in the gluon distribution.

The EMC group has inverted this argument by fixing $\hat{\Lambda}$ at 0.5 GeV and fitting for the gluon distribution in separate bins of $Q^2$ to the form $xG(x, Q^2) = A(1-x)^n$. They obtain an $n$ value consistent with the naive counting rule prediction $n = 5$ and roughly $Q^2$ independent.

The BPF group has studied the gluon distribution in the low-$Q^2$ limit through fits to the charm leptoproduction $v$ dependence. Their results prefer the counting rule glue $[xG(x) = (1-x)^5]$ but systematic uncertainties preclude exclusion of the other gluon distributions.

A final issue addressed by the leptoproduction experiments concerns the mechanism through which charm quarks hadronize into charm particles. The dressing process will primarily influence the energy spectrum of the daughter muons. Both EMC and BPF succeed in fitting the daughter energy spectrum, but they use significantly different dressing procedures. The BPF model hadronizes the charm quark in the quark-gluon center-of-mass system. The charmed mesons are produced along the direction of the original quarks and acquire an energy of $E_D = (\sqrt{s}/2)Z$ where $Z$ is drawn from the distribution $D(Z)$. Their data is most consistent with the form $D(Z) \propto (1-Z)^{0.4}$ first suggested by SPEAR data.

The EMC group has considered hadronization in both the photon-gluon center-of-mass and the lab frames. The choice of reference frame bears on the issue of whether the charm quarks hadronize by sharing light-quark pairs among themselves, or whether they share light-quark pairs with nucleon quarks. In the former case, the photon-gluon frame seems most appropriate; in the latter case, the lab frame is most appropriate. In either frame, the EMC group produces their D mesons with a fraction $Z$ of the charm quark momentum where $Z$ is drawn from $D(Z)$. We note that the EMC group uses $Z$ as a momentum-partitioning variable while BPF uses $Z$ as an energy-partitioning variable. Unfortunately, the energy and momentum in the photon-gluon center-of-mass frame are quite different near threshold, where the bulk of the cross section lies. Monte Carlo models are compared to the dimuon elasticity distribution ($z_\mu = E_\mu/v$) and the trimuon

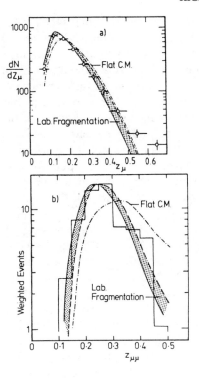

*Figure 30* EMC data on the dressing of charmed particles produced in virtual photoproduction. Figure (*a*) uses the EMC dimuon charm sample and histograms $Z_\mu$ or the fraction of $v$ carried by the single daughter muon. The curves are for flat dressing in the charm system center of mass and for dressing in the laboratory. Figure (*b*) uses the EMC trimuon charm sample and histograms $Z$ or the fraction of $v$ carried by both daughter muons. Again the histograms are compared to Monte Carlo calculations assuming charm system center-of-mass dressing and laboratory dressing.

elasticity distribution $[z_{\mu\mu} = (E_{\mu1} + E_{\mu2})/v]$ in Figure 30. The EMC group concludes their data are consistent with $D(Z) = \exp(1.6 \pm 1.6Z)$, where hadronization occurs in the lab frame.

## 7.  CONCLUDING REMARKS

Over the past decade photoproduction experiments have played a vital role in the development of our understanding of charmed particles. Ten years ago photoproduction experiments demonstrated conclusively that the then newly discovered $J/\psi$ particle was indeed a hadron. Since then, the observations of charmed particles produced in photon-nucleon interactions have shed light on the dynamics of heavy-quark pair formation, and provided tests of models describing the structure of the nucleon and the nature of the strong interaction. Recent measurements of the lifetimes of charmed particles produced in photon interactions have played a major role in determining the $D^+$ and $D^0$ lifetimes.

As we have indicated throughout this review, present heavy-flavor photoproduction data is consistent with either the PGF or VDM type of model. At higher energies, the underlying subprocess of the PGF model will become more apparent, and the predictions of the VDM and PGF models will start to diverge. It is thus of considerable interest to extend the charm photoproduction measurements to large energies, where threshold effects become less important. As more energetic photon beams become available, data on the photoproduction of still heavier flavors should augment the present data on charm photoproduction. Programs of measuring the yield, production mechanisms, and properties of beauty particles (b quarks) will be initiated in the near future at the Fermilab Tevatron using photons with energies up to 500 GeV. Of immediate interest will be a comparison of the observed yield to the theoretically expected yield based on the photon-gluon fusion model and present understanding of the gluon momentum distribution within the proton. Although the PGF model predicts that $\Upsilon$ production should be only a few tenths of a percent of open beauty production, the relative cleanliness of the dilepton decay mode should permit observation and measurement of $\Upsilon$ photoproduction. It should be possible to measure the $\Upsilon$-nucleon total cross section and check that it continues to follow the $M_v^{-2}$ mass scaling law that the presently measured vector mesons obey. High-resolution vertex chambers will be standard in the next generation of photoproduction experiments, and this will allow measurement of particle lifetimes as low as several times $10^{-14}$ s. The Tevatron experiments should thus provide new data on the properties and spectroscopy of beauty particles.

The photoproduction of particles containing the top quark (t quark), which present data indicate may be as massive as 30 to 40 GeV, awaits either the completion of an electron-proton collider or the construction of an external photon beam at an SSC scale hadron collider. The HERA electron-proton collider will be able to produce virtual photon beams with equivalent laboratory energies up to 51 TeV. The anticipated yield of 40-GeV top particles in a year's run of HERA ($L = 500$ pb$^{-1}$) should be in the thousands. HERA may become a fruitful hunting ground for some of the new states in the 100-GeV mass range that couple to gluons and photons, predicted by either the supersymmetric or technicolor theory. With the construction of an electron-proton collider built in association with either the proposed Superconducting Super Collider or with a hadron accelerator residing in the LEP tunnel, one could envision reaching equivalent laboratory photon energies of hundreds of TeV and extending sensitivity to new states with masses up to 500 GeV. We anticipate that high-energy photoproduction will continue to play an important role in spectroscopy, and in the understanding of the strong interaction far into the future.

ACKNOWLEDGMENTS

It is our pleasure to acknowledge the E516, E401, and SLAC Hybrid Facility Photon Collaborations for the use of some of their unpublished data. We also gratefully acknowledge Jean Therrien of Nevis Laboratory for her photographic skills, and Jeannine Adomaitis of the University of Illinois for her patience and skill in typing the many drafts of this manuscript. This work was supported in part by the Department of Energy and the National Science Foundation.

*Literature Cited*

1. Abe, K., et al. *Phys. Rev. Lett.* 48:1526 (1982)
2. Abe, K., et al. *Phys. Rev. Lett.* 51:156 (1983)
3. Abe, K., et al. *SLAC PUB-3271.* Stanford, Calif: SLAC (1983)
4. Abe, K., et al. Paper contributed to *XXII Int. Conf. on High Energy Phys.*, Leipzig, Germany (1984)
5. Abrams, G. S., et al. *Phys. Rev. Lett.* 33:1453 (1974)
6. Deleted in proof
7. Adamovich, M. I., et al. *Phys. Lett.* 99B:271 (1981)
8. Anderson, R. L., et al. *Phys. Rev. Lett.* 38:263 (1977)
9. Aston, D., et al. *Phys. Lett.* 94B:113–17 (1980)
10. Aston, D., et al. *Phys. Lett.* 100B:91 (1981)
11. Aston, D., et al. *Nucl. Phys. B* 189:205–11 (1981)
12. Atiya, M. S., et al. *Phys. Rev. Lett.* 43:414 (1979)
13. Aubert, J. J., et al. *Phys. Rev. Lett.* 33:1404 (1974)
14. Aubert, J. J., et al. *Phys. Lett.* 89B:267 (1980)
15. Aubert, J. J., et al. *Nucl. Phys. B* 213:30 (1983)
16. Aubert, J. J., et al. *Nucl. Phys. B* 213:31–64 (1983)
17. Augustine, J. E., et al. *Phys. Rev. Lett.* 33:1406 (1974)
18. Avery, P., et al. *Phys. Rev. Lett.* 44:1306–12 (1980)
19. Avery, P. 1980. *Photoproduction of the $D^{*\pm}$.* PhD thesis. Univ. Ill. 161 pp. (1980)
20. Barger, V., et al. *Phys. Lett.* 91B:253 (1980)
21. Barger, V., et al. In *XX Int. Conf. on High Energy Phys.*, Madison, Wisc., p. 271.

New York: Am. Inst. Phys. (1980)
22. Bauer, T. H., et al. 1978. *Rev. Mod. Phys.* 50:261 (1978)
23. Berger, E. L., Jones, D. *Phys. Rev. D* 23:1521 (1981)
24. Binkley, M., et al. *Phys. Rev. Lett.* 48:73 (1982)
25. Binkley, M., et al. *Phys. Rev. Lett.* 50:302 (1983)
26. Boyarski, A. M., et al. *Phys. Rev. Lett.* 34:1357 (1975)
27. Brandelik, R., et al. *Phys. Lett.* 80B:412 (1979)
28. Camerini, U., et al. *Phys. Rev. Lett.* 35:483 (1975)
29. Cazzoli, E. G., et al. *Phys. Rev. Lett.* 34:1125 (1975)
30. Chen, A., et al. *Phys. Rev. Lett.* 51:634 (1983)
31. Clark, A. R., et al. *Phys. Rev. Lett.* 43:187 (1979)
32. Clark, A. R., et al. *Phys. Rev. Lett.* 45:2092 (1980)
33. Drees, J., Montgomery, H. *Ann. Rev. Nucl. Part. Sci.* 33:385–452 (1983)
34. Denby, B. H., et al. *Phys. Rev. Lett.* 52:795 (1984)
35. Duke, D. W., Owens, J. F. *Phys. Lett.* 96B:184 (1980)
36. Duke, D. W., Owens, J. F. *Phys. Rev. D* 23:1671 (1981)
37. Duke, D. W., Owens, J. F. *Phys. Rev. D* 24:1403 (1981)
38. Egloff, R. M. 1979. *Measurements of elastic rho, omega and phi meson photoproduction cross sections on protons from 30 to 180 GeV.* PhD thesis. Univ. Toronto. 235 pp.
39. Feldman, G. J., Peruzzi, I., Piccolo, M., Abrams, G. S., Alam, M. S., et al. *Phys. Rev. Lett.* 38:1313–15 (1977)
40. Feynman, R. P. *Photon-Hadron Inter-*

*actions.* Reading, Mass: Benjamin (1972)
41. Fiorino, A., et al. *Lett. Nuovo Cimento* 30:166–70 (1981)
42. Fritzsch, H., Streng, K.-H. *Phys. Lett.* 72B:385 (1978)
43. Gittelman, B. In *Int. Symp. on Electron and Photon Interactions at High Energies,* p. 255. Stanford, Calif: SLAC (1975)
44. Gittelman, B., et al. *Phys. Rev. Lett.* 35:1616 (1975)
45. Glück, M. See Ref. 21, p. 268
46. Goldhaber, G., et al. *Phys. Rev. Lett.* 37:255 (1976)
47. Gollin, G. D., et al. *Phys. Rev. D* 24:559–89 (1981)
48. Halzen, F. *XXI Int. Conf. on High Energy Phys.,* p. 381. Paris: Ed. Physique (1982)
49. Hand, L. *Phys. Rev.* 129:1834 (1964)
50. Jones, L. M., Wyld, H. W. *Phys. Rev. D* 17:759–64 (1978)
51. Knapp, B., et al. *Phys. Rev. Lett.* 34:1040 (1975)
52. Knapp, B., et al. *Phys. Rev. Lett.* 37:882–84 (1976)
53. Lee, W. See Ref. 43, p. 213
54. Leveille, J. P., Weiler, T. *Nucl. Phys. B* 147:147 (1979)
55. Leveille, J. P., Weiler, T. *Phys. Lett.* 86B:377 (1979)
56. Leveille, J. P., Weiler, T. See Ref. 21, p. 263

57. Leveille, J. P., Weiler, T. *Phys. Rev. D* 24:1789 (1981)
58. Lüth, V., et al. *Phys. Rev. Lett.* 35:1124 (1975)
59. Nash, T., et al. *Phys. Rev. Lett.* 36:1233 (1976)
60. Nash, T. In *Int. Symp. on Lepton and Photon Interactions at High Energies,* p. 329. Ithaca, NY: Cornell Univ. Press (1983)
61. Particle Data Group. *Rev. Mod. Phys.,* Vol. 64 (1984)
62. Peruzzi, I., et al. *Phys. Rev. Lett.* 37:569 (1976)
63. Prepost, R. See Ref. 43, p. 241
64. Russell, J. J., et al. *Phys. Rev. Lett.* 46:799–802 (1981)
65. Schilling, K., Wolf, G. *Nucl. Phys. B* 61:381 (1973)
66. Sivers, D., et al. *Phys. Rev. D* 13:1234 (1976)
67. Strovink, M. In *Int. Symp. on Lepton and Photon Interactions at High Energies,* p. 594. Univ. Bonn, Germany (1981)
68. Tajima, T., Watanabe, T. *Phys. Rev. D* 23:1517 (1981)
69. Weiler, T. *Phys. Rev. Lett.* 44:304 (1980)
70. Wiss, J. In *VI Int. Conf. on Meson Spectroscopy,* Brookhaven Natl. Lab., pp. 257–84. New York: Am. Inst. Phys. (1980)
71. Silwa, K., et al. *Phys. Rev. D.* In press (1985)

*Ann. Rev. Nucl. Part. Sci. 1985. 35 : 455–99*

# INDUCED WEAK CURRENTS
# IN NUCLEI

*Laszlo Grenacs*

Université Catholique de Louvain, Institut de Physique, Chemin du Cyclotron 2, B-1348 Louvain-la-Neuve, Belgium

## CONTENTS

## 1. INTRODUCTION

"Induced weak currents" designate gradient-type contributions to the vector–axial vector (V-A) description of the charged hadronic current in

455

0163–8998/85/1201–0455$02.00

electroweak interactions. Their existence arises in general from the presence of the strong interaction, and is constrained by the symmetries governing any unified description of physical processes. In semileptonic reactions, these terms give rise to recoil-order departures of observables from their leading-order V-A prescriptions. The determination of these departures has historically played a critical role in the identification and verification of the symmetry structure of weak interactions. The important mechanisms for the induced effects are

1. a conserved vector current (CVC);
2. a partially conserved axial vector current (PCAC); and
3. a transformation of the currents under the combined operations of charge conjugation and charge symmetry (G-parity transformation), analogous to the strong interaction, suggesting the absence of the "second-class currents" (SCC).

The most celebrated among them, CVC, treats the strangeness-conserving weak vector currents and the isovector electromagnetic current as belonging to a single triplet of conserved currents. It constituted the first significant step in unifying weak and electromagnetic interactions and was the precursor of the $SU(2)_L \times U(1)$ gauge theory of electroweak interactions. Similarly, the existence of a (partially) conserved axial vector current and G-parity invariance had a historical role in encouraging the development of a unified $SU(3) \times SU(2)_L \times U(1)$ description embodying the strong interaction.

At present, the unified description of strong and electroweak processes is prescribed by local $SU(5)$ gauge invariance, in which the Standard Model of $SU(2)_L \times U(1)$ is embedded and its symmetries essentially preserved. With the direct verifications of the model through the discovery of predicted weak neutral currents and vector bosons, interest in the induced weak currents as low-energy signatures of the symmetry aspects of the Standard Model may appear to have waned. In fact, however, a new era in the investigation of these currents has emerged. The renewed interest lies in the fact that various low-energy experiments in beta decay, muon capture, and neutrino-induced reactions—formerly considered as a source of information on induced current–related symmetries—may now provide information on additionally hypothesized aspects of electroweak theories beyond the standard description. These aspects arise from larger unification schemes not yet accessible to precise, direct, higher-energy accelerator testings. They include

1. the Lorentz structure of the weak interaction beyond V-A (68, 104);
2. the possible existence of right-handed currents (70, 131);

3. the possible existence of massive neutrinos (28, 130);
4. additional tests of $\mu/e$ universality; and
5. the existence of the time reversal noninvariance (65).

Since accuracies of better than $10^{-3}$ are generally required in such tests, the contributions of induced weak currents cannot be neglected: they become input to the experimental analysis, and therefore significant constraints upon the uncertainty of the results and their interpretation. Unfortunately, the Standard Model prescription of these currents cannot be utilized without some caution because "imperfections" of the real world (e.g. isospin breaking) may cause departures of the actual induced contributions from their symmetry prescriptions at precisely this level; thus their accurate determination becomes even more important.

In general, the accurate measurement of induced contributions encounters several difficulties, and has only been achieved in nuclear and nucleon reactions. This is primarily because nuclear physics has some traditional advantages: availability of elaborate experimental techniques, and high statistics at low cost. Chief among the difficulties is that the effects due to these currents, being of recoil order, are small in most practical reactions: $\sim 10^{-2}$ in beta decay and $\sim 10^{-1}$ in muon capture. Moreover, these currents contribute to observables in different combinations, depending on the quantum numbers of the reactions considered and the choice of the observable. While this fact generally provides for the existence of several contributions simultaneously, it occasionally has the advantage of permitting a selective isolation of a specific contribution in special cases. Finally, additional complications arise from finite-size and Coulomb effects at the same level of the recoil contributions. The choice of nuclear transitions wherein the interpretation of the results is not further obscured by nuclear structure (many-body) effects is of crucial importance.

Present experimental information derives primarily from the $p \to n$ ($\mu$ capture) isodoublet transition, the $^8Li \to {}^8Be \leftarrow {}^8B(2^+ \to 2^+)$ and the $^{12}B \to {}^{12}C \leftarrow {}^{12}N(1^+ \to 0^+)$ isotriplet mirror beta decays, the $^{12}C \to {}^{12}B_{gs}$ capture reaction, the $^{19}Ne \to {}^{19}F$ isodoublet $\beta^+$ transition, and the $2^+ \to 2^+$ mirror transitions in $A = 20$ nuclei.

The most decisive results have been obtained in $A = 12$, in which the contributing induced weak currents preserve their essentially "nucleonic" character. Only in the mass triad $^{12}B$-$^{12}C$-$^{12}N$ does there exist a sufficient number of independent observables accessible to experiment to allow independent determinations of the induced weak form factor contributions. Measuring all of the relevant observables has become possible, however, only in recent years through the development of novel experimental techniques involving nuclear orientation of $^{12}B$ and $^{12}N$.

Despite the difficulties, experiments have yielded precise results in agreement with theory, essentially independent of possible contributions beyond the V-A structure of weak interactions. The CVC-implied induced current contribution is verified to within 6%. The PCAC-implied induced current contribution is verified to within 15%. The strength of a "second-class" current contribution resulting from a $G$-parity noninvariance is consistent with zero, and in any case is one order of magnitude smaller than that arising from CVC.

With both the theory of electroweak interactions and its extensions in mind, we herein review the critical determination of induced weak currents, and discuss their significance. Particular emphasis is given to $A = 12$ because of the decisiveness of the results.

Excellent theoretical reviews or works treating induced weak currents and symmetry properties of weak interactions are available in the literature, of which a few are Okun (109), Lee & Wu (89) in this *Annual Review* series, Marshak, Riazuddin & Ryan (93), Morita (101), Holstein (66), Primakoff (118), and Walecka (146). Considerations regarding induced weak currents in the gauge theoretical framework can be found in the review of Bég & Sirlin (10) in this series; see also conference reports by Weinberg (150) and Wolfenstein (156).

# 2.    INTERACTION SYMMETRIES AND INDUCED WEAK CURRENTS

## 2.1    *The V-A Description of Weak Interactions*

The weak reactions considered herein are treated within the framework of the current × current V-A interaction (24, 42, 135):

$$L = \frac{G_F}{\sqrt{2}} j_\lambda \bar{j}_\lambda + \text{h.c.} \qquad\qquad 1.$$

with

$$j_\lambda = j_\lambda^{(\ell)} + j_\lambda^{(h)} = \bar{v}_\ell \gamma_\lambda (1 + \gamma_5)\ell + [V_\lambda^{(0)} - A_\lambda^{(0)}] \cos \theta_c$$
$$+ [V_\lambda^{(\pm 1)} - A_\lambda^{(\pm 1)}] \sin \theta_c. \qquad\qquad 2.$$

The matrix elements of the left-handed lepton current are given in terms of the lepton fields $\ell = \text{e}, \mu$. The matrix elements of the hadronic vector ($V$) and axial vector ($A$) currents, where the superscripts indicate the strangeness-conserving ($\Delta S = 0$) or strangeness-changing ($|\Delta S| = 1$) parts, are constructed out of spinors, Dirac matrices, and the four-momentum transfer $q$. The terms involving $q$ are called *induced weak currents*. The constants of the description, $G_F$ and $\theta_c$, respectively the Fermi constant and the Cabbibo angle, are determined experimentally:

$$G_F = 1.16631 \pm 0.00002 \times 10^{-5} \text{ GeV}^{-2} \text{ (8)} \qquad \text{3a.}$$

$$\sin \theta_c = 0.229 \pm 0.003 \text{ (46).} \qquad \text{3b.}$$

The matrix elements of the weak nucleon n → p current, in the notations of Hwang & Primakoff (74), are (omitting superscripts)

$$\langle p|V_\lambda|n\rangle = \bar{u}(p_1)[f_V\gamma_\lambda + f_M\sigma_{\lambda\nu}q_\nu/2m_p + if_S q_\lambda(m_p + m_n)/m_\pi^2]u(p_2) \qquad \text{4a.}$$

$$\langle p|A_\lambda|n\rangle = \bar{u}(p_1)[f_A\gamma_\lambda\gamma_5 - f_T\sigma_{\lambda\nu}\gamma_5 q_\nu/3m_p - if_P q_\lambda\gamma_5(m_p + m_n)/m_\pi^2]u(p_2), \qquad \text{4b.}$$

where $u(p_1$ or $p_2)$ are free spinors, $q_\lambda = (p_1 - p_2)_\lambda$ is the four-momentum transfer, and $m_{n,p,\pi}$ are respectively the neutron, proton, and pion masses. Further, $f_{V,A}$ are identified as the leading-order or "conventional" vector and axial vector form factors, and $f_{M,S,T,P}$ are the weak magnetism, effective scalar, induced tensor (or "weak electricity"), and induced pseudoscalar form factors, respectively. The $f_i$'s, which are scalar functions of $q^2$, summarize the dynamical role of the corresponding terms of the current. In the $q^2 = 0$ limit, only $f_{V,A}$ contribute, and they are known experimentally:

$$f_V(0) = 1.000 \pm 0.0015 \text{ (46)} \qquad \text{5a.}$$

$$f_A(0) = 1.254 \pm 0.009 \text{ (119).} \qquad \text{5b.}$$

The above method of constructing baryonic matrix elements, constrained by Lorentz covariance only, is an extension of that used in the case of the electromagnetic baryonic current $j^{em}$. For nucleons, the latter is (93, 109)

$$\langle p|j_\lambda^{em}|p\rangle = \bar{u}(p_1)[F_1^p\gamma_\lambda + \mu_p^a F_2^p\sigma_{\lambda\nu}q_\nu/2m_p]u(p_2) \qquad \text{6a.}$$

$$\langle n|j_\lambda^{em}|n\rangle = \bar{u}(p_1)[F_1^n\gamma_\lambda + \mu_n^a F_2^n\sigma_{\lambda\nu}q_\nu/2m_n]u(p_2), \qquad \text{6b.}$$

where $\mu_p^a$ ($\mu_n^a$) is the anomalous magnetic moment of the proton (neutron), while $F_1^p$ ($F_1^n$) and $F_2^p$ ($F_2^n$) are respectively the Dirac and Pauli form factors normalized as $F_1^p(0) = F_2^p(0) = F_2^n(0) = 1$ and $F_1^n(0) = 0$.

To date, all known low-energy weak interaction phenomena, except for CP violation (89), are adequately described by the current × current V–A interaction. Neither the remaining basic Lorentz covariants (scalar, tensor, and pseudoscalar) nor the possibly existing right-handed currents (9) are known to contribute significantly. The question of these as yet undiscovered currents is of importance in the present context as their existence could obscure the identification of the a priori, not too-large, induced weak current contributions. There is, in principle, a scalar contribution of gauge-theoretical origin associated with the existence of the Higgs sector, but its contribution is estimated to be negligible (97). A tensor contribution is

precluded at present by the requirement of a renormalizable gauge theory
(11). In general, the pseudoscalar is a relativisitic effect and therefore
vanishing in the low-energy limit characterizing beta decay and muon
capture. With these considerations and above all with the existing upper
limit estimates for the contribution of any scalar, tensor, pseudoscalar (17,
67), and right-handed currents (70, 131), the adoption of the pure V–A
description is well enough justified. Further, $CP$-violating contributions to
the reactions considered are absent, in accordance with both the theory (82)
and experiment (60). Consequently, the $f_i$'s are all relatively real. Note that
since all of the observables considered herein are even under time reversal,
the significance of the resulting form factors remains valid in any case.

In general, the weak hadronic currents might be any vector and axial
vector operators: in a gauge theory they must be the currents associated
with the gauge symmetry, and must therefore be conserved except for the
effects of spontaneous symmetry breaking inherent to the theory (150). The
gauge symmetry determines the properties of the elementary weak currents
and hence restricts the $f_i$'s of the baryonic current, and to a large extent the
form factors of the weak nuclear current.

## 2.2   The Charged Weak Current in $SU(2)_L \times U(1)$

The Standard Model (SM) of strong and electroweak interactions (11),
which is broken down to $SU(3) \times SU(2)_L \times U(1)$, contains the gauge group
$SU(2)_L \times U(1)$ of the electroweak interaction (127, 149) as a subgroup. The
brief account here focuses on the $SU(2)_L \times U(1)$ aspect of SM, and is
intended to indicate the properties of the charged electroweak current of
elementary fermions evolving from the gauge symmetry.

The theory involves left-handed fermions forming SU(2) doublets and
right-handed fermions forming SU(2) singlets. The weak quark current
corresponding to the doublet (Q', Q) assumes the form $Q'\gamma_\lambda(1+\gamma_5)Q$, where
Q' is a linear combination of d, s, and b quarks

$$\begin{Bmatrix} d' \\ s' \\ b' \end{Bmatrix} = U \begin{Bmatrix} d \\ s \\ b \end{Bmatrix},$$

the (3 × 3) unitary matrix $U$ is the flavor-mixing matrix (82), and Q is one of
the u, c, and t (as yet unidentified) quarks. In practice, the admixture of b
into d' and s' is negligible, so that the mixing between d and s is adequately
described by the Cabbibo theory:

$$d' = d \cos \theta_c + s \sin \theta_c$$

$$s' = s \cos \theta_c - d \sin \theta_c.$$

The lepton current corresponding to the doublet ($\ell$, $v_\ell$) was given in

Equation 2 and is similar to the quark current. The difference is that the lepton families, according to current knowledge, do not mix.

The theory involves further the electromagnetic current, such as $\bar{u}\gamma_\lambda u$, $\bar{e}\gamma_\lambda e$, $\cdots$, and the neutral weak currents, but these have no direct relevance to our subject. The invariance of the electroweak interaction under the gauge group requires the universality of the coupling of both quark and lepton currents to the corresponding vector bosons. The nonlocal effects due to the propagator of vector bosons of $\sim 80$ GeV are negligible in beta decay and muon capture; the current $\times$ current description of the charged weak interactions is adequate. The appropriate current is

$$j_\lambda = \bar{v}_e\gamma_\lambda(1+\gamma_5)e + \bar{v}_\mu\gamma_\lambda(1+\gamma_5)\mu + \bar{d}\gamma_\lambda(1+\gamma_5)u\cos\theta_c + \bar{s}\gamma_\lambda(1+\gamma_5)u\sin\theta_c. \tag{7.}$$

The weak d-u currents comprise an isospin ($I$) subgroup,

$$[I^+, I^-] = 2I_3, \tag{8.}$$

such that CVC (42, 49), which identifies $V(n \to p)$, $V^+(p \to n)$ and the isovector electromagnetic current of nucleons with the charge-raising, charge-lowering, and third component of the isospin current (respectively $I^+$, $I^-$, and $I_3$), evolves naturally from the theory. Since the strong interactions do not affect the universality of the hadronic and leptonic electromagnetic currents, then via CVC the universality of the weak vector quark current is guaranteed. The quark-lepton universality required by the theory, reflected in Equation 7, holds in the presence of the strong interaction except for the axial vector quark current. The latter is renormalized as a result of the spontaneous symmetry breaking inherent to SM (11). A "measure" of this symmetry breaking is the pion mass, such that the PCAC hypothesis (48, 105), which relates the divergence of the axial current to the pion field times $m_\pi^2$, also has a natural place within the theory.

Gradient-type current contributions, such as $\bar{d}q_\lambda u$ and/or $\bar{d}\sigma_{\lambda v}q_v\gamma_5 u$, are forbidden. The former (latter) has the Lorentz covariance of the weak vector (axial vector) d-u current. Clearly, $\bar{d}q_\lambda u(\bar{d}\sigma_{\lambda v}q_v\gamma_5 u)$ and $\bar{d}\gamma_\lambda u(\bar{d}\gamma_\lambda\gamma_5 u)$ are uneven under charge ($C$) conjugation. Such derivative-type currents would lead to gauge noninvariant interaction densities, and are hence precluded. The inclusion of the strong interactions, which are invariant under $C$, does not modify the conclusion. In practice, the existence of the above gradient-type currents would give rise to SCC contributions to nucleon (and nuclear) currents, rendering for example $n \to p$ and $p \to n$ rates unequal.

## 2.3  Isospin Structure of the Weak Nucleon Currents, First- and Second-Class Currents

The question arises, at least at the stage of phenomenology, whether all of the terms of Equations 4a,b, allowed by Lorentz invariance, are dynamically allowed. It has been pointed out (147) that under the $G$-parity transformation the matrix elements of the conventional currents transform as

$$GV(A)G^{-1} = V(-A), \qquad\qquad 9a.$$

analogous to their strong interaction counterparts. These are by definition first-class currents. The weak magnetism and the induced pseudoscalar currents also belong to this class. The effective scalar and the induced tensor, which transform differently,

$$GV(A)G^{-1} = -V(A), \qquad\qquad 9b.$$

are by definition second-class currents.

Essentially, the $G$-operation transforms the matrix elements of n → p into those of the inverse p → n process. One may alternatively keep the sign of all transforms unchanged, while changing the sign of the corresponding form factors in accordance with Equations 9a,b:

$$f_{V,M,T} \to f_{V,M,T} \qquad\qquad 10a.$$

$$f_{A,P,S} \to -f_{A,P,S}. \qquad\qquad 10b.$$

The effective scalar and the induced tensor must vanish, as their quark equivalents are forbidden. The question of spurious SCC, due to isospin symmetry breaking, has been considered within standard theory (61). However, even if the d and u masses are quite different, the quark confinement to the spatial dimension of the nucleons makes the spurious SCC form factors much smaller than the form factors of first-class currents. Langacker and others (86, 133) have shown in fact that SCC may only exist under the constraints of anomalously large isospin breaking or dramatic modifications of standard theory.

## 2.4   The Isotriplet CVC and Weak Magnetism

The CVC-implied isospin algebra leads to the relation

$$\langle p | V_\lambda(0) | n \rangle = \langle p | j_\lambda^{em}(0) | p \rangle - \langle n | j_\lambda^{em}(0) | n \rangle \qquad\qquad 11.$$

between the weak vector and isovector electromagnetic currents, for which the substitution of Equation 4a and Equations 6a,b then requires the following relations between the weak and electromagnetic form factors:

$$f_V = F_1^p - F_1^n \qquad\qquad 12a.$$

$$f_M = \mu_p^a F_2^p - \mu_n^a F_2^n \qquad\qquad 12b.$$

$$f_S \equiv 0. \qquad\qquad 12c.$$

Thus, in the condition of the experiments considered, i.e. for $q^2 \approx 0$,

$$f_V(0) = 1 \qquad\qquad\qquad\qquad\qquad\qquad\qquad\qquad \text{13a.}$$

$$f_M(0) = \mu_p^a - \mu_n^a \equiv \mu_p - \mu_n - 1 \; (\approx 3.607) \qquad\qquad\qquad \text{13b.}$$

$$f_S(0) = 0, \qquad\qquad\qquad\qquad\qquad\qquad\qquad\qquad\qquad \text{13c.}$$

where $\mu_p$ ($\mu_n$) is the proton (neutron) magnetic moment.

The observed nonrenormalization of $f_V(0)$ in Equation 5a is compatible with the universality of $V_\lambda$ in Equation 12a. This condition holds, however, for any conserved vector current and is hence not an identifying mark of the isotriplet CVC condition (the so-called strong CVC), the signature of which is the weak magnetism (47) in Equation 12b.

Manifestly, the universality of the weak vector and isovector electromagnetic currents is most critically demonstrated by the equality of their magnetic anomalies, a demonstration reminiscent of the comparison of $g - 2$ lepton anomalies.

## 2.5 The PCAC Hypothesis and the Induced Pseudoscalar

Either $f_A(0) \neq 1$ in Equation 5b, or the existence of the $\pi \to e\nu$ decay channel precludes the conservation of $A_\lambda$. In the PCAC hypothesis, we find $\partial_\lambda A_\lambda = g_\pi f_\pi m^2 \phi_\pi$, where $g_\pi, f_\pi$, and $\phi_\pi$ are respectively the pion-decay constant, the pion-nucleon coupling constant, and the pion field operator. The pion-nucleon constant is assumed to be slowly varying in the interval $0 < q^2 < m_\pi^2$. The G parity of $A_\lambda$ is clearly required to be that of the pion (negative), such that only first-class axial vector currents may contribute. A full treatment of the matrix elements $\langle n | \partial_\lambda A_\lambda | p \rangle$ leads to two predictions:[1]

$$g_A(0) = 2g_\pi f_\pi / 2m_p; \qquad\qquad\qquad\qquad\qquad\qquad \text{14.}$$

$$g_P^\mu(q^2) = m_\mu g_P(q^2) = m_\mu 2m_p g_A(0)/(q^2 + m_\pi^2), \qquad\qquad \text{15.}$$

where $m_\mu$ is the muon mass. Equation 14, known as the Goldberger–Treiman relation (51), yields (160) $g_A(0) = 1.32 \pm 0.02$ in good agreement with the experimental result. The determination of $g_P$ is a test of PCAC at $q^2 \neq 0$. This form factor contributes essentially to muon capture only: it is negligible in beta decay, owing to the helicity suppression in the $\pi \to e\nu$ channel, i.e. $g_P^e/g_P^u \; m_e/m_\mu \approx 1/200$. For $q^2 \approx 0.88m_\mu^2$ prevailing in muon capture, the prediction is

$$g_P^\mu(q^2)/g_A(0) \approx 6.7, \qquad\qquad\qquad\qquad\qquad\qquad \text{16a.}$$

---

[1] This discussion follows Lee & Wu (89) in employing form factors related to those in Equation 4b as $g_A \equiv f_A$, $g_p \equiv -f_P 2m_p/m_\pi^2$.

or in terms of $f_A$ and $f_P$,

$$f_P(q^2) = -f_A(0)/(1+q^2/m^2) \approx -0.71f_A(0). \qquad \text{16b.}$$

## 3.  INDUCED WEAK CURRENTS IN NUCLEI AND THEIR OBSERVABLES

### 3.1  *Considerations in Tests of SCC, CVC, and PCAC*

Induced weak currents contribute in recoil order to the interaction density (Equation 1) resulting from the contraction of nucleon matrix elements (Equations 4a,b) with $j_\lambda^{(\ell)}$. Given $G_F$, $\cos\theta_c$, the $f_i$'s, and $q$, one may predict the observables in nucleon-nucleon reactions since the matrix elements of the currents involved are well defined. The leading-order contributions of induced currents to observables consists obviously of their interference with the conventional current contributions, e.g. $f_A f_M q/2m_p$, where in most cases $q$ is proportional to electron energy $E$ or neutrino energy $E_\nu$.

In nuclear reactions, however, the observables cannot be predicted exactly because the nuclear matrix elements of weak currents are not well known. Further, the relationship between matrix elements and form factors is a bit more complicated. For instance, a particular matrix element may have both first- and second-class contributions.

To illustrate the relation between nuclear matrix elements and form factors, and their connection with nucleon form factors, consider the representative $1^+ \leftrightarrow 0^+$ transitions in $A = 12$ nuclei, shown in Figure 1. The Gamow-Teller character of these weak reactions implies a potential contribution from only the axial vector, weak magnetism, induced tensor, and induced pseudoscalar form factors. The analogous electromagnetic

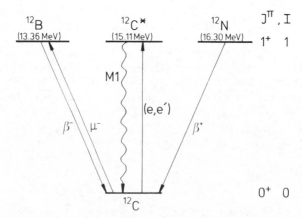

*Figure 1*    Reactions investigated to determine induced weak currents in $A = 12$ nuclei.

process is essentially a pure isovector M1 transition (27). In the "elementary particle treatment" (EPT) of nuclei (80), the matrix elements are constructed out of available four-vectors, $\xi$ (polarization vector of spin-one nuclei), $q_\lambda = (p_2 - p_1)_\lambda$, $Q_\lambda = (p_2 + p_1)_\lambda$, and the tensor $\varepsilon_{\lambda\chi\rho\eta}$. Following Hwang & Primakoff (74), we write

$$\langle {}^{12}C(p_1)|V_\lambda|{}^{12}B(p_2,\xi)\rangle = \sqrt{2}\varepsilon_{\lambda\chi\rho\eta}\xi_\chi \frac{q_\rho}{2m_p}\frac{Q_\eta}{2M} F_M^- \qquad 17a.$$

$$\langle {}^{12}C(p_1)|A_\lambda|{}^{12}B(p_2,\xi)\rangle = \sqrt{2}\left( \xi_\lambda F_A^- + q_\lambda \frac{q^2}{m_\pi^2} F_P^- - \frac{Q_\lambda}{2M}\frac{q\cdot\xi}{2m_p} F_E^- \right) \qquad 17b.$$

$$\langle {}^{12}C(p_1)|V_\lambda^+|{}^{12}N(p_2,\xi)\rangle = \sqrt{2}\varepsilon_{\lambda\chi\rho\eta}\xi_\chi \frac{q\xi}{2m_p}\frac{Q_\eta}{2M} F_M^+ \qquad 17c.$$

$$\langle {}^{12}C(p_1)|A_\lambda^+|{}^{12}N(p_2,\xi)\rangle = \sqrt{2}\left( \xi_\lambda F_A^+ + q_\lambda \frac{q\cdot\xi}{m_\pi^2} F_P^+ - \frac{Q_\lambda}{2M}\frac{q\cdot\xi}{2m_p} F_E^+ \right) \qquad 17d.$$

$$\langle {}^{12}C(p_1)|j_\lambda^{em}|{}^{12}C^*(p_2,\xi)\rangle = \sqrt{2}\varepsilon_{\lambda\chi\rho\eta}\xi_\chi \frac{q_\rho}{2m_p}\frac{Q_\eta}{2M} \mu, \qquad 17e.$$

where $F_{M,A,P,E}(q^2)$ are respectively the weak magnetism, axial, induced pseudoscalar, and induced tensor (weak electricity) nuclear form factors; $\mu(q^2)$ is the magnetic nuclear transition form factor and $M$ is the nuclear mass.

Nuclear and nucleon form factors are usually related via the impulse approximation (34). For instance, in the case of ${}^{12}B \to {}^{12}C$ (74, 102)

$$\sqrt{2}F_M^- \approx -(f_V + f_M)M_\sigma + \alpha_1 \qquad 18a.$$

$$\sqrt{2}F_A^- \approx -f_A M_\sigma \qquad 18b.$$

$$\sqrt{2}F_P^- \approx -f_P M_\sigma + \alpha_2 \qquad 18c.$$

$$\sqrt{2}F_E^- \approx f_T M_\sigma - f_A(1 + 2\alpha_3/M_\sigma)M_\sigma, \qquad 18d.$$

while in the case of ${}^{12}C^* \to {}^{12}C$,

$$\sqrt{2}\mu \approx -(\mu_p - \mu_n + 1)M_\sigma' + \alpha_1', \qquad 18e.$$

where $M_\sigma$, $M_\sigma'$ ($\approx M_\sigma$) are the leading single-particle ($\int\sigma$) matrix elements, and $\alpha_1$, $\alpha_2$, $\alpha_3$, $\alpha_1'$ ($\approx \alpha_1$) are recoil-order corrections terms [$\alpha_1 = \int r \times p$, $\alpha_2 = \int\sigma r^2$, and $\alpha_3 = \int r(\sigma\cdot p)$] that mimic induced contributions. The form factor $F_E$ contains both first(1)- and second(2)-class parts, $F_E^\mp =$

$F_E^{(1)\mp} \mp F_E^{(2)\mp}$. In other nuclear systems, the above relations are essentially the same except for spin-dependent factors.

In an experiment to determine the induced tensor $F_E^-$ form factor, the total recoil-order interference contribution contains both a first-class $\times$ first-class part ($\sim f_A \cdot f_A$) and a first-class $\times$ second-class part ($\sim f_A \cdot f_T$). The separation of them is possible only in mirror $\beta^{\pm}$ transitions as shown, for example, in Figure 1, because by virtue of Equations 10a,b the former (latter) is charge independent (dependent).

The determination of the weak magnetism form factor $F_M$, which also gives rise to a charge-dependent $\sim f_A \cdot f_M$ interference, also requires comparative measurements in mirror transitions. The nuclear matrix elements of both the isovector electromagnetic and weak vector currents may be developed in powers of $q$, with the terms of the same power of $q$ related by virtue of CVC. For a "critical" verification of CVC based on weak magnetism, one compares the transition matrix element of a pure M1 $1^+ \rightarrow 0^+$ isovector $\gamma$ transition, composed of the total (Dirac and Pauli) isovector magnetism $(\mu_p - \mu_n)q/2m_p$, with $(f_V + f_M)q/2m_p$ determined in analogue transitions. Such an experiment has been performed in mass multiplets $A = 8, 12$, and 20. In one case ($A = 8$), the CVC test involves the comparison of the $q^2$-order term in the isovector electromagnetic and weak vector currents.

One may note here that a determination of the induced tensor or weak magnetism form factor in mirror transitions is independent of a pseudo-scalar contribution, if there is any, since the latter gives rise to a charge-independent interference contribution. Moreover, a possible scalar contribution is suppressed by selection rules in appropriate $1^+ \rightarrow 0^+$ transitions.

The effective scalar, of which the quantum numbers are $1^-$ or $0^+$, may contribute either to parity-changing (forbidden) transitions or to allowed transitions without spin change ($\Delta J = 0$). In the former case it competes with a large number of forbidden terms of recoil order; in the latter, with the dominant matrix element(s). Its determination is therefore difficult in both cases. There is nevertheless an upper limit estimate for it (69, 138). In the description of experiments, $f_S = 0$ is assumed.

The induced pseudoscalar is determined in muon capture at fixed $q$, contrary to induced weak current determinations ($f_T, f_M$) in beta decay, where $q$ ($\sim E$, the beta-ray energy) is variable. The contribution of the induced pseudoscalar is best recognized by its scalar nature in spin space. For example, in the $^{12}C \rightarrow {}^{12}B(0^+ \rightarrow 1^+)$ capture reaction, it contributes to the population of only the $m = 0$ magnetic substate defined in the recoil frame. Were $F_p$ the only form factor contribution, the final nucleus would be unpolarized.

Let us note finally that the axial vector current is already modified in transfer from quark to nucleon description, and its properties may further be affected in the nuclear environment. The problem of the induced tensor is particularly complicated, since "cooperative" effects in nuclei (147) and exchange currents (84) may generate secondary effects.

## 3.2 Choice of Observables to Determine Induced Weak Currents

In beta decay, the general observable is the e-v correlation in oriented nuclei (67), which, after integration over the neutrino variables, reduces to (74)

$$d^3\Gamma^\mp/dE\,d\Omega = S_\mp(Z,E,E_0)(1+\eta_\mp+a_\mp E)\cdot W(\theta,E)_\mp, \qquad 19.$$

where $E$ ($E_0$) is the $\beta$ decay kinetic (maximum) energy.

The correlation function $W$, again for $1^+ \to 0^+$ transitions, has the form

$$W(\theta,E)_\mp = 1\mp P(1+\alpha_\mp E)P_1 + A\alpha_\mp EP_2. \qquad 20.$$

Here $P = p_1 - p_{-1}$ ($A = 1 - 3p_0 \equiv p_1 + p_{-1} - 2p_0$) is the nuclear polarization (alignment) of the parent and $p_m$ is the population of the magnetic substate $m$ (with normalization $\Sigma p_m = 1$); $\theta$ is the angle between the orientation axis and the electron momentum, and $P_{1,2}$ are Legendre polynomials.

The parameters $\eta_\mp \approx (\Delta/3M_p)\,[\mp\tilde{F}_M^\mp(0) + \tilde{F}_E^\mp(0)]$, where $\tilde{F}_i \equiv F_i/F_A$ and $\Delta$ is the energy release, are inaccessible directly to experiment. They contribute to the decay widths obtained by integrating Equation 19 over the electron variables. For $A = 12$ one has

$$\Gamma^- = 124.51\,|F_A^-(0)|^2\,[1+0.005\tilde{F}_E^-(0)]\;s^{-1} \qquad 21a.$$

$$\Gamma^+ = 251.28\,|F_A^+(0)|^2\,[1+0.006\tilde{F}_E^+(0)]\;s^{-1}. \qquad 21b.$$

As a consequence of the integration over both neutrino and electron variables, the beta decay rates do not depend upon $F_M$ (47, 147).

Observables of practical interest are the spectrum shape correction parameters $a_\mp$ and the correlation parameters $\alpha_\mp$, given to first-order in recoil by

$$a_\mp = (4/3m_p)\tilde{F}_M^\mp(0) \qquad 22a.$$

$$\alpha_\mp = (1/3m_p)[\pm\tilde{F}_M^\mp(0) - \tilde{F}_E^\mp(0)]. \qquad 22b.$$

The $a_\mp$ are the classic sources of CVC tests, and the $\alpha_\mp$ provide the best information on the second-class induced tensor $F_E^{(2)}$.

It seems appropriate at this point to comment on the principle of the methods employed to measure $a_\mp$ and $\alpha_\mp$ in the polarization and alignment terms, following a discussion of Telegdi (139). The leading $j = 1/2$ electron amplitude (Gamow-Teller interaction) leads to rank-zero (scalar) and rank-one (vector) observables essentially independent of $E$, i.e. to the first terms in Equations 19 and 20 and to $P \cos \theta$ in Equation 20. The electron amplitude corresponding to an induced interaction ($\sim E$) is characterized by $j = 1/2$ and/or $j = 3/2$, and its interference with the leading $j = 1/2$ amplitude leads to intensities, proportional to $E$, of rank zero ($a_{\mp} E$), of rank one ($P\alpha_{\mp} E P_1$), and of rank two ($\alpha_{\mp} E A P_2$). Clearly, the alignment term is composed entirely of couplings linear in $E$. Thus, measurement of the alignment term leads to a direct determination of induced form factors. Such an approach is reminiscent of direct determinations of $g$-factor anomalies. The "critical determination" of induced weak currents, referred to in the introduction, means precisely their determination through the alignment terms of the correlation functions.

Determinations of $a_{\mp}$ through the shape factor $(1 + a_{\mp} E)$, and $\alpha_{\mp}$ through the polarization term $P(1 + \alpha_{\mp} E)$ constitute indirect procedures: one has to subtract (or factorize) the dominant (Gamow-Teller) contribution from the measured quantity. Even a tiny error introduced by the subtraction procedure may have dramatic consequences with regard to $a_{\mp}$ or $\alpha_{\mp}$, since they contribute at the level of a few percent only. Past controversies regarding weak magnetism and second-class currents stemmed from the indirect nature of the methods employed to determine $a_{\mp}$ and $\alpha_{\mp}$.

In other reactions, information about induced weak currents derives from decay or capture rates, polarization- and alignment-correlation measurements in an analogous manner. Our description of the experimental results is thus conducted by analogy with $A = 12$.

# 4.   DETERMINATION OF INDUCED WEAK CURRENTS

## 4.1   The Induced Pseudoscalar in the $\mu^- + p \rightarrow \nu_\mu + n$ Reaction

This reaction is the ideal source of information for induced weak currents, since the recoil $q/2m_p$ is quite large ($\sim 5\%$) and the theoretical treatment of observables is accurate. However, the determination of a complete set of observables is a rather remote prospect, mainly because of the slowness of the reaction. In fact, only the singlet capture rate has been measured to date; this constitutes only an indirect approach to the determination of induced contributions.

The $\mu^-$ stopped in a hydrogen target form the quasi-stationary states indicated in Figure 2. The $\mu^- p$ atomic ground state is formed in either triplet (T) or singlet (S). In general, T converts into S, with a rate depending upon the hydrogen density. In gaseous hydrogen, for pressures above five atmospheres, the conversion rate is large compared to the disappearance rate $\lambda_0^-$ of $\mu^-$, so that the capture or disintegration of $\mu^-$ occurs in S. In liquid hydrogen, S converts rapidly (conversion rate $\lambda_{pp} \gg \lambda_0^-$) into orthomolecular, OM (paramolecular, PM) $p\mu^-p$ hydrogen with a branching ratio $0.997 \pm 0.003$. Further, the OM may convert into PM with a rate $\lambda_{OP}$.

The atomic capture rates $\Gamma_{T,S}$ are given in terms of the inverse beta decay rate and the nucleon weak form factors (118):

$$\Gamma_S = (29.2 \pm 0.3)(G_V - 3G_A + G_P)^2 \ s^{-1} \qquad \text{23a.}$$

$$\Gamma_T = (29.2 \pm 0.3)[(G_V + G_A)^2 - (2/3)(G_V + G_A)G_P + G_P^2] \ s^{-1}, \qquad \text{23b.}$$

where $G_V$, $G_A$, and $G_P$ are the so-called effective form factors. Assuming the weak form factors implied by the interaction symmetries evolving from SM,

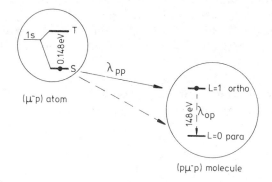

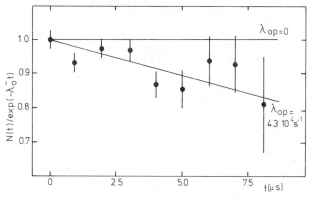

*Figure 2* (*Top*) Quasi-stationary states of $\mu^-$ stopped in hydrogen (159). (*Bottom*) Experimental demonstration of the ortho to para conversion in liquid hydrogen: the number of neutrons (N) decreases more rapidly vs time than that of decay electrons [$N_e \alpha \exp(-\lambda_0^- t)$].

one has

$$\Gamma_T \approx 12 \text{ s}^{-1} \qquad\qquad 24a.$$

$$\Gamma_S \approx 664 \text{ s}^{-1}. \qquad\qquad 24b.$$

The strong spin dependence of the atomic capture rates is a consequence of the V–A nature of the weak interaction. In $q = 0$ order, the rates are $\Gamma_T = (f_V - f_A)^2$ and $\Gamma_S = (f_V + 3f_A)^2$; in the absence of strong interactions $f_V/f_A = 1$ and $\Gamma_T$ vanishes. The prediction of $\Gamma_T/\Gamma_S \ll 1$ was put forward in the early days of the V–A theory (15) and it has been verified in the $\mu^- + {}^{19}F \to \nu_\mu + {}^{19}O$ reaction (87), where the unpaired proton of the initial nucleus (spin 1/2) plays the role of hydrogen.

The determination of $\Gamma_T$, which is composed essentially of recoil-order effects and hence reminiscent of the alignment term in $A = 12$, is obviously a difficult task and has yet to be done. The rate $\Gamma_S$, which is less dependent upon induced weak currents, has been determined by groups at CERN (3) and Dubna (23). In both of these experiments, $\Gamma_S$ was deduced from the measured ratio of neutrons vs $\mu^-$ stops. The results agree and their average,

$$\bar{\Gamma}_S = 661 \pm 47 \text{ s}^{-1}, \qquad\qquad 25.$$

is in agreement with that in Equation 24b.

The predicted molecular capture rates are (148)

$$\Gamma_{OM} = (2\gamma_O\Gamma_S/4)(3 + R) \quad [\approx 513 \text{ s}^{-1}] \qquad 26a.$$

$$\Gamma_{PM} = (2\gamma_P\Gamma_S/4)(3R + 1) \quad [\approx 276 \text{ s}^{-1}], \qquad 26b.$$

where $R = \Gamma_T/\Gamma_S$ and $\gamma_{O,P}$ are known mesomolecular factors.

As evident from these equations, the interpretation of the capture rate measured in liquid hydrogen depends upon $\lambda_{OP}$. A fair indication of the OM $\to$ PM conversion has existed for a long time (16) but until recently was generally overlooked. Obviously, the neglect of this conversion leads to an underestimate of $\Gamma_S$. The existence of the OM $\to$ PM conversion is now more firmly established (7) by a Saclay group (see Figure 2) and the result $\lambda_{OP} = 4.6 \pm 1.4 \times 10^4 \text{ s}^{-1}$ agrees with the recent theoretical estimate (5).

A new determination of the capture rate in liquid hydrogen, based on a novel approach that compares disappearance rates of $\mu^-$ and $\mu^+$, yields 460 $\pm 20 \text{ s}^{-1}$ (8). This is compatible with earlier determinations in counter experiments (16, 126), and taking into account the OM $\to$ PM conversion, leads to $\Gamma_{OM} = 531 \pm 33 \text{ s}^{-1}$.

With CVC and vanishing SCC assumed, the capture rate is a function of $f_P$ only. Analysis of gaseous and liquid hydrogen data yields the average (7)

$$f_P/f_P(\text{PCAC}) = 1.04 \pm 0.23. \qquad\qquad 27.$$

## 4.2    The Second-Class Induced Tensor in $^6He \rightarrow {}^6Li$ Beta Decay

Information on $F_E^{(2)}$ is obtained from the electron-antineutrino directional correlation in the $0^+ \rightarrow 1^+$ allowed $^6He \rightarrow {}^6Li$ beta transition ($E_0 = 3.1$ MeV). The measurement was performed 20 years ago (77) to clarify the nature of the Gamow-Teller interaction, as the axial vector (tensor) interaction implies $-1/3$ ($+1/3$) for the coefficient $a$ in the angular correlation $W(\theta_{ev}) = 1 + a\cos\theta_{ev}$. The correlation coefficient was deduced from the measured recoil spectrum of $^6Li$, and the result $a = -0.3343 \pm 0.0030$ clearly favored the axial vector interaction.

Beyond the leading-order V-A description, the correlation coefficient involves contributions of $F_E^{(1)}$,

$$a = a_{V-A} \ [1 - (8/3)E_0\tilde{F}_E^- / 2m_p], \qquad\qquad 28.$$

while the contribution of $F_M$ vanishes as a consequence of the integration over lepton variables in the measurement. By virtue of the allowed character of this transition, it seems reasonable to assume the relation between nuclear and nucleon form factors in Equations 18, with higher order correction terms neglected. The measured correlation coefficient $a$ then implies

$$\tilde{F}_E^{(2)} = +0.5 \pm 1.3, \qquad\qquad 29.$$

compatible with vanishing second-class currents.

## 4.3    The Second-Class Induced Tensor and Weak Magnetism in Beta Decay of $A = 8$ Nuclei

The low-energy level scheme of $A = 8$ nuclei together with the relevant weak and electromagnetic transitions is shown in Figure 3.

Ideally, the second-class current form factor $F_E^{(2)}$, which contributes to the $ft$ values of beta transitions, may be extracted from the charge-dependent fraction of mirror transitions (152)

$$\delta = ft^+/ft^- - 1 \approx (4/3)\tilde{F}_E^{(2)}(W_0^+ + W_0^-), \qquad\qquad 30.$$

where $W_0^\pm$ ($m_e c^2$ units) is the end-point energy of $\beta^\pm$. In practice, however, charge-dependent effects such as the $I_3$-dependent nuclear overlap integral, also contribute to $\delta$ and preclude the reliable determination of $F_E^{(2)}$.

In the case of mirror transitions feeding the broad $^8Be^*$ state (2.9 MeV), $\delta$ may be measured vs the excitation energy $E_{ex}$ of $^8Be^*$, i.e. vs the energy release $W_0^+ + W_0^-$ over 20 $m_e c^2$ units ($\sim 10$ MeV). Since the slope of $\delta$ is expected to be essentially independent of overlap integrals, it may serve to determine $F_E^{(2)}$ (154). The released energy in a particular beta decay event is determined from the kinetic energy of the coincident α particles resulting from the break-up of $^8$Be*. According to the experiment of Wilkinson & Alburger (154), δ is practically independent of $W_0^+ + W_0^-$ (Figure 3), which implies the absence of the second-class induced tensor:

$$|\tilde{F}_E^{(2)}| \leq 0.6. \qquad\qquad 31.$$

The CVC test in $A = 8$ nuclei is based on the measurement of the directional correlation between the β particle in $^8$Li → $^8$Be* ← $^8$B and the subsequent α particles. The $^8$Be* state is aligned because of the presence of recoil-order contributions to the beta decay amplitude. The degree of alignment is measured via the $\cos^2 \theta$ distribution of α particles. The β-α directional correlation and the alignment-electron momentum distribution in $A = 12$ nuclei are equivalent, with merely the role of initial/final states inverted; hence these experiments may be treated in the same terms (except for the traditionally different notations). The difference of the "reduced" $\beta^-(\beta^+)$-α correlation coefficients, $\delta^-/E$, corresponding to $\Delta\alpha = \alpha_- - \alpha_+$ in

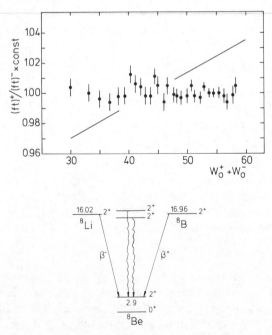

*Figure 3*    The charge-dependent fraction $ft^+/ft^- - 1$ vs the energy release $W_0^+ + W_0^-$ in mirror transitions shown in the bottom figure. The width ($\sim 1.5$ MeV) of the 2.9-MeV state in $^9$Be and its break-up into two α particles are not indicated. The straight line in the plot corresponds to $F_E^{(2)} \approx -2F_M$(CVC).

$A = 12$ (Equation 22b), is composed of SCC- and CVC-related contributions (18, 140).

The experiment in $A = 8$ has, however, qualitative differences with respect to that in $A = 12$. By virtue of the $2^+$-$2^+$ sequence, the analogue electromagnetic transition is not restricted to M1 : the electric quadrupolar (E2) matrix element may also contribute. Further, the analogue state in $^8$Be is a nearly degenerate doublet (16.6 and 16.9 MeV) of which both members are $I = 1$ and $I = 0$ mixtures of equal strength. The electromagnetic $2^+$-$2^+$ transitions may therefore have both isovector and isoscalar contributions. Finally, the large width of $^8$Be* leads to additional complications : the form factor ratios involved in $\delta^-$ have to be measured vs $E_{ex}$, then properly averaged.

The total radiative isovector M1 width, $\Gamma(M1)$, and the isovector E2/M1 mixing ratio, $\delta(E2/M1)$, were determined experimentally by Bowles & Garvey (18): $\Gamma(M1) = 6.10 \pm 0.53$ eV and $\delta(E2/M1) = 0.19 \pm 0.03$.

The CVC-implied correlation coefficient $M_n \delta^-/E$ (the role of the nucleon mass $M_n$ is to make the coefficient dimensionless), including the weak analogue of both the isovector M1 and E2 amplitudes and assuming vanishing SCC,

$$M_n \delta^- /E(\text{CVC}) = 8.5 \pm 1.1, \qquad\qquad 32.$$

compares well with the experimental result of Tribble & Garvey (140) in Figure 4,

$$M_n \delta^- /E(\text{exp}) = 7.0 \pm 0.5. \qquad\qquad 33.$$

The correlation coefficient calculated without the weak analogue of E2, 10.9 $\pm 1.4$, is not in agreement with the observation. The particular merit of this experiment is thus the verification of the isotriplet CVC condition up to E2 order of isovector currents.

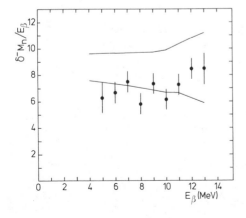

*Figure 4* $A = 8$ nuclei: the upper and lower limits of the CVC-implied difference of $\beta^- - \alpha$ and $\beta^+ - \alpha$ correlation coefficients (including the weak analogue of both M1 and E2 isovector strengths) and its experimental determinations. Data are taken from (140) and agree with the earlier results of Nordberg et al (106).

## 4.4 Weak Magnetism, Induced Tensor, and Induced Pseudoscalar Contributions in $A = 12$ Nuclei

Ground state to ground transitions of interest are indicated in Figure 1. Detailed nuclear structure calculations of these reactions were done in an impulse approximation both with (59) and without (59, 102) the inclusion of core polarization and first-class exchange-currents effects. From the general agreement between a number of calculated and measured quantities [rates, $F_E^{(1)}$, etc], one concludes that the microscopic structure of these transitions is well understood and the prediction of first-class current form factors in terms of nucleon form factors is trustworthy to within $\sim 5\%$. The relevant points are

1. The magnetic form factor $\mu$ is due almost entirely (98%) to nucleon magnetism, such that the comparison of $\mu$ with $F_M$ constitutes a CVC test as if it were done with free nucleons.

2. If PCAC is assumed, $F_P$ is composed chiefly ($\sim 75\%$) of $f_P$ and of a non-pole term ($\alpha_2$) in Equation 18c.

3. For the first-class component of $F_E$, which is merely an auxiliary parameter, one has the provision (59, 102)

$$F_E^{(1)}/F_A = 1 + 2\int r(\boldsymbol{\sigma} \cdot \mathbf{p})/M_\sigma \approx 3.6 \qquad 34.$$

4. The scaling of form factor ratios $F_i/F_A \equiv \tilde{F}_i$ vs $q^2$ is guaranteed, except for $\tilde{F}_E^{(1)}$ where a small correction is required (59). The scaling, which is an important aspect of EPT, allows for the combination of $\beta$ decay and $\mu$ capture results to form a larger basis of independent observables.

The test of CVC consists of the verification of the relation (74)

$$F_M^{\mp}(q^2) = \sqrt{2}\,\mu(q^2) \qquad 35.$$

implied by the conserved vector current of the electroweak interaction model (see Equations 8 and 17).

The M1 width $\Gamma_\gamma = 37.0 \pm 1.1$ eV (30), taking into account a small correction due to an $I = 0$ admixture in $^{12}C^*$ (27) and $\cos\theta_c$, yields $\sqrt{2}\mu(0)$ $= 2.00 \pm 0.06$. The axial vector form factor is obtained from $\Gamma$'s: $|F_A^-(0)|$ $= 0.516$ (74). The CVC-implied form factor ratio is (27, 74)

$$\tilde{F}_M^{\mp}(0) = \pm 3.88 \pm 0.12, \qquad 36.$$

where the sign is guaranteed by nuclear physics calculations referred to above. The shape correction factors are then (in $GeV^{-1}$),

$$a_-(\text{CVC}) = -a_+(\text{CVC}) = 5.5 \pm 0.16 \qquad 37a.$$

$$\Delta a(\text{CVC}) = a_- - a_+ = 11.0 \pm 0.32. \qquad 37b.$$

Higher order $(\sim q^2)$ contributions, which may also be viewed as manifestations of imperfect scaling, and Coulomb effects modify $a_\mp$ (12, 101). Following the elaborate calculations of Behrens & Szybisz (12), we have

$$a_- = 6.7 \qquad\qquad\qquad 38a.$$

$$a_+ = -2.7, \qquad\qquad\qquad 38b.$$

$$\Delta a = 9.4, \qquad\qquad\qquad 38c.$$

the difference less modified than the individual slope parameters. In the absence of SCC, and by adopting the calculated value of $\tilde{F}_E^{(1)}$ in Equation 34 (59, 74, 102), we find

$$\alpha_-(CVC) \approx 0.1 \qquad\qquad\qquad 39a.$$

$$\alpha_+(CVC) \approx -2.7, \qquad\qquad\qquad 39b.$$

$$\Delta\alpha(CVC) = 2.75, \qquad\qquad\qquad 39c.$$

which is obviously independent of $F_E^{(1)}$.

The correlation parameters $\alpha_\mp$ of the alignment term in Equation 20 are expected to be unaffected by higher-order and Coulomb effects (102), especially $\alpha_\mp(E_0)$ (66), i.e. they are practically constant as given in Equation 22b.

Experimental verification of Gell-Mann's prediction of weak magnetism was attempted in the early 1960s by observing a $(1 + a_\mp E)$ departure from the allowed shape in the beta spectra of mirror transitions $^{12}B(-)^{12}C(+)^{12}N$. The agreement between the theory and experiment (Table 1) seemed good, especially in the experiment of Lee et al (90).

The faith in CVC was so profound that a warning by Huffaker & Laird (72), who discovered errors in Fermi functions used in the experimental analyses (especially for the $e^+$ branch), went unnoticed until polarization-correlation experiments in $A = 12$ (137) and in $A = 19$ (26) nuclei seemed to show the presence of a significant second-class tensor contribution: $f_T \approx -f_M$. This prompted Calaprice & Holstein (27) to reanalyze the above experiments using corrected Fermi functions and end-point energies (which had also been incorrect). The reanalyzed results, shown in Table 1, weakened the experimental support for weak magnetism, especially in the experiment of Lee et al. The results of the latter were subsequently revised by Wu et al (158), including additional updated parameters such as branching ratios of internal transitions: their revised $a$'s are also given in Table 1.

A new measurement of $a_\mp$ was then performed by Kaina et al (78), using a NaI crystal as a spectrum analyzer instead of the magnetic spectrometer

**Table 1**    Slope parameters $a_\mp$ in $^{12}B(-)^{12}C(+)^{12}N$ mirror beta decays[a]

| Parameter (in GeV$^{-1}$) | Mayer-Kukuk & Michel (95) | Lee et al (90) | Wu et al (158) | Kaina et al (78) no $E^2$ term | Kaina et al (78) with $E^2$ term |
|---|---|---|---|---|---|
| $a_-$ | 18.2(0.9) | 5.5(1.0) | | 7.1(1.1) | 9.1(1.1) |
| $a_+$ | 6.0(0.8) | −5.2(0.6) | | −3.8(0.9) | −0.7(0.9) |
| $a_-$ | 19.2(0.9)[b] | 4.2(1.0)[b] | 4.1(1.0)[c] | | |
| $a_+$ | 10.7(0.8)[b] | −1.7(0.6)[b] | −4.5(0.9)[c] | | |

[a] CVC prediction, including higher-order and Coulomb effects: $a_- = 6.7$ and $a_+ = -2.7$ GeV$^{-1}$ (12).
[b] Revised by Calaprice & Holstein (27).
[c] Results of Lee et al corrected (158).

employed in the earlier measurements (see Figure 5). The crystal's response to electrons was calibrated with monoenergetic electrons obtained from an accelerator. The results are shown in Table 1 (last two columns). The $a_\mp$ parameters are not in agreement with previous (revised or not) determinations.

The cause of troubles lies in the indirect nature of the method employed. The conclusion of the experiment depends strongly upon the knowledge of

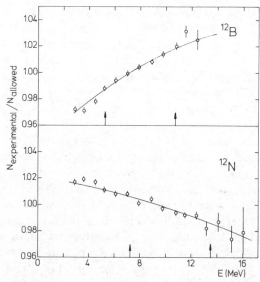

*Figure 5*    Shape correction factors in $A = 12$ nuclei (78). Solid curves are the CVC predictions (12). The $\beta$-ray energy domain exploited to determine $a_\mp$ slope parameters is delimited as indicated by the arrows.

the leading component(s) to be subtracted from the observed result. The role of the first internal positron transition (Table 2) is one example. A small $(-3 \times 10^{-3})$ modification of this branching $b_1^+$ produces a variation $(\delta a_+)$ in $a_+$ amounting to $\sim 25\%$ of the prediction in Equation 38b, $\delta a_+ = -0.7$ (see 158). In conclusion, the individual slope parameters are not trustworthy.

As seen in Table 2, the difference of slope parameters $\Delta a$ is more reliable, as it is less dependent upon, for example, common energy calibration errors. The results shown are obtained from Table 1 taking into account a small correction $\delta a_+$ for the internal branching $b_1^+$ (for which the "central" value taken by Wu et al is adopted). The weighted average $\Delta a = 9.4 \pm 0.7$ is in good agreement with the result of Kaina et al (78), which notes the existence of an additional correction for the energy dependence of positron annihilation in flight (37). By combining the above error and $\delta a_+ = \pm 0.7$ in quadrature, one gets

$$\Delta a = 9.4 \pm 1.0, \qquad\qquad 40.$$

which coincides with $\Delta a$ in Equation 38c, and corresponds to

$$\tilde{F}_M^- - \tilde{F}_M^+ = 7.76 \pm 0.85. \qquad\qquad 41.$$

The determination of the induced tensor form factor required oriented $^{12}$B and $^{12}$N nuclei. They are produced in nuclear reactions $^{11}$B(d,p)$^{12}$B and $^{10}$B($^3$He,n)$^{12}$N, respectively. Through the effect of $\ell \cdot s$ coupling, the recoils in these reactions are produced at certain momentum transfers with

Table 2  Difference of slope parameters $\Delta a = a_- - a_+$ in $^{12}$B($-$)$^{12}$C($+$)$^{12}$N mirror beta decays[a] and beta branches feeding the 4.4-MeV state in $^{12}$C assumed in the analysis of the corresponding experiment

| $\Delta a$ (in GeV$^{-1}$) (Ref.) | $^{12}$B $b_1^-$ (%) | $^{12}$N $b_1^+$ (%) |
|---|---|---|
| $9.2 \pm 1.2$  (95)[b] | 1.3 | 2.4 |
| $15.3 \pm 2.8$  (50)[b] | 1.3 | 2.4 |
| $8.6 \pm 2.4$ (158)[c] | 1.3 | 2.1[e] |
| $9.1 \pm 0.9$  (78)[d] | 1.14 | 1.8 |

[a] CVC prediction including higher-order and Coulomb effects: $\Delta a = 9.4$ GeV$^{-1}$ (12).
[b] Revised result by Calaprice & Holstein (27) and increased by 0.7.
[c] Corrected result of Lee et al (158).
[d] Result of Kaina et al (78) decreased by 0.7.
[e] Adopted herein.

sizeable polarization ($P$) and alignment ($A$) along the normal to the reaction plane. Such orientations can be exploited in beta decay correlation studies thanks to the remarkable discovery of the Johns Hopkins (13) and Osaka (136) groups that, when these nuclei are implanted into certain metals, their orientations relax with times considerably in excess of their mean lives. Collimated recoils (Figure 6a), implanted into a material in the presence of a holding field **B** (Figure 6b), which tends to preserve the nuclear orientation,

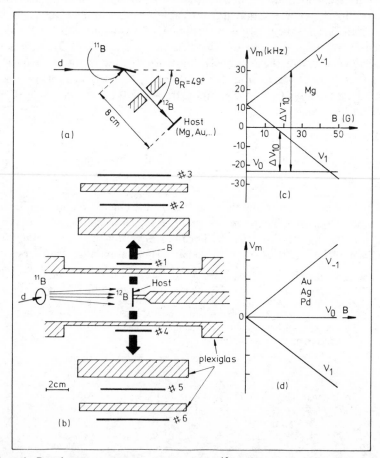

*Figure 6*    Beta decay asymmetry measurements with $^{12}$B. (*a*) Production target, $^{11}$B ($\sim 120\,\mu$g cm$^{-2}$) on inox backing. (*b*) Schematic side-view of the apparatus, 1, $\cdots$, 6 are scintillators. Coincidences are 123($\pi$), 456(0). Holding field is perpendicular to the reaction plane (coils not shown). (*c*) Frequencies of $m$ states of $^{12}$B in a single crystal of Mg (*c* axis parallel to *B*), $g > 0$ and $e^2Qq > 0$. Quadrupole frequency of $^{12}$B in Mg and Zeeman frequency from Haskell et al (63, 64). (*d*) $^{12}$B in face-centered cubic metals, Au, Ag, and Pd. Similar descriptions for measurements with $^{12}$N.

are viewed (along $\mathbf{B}$) by electron counters located above ($\theta = 180°$) and below ($\theta = 0°$) the reaction plane.

The indirect determination of $\alpha_\mp$ through the $E$ dependence of $P(1 + \alpha_\mp E)$, obtained by measuring the up/down beta decay asymmetry vs $E$, is even more difficult than that of the shape correction factor $a_\mp$ since the scale of $\alpha_\mp$ is four times smaller, and the nuclei are only partially polarized. The first indirect measurement produced a controversial result: $\Delta\alpha = 5.2 \pm 0.4$ GeV$^{-1}$ (137), in contrast with $\Delta\alpha(\text{CVC}) \approx 2.8$ GeV$^{-1}$, which implies a large second-class induced tensor form factor, $F_T^{(2)} \approx -F_M$. Accurate and definitive determinations of $\alpha_\mp$ were subsequently obtained by direct methods, via the alignment term in $W(\theta,E)$. A discussion of the experimental status of second-class currents at the moment of transition between indirect and direct measurements of correlation parameters can be found in a report by Telegdi (139).

To isolate the alignment term, the following (idealized) procedure was used (19, 20, 88): $^{12}$B ($^{12}$N) with *vanishing* initial alignment $A^0 = 0$ and *maximum* initial polarization $P^0$ were implanted, in the presence of a holding field $\mathbf{B}$, into a single crystal of Mg where the $|\Delta m| = 1$ Zeeman frequencies ($v_{10}$, $v_{\bar{1}0}$) are not degenerate. Selective *equalization* of populations ($p_m$) of $m = 1$ and $m = 0$ substates (situation 1), or of the $m = -1$ and $m = 0$ substates (situation $\bar{1}$), leads to a reduced polarization $P(1) = P(\bar{1}) = (3/4)P^0$ and to an *induced alignment* $A(1) = -A(\bar{1}) = -(3/4)P^0$.

The isolation of the $P_2$ term in $W(\theta,E)$ is based on this sign change of the induced $A$ for fixed $P^0$. Conditions 1 and $\bar{1}$ were alternated rapidly. In one implantation-counting cycle, the nuclei were subject to 1, in the next to $\bar{1}$. Denoting the corresponding electron rates (at fixed $\theta$) $N_1$ and $N_{\bar{1}}$, one can form a signal

$$S_\mp = 2(N_1 - N_{\bar{1}})/(N_1 + N_{\bar{1}})$$
$$= \tfrac{3}{2}P^0\alpha_\mp EP_2(\theta)/[1 \mp \tfrac{3}{4}P^0P_1(\theta)].$$   42.

In practice, the recoils are not produced with $A^0 = 0$, and an initial alignment leads to an induced polarization $P = A^0/2$; the signal is given in the general case by

$$S_\mp = \frac{\tfrac{3}{2}P^0\alpha_\mp EP_2 \mp A^0/2}{1 \mp \tfrac{3}{4}P^0P_1(\theta)}.$$   43.

Defining the numerator as the "reduced" signal $\tilde{S}_\mp(\theta,E)$, one clearly has

$$\alpha_\mp = |\tilde{S}_\mp(0,E) + \tilde{S}_\mp(\theta,E)|/3P^0E.$$   44.

It is essential to note here that the initial polarization $P^0$ is determined in a separate measurement by using the same apparatus and the same general conditions (background, etc), such that the signal $\alpha_\mp$ is calibrated in situ.

In order to obtain largest possible $S_\mp$ signals, the production conditions of $^{12}$B and $^{12}$N are so chosen as to maximize (minimize) $P^0$ ($A^0$). The selective equalization of populations is achieved by either NMR saturation techniques (88) or spontaneous level mixing at level crossing (19, 20). The level crossing of $|m| = 1$ and $m = 0$ substates of $^{12}$B in Mg, which occurs at $B^* \approx 45$ G, is indicated in Figure 6c (for $^{12}$N in Mg, $B^* \approx 120$ G). Conditions 1 and $\bar{1}$ are realized by alternating the polarity of $B^*$.

Figure 7, a plot of $\tilde{S}_+$ obtained in the decay of $^{12}$N by using spontaneous level mixing to induce alignment, illustrates well enough the sensitivity of the direct method to detect and measure induced weak currents.

Alignment induced by selective population *inversion*, achieved by using the adiabatic fast-passage technique, was also exploited to determine $\alpha_\mp$ (94). Figure 8 explains the conversion of the initial (ideally pure) polarization into a positive (HF) or a negative (LF) alignment. Clearly, the difference in count rates obtained with $A^H (A > 0)$ and $A^L (A < 0)$ leads to a signal similar to $S_\mp$ in Equation 43. The advantage of the method is that for a fixed $P^0$, the induced alignment is two times larger than that obtained by population equalization.

The correlation parameters $\alpha_\mp$ determined by direct methods are summarized in Table 3 and displayed vs $E$ in Figure 9. In those experiments where $\alpha_\mp$ is obtained by measurement of the slope only, the values are placed at $\bar{E}_\mp$. Also included is an early determination of $\alpha_- (E_0)$ obtained by exploiting the initial alignment in $^{11}$B(d,p)$^{12}$B (132). In this experiment, the production conditions were so tuned as to maximize (minimize) $A^0$ ($P^0$) of $^{12}$B implanted in polycrystalline Pd (level diagram of $m$ states is shown in Figure 6d). By comparing counting rates with ($B \neq 0$) and without ($B = 0$)

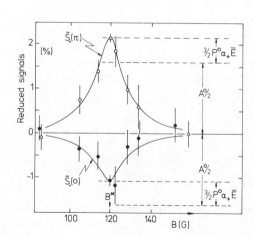

Figure 7  Reduced signals $\tilde{S}_+$ at and near level crossing (integrated over $E$, $\bar{E} \approx 9$ MeV). The sum $\tilde{S}_+(\pi) + \tilde{S}_+(0)$ is the induced alignment effect, $3P^0\alpha_+\bar{E}$, see Equation 44. [From Brändle et al (20) and unpublished material.]

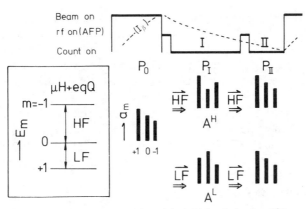

*Figure 8* Diagram to explain the conversion of the initial polarization $P^0$ into a positive or negative alignment $A$ by using adiabatic fast passage (AFP). A time-sequence program is shown for the $^{12}$N production, NMR transitions, and countings. The alignment correlation is measured in period $I$.

**Table 3** Correlation coefficients $\alpha_{\mp}$ in mirror transitions $^{12}$B$(-)^{12}$C$(+)^{12}$N measured through the alignment-electron momentum angular distribution (errors in parentheses)

| Alignment | $\alpha_-$ (GeV$^{-1}$) | $\alpha_+$ (GeV$^{-1}$) | Ref. |
|---|---|---|---|
| Initial | 0 (0.27)[a] | | 132[b] |
| Induced (NMR, $p_m$ equalization) | −0.07(0.20)[a] | | 88[c] |
| Induced (level crossing) | 0.24(0.40) | | 19[b] |
| Induced (NMR, $p_m$ inversion) | 0.25(0.34) | | 94[b] |
| Induced (NMR, $p_m$ inversion) | | −2.77(0.48) | 94[b] |
| Induced (level crossing) | | −2.73(0.39) | 20[b] |
| Induced (level crossing) | 0.1 (0.3) | | 20[b] |
| Induced (NMR, $p_m$ inversion) | | −2.67(0.56) | 94[b] |
| Induced (NMR, $p_m$ inversion) | −0.1 (0.21) | | 94[b] |
| | $\langle \alpha_- \rangle = 0.02(0.12)$ | $\langle \alpha_+ \rangle = -2.72(0.27)$ | |

[a] $\alpha_-(E_0)$.
[b] Electron energy is measured by using a plastic scintillator.
[c] Electron energy is measured by electron range.

nuclear orientation, the asymmetry $\varepsilon(\theta,E)$ was determined both at $\theta = 180°$ and $0°$. The coefficient $\alpha_-$ was deduced from the comparison of $\varepsilon(\pi,E) = P^0 + \alpha_- EA^0$ and $\varepsilon(0,E) = -P^0 + \alpha_- EA^0$. The estimate of $A^0$ was taken from measurements by Hori et al (71). The data indicate clearly that the $\alpha_\mp$ parameters are essentially constants, i.e. the effects contributing to the alignment term are of first-order in the recoil.

The sign of $\alpha_-$ ($\approx 0$) is irrelevant, but can be deduced from the nuclear physics experiment of Hori et al (71). That of $\alpha_+$ ($<0$) may be taken from the early (indirect) measurement of Sugimoto et al (137), which was confirmed by Masuda et al (94) through the angular distribution of the $\gamma$ ray in $^{12}$N $\rightarrow$ $^{12}$C$_1(\gamma)^{12}$C$_{gs}$.

The averages (in GeV$^{-1}$) are

$$\alpha_- = +0.02 \pm 0.12 \tag{45a}$$

$$\alpha_+ = -2.72 \pm 0.27. \tag{45b}$$

The sum, $\alpha_- + \alpha_- = 2.69 \pm 0.30$, leads to $\tilde{F}_E^{(1)} = 3.80 \pm 0.39$, in good agreement with the nuclear physics calculation given in Equation 34, so that $\alpha_\mp$ in Equations 39a,b are confirmed to be $\alpha_\mp$(CVC). The very good agreement between the above results and $\alpha_\mp$(CVC) shows the absence of a significant second-class induced tensor contribution. In order to reach a more quantitative conclusion, let us anticipate here the weak magnetism determination in $^{12}$C $\rightarrow$ $^{12}$B muon capture discussed just below (see Equation 60). That result, and the information obtained in beta decay

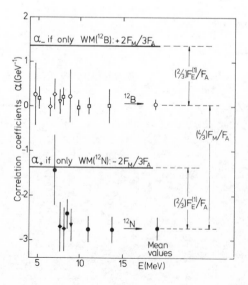

Figure 9   Alignment-electron momentum correlation coefficients $\alpha_\mp$ in $A = 12$ nuclei (data are taken from Table 3). Were weak magnetism the only induced contribution, we would have $\alpha_\mp = \pm 2F_M/3F_A$. Note that the deviation of the corresponding experimental coefficients from $\alpha_\mp$(CVC) is essentially charge independent $\approx 2F_E^{(1)}/3F_A$ and signals the absence of SCC. For definitions of form factors $F_i$'s, see text.

(Equation 41), yield the average

$$\tilde{F}_M = 4.05 \pm 0.34.$$  46.

This result combined with the difference of the correlation parameters determined experimentally, $\alpha_- - \alpha_+ = 2.75 \pm 0.30$, leads to

$$\tilde{F}_E^{(2)} = -0.20 \pm 0.50.$$  47.

The remainder of this section deals with the determination of weak magnetism and induced pseudoscalar form factors in the muon capture reaction $\mu^- + {}^{12}C \to \nu_\mu + {}^{12}B_{gs}$.

Since the helicity of the muon neutrino $h_{\nu_\mu}$ is $-1/2$, ${}^{12}B$ can have only $m = -1$ or $m = 0$ spin projections in the recoil frame. The helicity of the final state $({}^{12}B + \nu_\mu)$ is thus $-1/2$ or $+1/2$, for which the amplitudes are denoted by $M_{-1/2}$ and $M_{1/2}$, respectively (14). These amplitudes are composed of weak form factors (74):

$$M_{-1/2} = F_A(1 + \tilde{F}_A E_\nu/2m_p)\sqrt{2},$$  48a.

$$M_{1/2} = F_A(1 + \tilde{F}_P m_\mu E_\nu/m_\pi^2 - \tilde{F}_E E_\nu/2m_p),$$  48b.

where $E_\nu$ is the neutrino energy. Any observable in ${}^{12}C \to {}^{12}B_{gs}$ can be expressed as a function of a single dynamical parameter (116) $x = \sqrt{2}M_{1/2}/M_{-1/2}$, or

$$x = (1 + \tilde{F}_P m_\mu E_\nu/m_\pi^2 - \tilde{F}_E E_\nu/2m_p)/(1 + \tilde{F}_M E_\nu/2m_p).$$  49.

In the allowed limit $(E_\nu \to 0)$, there is but a single overall amplitude, and $x = 1$. Any deviation of $x$ from unity is thus indicative of recoil-order contributions. An independent determination of $F_M$ and $F_P$ is obtained by measuring the following three observables.

1. The capture rate, $\Gamma_c$, which is proportional to $(|M_{-1/2}|^2 + |M_{1/2}|^2)$. By virtue of $\mu$-e universality, $\Gamma_c$ can be expressed in terms of $\Gamma^-$, namely (74)

$$\Gamma_c = K |F_A(q^2)/F_A(0)|^2 (1 + \tilde{F}_M E_\nu/2m_p)^2 (2 + x^2).$$  50.

Taking into account the already known $F_E^-$ contribution to $\Gamma^-$, the constant $K$ amounts to $3.63 \times 10^3$ s$^{-1}$. The scaling factor $F_A(q^2)/F_A(0)$ is "known" from nuclear physics calculations referred to at the beginning of this section (59). It may also be obtained from $\mu(q^2)/\mu(0)$ determined in ${}^{12}C \to {}^{12}C^*$ (e, e') measurements (27, 30). Here the compromise between these (almost equivalent) alternatives is taken: $F_A(q^2)/F_A(0) = 0.765 \pm 0.015$. Insertion of $x$ into $\Gamma_c$ allows the determination of $(1 + \tilde{F}_M E_\nu/2m_p)$, i.e. weak magnetism in muon capture. We may remark here that the predicted weak magnetism, $\tilde{F}_M \approx 4$, multiplied by $E_\nu/2m_P = 0.0487$, increases $\Gamma_c$ by $\sim 40\%$!

Experimental inputs to the determination of $\Gamma_c$ are from the accurate determination of $\Gamma$ (all bound states) made by Maier, Edelstein & Siegel (92) and from three other (published) results taken from Reynolds et al (120). Their mean, corrected for the contribution $900 \pm 80$ s$^{-1}$ of excited bound states (57, 98, 123), yields $\Gamma_c(obs) = (6.18 \pm 0.26) \times 10^3$ s$^{-1}$.

2. The polarization of $^{12}$B in the recoil frame (the so-called longitudinal polarization $P_L = \langle \mathbf{J} \cdot \mathbf{p}_{recoil} \rangle / J$). Expressed vs $x$, this is

$$P_L = -|M_{-1/2}|^2/(|M_{-1/2}|^2 + |M_{1/2}|^2) = -2/(2+x^2).$$  51.

A parity-violating effect, it is essentially the recoil polarization measured in the Goldhaber–Grodzins–Sunyar experiment (52), which determined $h_{\nu_e}$, since the spin sequence is the same. The difference is that in $\mu$ capture the $q = 0$ approximation is not valid. With $E_\nu \to 0$ (i.e. $x = 1$), one recovers, of course, the classical e capture result, $P_L = -2/3$ (121).

3. The polarization of $^{12}$B in the $\mu^-$ frame (the so-called average polarization $P_{av} = \langle \mathbf{J} \cdot \mathbf{P}_\mu \rangle / J$, where $\mathbf{P}_\mu$ is the $\mu$ polarization). Expressed vs $x$, this is

$$P_{av} = (2/3)P_\mu(1 + 2x)/(2 + x^2).$$  52.

The interesting feature of $P_{av}$ is that it involves $x$ linearly, which can be exploited to determine $x$ by measuring the ratio $R$ of $P_{av}$ and $P_L$. Given Equations 51 and 52, one has

$$R = P_{av}(P_\mu = 1)/P_L = -(1/3)(1 + 2x).$$  53.

The dynamical parameter $x$ is determined from the ratio of $P_{av}$ and $P_L$ measured *simultaneously*. The experiment was performed (122; see also 142, 143) by using a recoil-direction-sensitive target, the principle of which is illustrated in Figure 10$a$. Muons from $\pi$ decay in flight are stopped in a carbon foil C sufficiently thin to let most of the $^{12}$B recoils emerge (recoil range in C is 180 $\mu$g cm$^{-2}$), which is sandwiched between a polarization-retaining (P) and a polarization-destroying (D) layer. According to the neutrino direction, the recoils will be stopped either in P or D. If a holding field **B** is applied along the axis of incidence of the muons, normal to foil C, polarization of $^{12}$B is retained along this axis in P, and is detected via the beta decay asymmetry of $^{12}$B with identical telescopes $T_1$ and $T_2$. The material of P is such that it also becomes depolarizing in the absence of **B**. If we assume $h_{\nu_\mu} < 0$, i.e. $P_L < 0$ (Equation 51) and disregard $P_{av}$, $T_1$ will count more beta rays than $T_2$ when the sandwich is oriented as in Figure 10$a$. When the sandwich is flipped by $\pi$ rad, the role of the telescopes is reversed. The beta decay asymmetry due to $P_{av}$ (a parity-conserving effect) is insensitive to the ordering of the P-C-D sandwich.

Defining as forward orientation (f) that direction where recoils along $P_\mu$ preserve their polarization, and as backward (b) the opposite, and integrating over all recoil directions, one has ideally for the total polarization (along **B**):

$$P^f = \tfrac{1}{2}(P_{av} + \tfrac{1}{2}P_L) \text{ and } P^b = \tfrac{1}{2}(P_{av} - \tfrac{1}{2}P_L),  \qquad 54.$$

so that (with $P_\mu = 2h_{\bar{v}_\mu} k$ where $k$ is the muon's kinematic and atomic depolarization factor)

$$R' = \tfrac{1}{2}(P^f + P^b)/(P^f - P^b)  \qquad 55.$$

$$= kP_{av}/P_L = -\tfrac{1}{3}k(1 + 2x),$$

$$R = R'/k = -\tfrac{1}{3}(1 + 2x).  \qquad 56.$$

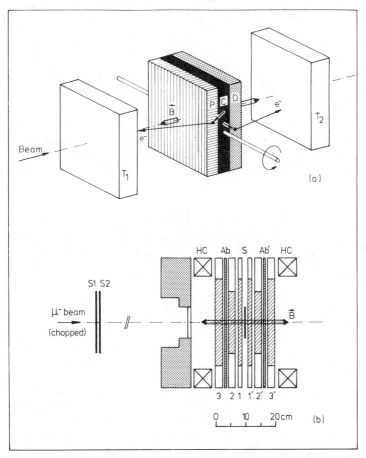

*Figure 10*   (*a*) Principle of the recoil-direction-sensitive PCD target and the experiment. (*b*) Actual setup: *S* (stack target), *Ab* (absorber), *HC* (Helmholtz coils), $S_1 S_2$ (beam monitor); 1, ···, 3′ are multiwire proportional chambers.

In practice the target S consisted of a stack of $N$ sandwiches as described, with for example D = 405 $\mu$g cm$^{-2}$ aluminum, C = 60 $\mu$g cm$^{-2}$ graphite, P = 1250 $\mu$g cm$^{-2}$ silver, and $N$ = 1000. Corresponding depolarizing $(CH)_2$, "dummy" (D-P), and recoil-direction-insensitive (P-C-P) targets were used for calibration purposes. The setup is shown schematically in Figure 10$b$. The muon depolarization factor was determined, using a separate setup (142), by the integral method (115). Corrected polarizations are shown in Figure 11.

Note here that the raw ratio $R$ is very close to the corrected one, as practically all of the corrections cancel. The final result (normalized to $P_\mu = 1$) is

$$R_{obs} = -0.516 \pm 0.041. \qquad 57.$$

The value of $P_L$ after corrections due to polarization losses is

$$P_L = -0.92 \pm 0.10. \qquad 58.$$

The average polarization (also a source of information for $x$) was determined in independent experiments using recoil-direction-insensitive targets, following a suggestion of Jackson, Treiman & Wyld (75) in determining the antineutrino helicity in $\pi$ decay. The principle of the measurement of $P_{av}$ is the following: negative longitudinally polarized muons from $\pi$ decay in flight are stopped in a bulk graphite target in the

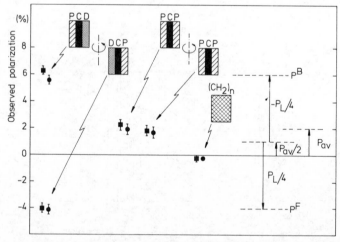

*Figure 11*  Polarizations of $^{12}$B observed with various stacks and stack orientations. P = Ag and D = Al; full squares (circles) correspond to polarizations measured with T$_1$ (T$_2$). On the right-hand side the measured polarizations are decomposed in terms of $P_{av}$ and $P_L$: $P^{f(b)}$ = $[P_{av} + (-)P_L/2]/2$. Note that the polarization obtained with the PCP target, namely $P_{av}$, gives, as expected, twice the "offset" of the polarizations of the PCD target.

presence of a holding field $\mathbf{B}$ along $P_\mu$; the polarization $P_{av}k$, of recoils produced in $^{12}C(\mu^-, v_\mu)^{12}B$ is determined again via the beta decay asymmetry of $^{12}B$; the muon depolarization factor $k$ is measured in a separate experiment. Two measurements of $P_{av}$ have been performed to date:

1. Possoz et al (116) used polycrystalline graphite. Measurement of $P_{av}$ was done at several field settings $[0.4 < B\,(kG) < 1.2]$. For the incomplete spin decoupling of $^{12}B$ in graphite, see (114, 116).

2. Kuno et al (85) used Grafoil (almost a single crystal of graphite). At $B = 3$ kG used in the experiments, the spin decoupling of $^{12}B$ is almost complete in this material.

These results (normalized to $P_\mu = 1$), $P_{av}^{(1)} = 0.452 \pm 0.042$ and $P_{av}^{(2)} = 0.456 \pm 0.047$, are in agreement.

Determinations of $R$, $P_L$, and $P_{av}$ are summarized in Table 4. They are corrected for the contribution of $^{12}B^*$. Note that in contrast with $P_L$ and $P_{av}$, $R$ is almost independent of that correction, too (122).

The dynamical parameter $x$ deduced from $R$ and $P_{av}$,

$$x = 0.26 \pm 0.06, \hspace{4cm} 59.$$

is rather far from unity, and evidence in itself for the presence of induced weak currents. As seen from Equation 49, this deviation is essentially due to the presence of $F_P^-$ and the known $F_E^-$.

The insertion of $x$ into $\Gamma_c$ yields $1 + \tilde{F}_M E_v/2m_p = 1.212 \pm 0.030$, i.e.

$$\tilde{F}_M = 4.27 \pm 0.60. \hspace{4cm} 60.$$

The insertion of the already known $\tilde{F}_M$ and $\tilde{F}_E$ into $x$ (Equation 49) yields

**Table 4**   Average polarization $(P_{av})$, longitudinal polarization $(P_L)$, and their ratio $(R = P_{av}/P_L)$ measured in $^{12}C(\mu,v_\mu)^{12}B$ (corrected results refer to $^{12}B_{gs}$)

| Observ-able | Observed | Corrected[c] (I) | Corrected[c] (II) | Deduced quantities | | | | |
|---|---|---|---|---|---|---|---|---|
| | | | | $X_I$ | $X_{II}$ | $F_P/F_A(I)$ | $F_P/F_A(II)$ | Ref. |
| $P_{av}$ | 0.435(40)[a] | 0.448(41) | 0.474(44) | | | | | 116 |
| | 0.466(47) | 0.480(48) | 0.508(51) | | | | | 85 |
| | 0.447(32)[b] | 0.460(32) | 0.487(34) | 0.21(5) | 0.25(7) | $-1.05(12)$ | $-0.95(17)$ | |
| $R$ | $-0.516(40)$ | $-0.500(40)$ | $-0.531(41)$ | 0.25(6) | 0.30(6) | $-0.95(15)$ | $-0.83(15)$ | 122 |
| $P_L$ | $-0.92\;(9)$[a] | $-0.98\;(10)$ | $-0.98\;(10)$ | | | | | 122 |

[a] Original result corrected for the directly determined $^{13}C$ contribution $\approx 1.1\%$ (85). Errors are in parentheses.
[b] Average.
[c] Corrected for $^{12}B^*$ contribution (123); the extrema are indicated by I and II.

the induced pseudoscalar form factor:

$$\tilde{F}_P = -0.93 \pm 0.15. \qquad\qquad 61.$$

This result corrected for the non-pole contribution (59, 73) leads finally to

$$f_P = -(0.70 \pm 0.15) f_A(0), \qquad\qquad 62.$$

in agreement with the hydrogen result.

Formulas of $P_L$, $P_{av}$, and $\Gamma_c$ and the discussion of their relevance may be found in many references: Foldy & Walecka (43), Devanathan et al (39), Subramanian et al (134), Holstein (68), Mukhopadhyay (103), Kobayashi et al (83), Guichon & Samour (59), O'Connell et al (107), Fukui et al (44), and many others.

## 4.5   The $0^+ \rightleftarrows 0^-$ Pseudoscalar Reactions in $A = 16$ Nuclei and the Induced Pseudoscalar

The comparison of the rates of $^{16}O(0^+) \rightleftarrows {}^{16}N^m(0^-)$ transitions represents a source of information for the induced pseudoscalar form factor. The argument of Shapiro & Blockhintsev (129) and Ericson, Sens & Rood (41) goes as follows. The amplitude of the $\mu$ capture in such a pseudoscalar transition $(a_\mu)$ is composed of the time component of $A_\lambda$ $(\sim g_A \int \boldsymbol{\sigma} \cdot \mathbf{V})$ and its space component times $g_P^\mu$ $(\sim g_P^\mu \int \boldsymbol{\sigma} \cdot \mathbf{r})$. The amplitude of the inverse beta decay $(a_\beta)$ is composed essentially of the first of these only. The ratio, $a_\mu/a_\beta \approx 1 + (g_P^\mu/g_A) \int \boldsymbol{\sigma} \cdot \mathbf{r} / \int \boldsymbol{\sigma} \cdot \mathbf{V}$, contains the desired information. The quotient of nuclear matrix elements, quasi-independent of the configuration mixing parameters of the involved wave functions, may be estimated in terms of single-particle matrix elements. This suggestion was followed in both $\mu$ capture and inverse beta decay rate measurements (Table 5). Comparison of the beta rate $\Gamma_\beta$ measured by Palffy et al (111) and the capture rate $\Gamma_c$ (the first three of the $\Gamma_c$ determinations were available at the

**Table 5**  Beta decay and muon capture rates in $^{16}O(0^+) \rightleftarrows {}^{16}N(0^-$ ; 120 keV) transitions (in s$^{-1}$)

| $^{16}N^m \rightarrow {}^{16}O_{gs}$ ($\Gamma_\beta$) | Ref. | $^{16}O_{gs} \rightarrow {}^{16}N^m$ ($\Gamma_c$) | Ref. |
|---|---|---|---|
| $0.43 \pm 0.10$ | 111 | $1100 \pm 200$ | 32 |
| $0.41 \pm 0.06$ | 45 | $1600 \pm 200$ | 4 |
| $0.60 \pm 0.07$ | 100 | $850 \pm 145$ $60$ | 38 |
| $0.45 \pm 0.10$[a] | | $1560 \pm 180$ | 79 |
| | | $1560 \pm 108$ | 55 |

[a] Private communication from Dr. L. Lessard.

time of the analysis) leads to either $g_P^\mu/g_A = 15 \pm 2$ or $19.0 \pm 1.5$. The former (latter) value corresponds to the highest (lowest) $\Gamma_c$. The latest determination of $\Gamma_c$ (55) favors $15.0 \pm 2$. The subsequent determinations of $\Gamma_\beta$ corroborated the first result, and it seemed that there was a serious indication of an upward renormalization of $g_P^\mu/g_A$ instead of the downward one expected for an "infinitely large" nuclear medium (53, 157).

However, subsequent calculations that go beyond the one-body impulse approximation descriptions show that one-pion exchange contribution enhances the time component of $A_\lambda$ by about 1.5 (56), which thus reduces $g_P^\mu/g_A$ to $10 \pm 2$. Further investigations revealed that a departure from the closed-shell hypothesis for $^{16}O$ mimics the role of one-pion exchange contribution (29, 58). In any case, the apparent upward renormalization of $f_P$ no longer exists (see also discussion in 76 and 81).

## 4.6  Weak Magnetism and Second-Class Current Form Factors

4.6.1  IN $A = 19$ NUCLEI    The super-allowed $^{19}Ne(1/2^+) \rightarrow {}^{19}F(1/2^+)$ $\beta^+$ transition is an important source of information for the weak magnetism and second-class current form factors. In such a transition, the observables may be treated in a model-independent way. For instance, the slope parameter of the up/down beta decay asymmetry, $dA/dE$, has the well-defined value (26)

$$dA/dE(CVC) = -3.4 \pm 0.1 \text{ GeV}^{-1}. \qquad 63.$$

The smallness of the $E$-independent contribution $A(0) \approx -0.040$, due to a destructive $f_V/f_A$ interference effect, is of course a fortunate aspect from an experimentalist's point of view.

In practice (26) the parent nucleus, produced in the $^{19}F(p,n)^{19}Ne$ reaction, is pumped into the source cavity of an atomic beam apparatus. Atoms emitted by the source traverse a Stern-Gerlach spin selector, which allows spin selection at will, and enter a storage cell viewed by scintillation counters located at $\theta = 0°$ and $180°$ with respect to the quantization axis. At this stage of the experiment, a systematic error characteristic of polarization-correlation measurements may be present: backscattering of beta rays from the detector at $0°$ ($180°$) into that at $180°$ ($0°$), which biases the asymmetry principally in the low-energy portion of the spectrum. The presence of a systematic error of this sort leads to a curvature of $A(E)$, and tends to increase (decrease) the absolute value of the actual slope parameter if the relative sign of $dA/dE$ and $A(0)$ is positive (negative). The first published result of Calaprice et al (26) indicates a curvature of $A$, i.e. the presence of backscattering. It is therefore not surprising that the slope parameter, $dA/dE = -6.5 \pm 1.5$, deviated substantially from the expected one. The "large SCC contribution," $f_T \approx -2f_M(\text{CVC})$ implied by the result disappeared, of course, when the experiment was repeated with minimized backscattering conditions. The improved result agrees with $dA/dE(\text{CVC})$, and if CVC is assumed, it yields the constraint (128)

$$|f_T/f_M(\text{CVC})| \lesssim 1/4. \qquad \qquad 64.$$

One must note here that the slope parameter is less dependent upon $f_T$ than $f_M$ (a matter of spin-dependent factors), so that one may infer an (almost) independent determination of the weak magnetism form factor,

$$f_M/f_M(\text{CVC}) = 1.00 \pm 0.15. \qquad \qquad 65.$$

4.6.2  IN $A = 20$ NUCLEI  The beta-gamma angular correlation measurements in $A = 20$ nuclei (Figure 12) are equivalent to the $\beta$-$\alpha$ correlation experiments in $A = 8$ nuclei described in Section 4.3, and their treatment is therefore quite analogous. The 90°–180° correlation asymmetry vs $\beta$ energy $(E)$ has the form $C_0 + C_1 E + C_2 E^2$. The parameter of interest is the difference between $\beta^-$-$\gamma$ and $\beta^+$-$\gamma$ angular correlation slope parameters, $\delta = C_1^- - C_1^+$.

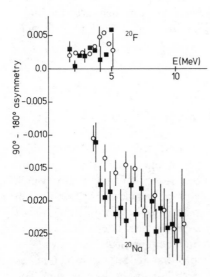

*Figure 12*  $\beta^- - \gamma$ and $\beta^+ - \gamma$ correlation asymmetries vs the beta energy $E$ in $A = 20$ nuclei, and the transitions involved in the experiment. $^{20}$F data from Dupuis-Rolin et al (36) (*open circles*) and Tribble & May (141) (*full squares*); $^{20}$Na data from (36) (*open circles*) and Tribble, May & Tanner (144) (*full squares*). Note that the $^{20}$F data (36) are in fact the corrected results of Ref. 124.

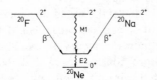

The $F_M/F_A$ contribution to $\delta$(CVC) is deduced from the $ft$ values of beta transitions ($\sim F_A$) and from the M1 width of the analogue $^{20}Ne^2$-$^{20}Ne^1$ gamma transition ($\sim F_M$). A calculation that takes into account various corrections (essentially of Coulomb origin) yields (36)

$$\delta(\text{CVC}) = 4.0 \pm 0.1 \text{ GeV}^{-1}. \tag{66.}$$

Data on the $\beta$-$\gamma$ asymmetries in the decays of $^{20}F$ and $^{20}Na$ are shown in Figure 12. In the $e^-$ branch, where $E$ is not too large ($E_0^- = 5.39$ MeV), the role of $C_2^- E^2$ is estimated to be negligibly small (25). A linear fit to the respective data yields the "direct" determinations $C_1^- = 0.41 \pm 0.18$ (Ref. 36), $C_1^- = 0.5 \pm 0.7$ (Ref. 141), and $C_1^- = 0.9 \pm 0.5$ by McKeown et al (96), which agree.

In the $e^+$ branch, where $E$ is larger ($E_0^+ = 11.76$ MeV), $C_2^+ E^2$ may have a nonnegligible contribution. Clearly, $C_2^+ \neq 0$ ($>0$), see Figure 12. Therefore, the determination of $C_1^+$ is an "indirect" one. One notes that the data of (36) and (144) almost agree, except for the curvature of the asymmetry vs $E$. Obviously, a smaller curvature observed by (36) [larger curvature by (144)] implies a smaller (larger) $-C_1^+$. Such a correlation is indeed reflected by the results: $C_1^+ = -2.93 \pm 0.32$ (Ref. 36) and $C_1^+ = -5.0 \pm 0.8$ (Ref. 144). Note that a three-free-parameter fit to the existing data yields poor accuracy for the $C_i$'s (see discussion by Tribble & May). Therefore, external constraints for either $C_0$ or $C_2$ (or for both of them) are imposed in data analyses. The error attributed to $C_1^+$ by Dupuis-Rolin et al (36) is smaller than that attributed to $C_1^\pm$ in the other experiments because constraints are more severe there than in the others. Application of more conservative constraints in the reanalysis of data of (36) might attenuate the discrepancy of the presently available $C_1^+$ coefficients.

An external consistency treatment of $C_1^+$ (36) and $C_1^+$ (144), combined with the average of $C_1^-$, yields $\delta = 4.6 \pm 1.1$. By comparing this figure with $\delta$(CVC) in Equation 66, one concludes that the CVC condition is verified to within about 25% uncertainty, or $|F_E^{(2)}| \lesssim F_M/4$ if CVC is assumed. A similar conclusion has been drawn by Tribble, May & Tanner (see discussion preceeding their Equation 11, Ref. 144).

## 4.7 The $\nu$-$\gamma$ Correlation in $A = 28$ Nuclei and the Induced Pseudoscalar Form Factor

Measurement of the angular correlation between the neutrino emitted in $\mu$ capture and a deexcitation gamma ray of the product nucleus, an approach analogous to the measurement of orientation parameters of $^{12}B$ discussed in Section 4.4, has been forward by Popov (113) and others (22, 110).

In practice, the angle between neutrino and gamma-ray momenta can be measured via the Doppler shift of the $\gamma$ ray (54). Since the recoil velocity of a not-too-heavy nucleus produced in $\mu$ capture is about 1% of the light velocity, the Doppler shift of the (prompt) $\gamma$ ray is quite large, and it can be determined by using commercially available high-resolution (Ge-Li) detectors. Obviously the low $\gamma$-detection efficiency of Ge(Li) detectors provides a statistical limitation to the method.

A $v$-$\gamma$ correlation experiment was performed in the $\mu^- + {}^{28}\text{Si} \to v_\mu + {}^{28}\text{Al}^*$ (2.2 MeV) reaction, between $v_\mu$ and the 1.23-MeV deexcitation $\gamma$ ray in the ${}^{28}\text{Al}^*$ (2.2 MeV) $\to {}^{28}\text{Al}^*$ (0.97 MeV) transition (99). The dynamical descriptions of this $0^+ \to 1^+$ and the ${}^{12}\text{C} \to {}^{12}\text{B}_{\text{gs}}$ capture reactions are identical. The orientation of ${}^{28}\text{Al}^*$, which is deduced from the Doppler pattern of the 1.23-MeV $\gamma$ ray, may be analyzed in terms of induced weak currents. Unfortunately, the available experimental result is inaccurate (low statistics) and the resulting estimate of the induced pseudoscalar (CVC and no SCC assumed) is therefore yet uncertain (31, 112).

## 4.8    Radiative Muon Capture in ${}^{40}Ca$ and the Induced Pseudoscalar Form Factor

The radiative muon capture $\mu^- + p \to v_\mu + \gamma + n$ and its nuclear analogue, $\mu^- + Z \to v_\mu + \gamma + (Z-1)$, are among the potential sources of information on induced weak currents (117, 125). An advantage of the radiative capture is that the momentum transfer is variable. The low branching ratio $(10^{-5})$ is, however, a serious disadvantage. A successful experiment has been performed in the $\mu^- + {}^{40}\text{Ca} \to v_\mu + \gamma + {}^{40}\text{K}$ reaction (62). The branching ratio for photon energies greater than 57 MeV was obtained as $(21.1 \pm 1.4) \times 10^{-6}$. Gamma rays were converted into electrons, such that the neutron background was practically eliminated. This method of neutron background suppression constitutes the major advance over earlier measurements (33, 35). By assuming CVC and no SCC, the above branching ratio determination yields the "indirect" result

$$f_P/f_P(\text{PCAC}) = 0.95 \pm 0.23. \qquad 67.$$

Experiments, including correlations (155), are in progress in several pion laboratories. The availability of high $\mu$ intensities will serve to compensate for the low branching yield. More direct and accurate determination of the induced pseudoscalar, resulting from both $v$-$\gamma$ and radiative capture measurements, may be expected in the near future.

## 5.    DISCUSSION OF EXPERIMENTAL RESULTS

Determinations of the induced weak currents discussed in Section 4 and additional information regarding SCC are summarized in Table 6.

**Table 6** Present status of experimentally determined induced weak form factors in nuclei[a]

| $A$ | Reaction | Weak magnetism $F_M/F_M(CVC)$ | Induced pseudoscalar $f_P/f_P(PCAC)$ | Induced tensor $F_E^{(2)}/F_A$ |
|---|---|---|---|---|
| 1 | $\mu$ capture | — | $1.04 \pm 0.23$ | — |
| 2 | $\nu$ scattering | — | — | $0.0 \pm 1.0$ |
| 6 | $\beta$ decay | — | — | $0.5 \pm 1.3$ |
| 8 | $\beta$ decay | $0.82 \pm 0.12$ | — | $0.0 \pm 0.6$ |
| 12 | $\beta$ decay | $1.00 \pm 0.11$ | — | $-0.2 \pm 0.5$ |
| 12 | $\mu$ capture | $1.10 \pm 0.15$ | $1.0 \pm 0.21$ | — |
| 19 | $\beta$ decay | $1.00 \pm 0.15$ | — | — |
| 20 | $\beta$ decay | $1.15 \pm 0.28$ | — | — |
| 40 | radiative $\mu$ capture | — | $0.95 \pm 0.23$ | — |

[a] In the case of first-class currents, the determination is compared with the corresponding theoretical prediction. The result for $A = 2$ is discussed in text.

## 5.1 First-Class Currents

From a conservative point of view, only the result in Equation 46, derived from the spectrum shape and muon capture measurements in $A = 12$, yields an independent determination of $\tilde{F}_M$. By comparing it with the CVC prediction, one has

$$F_M/F_M(CVC) = 1.04 \pm 0.09. \qquad 68.$$

But since the second-class currents are small compared to weak magnetism, the treatment on an equal ground with weak magnetism determinations in Table 6 is legitimate. The total average is then

$$F_M/F_M(CVC) = 1.02 \pm 0.06. \qquad 69.$$

According to the nuclear physics calculations of the $1^+ - 0^+$ transitions in $A = 12$ nuclei referred to in Section 4.4, the identification $F_M/F_A = (f_V + f_M)/f_A$ is well justified. By using $\tilde{F}_M$ in Equation 46 and $f_{V,A}$ in Equations 5a,b, one obtains $f_M = 4.0 \pm 0.4$. Including the determination of $\tilde{F}_M$ in the isodoublet transition $^{19}Ne - ^{19}F$, where the single-particle (valence-nucleon) description works well, the estimate of the weak magnetism nucleon form factor is then

$$f_M = 3.9 \pm 0.3. \qquad 70.$$

The independent estimate of $f_P$ in $A = 12$ nuclei (Equation 62) agrees with $f_P$ derived from the muon capture rate in hydrogen (Equation 27). Their mean yields the comparison

$$f_P/f_P(\text{PCAC}) = 1.01 \pm 0.15. \qquad 71.$$

It seems appropriate to comment here upon the question of renormalization of weak form factors in the nuclear context. For a nucleon inside the nuclear matter, the divergence condition of the axial vector current is expected to be modified (40, 53, 157; see also 2). Interaction with the surrounding nucleons modifies the pion propagator. For a nucleon inside an "infinite" nucleus, a reduction of $f_P/f_A$ by a factor of about 3 is expected (2, 40). While the result in $A = 12$ nuclei does not contradict this consideration, since the transforming nucleon is peripheral (1p shell), the result in $A = 40$ ("larger") nuclei seems to contradict the expected renormalization.

Renormalization of $f_A$ has also been considered (40). Among the cases discussed herein, the $A = 12$ and 19 nuclei allow the most quantitative conclusion. Equation 70 and the corresponding result for $A = 12$ nuclei show (indirectly) that the nuclear axial vector form factor in these nuclei is (within about 10%) the nucleon form factor $f_A$. Note, however, that this indication of a nonrenormalization of $f_A$ in the nuclear context (not so "nuclear" in fact, in these special cases) does not agree with a recent conclusion drawn from the comparison of magnetic moments and $ft$ values in mirror transitions: $f_A$ (in nucleus) $\approx 0.8 f_A$ (21).

## 5.2    Second-Class Currents

The connection between $F_E^{(2)}$ and $f_T$ ("primary" SCC) might be complicated by the possible existence of "secondary" SCC due to off-shell (84, 147) and exchange-current (84) effects. A model incorporating these current types has been worked out for the nuclear cases considered herein, and both $ft^+/ft^-$ ratios in mirror nuclei (153) and correlation experiments together with $ft^+/ft^-$ ratios (108) have been analyzed in the framework of this model. The analysis by Oka & Kubodera (108) included many of the results in Table 6, among them that of the alignment-correlation experiment in $A = 12$, which is particularly restrictive for $f_T + f'_T$, where the latter form factor characterizes off-shell-type currents. Numerically, the analysis provided $|f_T| \leq 0.4 f_M$. Unfortunately, however, the restrictive power of the alignment experiment in $A = 12$ assumed in this analysis was weaker than it is nowadays. Accordingly, the above limit reduces to

$$|f_T| \lesssim 0.20 f_M. \qquad 72.$$

According to this analysis, $f'_T$ and the strength of the exchange-type SCC vanish also, which is the trivial solution of the problem in the sense that the three SCC are expected to affect the observables differently from one case to an other. Since nowhere is there observed a positive signal that could be identified with an SCC, all of the current types must vanish.

The additional restrictions for the existence of SCC, not included in the analysis, derive from a survey of $e^+/K$ capture ratios by Vatai (145) and a $\nu_\mu + n \rightarrow \mu^- + p$ scattering experiment at $1 \lesssim q$ (GeV/c) $\lesssim 3$, using a deuterium target (6). At such a large $q$, the induced pseudoscalar is suppressed and the induced tensor may be identified and derived from, for example, the $E_\nu$ dependence of the cross section. A preliminary result obtained from the reanalysis of published and new data (private communication by Dr. M. Tanaka) is $|F_E^{(2)}| < 1$.

For the effective scalar there exists the estimate deduced from an analysis of superallowed $0^+ \rightarrow 0^+$ Fermi transitions (138):

$$|f_S/f_M| \lesssim 1. \qquad\qquad 73.$$

## 6. CONCLUSION

The effect of induced pseudoscalar and weak magnetism induced currents is identified and well determined in a variety of weak reactions, while the effect of the SCC-induced tensor is systematically absent (Table 6).

According to Equation 68, CVC is verified to within 9% accuracy, or to within 6%, by considering the total average in Equation 69 based on a less conservative criterion. By taking a vanishing induced tensor, the nuclear alignment–electron correlation measurements in $A = 12$ nuclei constitute the most direct verification of CVC. Figure 9, a display of the results of those experiments, exposes clearly the isotriplet CVC condition (keeping in mind the $G$-parity transformation of $F_A$, $F_A^- = -F_A^+$): $F_M^- = \sqrt{2}\mu = F_M^+$. It is well known that the PCAC-based estimate of many-particle contributions (essentially pions) to the equal-time commutator of quark axial vector charges, $[Q_A^+, Q_A^-] = 2I_3$, is in good agreement (to within a few percent) with experiment (1, 151). Since PCAC is rather accurately verified, to within 15% in Equation 71 and to within the same accuracy through the Goldberger–Treiman relation mentioned in Section 2.5, the universality condition of the fundamental fermion axial vector current, of which the above commutator is a formulation, is also well verified (91). One thus concludes that the weak hadronic currents behave as currents associated with the gauge symmetry. A recent survey (46) of weak reactions involving hyperons reached a similar (but more qualitative) conclusion.

Let us note finally that the present knowledge of induced weak currents in nuclei allows us to undertake investigations of aspects beyond the standard description. In beta decay, for instance, the observables may now be predicted down to $10^{-3}$ accuracy since weak magnetism contributes at the $10^{-2}$ level and is known to better than $10^{-1}$ (and $f_P \approx 0$). The corresponding limit in muon capture is $\sim 10^{-2}$.

ACKNOWLEDGMENTS

It is a pleasure to thank Professors M. Gell-Mann and V. L. Telegdi for advice and critical comments. I am indebted to Professor T. A. Girard for his constant assistance, to Professor A. Wetherell for enabling me to spend some time at CERN, and to Professor Gh. Grégoire for much encouragement.

*Literature Cited*

1. Adler, S. L. *Phys. Rev. Lett.* 14:1051 (1965)
2. Akhmedov, E. Kh. *JETP Lett.* 34:138 (1981)
3. Alberigi Quaranta, A., Bertin, A., Matone, G., Palmonari, F., Torelli, G., et al. *Phys. Rev.* 177:2118 (1969)
4. Astubry, A., Auerbach, L. B., Cuttz, D., Esterling, R. J., Jenkins, D. A., et al. *Nuovo Cimento* 33:1021 (1964)
5. Bakalov, D. D., Faifman, M. P., Ponomarev, L. I., Vinitsky, S. I. *Nucl. Phys.* A384:302 (1982)
6. Baker, N. J., Cnops, A. M., Conolly, P. L., Kahn, S. A., Kirk, H. G., et al. *Phys. Rev. D* 23:2499 (1981)
7. Bardin, G., Duclos, J., Magnon, A., Martino, F., Richter, A., et al. *Phys. Lett.* 104B:320 (1981)
8. Bardin, G., Duclos, J., Magnon, A., Martino, F., Richter, A., et al. *Nucl. Phys. A* 352:365 (1981)
9. Bég, M. A. B., Bundy, R. V., Mohapatra, R., Sirlin, A. *Phys. Rev. Lett.* 38:1252 (1977)
10. Bég, M. A. B., Sirlin, A. *Ann. Rev. Nucl. Sci.* 24:379 (1974)
11. Becher, P., Böhm, M., Joos, H. 1984. *Gauge Theories of Strong and Electroweak Interactions.* New York: Wiley
12. Behrens, H., Szybisz, L. *Z. Phys. A* 273:17 (1975)
13. Berlijn, J. J., Keaton, P. W. Jr., Madansky, L., Owen, G. E., Pfeiffer, L., Roberson, N. R. *Phys. Rev.* 153:1152 (1967)
14. Bernabeu, J. *Phys. Lett.* 55B:438 (1974)
15. Bernstein, J., Lee, T. D., Yang, C. N., Primakoff, H. *Phys. Rev.* 111:313 (1958)
16. Bleser, E. J., Lederman, L. M., Rosen, J. L., Rothberg, J. E., Zavattini, E. *Phys. Rev. Lett.* 8:288 (1962)
17. Boothroyd, A. I., Markey, J., Vogel, P. *Phys. Rev. C* 28:603 (1984)
18. Bowles, T. J., Garvey, G. T. *Phys. Rev. C* 18:1447 (1978)
19. Brändle, H., Grenacs, L., Lang, Z., Roesch, L. P., Telegdi, V. L., et al. *Phys. Rev. Lett.* 40:306 (1978)
20. Brändle, H., Miklos, G., Roesch, L. P., Telegdi, V. L., Truttman, P., et al. *Phys. Rev. Lett.* 41:299 (1978)
21. Buck, B., Perez, S. M. *Phys. Rev. Lett.* 50:1975 (1983)
22. Bukhvostov, A. P., Popov, N. P. *Nucl. Phys. A* 147:385 (1970)
23. Bystritskii, V. M., Dzhelepov, V. P., Ermolov, P. F., Oganesyan, K. O., Ormelyanenko, M. N., et al. *Sov. Phys. JETP* 39:19 (1974)
24. Cabbibo, N. *Phys. Rev. Lett.* 10:531 (1963)
25. Calaprice, F. P., Chung, W., Wildenthal, B. H. *Phys. Rev. C* 15:2178 (1977)
26. Calaprice, F. P., Freedman, S. J., Mead, W. C., Vantine, H. C. *Phys. Rev. Lett.* 35:1566 (1975)
27. Calaprice, F. P., Holstein, B. R. *Nucl. Phys. A* 273:301 (1976)
28. Calaprice, F. P., Millener, D. J. *Phys. Rev. C* 27:1175 (1983)
29. Cheng, W. K., Lorazo, B., Goulard, B. *Phys. Rev. C* 21:374 (1980)
30. Chertok, B. T., Sheffiled, C., Lightbody, J., Penner, S., Blum, D. *Phys. Rev. C* 8:23 (1973)
31. Ciechanowicz, S. *Nucl. Phys. A* 267:472 (1976)
32. Cohen, R. S., Devons, S., Kanaris, A. D. *Nucl. Phys.* 57:255 (1964)
33. Conversi, M., Diebold, R., di Lella, L. *Phys. Rev.* 136B:1077 (1964)

34. Delorme, J. *Nucl. Phys. B* 19 : 573 (1970)
35. di Lella, L., Hammerman, I., Rosenstein, L. *Phys. Rev. Lett.* 27 : 830 (1971)
36. Dupuis-Rolin, N., Deutsch, J. P., Favart, D., Prieels, R. *Phys. Lett.* 79B : 359 (1978)
37. Deutsch, J. P., Macq, P. C., van Elmbt, L. *Phys. Rev. C* 15 : 1587 (1977)
38. Deutsch, J. P., Grenacs, L., Lehmann, J., Lipnik, P., Macq, P. C. *Phys. Lett.* 28B : 66 (1969)
39. Devanathan, V., Parthasarathy, R., Subramanian, P. R. *Ann. Phys.* 73 : 291 (1972)
40. Ericson, M. 1978. *Prog. Part. Nucl. Phys.* 1 : 67
41. Ericson, T., Sens, J. C., Rood, H. P. C. *Nuovo Cimento* 34 : 51 (1964)
42. Feynman, R. P., Gell-Mann, M. *Phys. Rev.* 109 : 193 (1958)
43. Foldy, L., Walecka, J. D. *Phys. Rev. B* 133 : 1339 (1965)
44. Fukui, M., Koshigiri, K., Sato, T., Ohtsubo, H., Morita, M. *Phys. Lett.* 132B : 255 (1983)
45. Gagliardi, C. A., Garvey, G. T., Wrobel, J. R. *Phys. Rev. Lett.* 48 : 914 (1982)
46. Gaillard; J.-M., Sauvage, G. *Ann. Rev. Nucl. Part. Sci.* 34 : 351–402 (1984)
47. Gell-Mann, M. *Phys. Rev.* 111 : 362 (1958)
48. Gell-Mann, M., Lévy, M. *Nuovo Cimento* 16 : 705 (1960)
49. Gershtein, S. S., Zeldovich, J. G. *Sov. Phys. JETP* 2 : 576 (1956)
50. Glass, N. W., Peterson, R. W. *Phys. Rev.* 130 : 253 (1963)
51. Goldberger, M. L., Treiman, S. B. *Phys. Rev.* 111 : 354 (1958)
52. Goldhaber, M., Grodzins, L., Sunyar, A. W. *Phys. Rev.* 109 : 1015 (1958)
53. Green, A. M., Rho, M. *Nucl. Phys. A* 130 : 112 (1969)
54. Grenacs, L., Deutsch, J. P., Lipnik, P., Macq, P. C. *Nucl. Instrum.* 58 : 164 (1968)
55. Guichon, P., Bihoreau, B., Giffon, M., Gonçalves, A., Julien, J., et al. *Phys. Rev. C* 19 : 987 (1979)
56. Guichon, P. A. M., Giffon, M., Samour, C. *Phys. Lett.* 74B : 15 (1978)
57. Giffon, M., Gonçalves, A., Guichon, P. A. P., Julien, J., Roussel, L., Samour, C. *Phys. Rev. C* 24 : 241 (1981)
58. Guichon, P. A. M., Samour, C. *Phys. Lett.* 82B : 28 (1979)
59. Guichon, P. A. M., Samour, C. *Nucl. Phys. A* 382 : 461 (1982)
60. Hallin, A. L., Calaprice, F. P., MacArthur, D. W., Piilonen, L. E., Schneider, M. B., Schreiber, D. F. *Phys. Rev. Lett.* 52 : 337 (1984)
61. Halprin, A., Leen, B. W., Sorba, P. *Phys. Rev. D* 14 : 2343 (1976)
62. Hart, R. D., Cox, C. R., Dodson, G. W., Eckhause, M., Kane, J. R., et al. *Phys. Rev. Lett.* 39 : 399 (1977)
63. Haskell, R. C., Correll, F. D., Madansky, L. *Phys. Rev. B* 11 : 3266 (1975)
64. Haskell, R. C., Madansky, L. *J. Phys. Soc. Jpn. Suppl.* 34 : 167 (1973)
65. Holstein, B. R. *Phys. Rev. C* 5 : 1529 (1972)
66. Holstein, B. R. *Rev. Mod. Phys.* 46 : 789 (1974)
67. Holstein, B. R. *Phys. Rev. C* 16 : 753 (1977)
68. Holstein, B. R. *Phys. Rev. D* 17 : 2499 (1976)
69. Holstein, B. R. *Phys. Rev. C* 29 : 623 (1984)
70. Holstein, B., Treiman, S. B. *Phys. Rev. C* 13 : 3059 (1976)
71. Hori, M., Ochi, S., Minamisono, T., Mizobuchi, A., Sugimoto, K. *J. Phys. Soc. Jpn. Suppl.* 34 : 161 (1973)
72. Huffaker, J. N., Laird, C. E. *Nucl. Phys. A* 92 : 584 (1967)
73. Hwang, W.-Y. P. *Phys. Rev. C* 20 : 814 (1979)
74. Hwang, W.-Y. P., Primakoff, H. *Phys. Rev. C* 16 : 397 (1977)
75. Jackson, J. D., Treiman, S. B., Wyld, H. W. *Phys. Rev.* 107 : 137 (1957)
76. Jäger, H.-U., Kirchback, M., Truhlik, E. *Nucl. Phys. A* 404 : 456 (1983)
77. Johnson, C. H., Pleasonton, F., Carlson, T. A. *Phys. Rev.* 132 : 1149 (1963)
78. Kaina, W., Soergel, V., Thies, H., Trost, N. *Phys. Lett.* 70B : 411 (1977)
79. Kane, F. R., Eckhause, M., Miller, G. H., Roberts, R. L., Vislay, M. E., Welsh, R. E. *Phys. Lett.* 45B : 292 (1973)
80. Kim, C. W., Primakoff, H. *Phys. Rev.* 139B : 1447 (1965)
81. Krichback, M., Kamalov, S., Jäger, H.-U. *Phys. Lett.* 144B : 319 (1984)
82. Kobayashi, M., Maskawa, T. *Prog. Theor. Phys.* 49 : 652 (1973)
83. Kobayashi, M., Ohtuska, N., Ohtsubo, H., Morita, M. *Nucl. Phys. A* 312 : 377 (1978)
84. Kubodera, K., Delorme, J., Rho, M. *Nucl. Phys. B* 66 : 253 (1973)
85. Kuno, Y., Imazato, J., Nishiyama, K., Nagamine, K., Yamazaki, T., Minamisono, T. *Phys. Lett.* 148B : 270 (1984)
86. Langacker, P. *Phys. Rev. D* 14 : 2340 (1976)
87. Lathrop, J. L., Lundy, R. A., Telegdi, V. L., Winston, W., Yovanovitch, D. D. *Phys. Rev. Lett.* 7 : 107 (1961)

88. Lebrun, P., Deschepper, Ph., Grenacs, L., Lehmann, J., Leroy, C., et al. *Phys. Rev. Lett.* 30:302 (1978)
89. Lee, T. D., Wu, C. S. *Ann. Rev. Nucl. Sci.* 15:381 (1965)
90. Lee, Y. K., Mo, L. W., Wu, C. S. *Phys. Rev. Lett.* 10:258 (1963)
91. Llewellyn Smith, C. H. *Phys. Rep.* 3C:264 (1972)
92. Maier, E. J., Edelstein, R. M., Siegel, R. T. *Phys. Rev. B* 133:663 (1964)
93. Marshak, R. E., Riazuddin, Ryan, C. P. 1969. *Theory of Weak Interactions in Particle Physics*, ed. R. E. Marshak. New York: Wiley
94. Masuda, Y., Minamisono, T., Nojiri, Y., Sugimoto, K. *Phys. Rev. Lett.* 43:1083 (1979)
95. Mayer-Kukuk, T., Michel, T. C. *Phys. Rev.* 127:545 (1962)
96. McKeown, R., Calaprice, F. P., Alburger, D. E. *Bull. Am. Phys. Soc.* 22:28 (1977)
97. McWilliams, B., Li, L. F. *Nucl. Phys. B* 179:62 (1981)
98. Miller, G. H., Eckhause, M., Kane, F. R., Martin, P., Welsh, R. E. *Phys. Lett.* 41B:50 (1972)
99. Miller, G. H., Eckhause, M., Kane, F. R., Martin, P., Welsh, R. E. *Phys. Rev. Lett.* 29:1194 (1972)
100. Minamisono, T., Takeyama, K., Ishigai, T., Takeshima, H., Nojiri, Y., Asaki, K. *Phys. Lett.* 130B:1 (1983)
101. Morita, M. 1973. *Beta Decay and Muon Capture*. Reading, Mass: Benjamin
102. Morita, M., Nishimura, M., Shimizu, A., Ohtsubo, H., Kubodera, K. *Prog. Theor. Phys. Suppl.* 60:1 (1976)
103. Mukhopadhyay, N. C. *Phys. Rep.* Vol. 30 C, No. 1 (1977)
104. Mursula, K., Roos, M., Scheck, F. *Nucl. Phys. B* 219:321 (1983)
105. Nambu, Y. *Phys. Rev. Lett.* 4:380 (1960)
106. Nordberg, M. E. Jr., Morinigo, F. B., Barnes, C. A. *Phys. Rev.* 125:321 (1962)
107. O'Connell, J. S., Donnelly, T. W., Walecka, J. D. *Phys. Rev. C* 6:719 (1972)
108. Oka, M., Kubodera, K. *Phys. Lett.* 90B:45 (1980)
109. Okun, L. B. 1965. *Weak Interactions of Elementary Particles*. Int. Ser. Monogr. in Natural Philosophy, Vol. 5. Oxford: Pergamon
110. Oziewicz, Z., Pikulski, A. *Acta Phys. Pol.* 32:873 (1967)
111. Palffy, L., Deutsch, J. P., Grenacs, L., Lehmann, J., Steels, M. *Phys. Rev. Lett.* 34:212 (1975)
112. Parthasarathy, R., Sridhar, V. N. *Phys. Rev. C* 23:861 (1981)
113. Popov, N. P. *Sov. Phys. JETP* 17:1130 (1963)
114. Possoz, A. PhD thesis. Université Catholique de Louvain, Louvain-la-Neuve, 1978 (unpublished)
115. Possoz, A., Grenacs, L., Lehmann, J., Palffy, L., Julien, J., Samour, C. *Phys. Lett.* 87B:35 (1979)
116. Possoz, A., Deschepper, Ph., Grenacs, L., Lebrun, P., Lehmann, J., et al. *Phys. Lett.* 70B:265 (1977)
117. Primakoff, H. *Rev. Mod. Phys.* 31:802 (1959)
118. Primakoff, H. 1975. Elementary particle aspects of muon decay and muon capture. In *Muon Physics*, ed. V. W. Hughes, C. S. Wu 2:3–48. New York: Academic
119. Review of particle properties. *Rev. Mod. Phys.* Vol. 52, No. 2 (1980)
120. Reynolds, G. T., Scarl, D. B., Swanson, R. A., Waters, J. R., Zdanis, R. A. *Phys. Rev.* 129:1790 (1963)
121. Roesch, L. Ph., Telegdi, V. L., Truttmann, P., Zehnder, A., Grenacs, L., Palffy, L. *Am. J. Phys.* 50(10):931 (Oct. 1982)
122. Roesch, L. Ph., Telegdi, V. L., Truttman, P., Zehnder, A., Grenacs, L., Palffy, L. *Phys. Rev. Lett.* 46:1507 (1981)
123. Roesch, L. Ph., Schlumpf, N., Taqqu, D., Telegdi, V. L., Truttman, P., Zehnder, A. *Phys. Lett.* 107B:31 (1981)
124. Rolin, N., Deutsch, J. P., Favart, D., Lebrun, M., Prieels, R. *Phys. Lett.* 70B:23 (1977)
125. Rood, H. P. C., Yano, A. F., Yano, F. B. *Nucl. Phys. A* 228:333 (1974)
126. Rothberg, J. E., Anderson, E. W., Bleser, E. J., Lederman, L. M., Meyer, S. L., et al. *Phys. Rev.* 132:2664 (1963)
127. Salam, A. In *Elementary Particle Physics*, ed. N. Svartholm, p. 367. Stockholm: Almquist & Wiksells (1968)
128. Schreiber, D., Calaprice, F. P., Deway, M., Hallin, A., Kleppinger, W. E., et al. *Bull. Am. Phys. Soc.* 24:51 (1979)
129. Shapiro, I. S., Blokhintsev, L. D. *Sov. Phys. JETP* 12:775 (1961)
130. Shrock, R. E. *Phys. Lett.* 96B:159 (1980)
131. Skalsey, M., Girard, T. A., Newman, D., Rich, A. *Phys. Rev. Lett.* 49:708 (1982)
132. Steels, M., Grenacs, L., Lehmann, J., Palffy, L., Possoz, A. 1975. In *High Energy and Nuclear Structure*. AIP Conf. Proc. No. 26, ed. R. Mischke, C. Hargrove, C. Hoffman. New York: Am. Inst. Phys.
133. Stremnitzer, H. *Phys. Rev. D* 10:1327 (1974)

134. Subramanian, P. R., Parthasarathy, R., Devanathan, V. *Nucl. Phys. A* 262:433 (1976)
135. Sudarshan, E. C. G., Marshak, R. E. *Phys. Rev.* 109:1860 (1958)
136. Sugimoto, K., Misobuchi, A., Nakai, K., Matuda, K. *Phys. Soc. Jpn.* 21:213 (1966)
137. Sugimoto, K., Tanihata, I., Göring, J. *Phys. Rev. Lett.* 34:1533 (1975)
138. Szybisz, L., Silbergleit, V. M. *J. Phys. G* 7:L201 (1981)
139. Telegdi, V. L. 1977. *Int. Conf. on High-Energy and Nuclear Structure, 7th, Zürich, 1977*, ed. M. P. Locher, pp. 367–74. Basel/Stuttgart: Birkhauser
140. Tribble, R. E., Garvey, G. T. *Phys. Rev. C* 12:967 (1975)
141. Tribble, R. E., May, D. P. *Phys. Rev. C* 18:2704 (1978)
142. Truttman, P. Eidgenössiche Technische Hochschule Zürich dissertation, No. 6753, 1980 (unpublished)
143. Truttman, P., Brändle, H., Roesch, L. Ph., Telegdi, V. L., Zehnder, A., Grenacs, L. *Phys. Lett.* 83B:48 (1979)
144. Tribble, R. E., May, D. P., Tanner, D. M. *Phys. Rev. C* 23:2245 (1981)
145. Vatai, E. *Phys. Lett.* 34B:395 (1971)
146. Walecka, J. D. 1975. Semileptonic weak interaction in nuclei. See Ref. 118, 2:113–218
147. Weinberg, S. *Phys. Rev.* 112:1375 (1958)
148. Weinberg, S. *Phys. Rev. Lett.* 4:575 (1960)
149. Weinberg, S. *Phys. Rev. Lett.* 19:1264 (1967)
150. Weinberg, S. 1977. See Ref. 139, pp. 339–52
151. Weisberger, W. T. *Phys. Rev. Lett.* 14:1047 (1965)
152. Wilkinson, D. H. *Phys. Lett.* 31B:447 (1970)
153. Wilkinson, D. H. *Phys. Lett.* 48B:169 (1974)
154. Wilkinson, D. H., Alburger, D. E. *Phys. Rev. Lett.* 26:1126 (1971)
155. Wullschleger, A., Scheck, F. *Nucl. Phys. A* 326:325 (1979)
156. Wolfenstein, L. 1977. See Ref. 139, pp. 363–66
157. Wycech, S. *Nucl. Phys. B* 14:133 (1969)
158. Wu, C. S., Leen, Y. K., Mo, L. W. *Phys. Rev. Lett.* 39:72 (1977)
159. Zavattini, E. 1975. Muon capture. See Ref. 118, 2:219–61
160. Zavattini, E. Muon-electron universality and PCAC in muon capture in hydrogen. In *Proc. 4th Course Int. Sch. Intermediate Energy Phys., San Miniato, Italy*, ed. R. Bergere, S. Costa, C. Schaerf, pp. 385–95. Singapore: World Scientific (1983)

*Ann. Rev. Nucl. Part. Sci. 1985. 35 : 501–58*

# PARITY VIOLATION IN THE NUCLEON-NUCLEON INTERACTION[1]

## E. G. Adelberger

Department of Physics, University of Washington, Seattle, Washington 98195

## W. C. Haxton

Department of Physics, University of Washington, Seattle, Washington 98195; Theoretical Division, Los Alamos National Laboratory, Los Alamos, New Mexico 87545

CONTENTS

---

501

# 1. INTRODUCTION

In 1957, the same year that parity nonconservation was discovered in $\beta$ and $\mu$ decay, Tanner (1) reported the first search for parity violation in the N-N interaction. The following year Feynman & Gell-Mann (2) predicted, on the basis of their universal current-current theory of weak interactions, that in addition to the familiar weak processes of $\mu$, $\beta$, and hyperon decay, there should be a first-order weak (and parity-violating) interaction between two nucleons. The verification of this prediction proved to be a formidable experimental challenge. Preliminary evidence for this new form of weak interaction was presented by Abov et al (3) in 1964. However, the first persuasive result was obtained in 1967 by Lobashov et al (4), who by pioneering the technique of integral detection (as opposed to pulse counting) obtained a result consistent with the qualitative predictions of the current-current theory.

In the years following these experiments great progress has been made in understanding the weak interactions: the development of the "standard" $SU(2) \otimes U(1)$ electroweak theory, the discovery of the neutral weak current, and the recent experimental detection (5, 6) of the $W^{\pm}$ and $Z^0$ bosons, mediators of the weak force. Given the successes of the standard model, what is the present motivation for studying weak interactions between nucleons? Although elegant experiments have been carried out to test the leptonic and semileptonic weak interactions, the hadronic weak interactions have proven more elusive. They can be studied only when the strong and electromagnetic interactions between the hadrons are forbidden by a symmetry principle, such as flavor [i.e. strangeness (S) or charm (C)] conservation. However, according to the standard theory, neutral-current contributions to $\Delta S = 1$ and $\Delta C = 1$ weak processes are strongly suppressed by the GIM mechanism (7). Therefore the neutral-current weak interaction between quarks can only be studied in flavor-conserving processes, of which the N-N weak interaction is the single accessible example. In this review, we summarize the progress that has been made by exploiting parity nonconservation as a filter to isolate the weak component of the N-N interaction.

The standard theory successfully predicts the weak interactions of the "point" leptons and quarks, but does not fully determine the weak interactions of composite hadrons: the strong color interaction between quarks alters the hadronic weak interaction so that it is no longer simply related to the elementary weak interactions of the pointlike constituents. To compute the weak interactions of hadrons, one needs a reliable theory of strong interactions. Although quantum chromodynamics (QCD) may be the correct theory, it is a difficult and unsolved problem to work out its consequences in the low-$q^2$ (nonperturbative) limit. The nontrivial nature of the task is illustrated by the $\Delta I = 1/2$ rule governing the hadronic weak decays of strange particles, such as $\Lambda \to N\pi$. The standard model attributes $\Delta S = 1$ decays to the charged-current interaction among pairs of quarks $(s \to d)(d \to u)$. Under strong isospin ($I$) rotations the interaction should have the form $\Delta I = 1/2 \otimes 1 = 1/2 \oplus 3/2$. Empirically one finds that the $\Delta I = 1/2$ amplitudes are enhanced over the $\Delta I = 3/2$ amplitudes by a factor of $\sim 20$. This dominance of the $\Delta I = 1/2$ amplitudes is thought to be a dynamical symmetry, but its origin is not understood in detail.

Isospin symmetry is also an important aspect of the parity-nonconserving N-N interaction. In this case one expects the interaction to have the form $\Delta I = 0 \oplus 1 \oplus 2$. The relative strength of these three components depends, in part, on the interplay between the neutral and charged weak currents. Current research has focussed on determining the isospin dependence of the $\Delta S = 0$ hadronic weak interaction, thereby disentangling the neutral- and charged-current contributions. This goal motivated much of the recent progress in theory and experiment, especially the development of remarkably precise accelerator-based techniques for measuring small effects. Experimental technology is now refined to the point that parity-violating decay widths as small as $10^{-10}$ eV and parity-violating asymmetries as small as $10^{-7}$ can be measured with confidence.

In this review we concentrate on the parity-nonconserving (PNC) N-N interaction at low energies. Although our topic involves both nuclear and particle physics, we emphasize the nuclear aspects—the experimental techniques and theoretical analyses needed to extract the matrix elements of the N-N weak interaction from data in the two-nucleon system and in complex nuclei. Our treatment of the particle physics aspects of the subject—calculations of the PNC meson-nucleon couplings from theories of the weak interaction and models of hadrons—is less complete because a detailed discussion of this topic has recently appeared elsewhere (8).

## 2.  THE PARITY-NONCONSERVING NUCLEON-NUCLEON POTENTIAL

In this section we discuss the PNC N-N potential and its relation to the underlying theory of the weak interactions between quarks. In the low-energy regime of interest to us, the hadronic weak interaction can be described by a phenomenological current-current Lagrangian (see for example, 7): $L = G_F/(2)^{1/2}(J_W^\dagger J_W + J_Z^\dagger J_Z) + \text{h.c.}$, where $J_W$ and $J_Z$ are the charged and neutral weak currents respectively. We neglect the heavy quarks c, b, and t, retaining only those components of the currents that involve the light quarks u, d, and s. The charged current $J_W$ has two components: $J_W = \cos\theta_C J_W^0 + \sin\theta_C J_W^1$ where $\theta_C$ is a quark mixing angle (the Cabibbo angle), with $\sin\theta_C \approx 0.22$. The current $J_W^0$ drives the u → d transition and transforms as $\Delta I = 1$, $\Delta S = 0$, while $J_W^1$ drives the u → s transition and transforms as $\Delta I = 1/2$, $\Delta S = -1$. The neutral current $J_Z$ also has two components $J_Z^0$ and $J_Z^1$, which transform as $\Delta I = 0$, $\Delta S = 0$ and $\Delta I = 1$, $\Delta S = 0$, respectively. The weak $\Delta S = 0$ N-N interaction is then governed by the current-current Lagrangian:

$$G_F/(2)^{1/2}[\cos^2\theta_C(J_W^0)^\dagger J_W^0 + \sin^2\theta_C(J_W^1)^\dagger J_W^1$$

$$+(J_Z^0)^\dagger J_Z^0 + (J_Z^1)^\dagger J_Z^1 + (J_Z^0)^\dagger J_Z^1 + (J_Z^1)^\dagger J_Z^0] + \text{h.c.} \quad 1.$$

The symmetric product of two $J_W^0$ ($\Delta I = 1$) currents transforms as $\Delta I = 0, 2$, while the symmetric product of two $J_W^1$ ($\Delta I = 1/2$) currents transforms as $\Delta I = 1$. Thus the $\Delta I = 1$ component of the charged-current weak N-N interaction is suppressed by $\tan^2\theta_C$ compared to the $\Delta I = 0, 2$ components. On the other hand, the neutral-current contribution to the $\Delta I = 1$ weak N-N interaction is not suppressed. Therefore, on these simple grounds, we expect the neutral current to dominate the $\Delta I = 1$ component of the PNC N-N interaction. However, as illustrated by the $\Delta I = 1/2$ rule, the strong interaction between quarks can significantly alter weak matrix elements between hadrons, so this qualitative isospin argument may not be valid.

We now turn to more detailed (and model-dependent) considerations. In the low-energy regime, two descriptions of the PNC N-N interaction have proven useful. In the first, the matrix elements of the N-N interaction are expressed in terms of the five elementary S-P amplitudes (e.g. 9–11). In the second, the PNC N-N interaction is written as a potential arising from single-meson ($\pi, \rho, \omega$) and multiple-meson ($\pi\pi$) exchange, where one meson-nucleon vertex is governed by the weak interaction and the other(s) by the strong (e.g. 12–17).

The potential description of the low-energy PNC N-N interaction has several advantages over the S-P amplitude description. The potential predicts the energy dependence of the S-P amplitudes and the strengths of transitions between higher partial waves. The meson-exchange description is also convenient for estimating the effects of short-range N-N correlations (not included in conventional independent-particle models of nuclear structure) on nuclear matrix elements of the PNC interaction. Thus we emphasize this description in our numerical work in Section 6.

## 2.1 Meson-Exchange PNC Potentials

An obvious process contributing to the PNC N-N interaction is direct $W^\pm$ or $Z^0$ exchange between the nucleons. Because of the small Compton wavelengths of these bosons ($\sim 0.002$ fm), direct exchanges effectively occur only when two nucleons overlap. We do not yet have an adequate understanding of such short-range contributions to either the PNC or parity-conserving (PC) N-N interactions. Fortunately, for energies characteristic of bound nucleons, the N-N interaction takes place primarily at distances large compared to the nucleon size. This is due in part to the strong repulsion in the N-N interaction at short distances. Thus we expect long-range contributions, which can be described without explicit reference to the structure of the nucleon, to dominate the low-energy PNC interaction.

The strong PC N-N interaction at low energies ($E < 300$ MeV) has been quite successfully explained (18) in terms of meson-exchange potentials. This description can be enlarged to include the weak PNC interaction by replacing one of the strong meson-nucleon couplings by a weak coupling, as illustrated in Figure 1. All the physics of W and Z exchange between quarks is hidden inside the weak meson-nucleon vertex.

At small center-of-mass energies, exchanges of light mesons dominate the PNC potential because the meson mass and the momentum transfer between interacting nucleons appear in the static limit of the meson propagator $(\mathbf{p}^2 + m^2)^{-1}$. The meson-exchange description of the N-N interaction becomes increasingly cumbersome at higher energies as a result of the proliferation of mesons with $m^2 < \mathbf{p}^2$. In nuclei, the nucleons have $\langle \mathbf{p}^2 \rangle_{\text{average}} = 3/5 k_F^2 \approx (200 \text{ MeV})^2$ (we assume a Fermi energy of $\varepsilon_F = 37$ MeV and an effective mass $m_N^* = 0.7 m_N$), so that a description of the PNC interaction that includes only light mesons should be adequate (e.g. 10). We anticipate that exchanges involving the pseudoscalars $\pi(140$ MeV), $\eta(549$ MeV), and $\eta'(958$ MeV), the scalars S(975 MeV) and $\delta(983$ MeV), and the vector mesons $\rho(769$ MeV), $\omega(783$ MeV), and $\phi(1020$ MeV) along with two-pion exchange will determine the interaction of bound nucleons.

Fortunately, the possible scalar and pseudoscalar exchanges are greatly restricted by Barton's theorem, which states that $CP$ invariance forbids any coupling between neutral $J = 0$ mesons and on-shell nucleons. Therefore $\pi^0$, $\eta$, $\eta'$, S, and $\delta^0$ exchanges are excluded to the extent that $CP$ violation is negligible. McKellar & Pick (19) have argued that $\delta^{\pm}$ exchange can be

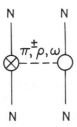

*Figure 1*  One-boson-exchange contribution to the PNC N-N interaction.

regarded as a form-factor correction to $\pi^{\pm}$ exchange. Furthermore, while the $\phi$ can be produced weakly by $Z^0$ exchange, the strong coupling of an $s\bar{s}$ to nucleons is greatly suppressed relative to that of the $\rho$ or $\omega$. As a result

$$V^{\text{PNC}} \approx V_{\pi^{\pm}} + V_{\text{vector meson}} + V_{2\pi}, \qquad 2.$$

where the vector meson potential is generated by $\rho^{\pm}$, $\rho^0$, and $\omega^0$ exchanges.

The nonrelativistic N-N potential corresponding to the first two terms in Equation 2 can be calculated directly from the strong and PNC weak meson-nucleon couplings (8):

$$H^{\text{PC}} = ig_{\pi\text{NN}}\bar{N}\gamma_5\tau \cdot \boldsymbol{\phi}_\pi N + g_\rho \bar{N}\left(\gamma_\mu + i\frac{\mu_v}{2M}\sigma_{\mu\nu}k^\nu\right)\tau \cdot \boldsymbol{\phi}_\rho^\mu N$$

$$+ g_\omega \bar{N}\left(\gamma_\mu + i\frac{\mu_s}{2M}\sigma_{\mu\nu}k^\nu\right)\phi_\omega^\mu N;$$

$$H^{\text{PNC}} = \frac{f_\pi}{2}\,\bar{N}[\tau \times \boldsymbol{\phi}_\pi)_z N$$

$$+ \bar{N}\left(h_\rho^0\tau \cdot \boldsymbol{\phi}_\rho^\mu + h_\rho^1\phi_{\rho z}^\mu + \frac{h_\rho^2}{2\sqrt{6}}(3\tau_z\phi_{\rho z}^\mu - \tau \cdot \boldsymbol{\phi}_\rho^\mu)\right)\gamma_\mu\gamma_5 N$$

$$+ \bar{N}(h_\omega^0\phi_\omega^\mu + h_\omega^1\tau_z\phi_\omega^\mu)\gamma_\mu\gamma_5 N - h_\rho'^1\bar{N}(\tau \times \boldsymbol{\phi}_\rho^\mu)_z \frac{\sigma_{\mu\nu}k^\nu}{2M}\gamma_5 N. \qquad 3.$$

This yields

$$V^{\text{PNC}}(\mathbf{r}) = \frac{iF_\pi}{M}[\tau(1) \times \tau(2)]_z[\boldsymbol{\sigma}(1) + \boldsymbol{\sigma}(2)] \cdot \mathbf{u}_\pi(\mathbf{r})$$

$$+ \frac{1}{M}\left(\left\{F_0\tau(1)\cdot\tau(2) + \frac{F_1}{2}[\tau(1)+\tau(2)]_z\right.\right.$$

$$\left. + \frac{F_2}{2\sqrt{6}}[3\tau(1)_z\tau(2)_z - \tau(1)\cdot\tau(2)]\right\}$$

$$\times \{(1+\mu_v)i[\boldsymbol{\sigma}(1)\times\boldsymbol{\sigma}(2)] \cdot \mathbf{u}_\rho(\mathbf{r}) + [\boldsymbol{\sigma}(1)-\boldsymbol{\sigma}(2)] \cdot \mathbf{v}_\rho(\mathbf{r})\}$$

$$+ \left\{G_0 + \frac{G_1}{2}[\tau(1)+\tau(2)]_z\right\}$$

$$\times \{(1+\mu_s)i[\boldsymbol{\sigma}(1)\times\boldsymbol{\sigma}(2)] \cdot \mathbf{u}_\omega(\mathbf{r}) + [\boldsymbol{\sigma}(1)-\boldsymbol{\sigma}(2)] \cdot \mathbf{v}_\omega(\mathbf{r})\}$$

$$+ \tfrac{1}{2}[\tau(1)-\tau(2)]_z[\boldsymbol{\sigma}(1)+\boldsymbol{\sigma}(2)] \cdot [G_1\mathbf{v}_\omega(\mathbf{r}) - F_1\mathbf{v}_\rho(\mathbf{r})]$$

$$+ H_1 i[\tau(1)\times\tau(2)]_z[\boldsymbol{\sigma}(1)+\boldsymbol{\sigma}(2)] \cdot \mathbf{u}_\rho(\mathbf{r})\Big), \qquad 4.$$

**Table 1**  Weak coupling constants from the "best value" and "reasonable range" results of Desplanques, Donoghue & Holstein (8) for the Glashow-Weinberg-Salam model[a]

| Coefficient | Equivalent | "Best value" ($\times 10^{-6}$) | "Reasonable range" ($\times 10^{-6}$) |
|---|---|---|---|
| $F_\pi$ | $g_{\pi NN} f_\pi / \sqrt{32}$ | 1.08 | 0 : 2.71 |
| $F_0$ | $-g_\rho h_\rho^0 / 2$ | 1.59 | $-1.59 : 4.29$ |
| $F_1$ | $-g_\rho h_\rho^1 / 2$ | 0.027 | 0 : 0.053 |
| $F_2$ | $-g_\rho h_\rho^2 / 2$ | 1.33 | $-1.06 : 1.54$ |
| $G_0$ | $-g_\omega h_\omega^0 / 2$ | 0.80 | $-2.39 : 4.29$ |
| $G_1$ | $-g_\omega h_\omega^1 / 2$ | 0.48 | 0.32 : 0.80 |
| $H_1$ | $-g_\rho h_\rho'^1 / 4$ | 0.0 | |

[a] We have taken $g_{\pi NN} = 13.45$, $g_\rho = 2.79$, and $g_\omega = 8.37$.

where $\mathbf{r} = \mathbf{r}_1 - \mathbf{r}_2$, $\mathbf{u} = [\mathbf{p}, e^{-mr}/4\pi r]$, $\mathbf{v} = \{\mathbf{p}, e^{-mr}/4\pi r\}$, and $\mathbf{p} = \mathbf{p}_1 - \mathbf{p}_2$. The coefficients in Equation 4, products of one strong and one weak meson-nucleon coupling, are given in Table 1. If vector dominance is assumed, the strong scalar and vector magnetic moments have the values $\mu_s = -0.12$ and $\mu_v = 3.70$. Note that $\pi$ exchange contributes only to the $\Delta I = 1$ N-N interaction. Because the pion is so light, we expect $\pi$ exchange to play a dominant role in the $\Delta I = 1$ interaction. Both $\rho$ and $\omega$ exchanges contribute to the $\Delta I = 0$ interaction, while the $\Delta I = 2$ interaction arises from $\rho$ exchange.

Although two-pion-exchange and other multi-meson-exchange diagrams contribute at some level to $V_{PNC}$, they are not considered here. The $2\pi$-exchange potential was discussed qualitatively by Fischbach & Tadic (16), and several explicit calculations have been reported (20, 21). These calculations generally predict a small $\Delta I = 0$ $2\pi$-exchange contribution and a negligible $\Delta I = 2$ contribution. The $\Delta I = 1$ component, corresponding to $2\pi$ exchange in a P state, is stronger but is, at least in part, already included in $\rho$ exchange. A similar problem of disentangling multi-meson exchanges from genuine single-meson exchanges is encountered for the strong interaction. However, the strong interaction double-counting problem with $\sigma$ exchange does not occur for the weak interaction because of Barton's theorem. Our single-meson-exchange potential $V_{PNC}$ probably incorporates the principal effects of $2\pi$ exchange, provided the couplings are regarded as effective.

## 2.2  Two-Nucleon S-P Amplitudes

At sufficiently low energies the two-nucleon matrix elements of the PNC interaction $V^{PNC} = V^{PNC}_{\Delta I = 0} + V^{PNC}_{\Delta I = 1} + V^{PNC}_{\Delta I = 2}$ can be written in terms of five elementary S-P amplitudes, summarized in Table 2. In addition to these

**Table 2**   Two-nucleon S-P weak amplitudes and meson-exchange equivalents

| Transition[a] $I \leftrightarrow I'$ | $\Delta I$ | n-n | n-p | p-p | N-N system exchanges |
|---|---|---|---|---|---|
| $^3S_1 \leftrightarrow {}^1P_1$   $0 \leftrightarrow 0$ | 0 | | x | | $\rho, \omega^0$ |
| $^1S_0 \leftrightarrow {}^3P_0$   $1 \leftrightarrow 1$ | 0 | x | x | x | $\rho, \omega^0$ |
| | 1 | x | | x | $\rho, \omega^0$ |
| | 2 | x | x | x | $\rho$ |
| $^3S_1 \leftrightarrow {}^3P_1$   $0 \leftrightarrow 1$ | 1 | | x | | $\pi^\pm, \rho, \omega^0$ |

[a] The states are labelled by $^{2s+1}L_J$.

zero-range amplitudes, it is customary to include a sixth parameter, the PNC pion-nucleon coupling constant $F_\pi$. The relatively long-ranged pion-exchange contribution to the $^3S_1 \to {}^3P_1$ amplitude can be distinguished because it generates significant amplitudes for higher partial waves at energies where the zero-range approximation remains valid for heavier meson exchanges.

A connection between the meson-exchange potential and the $S \to P$ amplitudes can be made by considering the weak N-N amplitudes near threshold:

$$f_{\left[\begin{smallmatrix}pp\\nn\end{smallmatrix}\right]}(\mathbf{k}',\mathbf{k}) = V_{\left[\begin{smallmatrix}pp\\nn\end{smallmatrix}\right]} [(\mathbf{k}'+\mathbf{k}) \cdot [\sigma_1 - \sigma_2] + i(\mathbf{k}'-\mathbf{k}) \cdot [\sigma_1 \times \sigma_2]] \qquad 5a.$$

$$f_{pn}(\mathbf{k}',\mathbf{k}) = \tfrac{1}{2} V_{pn}[(\mathbf{k}'+\mathbf{k}) \cdot [\sigma_p - \sigma_n] + i(\mathbf{k}'-\mathbf{k}) \cdot [\sigma_p \times \sigma_n]]$$

$$+ \tfrac{1}{2} U_{pn}[(\mathbf{k}'+\mathbf{k}) \cdot [\sigma_p - \sigma_n] - i(\mathbf{k}'-\mathbf{k}) \cdot [\sigma_p \times \sigma_n]]$$

$$+ W_{pn}(\mathbf{k}'+\mathbf{k}) \cdot [\sigma_p + \sigma_n], \qquad 5b.$$

where $\mathbf{k} = (\mathbf{k}_1 - \mathbf{k}_2)/2$ in Equation 5a and $(\mathbf{k}_p - \mathbf{k}_n)/2$ in Equation 5b; $\mathbf{k}' = (\mathbf{k}'_1 - \mathbf{k}'_2)/2$ and $(\mathbf{k}'_p - \mathbf{k}'_n)/2$, with $\mathbf{k}_1$ and $\mathbf{k}_2$ the initial nucleon momenta and $\mathbf{k}'_1$ and $\mathbf{k}'_2$ the final momenta. The symbol $\sigma_p(\sigma_n)$ denotes a matrix element of $\sigma$ between initial and final proton (neutron) spins. In the plane-wave Born approximation the $^1S_0 \to {}^3P_0$ amplitudes $V$ are given by

$$V_{\left[\begin{smallmatrix}pp\\nn\end{smallmatrix}\right]} \approx \frac{1}{m_\rho^2} [(F_0 \pm F_1 + F_2/\sqrt{6})(2+\mu_V) + (G_0 \pm G_1)(2+\mu_S)]$$

$$V_{pn} \approx \frac{1}{m_\rho^2} \left[ \left(F_0 - \sqrt{\frac{2}{3}} F_2\right)[2+\mu_V] + G_0[2+\mu_S] \right], \qquad 6a.$$

while the $^3S_1 \to {}^3P_1$ amplitude $W_{pn}$ is given by

$$W_{pn} \approx \frac{1}{m_\rho^2} \left[ 2F_\pi \left(\frac{m_\rho}{m_\pi}\right)^2 + 2H_1 + G_1 - F_1 \right]. \qquad 6b.$$

The $^3S_1 \rightarrow {}^1P_1$ amplitude is

$$U_{pn} \approx \frac{1}{m_\rho^2} [G_0\mu_S - 3F_0\mu_V].$$ 6c.

We have taken $m_\omega = m_\rho$.

While our PNC potential depends on seven independent couplings, only six weak amplitudes can be tested at low energies: the five couplings given by Equations 6 and $F_\pi$. On inspecting Equations 6, one sees that the following transformation leaves the zero-energy $S \rightarrow P$ amplitudes unchanged:

$$H_1 \rightarrow H_1 + \alpha$$

$$F_1 \rightarrow F_1 + \alpha \frac{[4+\mu_S]}{4+\mu_S+\mu_V} = F_1 + 0.496\alpha$$

$$G_1 \rightarrow G_1\alpha \frac{[4+2\mu_V]}{4+\mu_S+\mu_V} = G_1 - 1.504\alpha.$$ 7.

In the shell model calculations described below, these transformations leave the PNC matrix essentially unchanged ($\sim 0.1\alpha$) in agreement with the argument that the $S \rightarrow P$ approximation is appropriate for bound nucleons. It is thus impractical to determine separately $H_1$, $F_1$, and $G_1$ from low-energy data.

Ideally one would perform at least five low-energy experiments in the N-N system to determine independent combinations of the five partial-wave amplitudes, including one experiment in which the dependence of the observable on the center-of-mass energy could be exploited to separate the pion-exchange contribution to the $^3S_1 \rightarrow {}^3P_1$ amplitude from the shorter-range contributions. We discuss such a set of measurements below (omitting the completely impractical n-n observable). Unfortunately, the measurements are so difficult that a definite PNC effect has been seen in only one observable. As a result, one has to examine parity impurities in complex nuclei to obtain more complete information about the PNC N-N interaction.

## 2.3   Weak Meson-Nucleon Couplings

The PNC meson-nucleon couplings of Equation 4 and Table 1 can be predicted from the underlying theory of the weak interaction between quarks, given a reliable model for the strong dynamics. Thus they have become the common point for comparing particle physics and nuclear physics investigations of the weak hadronic interaction. Originally such comparisons were made simply to determine whether a set of empirical couplings could explain the nuclear PNC data while being consistent (within the broad ranges of theoretical uncertainties) with particle physics estimates. However, recent progress in experiment and in understanding the nuclear aspects of parity nonconservation has generated more optimism: empirical couplings may provide a detailed test of the effects of strong dynamics on the hadronic weak interactions. In this section we discuss the recent derivation of weak meson-nucleon couplings by Desplanques, Donoghue & Holstein (8) (DDH) to illustrate the present state of the particle theory.

DDH used many conventional techniques—current algebra, relations to known $\Delta S = 1$ hyperon-decay amplitudes, $SU(6)_w$ symmetry—in a calculation based on a specific description of the strong dynamics, the quark model, to work out the standard model estimates for the PNC nucleon-nucleon-meson couplings. The marriage of the quark model and such symmetry techniques removed some, though certainly not all, of the ambiguities that plagued earlier calculations.

DDH grouped the contributions to the weak meson-nucleon vertices into three classes.

1. Factorization terms: Most early work on the charged-current contribution to weak vector meson vertices used Michel's factorization approximation (22, 23):

$$\langle \rho^- p|^c H_w|n\rangle = G/\sqrt{2} \cos^2 \theta_c \langle \rho^- p|A^\mu_+(0)V^-_\mu(0)|n\rangle$$

$$\approx G/\sqrt{2} \cos^2 \theta_c \langle \rho^-|V^-_\mu(0)|0\rangle \langle p|A^\mu_+(0)|n\rangle, \qquad 8.$$

where only the vacuum intermediate states shown in Figure 2a are retained. (Here $V$ and $A$ are the charged vector and axial vector currents.) In quark model calculations the factorization term also generates, after a Fierz rearrangement of the four-quark operator, an amplitude for neutral vector meson production. The charged-current factorization contribution to pion production vanishes if $SU(3)$ breaking is ignored. (The matrix element $\langle \pi^-|A^\mu_-|0\rangle \approx q^\mu F_\pi$, and $q^\mu V_\mu = 0$ for a conserved vector current.) However, the product of left- and right-handed neutral currents generates, after Fierz transformation into a product of scalar and pseudoscalar densities, a factorization amplitude for $\pi^\pm$ emission proportional to $\sin^2 \theta_W$, where $\theta_W$ is the Weinberg angle. A term proportional to $(1 - 2 \sin^2 \theta_W)$ also appears when strong interaction enhancements are taken into account.

2. Quark model terms: The quark model terms are corrections to the factorization approximation corresponding to baryonic intermediate states, as shown in Figure 2b. Like the factorization terms, these terms cannot be related to the $\Delta S = 1$ charged-current amplitudes. Their

evaluation requires specific quark model wave functions for the baryons. As first stressed by McKellar & Pick (19), the quark model terms substantially alter the naive results based on factorization: in the $SU(6)_W$ limit the quark model charged-current term for vector meson emission is three times larger than the usual factorization term and opposite in sign.

3. Sum rule terms: These are contributions to the pion and vector meson emission amplitudes that can be related to experimental $\Delta S = 1$ hyperon PNC decay amplitudes via $SU(6)_W$ symmetry and PCAC. They are shown schematically in Figure 2c. The charged-current contributions to pion emission are entirely due to the sum rule terms. They are proportional to $\tan \theta_c$ times the s-wave hyperon-decay amplitudes and thus are suppressed. (The s-wave hyperon amplitudes are themselves proportional to $\sin \theta_c$ $\cos \theta_c$ so that the $\pi NN$ amplitude is proportional to $\sin^2 \theta_c$ as mentioned above.) Certain contributions to the neutral-current Hamiltonian belong to the same $SU(3)$ multiplets as the charged current, and thus can be treated similarly. These neutral currents produce a pion emission amplitude proportional roughly to $\cot \theta_c$, enhancing the pion emission by an order of magnitude.

DDH emphasized that there are substantial uncertainties in estimating the pion weak coupling. The DDH estimate of the sum rule contribution to $F_\pi$, based on the $SU(3)$ symmetry, could be uncertain by a factor of two. The quark model terms involve a ratio $R$ of reduced matrix elements that

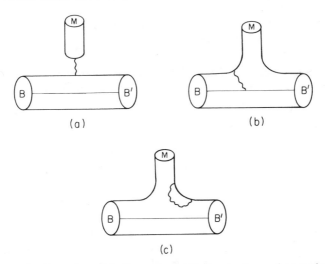

*Figure 2* Selected quark model diagrams contributing to the nucleon-nucleon-meson couplings: (a) "factorization" diagram; (b) diagram contributing to the "quark model" terms; (c) "sum rule" diagram.

depends on the quark wave functions ($R = 1$ in a nonrelativistic model, but $R \approx 2/3$ for the MIT bag). The factorization term is uncertain because it depends on the absolute masses of the u and d quarks. All of the terms depend on estimates of strong interaction enhancements based on the Wilson operator–product expansion and renormalization group arguments. These enhancements generate an additional factor-of-two uncertainty in $F_\pi$ (24).

The predictions for the vector meson couplings are also quite uncertain and there are no measured $\Delta S = 1$ vector meson decays to test the theory. The charged-current factorization, sum rule, and quark model terms are each large. In the $SU(6)_W$ limit the sum rule and quark model terms cancel against and dominate the factorization term. However, DDH estimate that the quark model contributions to the vector meson couplings could range from their $SU(6)_W$ values to near zero. Uncertainties in the sum rule contribution are similar to those discussed above for pions. Thus the vector meson couplings could vary between the factorization and $SU(6)_W$ limits.

These considerations led DDH to define broad "reasonable ranges" for the couplings appearing in Equation 4, as well as "best values" that reflected their theoretical preferences. These couplings are given in Table 1. Note that $h'^1_\rho$, which was not evaluated by DDH, was later considered by Holstein (25) and found to be small. One should recognize that the "reasonable range" uncertainties in the various couplings are strongly correlated. It makes no sense to compute a "theoretical error bar" for an observable by combining the "reasonable range" uncertainties in quadrature. Allowed variations in the vector meson couplings are essentially governed by three parameters, one describing the strong interaction enhancement effects and the other two weakening of the quark model and sum value terms from their $SU(6)_W$ values. The pion coupling can be treated as a fourth independent parameter. In Section 7 we discuss an attempt to extract these theoretical parameters from experiment.

## 3.  DETECTION OF PARITY VIOLATION IN STRONGLY INTERACTING SYSTEMS

What general principles govern experimental measurements of PNC perturbations in systems that have PC strong interactions? As the relative strength of the PNC and PC N-N interactions is given roughly by $4\pi G_F m_\pi^2/g_{\pi NN}^2 \approx 10^{-7}$, clearly PNC effects will be very small. Therefore, the dominant experimental consideration is minimizing the uncertainties from statistical and systematic errors.

Parity-violation experiments fall into three general classes (26):

1. Type I experiments determine rates of processes that would vanish if

parity were exactly conserved. The only practical examples involve $\alpha$ decays. Parity conservation forbids any unnatural parity state from decaying to a $0^+$ final state by alpha emission. An example is $^{16}O(2^-)$ $\rightarrow$ $^{12}C(0^+) + \alpha$, which competes with the PC $\gamma$ decay $^{16}O(2^-) \rightarrow$ $^{16}O + \gamma$. Type I experiments are sensitive to the *probability* of parity admixtures in nuclear eigenstates: $\Gamma_{PNC} \propto |\langle f|H_{PNC}|i\rangle|^2$.

2. Type II experiments determine pseudoscalar observables arising from the interference of PNC and PC matrix elements. One can probe pseudoscalar interactions in two ways. Usually one detects the longitudinal projection of a particle's spin along its momentum. A common example is the circular polarization $P_\gamma = \langle \sigma_\gamma \cdot \hat{p}_\gamma \rangle$ of $\gamma$ rays emitted by an unpolarized nucleus. One can also detect PNC interactions by orienting a particle's spin perpendicular to its motion. In this case $\sigma$ precesses around $p$ in the presence of a PNC interaction that breaks the degeneracy of the states with $\sigma \cdot \hat{p} = \pm 1$. This method has been exploited in reactions induced by cold neutrons. In all cases pseudoscalar measurements are proportional to the *amplitude* of the parity admixtures. For example, the circular polarization produced by the interference of a PNC E1 multipole with a dominant PC M1 multipole is

$$P_\gamma = 2\,\text{Re}\left[\frac{\langle f|E1|i\rangle \langle f|M1|i\rangle^*}{|\langle f|M1|i\rangle|^2 + |\langle f|E1|i\rangle|^2}\right] \approx 2\,\text{Re}\left[\frac{\langle f|E1|i\rangle}{\langle f|M1|i\rangle}\right]. \qquad 9.$$

3. Type III experiments also determine the interference of PNC and PC matrix elements but the observable is not a pseudoscalar. As a result the observable is *second order* in the weak interaction. An example of such an observable is an odd power of $\cos\theta$ in the angular distribution of $\gamma$ rays emitted by an isolated nuclear state. For states of good parity such terms must vanish. In the nuclear domain type III measurements are not competitive with type II experiments and are not discussed further here.

One might expect that, from statistical considerations alone, type I experiments (which are sensitive to the *square* of a very small PNC amplitude) would be greatly inferior to type II experiments (where the effects are *linear* in the PNC amplitude). However, this is not correct. Consider two simple hypothetical systems (see Figure 3). In the first, two competing processes lead to distinct final states with amplitudes $A_{PNC}$ and $A_{PC}$. For example, $A_{PNC}$ could correspond to $^{16}O(2^-) \rightarrow$ $^{12}C + \alpha$ and $A_{PC}$ to $^{16}O(2^-) \rightarrow$ $^{16}O + \gamma$. The rate $\Gamma_{PNC} \propto |A_{PNC}|^2$ is the observable. In the second system, two competing processes with amplitudes $A_{PNC}$ and $A_{PC}$ lead to a single final state, producing a pseudoscalar observable $P \propto 2A_{PNC}/A_{PC}$. For definiteness, we take $P$ to be a $\gamma$-ray circular polarization. Assume that in each case we can produce $N_0$ of the decaying states. For the first system the number of PNC decays is $N_{PNC} = N_0\Gamma_{PNC}/(\Gamma_{PC} + \Gamma_{PNC}) \approx N_0\Gamma_{PNC}/$

$\Gamma_{PC} \propto N_0 |A_{PNC}|^2/|A_{PC}|^2$. We infer $A_{PNC}$ from the relation $|A_{PNC}|^2 = |A_{PC}|^2 N_{PNC}/N_0$. The statistical uncertainty in $N_{PNC}$ leads to a corresponding statistical uncertainty in $A_{PNC}$ of

$$\Delta A_{PNC} = \frac{1}{2A_{PNC}} |A_{PC}|^2 \frac{\Delta N_{PNC}}{N_0} = \frac{A_{PC}}{2[N_0]^{1/2}}. \qquad 10.$$

For the second system assume that $P$ can be detected with 100% efficiency (not necessarily a realistic assumption!). The statistical error in $P$ (assuming $P \ll 1$) is $\Delta P = (N_0)^{-1/2}$. We obtain $\Delta A_{PNC}$ from the relation $\Delta A_{PNC} = A_{PC} \Delta P/2 = A_{PC}/(2N_0^{1/2})$. Therefore on statistical grounds neither class of experiment is preferred. In fact, if $P$ is a $\gamma$-ray circular polarization and one includes a realistic value for the analyzing power of the circular polarimeter (typically a few percent), the type I experiment is statistically superior. Of course, in practical cases systematic errors are as important as statistical uncertainties.

The chief distinction between type I and type II experiments is flexibility. Type I experiments are restricted to $\alpha$ decay of unnatural parity states in even-$A$ nuclei. The process typically involves highly excited states that are difficult to describe with confidence in nuclear models. Type II experiments (longitudinal analyzing powers, $\gamma$-ray circular polarizations and asymmetries, neutron spin rotations) have been carried out in the two-nucleon and few-nucleon systems, and in many complex nuclei. The observables, being linear in the weak amplitude, provide information on the sign of the weak amplitude. Type II experiments in complex nuclei usually involve $\gamma$ decay, a process that often can be analyzed more reliably than the strong interaction $\alpha$ decays of type I experiments.

The chief experimental problems in type I and type II experiments are also different. The dominant obstacle to type I experiments is finding a way to populate the unnatural-parity parent state with enough selectivity so that the very tiny PNC decay branch can be detected above the inevitable background. The chief problems in type II experiments are achieving statistical accuracy and reducing spurious contributions to the measured pseudoscalar. These points are discussed more fully below.

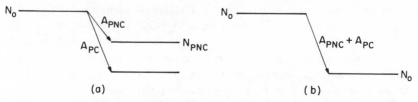

*Figure 3* Schematic representation of two classes of PNC experiments: (a) type I experiment; (b) type II experiment.

## 3.1   *Longitudinal Analyzing Powers*

The longitudinal analyzing power for reactions induced by polarized nucleons is defined by $A_L = (1/P_z)[\sigma^+(\theta) - \sigma^-(\theta)]/[\sigma^+(\theta) + \sigma^-(\theta)]$, where $P_z$ is the longitudinal polarization of the beam and $\sigma^+$ and $\sigma^-$ are cross sections for positive and negative helicity projectiles, respectively. $A_L$ is measured by repeatedly switching the beam helicity between right- and left-handed states and detecting the correlated change in the cross section. Measurements of $A_L$ are straightforward only if one can rapidly reverse the helicity of the beam without significantly changing any of the other parameters, such as intensity, position, and emittance. Therefore $A_L$ experiments place stringent demands on the polarization and spin-reversal apparatus.

In polarized proton experiments the dominant systematic errors are usually associated with residual transverse components of the polarization. The most difficult problems are not due to beam misalignment or to a nonzero mean transverse polarization, but to a second-order effect arising from variations in the residual transverse polarization across the beam profile. This is illustrated in Figure 4. Suppose that, for positive-helicity protons, there exists on the fringes of the beam a small component of transverse polarization directed clockwise (Figure 4a). In addition, suppose that the proton scattering has a strong transverse analyzing power, $A_T$, so that spin-up protons scatter preferentially to the left. Then the protons scatter preferentially toward the nearer counter. Assume, that, on reversing the helicity of the beam, the transverse polarization also reverses and is directed counterclockwise (Figure 4b). Because of the transverse analyzing power, the protons scatter preferentially toward the more distant counter.

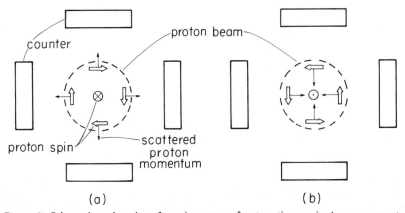

*Figure 4*   Schematic explanation of a major source of systematic error in $A_L$ measurements.

This second-order solid-angle effect is proportional to $\langle xp_y \rangle$ and $\langle yp_x \rangle$ and mimics a PNC helicity dependence. Although it vanishes for a line beam or for a perfectly efficient $4\pi$ counter, in realistic situations it is the dominant systematic effect once the mean transverse polarization, $\langle p_x \rangle$ and $\langle p_y \rangle$, has been minimized. Such "circulating transverse polarization" is produced, for example, whenever a polarized beam passes through a bending magnet asymmetrically with respect to its median plane.

## 3.2  $\gamma$-Ray Circular Polarizations

Unfortunately there are no efficient analyzers of $\gamma$-ray circular polarizations. Essentially all polarimeters are based on the spin dependence of Compton scattering by polarized electrons (27) in magnetized Fe. Because only $\sim 2/26$ of the electrons in Fe are polarized at saturation, the analyzing power of this polarimeter cannot exceed $\sim 8\%$. Two different geometries are employed. In the transmission geometry one looks for a small change in the transmitted $\gamma$-ray fraction when the magnetization of the Fe is reversed. Large solid angles and good $\gamma$-ray energy resolution (important for discrimination against backgrounds) can be obtained with this method. In the forward-scattering geometry one detects the scattered $\gamma$ ray. Energy resolution and solid angle are sacrificed for a higher analyzing power.

Transmission polarimeters have been used in most recent work. It is easy to compute the analyzing power for such a device if the $\gamma$-ray detector has sufficient energy resolution to identify the unscattered photons. For photons with circular polarization $P$ incident on magnetized Fe with the electron spins parallel ($S = +1$) or antiparallel ($S = -1$) to the incident photon momentum, the Compton cross section is $\sigma(E_\gamma) = \sigma_0(E_\gamma) + SP\sigma_1(E_\gamma)$ where $\sigma_1$ is the spin-dependent cross section. The polarimeter analyzing power is $\eta = (N_R - N_L)/(N_R + N_L)$, where $N_R(N_L)$ is the number of right-handed (left-handed) photons transmitted by the polarimeter when the electron spins are parallel to the incident photon momentum. Thus

$$\eta = \frac{\exp[-(\sigma_0 + f\sigma_1)nt] - \exp[-(\sigma_0 - f\sigma_1)nt]}{\exp[-(\sigma_0 + f\sigma_1)nt] + \exp[-(\sigma_0 - f\sigma_1)nt]}$$

$$= -\tanh(f\sigma_1 nt) \approx -f\sigma_1 nt, \qquad\qquad 11.$$

where $f$ is the number of polarized electrons per Fe atom, $n$ is the number of Fe atoms per volume, $t$ is the polarimeter thickness, and $\sigma_t$ is the total $\gamma$-ray attenuation cross section. One selects $t$ to optimize the signal-to-noise ratio.

Typically, experiments are limited by the strength of the $\gamma$-ray source. In this case (if background in the $\gamma$-ray detector is negligible) the optimum thickness is $t = 2/\sigma_t n$. The analyzing power for such an optimized polarimeter is simply $2f\sigma_1/\sigma_t$, as shown in Figure 5 (for Fe $f \approx 2.06$) at

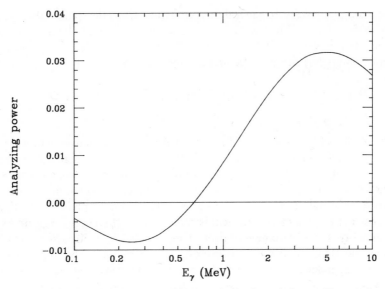

*Figure 5* Analyzing power for a transmission-mode circular polarimeter. The polarimeter thickness has been optimized as discussed in the text.

saturation). Notice that $\eta$ vanishes at low energies and at $E_\gamma \approx 0.64$ MeV, and has a broad maximum of $\eta \approx 0.03$ at $E_\gamma \approx 5$ MeV. Somewhat higher analyzing powers can be obtained from Fe-Co alloys.

### 3.3    $\gamma$-Ray Asymmetries

Asymmetry measurements are generally favored over circular polarization measurements whenever the initial state can be produced with a polarization greater than the analyzing power of the appropriate circular polarimeter. Two methods are used to polarize the initial, excited nuclear state: (*a*) production of the state in a reaction induced by polarized projectile or by using a polarized target (see, for example, 40, 52); and (*b*) "brute force" polarization generated by strong crystalline fields at very low temperatures (see, for example, 64).

The "brute force" technique is possible only for levels fed in $\beta$ decay of moderately long-lived activities because of the time required to cool a sample. On the other hand, the production of polarized, excited nuclei in reactions involving polarized particles is, in principle, widely applicable. Modern ion sources produce polarized protons with intensities of $\sim 10^{13}$ s$^{-1}$ and with polarizations up to 90%. Cold neutron beams from a modern reactor reach intensities of $6 \times 10^9$ s$^{-1}$ at 97% polarization. The difficulty lies in finding an efficient way to transfer the polarization from the projectile to the decaying state. In general this is maximal when the spins of the target and residual nuclei are low. Occasionally this efficiency can be surprisingly high: the $^{22}$Ne(p, $\alpha$)$^{19}$F* reaction (28) transfers $\sim 70\%$ of the proton transverse polarization to the first excited state of $^{19}$F.

# 4.    PARITY VIOLATION IN THE N-N SYSTEM

## 4.1    The p-p System

The obvious PNC observable in the p + p system is the analyzing power for longitudinally polarized protons. $A_L$ in the p + p reaction is sensitive only to the $\Delta I = 0$, 1, and 2 $^1S_0 \leftrightarrow {}^3P_0$ amplitudes in the low-energy limit. Therefore, for $E_p < 50$ MeV, $A_L$ is independent of $\theta$ and its energy dependence is determined by the known PC $^1$S and $^3$P phase shifts (if, as expected, the weak matrix element is constant over this energy region). As a result, any $A_L$ measurement at proton energies below $\sim 50$ MeV measures the same PNC quantity. At energies above $\sim 50$ MeV the angular and energy dependence of $A_L$ reflects the interplay of the S-P and higher partial-wave PNC amplitudes. The dependence of the low-energy $A_L$ upon the PNC N-N potential of Equation 4 is shown in Table 3.

One can determine $A_L$ either by detecting the scattered protons or by a transmission measurement. In either case, a large number of protons ($\sim 10^{15}$) must be detected to obtain statistical errors smaller than the expected effect ($10^{-7}$). Because it is difficult to count individual pulses accurately at rates much greater than $10^6$–$10^7$ s$^{-1}$, while ion sources can provide beams of $\sim 10^{12}$–$10^{13}$ s$^{-1}$, experimenters have chosen to integrate detector currents rather than count pulses. In this way the required statistical accuracy can be achieved after some tens of minutes of data collection rather than after several years. Transmission measurements are preferred at high energies where, by using thick targets, a significant fraction of the incident protons are scattered by the strong force.

At lower energies, where only a small fraction of the incident beam is scattered by the nuclear interaction, transmission experiments become impractical, so one must detect the scattered protons. The first result was reported by the Los Alamos group (29, 30), who employed the Lamb-shift polarized ion source of the Los Alamos tandem accelerator. The scattered protons were detected in plastic scintillators and the transmitted beam in a Faraday cup. They obtained a value $A_L = (-1.7 \pm 0.8) \times 10^{-7}$ at $E_p = 15$ MeV.

Recently a very detailed account was published of $A_L$ measurements at $E_p = 45$ MeV (31). The experiments were performed with the atomic beam polarized ion source of the SIN injector cyclotron. Scattered protons were detected in a cylindrical ion chamber surrounding a high-pressure gas

**Table 3** PNC observables in the N-N system: theory and experiment

| | Experiment | | Theory | | | | | | | | |
|---|---|---|---|---|---|---|---|---|---|---|---|
| Observable | Value $(10^{-7})$ | Ref. | Value$^a$ $(10^{-7})$ | $F_\pi$ | $F_0$ | $F_1$ | $F_2/\sqrt{6}$ | $G_0$ | $G_1$ | $H_1$ | Ref. |
| $\bar{p}+p$ | | | | | | | | | | | |
| $A_L^{tot}$ (45 MeV) | $-2.31\pm0.89$ | 31 | $-1.35$ | | $-0.053$ | $-0.053$ | $-0.053$ | $-0.016$ | $-0.016$ | | 31 |
| (15 MeV) | $-1.7\pm0.8$ | 30 | $-0.77$ | | $-0.030$ | $-0.030$ | $-0.030$ | $-0.011$ | $-0.011$ | | 31 |
| $np \rightarrow d\gamma$ | | | | | | | | | | | |
| $P_\gamma$ | $1.8\pm1.8$ | 38 | $0.56$ | | $0.022$ | $0.001$ | $0.043$ | $-0.002$ | | | 62 |
| $A_\gamma$ | $0.6\pm2.1$ | 41 | $-0.49$ | $-0.045$ | | | | | $-0.001$ | $-0.002$ | 62 |

$^a$ Obtained by combining the matrix elements in the columns to the right with the "best value" DDH weak meson-nucleon couplings of Table 1.

target. A schematic diagram of the experiment is shown in Figure 6. The proton helicity was reversed periodically (nominally every 30 ms). The first 20 ms of each period were devoted to measuring the scattered fraction of the beam (by the ratio of the ion chamber current to the Faraday cup current). During the last 10 ms scanners were activated to measure the intensity and transverse polarization profiles of the proton beam. The scanners consisted of carbon rods that were swept through the beam. Protons scattered from these rods were detected in plastic scintillators. Because of the large transverse analyzing power for $p + {}^{12}C$ ($A_T \approx 0.93$) the scanners gave a detailed map of the variation of the intensity and transverse polarization across the beam.

The most important and time-consuming aspect of the experiment was the study and minimization of possible sources of systematic errors. It was crucial to determine if reversing the helicity of the beam by switching between two ion source states changed other beam properties that could affect the measured $A_L$. Whenever possible the sensitivity to various sources

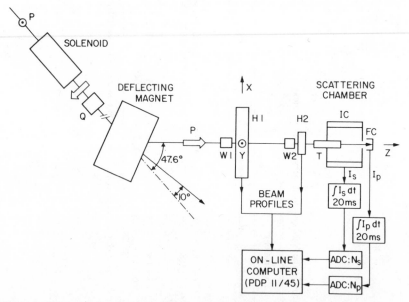

*Figure 6* Schematic diagram of the apparatus used in the SIN measurement of $A_L$ in $\vec{p} + p$ scattering. At the exit to the cyclotron the beam polarization P points upward (along $y$). The spin is precessed by the solenoid into the horizontal plane, and rotated into the $z$ direction by the deflecting magnets. The protons then pass through the $H_2$ target T and are stopped in the Faraday cup FC. Scattered protons are detected in the cylindrical ion chamber IC. The scanners H1 and H2 measure the intensity and polarization profiles of the beam. Items Q, W1, and W2 induce artificial modulations in the beam for assessing sensitivities to systematic errors.

of systematic error was measured directly. For example, during the experiment the scanners determined the intensity, position, and transverse polarization modulations that were coherent with the helicity reversal. Then in auxiliary measurements such modulations were deliberately induced by, for instance, altering the beam position with a magnet placed on the beam line upstream of the apparatus. The amplitude of the induced modulations was made much larger than the modulations monitored in the PNC experiment so that the spurious contributions to $A_L$ could be readily measured. The sensitivity of $A_L$ to these induced modulations, when multiplied by the actual modulations observed during PNC data runs, determined corrections that had to be applied to the measured $A_L$.

The only significant correction was that for the "circulating transverse polarization" effect discussed in Section 3.1. In spite of all precautions taken to reduce systematic errors in the SIN experiment, the corrections were roughly the size of the quoted asymmetry in the total cross section, $A_L^{tot}$ (45 MeV) $= (-2.31 \pm 0.89) \times 10^{-7}$. The error bar includes statistical and all identified systematic uncertainties. This result may be compared to preliminary results at essentially the same proton energy from other laboratories: $A_L$(46 MeV) $= (-1.3 \pm 2.3) \times 10^{-7}$ from Berkeley (32) and $A_L$(47 MeV) $= (-4 \pm 3) \times 10^{-7}$ from Texas A&M (33). From the known strong interaction phase shifts one expects that $A_L^{tot}$(45 MeV) $= (1.75 \pm 0.1) A_L^{tot}$(15 MeV). Scaling the Los Alamos results (30) up to 45 MeV we find $A_L^{tot}$(45 MeV) $= (-3.0 \pm 1.4) \times 10^{-7}$. The consistency of these four results is quite impressive considering the difficulty of the measurements.

At high energies, where one can use thick targets to scatter a significant fraction of the protons, $A_L$ is best measured in transmission experiments. A measurement at 800 MeV has been completed at LAMPF. The preliminary result (34) is $(+2.4 \pm 1.1) \times 10^{-7}$.

What are the theoretical expectations? Using nucleon-nucleon wave functions calculated from the Reid soft-core potential, the short-range correlations of Miller & Spencer (35), and the one-boson-exchange PNC interaction with the "best values" of DDH (8), the SIN group predicts $A_L^{tot}$(45 MeV) $= -1.35 \times 10^{-7}$. The measured results are roughly twice as large as the "best value" predictions. Oka (36) used the DDH "best value" potential to calculate $A_L$ for proton energies from 10 MeV to over 1 GeV. His unperturbed wave functions were obtained from phase-shift analyses and the resulting Born amplitudes of $V_{PNC}$ were "unitarized" to satisfy the generalized Watson theorem. However, Oka did not include the effects of short-range correlations in evaluating the $T$ matrix. In Figure 7 we compare the data to Oka's "best value" prediction. We have scaled his results to agree with the SIN prediction at 45 MeV. This should crudely approximate the correction for short-range correlations. The measured $A_L$

## p + p scattering

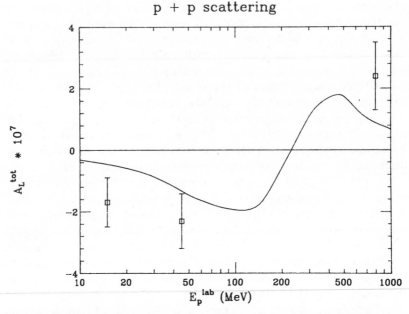

*Figure 7* Comparison of experimental $A_L$ to Oka's prediction. The calculation has been scaled as discussed in the text.

at 800 MeV is again roughly twice the "best value" prediction. As one expects the meson-exchange description of the PNC interaction to be valid up to energies of $\sim 1$ GeV, this result is not unexpected.

A measurement of $A_L$ in $\vec{p} + H_2O$ scattering at 6 GeV/$c$ has been reported (37). This higher-energy result is outside the scope of our review and is not discussed further.

### 4.2 *The n-p System*

Four different PNC observables are of interest in the n-p system.

1. $P_\gamma$: the circular polarization of the 2.2-MeV $\gamma$ ray emitted when unpolarized thermal neutrons are captured by protons. Alternatively one can study $A_L^\gamma$, the circular polarization dependence of the $\gamma d \to np$ cross section. These observables are sensitive to the $\Delta I = 0$ and $\Delta I = 2$ $^1S_0 \leftrightarrow {}^3P_0$ amplitudes and the $^3S_1 \leftrightarrow {}^1P_1$ amplitude.

2. $A_\gamma$: the asymmetry of the 2.2-MeV $\gamma$ ray emitted when polarized thermal neutrons are captured by protons. This observable is sensitive to the $^3S_1 \leftrightarrow {}^3P_1$ amplitude.

3. $\phi_{PNC}$: the spin rotation of polarized cold neutrons transmitted through a parahydrogen target. [Strong interaction spin-flip scattering is forbidden

from parahydrogen ($S = 0$) molecules so neutrons have long mean free paths and do not depolarize.] This observable is sensitive to the $^3S_1 \leftrightarrow {}^1P_1$, $^3S_1 \leftrightarrow {}^3P_1$, and $^1S_0 \leftrightarrow {}^3P_0$ amplitudes. It has not yet been studied experimentally.

4. $A_L^n$: the dependence of the capture cross section on the helicity of the neutrons. This observable and $\phi_{PNC}$ are sensitive to the same combination of amplitudes. $A_L^n$ also has not yet been measured.

The sensitivities of $P_\gamma$ and $A_\gamma$ to the PNC N-N potential (Equation 4) are shown in Table 3.

Recently a Soviet group (38) reported a new measurement of $P_\gamma$ at the Leningrad reactor. The major sources of systematic error in $P_\gamma$ are a circularly polarized $\gamma$-ray background from the bremsstrahlung of the left-handed electrons produced by fission product $\beta$ decays, higher-order effects in Compton scattering that vanish only for an axially symmetric polarimeter-detector geometry, and fluctuations in the reactor neutron flux.

Great care was taken with shielding so that the 2.2-MeV $\gamma$-ray beam from neutron capture in the $H_2O$ target was not significantly contaminated by bremsstrahlung from the reactor core. A transmission circular polarimeter was employed to exploit the sign change in the analyzing power at $E_\gamma \approx 0.64$ MeV (see Figure 5). This reduced the signal from circularly polarized bremsstrahlung because of the relatively low energy of the bremsstrahlung spectrum. To suppress reactor power-level fluctuations, a dual polarimeter system (shown in Figure 8) was employed. The two polarimeters were magnetized in opposite directions. When the signals from the two polarimeters were subtracted, fluctuations in the reactor flux tended to cancel while the effects of a $\gamma$-ray circular polarization added.

From three different measurements and various control experiments, Knyazkov et al (38) found $P_\gamma = (1.8 \pm 1.8) \times 10^{-7}$. This value supersedes an earlier result (39) that apparently was flawed by bremsstrahlung contamination from the reactor core. As the analyzing power of the polarimeter used in this experiment is 4.5%, the error corresponds to an experimental

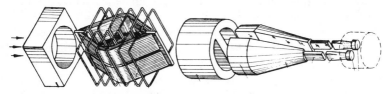

*Figure 8*   Dual circular polarimeter used in the Leningrad $P_\gamma$ measurement in np $\rightarrow$ d$\vec{\gamma}$. The 2.2-MeV photons enter from the left and are detected in the two small circular photodiodes shown at the right.

asymmetry of $8 \times 10^{-9}$! Unfortunately, even this very impressive precision is not sufficient to reach the level predicted by theory.

An experiment to study $A_L^\gamma$ in the $\gamma$d → np reaction is in progress at Chalk River. Although this observable and $P_\gamma$ are sensitive to the same combination of amplitudes, the $A_L^\gamma$ measurement is freer of background problems. Unfortunately, the statistical precision of the $A_L^\gamma$ measurement is expected to be at the $10^{-6}$ level (A. B. McDonald, private communication).

A measurement of $A_\gamma$ in np → d$\gamma$ at the ILL reactor has been reported (41) and a remeasurement is currently in progress (R. Wilson, private communication). The published limit (41) is roughly four times greater than the theoretical value, and does not provide an important constraint on the PNC amplitudes. However, the precision of the new result is expected to reach the level of theory.

## 5.   PARITY MIXING IN COMPLEX NUCLEI

This section is organized according to the mass numbers of the nuclei involved. The organization reflects the differences in experimental techniques and theoretical analyses employed in the different mass regions. In this section we address the following questions: What can be learned about the PNC N-N interaction from experiments in $A > 2$ nuclei? Are there mechanisms that amplify the PNC effects and make experiments more practical? How can we minimize the "theoretical error bars" when interpreting measured effects in terms of the PNC N-N interaction?

### 5.1   Few-Body Systems

In terms of PNC experiments, the few-body nuclei ($A = 3$ through $A = 5$) have much more in common with the N-N system than with heavier nuclei. Unlike the heavier nuclei, there are no bound excited states that can be studied by $\gamma$-ray techniques and usually no amplification factors, as discussed below, to increase the size of PNC effects. [Some amplification of the PNC effect in $^2$H(n, $\gamma$) at thermal energies does occur because the PC M1 capture rate is strongly suppressed.] For three-body systems "exact" Faddeev calculations are still possible, but for $A \geqslant 4$ calculations are dependent on models of the many-body dynamics.

The experimental results and the corresponding theoretical predictions in the few-body nuclei are shown in Table 4. The $A_L$ (46 MeV) measured (43) in $\vec{p} + {}^4$He is particularly noteworthy. In this case $A_L$ is a function of $\theta$ and, because of breakup processes, of the response of the detector to particles of different energies. To predict $A_L$ for a given PNC potential one must know the detector response function and have a good description of the strong p + $^4$He reaction. One might expect that the parity-conserving eigenstates

**Table 4** PNC observables in few-body systems: theory and experiment

| Observable | Experiment | | Theory | | | | | | | | |
| --- | --- | --- | --- | --- | --- | --- | --- | --- | --- | --- | --- |
| | Value ($\times 10^{-7}$) | Ref. | Value[a] ($\times 10^{-7}$) | $F_\pi$ | $F_0$ | $F_1$ | $F_2/\sqrt{6}$ | $G_0$ | $G_1$ | $H_1$ | Ref. |
| $\bar{p}+d$ | | | | | | | | | | | |
| $A_L$ (15 MeV) | $-0.35 \pm 0.85$ | 30 | $-1.4$ | $-0.097$ | $-0.017$ | $-0.007$ | $-0.000$ | $-0.004$ | $-0.003$ | $-0.007$ | 42[b] |
| $\bar{n}+d \rightarrow t+\gamma$ | | | | | | | | | | | |
| $A_\gamma$ | $78 \pm 34$ | 44 | $6.8$ | $0.29$ | $0.24$ | $-0.71$ | $-0.098$ | $0.053$ | $-0.013$ | $0.21$ | 44 |
| $\bar{p}+{}^4\text{He}$ | | | | | | | | | | | |
| $A_L$ (46 MeV) | $-3.3 \pm 0.9$ | 43 | $-3.3$ | $-0.14$ | $-0.10$ | $-0.034$ | | $-0.014$ | $-0.014$ | $-0.013$ | 43 |

[a] Obtained by combining the matrix elements in the columns to the right with the DDH weak meson-nucleon couplings of Table 1.
[b] We have used the Reid soft-core Faddeev result from the reference.

would be fixed by the known $p + {}^4He$ scattering phase shifts. But phase shifts do not determine the crucial short-distance behaviour. The calculations (43) shown in Table 4 employ the Jastrow factor of Miller & Spencer (35) to describe the short-range N-N correlations. (This same factor is used in the analyses of $\vec{p} + p$ discussed above and of the parity-mixed doublets described below.) The sensitivity of the $\vec{p} + {}^4He$ result to the strong interaction physics is apparent: if ${}^4He$ were a structureless, $I = 0$ particle, and breakup could be neglected, PNC $\pi$ exchange would not occur. In fact $\pi$ exchange and heavier meson exchanges contribute about equally to the calculated $A_L$ because of the important role of exchange (as opposed to direct) processes. The latest $\vec{p} + {}^4He$ result (43) shown in Table 4 agrees well with the "best value" predictions.

The measured $\vec{n}d \rightarrow t\gamma$ asymmetry (44) considerably exceeds the prediction based on "best value" couplings. In view of the large experimental error in this quantity, one should not yet place too much significance on this discrepancy.

## 5.2    Parity-Mixed Doublets in Light Nuclei

5.2.1    GENERAL CONSIDERATIONS    The pioneering nuclear parity violation experiments focussed simply on measuring effects in practical cases, such as circular polarization of $\gamma$ rays from radioactive sources. However, in general it proved very difficult to extract from such experiments quantitative information on the underlying PNC N-N interaction. The demands placed on nuclear structure theory are just too great. Consider, as an illustration, a $1^+ \rightarrow 0^+$ M1 $\gamma$-ray transition. The PNC N-N interaction induces small negative parity admixtures in the initial and final nuclear states and thus a small E1 component in the $\gamma$-ray transition. The interference of the E1 and M1 multipoles produces a PNC circular polarization, $P_\gamma$, of the $\gamma$ ray. In first-order perturbation theory the initial state $|i\rangle$ is

$$|i\rangle = |1^+\rangle + \sum_m \frac{\langle 1_m^- | V_{PNC} | 1^+ \rangle}{E_{1_+} - E_m} |1_m^-\rangle \qquad \text{12a.}$$

while the final state $|f\rangle$ is

$$|f\rangle = |0^+\rangle + \sum_n \frac{\langle 0_n^- | V_{PNC} | 0^+ \rangle}{E_{0_+} - E_n} |0_n^-\rangle. \qquad \text{12b.}$$

The sums on $m$ and $n$ run over a complete set of $1^-$ and $0^-$ states, respectively. One could evaluate $P_\gamma = 2\,\text{Re}\,(\langle f|E1|i\rangle / \langle f|M1|i\rangle)$ by calculating $\langle 0_n^- | E1 | 1^+ \rangle$, $\langle 0^+ | E1 | 1_m^- \rangle$, $\langle 0^+ | V_{PNC} | 0_n^- \rangle$, and $\langle 1_m^- | V_{PNC} | 1^+ \rangle$ for a complete set of $1_m^-$ and $0_n^-$ levels. However, it is usually easier to generate mixed-parity states $|i\rangle$ and $|f\rangle$ directly. In a shell model calculation this can be accomplished by adding $\alpha V_{PNC}$ to the shell model Hamiltonian. The scale factor $\alpha$ is a large number chosen to enhance numerical accuracy, though not so large that the response of the wave function to $\alpha V_{PNC}$ becomes nonlinear. Of course, the resulting $P_\gamma$ must be multiplied by $1/\alpha$ to recover the unscaled result. A similar approach has been quite successful in analyzing the parity mixing in heavy atoms induced by the neutral-current e-N interaction (45).

However, the predictive power of nuclear models is not sufficient to inspire great confidence in calculations of this complexity. Complete sets of states are required to saturate the sums over the $1_m^-$ and $0_n^-$ levels, so very large bases are required. These calculations must reproduce highly collective features of the nuclear response, such as the E1 giant resonance. Furthermore, because of the different spin dependence of the operators (the E1 operator has $\Delta S = 0$, while $V_{PNC} \approx \boldsymbol{\sigma} \cdot \mathbf{p}$ has $\Delta S = 1$), the sums over intermediate negative parity states are likely to be highly destructive.

For these reasons modern PNC experiments deal largely with special transitions involving parity doublets, closely spaced pairs of states having identical spins but opposite parities. Some examples are shown in Figure 9. In such cases the sums over $m$ and $n$ in Equations 12 are dominated by a single state, provided that the splitting between the doublet levels is sufficiently small. This greatly simplifies the task of the nuclear theoretician: he must provide accurate wave functions for the doublet states, and possibly for a third state to which they may decay, rather than tackle the enormously difficult "statistical" calculation described above. Typically these doublet states are well-studied levels near the ground state for which much experimental information is available. This provides the theorist with many checks on his structure calculations. Furthermore, the experimentalist benefits from the enhancement of the PNC observable due to the small energy denominator governing the mixing.

An important example of a parity doublet is the $0^+$; $I = 1$ and $0^-$; $I = 0$ levels at $E_{ex} \approx 1$ MeV in $^{18}$F. The splitting of the doublet is only 39 keV, while the energy denominator of the next nearest opposite parity $J = 0$ state is 3672 keV. We therefore tentatively assume that the parity impurities in the 1-MeV levels are predominantly due to an admixture of the other member of the doublet—i.e. that the two-state mixing approximation applies

$$|1081\rangle = |-\rangle + \varepsilon|+\rangle$$

$$\varepsilon = \frac{\langle +|V_{PNC}|-\rangle}{39 \text{ keV}} \qquad 13.$$

$$|1042\rangle = |+\rangle - \varepsilon|-\rangle.$$

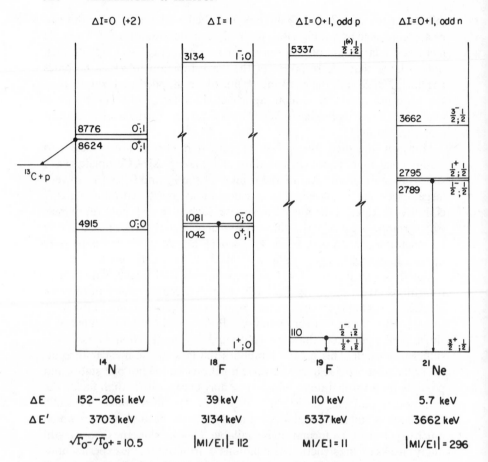

*Figure 9* Parity-mixed doublets in light nuclei. The transitions displaying the amplified PNC effect are indicated. The quantities $\Delta E$ and $\Delta E'$ are the smallest and next smallest energy denominators governing the parity mixing. The quantities shown in the bottom row are "amplification factors."

The $0^-$ and $0^+$ levels decay by dipole transitions to the $1^+$ ground state of $^{18}F$. Because the $\Delta I = 0$ E1 decay of the $0^-$ level is isospin forbidden, the $0^-$ level has a comparatively long lifetime, $\tau_- = 27.5 \pm 1.9$ ps. In contrast, the $\Delta I = 1$ M1 decay of the $0^+$ level is isospin favored and is observed to have a very short lifetime, $\tau_+ = 2.5 \pm 0.3$ fs, rendering it one of the strongest known M1 transitions ($10.3 \pm 1.5$ W.u.). (This further justifies the two-level mixing approximation: M1 transitions from other admixed $0^+$ levels are unlikely to compete favorably with this one.) Because of this large difference in lifetimes, a small admixture of the $0^+$ state into the $0^-$ level produces a sizeable circular polarization:

$$P_\gamma(1081) \approx 2\,\mathrm{Re}\left(\frac{\varepsilon\langle \mathrm{gs}|T_1^{\mathrm{mag};\gamma}|+\rangle}{\langle \mathrm{gs}|T_1^{\mathrm{el};\gamma}|-\rangle}\right)$$

$$= \frac{2}{39\,\mathrm{keV}}\,\mathrm{Re}\left(\frac{\langle +|V_{\mathrm{PNC}}|-\rangle\langle \mathrm{gs}|T_1^{\mathrm{mag};\gamma}|+\rangle}{\langle \mathrm{gs}|T_1^{\mathrm{el};\gamma}|-\rangle}\right). \qquad 14.$$

(Our $\gamma$-decay multipole operators are defined in Ref. 46.) The magnitude of the dipole matrix elements can be determined from the known lifetimes of the 1042- and 1081-keV states because

$$\frac{1}{\tau} \propto |\langle f|T_1^\gamma|i\rangle|^2 E_\gamma^3.$$

Thus we find (47)

$$\left|\frac{\langle \mathrm{gs}|T_1^{\mathrm{mag};\gamma}|+\rangle}{\langle \mathrm{gs}|T_1^{\mathrm{el};\gamma}|-\rangle}\right| = \left(\frac{\tau_-}{\tau_+}\right)^{1/2}\left(\frac{1081\,\mathrm{keV}}{1042\,\mathrm{keV}}\right)^{3/2} \approx 111 \pm 8.$$

The factor $\langle \mathrm{gs}|T_1^{\mathrm{mag};\gamma}|+\rangle/\langle \mathrm{gs}|T_1^{\mathrm{el};\gamma}|-\rangle$ can be regarded as an "amplifier" of the PNC effect. This amplification factor is determined up to a sign by measured lifetimes. The sign (once a convention for the phases of wave functions has been adopted) cannot be determined by experiment because the E1 and M1 matrix elements interfere only in PNC processes. Instead, it must be taken from a nuclear model calculation. (In the example chosen here, the isospin-forbidden E1 transition is so suppressed that it is probably impossible to calculate its sign reliably.)

To summarize, the power and utility of "two-level" PNC transitions follow from three considerations:

1. A measurement of a single PNC observable (such as $P_\gamma$) in conjunction with known lifetimes, energy splittings, etc determines a well-defined matrix element of $V_{\mathrm{PNC}}$ that can be compared to the predictions of theory.

2. Because only two levels are involved in the mixing, the transition "filters out" specific components of the PNC N-N interaction. In our example, where the doublet levels have $I = 0$ and $I = 1$, a measurement of $P_\gamma$ isolates the $\Delta I = 1$ component of the PNC N-N interaction and therefore probes the weak $\pi^\pm$ exchange amplitude.

3. If the members of the parity doublet have very different decay (or formation) amplitudes, the PNC observable can be enhanced significantly, making it practical to measure very small PNC matrix elements. In our example, the large amplification factor and the small energy splitting combine to produce an expected $P_\gamma$ of $\sim 1.5 \times 10^{-3}$ (47). This can be compared to the expected asymmetry $A_\gamma \approx -5 \times 10^{-8}$ for the $\vec{\mathrm{n}} + \mathrm{p} \to \mathrm{d}\gamma$ reaction. As $P_\gamma(^{18}\mathrm{F})$ and $A_\gamma(\mathrm{np} \to \mathrm{d}\gamma)$ are both sensitive primarily to the

weak $\pi^{\pm}$ amplitude, this dramatically illustrates the enhanced PNC observables one can achieve by selecting the appropriate nuclear transition.

### 5.2.2 PARITY-MIXED DOUBLETS IN $^{14}$N, $^{18}$F, $^{19}$F AND $^{21}$Ne

Salient properties of the parity-mixed doublets in $^{14}$N, $^{18}$F, $^{19}$F and $^{21}$Ne are shown in Figure 9. The doublets in $^{18}$F, $^{19}$F and $^{21}$Ne are bound states whose parity mixing can be determined by measuring $A_\gamma$ or $P_\gamma$. The doublet in $^{14}$N is unbound to proton emission so that its parity mixing is most easily probed by measuring a pseudoscalar associated with this decay. These four doublets are important because (a) the two-level approximation is valid in each case; (b) each has a substantial "amplification" factor; (c) each of the four admixtures has a different isospin character; and (d) they involve low-lying levels in simple nuclei whose structures are relatively well understood.

The transitions of interest are shown in Figure 9. The validity of the two-state approximation for each transition is plausible because the amplification factors range from 11 to 296, and because $\Delta E$ (the doublet splitting) is 14 to 642 times smaller than $\Delta E'$ (the next smallest energy denominator).

We stress the very different isospin properties of parity mixing in the four nuclei. The mixing of $I = 0$ and $I = 1$ levels in $^{18}$F is sensitive only to the $\Delta I = 1$ PNC N-N amplitudes. The mixing of the $I = 1$ levels in $^{14}$N ($I_3 = 0$) is sensitive to both the $\Delta I = 0$ and $\Delta I = 2$ PNC amplitudes but, as we discuss in Section 6, the ratio of the $\Delta I = 2$ to $\Delta I = 0$ matrix elements is expected to be small. The mixed $I = 1/2$ doublets in $^{19}$F and $^{21}$Ne are sensitive to both the $\Delta I = 0$ and $\Delta I = 1$ interactions. However, as $^{19}$F is an odd-proton nucleus and $^{21}$Ne and odd-neutron nucleus, the relative signs of the $\Delta I = 0$ and $\Delta I = 1$ contributions are different in the two nuclei. This will become apparent when we discuss an effective single-particle PNC potential in Section 6.1. Because of their complementary isospin characters and large amplification factors, the four parity doublets in the $A = 14$ through $A = 21$ nuclei provide a remarkable opportunity to measure the $\Delta I = 0$ and $\Delta I = 1$ components of the PNC N-N interaction.

The large amplification factors and small energy denominators in these nuclei lead us to expect PNC effects roughly $10^2$–$10^4$ times larger than those in the N-N and few-body systems, i.e. between $10^{-3}$ and $10^{-5}$. Therefore one can do experiments using pulse-counting techniques. In fact the pulse-counting technique is preferred in these cases because it provides the detector energy resolution needed to select only the transition of interest. In the remainder of this subsection we review the experimental techniques that have been developed to measure PNC observables in parity-mixed doublets. Although we compare measurements to theoretical predictions, detailed discussions of the nuclear theory are delayed until Section 6.

5.2.2.1  *The $0^-$, $I = 0 \leftrightarrow 0^+$, $I = 1$ doublet in $^{18}F$*    Because the doublet levels have $J = 0$, all asymmetries vanish and one must measure $P_\gamma$. The $^{18}F$ doublet is particularly important because the circular polarization of the 1091-keV $0^-$-ground state $\gamma$ ray is a very sensitive probe of $F_\pi$. Inserting the numerical values (47) into Equation 14, we obtain $|\langle 0^+|V_{PNC}|0^-\rangle| = (177 \pm 13)\text{eV}|P_\gamma(1081)|$.

Four different groups (48–51) have measured $P_\gamma(1081)$. The main difficulty is developing targets that can withstand the intense bombardments that are necessary to achieve statistical precision. All recent works produce the $^{18}F^*$ nuclei using the $(^3He, p)$ reaction on a flowing $H_2O$ target, as pioneered by Barnes et al (48). The ongoing experiments at Queen's University (51) and at Florence (P. G. Bizetti, private communication) use four transmission-mode circular polarimeters containing large-volume germanium detectors. The polarimeters are located at $\theta = 90°$ and spaced equally around the azimuth. The layout of the Queen's experiment is shown in Figure 10. The magnetizations in the polarimeters are directed as shown in Figure 11 in order to eliminate any significant dipole component of the fringing field along the beam axis. The magnetic fields are reversed periodically (approximately every 10 s in the Queen's experiment) between states denoted by $+$ and $-$, and digitized detector signals are accumulated into two sets of spectra according to the polarimeter state. The circular polarization can be obtained from the expression $P_\gamma = A/\eta$, where $\eta$ is the polarimeter analyzing power and $A$ is an asymmetry, $A = (R-1)/(R+1)$. The ratio $R$ is defined by $R = (N_L^+ N_R^+ N_U^- N_D^-/N_L^- N_R^- N_U^+ N_D^+)^{1/4}$ where, for example, $N_L^+$ is the integrated number of gamma rays detected in the left counter when the magnetization was in the $+$ state. A typical $\gamma$-ray spectrum from the Queen's experiment is shown in Figure 12.

The low polarimeter analyzing power at $E_\gamma \approx 1$ MeV ($\sim 1\%$ according to Figure 5) combined with the relatively long pulse widths required to maintain excellent energy resolution in the Ge detectors lead to fairly low rates for data collection and very long counting times. Fifteen hundred hours of data were collected in the original experiment (48). The total running time devoted to this measurement by all four groups is approximately 8000 hours and close to 9 mg of $^3He$ have been delivered to the targets! The published results for $P_\gamma(1081)$ are shown in Table 5 along with preliminary results from the Queen's University and Florence groups. The "world average" of $P_\gamma(1081)$ is significantly smaller than the "best value" prediction.

5.2.2.2  *The $1/2^-$, $I = 1/2 \leftrightarrow 1/2^+$, $I = 1/2$ doublet in $^{19}F$*    Because this doublet has $J = 1/2$, one can measure $A_\gamma$ in the decay asymmetry of

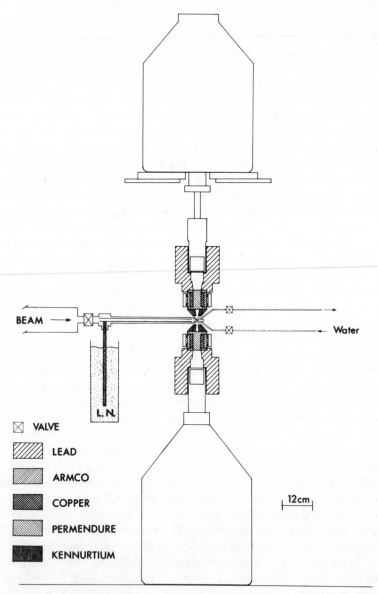

*Figure 10* Side view of the Queen's University $^{18}$F PNC experiment. Only two of the four Ge(Li) detectors are shown.

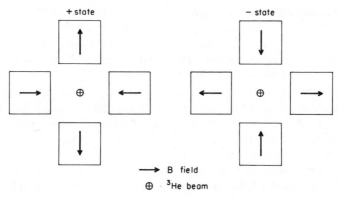

$\longrightarrow$ B field

$\oplus$  $^{3}$He beam

*Figure 11*  Magnetization scheme used in the four-counter polarimeters of Refs. 51 and 50.

polarized $^{19}$F$(1/2^-)$ nuclei. ($A_y$ is defined by $d\omega/d\Omega \approx 1 + A_y \mathbf{P}_F \cdot \hat{\mathbf{k}}_y$, where $\mathbf{P}_F$ and $\hat{\mathbf{k}}_y$ are the $^{19}$F$(1/2^-)$ polarization and $\gamma$-ray propagation vectors respectively.) In the two-state mixing approximation

$$A_y = \frac{2}{\Delta E} \, \mathrm{Re} \left( \frac{\langle 1/2^+ || T_1^{\mathrm{mag};\gamma} || 1/2^+ \rangle - \langle 1/2^- || T_1^{\mathrm{mag};\gamma} || 1/2^- \rangle}{\langle 1/2^+ || T_1^{\mathrm{el};\gamma} || 1/2^- \rangle} \right.$$

$$\left. \times \langle 1/2^+ | V_{\mathrm{PNC}} | 1/2^- \rangle \right). \qquad\qquad 15.$$

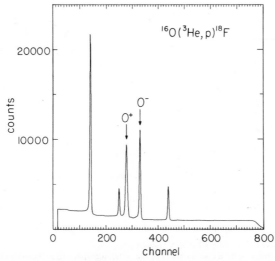

*Figure 12*  Spectrum of $\gamma$ rays transmitted by the circular polarimeter in the Queen's experiment. The need for good $\gamma$-ray energy resolution is apparent.

From the measured values of $\Delta E$ (110 keV), the lifetime of the $1/2^-$ level, and the magnetic moment of the $1/2^+$ state ($2.6289\mu_N$), we obtain $|\langle 1/2^+|V_{PNC}|1/2^-\rangle| = (5.2 \pm 0.4) \times 10^3$ eV $|A_\gamma|$. The magnetic moment of the $1/2^-$ state is not known, though various theoretical calculations give values between $-0.17$ and $-0.24\mu_N$. We have used $\mu(1/2^-) = (-0.21 \pm 0.21)\mu_N$; although this error is probably unrealistically large, it has little impact on the uncertainty of $|\langle 1/2 + |V_{PNC}|1/2^-\rangle|$ because $|\mu(1/2^+)/\mu(1/2^-)| \gg 1$. The sign of $A_\gamma$ depends on that of the E1 matrix element. Shell model calculations discussed in Section 6.3 predict a negative $A_\gamma$ for DDH couplings, in agreement with experiment, and magnitudes for the E1 matrix element somewhat larger than experiment.

Adelberger et al (52) discovered that the $^{22}$Ne$(\vec{p}, \alpha)$ reaction induced by transversely polarized protons is very efficient in producing polarized $^{19}$F* nuclei. (The $^{19}$F polarization was found by measuring the PC circular

**Table 5** PNC observables in parity doublets: experiment and theory

| Observable | Experiment | | | Theory[a] | | |
| | Value ($\times 10^{-5}$) | $-i\langle V_{PNC}\rangle$ | Ref. | Value ($\times 10^{-5}$) | $-i\langle V_{PNC}\rangle$ | Ref. |
|---|---|---|---|---|---|---|
| $^{14}$N | | | | | | |
| $A_L (0^+)$ | | | | $-3.1$ | 1.04 eV | |
| $^{18}$F | | | | | | |
| $P_\gamma$ (1081 keV) | $-70 \pm 200$ | | 48 | | | |
| | $-40 \pm 300$ | | 49 | | | |
| | $-100 \pm 180$ | | 50 | | | |
| | $15 \pm 55$ | | f | | | |
| | $20 \pm 60$ | | g | | | |
| | $8 \pm 39$ | $\leq 0.09$ eV | "best" | $208 \pm 49$[b] | 0.37 eV[d] | 68 |
| $^{19}$F | | | | | | |
| $A_\gamma$ (110 keV) | $-8.5 \pm 2.6$ | | 47 | | | |
| | $-6.8 \pm 1.8$ | | 28 | | | |
| | $-7.4 \pm 1.9$[c] | $0.38 \pm 0.10$ eV | "best" | $-8.9 \pm 1.6$ | 0.46 eV[d] | 47 |
| $^{21}$Ne | | | | | | |
| $P_\gamma$ (2789 keV) | $80 \pm 140$ | $\leq 0.029$ eV | 53 | $46$[b] | $-0.006$ eV[e] | 72 |

[a] Based on the "best value" DDH potential.
[b] Theory fixes only the absolute value of this quantity (see text).
[c] Includes a contribution from uncertainty in $P_F$ that is not incorporated in the individual values.
[d] Using the $\beta$-decay results of Table 6.
[e] Using the $0 + 1\hbar\omega - $ Kuo results of Table 6, scaled by an effective change of one third.
[f] H. B. Mak, private communication.
[g] M. Bini, private communication.

polarization correlation $\sigma_\gamma \cdot \hat{\mathbf{k}}_\gamma \; \mathbf{P}_F \cdot \hat{\mathbf{k}}_\gamma$ of the 110-keV $\gamma$ ray.) The measured polarization transfer coefficient of $-0.70 \pm 0.11$ (28) is close to the value $-1$ expected if the outgoing $\alpha$'s are purely s-wave and no depolarization occurs during the $\sim 1$-ns lifetime of the $1/2^-$ state. The $^{22}$Ne + p reaction also offers an automatic check on instrumental asymmetries. The $^{22}$Ne(p, n) reaction produces a 74-keV $\gamma$ ray from the decay of a $0^+$ state in $^{22}$Na. This 74-keV $\gamma$ ray must be isotropic and therefore provides a continuous monitor of instrumental asymmetries. Two different groups (28, 47) have measured $A_\gamma$ (see Table 5). A schematic diagram of the Seattle experiment is shown in Figure 13. The 74-keV $\gamma$-ray monitor was useful in establishing the reliability of the experimental results. For example, in the Seattle experiment (52) a 2-$\mu$m displacement of the beam accompanying the spin reversal would have produced a spurious asymmetry equal to the measured effect! The experiments of Refs. 47 and 28 agree very well and can be combined to yield $A_\gamma = (-7.4 \pm 1.9) \times 10^{-5}$. This corresponds to $|\langle 1/2^+ | V_{PNC} | 1/2^- \rangle| = 0.38 \pm 0.10$ eV.

5.2.2.3 *The $1/2^+$, $I = 1/2 \leftrightarrow 1/2^-$, $I = 1/2$ doublet in $^{21}$Ne*    The extreme retardation of the $1/2^- \rightarrow$ ground-state transition [$\tau_\gamma = 696 \pm 51$ ps, yielding $B(E1) \leqslant 1.3 \times 10^{-6}$ W.u.] and the very small energy denominator (5.74 $\pm 0.15$ keV) combine to make the 2789-keV ground-state $\gamma$ ray an extraordinarily sensitive probe of $\langle 1/2^+ | V_{PNC} | 1/2^- \rangle$. Two PNC observables are possible, $A_\gamma$ and $P_\gamma$. Lacking an efficient way to produce polarized $^{21}$Ne(2789), the experimenters (53) chose to measure $P_\gamma$ by populating the $1/2^-$ level in a $^{21}$Ne(p, p') resonance at $E_p \approx 4.08$ MeV. In

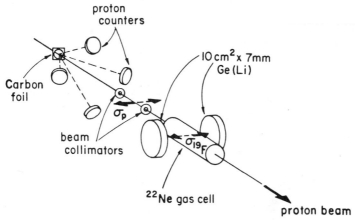

*Figure 13*   Schematic view of the Seattle $^{19}$F PNC experiment. The four proton counters view a thin carbon foil onto which a layer of Au has been evaporated. By comparing the scattering yields from C and Au, an on-line computer continuously monitors the transverse polarization. It also controls a spin precessor on the ion source to keep the spin "locked onto" the axis of the $\gamma$-ray counters.

the two-state mixing approximation

$$p_\gamma = -\frac{2}{\Delta E}\frac{1+\delta_-^*\delta_+}{1+|\delta_-|^2}\,\mathrm{Re}\left(\frac{\langle 3/2^+||T_1^{\mathrm{mag};\gamma}||1/2^+\rangle}{\langle 3/2^+||T_1^{\mathrm{el};\gamma}||1/2^-\rangle}\langle 1/2^+|V_{\mathrm{PNC}}|1/2^-\rangle\right),$$

16.

where $\delta_-$ is the M2/E1 mixing ratio for the $1/2^- \to$ ground-state $(3/2^+)$ transition and $\delta_+$ the E2/M1 mixing ratio for the $1/2^+ \to$ ground-state transition. A limit of $|\delta_+| < 0.39$ can be obtained from the measured lifetime of the $1/2^+$ level and from the extreme assumption that $B(\mathrm{E2}) = 30$ W.u., while a shell model calculation predicts $\delta_+ = 0.026i$. A recent measurement of the pair emission probability for the $1/2^- \to 3/2^+$ transition (54) determined $|\delta_-| < 0.6$. Adopting $|\delta_-| < 0.6$ and the theoretical prejudice that $\delta_+ \approx 0$ and inserting measured values for the lifetimes and energy splitting, we obtain $|\langle 1/2^+|V_{\mathrm{PNC}}|1/2^-\rangle| = (9.5^{+3.4}_{-0.6})$ eV $|P_\gamma|$. The absolute value occurs because theory cannot reliably predict the sign of the extremely hindered E1 matrix element. The measured value of $P_\gamma$ (see Table 5) corresponds to a matrix element $|\langle 1/2^+|V_{\mathrm{PNC}}|1/2^-\rangle| \leq 0.029$ eV, a value significantly smaller than that found in $^{19}$F. The shell model calculations described below predict that the $\Delta I = 0$ and $\Delta I = 1$ components of the matrix elements have similar magnitudes in $^{19}$F and $^{21}$Ne. We conclude that the total matrix elements differ by such a large factor because the $\Delta I = 0$ and $\Delta I = 1$ components interfere destructively in $^{21}$Ne and constructively in $^{19}$F. This is in accord with qualitative arguments we present in Section 6.1.

5.2.2.4   *The $0^+$, $I = 1 \leftrightarrow 0^-$, $I = 1$ doublet in $^{14}N$*   The obvious observable in this case is the circular polarization of the protons emitted in the decay of the longer-lived member of the doublet. This is equivalent to measuring $A_L$ for $\vec{p} + ^{13}$C scattering in the region of the $0^+$ resonance. In this case the amplification factor is roughly $[\Gamma_p(0^-)/\Gamma_p(0^+)]^{1/2} \approx 11$ (see Ref. 55 for numerical values). Even though this PNC decay is a hadronic

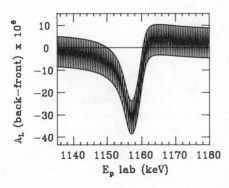

*Figure 14* Predicted signal (based on the DDH parameters) for the Seattle $^{14}$N PNC experiment. Scattered protons are detected in front and back counters. The signal, the difference in $A_L$ between the back and front counters, lies in the center of the shaded band. The shaded area denotes a $\pm 1\sigma$ error band, where $\sigma$ is the statistical error attained after 1 $\mu$A·day of integrated beam. A 25-$\mu$g cm$^{-2}$ $^{13}$C target is assumed.

process, it can be analyzed with almost as much confidence as the $\gamma$-ray decays discussed above because of the $0 \to 1/2 + 1/2$ spin structure and because elastic scattering is the only open hadronic channel (55). The predicted $A_L(0^+)$ signal is shown in Figure 14. From Table 6 it can be seen that $A_L(0^+)$ is sensitive almost exclusively to the $\Delta I = 0$ PNC N-N interaction. The magnitude and sign of $\langle 0^+ | V_{PNC} | 0^- \rangle$ can be obtained from $A_L(0^+)$ because theory makes a definite prediction for the signs of the proton-decay amplitudes of the $0^+$ and $0^-$ resonances. An experiment to measure $A_L$ is being prepared in Seattle.

## 5.3 PNC $\alpha$ Decay in Light Nuclei

Searches for PNC $\alpha$ decay have been reported for a number of transitions in $^6$Li, $^{16}$O, $^{18}$F, and $^{20}$Ne. We discuss two of the three cases (see Figure 15) where definite results have been observed: $^{16}$O$(2^-, 8.87 \text{ MeV}) \to {}^{12}$C$(0^+)$ $+ \alpha$ (56) and $^{20}$Ne$(1^+, 11.26 \text{ MeV}) \to {}^{16}$O$(0^+) + \alpha$ (57). A summary of other results is given by Desplanques (58). The PNC $\alpha$ decay of $^{16}$O$(2^-)$ is sensitive only to the $\Delta I = 0$ PNC interaction because both the initial and final states have $I = 0$. It was observed, after selectively populating the $2^-$ level in the $\beta$-decay of $^{16}$N, by counting the decay alphas in thin solid-state detectors. This impressive experiment determined the PNC $\alpha$ width $\Gamma_{PNC}$ $= (1.03 \pm 0.28) \times 10^{-10}$ eV! Unfortunately, nuclear structure theory is not adequate to take full advantage of this precision. The PNC $\alpha$ decay of the $2^-$

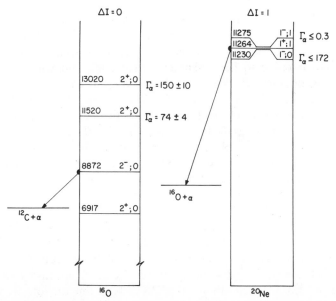

*Figure 15* PNC $\alpha$ decays in light nuclei. Nearby natural parity levels whose admixtures contribute to the PNC $\alpha$ widths are also shown.

level does not arise from the admixture of a single nearby $2^+$ level; rather one must include the effects of several interfering $2^+$, $I = 0$ levels (see Figure 15). As one of these states lies below the threshold for $\alpha$ decay, theory must provide both the magnitude and sign of the amplitude. The structure of low-lying states in $^{16}O$ is complicated by the coexistence of spherical and deformed states (59) so that an adequate shell model description of $^{16}O$ must include excitations up to $4\hbar\omega$. This has not yet been done reliably. Nevertheless, the measured $\Gamma_\alpha^{PNC}$ is in reasonable accord (within a factor of 2) with existing predictions (see 60).

The PNC $\alpha$ decay of the $1^+$, $I = 1$ level of $^{20}Ne$ at $E_x = 11.264$ MeV to the $I = 0$ ground state of $^{16}O$ is sensitive to $\Delta I = 1$ PNC interaction. This $\alpha$ width should be a very sensitive probe of the $\Delta I = 1$ PNC N-N interaction because an $I = 0$ level with $\Gamma_\alpha = 172$ keV lies only 34 keV away from the $1^+$ state. The PNC $\alpha$ decay was observed (57) by detecting the very weak $^{16}O(\alpha, \gamma_0)$ resonance corresponding to the 11.264-MeV $1^+$ state of $^{20}Ne$. The $1^+$ state could be populated with great selectivity because of the excellent excitation energy resolution ($\sim 1.4$ keV FWHM) achieved with a differentially pumped gas target.

Fifield et al (57) obtain a PNC $\alpha$ width of $(42 \pm 20) \times 10^{-6}$ eV. In the two-level approximation one has

$$\Gamma_\alpha^{PNC} = \left| \frac{\langle 1^+ | V_{PNC} | 1^- \rangle^2}{E_- - E_+ + (i/2)\Gamma_-} \right| \Gamma_\alpha[1^-].$$

Inserting the measured quantities, one finds $\langle 1^+ | V_{PNC} | 1^- \rangle \approx 1.5$ eV. Unfortunately, a third level complicates this picture. This level has $J^\pi = 1^-$ and $I = 1$, and thus undergoes $\alpha$ decay only via its isospin impurity. If it is allowed to interfere to the maximum extent consistent with its known width, one obtains $0.8$ eV $\leq |\langle 1^+ | V_{PNC} | 1^- \rangle| \leq 2.6$ eV where the range includes the possibility that the interference is constructive or destructive. How does this matrix element compare to theory? Unfortunately, the admixed $1^-$, $I = 0$ level is the fourth known $1^-$, $I = 0$ state, and a fifth such level lies only 0.17 MeV higher. As a result we do not have much confidence in the shell model identification of this $1^-$ state and do not quote a predicted matrix element.

## 5.4   Parity Nonconservation in Heavier Nuclei

The first observation of a parity impurity in a nuclear state was in the heavy nucleus $^{181}Ta$. Today the largest number of measured effects, the results with the smallest errors (expressed as a fraction of the effect), and the largest PNC effects are all found in heavy nuclei. However, in most cases it is difficult to learn much about the PNC N-N interaction from the experimental data because the relevant nuclear structure is not sufficiently well understood. Therefore, we discuss only a few examples of PNC in heavy nuclei—those cases that can be analyzed with some confidence or that have yielded some notable results. More comprehensive reviews are given by Boehm (61) and Desplanques (58, 62).

Formerly, most data in heavy nuclei came from circular polarization or asymmetry measurements of transitions fed in $\beta$ decay (58, 61). Most recent experiments use polarized low-energy neutron beams to measure longitudinal analyzing powers of PNC spin rotations (58). Only three of the many PNC transitions investigated in heavy nuclei can be used with some confidence to probe the PNC N-N interaction: $P_\gamma$ measurements for $^{41}$K $[7/2^- \rightarrow 3/2^+]$, $^{175}$Lu $(9/2^- \rightarrow 7/2^+)$, and $^{181}$Ta $[5/2^+ \rightarrow 7/2^+]$ (58). In these cases there are no substantial amplification factors so that the $P_\gamma$'s are quite small. Furthermore, almost all of the energy deposited in the $\gamma$-ray detector by the radioactive source comes from the transition of interest. Therefore, the measurements were made using the integral counting technique.

For these heavy nuclei, calculations can be done using a single-particle model for the nuclear states (63) and an effective single-particle PNC potential similar to that given in Equation 20 below. All three transitions are in odd-$Z$ nuclei and thus probe combinations of isoscalar and isovector PNC potentials similar to those tested in $^{19}$F and p + $^4$He. The results in p + $^4$He, $^{19}$F, $^{41}$K, $^{175}$Lu, and $^{181}$Ta can be explained consistently within the simple one-body model. However, the demands placed on nuclear models in order to interpret PNC measurements in the heavy nuclei are more severe than in cases of the light "two-level" systems. For example, in $^{181}$Ta the observable depends on the mixing of more than two levels. Furthermore, because the "wrong parity" multipole is E1 rather than M1, collective effects of the giant dipole resonance are presumably important in determining $P_\gamma$. Because of such complications we emphasize results in the lighter nuclei.

We conclude this section by mentioning the remarkably large PNC effects observed in certain heavy nuclei. The occurrence of a very large pseudoscalar effect is always associated with a retardation of the PC amplitude. One example is the highly retarded $\Delta K = 8$ transitions from the decay of the $8^-$ isomer $^{180}$Hf. The PNC asymmetry (64) of the 501-keV $\gamma$ ray is $A_\gamma = (-1.66 \pm 0.18) \times 10^{-2}$! An even more spectacular result was seen in low-energy neutron scattering from $^{139}$La, where a p-wave resonance occurs at $E_n = 0.73$ eV. On resonance, the longitudinal analyzing power of the total neutron cross section is $A_L = (7.3 \pm 0.5) \times 10^{-2}$ (65). This enormous effect presumably arises because the p-wave resonance is embedded in a dense "sea" of s-wave resonances. Thus there are many possibilities for parity admixing with very small energy denominators. In $^{139}$La + n the admixture happens to be unusually large, probably because the p-wave resonance is exceedingly narrow ($\Gamma_n \approx 7 \times 10^{-8}$ eV) and a strong s-wave state exists just 37 eV below threshold (see 80). S-wave admixtures into a p-wave resonance generally produce large effects because of a trivial kinematic enhancement. Near threshold, neutron widths are proportional to $(kr)^{2l+1}$. The PNC observable is proportional to $[\Gamma_n(s)/\Gamma_n(p)]^{1/2} \propto 1/kr$. For a 0.73-eV neutron incident on $^{139}$La, this amplification factor is approximately 760. Alfimenkov et al. (65) estimate that a PNC matrix element between the p-wave resonance and the subthreshold s-wave state of only $\sim 1 \times 10^{-3}$ eV could account for the phenomenal enhancement of $A_L$. Unfortunately it is highly unlikely that these remarkably large effects, which involve very complicated compound nuclear resonances whose wave functions are poorly known, can be exploited to probe the weak N-N interaction. It is easy to draw erroneous conclusions that a large effect is due to an exotic new weak interaction rather than to a chance suppression of the PC amplitudes. Such an unlikely possibility was considered by Stodolsky (66).

# 6.　EVALUATION OF PNC NUCLEAR MATRIX ELEMENTS

The crucial step in extracting PNC weak meson-nucleon couplings from parity-mixing measurements in nuclei is the evaluation of the nuclear matrix elements of the PNC potential. The problem is analogous to that encountered in computing PNC couplings from the elementary weak interactions between quarks. The accuracy of both calculations is limited by an incomplete understanding of the many-body strong interaction eigenstates. In this section we argue that the nuclear many-body calculations for certain "two-level" nuclei are more reliable than their quark model counterparts. Therefore studies of PNC effects in these nuclei can teach us about the particle physics governing the weak meson-nucleon couplings.

## 6.1　Gross Features of Nuclear PNC: The One-Body Approximation

Considerable insight into the nuclear physics of parity violation may be gained from a simple exercise, the derivation of effective one-body PNC potentials. This will clarify why different nuclei test different aspects of the PNC potential and will also demonstrate the intimate relationship between parity mixing and first-forbidden $\beta$ decay. This relationship can be exploited to provide essentially model-independent determinations of PNC matrix elements in $^{18}$F and $^{19}$F.

Effective one-body PNC potentials were first introduced by Michel (22) and have proved useful in many subsequent PNC investigations. We consider a simple case in which the relevant states of a nucleus are single-nucleon levels outside of a closed core. The nonzero matrix elements of the PNC two-body potential are then semidiagonal in one quantum number (corresponding to a core nucleon). By summing over the core nucleons an exactly equivalent one-body PNC potential $U_{PNC}$ can be defined:

$$\langle\alpha|U_{PNC}|\beta\rangle = \sum_{\delta < F} (\langle\alpha\delta|V_{PNC}|\beta\delta\rangle - \langle\alpha\delta|V_{PNC}|\delta\beta\rangle), \qquad 17.$$

where $\alpha$ and $\beta$ are the single-particle states available to the valence nucleon. We derive $U_{PNC}$ for a Fermi gas model, so that the core is spin symmetric, and we approximate $m_\omega \approx m_\rho$:

$$
\begin{aligned}
U_{PNC}(i) = \boldsymbol{\sigma}(i)\cdot&\frac{\mathbf{p}(i)}{M}\frac{2}{m_\rho^2}\\
&\times\Bigg(\rho_Z\left(\frac{1+\tau_3(i)}{2}\right)\{[(1+\mu_v)W_Z^\rho+1](F_0+F_1+F_2/\sqrt{6})\\
&+[(1+\mu_s)W_Z^\rho+1](G_0+G_1)\}+\rho_N\left(\frac{1-\tau_3(i)}{2}\right)\\
&\times\{[(1+\mu_v)W_N^\rho+1](F_0-F_1+F_2/\sqrt{6})\\
&+[(1+\mu_s)W_N^\rho+1](G_0-G_1)\}+\rho_N\left(\frac{1+\tau_3(i)}{2}\right)\\
&\times\Big\{2\left(\frac{m_\rho}{m_\pi}\right)^2 W_N^\pi F_\pi + 2W_N^\rho H_1 + G_1 - F_1\\
&+[(1+\mu_v)W_N^\rho+1](F_0/2-F_2/\sqrt{6})+[(1+\mu_s)W_N^\rho+1]G_0/2\\
&+3[(1+\mu_v)W_N^\rho-1]F_0/2-[(1+\mu_s)W_N^\rho-1]G_0/2\Big\}\\
&+\rho_Z\left(\frac{1-\tau_3(i)}{2}\right)\Big\{-2\left(\frac{m_\rho}{m_\pi}\right)^2 W_Z^\pi F_\pi - 2W_Z^\rho H_1 - G_1 + F_1\\
&+[(1+\mu_v)W_Z^\rho+1](F_0/2-F_2/\sqrt{6})\\
&+[(1+\mu_s)W_Z^\rho+1]G_0/2+3[(1+\mu_v)W_Z^\rho-1]F_0/2\\
&-[(1+\mu_s)W_Z^\rho-1]G_0/2\Big\}\Bigg). \qquad 18.
\end{aligned}
$$

The proton and neutron densities are determined by the Fermi momenta, e.g. $\rho_Z = (k_F^Z)^3/3\pi^2$. The functions $W$ are given by

$$W_Z^\pi \equiv W(k_F^Z, m_\pi, p) = \left(\frac{3\tilde{m}_\pi^2}{32\tilde{p}^3}\right)\left\{4\tilde{p}(1+\tilde{p}^2+\tilde{m}_\pi^2)\right.$$

$$\left.-\left[(1-\tilde{p}^2)^2+2\tilde{m}_\pi^2(1+\tilde{p}^2)+\tilde{m}_\pi^4\right]\ln\left(\frac{\tilde{m}_\pi^2+(1+\tilde{p})^2}{\tilde{m}_\pi^2+(1-\tilde{p})^2}\right)\right\}$$

$$\rightarrow 1 \text{ as } \tilde{m}_\pi \rightarrow \infty, \qquad\qquad\qquad 19.$$

where $\tilde{m} = m/k_F^Z$ and $\tilde{p} = |\mathbf{p}|/k_F^Z$. The functions $W_N^\pi$, $W_Z^\rho$, and $W_N^\rho$ are defined similarly. Although $W$ is sensitive to the mass of the exchanged particle, it is relatively insensitive to $\tilde{p}$, varying smoothly from 0.20 to 0.14 over the range $0 < \tilde{p} < 1$ for $m_\pi = 140$ MeV and from 0.88 to 0.79 for $m_\rho = 770$ MeV. Thus, to a good approximation, we can regard $U_{PNC}$ as a density-independent potential. As Equation 18 was derived for an independent-particle model, the effects of short-range correlations have been neglected.

Equation 18 has a simple interpretation. The interaction of the valence nucleon with the core is identical in form to the N-N interaction we derived in the S-P plane-wave Born approximation of Section 2.2, apart from the reduction of the exchange terms by $W$. This reduction is due to the decreased range of these terms because of the average momentum transfer appearing in the propagators $(m^2 + \langle p^2\rangle)^{-1}$. If $m$ is large compared to $k_F$, $\langle p^2\rangle$ is unimportant and $W \approx 1$, as indicated in Equation 19. Effectively, $W$ reduces the magnetic moments $(1+\mu)$ and the strengths of $F_\pi$ and $H_1$.

We can rewrite Equation 18 to illustrate how the various weak couplings probe different nuclear densities. Defining the isoscalar and isovector direct and exchange nuclear densities by

$$\rho_S = \rho_N + \rho_Z \qquad \rho_S^{\pi ex} = \rho_N W_N^\pi + \rho_Z W_Z^\pi$$

$$\rho_V = \rho_N - \rho_Z \qquad \rho_V^{\pi ex} = \rho_N W_N^\pi - \rho_Z W_Z^\pi,$$

we find

$$U_{PNC}(i) = \sigma(i)\cdot\frac{\mathbf{p}(i)}{M}\frac{2}{m_\rho^2}\left(F_\pi\left(\frac{m_\rho}{m_\pi}\right)^2[\rho_V^{\pi ex}+\tau_3(i)\rho_S^{\pi ex}]\right.$$

$$+F_0\{\tfrac{3}{2}(1+\mu_v)\rho_S^{\rho ex}+\tau_3(i)[\tfrac{1}{2}(1+\mu_v)\rho_v^{\rho ex}-\rho_v]\}$$

$$+F_1[-\rho_v-\tfrac{1}{2}(1+\mu_v)\rho_v^{\rho ex}+\tfrac{1}{2}\tau_3(i)(1+\mu_v)\rho_S^{\rho ex}]$$

$$-F_2/\sqrt{6}\tau_3(i)[\rho_v+(1+\mu_v)\rho_v^{\rho ex}]$$

$$+G_0[\rho_s+\tfrac{1}{2}(1+\mu_s)\rho_S^{\rho ex}-\tfrac{1}{2}\tau_3(i)(1+\mu_s)\rho_v^{\rho ex}]$$

$$-G_1\{\tau_3(i)[\rho_s+\tfrac{1}{2}(1+\mu_s)\rho_S^{\rho ex}]-\tfrac{1}{2}(1+\mu_s)\rho_v^{\rho ex}\}$$

$$\left.+H_1[\rho_v^{\rho ex}+\tau_3(i)\rho_S^{\rho ex}]\right). \qquad\qquad 20.$$

Even in heavy nuclei we expect the isoscalar density contributions to $U_{\text{PNC}}$ to dominate over the isovector contributions. Therefore we can make the following qualitative observations: (a) According to the DDH estimates, $F_1$ and $H_1$ are small, so that nuclear PNC effects are dominated by couplings $F_\pi$ and $F_0$. (The $\rho$-exchange parameters are enhanced relative to $\omega$ exchange because of the large isovector magnetic moment.) (b) The contributions of $F_\pi$ and $F_0$ have the same sign for odd-proton nuclei and opposite signs for odd-neutron nuclei. (c) Terms proportional to the isotensor coupling $F_2$ are suppressed because $F_2$ multiplies isovector densities only.

Before one can use Equation 20 to estimate the sensitivity of a transition to the weak couplings, a correction must be introduced to account for short-range correlations. The correlation function used in the shell model calculation of Section 6.3 reduces the pion-exchange contribution by a factor of $\sim 1.4$ and the vector meson-exchange contributions by $\sim 3.6$.

## 6.2  Detailed Features of Nuclear PNC: Matrix Elements of the One-Body Axial Charge Operators

According to Equation 20 the detailed nuclear physics of parity mixing is governed by two one-body operators, the isovector and isoscalar axial charge operators $\sigma(i) \cdot \mathbf{p}(i)\tau_3(i)$ and $\sigma(i) \cdot \mathbf{p}(i)$. The isovector operator is familiar from first-forbidden $\beta$ decay: it is the dominant term driving parity-changing $\Delta J = 0$ transitions (46, 67). The isoscalar operator is "new" in the sense that it is not known from other contexts. Particularly in the light nuclei, where $\rho_N \approx \rho_Z$, the matrix elements of these two operators determine the relative importance of the isovector ($F_\pi$, $F_1$, $G_1$, $H_1$) and isoscalar ($F_0$, $G_0$) couplings.

Unfortunately, it is not easy to compute matrix elements of the axial charge operators. Simple models of nuclear structure overlook a number of effects that suppress axial charge matrix elements and thus often grossly overestimate these quantities. To illustrate some of the nuclear structure issues, we discuss the isoscalar and isovector axial charge transitions in $^{19}$F. A similar discussion for $^{18}$F was given by Haxton (68). The $^{19}$F many-body matrix elements of these operators are

$$\left\langle 1/2^- 1/2 \left\| \sum_{i=1}^{A} \sigma(i) \cdot \frac{\mathbf{p}(i)}{M} I_T(i) \right\| 1/2^+ 1/2 \right\rangle = \sum_{\alpha\beta} \psi_{\alpha\beta}^{T} \left\langle \alpha \left\| \sigma \cdot \frac{\mathbf{p}}{M} I_T \right\| \beta \right\rangle, \quad 21.$$

where $I_{T=0}(i) = 1$ and $I_{T=1}(i) = \tau_3(i)$, and the sum over the one-body density matrix elements $\psi_{\alpha\beta}^{T}$ extends over a complete set of single-particle quantum numbers $\alpha$ and $\beta$. The notation $\|$ indicates a matrix element reduced in angular momentum and isospin.

Equation 21 is exact. The nuclear theorist must find a reasonable approximation to the unknown nuclear structure coefficients $\psi_{\alpha\beta}$. We consider a shell model approximation to $\psi_{\alpha\beta}$ in which wave functions for $^{19}$F are generated in the 1s-1p-2s1d-2p1f space. We adopt a harmonic oscillator single-particle basis; improvements in the basis wave function are considered later. Regardless of the complexity of the many-body wave functions, the matrix elements of the axial charge operators have the form

$$-\frac{i}{Mb}\sqrt{2(2T+1)}\,(\sqrt{3}\bar{\psi}_{1p_{1/2}\,1s_{1/2}}-\sqrt{2}\bar{\psi}_{2s_{1/2}\,1p_{1/2}}+\sqrt{10}\bar{\psi}_{1d_{3/2}\,1p_{3/2}}$$

$$+\sqrt{5}\bar{\psi}_{2p_{1/2}\,2s_{1/2}}-2\bar{\psi}_{2p_{3/2}\,1d_{3/2}}+\sqrt{21}\bar{\psi}_{1f_{5/2}\,1d_{5/2}}),\qquad 22.$$

where $b$ is the oscillator parameter and $\bar{\psi}_{\alpha\beta}=\psi_{\alpha\beta}-\psi_{\beta\alpha}$.

One relatively simple approximation to Equation 22 is provided by the single-particle Nilsson model. In this model the low-lying positive and negative $J=1/2$ states in $^{19}$F are proton holes in a $^{20}$Ne core. The holes have asymptotic quantum numbers $[Nn_3\Lambda K]=[220\ 1/2]$ and $[101\ 1/2]$, respectively. Then the nonzero density matrix elements in Equation 22 are $\bar{\psi}_{2s_{1/2}\,1p_{1/2}}=\psi_{2s_{1/2}\,1p_{1/2}}$ and $\psi_{1d_{3/2}\,1p_{3/2}}$.

In the spherical limit the $[220\ 1/2]$ orbital is in the $1d_{5/2}$ shell. Therefore the axial charge matrix elements vanish. Likewise, in the limit of large deformations axial change transitions are forbidden because the states differ by $\Delta n_3=2$. The matrix elements obtain their peak values (about 0.3 of the "single-particle value" obtained with $\psi_{2s_{1/2}\,1p_{1/2}}=1$) at $\delta\approx0.2$, while $^{19}$F is well described by $\delta\approx0.3$ (69). It is apparent that the axial charge matrix elements in $^{19}$F are quite sensitive to the description of the nuclear quadrupole mean field. Now consider more "realistic" wave functions, those generated by the popular Zuker-pds effective shell model Hamiltonian for the space $1p_{1/2}-2s_{1/2}-1d_{5/2}$. While such wave functions may appear to be more sophisticated than those of the single-particle Nilsson model, they lack essential physics. Omitting the $\psi_{1d_{3/2}\,1p_{3/2}}$ density matrix element destroys the $\Delta n_3=2$ forbiddenness, which occurs because of an exact cancellation between $\psi_{1d_{3/2}\,1p_{3/2}}$ and $\psi_{2s_{1/2}\,1p_{1/2}}$ in the asymptotic limit. Therefore, for a well-deformed nucleus such as $^{19}$F, the Zuker-pds wave functions will overestimate matrix elements of the axial charge. On the other hand, a full $0\hbar\omega+1\hbar\omega$ shell model calculation, as described in the next section, does not suffer from this defect.

The pairing force also suppresses matrix elements of the axial charge. The effect of the pairing force on operator matrix elements depends on the behavior of the operator under particle-hole conjugation. Axial charge multipole operators and the more familiar transverse electric multipoles $T_{JM}^{\text{el};\gamma}$ are both odd under particle-hole conjugation. The pairing force suppresses matrix elements of odd operators (70). In our density matrix formalism, the pairing suppression is embodied in the coefficients $\psi_{\beta\alpha}$ that, having the same sign as $\psi_{\alpha\beta}$, generate cancellations in $\psi_{\alpha\beta}^- = \psi_{\alpha\beta} - \psi_{\beta\alpha}$. Unfortunately, in the shell model the $\psi_{\beta\alpha}$ enter only in very large-basis calculations. For instance, $\psi_{1p_{1/2} 2s_{1/2}}$ is produced only if $2\hbar\omega$ "pairing excitations" are included in the $1/2^+$ wave function (corresponding to the promotion of two p-shell nucleons to the 2s 1d shell). Because of the similar behavior of the axial charge and electric multipole operators under particle-hole conjugation, predictions of low-lying E1 transition strengths are an important test of the pairing properties of wave functions used in PNC calculations.

A third important aspect of the axial charge operators is their dependence, via the derivative $\mathbf{p}(i)$, on the shape of the nuclear surface. Our harmonic oscillator, single-particle, radial wave functions fall off too sharply at the surface. In the shell model the nuclear surface is "softened" by including the $2\hbar\omega$ 1p1h excitations that generate the coefficients $\psi_{1p_{1/2}1s_{1/2}}^-$, $\psi_{2p_{1/2} 2s_{1/2}}$, etc. Haxton showed (68) that including $2\hbar\omega$ 1p1h excitations further suppresses matrix element of the axial charge. By diagonalizing the one-body density matrix one can demonstrate that the principal effect of these excitations is to redefine the single-particle basis. Alternatively, one can adopt a more realistic single-particle basis at the outset. Millener & Warburton (71) have shown that Woods-Saxon calculations of isovector axial charge matrix elements frequently give values significantly smaller than those for harmonic oscillator wave functions.

## 6.3 Shell Model Studies of PNC Matrix Elements in Light Nuclei

It is apparent that the nuclear PNC operators must be evaluated with care. In this subsection we discuss calculations we performed in the light nuclei $^{18}$F, $^{19}$F, $^{21}$Ne, and $^{14}$N, the most tractable cases for shell model studies. The $^{18}$F, $^{19}$F, and $^{21}$Ne calculations are described in more detail elsewhere (47, 68, 72).

Two technical points should be raised at the outset. First, because the nuclear shell model binds nucleons with a fictitious central potential, the wave functions may contain spurious center-of-mass excitations that create difficulties in PNC calculations. Spurious components are troublesome in calculations that employ translationally noninvariant two-body PC potentials (two-body potentials derived from the N-N interaction are frequently not translationally invariant because of approximations made in evaluating the $g$ matrix). Spurious wave functions also must be avoided when evaluating matrix elements of effective one-body PNC potentials: operators proportional to $Y_1(\hat{r})$ strongly couple spurious and nonspurious components. Spurious components can be removed exactly in calculations employing a harmonic oscillator single-particle basis, if the model space contains a full $n\hbar\omega$ basis. This has been done in all such cases reported below.

Second, as short-range two-nucleon correlations are almost totally neglected in the shell model, the shell model two-nucleon densities must be modified by a correlation function. Clearly the relative strengths of pion and vector meson PNC couplings extracted from experiment depend on the treatment of short-range correlations. We have consistently used the correlation function (35)

$$f(r_{12}) = 1 - \exp(-ar_{12}^2)(1 - br_{12}^2),$$

with $a = 1.1$ fm$^{-2}$ and $b = 0.68$ fm$^{-2}$, in our analysis of data from the two-nucleon, few-nucleon, and light nuclear systems. The reduction of the pion and vector meson PNC matrix elements by this simple function is very similar to that obtained (73) by solving the coupled Brueckner-Goldstone equations for Hamada-Johnston and Reid soft-core potentials. The reader should be aware that supersoft-core potentials have been used in a number of PNC investigations. PNC couplings extracted from experimental data using supersoft-core potentials will be smaller than those we deduce.

To illustrate some of the nuclear structure issues discussed in Section 6.2, the results of a series of $^{18}$F PNC matrix element calculations are given in Table 6. The first, denoted $0 + 1\hbar\omega - $ Kuo, allows all $0\hbar\omega$ and $1\hbar\omega$ configurations to contribute to the positive and negative parity bases. Thus a large model space (1p-2s1d-2p1f) and nonspurious wave functions are employed, but the effects of multiparticle-hole correlations are omitted entirely. The effective interaction for the 2s1d shell is taken from Kuo & Brown (74), while cross-shell matrix elements are taken from Millener & Kurath (75) and from the bare-$g$-matrix results of Kuo & Brown (74). We can study the qualitative features of the $0 + 1\hbar\omega - $ Kuo calculation by examining the matrix element of the effective operator $\sigma(i) \cdot \mathbf{p}(i)\tau_3(i)$

$$\langle 0^- 0 | U_{\text{PNC}}^{\text{1-body}} | 0^+ 1 \rangle_{0+1\hbar\omega} \propto -\sqrt{2}\psi_{2s_{1/2}\,1p_{1/2}} + \sqrt{10}\psi_{1d_{3/2}\,1p_{3/2}}$$
$$+ \sqrt{5}\psi_{2p_{1/2}\,2s_{1/2}} - 2\psi_{2p_{3/2}\,1d_{3/2}} + \sqrt{21}\psi_{1f_{5/2}\,1d_{5/2}} \equiv B. \quad 23.$$

(In the $0 + 1\hbar\omega$ approximation, $\psi_{\alpha\beta}^- = \psi_{\alpha\beta}$ and $\psi_{1p_{1/2}\,1s_{1/2}} = 0$ in Equation 22.) The principal component in this expression, $\psi_{2s_{1/2}\,1p_{1/2}}$, yields $B_{2s_{1/2}\,1p_{1/2}} = 0.55$, but the minor components all conspire to cancel against this contribution to yield $B = 0.28$. We argued in Section 6.2 that this cancellation arises from deformation and from corrections to the shape of the 1p radial wave function from 2p orbitals.

The second $^{18}$F calculation, which generated the two-body PNC matrix elements shown in the second row of Table 6, uses Zuker-pds wave functions. The basis consists of all configurations within the restricted (and

**Table 6** Shell model and $\beta$-decay scaling results for the nuclear matrix elements of $V_{PNC}$ given in Equation 4 (results are in MeV)[a]

| Calculation | $F_\pi$ | $F_0$ | $F_1$ | $F_2/\sqrt{6}$ | $G_0$ | $G_1$ | $H_1$ | $(V_{PNC}^{DDH})_{best}$ |
|---|---|---|---|---|---|---|---|---|
| $^{18}$F: $-i\langle 0^- 0|V_{PNC}|0^+ 1\rangle$ | | | | | | | | |
| Z-pds | 0.722 | | 0.088 | | | 0.053 | 0.048 | 0.81 |
| $0+1\hbar\omega$–Kuo | 0.953 | | 0.160 | | | 0.090 | 0.069 | 1.08 |
| $1+2\hbar\omega$ | 0.247 | | 0.047 | | | 0.027 | 0.020 | 0.28 |
| $\beta$ decay | 0.324 | | 0.062 | | | 0.035 | 0.026 | 0.37 |
| $^{19}$F: $-i\langle\frac{1}{2}^- \frac{1}{2}|V_{PNC}|\frac{1}{2}^+ \frac{1}{2}\rangle$ | | | | | | | | |
| Z-pds | 0.825 | 0.415 | 0.162 | | 0.085 | 0.086 | 0.062 | 1.67 |
| $0+1\hbar\omega$–Kuo | 0.606 | 0.373 | 0.112 | | 0.070 | 0.063 | 0.048 | 1.34 |
| $\beta$ decay | 0.206 | 0.127 | 0.038 | | 0.024 | 0.021 | 0.016 | 0.46 |
| $^{21}$Ne: $-i\langle\frac{1}{2}^- \frac{1}{2}|V_{PNC}|\frac{1}{2}^+ \frac{1}{2}\rangle$ | | | | | | | | |
| $0+1\hbar\omega$–Kuo | 0.344 | −0.234 | 0.067 | | −0.046 | 0.032 | 0.024 | −0.018 |
| $^{14}$N: $-i\langle 0^+ 1|V_{PNC}|0^- 1\rangle$ | | | | | | | | |
| $1+2\hbar\omega$ | | 0.648 | | −0.168 | 0.123 | | | 1.04 |

[a]Combining these with the DDH "best value" couplings gives the mixing matrix elements (in eV) shown in the last column.

spurious) $1p_{1/2}$-$2s_{1/2}$-$1d_{5/2}$ space. As the only nonzero $\psi_{\alpha\beta}$ are $\psi_{2s_{1/2} 1p_{1/2}}$ and $\psi_{1p_{1/2} 2s_{1/2}}$, the calculation demonstrates the effects of multiparticle-hole configurations that govern the pairing-force suppression. Including the contribution of $\psi_{p_{1/2} 2s_{1/2}}$ reduces $B_{2s_{1/2} 1p_{1/2}}$ from 0.57 to 0.20. [The Zuker-pds wave functions are known to overestimate the strengths of multiparticle-hole corrections (81) but have been used in previous studies of PNC.]

These results confirm our general arguments that a realistic calculation must include both a large model space and multiparticle-hole correlations, principally the 4p2h admixtures in the $0^+$ 1.04-MeV level in $^{18}$F. This can be accomplished by expanding the first calculation described above to include all $2\hbar\omega$ configurations, as described by Haxton (68). The resulting value for $B$ is 0.054. Correlations and the minor components of the density matrix conspire to yield a value much smaller than one would expect from either of the simpler calculations discussed above. The PNC two-body matrix elements are similarly reduced, as shown in the third row of Table 6. These results agree quite well with the essentially model-independent values extracted from the first-forbidden $\beta$-decay rate of $^{18}$Ne (see Section 6.4).

Also shown in Table 6 are the $0+1\hbar\omega-$Kuo results for $^{19}$F and $^{21}$Ne. Based on our qualitative arguments of Section 6.2 and on the $^{18}$F calculations discussed above, we expect these calculations to overestimate the PNC matrix elements. Extensive comparisons of the electromagnetic transition strengths in $^{18}$F, $^{19}$F, and $^{21}$Ne to calculated values were made by Adelberger et al (47). The $0+1\hbar\omega-$Kuo wave functions reproduce measured M1 and E2 strengths quite successfully, but overestimate the low-lying E1 transition rates. However, the improved $2\hbar\omega$ $^{18}$F calculations predict weaker E1 and PNC matrix elements, in better agreement with experiment. This is consistent with our earlier observations that matrix elements of the axial charge and electric multipole operators are suppressed by pairing correlations.

Although an experimental result is not yet available for the parity mixing in $^{14}$N, a full $1+2\hbar\omega$ calculation of the PNC matrix element has been performed. The wave functions were taken from a study of 1p-shell nuclei by Dubach & Haxton (81) that has enjoyed considerable success. The calculated level spectrum of $^{14}$N is shown in Figure 16. Our confidence in these calculations is due in part to the close correspondence between the theoretical and experimental energies of several low-lying levels, including the $0^+$, $I = 1$ member of the 8.6-MeV parity doublet, that are predominantly $2\hbar\omega$ in character. Because of the 2p4h character of the $0^+$, $I = 1$ state, our general arguments about the suppression of PNC matrix elements by pairing correlations do not apply to $^{14}$N. (In $^{18}$F the pairing

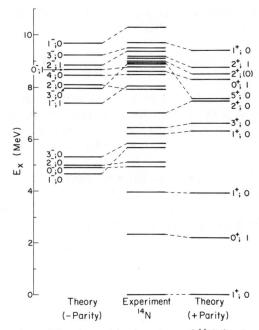

*Figure 16*   Comparison of the observed level spectrum of $^{14}$N (for $E_x \leq 10$ MeV) to the $0+2\hbar\omega$ calculation by Dubach & Haxton (81).

suppression is due to interference between $0\hbar\omega$ and $2\hbar\omega$ configurations that connect to the $1\hbar\omega$ $0^-0$ state via the PNC operators. In $^{14}$N the absence of such interference leads to an exceptionally strong PNC matrix element.) Although the axial charge $\beta$-decay calibration of Section 6.4 cannot be exploited in this case, the parity mixing in $^{14}$N is probably the most reliably calculated PNC quantity in any complex nucleus.

### 6.4   PNC in $^{18}F$ and $^{19}F$ and the First-Forbidden Beta Decays of $^{18}Ne$ and $^{19}Ne$

We now discuss the relationship between parity mixing of "two-level" systems and the corresponding first-forbidden $\beta$-decay rates that connect one state to the isospin analogue of the second state. This relationship is important because recent measurements (47, 76) of the $\beta$-decay rates of $^{18}$Ne $(0^+, 1) \rightarrow$ $^{18}$F $(0^-, 0)$ and $^{19}$Ne $(1/2^+, 1/2) \rightarrow$ $^{19}$F $(1/2^-, 1/2)$ have played a decisive role in interpreting the parity mixing in the daughter nuclei (see Figure 17).

In general, six independent operators contribute to first-forbidden $\beta$ decay (we use the notation of Refs. 46 and 47). However, for odd-parity

$J = 0 \to J = 0$ transitions [such as $^{18}\text{Ne}(0^+) \to {}^{18}\text{F}(0^-)$], spin and parity selection rules eliminate all but two of these,

$$M_{00}^5 = \frac{1}{\sqrt{4\pi}} \int d\mathbf{x}\, J_0^5(\mathbf{x}) \qquad L_{00}^5 = \frac{iq}{6\sqrt{4\pi}} \int d\mathbf{x}\, \mathbf{x}^2 \mathbf{V} \cdot \mathbf{J}^5(\mathbf{x}), \qquad 24.$$

where $J_\mu^5$ is the axial vector charge-changing hadronic current and $q$ is the three-momentum transfer from the nucleus. The second operator in Equation 24 vanishes in the long wavelength limit. Since the momentum transfer in the $^{18}\text{Ne}$ decay is small ($qR \approx 0.02$), the decay rate is determined primarily by $|\langle 0^- | M_{00}^5 | 0^+ \rangle|^2$.

The dominant $M_{00}^5$ operator contains, in principle, one- through $A$-body axial currents. The most important of these are the one- and two-body currents. The one-body contribution, $M_{00}^5(1)$, is obtained from the nonrelativistic reduction of the axial current for a free nucleon. To $O(1/M)$ one finds (46)

$$J_{0(1)}^{5\pm} = F_A \sum_{i=1}^{A} \tau_\pm(i) \left[ \sigma(i) \cdot \frac{\mathbf{p}(i)}{2M} \delta[\mathbf{x} - \mathbf{x}_i] + \delta[\mathbf{x} - \mathbf{x}_i] \frac{\mathbf{p}(i)}{2M} \cdot \sigma(i) \right]$$

$$- i \left[ \hat{H}, \frac{F_p}{2M} \mathbf{V} \cdot \sum_{i=1}^{A} \sigma(i)\tau_\pm(i)\delta[\mathbf{x} - \mathbf{x}_i] \right], \qquad 25.$$

where $\hat{H}$ is the nuclear Hamiltonian and $F_A = -1.25$. We use partial conservation of the axial vector current (PCAC) to evaluate the pseudoscalar coupling constant, $F_p = 2MF_A/m_\pi^2$. The first term in Equation 25,

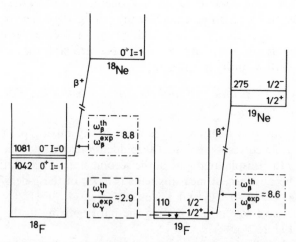

*Figure 17*    First forbidden $\beta$-decay analogues of the parity mixing in $^{18}\text{F}$ and $^{19}\text{F}$. The ratios $\omega^{\text{th}}/\omega^{\text{exp}}$ refer to the $0 + 1\hbar\omega - \text{Kuo}$ calculation.

although forbidden in the sense that $|\mathbf{p}(i)|/M \approx 1/5$, is finite as $q \to 0$, while the second term does not contribute to $M_{00}^5$ in the long wavelength limit.

The two-body contribution to the axial charge operator, $M_{00}^5$ (2), is dominated by pion exchange (see Figure 18) and can be evaluated from a low-energy theorem based on PCAC and current algebra (77, 78). The result, to leading order in $(1/M)$, is given entirely by the seagull term

$$J_{0(2)}^{5\pm} = \frac{m_\pi^2 g_{\pi NN}^2 F_1^V}{8\pi M^2 F_A} \frac{1}{2} \sum_{\substack{i,j \\ i \neq j}} [\tau(i) \times \tau(j)]_\pm [\sigma(i) \cdot \hat{\mathbf{r}}_{ij} \delta[\mathbf{x} - \mathbf{x}_j]$$
$$+ \sigma(j) \cdot \hat{\mathbf{r}}_{ij} \delta[\mathbf{x} - \mathbf{x}_i]] \phi[m_\pi r_{ij}], \qquad 26.$$

where $\mathbf{r}_{ij} = \mathbf{r}_i - \mathbf{r}_j$, $[\ ]_\pm \equiv [\ ]_1 \pm i[\ ]_2)/2$, $\phi(x) = (e^{-x}/x)(1 + 1/x)$, and $F_1^V = 1$. The one-pion-exchange current plays an important role in axial charge transitions because it is of the same order ($|\mathbf{p}|/M \approx v/c$) as the one-body contribution.

Repeating the discussions of Section 6.1 one finds that, for a nucleus with $\rho_N \approx \rho_Z$, $M_{00}^5(2)$ can be represented by an effective one-body operator proportional to $\sigma \cdot \mathbf{p}\tau_\pm$. To an excellent approximation, the exchange current merely renormalizes the one-body axial charge. The Fermi gas prediction for the ratio $\alpha$ of two-body to one-body matrix elements is

$$\alpha \equiv \frac{\langle \|M_0^5(2)\| \rangle}{\langle \|M_0^5(1)\| \rangle} = \frac{g_{\pi NN}^2 \rho_A W[\tilde{p}, \tilde{m}_\pi]}{2m_\pi^2 M F_A^2} \qquad 27.$$

where $\rho_A = \rho_N + \rho_Z$. We argued before that the function $W$ is approximated well by its average value, $\langle W \rangle \approx 0.17$. For a nuclear matter density of $\rho_A = 0.195$ nucleons per fm³, we find $\alpha \approx 0.84$. This value must be corrected for short-range correlations ignored in the Fermi gas model. Our correlation function (Equation 23) reduces the shell model $\Delta I = 1$ pion-exchange matrix elements to $0.74 \pm 0.03$ of their uncorrelated values. Thus we expect $\alpha^{cor} \approx (0.84)(0.74) \approx 0.62$.

These ideas are supported by detailed shell model calculations of the $^{18}$Ne $0^+ \to 0^-$ decay. In Table 7 we present the various predictions for this decay rate. Although the different structure assumptions lead to $\beta$-decay rates that differ by as much as a factor of 16, the calculated strengths of the

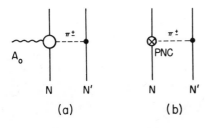

Figure 18   Schematic representation of the relation between semileptonic axial charge and isovector parity-mixing operators.

**Table 7**    PNC analogue $\beta$-decay rates in units of $10^{-6}$ s for various model calculations[a]

| Transition | Calculation | $\omega_{1+2}$ | $\omega_1$ | $\omega_1 (q = 0)$ | $\alpha$ | $\omega^{exp}$ |
|---|---|---|---|---|---|---|
| $^{18}Ne \rightarrow {}^{18}F(0^-)$ | Z-pds | 52.2 | 23.0 | 14.0 | 0.65 | |
| | $0 + 1\hbar\omega - Kuo$ | 75.3 | 29.5 | 26.2 | 0.63 | $8.6 \pm 1.2$ |
| | $1 + 2\hbar\omega$ | 4.80 | 1.84 | 1.00 | 0.83 | |
| $^{19}Ne \rightarrow {}^{19}F(1/2^-)$ | Z-pds | 104.0 | 50.2 | 42.4 | 0.48 | $4.8 \pm 0.8$ |
| | $0 + 1\hbar\omega - Kuo$ | 41.3 | 17.1 | 15.3 | 0.58 | |

[a] The one-body and one-body-plus-exchange-current results are $\omega_1$ and $\omega_{1+2}$, while $\omega_1 (q = 0)$ denotes the one-body result with no $q$-dependent corrections. See Section 6.4 for the definition of $\alpha$.

exchange-current contribution *relative* to the one-body amplitude are remarkably constant, $\alpha = 0.73 \pm 0.10$, in accordance with the Fermi gas prediction that this ratio is determined by coupling constants, the pion mass, and gross nuclear properties.

A comparison of Equations 26 and 3 reveals that, apart from a rotation in isospin, the pion-exchange contribution to $V_{PNC}$ is proportional to $M_{00}^5 (2)$, as shown schematically in Figure 18. This was observed first by Bennett, Lowry & Krien (79) and later independently by one of us and, when combined with the virtual model independence of $\alpha$, makes possible a $\beta$-decay "calibration" of the nuclear matrix elements needed to analyze the parity mixing in $^{18}F$ and $^{19}F$ (47, 68).

The measured $^{18}Ne$ $\beta$-decay rate (68) effectively determines the matrix element of $M_{00}^5 = M_{00}^5(1) + M_{00}^5(2)$. The scaling relation, Equation 27, allows us to place stringent and *largely model-independent* constraints on the matrix element of $M_{00}^5(2)$ and thus, after an isospin rotation, on the strength of the pion-exchange contribution to the $\Delta I = 1$ PNC mixing matrix elements. Applying this to $^{18}F$ we find in the long wavelength approximation

$$\langle 0^- 0 | V_{PNC}^\pi | 0^+ 1 \rangle \approx (0.366 \pm 0.055 \text{ MeV}) F_\pi. \qquad 28.$$

The $O(q)$ contributions to the $0^+ \rightarrow 0^-$ $\beta$-decay amplitude are not completely negligible because the matrix elements of $M_{00}^5(q = 0)$ are so suppressed. However, because of the particle-hole conjugation properties of $L_{00}^5$, the $O(q)$ corrections depend on the density matrix element combination $\psi_{\alpha\beta}^+ \equiv \psi_{\alpha\beta} + \psi_{\beta\alpha}$. This combination is enhanced by the multiparticle-hole correlations and thus is not subject to the delicate cancellations occurring for $M_{00}^5$. Therefore, the $O(q)$ corrections should be predicted reliably by the shell model. Indeed, our various calculations yield quite consistent predictions of these corrections,

$$\langle 0^- 0 | V_{PNC}^\pi | 0^+ 1 \rangle = \begin{cases} (0.326 \pm 0.051 \text{ MeV}) F_\pi, & 0 + 1\hbar\omega - Kuo, \\ (0.322 \pm 0.51 \text{ MeV}) F_\pi, & 1 + 2\hbar\omega, \end{cases} \qquad 29.$$

and we take as our final result

$$\langle 0^- 0|V^\pi_{\text{PNC}}|0^+ 1\rangle = (0.324 \pm 0.053 \text{ MeV})F_\pi. \qquad 30.$$

The $^{18}$F PNC matrix element also involves isovector heavy meson exchanges. The vector meson ($F_1$, $G_1$, and $H_1$) and pion-exchange matrix elements are governed by the same effective one-body operator, $\boldsymbol{\sigma} \cdot \mathbf{p}\tau_3$, so their relative strengths should not depend on the details of the shell model calculation. This expectation is borne out by the numerical results in Table 6. Therefore we estimate the heavy meson-exchange contribution to $P_\gamma$ by scaling the $1+2\hbar\omega$ calculation to reproduce Equation 32 (Table 6). We find $|P_\gamma| = (1.83 \pm 0.43) \times 10^{-3}[F_\pi + 0.19F_1 + 0.11G_1 + 0.08H_1]$. For "best value" couplings the vector meson contribution to $|P_\gamma|$ is approximately 5% that of $F_\pi$ and adds constructively. We ignore these contributions in setting an upper bound on $F_\pi$ from the measured polarization, $P_\gamma = (0.8 \pm 5.2) \times 10^{-4}$. We obtain $|F_\pi| \leq 0.34 \times 10^{-6}$. This upper bound is essentially independent of nuclear structure assumptions and rules out a large part of the DDH reasonable range, $0 < F_\pi < 2.7 \times 10^{-6}$. The DDH best value of $F_\pi = 1.1 \times 10^{-6}$ exceeds the upper bound by more than a factor of three.

The extension of these arguments to $^{19}$F is somewhat more complicated. For a $J = 1/2 \rightarrow J = 1/2$ odd-parity transition [such as $^{19}$Ne$(1/2^+)$ $\rightarrow$ $^{19}$F$(1/2^-)$] three additional operators satisfy the spin and parity selection rules:

$$M_{1\text{M}} = \frac{q}{3}\int d\mathbf{x}\, xY_{1\text{M}}[\Omega_x]J_0(\mathbf{x}), \qquad T^{\text{el}}_{1\text{M}} = \frac{q_0}{q}\sqrt{2}M_{1\text{M}},$$

$$T^{\text{mag}5}_{1\text{M}} = \frac{iq}{2\sqrt{6\pi}}\int d\mathbf{x}\,[\mathbf{x} \times \mathbf{J}^5(\mathbf{x})]_{\text{M}},$$

where $q_0 = E_i - E_f$ is the energy release and $J_\mu$ the vector current. In Equation 31 we have used current conservation to relate $T^{\text{el}}_{1\text{M}}$ to the vector charge operator $M_{1\text{M}}$. Note that all three of the additional operators vanish in the long wavelength limit ($q_\mu \rightarrow 0$). By the conserved vector current (CVC) hypothesis the matrix elements of the two vector operators can be determined from the known E1 $\gamma$-decay rates for the analog transition in $^{19}$Ne and $^{19}$F, assuming the $1/2^-$ and $1/2^+$ states have good isospin (47). These vector operators account for only a very small fraction of the measured $^{19}$Ne $\beta$-decay rate. Matrix elements of the axial vector operator, $T^{\text{mag}5}_{1\text{M}}$, must be taken from a nuclear model calculation.

The results in Table 7 demonstrate that, as expected, the $O(q)$ contributions to the $^{19}$Ne decay rate are small. Thus, as in $^{18}$Ne, this decay measures the matrix element of the axial charge. The difficult $1+2\hbar\omega$ shell model calculation has not been done for $A = 19$. However, the ratio $\alpha$ for the two simpler calculations of Table 7 is $0.53 \pm 0.05$, again in good agreement with the Fermi gas estimate. The ratio of the observed $^{19}$Ne first-forbidden beta-decay rate to the $0 + 1\hbar\omega - $Kuo prediction, $\beta \equiv (\omega^{exp}/\omega^{th})^{1/2} = 0.34 \pm 0.03$, is almost identical to the ratio $0.33 \pm 0.03$ for the corresponding wave functions in $A = 18$. As the $1 + 2\hbar\omega$ calculation of the $A = 18$ $\beta$-decay rate is quite close to the experimental value, we conclude that the omission of 5p2h pairing correlations in the $A = 19$ wave functions accounts for the discrepancy between the calculated and experimental $\beta$-decay rates.

The contributions of the isovector couplings $F_\pi$, $F_1$, $G_1$, and $H_1$ to the $^{19}$F PNC matrix element are also governed by the effective one-body operator $\sigma \cdot \mathbf{p}\tau_3$. Thus, as in $^{18}$Ne, the $^{19}$Ne $\beta$-decay calibration of the pion-exchange matrix element also determines the scale of $\Delta I = 1$ vector meson matrix elements. (For "best value" couplings, the vector mesons make only a small contribution to the $\Delta I = 1$ parity mixing.) For $^{19}$F, however, the isoscalar couplings $G_0$ and $F_0$, which multiply $\sigma \cdot \mathbf{p}$ in our effective one-body potential, also contribute to the parity mixing. Although our qualitative arguments about the effects of deformation, pairing, and the single-particle basis do not depend on isospin, we have no experimental test of the isoscalar axial charge. Thus we must rely more heavily on numerical calculations.

Unfortunately, because a $1 + 2\hbar\omega$ shell model calculation for $^{19}$F has not yet been performed, it is difficult to draw firm conclusions. In the $0 + 1\hbar\omega$ $-$Kuo and Zuker-pds calculations, the isovector and isoscalar heavy meson matrix elements track the pion-exchange matrix elements reasonably well. Thus we assume that the isoscalar "effective charge" needed to correct the $0 + 1\hbar\omega -$Kuo calculation for multiparticle-hole correlations is the same as the isovector "effective charge" determined by the $^{19}$Ne $\beta$-decay rate. (This conjectured equality of the isoscalar and isovector effective charges should be tested in future shell model studies of light nuclei.) Under this assumption we then should scale the $A = 19$ matrix elements by that factor needed to reproduce the measured $\beta$-decay rate. This procedure and the DDH best value couplings then yield $|\langle |V_{PNC}| \rangle| = 0.46 \pm 0.08$ eV and $A_\gamma = (-8.9 \pm 1.6) \times 10^{-5}$, in good agreement with the experimental result $A_\gamma = (-7.4 \pm 1.9) \times 10^{-5}$.

# 7.   EXPERIMENTAL CONSTRAINTS ON THE PNC N-N INTERACTION

In this section we extract, from the available data, a "best set" of PNC N-N couplings. The data we fit are $A_L$ in $\vec{p} + p$, $\vec{p} + d$, and $\vec{p} + \alpha$ and the parity-mixing matrix elements in $^{18}$F, $^{19}$F, and $^{21}$Ne. Other PNC observables are excluded for reasons detailed previously. Clearly the data set is not complete. Two of the results (the "odd-proton" systems $\vec{p} + \alpha$ and $^{19}$F)

**Table 8**  Weak meson-nucleon coupling constants

|  | Range ($\times 10^{-6}$)[a] | Fitted values ($\times 10^{-6}$)[b] |
|---|---|---|
| $F_\pi$ | $0:2.7$ | $0.5$ |
| $F_0$ | $-1.5:4.3$ | $0.8$ |
| $F_1$ | $-0.06:0.07$ | $0.03$ |
| $F_2/\sqrt{6}$ | $0.36:0.59$ | $0.4$ |
| $G_0$ | $-1.1:5.1$ | $2.7$ |
| $G_1$ | $0.46:1.3$ | $1.0$ |

[a] For the vector meson couplings, these are determined from Ref. 62 allowing $1 \leq K \leq 7$ and $0 < \eta', \eta'' < 1$. The range for $F_\pi$ is that of DDH.
[b] Corresponding to $K = 4$ and $\eta = \eta' = 0.5$.

probe very similar combinations of the PNC N-N amplitudes, while other combinations are essentially unconstrained by experiment. Furthermore, three of the results are only sensitive upper limits. As a result, it is not presently possible to determine the six independent parameters of $V_{\text{PNC}}$ or, equivalently, $F_\pi$ and the five S-P amplitudes.

Instead we must impose certain theoretical constraints to make our task more manageable. We use as our guide the $SU(6)_W$ expressions for the vector meson amplitudes of Desplanques (62). (These expressions are virtually identical to those of DDH.[2]) The strong interaction enhancement of the vector meson amplitudes is governed by the parameter $K$, $K \geq 1$. In accordance with our earlier discussions of the uncertainties in the $SU(6)_W$ predictions, we multiply the quark model and sum rule terms by scale factors $\eta'$ and $\eta''$, where these quantities run between one and zero. This permits us to dial between the $SU(6)_W$ and factorization predictions for the vector meson couplings. The vector meson amplitudes are then determined by $K$, $\eta'$, and $\eta''$. If we require $K$ to lie in the theoretically reasonable (24) range $1 \leq K \leq 7$, the resulting vector meson couplings can have the values shown in Table 8. The ranges are similar to the reasonable ranges defined by DDH.

Unfortunately, even a four-parameter fit ($F_\pi$, $\eta'$, $\eta''$, and $K$) is too ambitious. Thus we adopt the theoretical prejudice that $K \approx 4$. Furthermore, as the predicted vector meson couplings are determined primarily by the value of $\eta' + \eta''$, we take $\eta' = \eta''$. We then fit the PNC observables in order to determine $F_\pi$ and $\eta'$. The theoretical expressions for these observables (see Tables 3, 4, and 6) were obtained with soft core (rather than supersoft core) strong potentials, or with correlation functions

[2] The differences are Desplanque's choice of $\xi = 1$, where $\xi$ is a reduced matrix element set equal to zero in DDH, and a slight change in the strong interaction enhancement of the sum rule term.

compatible with such potentials. For consistency with the $^{18}$F and $^{19}$F effective charges deduced in Section 6, we have scaled the $^{21}$Ne $0+1\hbar\omega$ results by one third.

The results of our least-square fit are $F_\pi = 0.5 \times 10^{-6}$ and $\eta' = 0.5$. Thus the data require $F_\pi$ to be weaker than the DDH best value, $1.1 \times 10^{-6}$. The vector meson amplitudes are midway between the SU(6)$_W$ and factorization predictions. The corresponding weak couplings constants are given in Table 8. The calculated PNC observables are

$\vec{p}+p$, $A_L^{tot}$ (45 MeV):    $-1.3 \times 10^{-7}$ $[(-2.31 \pm 0.89) \times 10^{-7}]$

$\vec{p}+p$, $A_L^{tot}$ (15 MeV):    $-0.79 \times 10^{-7}$ $[(-1.7 \pm 0.8) \times 10^{-7}]$

$\vec{p}+{}^4$He, $A_L$ (46 MeV):    $-2.6 \times 10^{-7}$ $[(-3.3 \pm 0.9) \times 10^{-7}]$

$\vec{p}+d$, $A_L$ (15 MeV):    $-0.77 \times 10^{-7}$ $[(-0.35 \pm 0.85) \times 10^{-7}]$

$^{18}$F, $-i\langle V_{PNC}\rangle$:    $0.20$ eV $[|\langle V_{PNC}\rangle| < 0.090$ eV$]$

$^{19}$F, $-i\langle V_{PNC}\rangle$:    $0.29$ eV $[-i\langle V_{PNC}\rangle = (0.38 \pm 0.10)$ eV$]$

$^{21}$Ne, $-i\langle V_{PNC}\rangle$:    $-0.036$ eV $[|\langle V_{PNC}\rangle| < 0.029$ eV$]$

with the experimental values given in brackets. The experimental and theoretical values are in reasonable agreement. The calculated $^{18}$F matrix element is too large by a factor of two and accounts for nearly half of the mean-square error in our fit.

We regard the couplings in Table 8 only as present benchmark values. Changes in experimental results or in the interpretation of those results can substantially alter our conclusions. In particular, we stress that the $^{21}$Ne limit exerts considerable influence on our conclusions. If we ignore the $^{21}$Ne result, a much improved fit to the remaining data can be achieved with $F_\pi = 0$. Effectively, the evidence for *any* isovector component in the PNC interaction rests entirely on the shell model prediction that the isoscalar matrix element in $^{21}$Ne is not small. (In $^{21}$Ne, unlike $^{18}$F and $^{19}$F, there is no $\beta$-decay test of the shell model calculation.) Regardless of the weight given to the $^{21}$Ne calculation, it is clear that the neutral current enhancement of $F_\pi$ is less than that expected theoretically. Perhaps the isospin dependence of $V^{PNC}$ will prove as surprising as the $\Delta I = 1/2$ rule in $\Delta S = 1$ interactions. A definitive measurement of $F_\pi$ is needed. The continuing $^{18}$F experiments or a considerably more precise measurement of $A_\gamma$ in $\vec{n}p \rightarrow d\gamma$ could yet provide this. Alternatively, results from the upcoming $^{14}$N experiment could be combined with the $^{19}$F results to separately determine $F_\pi$ and $F_0$.

In summary, the existing experimental data are roughly consistent with the PNC N-N couplings expected for the standard electroweak theory and the quark model of hadron structure. Virtually all of the glaring discrepancies claimed previously have been resolved. However, the program of determining the full low-energy PNC N-N interaction is still incomplete. We look forward to the next few years, when new results in light nuclei, a new generation of precise n + p experiments, and our continuing progress in understanding the nuclear structure of PNC will greatly improve our knowledge of the flavor-conserving hadronic weak interaction.

We thank J. Donoghue and B. Holstein for helpful discussions, and Barbara Fulton for her cheerful and careful typing. This work was supported in part by the US Department of Energy.

*Literature Cited*

1. Tanner, N. *Phys. Rev.* 107:1203 (1957)
2. Feynman, R. P., Gell-Mann, M. *Phys. Rev.* 109:193 (1958)
3. Abov, Yu. G., Krupchitsky, P. A., Oratovsky, Yu. A. *Phys. Lett.* 12:25 (1964)
4. Lobashov, V. M., Nazarenko, V. A., Saenko, L. F., Smotriskii, L. M., Kharkevitch, O. I. *JETP Lett.* 5:59 (1967); *Phys. Lett. B* 25:104 (1967)
5. Arnison, G., et al, UA-1 Collaboration. *Phys. Lett. B* 122:103 (1983)
6. Banner, M., et al, UA-2 Collaboration. *Phys. Lett. B* 122:476 (1983)
7. Commins, E. D., Bucksbaum, P. H. *Weak Interaction of Leptons and Quarks.* Cambridge Univ. Press (1983)
8. Desplanques, B., Donoghue, J. F., Holstein, B. R. *Ann. Phys.* 124:449 (1980)
9. Danilov, G. S., *Phys. Lett.* 18:40 (1965); *Soc. J. Nucl. Phys.* 14:443 (1972)
10. Box, M. A., McKellar, B. H. J., Pick, P., Lassey, K. R. *J. Phys. G* 1:493 (1975)
11. Desplanques, B., Missiner, J. *Nucl. Phys. A* 300:286 (1978)
12. Blin-Stoyle, R. J. *Phys. Rev.* 118:1605 (1960); 120:181 (1960)
13. Barton, G. *Nuovo Cimento* 19:512 (1961)
14. McKellar, B. H. J. *Phys. Lett. B* 26:107 (1967); *Phys. Rev. Lett.* 20:1542 and 21:1822 (1968); *Phys. Rev.* 178:2160 (1969)
15. Henley, E. M. *Ann. Rev. Nucl. Sci.* 19:367 (1969)
16. Fischbach, E., Tadic, D. *Phys. Rep.* 6C:123 (1973)
17. Gari, M. *Phys. Rep.* 6C:318 (1973)
18. Vinh Mau, R. *Nucleon-Nucleon Interactions—1977, AIP Conf. Proc.* 41:140 (1978)
19. McKellar, B. H. J., Pick, P. *Phys. Rev. D* 6:2184 (1972); 7:260 (1973)
20. Desplanques, B. *Phys. Lett. B* 41:461 (1972)
21. Pirner, H. J., Riska, D. O. *Phys. Lett. B* 44:151 (1973)
22. Michel, F. C. *Phys. Rev. B* 133:329

(1964)
23. Fischbach, E., Tadic, D., Trabert, K. *Phys. Rev.* 186:1688 (1969)
24. Donoghue, J. F., Holstein, B. R. *Phys. Rev. Lett.* 46:1603 (1981)
25. Holstein, B. R. *Phys. Rev. D* 23:1618 (1981)
26. Wilkinson, D. H. *Phys. Rev.* 109:1603, 1610, 1614 (1958)
27. Schopper, H. *Nucl. Instrum.* 3:158 (1958)
28. Elsener, K., Grüebler, W., König, V., Schmelzbach, P. A., Ulbricht, J., et al. *Phys. Rev. Lett.* 52:1476 (1984)
29. Potter, J. M., Bowman, J. D., Hwang, C. F., McKibben, J. L., Mischke, R. E., et al. *Phys. Rev. Lett.* 33:1307 (1974)
30. Nagle, D. E., Bowman, J. D., Hoffmann, C., McKibben, J. L., Mischke, R. E. *Proc. 3rd Int. Symp. on High Energy Physics with Polarized Beams and Polarized Targets*, Argonne, 1978. *AIP Conf. Proc.* 51:24 (1979)
31. Balzer, R., Henneck, R., Jacquemart, Ch., Lang, J., Nessi-Tedaldi, F., et al. *Phys. Rev. C* 30:1409 (1984)
32. von Rossen, P., von Rossen, U., Conzett, H. E. *Proc. 5th Int. Symp. on Polarization Phenomena in Nuclear Physics*, Sante Fe. *AIP Conf. Proc.* 69:1442 (1981)
33. Tanner, D. M., Mihara, Y., Tribble, R. E., Gagliardi, C. A., Nesse, R. E., Sullivan, J. P. *Proc. Int. Conf. on Nuclear Physics*, 1, Florence, 1:697: Tipografia Compositori (1983)
34. Yuan, V. *Proc. 18th LAMPF Users Group Meeting.* To be published (1985)
35. Miller, G. A., Spencer, J. E. *Ann. Phys.* 100:562 (1976)
36. Oka, T. *Prog. Theor. Phys.* 66:977 (1981)
37. Lockyer, N., Romanowski, T. A., Bowman, J. D., Hoffman, C. M., Miscke, R. E. *Phys. Rev. Lett.* 45:1821 (1980)
38. Knyazkov, V. A., Kolomenskii, E. A., Lobashov, V. M., Nazarenko, V. A., Pirozhov, A. N., et al. *Nucl. Phys. A* 417:209 (1984)
39. Lobashov, V. M., Kaminker, D. M.,

Kharkevich, G. I., Knizkov, V. A., Lozovoy, N. A., et al. *Nucl. Phys. A* 197:241 (1972)

40. Abov, Y. G. *Zh. Eksp. Teor. Fiz.* 65:1738 (1973)
41. Caviagnac, J. F., Vignon, B., Wilson, R. *Phys. Lett. B* 67:148 (1977)
42. Desplanques, B., Benayoun, J. J., Gignous, C. *Nucl. Phys. A* 324:221 (1979)
43. Lang, J., Maier, Th., Müller, R., Nessi-Tedaldi, F., Roser, Th., et al. *Phys. Rev. Lett.* 54:170 (1985)
44. Avenier, M., Cavaignac, J. F., Koang, D. H., Vignon, B., Hart, R., Wilson, R. *Phys. Lett. B* 137:125 (1984)
45. Fortson, E. N., Lewis, L. L. *Phys. Rep.* 113:289 (1984)
46. Walecka, J. D. In *Muon Physics*, ed. V. W. Hughes, C. S. Wu, Vol. 2. New York: Academic (1975)
47. Adelberger, E. G., Hindi, M. M., Hoyle, C. D., Swanson, H. E., Von Lintig, R. D., Haxton, W. C. *Phys. Rev. C* 27:2833 (1983)
48. Barnes, C. A., Lowry, M. M., Davidson, J. M., Marrs, R. E., Morinigo, F. B., et al. *Phys. Rev. Lett.* 40:840 (1978)
49. Bizetti, P. G., Fazzini, T. F., Maurenzig, P. R., Perego, A., Poggi, G. *Lett. Nuovo Cimento* 29:167 (1980)
50. Ahrens, G., Harfst, W., Kass, J. R., Mason, E. V., Schober, M. *Nucl. Phys. A* 390:486 (1982)
51. Queen's Univ. Nucl. Phys. Group, and McDonald, A. B., Barnes, C. A., Alexander, T. K., Clifford, E. T. H. *Rep. on Res. in Nucl. Phys. at Queen's Univ.*, p. 14 (1985)
52. Adelberger, E. G., Swanson, H. E., Cooper, M. D., Tape, W., Trainor, T. A. *Phys. Rev. Lett.* 34:402 (1975); Adelberger, E. G. In *Polarization Phenomena in Nuclear Physics—1980*, ed. G. G. Ohlsen, R. E. Brown, N. Jarmie, W. W. McNaughton, G. M. Hale, *AIP Conf. Proc.* 69:1367 (1981)
53. Snover, K. A., Von Lintig, R., Adelberger, E. G., Swanson, H. E., Trainor, T. A. *Phys. Rev. Lett.* 41:145 (1978); Earle, E. D., McDonald, A. B., Adelberger, E. G., Snover, K. A., Swanson, H. E., et al. *Nucl. Phys. A* 396:221c (1983)
54. McDonald, A. B., Earle, E. D., Simpson, J. J., Robertson, R. G. H., Mak, H. B. *Phys. Rev. Lett.* 47:1720 (1981)
55. Adelberger, E. G., Hoodbhoy, P., Brown, B. A. *Phys. Rev. C* 30:456 (1984); Recent unpublished work gives $\Gamma(0^+) = 3.8$ keV.
56. Neubeck, K., Schober, H., Wäffler, H.

*Phys. Rev. C* 10:320 (1974)

57. Fifield, L. K., Catford, W. N., Chew, S. H., Garman, E. F., Pringle, D. M. *Nucl. Phys. A* 394:1 (1983)
58. Desplanques, B. *Proc. 8th Int. Workshop on Weak Interactions and Neutrinos*, Javea, Spain, p. 515. Singapore: World Scientific (1983)
59. Erikson, T., Brown, G. E. *Nucl. Phys. A* 277:1 (1977) and references therein; see also Engeland, T. *Nucl. Phys.* 72:68 (1965)
60. Brown, B. A., Richter, W. A., Godwin, N. S. *Phys. Rev. Lett.* 45:1681 (1980)
61. Boehm, F. *Proc. 6th Int. Conf. on High Energy Physics and Nuclear Structure*, ed. D. E. Nagle, p. 488. New York: AIP (1975)
62. Desplanques, B. *Nucl. Phys. A* 335:147 (1980)
63. Desplanques, B. *Nucl. Phys. A* 316:244 (1979)
64. Krane, K. S., Olsen, C. E., Sites, J. R., Steyert, W. A. *Phys. Rev. C* 4:1906 (1971)
65. Alfimenkov, V. P., et al. *Pis. Zh. Eksp. Teor. Fiz.* 35:42 (1982) [*JETP Lett.* 35:51 (1982)]
66. Stodolsky, L. *Phys. Lett.* 96B:127 (1980)
67. Donnelly, T. W., Haxton, W. C. *At. Data Nucl. Data Tables* 23:103 (1979)
68. Haxton, W. C. *Phys. Rev. Lett.* 46:698 (1981)
69. Haxton, W. C. Ph.D. thesis, Stanford Univ. (1975)
70. Bohr, A., Mottelson, B. R. *Nuclear Structure, II*, pp. 650–51. New York: Benjamin (1975)
71. Millener, D. J., Warburton, E. K. In *Proc. Talmi Symp. on the Nuclear Shell Model*. To be published (1985)
72. Haxton, W. C., Gibson, B. F., Henley, E. M. *Phys. Rev. Lett.* 45:1677 (1980)
73. Gari, M., Huffman, A. H., McGrory, J. B., Offermann, R. *Phys. Rev. C* 11:1485 (1975)
74. Kuo, T. T. S., Brown, G. E. *Nucl. Phys. A* 114:241 (1968)
75. Millener, D. J., Kurath, D. *Nucl. Phys. A* 255:315 (1975)
76. Hernandez, A. M., Daehnick, W. W. *Phys. Rev. C* 25:2957 (1982); Bowles, T. Private communication
77. Kubodera, K., Delorme, J., Rho, M. *Phys. Rev. Lett.* 40:755 (1978)
78. Guichon, P. A. M., Griffon, M., Samour, C. *Phys. Lett. B* 74:15 (1978)
79. Bennett, C., Lowry, M. M., Krien, K. *Bull. Am. Phys. Soc.* 25:486 (1980)
80. Shwe, H., Cote, R. E., Prestwich, W. W. *Phys. Rev.* 159:1050 (1967)
81. Dubach, J., Haxton, W. C. Unpublished work (1980)

*Ann. Rev. Nucl. Part. Sci. 1985. 35 : 559–604*
*Copyright © 1985 by Annual Reviews Inc. All rights reserved*

# LATTICE GAUGE THEORIES

## *Anna Hasenfratz‡*

SCRI, The Florida State University, Tallahassee, Florida 32306

## *Peter Hasenfratz*

Physics Department, University of Bern, Bern, 3012, Switzerland

CONTENTS

‡ On leave from CRIP, Budapest.

0163–8998/85/1201–0559$02.00

# 1. INTRODUCTION

Quantum chromodynamics (QCD) and the Weinberg-Salam model, the successful theories of strong and electroweak interactions, are quantum field theories (QFT). In all recent attempts for unification, including gravity or new symmetries, although some aspects of the ideas might be very unconventional, the language and notions of QFT are used. In all cases, a significant part of the dynamics is beyond the reach of perturbation theory. In a nonasymptotically free theory the very definition of the model requires nonperturbative methods. Even in asymptotically free theories, like QCD, the discussion of spectroscopy and other low-energy phenomena must go beyond perturbation theory.

Conventional analytic techniques expand around some (trivial or nontrivial) classical solution and remain always essentially perturbative. Although phenomenological models might reflect certain nonperturbative aspects properly, there is a growing need for a quantitative understanding of these theories, one that starts from basic principles. In recent years the lattice regularization of field theories, with the advent of new and powerful computers, became an effective method for nonperturbative studies.

This paper is an introduction to lattice gauge theories, reviewing mainly those aspects that are closely related to SU(3) Yang-Mills gauge theory and QCD.

QCD, which is believed to be the fundamental theory of strong interactions, is a field theory of basic quark and gluon constituents. The interaction of these constituents is specified by an intriguingly simple Lagrangian

$$\mathcal{L} = -\tfrac{1}{4}F^a_{\mu\nu}F^a_{\mu\nu} - \sum_f [i\bar{\psi}^{if}\gamma_\mu D^{ij}_\mu\psi^{jf} + m_f\bar{\psi}^{if}\psi^{if}],$$

where

$$F^a_{\mu\nu} = \partial_\mu A^a_\nu - \partial_\nu A^a_\mu - gf^{abc}A^b_\mu A^c_\nu,$$

$$D^{ij}_\mu = \partial_\mu\delta^{ij} + ig(\tfrac{1}{2}\lambda^a A^a_\mu)^{ij}.$$

1.

The index $f$ runs over the different quark flavors u,d,s,···. Simple as it is, this interaction should account for several puzzling phenomena. Experimentally, mesons and baryons are observed rather than quarks and gluons. Quarks seem to be permanently bound inside hadrons ("confinement"). Additionally, the hadron spectrum reflects the (approximate) symmetries of $\mathscr{L}$ in a nontrivial way: chiral symmetry (present in Equation 1 for zero quark masses) is broken spontaneously, producing light pseudoscalar mesons, while the flavor singlet part of the chiral symmetry should be broken explicitly by some, not completely understood, nonperturbative effects ["U(1) problem"]. All these problems require a quantitative, nonperturbative understanding of QCD.

Lattice regularization of gauge theories and its consequent nonperturbative techniques were discussed first by Wilson in 1974 (1). The standard techniques of statistical physics, such as mean field and variational methods, strong coupling expansion and Monte Carlo (MC) simulations, became available in high-energy physics.

The first results on the confining part of the potential (string tension) and on hadron spectroscopy were derived by strong coupling expansions (2). These results were very encouraging (although not really quantitative in the continuum limit) and stimulated interest in the subject. After the first pioneering works (3), MC simulation and the techniques combined with it rapidly became the most powerful methods in lattice QCD. This was partly due to the recent, rather dramatic, increase of computer resources available in this field. The quality of the results and the control over the systematical and statistical errors increased in proportion.

Most of the recent efforts went into the study of pure SU(3) gauge theory with static quark sources only. All the available results confirm that this theory confines. The quark-antiquark potential is predicted with reasonable precision. The temperature of the deconfining phase transition, the string tension, and the mass gap (in this order of reliability) are predicted also.

The first exploratory and feasibility studies on hadron spectroscopy (4) confirmed that the lattice methods are powerful enough to provide quantitative information on the low-lying part of the spectrum—at least in the approximation where the virtual quark contribution is suppressed ("quenched approximation"). This subject is under active investigation; new and much improved results are expected soon.

This review attempts to be self-contained (5). No previous knowledge in lattice gauge theories is assumed. Readers who are familiar with the subject are advised to move directly to Sections 8 and 11, where the results are discussed.

## 2.  DEFINING QUANTUM FIELD THEORIES ON THE LATTICE

### 2.1  *Euclidean Quantum Field Theory as a Classical Statistical System*

In the path integral formulation of QFT, the vacuum functional is given by

$$Z = \sum_{\substack{\text{field} \\ \text{configurations}}} \exp(i \int dt \int d^3x \mathscr{L}), \qquad\qquad 2.$$

where $\mathscr{L}$ is the classical Lagrange density depending on the fields and their derivatives. The vacuum expectation value of the time ordered product of fields (Green's functions) is obtained as the expectation value of the product of the fields, using the measure

$$\sim \frac{1}{Z} \exp(i \int dt \int d^3x \mathscr{L}). \qquad\qquad 3.$$

These expressions resemble the corresponding formulas in classical statistical physics ($Z$ being the partition function there), except that the measure in Equation 3 is complex. A full analogy between QFT and classical statistical physics is established in Euclidean (rather than Minkowski) space, however. After the analytic continuation $x_0 \to ix_4$ ($x_4$ is real), a normalized, real measure is obtained:

$$\sim \frac{1}{Z} \exp(-\int d^4x \mathscr{L}_E), \qquad\qquad 4.$$

where $\mathscr{L}_E$ is the Euclidean Langrangian and space-time indices enter symmetrically with positive metric diag. (1,1,1,1). Lorentz symmetry is replaced by four-dimensional Euclidean rotation symmetry.

The possibility of the analytic continuation to imaginary time can be established easily in perturbation theory (6). Although it is sometimes questioned whether this step is valid in the full nonperturbative theory (especially in confining models, where the final spectrum is far removed from the original fields), we take the Euclidean formulation as our starting point. For all the questions discussed in this paper, the answer can be found directly within the Euclidean formulation; an explicit continuation back to Minkowski space is not necessary.

### 2.2  *Lattice Regularization*

One must define the summation in Equation 2 in some sensible way; without such a definition Equation 2 is formal. The path integral representation of the propagation kernel in quantum mechanics (7) is defined usually by discretizing the time variable: the continuous path $x(t)$ is replaced by the discrete set of points $x_k = x(t_k)$ $(k = 0, 1, \cdots, n+1)$ and the summation is represented by ordinary integrals over the variables $x_k$ $(k = 1, 2, \cdots, n)$. At the end the limit, $n \to \infty$ is taken.

In field theory, where the dynamical variables (fields) are labeled by the points of the four-dimensional Euclidean space, the natural generalization of these steps is to introduce a four-dimensional, hypercubic lattice. By defining the fields on the elements of this lattice, one obtains a discrete set of variables, and the summation over the field configurations will be represented by ordinary integrals over them.

The lattice introduced this way not only defines the path integral representation of the vacuum functional (partition function), but gives an ultraviolet regularization as well. Let us denote the lattice spacing by $a$ and a lattice point by four integers $n = (n_1, n_2, n_3, n_4)$. The Fourier transform of a function $f(n)$ and its inverse are given by

$$\tilde{f}(p) = a^4 \sum_n \exp[i(pn)a] f(n),$$

$$f(n) = \frac{1}{(2\pi)^4} \int\!\!\!\int\!\!\!\int\!\!\!\int_{-\pi/a}^{\pi/a} d^4p \, \exp[-i(pn)a] \tilde{f}(p). \qquad 5.$$

Therefore the lattice provides a cutoff in momentum space, $\Lambda^{\text{cut}} \sim 1/a$.

There is a natural way to define scalar or vector fields on the lattice: scalar fields are defined on the sites, vector fields (characterized by a position and a direction) on the bonds of the lattice. Replacing the derivatives of the continuum action by the discrete difference ($\hat{\mu}$ is the unit vector along the $\mu$th direction),

$$\partial_\mu f(x) \to \Delta_\mu f(n) = \frac{1}{a} [f(n+\hat{\mu}) - f(n)], \qquad 6.$$

yields a lattice-regularized version of the model. The internal, global symmetries of the model are obviously preserved by this regularization.

The requirement of a local gauge symmetry, or the presence of fermions necessitate special considerations, however. The first problem has a beautiful and simple solution. Actually, the way the lattice incorporates gauge symmetries is more satisfactory than that of the continuum formulations. On the other hand, as we discuss below, the fermions are a permanent headache, without a fully satisfactory solution at present.

## 2.3    *Gauge Fields on the Lattice*

We want to preserve explicit gauge symmetry at any value of the lattice spacing $a$. Then the final continuum theory obtained in the limit $a \rightarrow 0$ will also have this property.

Consider a matter field $\phi_i(x)$ belonging to some irreducible representation of the gauge group $G$. The index $i$ refers to this representation, other possible indices are suppressed. The generators in this representation are denoted by $\lambda^a$, the gauge potentials by $A_\mu^a(x)$.[1] The field and its covariant derivative

$$[D_\mu \phi(x)]_i = [\partial_\mu \delta_{ij} + ig(\lambda^a)_{ij} A_\mu^a(x)] \phi_j(x) \qquad 7.$$

transform identically under the local gauge transformation

$$\phi_i(x) \rightarrow V_{ij}(x) \phi_j(x)$$

$$A_\mu^a(x) \lambda^a \rightarrow V(x) A_\mu^a(x) \lambda^a \cdot V^{-1}(x) - \frac{i}{g} \partial_\mu V(x) \cdot V^{-1}(x). \qquad 8.$$

Therefore the combinations $\phi_i^+(x)\phi_i(x)$, $\phi_i^+(x)(D_\mu\phi)_i(x), \cdots$ are gauge invariant and can be used to construct a gauge-invariant Lagrangian. In Equation 8, $V(x)$ is an element of the gauge group (in the representation of $\phi$) and summation over repeated indices is understood.

A naive generalization of these equations on the lattice—by associating the vector potentials with bonds and replacing the derivatives by finite differences—does not work. The discretized covariant derivative will have the correct transformation properties only in the limit when $a$ becomes infinitesimal. For finite lattice spacings $a$, gauge invariance will be violated.

How does one construct a gauge-invariant expression from $\phi^+(x)\phi(y)$ when $x - y$ is finite? A possible solution is to take

$$\phi^+(x)_i U(x,y)_{ij} \phi_j(y), \qquad 9.$$

where

$$U(x,y) = P\left\{\exp\left[ig \int_x^y \lambda^a A_\mu^a(z)\,dz^\mu\right]\right\} \qquad 10.$$

The line integral in Equation 10 is taken along some path connecting the points $x$ and $y$, and $P$ defines path-ordering.[2] Under the gauge trans-

---

[1] In QCD $G = SU(3)$, and we might think of $\phi_i$ as the quark fields in the fundamental representation. In this case, $i = 1,2,3$, while $a = 1,2,\cdots,8$.

[2] $\lambda^a A_\mu^a(z_1)$ is to the left of $\lambda^a A_\mu^a(z_2)$ if, along the path, $z_1$ is closer to $x$ than $z_2$.

formation Equation 8, $U(x,y)$ has the simple transformation property

$$U(x,y) \rightarrow V(x)U(x,y)V^{-1}(y),$$    11.

which assures that the expression in Equation 9 is gauge invariant.

If we put $y_\mu = x_\mu + \varepsilon\hat{\mu}$ in Equation 10, and expand the exponential, we get

$$\lim_{\varepsilon \to 0}\frac{1}{\varepsilon}[\phi_i^+(x)U_{ij}(x,y)\phi_j(y) - \phi_i^+(x)\phi_i(x)] = \phi_i^+(x)(D_\mu\phi)_i(x),$$    12.

but for finite $(x-y)$ the full expressions should be kept in order to preserve gauge invariance.

These considerations suggest that we use the matrix $U$ rather than the vector potential $\lambda^a A_\mu^a$ as the basic quantity for constructing an invariant action and path integral. To every directed bond connecting the points $n$ and $n+\hat{\mu}$, a gauge variable $U_{n\mu}$ is associated:

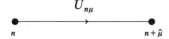

$$U_{n\mu}$$

$$n \qquad\qquad n+\hat{\mu}$$

Here $U_{n\mu}$ is an element of the gauge group $G$. The oppositely directed link is associated with the inverse

$$U_{n+\mu,-\mu} = U_{n\mu}^{-1}$$

$$n \qquad\qquad n+\hat{\mu}$$

The link variable $U_{n\mu}$ is defined to have the transformation property

$$U_{n\mu} \rightarrow V_n U_{n\mu} V_{n+\mu}^{-1},$$    13.

where $V_n \in G$ is the gauge transformation at the point $n$. Equation 13 assures that $\phi_n^+ U_{n\mu}\phi_{n+\mu}$ is gauge invariant and gives a gauge-invariant, discretized covariant derivative even for finite lattice unit $a$. Although $U_{n\mu}$ is the primary dynamical variable of the lattice formulation, the vector potential can also be defined via the equation

$$U_{n\mu} = \exp(igaA_{n\mu}^a\lambda^a).$$    14.

Equations 13 and 14 correspond to the classical expressions in Equations 11 and 10 respectively and reduce to them in the formal $a \rightarrow 0$ limit.

Our first task is to construct an action that describes the dynamics of the gauge variables $U_{n\mu}$. The next problem is to introduce matter fields (quarks in QCD) and define their interaction with the gauge fields. As an example, consider the case where $G = \mathrm{SU}(N)$ and $U_{n\mu}$ is an element of the defining representation.

## 2.4   The Gauge Field Action

Apart from the requirement of gauge symmetry, the guiding principles of constructing an action are locality and the wish to keep the global symmetries (parity, charge conjugation, translation, etc) of the continuum SU(N) Yang-Mills action—at least to the extent the lattice structure allows us to keep them.

The transformation property Equation 13 implies that the trace of the product of $U$ matrices along a closed loop is gauge invariant. The simplest gauge-invariant, local interaction can be defined as (1)

$$S_W = \text{const.} \sum_{\text{plaquettes}} (\text{Tr } U_p + \text{Tr } U_p^+), \qquad\qquad 15.$$

where

$$U_p = U_{n\mu} U_{n+\mu,\nu} U_{n+\nu,\mu}^+ U_{n\nu}^+, \qquad P: \quad \begin{array}{c} n+\hat{\nu} \quad\quad n+\hat{\nu}+\hat{\mu} \\[2pt] \square \\[2pt] n \quad\quad n+\hat{\mu} \end{array} \qquad 16.$$

In the classical $a \to 0$ limit, the gauge variable $U_{n\mu}$ and therefore the action $S_W$ can be expanded in increasing powers of $a$. It is a simple algebraic exercise to show that the leading terms give the usual continuum Yang-Mills action if const. $= -1/g^2$ is chosen (5, 8):

$$S_W = -\frac{1}{g^2} \sum_p (\text{Tr } U_p + \text{Tr } U_p^+) \xrightarrow[\substack{a \to 0 \\ \text{formal limit}}]{} -\frac{1}{4} \int d^4x F_{\mu\nu}^a F_{\mu\nu}^a + \text{const.} \qquad 17.$$

The partition function of the pure SU(N) gauge theory on the lattice has the form

$$Z = \left( \prod_{\text{links}(n,n+\hat{\mu})} \int dU_{n\mu} \right) \exp(-S_W), \qquad\qquad 18.$$

where $dU$ is the invariant Haar measure with properties

$$\int dU = 1, \qquad \int dU f(U) = \int dU f(VU), \qquad\qquad 19.$$

where $V$ is an arbitrary element of the group $G$. The action $S_W$ is gauge invariant, which, together with the property in Equation 19, assures that the expectation values

$$\langle O \rangle = \frac{1}{Z} \prod_{\text{links}} \int dU_{n\mu} O(U) \exp(-S_W) \qquad\qquad 20.$$

are also gauge invariant under the local gauge transformation Equation 13.

On a four-dimensional hypercubic lattice of linear dimension $L$ there are $4L^4$ bonds. The partition function in Equation 18 is defined by $4L^4$ integrals over the group space. The group integral itself is an integral over a compact space, which, in principle, can be expressed as an integral over $N^2 - 1$ generalized Euler angles [in the case of SU($N$). This explicit representation we do not need here, however. We want to emphasize only that Equations 18 and 20 are completely well defined, finite expressions. Unlike in the continuum formulation, no gauge fixing is needed.

In the continuum formulation one works with the vector potential $A_\mu^a \lambda^a$, which takes its values in the Lie algebra of the group. On the lattice one works with the gauge variables $U_n$, which take their values in the group $G$ itself. This permits defining and investigating local gauge theories defined on discrete groups that do not even have a Lie algebra (9).

## 2.5  Matter Fields and Their Interactions

We consider here fermions in the fundamental representation of the SU($N$) gauge group (quarks). The introduction of scalar fields and the generalization to other gauge groups is rather straightforward.

The quark field is denoted by $\psi_\alpha^{if}(x)$, where $i = 1, 2, \ldots, N$, $\alpha = 1, \ldots, 4$, and $f = $ u, d, s, c, $\ldots$ are the color, Dirac, and flavor indices respectively. Following the usual recipe, the lattice form of the action of free fermions becomes

$$S = a^4 \sum_n \left\{ \sum_{\mu=1}^{4} \frac{1}{2a} [\bar{\psi}_\alpha^{if}(n)(\gamma_\mu)_{\alpha\beta} \psi_\beta^{if}(n+\hat{\mu}) - \bar{\psi}_\alpha^{if}(n+\hat{\mu})(\gamma_\mu)_{\alpha\beta} \psi_\beta^{if}(n)] \right.$$

$$\left. + m_f \bar{\psi}_\alpha^{if}(n) \psi_\alpha^{if}(n) \right\}, \qquad 21.$$

where $m_f$ are the flavor-dependent bare quark masses and $\gamma_\mu$ are the Dirac matrices satisfying $\{\gamma_\mu, \gamma_\nu\}_+ = 2\delta_{\mu\nu}$.

The discussion in Section 2.3 leads to the following gauge-invariant form of the quark–gauge field interaction

$$S_{\mathrm{M}} = a^4 \sum_n \left\{ \sum_{\mu=1}^{4} \frac{1}{2a} [\bar{\psi}^{if}(n)\gamma_\mu U_{n\mu}^{ij} \psi^{jf}(n+\hat{\mu}) - \bar{\psi}^{if}(n+\hat{\mu})\gamma_\mu U_{n\mu}^{+ij} \psi_{(n)}^{if}] \right.$$

$$\left. + m_f \bar{\psi}^{if}(n) \psi^{if}(n) \right\}, \qquad 22.$$

where the Dirac indices are not written out explicitly. After the replacement

$$\psi^{if}(n) \to (a^4 m_f)^{-1/2} \psi^{if}, \qquad 23.$$

one obtains

$$S_M = \sum_n \left\{ \bar{\psi}^{if}(n)\psi^{if}(n) + \sum_{\mu=1}^{4} \frac{1}{2m_f a} [\bar{\psi}^{if}(n)\gamma_\mu U_{n\mu}^{ij}\psi^{jf}(n+\hat{\mu}) \right.$$

$$\left. - \bar{\psi}^{if}(n+\hat{\mu})\gamma_\mu U_{n\mu}^{+ij}\psi^{if}(n)] \right\}. \qquad 24.$$

Together with the gauge field action in Equation 15, we now have an explicitly gauge-invariant, lattice-regularized version of QCD.

Unfortunately, as it stands, the action $S_M$ is not appropriate to describe QCD. The regularization process followed above introduced new, unwanted fermionic degrees of freedom. There is a hidden, 16-fold degeneracy in the spectrum of $S_M$ that renders it useless in its present form. We postpone the discussion of this problem to Section 9.

In the path integral formulation, the fermion fields are represented by anticommuting $c$ numbers, called Grassmann variables (10):

$$\{\psi_\alpha^{if}, \psi_\beta^{jf'}\}_+ = \{\psi_\alpha^{if}, \bar{\psi}_\beta^{jf'}\}_+ = \ldots = 0. \qquad 25.$$

Because of the anticommutation property, we have $(\psi_\alpha^{if})^2 = (\bar{\psi}_\alpha^{if})^2 = 0$. The integral over these Grassmann variables is defined to vanish except when the integrand is a product of all the Grassmann variables, each variable occurring once and only once. For a simple pair of variables $\psi, \bar{\psi}$ this rule reads

$$\int d\bar{\psi}\, d\psi = 0, \qquad \int d\bar{\psi}\, d\psi\, \psi = \int d\bar{\psi}\, d\psi\, \bar{\psi} = 0$$

$$\int d\bar{\psi}\, d\psi\, \bar{\psi}\psi = -\int d\bar{\psi}\, d\psi\, \psi\bar{\psi} = 1 \qquad 26.$$

with trivial generalization to many variables. For later use let us mention a simple consequence of the rules defined above. Let $A$ be an $n \times n$ matrix. It is an easy exercise to show that

$$\int d\bar{\psi}_1\, d\psi_1\, d\bar{\psi}_2\, d\psi_2 \ldots d\bar{\psi}_n\, d\psi_n \exp(\bar{\psi}_n A_{nm}\psi_m) = \det A. \qquad 27.$$

## 3.   OBSERVABLES IN A YANG-MILLS THEORY

As discussed above, the trace of the product of gauge variables along a closed loop $C$ on the lattice (called a Wilson loop)

$$W(C) = \text{Tr} \prod_C U_{n\mu}, \qquad \underset{n}{\bullet}\!\!\!-\!\!\!-\!\!\!-\!\!\!-\!\!\!\underset{n+\hat{\mu}}{\bullet} \in C \qquad 28.$$

is gauge invariant. Wilson loop expectation values and correlations between Wilson loops are the most important observables in a pure gauge theory. They are directly related to the static potential between external, colored sources and to the glueball spectrum, respectively. In both cases the connection comes from the following basic observation. Consider an operator $O(t)$ in Minkowski space, which is constructed from the basic variables on the time $= t$ hyperplane. By inserting a complete set of energy eigenstates into the connected correlation

$$\langle 0|O(t)O(0)|0\rangle_C \equiv \langle 0|O(t)O(0)|0\rangle - \langle 0|O(t)|0\rangle\langle 0|O(0)|0\rangle, \qquad 29.$$

one obtains

$$\langle 0|O(t)O(0)|0\rangle_C = \sum_{n\neq 0} \exp[-it(E_n-E_0)]|\langle n|O(0)|0\rangle|^2. \qquad 30.$$

By going to Euclidean space and taking the limit $it = x_4 \to \infty$, we get

$$\langle O(x_4)O(0)\rangle_C = \sum_{n\neq 0} \exp[-x_4(E_n-E_0)]|C_n|^2 \xrightarrow[x_4 \to \infty]{}$$

$$\sim \exp[-x_4(E_1-E_0)] + (\text{exp. small corrections}), \qquad 31.$$

where $E_1$ is the first excited state above the ground state with the quantum numbers carried by the variable $O$.[3] If the spectrum is continuous, there might occur some power correction $x_4^{-n}$ in front of the exponential, but the main conclusion remains unchanged: the long-distance correlations provide information on the spectrum of low-lying excitations.

## 3.1 Wilson Loop Expectation Values and the Potential

A Wilson loop expectation value describes the creation, propagation, and annihilation of a static quark-antiquark source.

In order to illustrate this property, consider the following expectation value

$$\langle \bar{\psi}_\alpha^i(0,T)\left(\prod_C U_{n\mu}\right)^{ij} \psi_\alpha^j(\mathbf{R},T)\bar{\psi}_\beta^k(\mathbf{R},0)\left(\prod_{C'} U_{n\mu}\right)^{kl} \psi_\beta^l(0,0)\rangle_C, \qquad 32.$$

Where $\mathbf{R} = (R,0,0)$ and the paths $C$ and $C'$ are straight lines connecting the point $(0,T)$ with $(\mathbf{R},T)$ and $(\mathbf{R},0)$ with $(0,0)$ respectively (Figure 1).

As discussed above (Equations 9–12), the combination $\bar{\psi}(\Pi_C U)\psi$ describes a q-q̄ pair in a gauge-invariant way; therefore Equation 32 gives the propagation of this gauge-invariant object over a "time" interval $T$.

In the limit of static sources the matrix element in Equation 32 should be

---

[3] Asuming that $O|0\rangle$ is not orthogonal to $|1\rangle$.

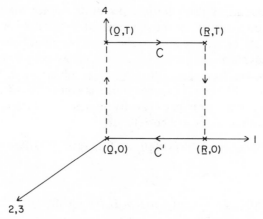

*Figure 1* The connected correlation function in Equation 32 is proportional to the expectation value of the Wilson loop in this figure (in the $m \to \infty$ static limit).

evaluated in the $m \to \infty$ limit, where $m$ is the mass of the quark. Consider therefore

$$
\lim_{m \to \infty} \frac{1}{Z} \int DUD\bar{\psi}D\psi \, \bar{\psi}^i_\alpha(0, T) \left( \prod_C U_{n\mu} \right)^{ij} \psi^j_\alpha(\mathbf{R}, T) \bar{\psi}^k_\beta(\mathbf{R}, 0) \left( \prod_{C'} U_{n\mu} \right)^{kl} \psi^l_\beta(0, 0)
$$

$$
\times \exp\left( -S_W - \sum_n \left\{ \bar{\psi}^i_\alpha(n)\psi^i_\alpha(n) + \sum_{\mu=1}^4 \frac{1}{2ma} [\bar{\psi}^i_\alpha(n) (\gamma_\mu)_{\alpha\beta} U^{ij}_{n\mu} \psi^j_\beta(n + \hat{\mu}) \right. \right.
$$

$$
\left. \left. - \bar{\psi}^i_\alpha(n + \hat{\mu}) (\gamma_\mu)_{\alpha\beta} U^{+ij}_{n\mu} \psi^j_\beta(n)] \right\} \right).
\qquad 33.
$$

According to the rules of Grassmann integrals (Section 2.5), $\psi^l_\beta(0, 0)$ requires a $\bar{\psi}^l_\beta(0, 0)$ from the exponent in order to produce a nonzero integral. This can come only from the interaction part that is multiplied by $1/(2ma)$ since the first term is diagonal and $\psi^l_\beta(0, 0)\bar{\psi}^l_\beta(0, 0)\psi^l_\beta(0, 0) = 0$. When, however, we get $\bar{\psi}^l_\beta(0, 0)$ by expanding the interaction part of the exponent, we bring down at the same time a gauge matrix $U$, a Dirac matrix, and another paired $\psi$ fermion field. This field again needs a pair from the exponent, and so on. Since the interaction connects nearest-neighbor points, a continuous line is produced this way: the fermion repeatedly "hops" from one point to its nearest neighbor until it arrives at the point $(0, T)$, where it is eaten by $\bar{\psi}^i_\alpha(0, T)$ and the process ends.[4] With every step we acquire a factor $1/(2ma)$. It is clear that for $m \to \infty$, the leading contribution comes when the quark

---

[4] The possibility that it arrives at $(\mathbf{R}, 0)$ and is eaten by $\bar{\psi}^k_\beta(\mathbf{R}, 0)$ is excluded by considering the connected correlation function.

hops from $(0, 0)$ to $(0, T)$ and from $(\mathbf{R}, T)$ to $(\mathbf{R}, 0)$ along the shortest straight lines depicted in Figure 1.

The result is proportional to the trace of the product of gauge variables along the rectangular path [i.e. the Wilson loop $W(R, T)$] multiplied by irrelevant constants. Therefore the matrix element of Equation 31 is proportional to the Wilson loop expectation value $\langle W(R, T) \rangle$:

$$\langle \bar{\psi}(0, T) \prod_C U_{n\mu} \psi(\mathbf{R}, T) \bar{\psi}(R, 0) \prod_{C'} U_{n\mu} \psi(0, 0) \rangle_C \sim \langle W(R, T) \rangle, \qquad 34.$$

what we wanted to show.

Using this interpretation of the Wilson loop expectation value, on the basis of Equation 31 one obtains

$$\langle W(R, T) \rangle \xrightarrow[T \to \infty]{} C \exp[-T \cdot V(R)], \qquad 35.$$

where the energy difference between the lowest state of a source at a distance $R$ and the vacuum is denoted by $V(R)$. The function $V(R)$ is the potential energy of a q-q̄ source. From Equation 35 we get

$$V(R) = \lim_{T \to \infty} V_T(R) \equiv \lim_{T \to \infty} \left( -\ln \frac{\langle W(R, T+1) \rangle}{\langle W(R, T) \rangle} \right). \qquad 36.$$

In a confining theory, $V(R)$ increases without bound as $R$ increases. Based on model studies and on the physical picture that a color electric flux tube (a string with finite width) is formed between the sources, one expects

$$V(R) \xrightarrow[R \to \infty]{} \sigma R, \qquad 37.$$

where $\sigma$ is called the string tension. In Equation 37 and elsewhere, the physical dimensions should be restored with the lattice unit $a$. This is not always done explicitly in our equations. The function obtained in Equation 36 is dimensionless, $a^{-1} V(R)$ is the physical potential energy at distance $r = Ra$. Equation 37 reads then

$$a^{-1} V(R) \xrightarrow[Ra \to \infty]{} (\sigma/a^2) Ra, \qquad 38.$$

where $\sigma/a^2$ and $Ra$ are the physical string tension and distance, respectively.

In a nonconfining theory (like QED) we expect

$$V(R) \xrightarrow[R \to \infty]{} \text{const.} \qquad 39.$$

Equations 36, 37, and 39 imply that large Wilson loops behave qualitatively differently depending on whether the sources are confined or not

$$\langle W(R, T)\rangle \underset{\substack{T \to \infty \\ R \to \infty \\ T \gg R}}{\propto} \begin{cases} \exp-(\text{area of the loop}), \text{confinement} \\ \\ \exp-(\text{perimeter of the loop}), \text{nonconfinement,} \end{cases} \qquad 40.$$

i.e. the Wilson loop expectation value is an order parameter for confinement.

One should note that since continuous rotation symmetry is broken by a hypercubic lattice, in general the potential energy between sources will depend on $\mathbf{R}$ rather than on $|\mathbf{R}|$ only; we have $V = V(\mathbf{R})$. In the equations above $V(R) = V(R, 0, 0)$ is meant. The important question of restoration of rotation symmetry is discussed in Section 8.

## 3.2  Loop-Loop Correlations and the Glueball Spectrum

The glueballs are hypothetical particles composed of gluons (and virtual q-q̄ pairs in the full theory). They are colorless, flavorless objects. A rich spectrum might exist, however, with different $J^{PC}$ quantum numbers and with possible recurrences ("radial excitations"). Although their existence is predicted by QCD, there is no experimentally firmly established glueball yet (11). The glueball with smallest mass defines the mass gap of the theory.

In searching for the smallest mass with given quantum numbers, the usual procedure is to create an excitation with these quantum numbers and—as suggested by Equation 31—follow it over large distances. Let $\theta = \theta(\mathbf{n}, T)$ be a local, gauge-invariant operator having the required quantum numbers. According to Equation 31 we get

$$\langle \theta(\mathbf{0}, T)\theta(\mathbf{0}, 0)\rangle_C \underset{T \to \infty}{\propto} \exp(-mT), \qquad 41.$$

where $m$ is the lowest mass in the channel. The variable $\theta(\mathbf{n}, T)$ is composed of Wilson loops lying in the $n_4 = T$ hyperplane, and localized around the point $\mathbf{n}$. Therefore the expectation value in Equation 41 is a linear combination of long-distance loop-loop correlations.

It is not completely trivial to construct an operator for given $J^{PC}$ quantum numbers. Parity and charge conjugation transformations can easily be defined on the lattice (12); angular momentum requires special considerations, however (13). On the lattice the full rotation symmetry is reduced to the cubic group with 24 elements. The cubic group has five irreducible representations denoted by $A_1$, $A_2$, $E$, $T_1$, and $T_2$. Local operators can be constructed on the lattice that transform according to these representations. These operators create excitations, which, from the point of view of the full rotation group, contain (infinitely) many different $J$

values in the continuum limit. There exists a standard way to determine this decomposition. Consider a representation $J$ of the rotation group. Identify the 24 elements corresponding to the cubic group. The character of these elements is a function (defined over five points) that can be expanded in terms of the characters of the cubic group representations. This gives the decomposition of the representation $J$ from the point of view of the cubic group and vice versa. As an example, the lowest $J$ occurring in $T_1$ is $J = 1$, but it contains $J = 3$ and higher angular moments also. It is expected, however, that the smallest mass in the $J = 1$ channel will be lighter than the masses in the $J = 3, \ldots$ channels; therefore at large distances $T_1$ singles out the $J = 1$ channel. In a similar way $A_1$ is associated with $J = 0$, $A_2$ with $J = 3$, while $E$ and $T_2$ with $J = 2$.

# 4. THE CONTINUUM LIMIT

The regularization has a temporary role only. At the end, the regularization should be removed and, in a renormalizable theory, the final predictions will not depend upon the specific regularization used. In lattice regularization, this is the process of taking the continuum limit.

## 4.1 Continuum Limit of Lattice Gauge Theories

Removing the regularization is a subtle problem. In a local field theory the elementary interactions are expressed via derivatives, and therefore connect variables that are separated over infinitesimal distances. After lattice regularization the elementary interactions extend over $O(a)$ distances, as illustrated by Equations 15 and 24. On the other hand, the typical distance over which events are correlated, for instance in strong interactions, is $O(1 \text{ fm})$. When $a \to 0$, 1 fm corresponds to a large number ($\to \infty$) of lattice units; therefore the final solution of the field theory should have nonzero correlations over many lattice units. It requires a highly nontrivial collective behavior to get correlations over finite distances starting from elementary interactions extending over infinitesimal distances. Achieving this collective behavior is the main difficulty of a QFT.

For simplicity, in the following arguments consider a field theory without dimensionful parameters (like a Yang-Mills model, or QCD with massless quarks). The only parameter available to carry the dimensions is the lattice unit $a$. Therefore a mass prediction should have the form

$$m = a^{-1} f(g_1, g_2, \ldots),$$ 42.

where $g_1, g_2, \ldots$ are the dimensionless couplings of the theory. In order to obtain finite mass predictions when $a \to 0$, the couplings $g_1, g_2, \ldots$ should also be tuned toward some $g_1^*, g_2^*, \ldots$, where

$$f(g_1^*, g_2^*, \ldots) = 0. \qquad 43.$$

Looking at the lattice system as a system in classical statistical mechanics, the point $g_1^*, g_2^*, \ldots$ corresponds to a continuous phase transition point. Really, at this point the dimensionless correlation length (i.e. the correlation length measured in lattice units) $\xi \approx (am)^{-1}$ approaches infinity.

Therefore the continuum limit of lattice field theories involves the tuning procedure

$$\begin{pmatrix} a & \to 0 \\ g_1 & \to g_1^* \\ g_2 & \to g_2^* \\ & \vdots \end{pmatrix}, \qquad 44.$$

where $g_1^*, g_2^*, \ldots$ is a critical point of the corresponding statistical physical problem. Close to this critical point there will be large-scale fluctuations in the system that tend to wash away the underlying lattice structure, which leads to the possibility of a Euclidean rotationally symmetric continuum theory.

## 4.2   Renormalizability, the $\beta$ Function, and the Mass Scale in Yang-Mills Theory

A pure Yang-Mills gauge theory has a single dimensionless coupling constant $g$. As discussed, $g$ is a function of $a$: $g = g(a)$, which should be chosen in such a way that the physical predictions become independent of the lattice spacing (and stay finite) when $a$ is small. This is the usual requirement of renormalizability.

What is $g^* \equiv g(a = 0)$? There are several arguments leading to the conclusion that $g^* = 0$. The simplest of them observes that $g^*$ is the bare coupling constant of the resulting continuum Yang-Mills theory. On the other hand, this theory is asymptotically free; therefore the bare coupling, which describes the interactions at the scale of the cutoff, goes to zero when the cutoff ($\sim 1/a$) goes to infinity.

Consider a physical quantity $X$, which has a dimension (mass)$^l$. Then we have

$$X = a^{-l} f(g). \qquad 45.$$

The requirement of renormalizability, i.e. that all the physical predictions must become independent of the cutoff when the cutoff is large, fixes uniquely the $g$ dependence of the function $f$. Cutoff independence implies

$$a \frac{\mathrm{d}}{\mathrm{d}a} X \bigg|_{a \to 0} = 0, \qquad 46.$$

or

$$\left[ a\frac{\partial}{\partial a}\bigg|_g + \left( a\frac{\partial g}{\partial a} \right)\frac{\partial}{\partial g}\bigg|_a \right] X_{a\to 0} = 0,$$    47.

which leads to the following first-order differential equation for $f$:

$$lf(g) + B(g)\frac{\mathrm{d}}{\mathrm{d}g}f(g) = 0.$$    48.

Here $B(g) = -a\,\mathrm{d}/\mathrm{d}a\,g(a)$ is the $\beta$ function, which specifies how the bare coupling should be tuned as a function of $a$. The leading behavior of $B(g)$ for small $g$ is known from perturbation theory (14)

$$B(g) = -\beta_0 g^3 - \beta_1 g^5 + \ldots,$$    49.

where

$$\beta_0 = \frac{11}{3}\frac{N}{16\pi^2}, \qquad \beta_1 = \frac{34}{3}\left( \frac{N}{16\pi^2} \right)^2.$$    50.

The solution of Equation 48 gives

$$X = c_x(\Lambda^{\text{latt}})^l,$$    51.

where

$$\Lambda^{\text{latt}} \equiv \frac{1}{a}\exp\left[ -\int^g \frac{1}{B(g')}\,\mathrm{d}g' \right]$$    52a.

$$= \frac{1}{a}\exp\left( -\frac{1}{2\beta_0 g^2} \right)(\beta_0 g^2)^{\beta_1/2\beta_0^2}[1 + O(g^2)].$$    52b.

$\Lambda^{\text{latt}}$ is a cutoff-independent mass parameter. (It satisfies Equation 46.) This mass parameter sets the scale for QCD. It is an external parameter, like the fine structure coupling $\alpha \sim 1/137$ in QED. It is defined in an analogous way to the $\Lambda$ parameters of the continuum regularization/renormalization schemes (like $\Lambda^{\text{MS}}$ or $\Lambda^{\text{MOM}}$). Classical Yang-Mills theory does not have a scale. It has a single dimensionless coupling constant $g$. In the quantized theory this $g$ becomes the unphysical bare coupling, while the regularization process brings in an unphysical dimensionful parameter "$a$". A special combination (Equation 52) of these unphysical parameters gives, however, a cutoff-independent, physical mass $\Lambda^{\text{latt}}$, which carries the dimension of all the physical predictions. This beautiful process by which a renormalizable field theory creates a physical scale from "nothing" is called dimensional transmutation (15).

The constant $c_x$ in Equation 51 is an integration constant, which cannot be predicted from these considerations alone. One of the main tasks of the nonperturbative techniques is to dig out $c$ from the full theory for different physical quantities.

## 4.3 *Scaling*

In the coupling constant region where cutoff independence applies, the $g$ dependence of $a\Lambda^{\text{latt}}$ or $a^l X$ is completely fixed by Equations 52. This $g$ dependence is governed by the $\beta$ function and is called scaling. For small $g^2$, where the $O(g^2)$ corrections are small in Equation 52, this functional dependence is known. This limiting behavior is called asymptotic scaling.

## 5.  WEAK AND STRONG COUPLING EXPANSIONS

There exist two analytic expansion techniques in the limits $g \to 0$ and $\infty$. We do not discuss these methods in detail. This section serves only to orient the reader on the role they now play in nonperturbative lattice QCD studies.

In the limit $g \to 0$, the factor $1/g^2$ in Equation 17 becomes very large, which suppresses the fluctuations in the plaquette variable: $1/N \operatorname{Tr} U_p \to 1$ in those configurations that give a nonzero contribution to the partition function. The fluctuations around this trivial limit can be treated in perturbation theory. Apart from technical complications, this expansion is completely analogous to the perturbation theory of the continuum formulation. After gauge fixing (16) (the usual Fadeev-Popov procedure can be followed, for instance), the gauge variable $U_{n\mu}$ can be expanded

$$U_{n\mu} = \exp(igaA_{n\mu}) = 1 + igaA_{n\mu} + \dots, \qquad 53.$$

the measure $DU$ and the action are rewritten in terms of the vector potential $A_{n\mu}$, and an ordinary (though rather complicated) perturbation theory can be set up. In this case the lattice is just one of the possible regularizations. The renormalized perturbation theory obtained this way is identical to that obtained by, say, dimensional regularization.

In Section 4 we argued that the continuum limit of a lattice Yang-Mills theory is the $g \to 0$ limit. Does it mean that in the continuum limit the model becomes perturbative? Of course not. In calculating an $R \times R$ Wilson loop expectation value in perturbation theory, the correction to the leading result typically has the form

$$\sim g^2 \ln R = g^2 \ln(r/a) \qquad 54.$$

where $r = Ra$ and $a$ are the size of the Wilson loop and the lattice unit respectively, both measured in fm. When $g^2 \to 0$, $a \to 0$ also in such a way as to keep $\Lambda^{\text{latt}}$ (which is cutoff independent) constant. By using the explicit form of $\Lambda^{\text{latt}}$ in Equation 52b, it is easy to show that $g^2 \ln(r/a)$ becomes $O(1)$ in the $\{{}^{g^2 \to 0}_{a \to 0}\}$ limit, if $r$ is large. Perturbation theory breaks down at large distances, as is well known from the continuum formulations.

Since lattice perturbation theory is equivalent to, but technically more involved than, continuum perturbation theory, it would not make sense to repeat or to attempt extending the continuum perturbative calculations on the lattice. It has a different role, as we summarize next.

1. First, the perturbative results obtained in the continuum formulation are connected with the nonperturbative predictions obtained on the lattice. For this purpose we need the connection between the respective scale parameters. This relation can be obtained exactly by a one-loop perturbative calculation on the lattice, which gives (17)

$$\Lambda^{\text{latt}}_{\text{W}} = \frac{1}{57.5} \Lambda^{\text{MOM}} \qquad \text{SU(2)}$$

$$= \frac{1}{83.5} \Lambda^{\text{MOM}} \qquad \text{SU(3)}, \tag{55.}$$

where $\Lambda^{\text{latt}}_{\text{W}}$ is the scale parameter of the Wilson action (Equation 17). These results have been extended to other lattice actions and to QCD with fermions (18).

2. Lattice artifacts are systematically eliminated in perturbation theory. The general aim is to improve the rate at which the lattice-regularized model approaches the continuum theory in the $g \to 0$ limit. The improvement program depends on the application: significant effort has been spent on improving actions for numerical studies (19), or on improving renormalization group schemes in MC renormalization group (MCRG) studies (20).

3. Next, unwanted perturbative tails are subtracted. The lattice definition of certain nonperturbative quantities is contaminated by unwanted perturbative (usually divergent) contributions. This problem is present in the case of the vacuum condensate $\langle FF \rangle$ and also for certain definitions of the topological charge.

4. Finally, a possible direct role of perturbation theory in spectrum calculations is sought. Lüscher suggested calculating perturbatively the mass gap of an asymptotically free system in a small $L^d$ volume (21). By adding information obtained from the renormalization group, this expansion is extrapolated to the large volume limit.

Apart from the last application, weak coupling perturbation theory plays only an indirect role in obtaining nonperturbative results. On the other hand, the strong coupling expansion is a full-fledged nonperturbative method.

Lattice gauge theories can be solved exactly in the $g \to \infty$ strong coupling limit. Because the factor $1/g^2$ in front of Equation 17 is small, the Boltzmann factor can be expanded, and a systematic expansion in $1/g^2$ can be constructed. The strong coupling expansion is in complete analogy with the high-temperature expansion in statistical physics.

For large $g^2$ the expansion is convergent (22) and describes a confining system. Specifically, in the strong limit of a SU($N$) gauge theory, sources in the fundamental representation are confined. In fact, by using Equation 20 and the group integrals [in SU(3)]

$$\int dU\, U = \int dU\, U^+ = 0$$

$$\int dU\, U_{ij}U_{kl}^+ = \tfrac{1}{3}\delta_{il}\delta_{jk}, \qquad\qquad 56.$$

it is easy to show that the leading contribution to the expectation value of a Wilson loop $R \times T$ is

$$\langle W(R, T)\rangle \underset{g^2 \to \infty}{=} 3\frac{1}{(3g^2)}RT = 3\exp(-RT\cdot\ln 3g^2). \qquad 57.$$

According to Equation 40, the area behavior of Equation 57 implies confinement with a string tension

$$\sigma \underset{g^2 \to \infty}{=} a^{-2}\ln 3g^2. \qquad\qquad 58.$$

Unfortunately, the continuum limit is defined by the opposite, $g \to 0$ limit. The idea of the strong coupling expansions is to derive a (long) series for the physical quantity in question, and then extrapolate this power series toward the continuum point $g = 0$.

For many classical statistical systems the strong coupling expansion provides precise information on the critical behavior. In asymptototically free models, the results obtained so far are much less spectacular. No series exists for the potential $V(R)$ itself. For the string tension, the extrapolation of the available series (2, 23) toward the continuum seems to be hindered by a special lattice singularity.[5] The strong coupling expansion for the glueball spectrum (25) is technically difficult, and the extrapolation of the available, rather short series is plagued by ambiguities (26).

On the other hand, the glueball spectrum is also a very difficult problem within the Monte Carlo technique, and much less effort has been spent on deriving long series than on running the Monte Carlo simulation to exhaustion. Nevertheless, it is difficult to see how the strong coupling

---

[5] This singularity is due to the roughening transition. At this point, strong, long-wavelength fluctuations "roughen" the flux tube connecting the sources (24).

expansion can become competitive without some new technical developments.

# 6.  MONTE CARLO SIMULATION

As discussed in Section 3, the relevant physical quantities can be extracted from appropriate gauge-invariant expectation values. The main problem is therefore to calculate expectation values like[6]

$$\langle A \rangle = \frac{(\prod_l \int dU_l) A(U) \exp[-S(U)]}{(\prod_l \int dU_l) \exp[-S(U)]}.$$   59.

On a finite lattice, Equation 59 is a well-defined, multidimensional integral. In the case of SU(N) on an $L^4$ lattice, $\prod_l \int dU_l$ is given by $4L^4(N^2-1)$ integrals over the generalized Euler angles. Monte Carlo (MC) simulation is a numerical method to evaluate these integrals (27). With the advent of powerful computers this method became practical for investigating relevant quantum field theories.

Representing the integrals in Equation 59 by a sum over sufficiently many points, the multidimensional integral becomes a summation. Each term is represented by a set of gauge variables on the links $\{U_l\}$, called a gauge field configuration. The contribution of a given configuration to $\langle A \rangle$ is

$$\sim A(\{U_l\}) \exp[-S(\{U_l\})].$$   60.

The number of configurations is so large (for continuous gauge groups it is infinite) that it is impossible to sum over all the configurations exactly. This would not be a wise procedure anyhow: most of the configurations have a very small Boltzmann factor; therefore their contribution to the total sum is negligible. Equation 60 suggests sampling over the configurations in a biased way: sum over a set of configurations, where the probability that a given configuration $\{U_l\}$ is included in this set is proportional to the Boltzmann weight $\exp[-S(\{U\})]$. This procedure is called importance sampling.

Let us assume that a sequence of $M$ configurations is generated with the equilibrium probability

$$P_{eq}(\{U_l\}) \sim \exp[-S(\{U_l\})].$$   61.

---

[6] In this section, only pure gauge theories are considered. Extending these considerations to scalar fields is simple, while fermion fields (represented by Grassmann variables) create special problems (see Section 10).

In this case the expectation value $\langle A \rangle$ can be approximated by

$$\langle A \rangle \approx \bar{A} = \frac{1}{M} \sum_{\nu=1}^{M} A(\{U_l\}).$$  62.

Configurations with the equilibrium distribution Equation 61 can be generated by a Markov process. The elements of the Markov chain are configurations. These configurations are generated subsequently, each configuration from the previous one. The transition probability $\nu \to \nu'$ of creating the configuration $\{U\}_{\nu'}$ from $\{U\}_\nu$ in a step is given by $W(\nu \to \nu')$. The transition probability must satisfy the following properties: (a) $\sum_{\nu'} W(\nu \to \nu') = 1$; (b) any finite action configuration should be reachable in a finite number of steps; and (c) a detailed balance condition must hold:[7]

$$P_{eq}(\nu)W(\nu \to \nu') = P_{eq}(\nu')W(\nu' \to \nu).$$  63.

It is easy to show (at least on a nonrigorous level) that a transition probability with properties $a$, $b$, and $c$ drives the system toward equilibrium, i.e. starting from an arbitrary configuration, after many steps along the chain, the probability that a given configuration is created will be equal to $P_{eq}$.

There are many different ways to construct $W(\nu \to \nu')$ satisfying the constraints above. In most of the lattice gauge theory applications, some version of the Metropolis or the heatbath methods were used.

### 6.1    Metropolis and the Heatbath Method

Let us create a new configuration ($\nu'$) from the old one ($\nu$) by some random process. This transition probability must satisfy $\omega(\nu \to \nu') = \omega(\nu' \to \nu)$. If the configuration $\nu'$ has a smaller action than $\nu$, then the new configuration is accepted as the next member of the Markov chain. If, however $S_{\nu'} - S_\nu > 0$, the change is accepted with the conditional probability $\exp(S_\nu - S_{\nu'})$. For that, a random number $r \in (0, 1)$ is drawn and the next element of the Markov chain is taken to be $\nu'$ if $r < \exp(S_\nu - S_{\nu'})$, otherwise it remains $\nu$. The transition probability $W(\nu \to \nu')$ corresponding to this procedure satisfies the detailed balance condition Equation 63. When $S_{\nu'} < S_\nu$, the change $\nu \to \nu'$ is accepted with probability 1, and therefore

$$W(\nu \to \nu') = \omega(\nu \to \nu').$$  64.

On the other hand, for the opposite process $\nu' \to \nu$, $\delta S_{\nu' \to \nu} = -\delta S_{\nu \to \nu'}$ and we have

$$W(\nu' \to \nu) = \exp(S_{\nu'} - S_\nu)\omega(\nu' \to \nu).$$  65.

---

[7] This condition is sufficient, but not necessary. The conventional implementations satisfy this condition.

Since $\omega(v \rightarrow v') = \omega(v' \rightarrow v)$, Equations 64 and 65 imply Equation 63.

In most of the applications the configuration is changed locally. The configuration $v'$ is the same as $v$ except on a given link, where $U_\ell \rightarrow U'_\ell$. A possibility is to define this local change as follows. Let $V_1, V_2, \ldots, V_k$ be a set of group elements. If $V$ is an element of this set, $V^{-1}$ should be an element also. Choose randomly one of the elements $V_i$ and define $U'_\ell = V_i U_\ell$. The probability $W$ defined this way clearly satisfies $\omega(v \rightarrow v') = \omega(v' \rightarrow v)$. The Metropolis algorithm requires that we calculate $(S_{v'} - S_v)$. Since $v$ and $v'$ differ only along a link, only a few terms of the local action contribute to this difference. Usually a given gauge variable is updated several times before moving to the next variable. The reason is that, in subsequently updating the same variable several times, many of the numerical manipulations (which are needed for the first update) need not be repeated and the method becomes more effective. This local sequence $U_\ell \rightarrow U'_\ell \rightarrow U''_\ell \rightarrow \ldots$ is a Markov chain itself, which, after many steps, will have an equilibrium distribution. It is like attaching a local heatbath to the link that is updated (all the other links are fixed). For certain systems an updating procedure can be devised that creates this one variable-limiting distribution directly, without going through this local Markov chain. This is called the heatbath method (28).

## 6.2 Statistical Errors

The expectation value of $A$ is approximated by the average over the generated configuration (Equation 62). If the configurations over which $A$ is calculated are statistically independent, the statistical error is given by

$$\delta A = \left[ \frac{1}{M^2} \sum_{v=1}^{M} (A_v - \langle A \rangle)^2 \right]^{1/2}. \qquad 66.$$

The statistical error decreases as $M^{-1/2}$ for large $M$, which is a very slow decrease—a notorious problem in Monte Carlo studies.

The error estimate Equation 66 is true only if the configurations are independent. Since the update changes only one variable at a time, it requires many updating steps to obtain (quasi) independent configurations. Even if the configurations are well separated, there is always some remaining correlation between them. It is an important part of any MC simulation to determine this correlation and to increase the statistical error estimate accordingly.

## 6.3 Finite Size Errors and Boundary Conditions

Monte Carlo simulations are performed on systems with finite, fixed volume (measured in lattice units). The maximum volume in a MC study is always limited by the size of the computer memory and the available computer time.

On the other hand, quantum field theories are defined in the infinite volume limit. Any distortive effect coming from the finite size should be considered as a systematical error. These systematical errors can be checked and controlled by comparing results obtained on systems with different volumes, and also by theoretical considerations (29, 30).

In most of the applications, periodic boundary conditions were used, since they are easy to implement and result in smaller finite size errors than other obvious choices.

# 7. MONTE CARLO RENORMALIZATION GROUP AND THE $\beta$ FUNCTION

The Monte Carlo renormalization group (MCRG) method combines RG ideas (31) with the power of MC simulations. It is a well-known, powerful method for spin systems in statistical physics. The basic idea, raised by S. K. Ma (32), was later significantly developed and turned into a practical scheme by R. H. Swendsen (33). Although technical and theoretical complications delayed somewhat its application in gauge theories, recent results indicate that MCRG is well suited for these kind of theories. Presumably, this method will play an even greater role in the future in defining and studying nonasymptotically free theories.

As discussed in Section 4, the lattice spacing $a$ and the bare coupling $g$ should be tuned together in order to arrange a cutoff-independent, sensible limit. The existence and properties of the function $g = g(a)$, or the $\beta$ function

$$B(g) = -a\frac{\mathrm{d}}{\mathrm{d}a}g(a) \qquad\qquad 67.$$

are basically important. One should confirm that the $\beta$ function obtained on the lattice approaches the expected behavior (Equations 49 and 50) for small $g$. Additionally, the knowledge of the $\beta$ function allows us to establish the relation between the measured dimensionless numbers for physical quantities and the $\Lambda^{\mathrm{latt}}$ parameter via Equation 52a.

A possible way to find the $\beta$ function might be to determine a mass gap, or the string tension on the lattice at different, small $g$ values. In a lattice calculation the dimensionless quantity $am$ (or $\sigma^{1/2}a, \ldots$) is obtained. Knowing this quantity as a function of $g$, and using the relation

$$ma(g) = c\, \exp- \int^{g} \frac{1}{B(g')}\, \mathrm{d}g',$$

one can find the $\beta$ function.

This is, however, not the best way to proceed. The physical properties of the continuum field theory are reflected by the long-distance properties of the lattice-regularized system in the $a \to 0$ limit. Consequently, a quantity that describes long-distance physics even in the continuum (like mass gap or string tension, see Section 3) is a long-distance feature in a double way on the lattice. Therefore, these quantities are difficult to determine.

The $\beta$ function entered our discussion via the homogeneous RG Equation 47, which expressed the fact that $X$ is cutoff independent and physical. Physical quantities corresponding to short-distance continuum properties also satisfy this equation. They might be easier to calculate.

Actually, the freedom in choosing quantities for the study of the $\beta$ function is even greater. There are many ways to combine basic observables (Wilson loop expectation values, for instance) in such a way that the combination satisfies the homogenous RG Equation 47 without this quantity being directly related to any measurable physical property of the system.

In a MCRG study, those variables related to short-distance (on the scale $a$) interactions and heavily contaminated by lattice artifacts are integrated out systematically in a step-by-step process. Starting from the original configurations generated by MC simulation, block variables and blocked configurations are constructed by averaging over the irrelevant short-distance properties, but keeping the long-distance, physical content unchanged. In this case the block loop expectation values provide the physical quantities needed to determine the $\beta$ function (20, 34).

In another method (the ratio method), which was tested extensively recently (35), appropriate ratios of Wilson loop expectation values were formed; these satisfy Equation 47 (36). Although the ratio method does not have the theoretical appeal of the MCRG blocking method, for the problems discussed here it performs remarkably well and it is easy to generalize, when fermions are present (37).

In the MCRG approach, not the $\beta$ function itself but a related quantity $\Delta\beta(\beta)$ is determined. This function carries the same information as the $\beta$ function and is defined in the following way.

Let us introduce the notation

$$\beta = \frac{2N}{g^2}\left[ = \frac{6}{g^2} \text{ for SU(3)} \right].$$

68.

In the continuum limit we have $\beta = \beta(a) \xrightarrow[a \to 0]{} \infty$. The function $\Delta\beta(\beta)$ is defined as

$$\Delta\beta(\beta) = \beta - \beta',$$

69.

where

$$\beta = \beta(a) \qquad \beta' = \beta(2a).$$

70.

The function $\Delta\beta(\beta)$ gives the change in the coupling that occurs when the lattice spacing $a$ is increased by a factor of two. By integrating Equation 67 between $a$ and $2a$ one obtains

$$-\int_a^{2a} \frac{da}{a} = \int_{g(a)}^{g(2a)} \frac{dg}{B(g)}$$

71.

or (in the case of $N = 3$)

$$\int_{\beta-\Delta\beta(\beta)}^{\beta} \frac{dx}{x^{3/2}B(\sqrt{6/x})} = -\frac{2\ln 2}{\sqrt{6}}.$$

72.

If $\Delta\beta(\beta)$ is known, Equation 72 gives the $\beta$ function and vice versa.

# 8.   RESULTS IN SU(3) GAUGE THEORY

During the last few years a large amount of information has been collected on discrete Abelian and non-Abelian gauge theories, or gauge–scalar matter interaction. Here we consider only some of the results obtained in SU(3) gauge theory by MC simulation or by MCRG techniques.

## 8.1   *Confinement*

The most important qualitative question is whether or not quark sources are confined in an SU(3) gauge theory. In the lattice formulation, in the strong coupling $g \to \infty$ limit sources in the fundamental representation are confined. The question is, does this property survive the $g \to 0$ continuum limit?

All the available numerical results indicate that SU(3) lattice gauge theory remains in the same confining phase for all $g$ values. There is increasing numerical evidence that a sensible continuum theory is defined by the $g \to 0$ continuum limit, and the limiting theory confines at large distances (it has a nonzero string tension), while remaining asymptotically free at short distances. The following results should serve as a quantitative illustration of this statement.

## 8.2   *Numerical Results on the $\beta$ Function*

As discussed in Sections 4 and 7, the existence and functional form or the $\beta$ function [or the related function $\Delta\beta(\beta)$] are basically important for the existence of a renormalizable continuum theory and for the correct interpretation of the lattice results.

The behavior of $\Delta\beta(\beta)$ for large $\beta$ is fixed by Equation 72 through the known, universal, perturbative terms of the $\beta$ function (Equation 50). One obtains

$$\Delta\beta(\beta) \xrightarrow[\beta \to \infty]{} 0.579 + 0.204/\beta + O(\beta^{-2}). \qquad\qquad 73.$$

The function $\Delta\beta(\beta)$, as determined on the lattice by nonperturbative methods, should approach this form for large $\beta$.

In the continuum limit there must exist a well-defined unique function $\Delta\beta(\beta)$—this is the requirement of renormalizability. It is unique in this sense but not universal: different schemes and formulations give different $\beta$ functions. Only the first two terms in Equation 49 are universal. Since in SU(3) most of the calculations used Wilson's Equation 15, we concentrate on results related to this action.

The limit $\beta \to \infty$ is the continuum limit. What is the coupling constant value beyond which continuum behavior is expected?[8] The first measurements (38) of the SU(3) string tension already revealed a sharp change in the $\beta$ dependence of $\sigma^{1/2}a$: in the strong coupling region $\sigma^{1/2}a$ decreases slowly with increasing $\beta$ (see Equation 58), while somewhere between $\beta = 5$ and 6 this decrease turns into a fast, exponential decay, which is in qualitative agreement with Equation 52b, dictated by RG.

Since Creutz's first calculation of the SU(2) string tension, the common practice has been to assume that continuum limit starts at the $\beta$ value where this qualitative change in the $\beta$ behavior of masses occurs. The data were consistent with the asymptotic form Equation 52b, at least within the rather large statistical and systematical errors of the first MC measurements.

Recent, more precise, numerical data required the revision of these assumptions. The available information on $\Delta\beta(\beta)$ as obtained by different MCRG methods is summarized in Figure 2 (35, 39–42). The function $\Delta\beta(\beta)$ gives the change in $\beta$ that results in a change of a factor of two in the lattice unit: $a[\beta - \Delta\beta(\beta)]/a(\beta) = 2$. The expected asymptotic form (Equation 73) is also plotted in this figure.

Around $\beta \approx 6.0$, there is a pronounced dip in $\Delta\beta(\beta)$, and the asymptotic form is approached from below, rather slowly. The $\beta$ function cannot be described by the first two perturbative terms in this region; there is a significant nonperturbative contribution.[9] At $\beta = 6.0$, $\Delta\beta \approx 0.35$, which

---

[8] Of course, for any finite $\beta$ (and correspondingly, any finite lattice unit $a$) there will be nonzero distortive lattice artifacts in the results. Instead of perfect cutoff independence, there will be $O(a^2)$ corrections to the right-hand side of Equation 46. Since, however, $a \sim \exp(-c/\beta)$, the onset of the continuum behavior is expected to occur in a narrow $\beta$ region, beyond which the cutoff dependence is negligible for all the low-energy observables.

[9] It is improbable that higher-order perturbative terms can account for this behavior. There exists an estimate for the third, perturbative term of the $\beta$ function (43), which gives a very small contribution to Equation 73.

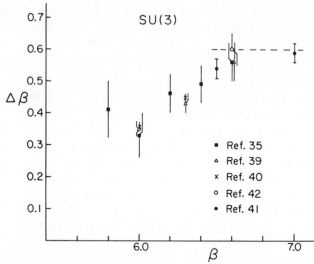

*Figure 2*    MCRG results on $\Delta\beta(\beta)$. The function $\Delta\beta(\beta)$ reflects the behavior of the $\beta$ function between $\beta$ and $\beta' = \beta - \Delta\beta(\beta)$.

implies that $a$ is decreased by a factor of two between $\beta \approx 5.65$ and 6.0. On the other hand, $\Delta\beta(\beta = 6.6) \approx 0.55$, therefore $a \rightarrow a/2$ between $\beta \approx 6.05$ and 6.6. This is quite close to the asymptotic prediction, Equation 73, which would give $\Delta\beta(\beta = 6.6) = 0.61$.

Different methods give consistent predictions for $\Delta\beta(\beta)$ at $\beta = 6.0$ with rather small estimated systematical error. This suggests that a $\beta$ function can be defined; continuum behavior sets in for $\beta \gtrsim 5.7$. Therefore scaling is expected for $\beta \gtrsim 5.7$ (perhaps even earlier); the coupling constant dependence is dictated by Equation 52a with a nontrivial $\beta$ function. On the other hand, according to Figure 2, beyond $\beta \sim 6.0$ the coupling constant dependence is expected to be close to the asymptotic form Equation 52b.

As we shall see, all these predictions are in agreement with recent high-statistics string tension, potential, and critical temperature calculations. This consistency increases our confidence that the continuum limit of SU(3) lattice gauge theory is getting under control. It is important to check, however, with further MCRG studies in the future, that $\Delta\beta(\beta)$ really follows the asymptotic line in Figure 2 as the system is pushed deeper into the continuum limit.

## 8.3    The q-q̄ Potential, Tension

The potential energy of a static q-q̄ pair is determined by the ratio of elongated Wilson loops. In Equation 36 the "time" $T$ must be large enough to suppress the contribution of excited states, which might be flux-tube vibrations, breathing modes, gluon excitations, or other excitations. Finite $T$ always overestimates the ground-state energy $aV(R)$.

In any MC investigation the maximum value of $T$ is limited by the size of the lattice and by the fact that large loop expectation values are very small (Equation 40); this requires small statistical errors, therefore very long runs. A large variety of extrapolation procedures are used in different works to estimate $V(R)$ from the measured values of $V_T(R)$ ($T = 1, 2, \ldots, T_{max}$). A possibility is to include corrections in Equation 31 and fit the parameters, or to use some simple model (like a vibrating string) to guess the nonleading $T$ dependence (44). Using different methods helps in obtaining an estimate on the systematical error in $V(R)$.

In principle there is a cleverer way to overcome this problem. In the ground state of the q-q̄ pair the color flux propagates between the sources in a definite, dynamically determined way. The space-like part of a single Wilson loop represents this flux distribution rather poorly. In other words, it has a projection not only to the ground state, but to all the excited states. It requires a long "time" $T$, until these unwanted states die out from the system. One might try to start with an operator in Equation 32, where the single space-like path $C$ is replaced by a superposition of several paths connecting the sources at $\mathbf{0}$ and $\mathbf{R}$. The unknown coefficients of this wave function are determined variationally, by minimizing $V(R)$—as in ordinary problems in quantum mechanics. The results obtained in exploratory studies are promising (45). Unfortunately, no large-scale MC calculation has yet been performed to explore this possibility.

Having the potential $V(R)$, the string tension is obtained from its asymptotic behavior for large $R$. Again, in any MC study $R$ is limited, and some extrapolation method is needed. In many recent works the long-distance part of the potential is fitted with the form

$$V(R) \xrightarrow[R \to \infty]{} C/R + \sigma R. \qquad 74.$$

The $C/R$ correction is predicted by a flux-tube picture with a definite constant $C = -\pi/12$ (46). The results are in surprising agreement with this form.

The final estimates on $\sigma^{1/2}a$ as a function of $\beta$ are summarized in Table 1, which quotes only those recent results obtained on large lattices (ranging from $12^3 \times 16$ up to $24^3 \times 48$) with high statistics. The string tension estimates of de Forcrand et al (53, 54) were obtained by a different technique. By measuring special loop-loop correlations, the tension is determined at a finite physical temperature. The tension $\sigma$ is expected to decrease with increasing temperature. Unfortunately, the functional relation is not known. The quoted number includes a $(\pi/3)$(temperature)$^2$ correction, which again is guessed from flux-tube models.

By measuring the potential energy between on- and off-axis sources, strong violation of rotation symmetry was observed at $\beta = 5.4$ (47). The tension measured along the axes is smaller than that measured off the axes. Rotation symmetry is essentially restored beyond $\beta = 5.7$ (47, 52). For $\beta \gtrsim 5.7$, the distortive effects are reduced to the level of a few percent.

The numbers in Table 1 show the same trend as Figure 2 on $\Delta\beta(\beta)$. Although scaling might start around $\beta = 5.7$, asymptotic scaling (Equation 52b) certainly does not. In fact, assuming asymptotic scaling, strong $\beta$ dependence is observed in the ratio $\sigma^{1/2}/\Lambda^{\text{latt}}$ (Figure 3). The tension $\sigma$ and $\Lambda^{\text{latt}}$ are physical quantities, their ratio should be $\beta$ (cutoff) independent in the continuum limit. It is clear from Figure 3 that the function $\beta - a(\beta)$ and therefore $\sigma^{1/2}a$ drops more rapidly than the asymptotic form (Equation 52b) demands. This is consistent with the MCRG result discussed before. Above $\beta = 6.0$, the $\beta$ dependence in Figure 3 becomes weak, the calculations are consistent with asymptotic scaling. One might quote as the present best estimate

$$\sigma^{1/2} = (95 \pm 10)\Lambda^{\text{latt}}. \qquad\qquad 75.$$

**Table 1**  Monte Carlo-calculations of the string tension[e]

| $\beta$ | 5.4 | 5.5 | 5.6 | 5.7 |
|---|---|---|---|---|
| $\sqrt{\sigma}a$ | 0.75 (47)[a] | 0.583 ± 0.013 | 0.53 ± 0.01 (48) | 0.442 ± 0.011 (47) |
| | | | | 0.367 ± 0.007 (53) |

| $\beta$ | 5.8 | 5.9 | 6.0 | |
|---|---|---|---|---|
| $\sqrt{\sigma}a$ | 0.332 ± 0.003 (48) | 0.253 ± 0.009 (53) | 0.221 ± 0.018 (40) | 0.21 ± 0.07 (50)[c] |
| | 0.32 ± 0.03 (52) | | 0.246 ± 0.003 (48) | 0.214 ± 0.004 (54) |
| | | | 0.250 ± 0.008 (51)[b] | 0.205 ± 0.008 (53) |

| $\beta$ | 6.2 | 6.3 |
|---|---|---|
| $\sqrt{\sigma}a$ | 0.189 ± 0.002 (48) | 0.150 ± 0.020 (40) |
| | | 0.155 (49)[d] |

[a] Rotation symmetry is violated; effective tension is quoted.
[b] Obtained from a fit in $\beta \in (6.0, 6.4)$.
[c] Obtained from a fit in $\beta \in (6.0, 7.2)$.
[d] Analyzed in (44).
[e] Numbers in parentheses are references.

This is the first point in our discussion where abstract characteristics of the lattice calculations can be filled with life. The value of the tension is fixed by different phenomenological studies to be $\sigma_{\exp}^{1/2} \sim 0.42$ GeV. Then Equations 75 and 52b predict

$$\Lambda^{\text{latt}} = 4.4 \pm 0.4 \text{ MeV} \quad \text{or} \quad \Lambda^{\text{MOM}} = 370 \pm 34 \text{ MeV}. \qquad 76.$$

Using Equations 76 and 52b, one obtains $a = a(\beta)$, specifically

$$a(\beta = 6.0) = 0.53 \pm 0.05 \text{ GeV}^{-1} \approx 0.11 \pm 0.01 \text{ fm}. \qquad 77.$$

Therefore, a periodic $16^4$ lattice at $\beta = 6.0$ represents a $\sim 1.76$ fm periodic box with a short-distance resolution of $\sim 0.1$ fm. The same lattice at $\beta = 6.6$ is only $\sim 0.88$ fm in size and has a resolution of $\sim 0.05$ fm.

The results on the potential energy are plotted in Figure 4. In this figure the potential energy is measured in $\sigma^{1/2}$ ($\approx 0.42$ GeV) while the distance is in $1/\sigma^{1/2}$ ($\approx 0.48$ fm) units. An overall, additive constant in the potential energy remains undetermined; the zero level in Figure 4 was chosen arbitrarily. The shape of the curve and the scale are completely fixed, however. In Figure 4, there are more than 60 points coming from different works and from 10 different $\beta$ values between 5.7 and 7.2. When the numbers are expressed in physical units, these points form a smooth curve determining the heavy quark potential from first principles with a good precision up to

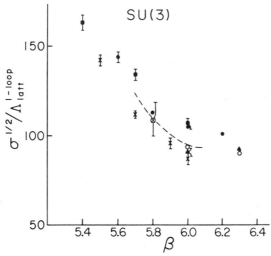

*Figure 3* The results of Table 1 are plotted assuming asymptotic scaling. The dotted line of this figure was used in the ratios of Figure 5.

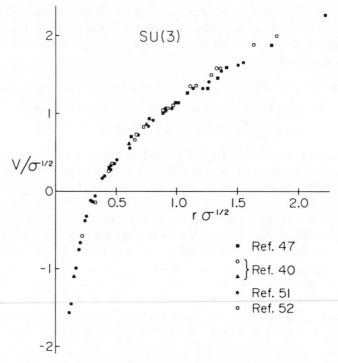

*Figure 4*  The quark-antiquark potential as obtained from potential calculations at different $\beta$ values.

$r \sim 1$ fm. The consistency of these points is a strong support for scaling. Figure 4 presumably represents the most truthworthy prediction available up to now in SU(3) and illustrates nicely the power of lattice methods. These techniques should be able to predict a curve of similar quality on the spin-dependent part of the potential in the near future (55).

## 8.4  *Glueballs*

As we discussed in Section 3.2, the glueball spectrum is related to the long-distance decay of appropriately connected correlations. It is a natural expectation that the lowest-lying glueball (mass gap) is in the $0^{++}$ channel. An appropriate operator whose correlation is controlled by the $0^{++}$ excitation is

$$\theta(n_4) = \sum_n \sum_{\substack{ij \\ i \neq j}} \left( \text{Tr} \, \begin{matrix} {\scriptstyle (n+\hat{j},n_4)} & \quad & {\scriptstyle (n+\hat{i}+\hat{j},n_4)} \\ & \square & \\ {\scriptstyle (n,n_4)} & \quad & {\scriptstyle (n+\hat{i},n_4)} \end{matrix} + \text{c.c.} \right), \qquad 78.$$

where $\square$ is the product of $U$ matrices along the plaquette and $\hat{i}, \hat{j}$ are the spatial unit vectors. Summing over **n** assures that the excitation created by $\theta(n_4)$ is in the rest frame ($\mathbf{p} = 0$).

The main difficulty of glueball mass calculations is that it is statistically very difficult to determine the connected correlations over distances where the lowest-lying mass already dominates. A possibility is to construct operators that have a large projection to the lowest-lying glueball state: $\theta$ is taken to be the combination of different loop products with unknown coefficients that are fixed by a variational calculation:

$$\min\left[ -\frac{1}{T} \ln \frac{\langle \theta(0)\theta(T)\rangle_c}{\langle \theta(0)\theta(0)\rangle_c} \right]. \qquad 79.$$

This method is not without problems. The presence of statistical errors requires special care in doing the variational calculations. Additionally, it is difficult to increase the volume of the system and keep the statistical errors under control. Since $\theta$ is a sum over **n**, the correlation contains a double sum $\sum_{\mathbf{n}, n_4 = T}$. In a large volume, for most of the terms $|\mathbf{n}\text{-}\mathbf{m}|$ is large, the corresponding contribution to the signal is negligible, and only the noise is increased.

There exist suggestions to improve on this situation, for example using $p \neq 0$ or replacing one of the operators by an extended, nonfluctuating source (source method). Nevertheless, the glueball mass calculations remain notoriously difficult to do on a level where the systematical errors can be efficiently controlled.

The recent situation on $ma$ in the $0^{++}$ channel is summarized in Table 2. The calculations of (56) and (57) were performed on small $5^3 \times 8$ and $4^3 \times 8$ lattices, respectively. The leading large-volume correction to the glueball mass is predicted to be (30)

$$m(L) = m(\infty)\left\{1 - G \frac{\exp[-\sqrt{3/2}m(\infty)aL]}{m(\infty)aL}\right\}, \qquad 80.$$

**Table 2**   Monte Carlo calculations of the mass gap[a]

| $\beta$ | 5.4 | 5.5 | 5.6 | 5.7 | 5.8 |
|---|---|---|---|---|---|
| $ma$ | $1.30 \pm 0.19$ (56) | $1.09 \pm 0.03$ (57) | $1.07 \pm 0.13$ (56) | $1.02 + 0.13$ (56) | $0.87 \pm 0.18$ (57) |
| | | $1.14 \pm 0.03$ (58) | | $1.02 \pm 0.05$ (57) | $0.67 \pm 0.04$ (58) |
| | | | | $0.90 \pm 0.04$ (58) | |

[a] Numbers in parentheses are references.

where $L$ is the linear spatial size of the lattice; $G$ is related to the triple glueball coupling and is unknown. A crude estimate derived from a low-order strong coupling expansion gives $G \times 100$ (59), which would indicate that the finite size effects on $5^3 \times 8$ and $4^3 \times 8$ lattices are large for the higher $\beta$ values considered. This prediction was confirmed by the latest MC simulation (58) in which the mass gap predictions obtained on $6^3 \times 16$ and $8^3 \times 16$ lattices were significantly different and gave $G \approx 155 \pm 45$. Strangely enough, the large systematical error of the earlier calculations was mostly cancelled by another systematical error: the mass on the actual lattice was overestimated.

In the continuum limit the ratio of physical masses should become cutoff (or $\beta$) independent. From the potential and MCRG calculations it is expected that the scaling region starts somewhere around $\beta = 5.7$. In Figure 5 the mass ratios are plotted as a function of $\beta$ for $\beta \geq 5.7$. For the mass gap the extrapolated results of de Forcrand et al (58) were used, while for the string tension we took the points of the dotted curve of Figure 3. The error bars reflect not only the statistical errors but also the uncertainty coming from the string tension estimates. For completeness, the ratio $T_C/\sqrt{\sigma}$, where $T_C$ is the critical temperature, is also given in Figure 5 (60).

It is difficult to draw firm conclusions on scaling from this figure. Within the statistical and (estimated systematical errors) the data points are consistent with scaling. Further work is needed. Figure 5 gives the

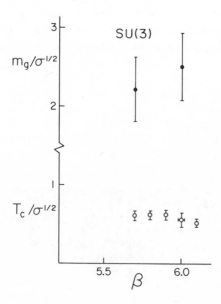

*Figure 5*  Monte Carlo predictions for the ratios $m_g/\sqrt{\sigma}$ and $T_C/\sqrt{\sigma}$.

estimates:

$$m_g = (2.35 \pm 0.40)\sigma^{1/2} \approx 1.0 \pm 0.15 \text{ GeV}$$

$$T_c = (0.57 \pm 0.05)\sigma^{1/2} \approx 0.24 \pm 0.02 \text{ GeV}. \qquad 81.$$

There are exploratory studies on higher-lying glueball states with different $J^{PC}$ quantum numbers. In the light of the difficulties discussed above, we avoid going into the details here.

# 9. FERMIONS ON THE LATTICE

In previous sections, quarks were present only as external, static sources. The study of hadron spectroscopy, chiral symmetry (breaking), hadronic matrix elements, and so on requires the introduction of dynamical quarks into the theory. This leads to theoretical problems and new tough challenges for the numerical methods.

## 9.1 Species Doubling

As mentioned in Section 2.5, Equation 24 has a 16-fold hidden fermion degeneracy. In order to illustrate this problem, consider a free Dirac fermion as defined by Equation 21. For $m = 0$, the propagator in Fourier space has the form

$$G(p) = 1 \left/ \frac{1}{a} \sum_{\mu=1}^{4} \gamma_\mu \sin(p_\mu a), \right. \qquad 82.$$

which for small $p_\mu a$ reproduces the correct, continuum propagator $(p_\mu \gamma_\mu)^{-1}$ and has a pole at $p_\mu = 0$. The problem is, however, that $G(p)$ has additional poles in the Brillouin zone $p_\mu a \in (-\pi, \pi)$. The propagator $G(p)$ has 16 poles at the points $p_\mu = (0,0,0,0), (\pi,0,0,0), \ldots (\pi,\pi,\pi,\pi)$ (species doubling).

Is this property generic? The lattice action, Equation 21, is not unique, there are other possible ways to discretize a continuum action. It turns out, however, that the species doubling is a general problem for all lattice actions that preserve the chiral symmetry of the massless continuum action. When $m = 0$, the continuum action and its lattice version (Equation 21) are symmetric under the global transformation

$$\psi \to \exp(i\alpha\gamma_5)\psi, \qquad \bar\psi \to \bar\psi \exp(i\alpha\gamma_5). \qquad 83.$$

A more general chiral symmetric (i.e. free of mass-like terms) lattice action will lead to the propagator

$$G(p) = 1 \left/ \frac{1}{a} \sum_{\mu} \gamma_\mu F_\mu(pa). \right. \qquad 84.$$

By demanding that $G \to (\gamma_\mu p_\mu)^{-1}$ for small $(pa)$ and remembering that $F_\mu$ is a periodic function of the momenta, we conclude that—if it is a continuous function—it must have an extra zero between $ap_\mu = 0$ and $2\pi$ (Figure 6). Chiral symmetry implies species doubling on the lattice. Rigorous theorems establish that there is a deep connection between chiral symmetry, species doubling, and the topology (in Fourier space) of the lattice.

Different ways of avoiding or reducing the number of unwanted fermion species have been suggested, but none of them is completely satisfactory. We face the choice: either a nonchiral symmetric lattice regularization is defined or doubling is accepted. The procedure suggested by Wilson opts for the first, while the Kogut-Susskind (staggered) fermion method chooses the second.

## 9.2  Wilson Fermions

Consider the naively discretized Equation 21 and add the term (61, 62)

$$a^4 \sum_n \sum_{\mu=1}^4 \frac{1}{2a} [\bar\psi_\alpha^{if}(n+\hat\mu) - \bar\psi_\alpha^{if}(n)][\psi_\alpha^{if}(n+\hat\mu) - \psi_\alpha^{if}(n)]. \qquad 85.$$

It is easy to check that this extra term goes to zero in the formal classical continuum limit; therefore the new action is an a priori legitimate choice. Introducing new fields

$$\psi \to [a^3(ma+4)]^{-1/2}\psi, \qquad 86.$$

one obtains

$$-S = \sum_n \left\{ \sum_{\mu=1}^4 K_f \bar\psi_\alpha^{if}(n)(1-\gamma_\mu)_{\alpha\beta}\psi_\beta^{if}(n+\hat\mu) \right.$$
$$\left. + \sum_{\mu=1}^4 K_f \bar\psi_\alpha^{if}(n+\mu)(1+\gamma_\mu)_{\alpha\beta}\psi_\beta^{if}(n) - \bar\psi_\alpha^{if}(n)\psi_\alpha^{if}(n) \right\}$$

where

$$K_f = (8+2m_f a)^{-1}. \qquad 87.$$

The propagator now has the form

$$G(p) \sim 1 \left/ \left[ 2K_f \sum_{\mu=1}^4 (\gamma_\mu \sin p_\mu a - \cos p_\mu a) + 1 \right] \right., \qquad 88.$$

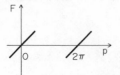

Figure 6   The function $F(pa)$ of Equation 84 near $ap = 0$ and $ap = 2\pi$.

which has only one low-momentum pole [for $m = 0$ ($K = \frac{1}{8}$) at $p = 0$). The poles at the edges of the Brillouin zone are suppressed by the extra term of Equation 85. There is no species doubling. The corresponding equation in QCD has the form

$$-S = \sum_n \left\{ \sum_\mu K_f [\bar{\psi}_\alpha^{if}(n)(1 - \gamma_\mu)_{\alpha\beta} U_{n\mu}^{ij} \psi_\beta^{jf}(n + \mu) + \bar{\psi}_\alpha^{if}(n + \hat{\mu})(1 + \gamma_\mu)_{\alpha\beta} \right.$$

$$\left. \times\, U_{n\mu}^{+ij} \psi_\beta^{jf}(n)] - \bar{\psi}_\alpha^{if}(n)\psi_\alpha^{if}(n) \right\} + \frac{1}{g^2}\sum_p (\mathrm{Tr}\, U_p + \text{c.c.}), \qquad 89.$$

where $K_f$ (f $= u, d, s, c, b$) is the so-called hopping parameter for the different quark flavors; it is related to the corresponding quark mass as given above.

Equation 85 explicitly breaks chiral symmetry. For massless quarks the action in Equation 22 has a $U(n_f)_V \times U(n_f)_A = U(1)_V \times U(1)_A \times SU(n_f)_V \times SU(n_f)_A$ symmetry where $n_f$ is the number of massless flavors. The axial $U(1)$ is expected to be broken by quantum effects while the $SU(n_f)$ chiral symmetry is broken spontaneously and produces Goldstone bosons. Wilson's action, Equation 89, explicitly breaks chiral symmetry even if $K = \frac{1}{8}(m = 0)$ is taken. Since the breaking terms are proportional to $a$, chiral symmetry is expected to be recovered in the continuum limit (and broken only spontaneously).

## 9.3 Kogut-Susskind (Staggered) Fermions

Using a clever trick, one can reduce the 16-fold degeneracy by a factor of four without losing completely the original chiral symmetry: a $U(1)_V \times U(1)_A$ symmetry survives, which offers important advantages (63).

Consider the following transformation of the quark fields (64):

$$\psi_\alpha^{if}(n) = T_{\alpha\beta}(n)\chi_\beta^{if}(n), \qquad \bar{\psi}_\alpha^{if}(n) = \bar{\chi}_\beta^{if} T_{\beta\alpha}^+(n), \qquad 90.$$

where

$$T(n) = \gamma_1^{n_1}\gamma_2^{n_2}\gamma_3^{n_3}\gamma_4^{n_4}. \qquad 91.$$

Rewriting the equation in terms of the $\chi$ fields, one obtains a spin-diagonal expression

$$S = \sum_n \left\{ \sum_\mu \tfrac{1}{2}\eta_\mu(n)[\bar{\chi}_\alpha^{if}(n)U_{n\mu}^{ij}\chi_\alpha^{jf}(n + \mu) - \bar{\chi}_\alpha^{if}(n + \hat{\mu})U_{n\mu}^{+ij}\chi_\alpha^{jf}(n)] \right.$$

$$\left. + \bar{\chi}_\alpha^{if}(n) \cdot ma \cdot \chi_\alpha^{if}(n) \right\} + S_g. \qquad 92.$$

Here $\eta_1(n) = 1$, $\eta_2(n) = (-1)^{n_1}$, $\eta_3(n) = (-1)^{n_1 + n_2}$, and $\eta_4(n) = (-1)^{n_1 + n_2 + n_3}$; this form is diagonal in the Dirac indices. One might consider one component only and drop the other three. Then $\chi$ becomes a one-component spinor field, and the four-component quark fields are constructed from the $\chi$ spinors at different corners of the $2^4$ hypercube of the lattice. By this trick the number of quark fields is reduced to four, which might be interpreted as describing four different flavours.

An important property of the reduced action is that it retains the $U(1) \times U(1)$ part of the original symmetry:

$$\left.\begin{array}{l} \chi(n) \to \exp(i\alpha)\chi(n) \\ \bar{\chi}(n) \to \exp(-i\beta)\bar{\chi}(n) \end{array}\right\} \quad \text{if} \, (-1)^{n_1+n_2+n_3+n_4} = 1$$

and                                                                                              93.

$$\left.\begin{array}{l} \chi(n) \to \exp(i\beta)\chi(n) \\ \bar{\chi}(n) \to \exp(-i\alpha)\bar{\chi}(n) \end{array}\right\} \quad \text{if} \, (-1)^{n_1+n_2+n_3+n_4} = -1,$$

where $\alpha$ and $\beta$ are two independent phases. This symmetry is enough to prevent the occurrence of mass counterterms in the renormalization process (therefore $m_{\text{bare}} = 0$ implies $m_{\text{renormalized}} = 0$) and the possible spontaneous breakdown of the $U(1)_A$ produces a Goldstone boson for any value of the lattice spacing $a$ (65).

The staggered fermion prescription has, however, some unpleasant features as well. Flavor symmetry is explicitly broken and hoped to be recovered only in the continuum limit. Additionally it is highly nontrivial to construct baryon operators with definite quantum numbers. It was observed only recently for instance, that all the local baryon operators couple to the spin-$\frac{1}{2}$ states only and the study of spin-$\frac{3}{2}$ exitations requires nonlocal combinations (66). This observation required the reconsideration of some earlier calculations. It is far from certain that all the tricky properties of the staggered fermions are understood.

# 10.   NUMERICAL METHODS FOR FERMIONIC SYSTEMS

In a path integral formulation the fermion fields are described by anticommuting $c$ numbers, Grassmann variables. In principle, they can be represented by appropriate anticommuting matrices; in practice, however, no computer memory could cope with this problem.

Since the fermions enter the action only quadratically, they can be integrated over. In a short notation

$$S = \sum \bar{\psi}_i \Delta_{ij}[U]\psi_j + S_g, \tag{94.}$$

where $i, j$ represent all kinds of indices. By integrating over the fermion variables in the vacuum functional or in any expectation value of fermion fields, one obtains (Section 2.5)

$$\int D\psi D\bar{\psi} e^{-S} = \det[\Delta(U)]e^{-S_g}$$

$$\int D\psi D\bar{\psi}\bar{\psi}_i\psi_j e^{-S} = \Delta_{ij}^{-1}(U)\det[\Delta(U)]e^{-S_g}. \qquad 95.$$

## 10.1   Fermion Determinant and the Quenched Approximation

After integrating over the fermion fields, the remaining gauge field dynamics is governed by the effective action

$$S_{\mathrm{eff}} = S_g - \ln\det[\Delta(U)]. \qquad 96.$$

The second term represents the effect of virtual quark loops. It is a highly nonlocal expression: a given link variable $U_l$ interacts directly not only with its neighbors but with distant gauge variables as well. This is an unpleasant property in MC simulation. At the update of every link the calculation of the determinant requires a large number of operations. Although several suggestions exist to attack this problem (67), it is clear that full QCD calculations require much more computing power than those in a pure gauge theory.[10] The first serious attempts are related to finite temperature QCD studies (69).

In calculating an expectation value, the fermionic determinant enters both the numerator (see Equation 95) and the denominator (via the normalizing vacuum functional). Replacing it by its expectation value (i.e. suppressing its fluctuation) the determinant drops out from the expressions. This approximation is called the quenched approximation. The effect of virtual quark loops is thus neglected. In the quenched approximation, the quark propagates in a background gauge field generated by the pure gauge field action. All the existing QCD spectrum calculations rely on this approximation. For many spectroscopic questions, it is expected (on the basis of qualitative arguments and indirect numerical studies) (70) that the effect of virtual quarks is small. Even if it is true, there are simple features like the decay of hadrons that cannot be discussed within the quenched approximation.

If $\det\Delta = 1$ is put into Equation 95, the calculation of the quark propagator (from which the hadron propagators are constructed) is reduced to the inversion of the matrix $\Delta[U]$ defined on the generated gauge configurations.

---

[10] It is not even clear, how much more (68). The suggested methods have not been checked carefully yet.

## 10.2   Calculating the Quark Propagator in a Background Gauge Field Configuration

The matrix $\Delta[U]$ is a large matrix with indices in configuration, color, and Dirac space. There are several suggestions for the inversion problem.

10.2.1   ITERATIVE METHODS   For concreteness, consider Wilson fermions. The matrix $\Delta$ has the form (Equation 89)

$$\Delta(U) = \mathbb{1} - KB(U). \qquad 97.$$

The equation

$$\chi = [1 - KB(U)]^{-1}\phi \qquad 98.$$

where $\phi$ is some initial vector, can be rearranged as

$$\chi = KB(U)\chi + \phi. \qquad 99.$$

Equation 99 can be solved by iterative methods. The simplest is the Jacobi iteration,

$$\chi^{(n+1)} = KB(U)\chi^{(n)} + \phi, \qquad 100.$$

where $\chi^{(n)}$ is the $n$th approximation. Different, improved versions of this iterative method exist, like the Gauss-Seidel iteration and the "second-order" method. In many recent applications, the conjugate gradient method was used (71).

10.2.2   HOPPING PARAMETER EXPANSION   The previous methods numerically invert the matrix for a fixed value of the hopping parameter, $K$. In the hopping parameter expansion method (61, 72), the inverse is expanded in powers of $K$ and the expansion coefficients are determined numerically. This method gives an approximation to the propagator for all values of $K$ at once. Also, by having the power series explicitly and adding some information on the singularity structure of the hadron propagators in the complex $K$ plane, different extrapolation methods (like Pade analysis) can be used (73). On the other hand, a high-order expansion requires extensive data handling and it remains to be seen whether this method can be competitive on parallel machines.

# 11.   RESULTS ON SPECTROSCOPY IN QUENCHED QCD

In the quenched approximation the quark propagates over a background gauge field configuration generated by the pure gauge field action. The correlation length (which might be identified with the inverse dimensionless mass gap) and the renormalization group behavior in the continuum limit are fixed by the gauge field dynamics alone. It follows that the basis of a reliable hadron spectroscopy is a reasonable understanding of the pure gauge sector. The increasingly consistent picture obtained in the gauge sector by recent investigations should provide a good starting point for more advanced hadron spectroscopy.

Part of the present situation is summarized in Tables 3 and 4 for Wilson and Kogut-Susskind fermions, respectively (67, 74–79). The basis of the mass determination is again Equation 31. The decay of the correlation function at large distances is calculated. The only free parameters that enter are the quark masses. In the case of hadrons containing u and d quarks only, usually $m_u = m_d = m$ is taken and the single parameter $m$ is fixed by the $m_\rho/m_\pi$ mass ratio. Since the pion is very light, the bare quark mass obtained this way is very close to $m_{cr}$, where $m_\pi = 0$. For staggered fermions, where the remaining $U(1) \times U(1)$ symmetry prevents the occurrence of mass counterterms, $m_{cr} = 0$. In the case of Wilson fermions, $m_{cr}$ (or equivalently, $K_{cr}$) is not known a priori, $K_{cr} = K_{cr}(\beta)$, and it is determined numerically by the condition $m_\pi(K = K_{cr}) = 0$. After the quark mass $m$ (or $K$) is fixed, the corresponding measured $m_\rho \cdot a$ fixes the lattice unit $a$ in physical units. The lattice unit $a = a(\beta)$ obtained this way should agree with predictions given by the pure gauge theory results. Of course, no physical prediction should depend on which fermion prescription, Wilson or Kogut-Susskind, is used.

There are many further predictions of hadron masses where strange or heavier quarks enter, of excited states, and of other static properties (like anomalous magnetic moments). However, a few conclusions on the present situation can already be drawn from Tables 3 and 4.

**Table 3**  Spectrum calculations with Wilson fermions[a]

| | $\beta = 5.7$ | | | $\beta = 6.0$ | | |
|---|---|---|---|---|---|---|
| | $8^4$ (74) | $8^4$ [b] | $16^4$ (75) | $8^4$ (74) | $10^3 \times 20$ (76) | $16^3 \times 28$ (77) |
| $K_{cr}$ | $0.1695 \pm 0.0007$ | $0.1696 \pm 0.0016$ | $0.1690 \pm 0.0005$ | $\sim 0.1560$ | $\sim 0.157$ | — |
| $m_\rho a$ | $0.53 \pm 0.03$ | $0.57 \pm 0.01$ | $0.58^{+0.12}_{-0.06}$ | $0.37 \pm 0.07$ | $\sim 0.35$ | $0.37 \pm 0.03$ |
| $a$ (fm)[c] | $0.136$ | $0.150 \pm 0.003$ | $0.15^{+0.03}_{-0.01}$ | $0.094 \pm 0.018$ | $\sim 0.079$ | $0.097 \pm 0.008$ |
| $m_p$ (GeV) | $\sim 1.6$ | $\sim 1.29$ | $\sim 1.44$ | — | $\sim 0.97$ | $\sim 1.4$ |
| $m_\Delta$ (GeV) | $\sim 1.6$ | $\sim 1.63$ | $\sim 1.6$ | — | $\sim 1.1$ | $\sim 1.6$ |

[a] Numbers in parentheses are references.
[b] Langguth & Montvay in (67).
[c] From tension.

Spectroscopical calculations with Wilson fermions require at least $\beta = 6.0$ or beyond. Results at lower $\beta$ values (e.g. $\beta = 5.7$) might reflect correctly the content of the model at this $\beta$, but seem to be inconsistent with continuum physics.

At $\beta = 5.7$, the excellent agreement between the results of References (67, 74, 75), where completely different methods were used, suggests that the predicted numbers are correct, in the sense that they reflect the property of the model at $\beta = 5.7$. However, there is a definite discrepancy between the lattice unit values obtained from spectroscopy and from the string tension (i.e. the lattice spacing fixed by the tension would give far too heavy a $\rho$ meson). The scaling behavior between $\beta = 5.7$ and 6.0 is also inconsistent with Figure 2, which indicates that suppressing lattice artifacts coming from Wilson fermions requires higher $\beta$ values than those coming from the gauge sector. As a whole, at $\beta = 5.7$ the hadron spectrum is very similar to that obtained in the $\beta = 0$ strong coupling limit. The later onset of continuum behavior might be related to the explicit chiral symmetry breaking of the Wilson formulation.

The continuum limit of the Kogut-Susskind fermions might start earlier. The scaling behavior of $a(\beta)$ between $\beta = 5.7$ and 6.0 is consistent with the pure gauge theory prediction.

A lattice spacing of $a(\beta = 6.0) \approx 0.1$ fm, which seems to occur consistently from different calculations, puts a constraint on the minimum size of the lattice that can be sensibly used in spectroscopy (80). A $16^3 \times N_t$ lattice, for instance, defines a periodic spatial box with linear size $\sim 1.6$ fm. We do not know how large the hadrons are in the quenched approximation. The experimental charge radius of the proton giving a proton diameter of the order of 1.6–2 fm might serve as an indicator.

Spontaneous breakdown of chiral symmetry and the occurrence of

**Table 4**  Spectrum calculations with Kogut-Susskind fermions[a]

|  | $\beta = 5.7$ | | $\beta = 6.0$ | |
|---|---|---|---|---|
|  | $8^4$ (74) | $10^3 \times 16$ (78) | $10^3 \times 20$ (76) | $10^3 \times 20$ (79) |
| $m_\rho a$ | $0.88 \pm 0.06$ | $\sim 0.98$ | $\sim 0.37$ | — |
| $a$ (fm) | $0.255 \pm 0.016$ | $\sim 0.25$ | $\sim 0.1$ | — |
| $m_p$ (GeV) | $\sim 0.91$ | $\sim 0.94$ | $\sim 1.0$ | $\sim 1.1$ |
| $m_\Delta$ (GeV) | — | — | — | $\sim 1.6$ |
| $f_\pi$ (MeV) | — | 134 | $\sim 160$ | — |

[a] Numbers in parentheses are references. In References 74, 76, and 78, the proton mass was fitted with an incorrect ansatz (66).

Goldstone bosons in the spectrum are characteristic features of the observed hadron spectrum. For staggered fermions, $\langle \bar{\psi}\psi \rangle$ is a good order parameter of chiral symmetry, as in the continuum formulations. A nonzero $\langle \bar{\psi}\psi \rangle$, indicating a spontaneous breakdown of chiral symmetry, is observed rather convincingly (81). Further information comes from finite temperature studies where the restoration of chiral symmetry at some finite temperature is investigated in detail (60, 82).

## 12. OUTLOOK

This paper deals almost exclusively with applications in QCD. Presumably QCD will remain in the center of lattice calculations in the near future. The existing techniques and the available computer resources should be able to produce trustworthy results in pure SU(3) gauge theory and in quenched hadron spectroscopy. Going beyond the quenched approximation might require some technical breakthrough or exceptional computer resources, or both.

Computational physics has entered high-energy physics. From this point of view, lattice QCD is only one (although the most important, at present) of the research fields. Increasing attention is devoted to the study of other QFTs. It is certain that the investigation of nonasymptotically free theories, the Higgs phenomenon, or field theories that are not perturbatively renormalizable will be important research areas in the future.

ACKNOWLEDGMENT

We are indebted to our colleagues, especially to Urs Heller and Frithjof Karsch, for the valuable discussions. One of us (P.H.) would like to thank the kind hospitality of the members of SCRI and Physics Department at Tallahassee, where the manuscript was completed.

*Literature Cited*

1. Wilson, K. G. *Phys. Rev. D* 10:2445 (1974)
2. Kogut, J. B., Sinclair, D., Susskind, L. *Nucl. Phys.* 114B:199 (1976); Kogut, J. B., Pearson, R. B., Shigemitsu, J. *Phys. Rev. Lett.* 43:484 (1979); Banks, T., Raby, S., Susskind, L., Kogut, J. B., Jones, D., et al. *Phys. Rev. D* 15:1111 (1977)
3. Creutz, M., Jacobs, L., Rebbi, C. *Phys. Rev. Lett.* 42:1390 (1979); Creutz, M. *Phys. Rev. Lett.* 43:553 (1979)
4. Hamber, H., Parisi, G. *Phys. Rev. Lett.* 47:1792 (1981); Marinari, E., Parisi, G., Rebbi, C. *Phys. Rev. Lett.* 47:1795 (1981); Weingarten, D. H. *Phys. Lett. B* 109:57 (1982); Hamber, H., Marinari, E., Parisi, G., Rebbi, C. *Phys. Lett. B* 108:314 (1982); Hasenfratz, A., Hasenfratz, P., Kunszt, Z., Lang, C. B. *Phys. Lett. B* 110:282; B117:81 (1982); Fucito, F., Martinelli, G., Omero, C., Parisi, G., Petronzio, R., Rapuano, F. *Nucl. Phys. B* 210:407 (1982); Martinelli,

G., Omero, C., Parisi, G., Petronzio, R. *Phys. Lett.* 117B : 434 (1982); Martinelli, G., Parisi, G., Petronzio, R., Rapuano, F. *Phys. Lett.* 116B (1982); Weingarten, D. H. *Nucl. Phys. B* 215 [FS7] : 1 (1983); Bernard, C., Draper, T., Olynyk, K. *Phys. Rev. Lett.* 49 : 1076 (1982); Hamber, H., Parisi, G. *Phys. Rev. D* 27 : 208 (1983)

5. A partial list of summary papers on the subject where many additional details and further references can be found: Drouffe, J. M., Itzykson, C. *Phys. Rep.* 38C : 133 (1978); Kadanoff, L. P. *Rev. Mod. Phys.* 49 : 267 (1977); Kogut, J. B. *Rev. Mod. Phys.* 51 : 659 (1979); 55 : 775 (1983); Drouffe, J. M., Zuber, J. B. *Phys. Rep.* 102 : 1 (1983); Creutz, M., Jacobs, L., Rebbi, C. *Phys. Rep.* 95 : 201 (1983); Hasenfratz, P. Lattice Quantum Chromodynamics, NATO Adv. Summer Inst., Munich, 1983. *CERN preprint TH : 3737-CERN* (1983); Monte Carlo Renomalization Group Methods and Results in QCD, Int. Sch. Subnucl. Phys., Erice, 1984. *CERN preprint TH : 3999-CERN* (1984); Montvay, I. Numerical Calculation of Hadron Masses in Lattice Quantum Chromodynamics, Aspen Ctr. for Phys., 1984. *Rev. Mod. Phys.* In press (1985); Berg, B. Cargese Lectures 1983. *DESY Preprint 84-012* (1984)

6. Schwinger, J. *Phys. Rev.* 115 : 721 (1959); Wick, G. C. *Phys. Rev.* 96 : 1124 (1954)

7. Feyman, R. P. *Rev. Mod. Phys.* 20 : 367 (1948); Feyman, R. P., Hibbs, A. R. *Quantum Mechanics and Path Integrals.* New York : McGraw-Hill (1965)

8. Balian, R., Drouffe, J. M., Itzykson, C. *Phys. Rev. D* 10 : 3376 (1974)

9. The first discussion on a discrete lattice gauge theory was given by Wegner, F. J. *Math. Phys.* 12 : 2259 (1971)

10. Berezin, F. *Method of Second Quantization.* New York : Academic (1966)

11. Bloom, E. D. *J. Phys. C* 43(3) : 403 (1982); Hitlin, D. *Proc. Int. Symp. on Lepton and Photon Interactions of High Energies,* Ithaca, 1983, p. 746. Cornell Univ. Press (1983)

12. See for instance the 7th reference in Ref. 5

13. Johnson, R. C. *Phys. Lett.* 114B : 147 (1982); Berg, B., Billoire, A. *Nucl. Phys. B* 221 : 109 (1983)

14. Gross, D., Wilczek, F. *Phys. Rev. D* 8 : 3633 (1976); Politzer, H. *Phys. Rev. Lett.* 30 : 1346 (1973); Jones, D. R. T. *Nucl. Phys. B* 75 : 531 (1974); Caswell, W. E. *Phys. Rev. Lett.* 33 : 244 (1974)

15. Coleman, S., Weinberg, E. *Phys. Rev. D* 7 : 1888 (1973)

16. Baaquie, B. E. *Phys. Rev. D* 16 : 2612 (1977)

17. Hasenfratz, A., Hasenfratz, P. *Phys. Lett.*

93B : 165 (1980); Dashen, R., Gross, D. *Phys. Rev. D* 23 : 2340 (1981); Kawai, H., Nakayama, R., Seo, K. *Nucl. Phys. B* 189 : 40 (1981); Hasenfratz, A., Hasenfratz, P. *Nucl. Phys. B* 193 : 210 (1981)

18. Weisz, P. *Phys. Lett.* 100B : 331 (1981); Gonzales-Arroyo, A., Korthals Altes, C. P. *Nucl. Phys. B* 205 (FS5) : 46 (1982); Scharatchandra, H. S., Thun, H. J., Weisz, P. *Nucl. Phys. B* 192 : 205 (1981); Karsch, F. *Nucl. Phys. B* 205 (FS5) : 285 (1982)

19. Symanzik, K. *Nucl. Phys. B* 226 : 187, 205 (1983); Weisz, P. *Nucl. Phys. B* 221 : 1 (1983); Weisz, P., Wohlert, R. *Nucl. Phys. B* 236 : 397 (1984); Curci, G., Menotti, P., Paffuti, G. P. *Phys. Lett.* 130B : 205 (1983)

20. Hasenfratz, P., Hasenfratz, A., Heller, U., Karsch, F. *Phys. Lett.* 140B : 76 (1984); Heller, U., Karsch, F. *Nucl. Phys. B* 251 (FS13) : 254 (1984)

21. Luscher, M. *Phys. Lett.* 118B : 391 (1982); *Nucl. Phys. B* 219 : 233 (1983); Floratos, E., Petcher, D. *Phys. Lett.* 133B : 206 (1983)

22. Osterwalder, K., Seiler, E. *Ann. Phys.* 110 : 440 (1978)

23. Munster, G. *Phys. Lett.* 95B : 59 (1980); Munster, G., Weisz, P. *Phys. Lett.* 96B : 119 (1980); Errata, 100B : 519 (1981)

24. Hasenfratz, A., Hasenfratz, E., Hasenfratz, P. *Nucl. Phys. B* 181 : 353 (1981); Itzykson, C., Peskin, M., Zuber, J. B. *Phys. Lett. B* 95 : 259 (1980); Luscher, M., Munster, G., Weisz, P. *Nucl. Phys. B* 180 : 1 (1980)

25. Munster, G. *Nucl. Phys. B* 190 (FS3) : 439 (1981); Errata, 205 (FS5) : 648 (1982)

26. Smit, J. *Nucl. Phys. B* 206 : 309 (1982)

27. Binder, K. In *Phase Transitions and Critical Phenomena,* ed. C. Domb, S. Green. New York : Academic (1976); Binder, K., ed. *Monte Carlo Methods in Statistical Physics,* Vol. 5B. New York : Springer Verlag (1979)

28. Creutz, M. *Phys. Rev. D* 21 : 2308 (1980); Pietarinen, E. *Nucl. Phys. B* 190 (FS3) : 349 (1981); Marinari, E., Cabibbo, N. *Phys. Lett.* 119B : 387 (1982)

29. Fisher, M. E. In *Proc. Int. Sch. Phys. Enrico Fermi Varenna, 1970.* New York : Academic (1971)

30. Luscher, M. DESY preprint 83-116 (1983)

31. Wilson, K., Kogut, J. B. *Phys. Rev.* 12C : 75 (1974)

32. Ma, S. K. *Phys. Rev. Lett.* 37 : 461 (1976)

33. Swendsen, R. H. *Phys. Rev. Lett.* 42 : 859 (1979); in *Statistical and Particle Physics, Proc. Scot. Univ. Summer Sch. in Phys.,* ed. K. C. Bowler, A. J. McKane. SUSSP (1984)

34. Wilson, K. G. In *Recent Developments in Gauge Theories,* ed. G. 't Hooff, et al.

New York: Plenum (1980); Shenker, S., Tobochnik, J. *Phys. Rev. B* 22:4462 (1980); Hasenfratz, A., Margaritis, T. *Phys. Lett.* 133B:211 (1983)

35. Hasenfratz, A., Hasenfratz, P., Heller, U., Karsch, F. *Phys. Lett.* 143B:193 (1984)

36. Creutz, M. *Phys. Rev. D* 23:1815 (1981)

37. Heller, U., Karsch, F. *Illinois preprint P/85/342* (1985)

38. Creutz, M. *Phys. Rev. Lett.* 45:313 (1980); see also Pietrinen in Ref. 28

39. Bowler, K. C., Hasenfratz, A., Hasenfratz, P., Heller, U., Karsch, F., et al. *Nucl. Phys. B* 257:155 (1985)

40. Bowler, K. C., Gutbrod, F., Hasenfratz, P., Heller, U., Karsch, F., et al. *Amsterdam preprint ITFA-85-07* (1985)

41. Gupta, R., Patel, A. *Caltech preprint CALT-68-1142* (1984)

42. Kennedy, A. D., Kuti, J., Meyer, S., Pendleton, B. J. *Santa Barbara preprint NSF-ITP-84/61*; *NSF-ITP-85/11*

43. Ellis, R. K., Martinelli, G. *Frascati preprint LNF-84/1 (P)* (1984); Ellis, R. K. *Fermilab preprint FERMILAB-CONF-84/41-T* (1984)

44. Ambjorn, J., Olesen, P., Peterson, C. *Nucl. Phys. B* 240 (FS12):189 (1984); Flensburg, M., Peterson, C. *Lund preprint LU TP 84-21* (1984)

45. Griffiths, L. A., Michael, C., Rakow, P. E. L. *Phys. Lett.* 150B:196 (1985)

46. Luscher, M. *Nucl. Phys. B* 180:317 (1981)

47. Hasenfratz, A., Hasenfratz, P., Heller, U., Karsch, F. *Z. Phys. C* 25:191 (1984)

48. Barkai, D., Moriarty, K. J. M., Rebbi, C. *Phys. Rev. D* 30:1283 (1984)

49. de Forcrand, P. Unpublished (1985)

50. Stack, J. *Phys. Rev. D* 29:1213 (1984)

51. Otto, S., Stack, J. D. *Phys. Rev. Lett.* 52:2328 (1984); Errata, 53:1028 (1984)

52. Sommer, A., Schilling, K. *Z. Phys.* In press (1985)

53. de Forcrand, P., Schierholz, G., Schneider, H., Teper, M. *DESY preprint 84-16* (1984)

54. de Forcrand, P., Roisnel, C. *Phys. Lett.* 137B:213 (1984)

55. Michael, C. *Liverpool preprint* (1985)

56. Berg, B., Billoire, A. *Nucl. Phys. B* 226:405 (1983); see also Berg & Billoire in Ref. 13

57. Ishikawa, K., Schierholz, G., Teper, M. *Z. Phys. C* 19:327 (1983); 21: 167 (1984)

58. de Forcrand, P., Schierholz, G., Schneider, H., Teper, M. *Preprint DESY/LAPP-TH-119* (1984); *Phys. Lett. B* 152:107 (1985)

59. Munster, G. *DESY preprint* (1985)

60. Celik, T., Engels, J., Satz, H. *Phys. Lett.* 125B:411 (1983); Kogut, J., Stone, M.,

Wyld, H. W., Gibbs, W. R., Shigemitsu, J., et al. *Phys. Rev. Lett.* 50:393 (1983); Karsch, F., Petronzio, R. *Phys. Lett.* 136B:403 (1984); Kennedy, A. D., Kuti, J., Meyer, S., Pendleton, B. J. *Phys. Rev. Lett.* 54:87 (1985); For a summary of recent status of finite temperature QCD calculations, see Satz, H. *Ann. Rev. Nucl. Part. Sci.* 35:245–70 (1985)

61. Wilson, K. In *New Phenomena in Subnuclear Physics, Erice 1975*, ed. A. Zichichi. New York: Plenum (1977)

62. Karsten, L. H., Smit, J. *Nucl. Phys. B* 183:103 (1981)

63. Susskind, L. *Phys. Rev. D* 16:3031 (1977); Kogut, J. B., Susskind, L. *Phys. Rev. D* 11:395 (1975)

64. Kawamoto, N., Smit, J. *Nucl. Phys. B* 192:100 (1981)

65. Kluberg-Stern, H., Morel, A., Napoly, D., Peterson, B. *Nucl. Phys. B* 190 (FS3):504 (1981); Kluberg-Stern, H., Morel, A., Peterson, B. *Nucl. Phys. B* 215 (FS7):527 (1983)

66. Morel, A., Rodrigues, J. P., *Saclay preprint SPH-T/84/41* (1984)

67. Fucito, F., Marinari, E., Parisi, G., Rebbi, C. *Nucl. Phys. B* 180 (FS2):369 (1981); Weingarten, D. H., Petcher, D. N. *Phys. Lett.* 99B:333 (1981); Scalapino, D. J., Sugar, R. L. *Phys. Rev. Lett.* 46:519 (1981); Kuti, J. *Phys. Rev. Lett.* 49:183 (1982); Berg, B., Foerster, D. *Phys. Lett.* 106B:323 (1981); Montvay, I. *Phys. Lett.* 139B:70 (1984); Langguth, W., Montvay, I. *Phys. Lett.* 145B:261 (1984); Polonyi, J., Wyld, H. W. *Phys. Rev. Lett.* 51:2257 (1983)

68. Weingarten, D. Algorithms for Monte Carlo calculations with fermions. *IBM preprint* (1985)

69. Fucito, F., Solomon, S. *Cal tech Rep. CALT-68-1084, 1127* (1984); Gavai, R. V., Lev, M., Peterson, B. *Phys. Lett.* 140B:397 (1984); 149B:492 (1984); Polonyi, T., Wyld, H. W., Kogut, J. B., Shigemitsu, J., Sinclair, D. K. *Phys. Rev. Lett.* 53:644 (1984)

70. Weingarten, D. *Preprint IUHET-82* (1982)

71. For a recent brief summary on these methods, see Montvay in Ref. 5

72. Hasenfratz, A., Hasenfratz, P. *Phys. Lett.* 104B:489 (1981); Lang, C. B., Nicolai, H. *Nucl. Phys. B* 200 (FS4):135 (1982); Stamatescu, I. O. *Phys. Rev. D* 25:1130 (1982)

73. See Hasenfratz et al in Ref. 4

74. Bowler, K. C., Chalmers, D. L., Kenway, A., Kenway, R. D., Pawley, G. S., Wallace, D. J. *Nucl. Phys. B* 240 (FS12):213 (1984)

75. Hasenfratz, P., Montvay, I. *Nucl. Phys. B*

237:237 (1984); Kunszt, Z., Montvay, I. *Phys. Lett.* 139B:195 (1984)

76. Billoire, A., Marinari, E., Petronzio, R. *Nucl. Phys. B* 251 (FS13):141 (1985)

77. Konig, A., Mutter, K. H., Schilling, K. *Wuppertal preprint WUB 84-14* (1984)

78. Gilchrist, J. P., Schneider, H., Schierholz, G., Teper, M. *Phys. Lett.* 136B:87 (1984)

79. Billoire, A., Marinari, E., Morel, A., Rodrigues, J. P. *Phys. Lett. B* 148:166 (1984)

80. For discussions on finite size problems, see Hasenfratz, P., Montvay, I. *Phys. Rev. Lett.* 50:309 (1983); Bernard, C., Draper, T., Olynyk, K. *Phys. Rev. D* 27:227 (1983); Gupta, R., Patel, A. *Phys. Lett.* 124B:94 (1983); Martinelli, G., Parisi, G., Petronzio, R., Rapuano, F. *Phys. Lett. B* 122:283 (1983)

81. Barbour, I. M., Gibbs, P., Gilchrist, J. P., Schneider, H., Schierholz, G., Teper, M. *Phys. Lett.* 136B:80 (1984)

82. Engels, J., Karsch, F., Montvay, I., Satz, H. *Nucl. Phys. B* 205 (FS5):545 (1982)

*Ann. Rev. Nucl. Part. Sci. 1985. 35 : 605–60*

# THE TEVATRON ENERGY DOUBLER:
# A Superconducting Accelerator

*Helen T. Edwards*

Fermi National Accelerator Laboratory,[1] Batavia, Illinois 60510

CONTENTS

## 1. INTRODUCTION

### 1.1 *The Facility*

The Tevatron project is a massive upgrade of the original Fermilab accelerator complex, which became operational in 1972 and operated from

[1] Operated by Universities Research Association, Inc. under contract with the US Department of Energy. The US Government has the right to retain a nonexclusive royalty-free license in and to any copyright covering this paper.

1973 until mid-1982 at typically 400 GeV and $2 \times 10^{13}$ protons per pulse (ppp). The three parts of the Tevatron project when complete will allow for both fixed-target and collider hadron physics using primary beam energies in the 800–1000-GeV range. The Fermilab facility should thus be able to maintain its position at the forefront of high-energy physics research facilities (1). This paper describes the superconducting accelerator phase of the program. Because the new synchrotron provides protons with twice the energy of the "old" Main Ring, it is frequently called the Energy Doubler. It first accelerated a beam in July 1983 and is presently operating at 800 GeV with intensities above $1 \times 10^{13}$ ppp.

During Main Ring operation from 1972 to 1982, maximum accelerator energies of 500 GeV and intensities of $3 \times 10^{13}$ ppp were obtained (not simultaneously). The normal operating energy was 400 GeV. The accelerator chain consisted of a 200-MeV $H^-$ Linac, an 8-GeV Booster Accelerator, and the Main Ring. The beam was resonantly extracted and split for distribution to numerous targets and secondary beam lines for fixed-target physics (2). The new project consists of the following additions or upgrades: (a) The Energy Doubler (3), which is the first large accelerator to use a superconducting magnet system for the main guide field. This technical achievement is the key to the Tevatron and for that matter to future higher energy proton accelerators. Excitation should be sufficient for 800–1000-GeV beam energies. (b) Beam lines to handle the increased energy capability of the accelerator for fixed target physics experiments. (c) The $\bar{p}$ Source—two 8-GeV rings the size of the Booster that accumulate anti-protons to be used for $\bar{p}$-p collisions in the Energy Doubler for colliding-beam experiments.

The Energy Doubler uses the Main Ring operating at 150-GeV peak excitation as an injector. The Doubler beam is located beneath that of the Main Ring in the same tunnel. The Main Ring is also operated at 120-GeV excitation in order to accelerate and extract protons to the $\bar{p}$ production target for the Source.

Operation with the Main Ring at 400 GeV typically was at a cycle time of 7 to 16 s. The slow 16-s limit was imposed to limit power demand during the daytime. The guide field ramp was such as to allow 1 to $1\frac{1}{4}$ s of 400-GeV flattop excitation during which the beam was slowly extracted from the ring to the various experimental areas. In addition, neutrino experiments required a few fast pulses of beam ($\sim 1$ ms) with intensities up to $\sim 10^{13}$ ppp.

Cycle time with the superconducting magnets is of the order of once per minute, with slow extraction over 20 s. The beam is split in the switchyard area using electrostatic septa and directed to a number of primary targets. This very long time for slow extraction gives a 1/3 duty factor and provides for very good data-taking rates in experiments requiring low intensity. A number of pulses of fast resonantly extracted beam are produced during the excitation flattop along with the slow spill. These pulses of beam typically contain up to $2 \times 10^{12}$ particles and last about 2 ms. The pulsed beam is diverted around the splitting septa and goes only to the neutrino production target. Though the long cycle time with limited intensity per cycle is not so favorable for neutrino experiments, the increase in the neutrino flux and interaction cross section with primary energy compensates for the slow repetition rate. Thus, in the Energy Doubler with a 60-s cycle time at 1000 GeV, one can expect the same average neutrino event rate as with the Main Ring at 400 GeV and a 10-s cycle time, assuming equivalent numbers of protons targeted per cycle for neutrino production.

The third part of the Tevatron project is the construction of the p̄ Source rings and the adaptation of the Energy Doubler to p̄-p storage operation (4, 5). In order for the superconducting ring to do colliding-beam physics, protons and antiprotons will be accelerated and stored in it while circulating in opposite directions. Initial operation calls for three bunches of beam in each direction. Collisions will take place at each of the six long straight sections. Major detectors will be constructed at two of the collision points.

The antiprotons are produced by extracting and targeting 120-GeV beam from the Main Ring. Resultant 8-GeV antiprotons are collected first in the debuncher source ring at an approximately 3-s cycle rate and then are transferred to the accumulator source ring where they are allowed to build up for a number of hours. During the accumulation time the p̄ beam phase space is reduced or "cooled" so that the required emittance can be obtained for efficient injection into and acceleration in the Main Ring. Three equally spaced bunches of p and p̄ are to be injected into the Energy Doubler and accelerated to full field, where they will be stored and collide for a number of hours.

Table 1 lists the general parameters of the Tevatron for both fixed-target and collider operation. Figure 1 illustrates the present operating cycle for the Main Ring and the Energy Doubler during fixed-target operation, with the Main Ring also being used for p̄ production. Figure 2 is a schematic of the accelerator complex, and Figure 3 illustrates the beam lines available for fixed-target operation.

## 1.2   Historical Motivation

The original impetus for the Energy Doubler in the early 1970s was just what the name implies; to double the magnetic field and consequently the beam energy without constructing a new tunnel and support facilities. The physics justification was basically just higher energy to explore new regions. The Main Ring and its service buildings were originally laid out with the

**Table 1**    Tevatron parameters

| | |
|---|---|
| **General** | |
| Accelerator radius | 1 km |
| Peak beam energy | 800–1000 GeV |
| Injection energy | 150 GeV |
| Bend magnetic field at 1000 GeV | 44 kG at 4400 amp |
| Beam emittance (95% normalized) | $24\pi$ mm mr |
| **Fixed-target** | |
| Intensity | $\sim 2 \times 10^{13}$ protons/cycle |
| Acceleration rate | 50 GeV s$^{-1}$ |
| Cycle time | 60 s |
| Slow spill duration (flattop time) | 20 s |
| Fast spill | 5 pulses at $2 \times 10^{12}$ protons/cycle expected |
| **Collider** | |
| Intensity per bunch | $0.6 \times 10^{11}$ protons (antiprotons) expected |
| Number of bunches | $3p, 3\bar{p}$ |
| Luminosity | $1 \times 10^{30}$ cm$^{-2}$ s$^{-1}$ |
| Storage time between fills | $\sim 4$ hr |
| Interaction point amplitude function ($\beta$) | 1 meter ($x, y$) |

idea that there would someday be a superconducting ring in the same tunnel, and the service buildings had room for the additional electronics of the superconducting ring (6).

Between 1973 and 1978 total power usage on site doubled, and the cost of power per kilowatt hour also doubled. It was clear that power costs would continue to rise and, in fact, over the ten-year period from 1974 to 1984 they have increased sixfold. The Main Ring magnet system alone at 400-GeV operation was by far the largest single user: over 50% of the total usage. The use of superconducting magnets could substantially reduce usage; the total power requirements for all Main Ring and Energy Doubler systems together at either 400- or 800-GeV operation uses half of what the Main Ring required when running with a 70% utilization factor at 400 GeV. As a bonus, the flattop duty factor relative to the whole magnetic cycle has increased from about 1/10 to 1/3, though obviously the number of protons accelerated per second has decreased. Details of cryogenic usage, manpower requirements, and equipment complexity all enter into the cost-benefit analysis, but the fact remains that the superconducting ring does operate at twice the energy with much better flattop duty factor for substantially less power than the old Main Ring.

It became apparent, as research and development of superconducting

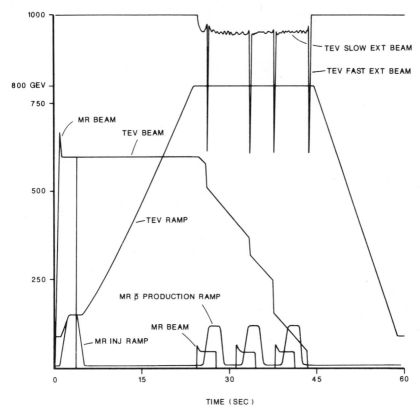

*Figure 1*  Main Ring and Energy Doubler cycle for 800-GeV fixed-target operation with three additional Main Ring p̄ production ramps included in the super cycle. (Vertical scale for beam intensity not shown.)

magnets proceeded, that collider physics was indeed the exciting new field with its great potential to reach high center-of-mass energies. The Main Ring would be extremely expensive to operate above 200 GeV for storage, and it probably would not make a good storage ring in any case. The tremendous advantage of superconducting ring magnets for beam storage at high excitation for hours with no additional operational cost became apparent. In fact, though there was talk of storage in the Main Ring or construction of other conventional rings for various colliding-beam scenarios, the only thing that really made sense was the development of the p̄ Source and the use of the Energy Doubler as a storage ring for p-p̄ collision experiments at 2-TeV center-of-mass energy.

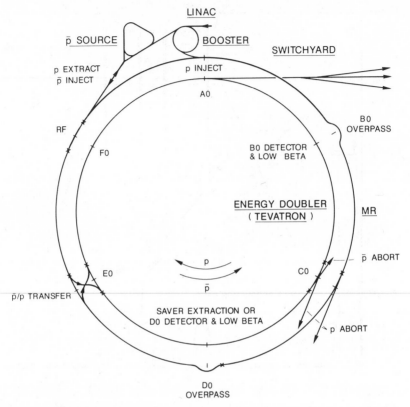

*Figure 2*    Tevatron Accelerator including the new p̄ Source rings and the Energy Doubler.

## 1.3   *Design Issues*

The design of a superconducting-magnet accelerator presented many real engineering challenges and raised several accelerator physics questions.

The magnets had to have a small cross section in order to fit under the Main Ring, and the superconducting cable could not be cryostatically stabilized as in the large bubble chamber magnets of the 1970s. This meant that the magnets had to be protected from their own stored energy in the event they stopped superconducting or "quenched." The magnet coils had to be thermally insulated from room temperature and cooled to 4.6 K with sufficient refrigeration to maintain their low temperature in the face of cryostat heat leaks, ramping eddy currents, and field hysteresis energy losses. The big question was how to wind and clamp the coils so they would not move under ramping field forces, and so that they had highly linear and reproducible magnetic fields throughout their excitation cycle.

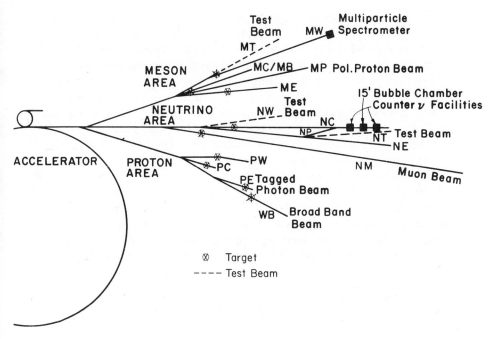

*Figure 3*    Expected layout of fixed-target physics beam lines upon completion of construction (1987).

The major accelerator physics questions were threefold. First, how good did the magnetic fields have to be over what region of the magnet aperture? Second, could beam losses be kept low enough so as not to quench the magnets during normal operation? Third, what sort of accelerator adjustment and control, instrumentation, and diagnostics would be required to aid in smooth commissioning and operation of an accelerator in which it was proposed to challenge the laws of nature that require an environment a few degrees above the absolute zero of temperature.

The first question tends to be one of the hardest to quantify in accelerator design, especially for proton accelerators where the memory of the particles is very long in the absence of synchrotron radiation. Designers really do not know what field quality is adequate to make a storage ring with long lifetime, for instance. An equally difficult problem is assessing the importance of various nonoptimized operational conditions, how often and to what extent they are likely to occur, and to what extent the design should allow for their occurrence. It is obvious, for instance, that the "good" magnetic field region should be as large as the beam size, but what about beam steering? In principle one can steer the beam down the center of the beam pipe within the resolution of the detectors and power supplies, but how do you get it there to begin with? How much will it drift with time? How much do you want to move it to explore its optical properties, etc? Taken in total, what is the difference between a good accelerator and a bad one? Why do some just run and others require continuous attention?

For the Energy Doubler, addressing these imponderables was cut short by evaluating the demands of the resonant extraction process. This process is precisely defined in terms of the amount of magnetic aperture it requires and the percentage of beam particles that under optimum conditions will hit the extraction septum. Thus, for the Energy Doubler beam-tracking studies were performed in detail for the few hundred turns required for extraction particle trajectories. Inside this required aperture, field quality was made as good as was reproducibly achievable by the manufacturing techniques and consistent with small resonant bands and realistically sized correction magnets. The losses from fast extraction were modeled by tracking particles after their interaction with the electrostatic septa. Particles were tracked either until they were extracted as halo or until they struck magnets or collimators and produced cascade showers. Energy deposition of the cascades in the superconducting magnets was calculated and could be compared with expected superconductor tolerance levels, and then allowed extraction beam intensities could be estimated.

The third question of control and diagnostics was confronted by realizing that the superconducting magnet system would not be forgiving, that magnets would quench from beam or refrigeration problems, that quench recovery would not be rapid, that cycle times would be long, and that a massive amount of hardware would be required for the refrigeration and power supply quench protection systems. Thus large amounts of distributed control and data processing capability were envisaged, along with diagnostics capable of recording the sequence of events during an acceleration cycle and leading up to a quench. The quench is prohibited, if possible, by "aborting" or removing the beam from the ring within the time duration of one turn, if for instance beam loss levels approach those that might produce a quench.

Finally, the problems of installation, commissioning, and operation had to be considered. The question of whether or not the vacuum system would ever become leak tight caused serious concern during initial installation. Reliability of the plant was not expected to be high during commissioning; accordingly, beam instrumentation and diagnostics were designed to provide information as efficiently as possible. Operating experience and operational reliability can now begin to be analyzed and the design and implementation judged in retrospect. It is possible to speculate upon changes and differences in approach for future accelerators.

# 2.  SPECIAL DESIGN FEATURES

## 2.1  *Overview*

The main special design features of a superconducting accelerator are, of course, the magnets themselves and the cryogenic system to cool them. The magnets are packaged in their own cryostats, which provide a reservoir for the helium cooling and thermal isolation between liquid helium temperature and room temperature. Vacuum systems are required not only for the tube in which the beam circulates but also for thermal insulation in the cryostat. In the Energy Doubler, the magnet cryostats themselves also provide a transfer line and heat exchanger system for the helium cooling as well as a nitrogen heat shield to reduce the refrigeration requirements at helium temperatures (7).

Second only to these requirements are the questions of protecting the magnets when they quench and of determining what level of beam intensity can be tolerated without quenching. The power supply and quench protection system for the Energy Doubler main bend and quad magnet string is a large distributed "active" system in which resistive voltage development is monitored continuously across > 200 separate segments of the ring in order to divert current and extract electrical energy from the magnets in the event of a quench. Regarding beam loss, arduous calculations are necessary to convince oneself that a superconducting accelerator is able to operate without quenching.

The system required to control and monitor the refrigeration, vacuum, power supply and quench protection systems, and special beam diagnostics must acquire and process far more data than its predecessors, even before considering the requirements of the more conventional accelerator systems. In order to obtain sufficient support, distributed process control and monitoring have been provided in the 30 remote service buildings around the ring.

## 2.2  *Magnet Cross Section and Accelerator Layout*

The cross section of the magnet-cryostat illustrated in Figure 4 shows the concentric design. Starting with the inside is the vacuum space for the beam. This is followed by the single-phase (liquid helium) cryostat, which contains the magnet coil clamped by stainless steel collars in a highly reproducible, accurate configuration that does not distort during magnetic excitation. Though most of the single-phase volume is filled by the collared coil assembly, small annular regions between the beam tube and coil, and between the collars and the outer single-phase pipe, allow helium to flow along the length of the magnet. The next concentric pipe encloses a volume for the two-phase (liquid and gas) helium, which returns along the length of the magnet in the opposite direction and thereby allows for counterflow heat exchange at the surface of the single-phase tube. Outside of the helium container is an insulating vacuum space and then two concentric pipes of only slightly different diameter; these contain nitrogen liquid and intercept heat flow from room temperature to liquid helium temperature (4.6 K). The insulating vacuum region between the nitrogen shield and the room temperature outer cryostat tube contains superinsulation (aluminized Mylar) as an additional radiation shield. This whole magnet-cryostat assembly is vacuum tight. It is held in a laminated iron yoke that contributes $\sim 18\%$ to the total magnetic field. The assembly is precisely ($\sim 1$ mil) adjusted relative to center with G10 suspension blocks and pre-loaded suspension cartridges that allow for contraction and expansion during the thermal cycle. The dipole (21 ft in total length) requires nine sets of suspension points.

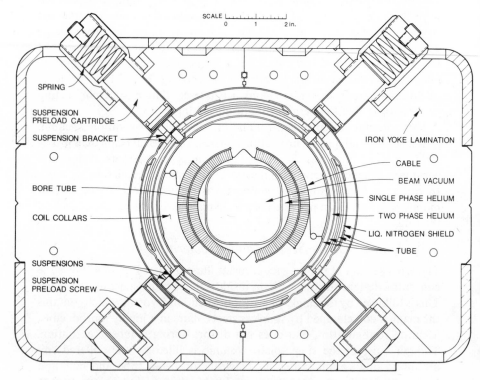

*Figure 4* Cross section of the Energy Doubler dipole magnet showing the collared coil assembly, the cryostat, and the warm iron yoke.

*Figure 5*    The Main Ring tunnel with the superconducting Energy Doubler magnets installed underneath the conventional Main Ring magnets.

The magnet layout for the Energy Doubler follows closely that of the Main Ring, constrained as it is by the requirement that the Doubler fit under the Main Ring magnets with their center lines separated by $25\frac{1}{2}$ inches (Figure 5). Both rings have six sectors and six long straight sections. In addition, each sector has two additional drift regions of 6 and 12 meters where kickers, special magnets, and detectors for injection, extraction, etc can be placed.

Each sector has 31 half-cells as well as cells with quadrupole doublets at the ends of the straight sections. A standard half-cell includes one quadrupole, one correction coil unit ("spool piece"), and four bend magnets. The sector is divided into four magnet string units or cryo sections, which typically contain eight half-cells. These 1/4 sector units are sometimes called "houses" because cryogens, power, vacuum, control, and monitoring for each unit are provided by a service building located above the tunnel. There are 24 houses for the support of the six ring sectors and an additional six for the support of the six straight sections.

## 2.3  *Magnet*

Superconducting accelerator magnet design was discussed in a prior review article (8). Here we only review some of the more basic features of the Energy Doubler magnets.

The desired $\cos \theta$ current distribution for the coil is approximated by two layers of NbTi Rutherford-style cable with the inner and outer layers extending to $\pm 72°$ and $\pm 36°$ respectively. The 23-strand cable is keystone shaped in cross section to form a self-supporting semicircular Roman arch. Cable dimensions are 0.044 to 0.055 keystone thickness $\times 0.308$ inch width. Twenty-three strands of 0.0268-inch diameter wire with a copper-to-superconductor-area ratio of 1.8 : 1.0 are twisted flat to make up the cable. Each strand has 2050 filaments of NbTi alloy (53.5 to 46.5% by weight) that average 8.7 $\mu$m in diameter. The small filaments are necessary for prevention of flux jumping in the superconductor (9), and also play a role in minimizing persistent current effects. To avoid inductive loops within the strands themselves, which would lead to interfilament coupling and large ac losses, the strands are twisted twice per inch. The cable is insulated with 1-mil thick double-lapped Kapton. Outside of the Kapton wrap is a layer of spirally wrapped fiberglass tape impregnated with epoxy that flows during the coil-molding process to provide a reasonably solid coil package. Since a typical cable in the magnet is subject to very large Lorentz forces ($\sim 100$ lb per linear inch), laminated stainless steel collars are press-fitted around the insulated coil to prevent any motion that could cause quenching. The magnets must operate to $\pm 500$ volts to ground under normal operation and 2 kV during failure mode, when the coils may be surrounded by up to 180-K helium gas at 40 pounds per square inch absolute (psia), i.e. relative to vacuum. The magnets are hi-potted to 2 kV at 4-atm room temperature helium prior to installation. (Note that while helium liquid is a good insulator, helium gas has a low breakdown threshold.)

Tests on superconducting cable for the Doubler magnets give an average $J_c$ of 1800 amp mm$^{-2}$ at 5 T and 4.2 K. Taking into account the magnet geometry, field, and operating point of 4.6 K, one finds the average magnet current that this $J_c$ would allow is 4.6 kA (10). All magnets prior to their installation in the ring were measured at the Magnet Test Facility under two different excitation sequences: "quench" and "cycle." In the first test, magnets were ramped at 200 A s$^{-1}$ until a quench occurred. In the second test repetitive ramps approximating the accelerator cycle were used and the flattop current increased until the quench occurred. The results from these tests for the magnets installed in the ring (see Figure 6) indicate that ring excitations of 900–950 GeV should be possible with only a few magnet replacements, whereas to reach 1000 GeV will probably require either very slow ramps or cryogenic modifications, such as "cold compressors" to lower the operational pressure (temperature).

Magnetic field quality is given by the multipole coefficients in the expansion

$$B_y + iB_x = B_0 \sum_{n=0}^{\infty} (b_n + ia_n)(x + iy)^n,$$

where the pole number is $2(n+1)$ and $b_n (a_n)$ is the normal (skew) multipole coefficient, and $b_0$ is unity. The multipoles allowed by dipole symmetry, $b_2, b_4, b_6, \ldots$, are designed to be small and would be zero for a pure cos $\theta$ winding. The Energy Doubler magnets were designed assuming the conductor could be placed accurately enough ($\sim 1$ mil) (8) to give high field multipoles a few times $10^{-4}$ at a reference radius of one inch. Generally speaking the measured values meet those expectations (11–13). An important feature of the magnet is the ability to null the normal and skew quadrupole moments by off-centering the collared coil in the external iron with the suspension cartridges. In this way the $b_1$ and $a_1$ rms widths for collared coils of $1.9 \times 10^{-4}$ and $2.9 \times 10^{-4}$ were both reduced to $\sim 0.7 \times 10^{-4}$ for the finished magnets.

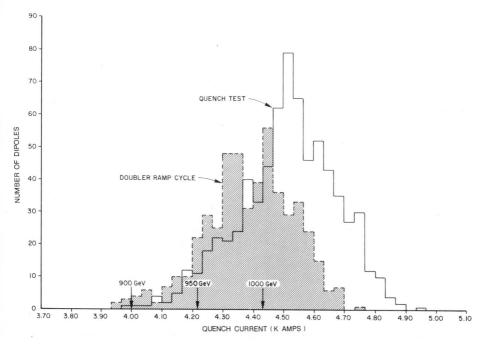

*Figure 6*  Quench current distribution for dipoles installed in the Energy Doubler.

Multipole coefficients that result from the position of conductors are expected to be independent of current as long as the coil assembly is rigid. "Persistent currents" in the superconducting filaments, resulting from eddy current shielding effects (Meisner effect), produce a hysteresis behavior in the $b_0, b_2, b_4$ coefficients that is dependent on whether the field is increasing or decreasing as well as on the size of the filament, the $J_c(B)$ of the filament, and the coil radius (14). This results in magnetization that persists even when the power supply current is reduced to zero (Figure 7). For the Energy Doubler, the persistent dipole and sextupole fields at one inch are about ~6 gauss; this means that the sextupole moment $(b_2)$ is $-5 \times 10^{-4}$ at injection whereas the average high field value is $1 \times 10^{-4}$. The standard deviation from the mean is $3.6 \times 10^{-4}$, independent of excitation (15).

The filament-induced magnetization exhibits hysteresis that in turn leads to cyclic energy losses as the magnets are ramped from minimum to maximum excitation and back. Additional eddy currents induced between

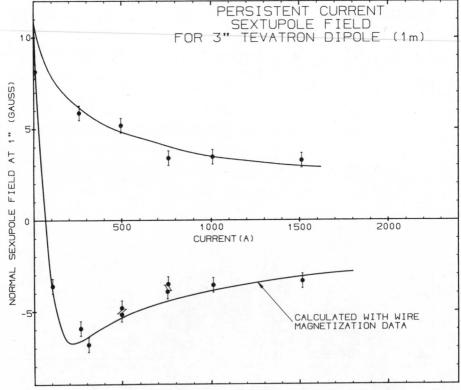

*Figure 7*   Persistent current sextupole field of an Energy Doubler dipole.

strands and in the Cu matrix surrounding the filaments also cause ac losses that are proportional to both $dB/dt$ and $B_{max}$-$B_{min}$. For a typical 60-s cycle, the total ac loss is $\sim 500$ joules per dipole, which results in an average power consumption of 8.4 W (13). Superconductor hysteresis is responsible for about 50% of this value.

Average values of the low order multipole coefficients ($b_1$, $b_2$, $b_3$, $a_1$, etc) produce tune shifts, chromaticity (momentum-dependent tune), amplitude dependence of tune, and horizontal-to-vertical coupling in the accelerator that can be compensated for with distributed correction elements. The rms fluctuations of the multipoles in the magnets will produce resonant bands and make the accelerator unstable at certain tunes. These also can be compensated by corrections, but their effect can be minimized if the magnets are installed in an order that cancels the magnetic imperfections. This "shuffling procedure" (16) was performed on groups of $\sim 32$ bend magnets (1/24th of the ring) and was designed to minimize sextupole and skew octupole resonance driving terms. The driving term for the $3v_x$ resonance was reduced by over a factor of 30; more typically other terms were reduced by factors of 2 to 3 (17). Third-integer resonance (full) widths of $\Delta v = 0.003$ to 0.004 would have been expected at injection (for $0.15\pi \times 10^{-6}$ m emittance) from the measured $b_2$ rms value (18). Half-integer full widths of $\Delta v = 0.006$ were expected from the $b_1$ rms of 0.7 and $\Delta v = 0.007$ was measured at 500 GeV (19).

## 2.4  Cryogenic System

The Energy Doubler helium refrigeration system is the world's largest. It is a hybrid system and contains a large helium liquefier, a nitrogen reliquefier, a distribution system around the ring for liquid helium and nitrogen (transfer line), as well as header pipes for room temperature high- and low-pressure gases, and a distributed satellite refrigerator system that supplies the 24 individual magnet strings with helium and nitrogen cooling (20). These components (Figure 8) can provide a total of 24 kW of cooling at 4.7 K for the magnets as well as 600 liters per hour ($\ell$ h$^{-1}$) of liquid helium for power lead cooling and 1000 $\ell$ h$^{-1}$ of nitrogen for the magnet heat shields.

The large helium liquefier can supply over 4500 $\ell$ h$^{-1}$ of liquid and uses 3.6 MW of power. It consists of two 2000-hp, three-stage compressors (and a spare) that operate at 12 atm and supply 1200 g s$^{-1}$ of helium gas to a heat exchanger "cold box" that operates with three turbine expanders. The first stage of cooling is provided by a liquid nitrogen counter flow heat exchanger that uses 0.64 liters of nitrogen per liter of helium produced.

The nitrogen reliquefier, which will become operational in 1985, can produce 85 tons of liquid per day and will use 3.0 MW. For operation without the reliquefier, liquid nitrogen must be delivered by 5000-gal commercial tankers (5000 gal = 17 tons). During full operation one tanker must arrive on the average of every four hours.

The distribution system around the ring is 6.5 km long and consists of a "transfer line" for liquid nitrogen and liquid helium and header pipes for high- and low-pressure helium and low-pressure nitrogen. The transfer line is made of four nested stainless steel pipes containing (from the inside out) supercritical helium at 5 K and 3 atm, an insulating vacuum space, 3-atm subcooled nitrogen that acts both as a radiation shield for the helium and as a nitrogen supply for the satellite refrigerators and magnets, and, finally, an

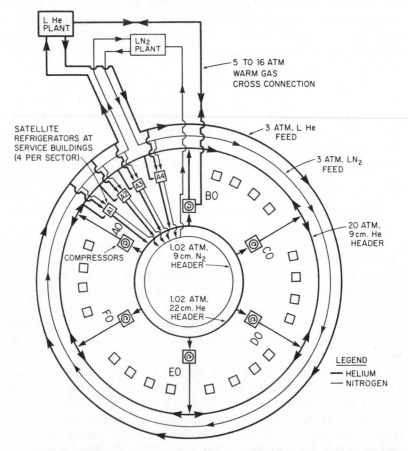

*Figure 8*  Layout of the helium and nitrogen refrigeration system, comprised of the central helium and nitrogen plant, the transfer line and header distribution system, and the satellite refrigerators and compressors.

outside vacuum insulation space. This transfer line is mounted on the ground above the tunnel and interrupted at each service building for branch taps to the satellite refrigerators and magnet strings. The transfer line can distribute 5000 $\ell$ h$^{-1}$ of helium and 3000 $\ell$ h$^{-1}$ of nitrogen. Estimated heat leak for the total line is very small: of the order of 3.6 kW (80 $\ell$ h$^{-1}$) at nitrogen temperature and 240 W at helium temperature.

Adjacent to the transfer line is a high-pressure (20 atm) helium header that supplies gas from the satellite compressors located at the straight-section service buildings (zero buildings) to the satellite refrigerators located at the sector service buildings (1–4 buildings). In addition, two low-pressure collection headers are located in the tunnel adjacent to the magnets. These headers collect helium and nitrogen from the satellite refrigerators and magnet system. Helium is vented through valves into the header from the magnets only during cooldown or during a quench. These safety valves are mounted on every magnet. Nitrogen from the magnet shield is vented at the end of every cryo string during normal operation. It also can be vented at every spool piece if pressure gets too high.

In principle no large quantities of helium or nitrogen should be vented into the tunnel except under extreme quench conditions (e.g. the simultaneous quenching of every magnet in a sector) or from the rupture of a magnet cryostat. Standard safety procedures do not allow personnel in the tunnel when the magnets are excited, and personnel entering the tunnel (whenever there is liquid in the magnets) are required to carry oxygen monitors and air escape breathing devices.

The satellite refrigerator system consists of 31 screw compressors, each of which requires 350 hp and can deliver 58 g s$^{-1}$ at 20 atm, and 24 refrigerators, each of which contains a cold box, a liquid reciprocating expansion engine, a standby gas expansion engine, and a pair of subcoolers. The refrigerators are located directly above the tunnel and feed helium and nitrogen directly into the magnet string below. The flow is split and goes upstream and downstream through the magnets for typically four half-cells (16 bends, 4 quads, 4 spools) in each direction. At the ends of the string helium flow is controlled by Joule-Thompson valves that connect the single-phase region with the two-phase return region of the cryostat. The two-phase helium is returned along the length of the magnets to the shell side of the refrigerator heat exchanger. Nitrogen for the magnet shields is supplied at the refrigerator feed point and makes a single pass to the ends of the strings where it is discharged into the nitrogen header as 92-K gas.

Under normal (satellite) operation, the satellite refrigerator system uses about 7.0 MW for compressor power, 1500 $\ell$ h$^{-1}$ (32 tons per day) of nitrogen (including magnet shields), and 4000 $\ell$ h$^{-1}$ of helium from the Central Helium Liquefier. The Central Helium Liquefier in turn uses

~ 2600 $\ell$ h$^{-1}$ (56 tons per day) of nitrogen and 3.6 MW of power (including all plant usage). This configuration results in 17 kW of refrigeration at 4.6 K and 600 $\ell$ h$^{-1}$ of helium for lead cooling. As much as 24 kW is available under ideal conditions.

Refrigeration at 100% efficiency would require 66 watts at room temperature for every watt at 4.5 K, and 234 watts at room temperature for every liter per hour of liquid helium produced. In addition 100% efficient nitrogen production would require 161 watts per liter per hour. Typically nitrogen is produced at 30% efficiency and so requires 540 W $\ell^{-1}$ h. With the above numbers, the Energy Doubler helium cryogenic system is of the order of 10–14% efficient (21).

Helium inventory in the system is 35,000 liters and includes 24,000 $\ell$ in the magnets and 11,000 $\ell$ in the transfer line plus liquid storage. Most of the magnet inventory is lost in less than one hour when a site power failure shuts down all compressors. Typical loss rates are 40,000 to 60,000 liters per month, of which ~ 60% is estimated to be steady-state leakage, ~ 15% is due to magnet quenches, and the remaining 25% is due to transient small leaks as well as maintenance decontamination and repair.

The cryogenic system has a broad spectrum of operating conditions. In "satellite mode," the central plant supplies large amounts of cold helium to the magnet strings and thus to the return side of the satellite heat exchangers. This excess flow imbalance in the exchangers results in 1000 W of refrigeration without use of the gas expansion engine and with minimal liquid nitrogen (12 $\ell$ h$^{-1}$) to do initial cooling of the warm high-pressure helium gas at the input to the cold box. At the other end of the spectrum in "stand-alone mode" no cold helium is used but the gas expansion engines must operate and over 60 $\ell$ h$^{-1}$ of liquid nitrogen is required in the first heat exchanger. Of the order of 450–500 W of refrigeration plus 25 $\ell$ h$^{-1}$ of liquid results from this mode and is sufficient to handle only the static heat load of the magnets. A variety of intermediate cases are possible depending on the availability of helium from the central plant.

The magnet strings are cooled by the force-fed single-phase subcooled liquid helium driven by the wet expansion engines. The single-phase liquid is continuously cooled along the length of the string by the return two-phase counter flow heat exchanger. This cooling balances the heat deposited in the string by the static heat load (thermal radiation and conduction through the magnet supports) and the ramping heat load (hysteresis and eddy currents in the magnets). The two-phase return helium changes from mostly liquid at the far end of the string to mostly gas near the refrigerator return, and it is this evaporation that supplies the required cooling. The flow control valves at the ends of the magnet strings are adjusted so that only a small amount of liquid (10%) remains in the end of the two-phase system independent of heat load. Typical flow in the magnet string is 21 g s$^{-1}$.

Operating temperature of the superconducting magnets is of critical importance as the allowable peak excitation varies by $\sim 14\%$ per degree Kelvin (22). In the Energy Doubler, the operating temperature is set by the requirement that all parts of the refrigerator system operate at positive pressure with respect to the atmosphere so as to reduce the possibility of contamination of the helium gas by leakage of air into the system (air freezes and blocks the cold piping). Because of this, the two-phase region of the magnets operates at 5 pounds per square inch "gauge" pressure (psig), i.e. above atmospheric pressure, and the helium boils at 4.5 K. A $\Delta T$ of $\sim 0.4$ K is required to carry the heat from the magnet conductor through the insulation, to the single-phase liquid, from the liquid to the stainless single-phase/two-phase tube, and through the tube to the two-phase flow. Thus the magnets are probably operating at $\sim 4.9$ K. Lowering the temperature at fast cycle rates more than 0.2 K may be difficult, but 0.4 K may be possible with very slow ramps.

Static heat leak measurements on the magnet strings give $430 \pm 60$ W for eight half-cells (32 dipoles, 8 quads, 8 spools) (23). The heat load due to ramping is an additional 270 W (500 J $\times$ 32 dipoles at 60-s cycle time), making a total of 700 W heat load per refrigerator. The measured static load ($\sim 9.5$ W per bend magnet) (24) is typically 50–100% greater than expected from calculations.

## 2.5    Vacuum System

The vacuum system for the Energy Doubler consists of three separate subsystems with different characteristics and requirements (25). The cryostat insulating vacuum system is the most complex and is completely isolated from the high-vacuum cold beam tube system inside the magnets. The straight sections and other noncryogenic regions have warm beam tube, bakeable, conventional vacuum systems.

In total the three systems employ of the order of 700 vacuum gauges (Pirani, cold cathode, and ion gauges), 200 ion pumps, 50 turbo roughing pump stations, and 250 electropneumatic valves, all with computer control and readout. The system is divided into 30 individual units corresponding to the 24 cryogenic refrigeration loops and the six warm long straight sections.

All in all there are about 1300 cryogenic interfaces between magnets or between magnets and spool pieces (correction coil modules) or other special purpose modules. A magnet-to-magnet interface (Figure 9) consists of a beam tube seal, two liquid helium connections, one liquid nitrogen connection, and a large external room temperature insulating vacuum seal.

Solder splice joints between the pairs of high-current superconducting leads of the magnets must be made up and insulated inside the single-phase connection. Each of the cryogenic seals must be able to be verified at room temperature with sufficient sensitivity to assure that it will not leak liquid helium. By far the most time-consuming aspect of installation is the interface connection and leak checking. During initial installation each interface took on the average one man-week; subsequent work has been done in about half this time.

The leak-checking procedure is done in three steps. First each device,

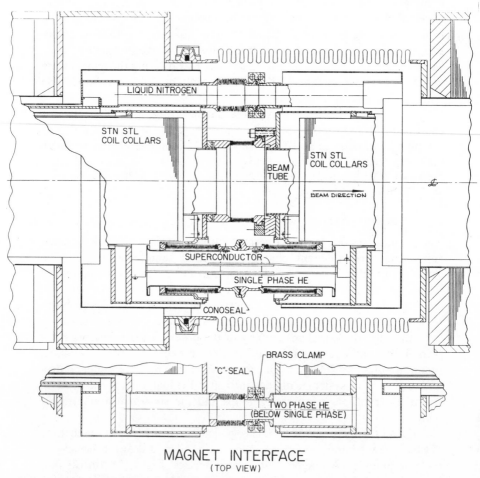

LIQUID NITROGEN

STN STL
COIL COLLARS

BEAM
TUBE

STN STL
COIL COLLARS

BEAM DIRECTION

SUPERCONDUCTOR

SINGLE PHASE HE

CONOSEAL

BRASS CLAMP

"C"-SEAL

TWO PHASE HE
(BELOW SINGLE PHASE)

## MAGNET INTERFACE
(TOP VIEW)

*Figure 9*   Magnet-to-magnet connecting interface with beam tube, superconductor, single-phase helium, two-phase helium, nitrogen, and external room temperature bellows connections.

during its manufacture, is checked for leaks after each major assembly and welding procedure. Finished assemblies are checked for leaks and all devices are cooled to at least liquid nitrogen temperatures (all magnets and spools are cooled to helium temperatures) and rechecked. To pass requires that no leaks be detected, with leak detector sensitivities of better than $2 \times 10^{-10}$ atm cm$^3$ s$^{-1}$. The second step is to install, connect, and test a complete half-cell (4 bends and 1 quad) in the tunnel. At this stage spool pieces are left out so that the ends of the half-cell can be capped off, with each cryogenic circuit brought through vacuum feedthroughs in the temporary end caps. The beam tube connections are first checked by pumping down the beam tube and checking the connections with helium in the usual fashion. (At this stage the large room temperature bellows interface has not yet been closed.) Next, the rest of the interface connections are made, the end caps installed, and the insulating vacuum space of the cryostat pumped down with leak detectors on pumpout ports adjacent to each interface region. After about three hours, leak checking can begin at $2 \times 10^{-10}$ atm cm$^3$ s$^{-1}$. The leak detector output signals are connected to multichannel chart recorders for comparison of responses and permanent record. Each cryogenic circuit (single-phase, two-phase, and nitrogen) is pressurized in turn to 2 atm (absolute) of helium gas and held for 15 min. If a leak is detected it can be located by using the time and magnitude response of the various detectors on the chart recorder. If any leaks are detected, seals are replaced and the string is rechecked. The bore tube vacuum is also checked when the single-phase system is pressurized. This procedure is repeated for each of the eight or nine half-cells in a house (or quarter sector). When all half-cells have been certified, the interconnecting spool pieces are installed and a third and final leak check is performed with a leak detector now on each half-cell. Since there are insulating vacuum barriers in each spool piece, leaks can still be isolated.

The static heat leak of the cryostat-magnet system is due to thermal radiation and heat conduction through magnet supports and other structural elements as well as through residual gas in the insulating space of the cryostat. We find that for our geometry the static heat load doubles at a pressure of $2 \times 10^{-5}$ torr of helium. We require a pressure below $10^{-5}$ for operation (which reads $3 \times 10^{-6}$ on nitrogen-calibrated cold cathode gauges). Typically, vacuum reads $10^{-7}$ torr (nitrogen-calibrated). The insulating vacuum is pumped with turbo molecular pumps. Additional mobile pump stations can be added at the location of leaks to permit continued operation of the accelerator.

The beam tube vacuum is divided, by isolation valves, into 30 major sections. These consist of 24 250-m cryogenic sections and six 50-m warm straight sections. In addition there are some 23 short warm sections.

The pressure in the cold beam tube is expected to be very low if helium leaks are absent. Pressures of $5 \times 10^{-11}$ torr cold ($5 \times 10^{-10}$ torr as measured warm) are typical. The cold beam tube provides an economical practical way of obtaining a very high vacuum (required for long beam storage times) over the major fraction of the ring circumference. A warm beam tube inside the superconducting magnets would have required considerable complexity and larger magnet coil diameter to accommodate insulation, in situ high-temperature baking, and conventional pumping. Beam current thresholds for pressure instability cascades induced by beam ionization of residual gas molecules have been estimated to be more than a factor of 30 above expected operating currents (25). In the cascade process the ions are accelerated to the vacuum wall and desorb more molecules.

Helium leaks to the beam tube are extremely difficult to detect or pump without warming the system up because the gas will condense on the beam tube surface and only slowly migrate along the pipe. Because of this, the cryostat was designed so that the beam tube seam weld is the only weld between the helium space and bore tube vacuum.

The warm sections of the beam tube vacuum contain not only pipe but also devices such as rf cavities, kicker magnets, electrostatic septa, magnets with vacuum tubes, and "Lambertson septum magnets," the last of which have their steel laminations exposed directly to the vacuum. Because the Lambertson magnets have the potential for outgassing from large virtual leaks, great care is taken in cleaning and baking them during assembly. Their outgassing rate has thus been reduced to a few times $10^{-8}$ torr $\ell$ s$^{-1}$ per meter of exposed laminations.

The fact that most of the circumference of the ring (93%) is cold and has very good vacuum means that the pressure in the warm regions (7%) need not be particularly low. With $5 \times 10^{-11}$ torr in the cold regions and $10^{-8}$ torr in the warm regions, the reduction of luminosity during storage due to Coulomb and nuclear scatterings of the beam from residual gas is expected to be $\sim 23\%$ after 20 hours (4). The warm region gas contributes about one-half of this reduction.

The vacuum systems, though large and containing numerous flanges, seals, pumps, valves, and pressure-monitoring devices, have been remarkably trouble free and reliable. This must, to a large extent, be due to the substantial cryo-pumping ability of the refrigerated surfaces.

## 2.6  Power Supply and Quench Protection System

This system (26) has two opposing goals; namely, powering the main bend and quadrupole magnet string and protecting these same magnets from the stored energy in the magnetic field should any fraction of the superconductor in the whole ring become normal (resistive) for any reason. In addition to the power supplies, the power system includes dump resistors to dissipate rapidly the 350 MJ of stored energy. The quench protection system consists of detection units, heater-firing units to distribute the normal region in quenched magnets and thus reduce the energy dissipated in any section of conductor, and current bypass units that direct current around the quenched magnets. Because this system requires rapid detection of quenches and action of subsidiary electrical components in order to save the magnets from self-destruction, it is called "active" as opposed to one that might require little or no external action, i.e. "passive" (diode protection).

In order to understand the protection system it is necessary first to consider what happens in the superconducting cable when it becomes normal and current transfers from the NbTi to the copper (27). The copper, which now conducts most of the current, is generally insufficient to prevent further heating, and the cable will melt unless some means is found to remove the current. The rate at which the cable temperature rises is difficult to calculate because of the nonlinear behavior of the parameters (specific heat, resistivity, thermal conductivity, etc) that describe the cable constituents at cryogenic temperatures. Instead, the cable temperature, after the initiation of a quench, has been measured at discrete constant currents $(I)$ as a function of $\int I^2 \, dt$. In the Energy Doubler the peak temperature limit allowed was set at 460 K in order to protect the numerous soft-solder splices and the silver-tin coating on cable strands. At its maximum operating current of 4.4 kA, the cable reaches 460 K after $7 \times 10^6$ A$^2$ s ($10^6$ A$^2$ s = 1 MIIT), i.e. there is less than one-half second available for removing the magnet current to prevent permanent damage. At a current of 1 kA, the thermal damage threshold is reached after 12 MIIT.

A schematic of the Energy Doubler magnet power supply system is shown in Figure 10. The 774 dipoles, 216 quadrupoles, and 12 power supplies form a single series circuit with "upper" and "lower" busses connected at a "fold" in the B0 straight section. Each power supply is capable of ramping to 4500 A at 1 kV. Eleven of the power supplies operate in a voltage regulation mode to ramp the current up and down, with either the A2 or A3 supply acting as the system current regulator. Since the resistance of the system is small, the current regulator can provide the required voltage during flattop; hence only this power supply must be capable of conducting continuous flattop current.

The magnets and their interconnection leads are continuously monitored for a resistive voltage component. Once detected, the power supplies are turned off and 0.25-ohm resistive loads are inserted in their place by thyristor switches and a backup dc breaker. The resulting exponential current decay (12-s time constant) is too slow to protect the "normal zone" in the magnet that has quenched. To ensure removal of the current from a quenching element, the magnet circuit is configured into quench protection units, as shown in Figure 11. Eight dipoles and two quadrupoles (one cell) form a typical unit with half of the magnet coils connected into each of the upper and lower bus circuits. A "safety lead" connects the superconducting bus to a room temperature bypass circuit at the ends of each unit. This lead cannot carry steady-state operating currents, but can carry the decaying magnet current around a quenching section of magnets during the dump. Thyristors in each of two redundant external bypass circuits are triggered to allow current to pass as soon as the resistive voltage in the unit exceeds the inductive voltage arising from the decaying current. The fate of the cable depends on the outcome of a race between the cable temperature (function of $\int I^2 \, dt$) and the current bypass rate, which is determined by the total resistance of the normal zone. [Longitudinal and transverse normal zone velocities are proportional to $I^2$ for currents between 1 and 4 kA (28).]

The problem is further complicated in the Energy Doubler by the requirement to protect intermagnet connections, which are not fully stabilized because of geometric restrictions. Quenches originating at these

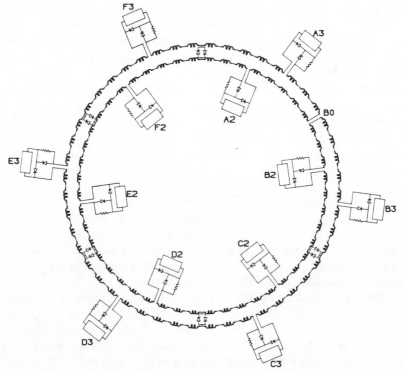

*Figure 10*    Energy Doubler Magnet Power Supply and Dump System configuration.

locations can only propagate longitudinally and are detected much later than quenches that start within a magnet coil. This uncertainty in quench resistance growth is overcome with resistive heater strips in each dipole. When a quench is detected in a protection unit, energy stored in capacitor banks (heater-firing units) is dumped into the dipole heaters of the protection unit. The resulting rapid resistance growth drives the current into the bypass thyristors and also yields a uniform voltage distribution within the bypassed unit.

The system is controlled by a network of 25 special purpose microcomputers [24 Quench Protection Monitors (QPMs) and one Tevatron Excitation Control and Regulation (TECAR) processor] connected by a serial communication link and redundant hardwire loop cable. The TECAR processor, in addition to controlling the power supply program-

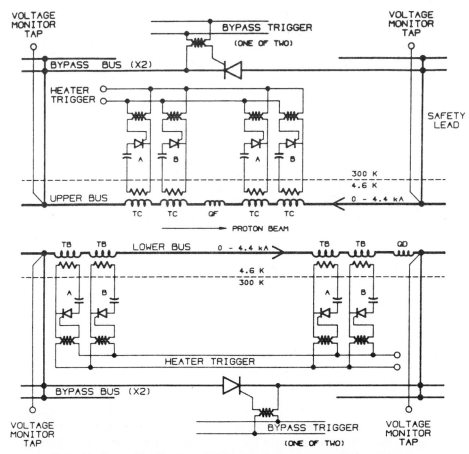

*Figure 11* Energy Doubler Quench Protection System for one cell of magnets.

ming and current regulation, coordinates quench protection actions. The system is fail-safe in the sense that element failure results in a shutdown in a manner that protects the magnets from damage. Critical elements and actions are made redundant, and there is continual or frequent verification of the redundancy.

The QPM samples the voltage across each cell (a total of about 200 points) at a rate of 60 Hz; it then calculates the average $dI/dt$ for the cells using the inductance values stored in the memory for each cell. For each cell, the "resistive voltage," equal to the measured voltage minus the voltage due to $L \, dI/dt$, is compared against the quench limits. If the resistive voltage is outside the limits ($\sim 1/2$ V), the protective actions described above are taken. This method of determining $dI/dt$ was selected because it had less electronic noise than did methods that differentiated the current signal from a transductor, and because it avoids heavy reliance upon a communication system. Further checks are made, however, between the "magnet" $dI/dt$ and the $dI/dt$ derived from a transductor.

The QPM also communicates to the refrigerator microprocessor that a quench has occurred and which cell(s) are involved, so that the quench recovery cooldown procedures can be initiated. The recovery time can be as little as 20 minutes if the cooldown begins promptly. Even short delays result in hot gas returning to the refrigerator, which makes the cooldown much more difficult. The refrigeration status is also checked by the QPM so that the current can be removed from the magnets before they quench from insufficient cooling.

In addition to its protective role, the QPM also maintains a record of the event. This is accomplished by continuously storing data in a "circular buffer" six seconds long and then freezing the buffer when an event occurs. The data record includes the cell voltages, the calculated resistive voltages, $dI/dt$, the current, voltages-to-ground, equipment status, and the voltages across the vapor-cooled current leads (which the QPM also monitors and protects). This data record is essential for properly understanding the system behavior.

The QPMs normally monitor the differential voltages across each cell. In the event of repeated quenches or other anomalous behavior, a special QPM can be installed to monitor the voltages within a cell, typically on every dipole. Faults can usually be localized to a particular device or interface.

## 2.7    Control System

The Accelerator Central Control System (29, 30) for the whole accelerator complex, including the superconducting ring, was redesigned and partially implemented in parallel with the design and construction of the Energy

Doubler. Further implementation is ongoing, along with the p̄ Source construction. The new system thus has diverse requirements.

1. It is an upgrading and conversion of the original Linac, Booster, Main Ring, and Switchyard central controls. It has to be compatible with and adaptable to the interfaces of existing equipment. The conversion has had to take place over a time scale of years, with minimal interference to operation.

2. The Energy Doubler Accelerator requires remotely distributed processors operating independently of the central system. Over 500 microprocessors are in the Doubler system and there are now 1000 in the entire complex.

3. The control system must also be suitable for control of the new p̄ Source rings: the debuncher and accumulator.

4. The system must support not only fixed-target operation as in the past but also simultaneous or sequenced operation for (a) p̄ production, stacking, and accumulation in conjunction with fixed-target physics or beam storage; (b) injection and storage of protons in the Energy Doubler; and (c) unstacking, injection, and storage of antiprotons in the Energy Doubler.

5. The system must provide console (operator station) support to remote facilities such as the Central Helium Liquefier, the Colliding Detector Facility at B0 (and eventually D0), and the RF Building at F0.

The control system can be seen as two separate parts: the central "Host" system and the remotely distributed processors and interfaces for equipment in service buildings around the rings. Communication between the central and remote parts (Figures 12a,b) takes place via a serial CAMAC link for the Energy Doubler. General specifications for the system include the following features.

2.7.1.  CONSOLES    All consoles are identical and are able to control and monitor all parts of the accelerator.

2.7.2  DATA RETRIEVAL    Selective retrieval is required so that only data required at a specific time is brought back to the central system and only as often as required. Thus instead of an indiscriminate central data pool where all monitor points are updated at a specified repetition rate, as in the old system, now a Data Pool Manager allows for discriminate data retrieval. This selective system was necessary because possible control/readback points increase from 6000 in the old accelerator to over 100,000 with the new Tevatron systems.

Data required at any time include those necessary for applications programs requested by the operators, by data-logged devices, and by alarms scans, though alarms consolidation is done as much as possible at

**CONSOLES**

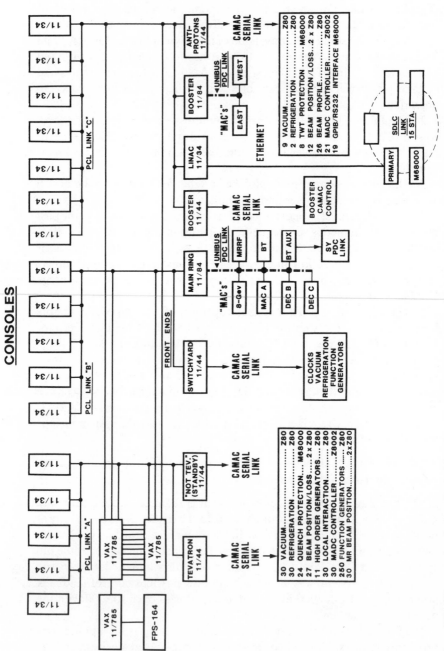

*Figure 12a* Tevatron Accelerator Control System: Central Host System including console computers, central VAX computers, and front-end computers that interface to the remote processors and discrete accelerator components.

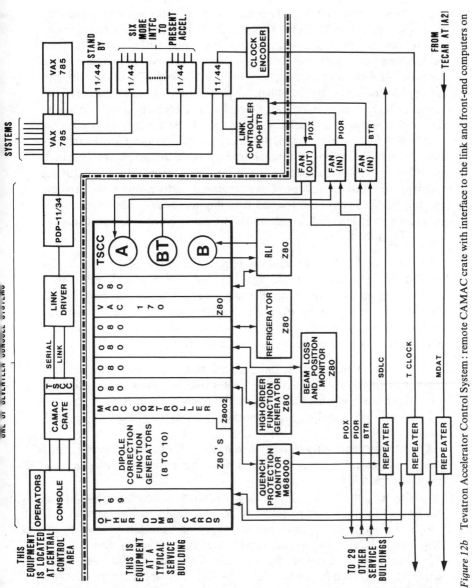

*Figure 12b* Tevatron Accelerator Control System: remote CAMAC crate with interface to the link and front-end computers on the one hand, and to stand-alone multibus microprocessors and accelerator components on the other hand.

the remote systems. Only small quantities of the data generated and used by the remote processors are ever brought back to the central system.

2.7.3  REMOTE PROCESSORS  The remote processors operate separately from the central system either as quasi stand-alone units (refrigerator control processors) or in constant communication with each other (quench protection monitors). These remote systems act as servos in real time and in conjunction with the power supply ramp cycle. Thus each of the refrigeration processors manage 15 control loops and monitor of the order of 100 points at each satellite refrigerator location. The correction coil processors have full adjustment as a function of ramp excitation energy and cycle time ($V_{out} = E^*[f(E) + g(t)]$, where $E$ is the ramp excitation and $f, g$ are arbitrary line segment functions). Additional distributed microcomputers have been provided for vacuum scanning, beam position/intensity/loss monitoring, quadrupole excitation control for resonant extraction, and fast-time plotting of ADC channels. The central computer does *no* real-time control or precise real-time data collecting and thus avoids placing excessive demands on the links.

2.7.4  SNAPSHOT  A number of the remote processors have been implemented with circular buffer storage that can save data recorded at time intervals preceding or following beam aborts or quenches. Any part of these data can be retrieved by the central system on operator demand for analysis. No attempt is made to retrieve all buffer data, as the total amount of the remote computer memory is about two million bytes.

2.7.5  LINKS  Aside from the three control links from the central control computers to the remote CAMAC crates and multibus microprocessor units, there are five other links around the Energy Doubler ring.
  1. Time clock—this link has a real-time clock signal along with "events" encoded on the link.
  2. Tevatron beam sync—this is a beam sync clock to track the position of beam bunches in the ring. It allows for precise kicker timing or gate triggering relative to the bunches.
  3. Magnet data clock—this link carries information on the Main Ring and Energy Doubler guide-field excitation encoded at a 720-Hz rate.
  4. SDLC—this link carries power supply and quench protection information only. It is an integral part of the magnet safety system.
  5. Abort link—this link is a fail-safe link. When interrupted it causes the beam to be extracted in one turn from the ring to an abort dump. This link can be deactivated by any one of a number of systems from any of the service buildings. For example, if beam losses above a given threshold are detected or if a quench is detected, the abort link is deactivated. The time

sequence in which different inputs pull on the link during an abort is recorded.

In the Central System there are two VAX 11/785 mainframe processors linked via DEC HSC units, one designated the "Operational VAX" and the other the "Development VAX." The Operational VAX supports a large central data base that contains addressing information for all "control points" in the accelerator system. It also supports certain "central applications programs" not run from the consoles. The Development VAX is used to support software development for computers throughout the control system. Examples of "Central Applications" are logging data for latter analysis, sending alarm messages on to the console computers, and gathering data from the beam position processors as input to the orbit-smoothing program.

Seven "Front End" PDP-11/44-style computers are used to interface the "Central Host" (Consoles and VAXs) to the various discrete accelerator components. A "Front End" computer attempts to hide from the Host computer the individual differences of the various subsystems. This allows the "Host" to treat all requests for receiving or sending data identically. Individual software drivers, such as that for CAMAC and those for communicating with the "smart" subsystems, are located in the appropriate Front End. The Front End also has the task of scanning for any alarms and then alerting the Host to devices that are at values outside their nominal limits. This includes the unsolicited messages generated by the "smart" subsystems as well as LAMs ("look-at-me" signals), which can be generated by "dumb" CAMAC modules such as power supply controllers.

There are 17 identical control consoles. Each console is supported by a DEC DP-11/34 with RSX/11 operating system, and up to four simultaneous user tasks. Each console is interfaced via a CAMAC crate to the PDP-11 and can be an arbitrary distance from it. Console equipment includes a keyboard, track ball, interrupt button, two Hitachi color monitors, a Lexidata precision color graphics monitor, and a Tektronix 613 storage scope. All PDP-11s and the two VAXs are interconnected with DEC PCL (Parallel Communications Link) busses. Three such busses are used since the fixed-time slice architecture does not permit a very high multiplicity on any one bus.

## 2.8  Beam Loss and Resonant Extraction

As discussed in the cryogenic section, the maximum operating field of the superconducting magnets is critically dependent on the temperature of the coils. Loss of even a small fraction of the circulating beam into the magnets will result in cascade showers, energy deposition, and heating in the superconductor. This heating may be sufficient to cause the magnets to quench even at low excitation; at high excitation where there is little $\Delta T$ margin, the peak accelerator energy may have to be reduced for reliable operation. In a system where quench recovery typically takes an hour, one must try to understand and reduce the effects of beam loss.

Beam loss in the accelerator fits into one of two categories: accidential and unavoidable. By unavoidable we mean loss produced during standard operation of the accelerator; by accidental we mean loss that can be minimized by being careful and by reducing failures or mistakes. The resonant extraction process of the beam from the accelerator to the fixed-target areas is an example of a standard operational mode that produces unavoidable losses. In fact, the fast resonant extracted beam pulses, which produce energy deposition over a short time interval (2 ms), must be carefully analyzed to determine if they are feasible at all.

Calibration studies with simple geometries have been carried out to determine how sensitive the magnets are as a function of excitation and time duration of energy deposition (3, 31). These studies consist of measuring the beam level at which a magnet quenches and comparing that to the calculated deposition per incident particle. In this way a design guideline of less than 4 mJ $g^{-1}$ at 800 GeV was established for millisecond-like beam pulses. Modeling of the extraction process then allowed us to determine what intensity is possible and how to design straight-section components that maximize this intensity.

Resonant extraction is accomplished by producing a half-integer stop-band with quadrupoles distributed on the 39th harmonic of the accelerator circumference ($v = 19\frac{1}{2}$). Octupoles with the same phase relationship are used to provide a nonlinear amplitude-dependent tune so that large amplitude oscillations are unstable. As the quadrupole strength is slowly increased, the stable region shrinks down on the beam and particles stream out along specific trajectories (separatrices) of the phase space plot. The oscillation amplitude of particles grows as they progress away from the stable region; when this growth becomes of the order of 1 cm every two turns, it is possible to "split" the beam with an electrostatic septum. Particles that have "jumped" from the field-free region to the high-field gradient region are deflected sufficiently then to jump a magnetic septum on the other side of the accelerator ($9\frac{3}{4}$ oscillations away) and be deflected into the extraction beam line.

Statistically it is not possible for all particles to jump the electrostatic septum without any hitting the 2-mil wires that make up the anode plane. Of the order of $1\frac{1}{2}\%$ of the beam hits the wires, and it is this beam and its resultant interaction particles that must be tracked.

Monte Carlo analysis using CASIM (32, 33) and similar programs is done in four steps. First the elastically and inelastically scattered particles from the septa are calculated and their energy and direction determined. The particles are then tracked through a geometry that models the accelerator configuration. Inelastics need only be tracked through the first few downstream accelerator magnets. Elastics must be tracked through the machine until they hit a magnet or are extracted. This tracking is followed for up to two turns. Once the locations of hits in the superconducting magnets are recorded, energy deposition in regions with a large "hit probability" can be calculated. Finally, the expected response of loss monitors in the region is also calculated.

Results show that of the particles that hit the septa wires, 20% are inelastics deposited locally downstream of the septa. Because of the magnetic shielding design (resulting from the calculations) in the straight section, only 7% of these actually enter the superconducting magnets. The other 80% of the beam hitting the septa are elastically scattered; 10% are lost in the half-turn between the septa and the extraction channel; 5% are lost in the extraction channel; and 65% leave the machine via the extraction channel on the first or third turn.

It is the $1\frac{1}{2}\%$ inelastics and the 10% elastics lost in the half-ring, or less than $2 \times 10^{-3}$ of the total beam intensity, that must be considered in detail. Calculations show that beam should hit in about a dozen locations, in good agreement with what is observed. Exact predictions are sensitive to the local closed-orbit beam position relative to the magnet bore tube and vary by about a factor of 3 for 1-mm changes. Particles hitting the magnets deposit about 40% of their energy in the superconducting coils.

Comparison of calculations with actual measurements of beam quench levels predicts that 4–15 mJ g$^{-1}$ of energy is deposited locally in the coil to produce the quench when of the order of $2.5 \times 10^{12}$ protons are extracted in a fast pulse. Considering the overall complexity of the calculations, the good consistency of the results with early calibration studies leads one to believe that these calculations are a viable design tool and can be used for further shielding and scraper design improvements (34).

## 3.   ACCELERATOR LATTICE AND CONVENTIONAL SYSTEMS

In the previous section, aspects of the accelerator that are specifically associated with or strongly impacted by the superconducting nature of the magnets were discussed. In this section we describe other systems of the Tevatron, which, though impacted by the superconducting design, are basically required in any accelerator. The lattice, correction coil configuration, diagnostics, and radiofrequency acceleration system are some of these. The accelerator parameters are given in Tables 1 and 2 and the general layout was illustrated in Figure 2.

## 3.1  *Lattice*

As already described, the lattice for the Energy Doubler follows that of the Main Ring closely. The $15\frac{1}{2}$ normal cells per sector follow the pattern of focussing quad, 4 bends, defocussing quad, 4 bends. The bend magnets have a magnetic length of 6.12 m and an overall length of 6.40 m; the quadrupoles have a magnetic length of 1.68 m and a focal length of 26 meters. Adjacent to each quad is a 2.16-m space for the correction coil spool piece. Spacing between quads (half-cell length) is 30 meters and the betatron phase advance per cell is 68°. One of the half-cells is exceptional in that two of the bending magnets are omitted as in the Main Ring to provide space for special warm components.

Generally the fact that the Energy Doubler must follow the Main Ring geometry so closely is very constraining, especially in the straight-section regions for injection, extraction, minimization of the beam loss on the superconducting magnets, and in the design of the low-$\beta$ regions for collisions. A number of compromises in design have been made, which would not have been necessary if the lattice had been designed independent of the Main Ring. Remember that beam energies of 150–1000 GeV instead of 8–400 GeV must still be handled by conventional iron magnets and electrostatic devices in spaces designed for the lower energy. Deviations from the Main Ring lattice do take place at the six 50-meter straight sections (A0–F0). Figure 13 illustrates the layout of these regions.

Two of the straight sections, E0 and F0, where injection and rf acceleration take place, are optically very similar to the Main Ring; the only difference is that the insertion quadrupole doublet on either side of the straight section is made up of two quadrupoles instead of four. At C0 where the beam is extracted to a dump when aborted by a one-turn kicker, superconducting bending magnets have been shortened on either side of the straight section, and conventional extraction magnets with equivalent $\int B\, dL$ have been substituted in the center of the straight section. Not only does this make space for the abort kicker, but it provides a certain amount of shielding for the superconducting magnets at the downstream end of the straight section.

The straight section A0 contains the extraction magnets to divert the beam into the switchyard for fixed-target operation. In both A0 and D0 the lattice quadrupoles on either side of the straight section have been reconfigured so as to produce a large horizontal amplitude function ($\beta$), at the upstream end of the straight section, suitable for fast extraction. This

**Table 2**  The accelerator parameters

General

| | |
|---|---|
| Number of dipoles | 774 |
| Number of quadrupoles (without low-$\beta$) | 216 |
| Number of sectors | 6 |
| Good field region | $\pm 2$ cm horizontal |
| | $\pm 1.5$ cm vertical |

Tune ($\nu$)

| |
|---|
| 19.41 horizontal |
| 19.38 vertical |

Natural chromaticity ($\xi$)    $-22$

Lattice

Lattice for sector arc: 2(F, 4B, D, 4B) F, $S_1$2B, D, 4B, 12(F, 4B, D, 4B)

Warm length $S_1 = 12.5$ m

$\beta_{max}$ 100 m    $\beta_{min}$ 29 m

$\eta_{max}$  6 m    $\eta_{min}$ 1.2 m

Lattice for normal straight section:

F, $S_2$, 3B, F, D, LS, F, D, 4B, D, 4B

Warm length long straight $LS = 51$ m

Warm length $S_2 = 6$ m

$\beta_{max}$ 110 m, $\beta_{min}$ 60 m in straight section

Lattice for high-$\beta$ straight section:

F, $S_2$, 3B, D, F, LS, D, F, 4B, D, 4B

$\beta_{max}$ 243 m, $\beta_{min}$ 34 m in straight section

Lattice for low-$\beta$ straight section: F, $S_2$, 3B, F, D, F, D, F, CS, D, F, D, F, 4B, D, 4B

Warm length collider straight $CS = 13$ m

$\beta_{max}$ 900 m, $\beta_{min}^{*}$ 1 m

Radio frequency

Longitudinal phase space:

0.3 eV-s fixed-target, 3.0 eV-s collider

rf frequency 53 MHz

Harmonic number 1113

rf voltage, 8 cavities at 300 kV each

Power supply

System stored energy 350 MJ

Total inductance of ring 36 h

Dipole inductance 0.045 h

Total resistance of ring:

90 m ohm (ring) + 90 m ohm (filter chokes)

Power supplies, 12 at 1 kV, 4500 A

Volts to ground (normal) 500 V peak

Volts to ground (fault conditions) 1.5 kV maximum

Volts bus-to-bus 1–1.5 kV peak

Injection

Single turn (single bunch for collider)

Abort

Single-turn extraction

Extraction

Half-integer resonant

lattice change does not affect the optics elsewhere in the machine but magnifies the horizontal beam size at the position of the electrostatic septa (D0) and magnetic Lambertson septa (A0). This magnification is crucial for obtaining efficient extraction and yet maintaining a reasonably small magnetic aperture in the superconducting arcs of the ring.

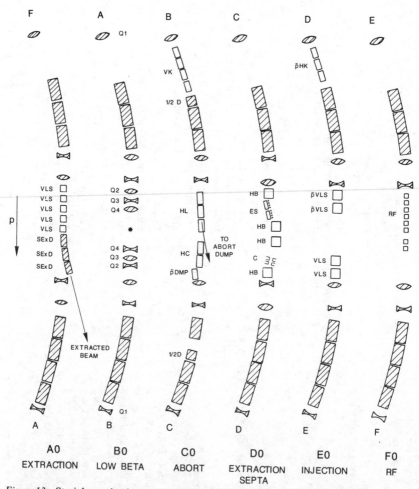

*Figure 13* Straight-section layout A0–F0. Cross hatching indicates superconducting elements. VLS = Vertical bending Lambertson Septum magnet; SExD = Skew Extraction Dipole; VK = Vertical Kicker; 1/2D = half-length Doubler dipole; HL = Horizontal Lambertson septum magnet; HC = Horizontal C magnet; HB = Horizontal Bend; ES = Electrostatic Septum; C = Collimator; and HK = Horizontal Kicker.

The straight section D0 will be reconfigured depending on whether fixed-target or collider operation is underway. For the fixed-target operation, a high-$\beta$ lattice configuration identical to that at A0 is used. In addition to the standard superconducting magnets on either side of the straight section, a chevron arrangement of conventional magnets is installed in the straight section, with the extraction electrostatic septa located after the first of these magnets. This configuration was devised in order to protect the downstream superconducting magnets from beam spray off the septa, as discussed in a prior section. For collider physics, straight sections B0 and D0 will be used for large solid angle detectors. In the case of D0, the extraction bends and electrostatic septa will be removed and the low-$\beta$ quads and experimental detector installed. Changeover time will be less than two weeks.

The B0 straight section has been modified in order to produce a low-$\beta^*$ of 1 m at the collision point. The quadrupole layout is such that the optics can be adjusted between the standard fixed-target optics (like E0 and F0) and the low-$\beta$ optics by adjusting four quadrupole circuits (35).

In addition to the standard Collins quadrupole doublets at either end of the straight section, which are powered in series with the whole super-conducting ring, triplets have been added inboard. Pairs of elements of the triplet (either side of the collision point) are powered in series by separate supplies. In addition, the quads a half-cell removed from the straight section need to be separately powered as a pair. During normal ramping these latter must track the normal excitation ramp and the other three supplies are off. During the process of squeezing the $\beta^*$ from 70 m to 1 m with the energy held constant, all four of these supplies must simultaneously follow prescribed excitation curves in order not to perturb unduly the optics outside of the interaction straight-section region. Magnetic gradients of 25 kG per inch are required in the low-$\beta$ quads. These quads have a 3-inch inside coil diameter, a copper-to-superconductor ratio of 1.3 : 1 and a 20-$\mu$m filament size.

In this design we have placed a minimum of reliance on the operation of the special low-$\beta$ quads during fixed-target physics (one circuit). We have also allowed for continuous tuning from the fixed-target lattice to the low-$\beta$ lattice, whereas in reality injection and acceleration for storage could probably take place at an intermediate $\beta^*$ value. We have also chosen to minimize the number of special quads and power supplies for economy of high-current cryogenic power leads and separate high-current supplies and quench protection. Because of this simplicity the amplitude function, $\beta(z)$, and dispersion function, $\eta(z)$, [$\Delta x(z, \Delta p/p) = \eta(z)\Delta p/p$] are not completely matched to fixed-target values around the ring. In particular, the dispersion is $\eta = 0.2$ m at the collision point and 60% larger in the arcs than for fixed-target operation.

Part of the preparation for collider operation is a reconfiguration of the Main Ring at B0 and D0 (36). Because the Main Ring will be accelerating and extracting beam for p̄ production during the time the Energy Doubler is in storage mode, it is desirable not to have the Main Ring beam going through the center region of the detectors. Otherwise there would be difficulties in Main Ring acceleration with the detector magnetic fields turned on as well as background and radiation damage in the detector, produced by the Main Ring beam. The B0 overpass allows the Main Ring to bypass the detector completely; the D0 overpass bypasses the central region and argon calorimeter of the D0 detector.

The Main Ring is arched vertically away from the Energy Doubler in these regions in so-called overpasses; this allows for a distance between beams of 21 feet (B0) and $6\frac{1}{4}$ feet (D0). In order to generate the necessary deflection, vertical bending magnets will be inserted in the Main Ring lattice 250 meters either side of B0 and 105 meters either side of D0 straight-section centers. These magnets and the two adjacent horizontal bends, along with similar down bends near the straight sections, are operated at twice the field of the regular Main Ring magnets and limit the Main Ring to a maximum of 200-GeV operation. Separate tunnel enclosures for the Main Ring near B0 are also required.

## 3.2  Correction and Adjustment Magnets

Correction magnets are those required to correct field imperfections and alignment errors of the main quadrupoles and bend magnets. Adjustment magnets are those required to tune the optics depending on the desired operating conditions. Often the same "correction" magnets perform both functions. Thus, dipole steering magnets are necessary to compensate for alignment errors and put the beam in the center of the aperture, but they are also necessary to bump the beam away from the center where this is required to avoid certain restrictions like injection or extraction magnets. Chromaticity sextupole circuits adjust the variation of the tune of the machine (number of betatron oscillations per turn) with momentum, $\Delta\nu(\Delta p/p) = \xi\Delta p/p$. These circuits are necessary as adjustments, but they are also necessary as corrections to the dipole magnet persistent current sextupole error at low excitation.

In the Energy Doubler most of the correction magnets are superconducting and are designed with strengths sufficient for use at full excitation (37). Their power supplies can be programmed to different values continuously throughout the acceleration cycles. The correction magnets are located in "spool pieces" adjacent to each of the more than 200 quadrupole magnets in the lattice. Most spool pieces contain two "packages" of three concentrically wound correctors with six pairs of leads coming from cryo temperature to room temperature. In principle all correctors (over 800) could be powered independently but, in fact, only the dipole steering elements need independent control—the rest are connected in series combinations to produce "correction coil families" powered by single supplies. Originally, putting the correctors inside the quadrupole magnet was considered; however, it was expedient not to complicate the quadrupole design by having a separate spool with its quench protection safety leads and correction elements.

The first correction package (DSQ) contains a dipole (horizontal or vertical), normal sextupole, and normal quadrupole for steering, chromaticity adjustment, and tune adjustment. The second package (OSQ) contains an octupole, a sextupole (normal or skew), and a quadrupole (normal or skew). Not all of these units are connected and powered. (See Table 3.)

Power supplies for the correction elements are of two types: one for single elements (dipoles), or possibly two to three elements in series, with $\pm 15$ V, $\pm 50$ A output, and a current stability and ripple limit of 0.1% of full scale; a second high-precision supply (38) used for 4 to 90 elements in series strings. These precision supplies are rated at $\pm 500$ V, $\pm 50$ A with a stability of better than 2.5 ppm per °C and output voltage ripple less than 30 mV peak to peak. Reference voltage to these supplies is provided by an analog-smoothed high-precision 16-bit D/A (39). Every effort to minimize ripple has been made in order to minimize intensity modulation of the slow extracted beam. This requirement of $dv/dt < 0.0001$ s$^{-1}$ is expected to be more demanding than the requirements of storage.

Desired excitation of the circuits typically is derived from formulas of the form given below for the tune chromaticity adjustment circuits at the focussing and defocussing quadrupoles (40):

$$I_{QF} = E(39\Delta v_x + 11\Delta v_y - 3.3b_1)$$

$$I_{QD} = -E(12\Delta v_x + 43\Delta v_y + 3.3b_1)$$

$$I_{SF} = E(0.17\Delta\xi_x + 0.05\Delta\xi_y - 3.3b_2)$$

$$I_{SD} = -E(0.09\Delta\xi_x + 0.30\Delta\xi_y + 5.1b_2)$$

where $I$ is in amps, $E$ is in TeV, and $\Delta v_{x,y} = v_{x,y} - 19.4$, $\Delta\xi_{x,y} = \xi_{x,y} + 22.5$. Here 19.4 is the natural tune of the accelerator and $-22.5$ is the natural chromaticity. Coefficients $b_1$, $b_2$ are the average bend magnet multipole coefficients in units of $10^{-4}$ at one inch.

**Table 3**    Correction coil circuits and their uses

| | Number of circuits/ number of elements per circuit | Strength each element (kG-in. at 1 in., 50 A) |
|---|---|---|
| **Steering dipole** | | |
| Horizontal and vertical | 224/1 | 181 |
| **Quad tune adjustment** | | |
| Horizontal and vertical | 2/90 | 75 |
| **Chromaticity sextupole** | | |
| Horizontal and vertical | 2/90 | 57 |
| **Skew quad coupling adjustment** | | |
| 0 harmonic $(v_x - v_y = 0)$ Horizontal-to-vertical coupling compensation | 1/48 | 75 |
| **39th harmonic quad** | | |
| Horizontal sin and cos terms for 39th harmonic $(2v_x = 39)$ 1/2-integer stopband adjustment for resonant extraction | 2/4 | 75 |
| **39th harmonic octupole** | | |
| Horizontal sin and cos terms 1/2-integer resonant extraction adjustment | 2/16 | 33 |
| **0 harmonic octupole** | | |
| Horizontal and vertical average octupole compensation | 1/24 | 33 |
| Extraction adjustment | 1/12 | — |
| **Circuits below not powered** | | |
| **39th harmonic quad** | | |
| Vertical sin and cos terms allow for simultaneous $2v_x$, $2v_y = 39$ compensation | 8/2 | 75 |
| **39th harmonic skew quad** | | |
| Coupling resonance sin and cos terms compensation for $v_x + v_y = 39$ | 4/2 | 75 |
| **58th harmonic normal sextupole** | | |
| Sin and cos terms for $3v_x$, $v_x + 2v_t = 58$ | 14/2 | 44 |
| 1/3-integer resonance compensation | 4/1 | — |
| **58th harmonic skew sextupole** | | |
| Sin and cos terms for $3v_y$, $2v_x + v_y = 58$ | 14/2 | 44 |
| 1/3-integer resonance compensation | 4/1 | — |
| **0th harmonic skew quadrupole** | 12/1 | 75 |

## 3.3    Beam Position and Beam Loss Monitors

Beam diagnostics in the Energy Doubler are still under development and rapid evolution, especially for storage operation. However, considerable effort has been taken with recording and displaying data from the more basic diagnostic devices so that information is available for analyzing causes of beam-induced quenches, aborts, and losses. The fact that recovery from quenches can take an hour or more makes it imperative that reasons for erratic beam behavior be found with as few beam pulses as possible, and at best with just one.

The basic diagnostic devices are the beam position and loss monitor systems (41). In general large amounts of memory have been incorporated into the electronics associated with the devices so that information related to first-turn injection and to 1000 consecutive turns, or to 500 samples at 15-ms intervals prior to an abort, can be automatically recorded, as well as data at other desired times. The Beam Position Monitor (BPM) system consists of about 200 position detectors and associated rf, analog, and digital circuitry. The detectors are a pair of directional strip line plates 20 cm long and can independently measure the beam in two directions. A detector is mounted at the end of a superconducting quadrupole inside the cold beam tube. Signals from the detectors go to the nearest service building for processing. They first pass through a 53-MHz resonant filter to an amplitude-to-phase modulation unit that produces an analog voltage proportional to the ratio of signals from the two plates. The resonant filter is necessary for isolated single bunches in collider operation. In addition to the AM/PM processing, the two-plate signals are added to provide intensity information. Position and intensity signals are then processed and digitized. There are of the order of 8 to 10 detectors per service building, with parallel analog processing for each detector. The digitized position and intensity information is read into a microprocessor, which can store either the single-turn information or average information over a selected time period in order to find the equilibrium orbit independent of coherent betatron oscillations.

The position system has a least count resolution of 0.15 mm. The dynamic range of the system is dependent on the number of rf buckets around the ring circumference filled with beam; its minimum threshold is $5 \times 10^9$ protons in less than 50 consecutive buckets, its maximum is $3 \times 10^{13}$ protons in a full ring ($\sim$ 1000 buckets). This wide dynamic range is required in order to avoid quenches during initial turn on, as well as for collider operation.

Data from the beam position monitors are used in various ways. One can plot beam position as a function of location around the ring for one-turn information and for closed-orbit information. This allows one to make corrections with the dipole correctors over the whole accelerator cycle. On any two detectors one can obtain position information on 1000 consecutive turns. This allows one to observe and correct injection errors in either the transverse or longitudinal phase space by observing coherent betatron and synchrotron oscillations. By using a fast Fourier transform routine with these data, one can obtain coherent tune measurements either at injection time or later in the cycle by inducing a coherent oscillation. By looking at the turn-by-turn information during times of erratic operation, one can see if coherent oscillations have been induced by some unknown mechanism. If two detectors in the same plane (horizontal or vertical) with ~90° phase advance between them are used for the turn-by-turn information, it is possible to plot the phase space $(x, x')$ mapped by the coherent oscillation. In Figure 14 a closed-orbit position plot around the ring taken during beam extraction time shows a large deviation from center line to avoid injection devices at E0. The small horizontal deviations in D and F sectors and in the D0 and A0 straight sections are made so as to reduce beam loss from particles scattered off the electrostatic septum during extraction. A turn-by-turn plot at injection time of a horizontal detector (Figure 15) illustrates coherent betatron and synchrotron motion with and without the use of the fast beam damper. Data are displayed for both 128 turns and 1024 turns.

Beam loss monitor (BLM) detectors are located near each quadrupole in the ring. Their electronics are packaged along with the BPM electronics.

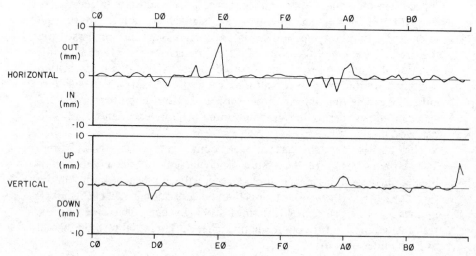

*Figure 14*  Horizontal and vertical closed orbit measured by the beam detector system at 800 GeV during resonant extraction.

Additional BLMs are installed in straight-section regions associated with extraction or abort. The BLM detector consists of a sealed glass argon ion chamber with a sensitivity of $2 \times 10^{-8}$ coulombs per rad (C rad$^{-1}$) and linearity to 100 rad of instantaneous loss. The electronics for each detector consists of a 4-decade integrating logarithmic amplifier with an integrating time of 1/16 second to try to match the quench properties of the superconducting magnets so that the output voltage is related to the probability of quenching the magnets. Outputs of the loss monitors are continuously checked so as to abort the beam if the abort threshold is exceeded. Signals to abort the beam can be generated within 200 $\mu$s. Though abort triggers can be generated by numerous other signals, such as the kicker not being ready or the beam position being out of tolerance, the loss monitor system is by far the most used for preventing beam-induced quenches. Figure 16 illustrates loss patterns obtained during a quench induced by too much fast extracted beam; it also indicates abort thresholds. Loss readings are higher in the straight section because the detectors are not shielded as well by the magnets. Loss points in sectors D, E, F are typical of those caused by extraction beam spray.

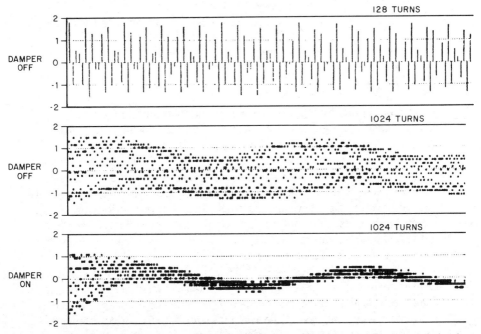

*Figure 15* Turn-by-turn plot illustrating coherent betatron and synchrotron oscillations and showing betatron oscillations being damped by the beam damper system.

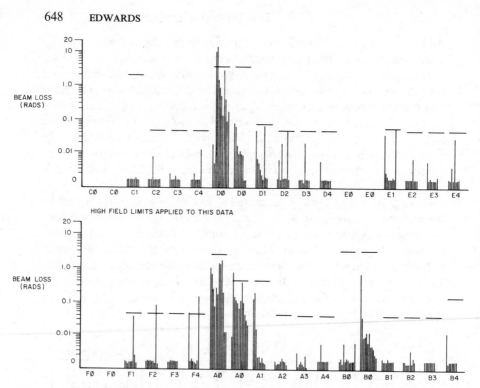

*Figure 16*  Beam loss reading during a quench induced by fast extraction losses.

## 3.4  *Beam Damper System*

The beam damper system (42) is used to inhibit beam blow up caused by electromagnetic coupling between the beam and the vacuum chamber wall or other devices installed in the ring. The damper can work on any collective betatron motion that the beam in individual rf buckets undergoes as a whole. The damper system is also a diagnostic device that can measure betatron motion of individual buckets, can damp or antidamp individual buckets selectively, or can antidamp small amplitude oscillations and damp larger amplitudes; this allows for continuous tune measurements on individual buckets during acceleration and extraction. We have just begun to explore the potential of this device.

The damper consists of two separate parts. The first is the low-level measurement and storage electronics, which also incorporates analysis and display electronics; the second is the high-level power system and deflector plates that act on the beam. The system is similar to that used in the Main Ring but it is considerably more powerful. The principle is very simple. The position of the beam in each rf bucket is measured as it goes by pickup electrodes. On the next turn it passes damper electrodes after $(n+\frac{1}{4})$ betatron oscillations and it receives a transverse deflection proportional to its measured displacement from center. Thus, after successive turns a coherent oscillation can be reduced to zero.

The low-level electronics consists of flash ADCs followed by a specialized "pipe line" computer for computing the beam position and multiplying it by the proper gain transfer function. The flash ADCs and computer are clocked at the 53-MHz frequency of the rf system, and the delay in the electronics is just equal to the time of flight of the beam around the ring so that the correct beam bucket is acted on.

The high-level electronics for each plate consists of a 24-tube (4CW800F tetrode) distributed amplifier feeding 50-ohm lines to the deflector strip line. Output to each plate is zero to $+1400$ V, with a 10–90% response of $3\frac{1}{2}$ ns. The gap between the plates is $2\frac{1}{2}$ inches and the length is $\lambda/4$ (of 53 MHz). The filling direction is opposite to the beam direction and a pulse length of the order of 12 ns is required.

## 3.5   Flying Wire

The flying wire (43) is a device to measure the profile of the beam and thus deduce the beam emittance if the amplitude function $(\beta)$ is known or to deduce relative amplitude functions at different locations with simultaneous measurements. The principle is very simple: A thin wire is quickly rotated through the beam in either the horizontal or vertical direction and losses from the beam hitting the wire are picked up in a photomultiplier counter. (Alternatively, charge depletion from the wire can be measured directly.) The signals are proportional to the beam intensity as a function of wire position.

This very simple concept has some interesting engineering problems associated with it. Namely, the wire must go fast enough so that it does not burn up or unduly scatter the beam (carbon or beryllium 4-mil wires are used). The data-digitizing rate must be fast enough to obtain sufficient data points over the beam profile, and the wire must move in a reasonably continuous linear fashion when in the beam and stop after one transversal. Typically, velocities of 5–10 m s$^{-1}$ and a digitizing rate of 47 kHz (revolution frequency of the beam) are used. Figure 17 illustrates wire profile measurement at $\beta = 100$ m, 1 m for 800-GeV beam.

## 3.6   Radiofrequency Accelerating System

The rf accelerating system (44) consists of eight 53-MHz cavities driven by pulse amplifiers. The cavities each produce a peak accelerating voltage of 1/3 MV and are coaxial, two-gap resonators, 12 inches in diameter and 160° total electrical drift tube length, or 108 inches total length. The cavities are

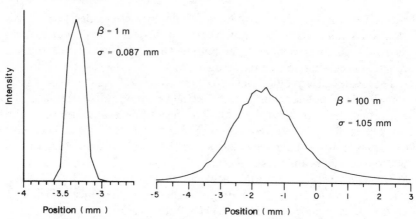

*Figure 17*   Beam profiles measured by the flying wire system at the collider low-$\beta$ point and at a standard-$\beta$ point.

excited from their center point in a $\lambda/4$ resonant mode. The small diameter unfolded structure was based on the geometrical requirement that the cavities fit between the Main Ring cavities and the floor. They can be mounted end to end with 180° electrical spacing if desired. The pulse amplifiers for the cavities are 200-kW units and produce $\pm 27$ kV peak to peak and 10 amps rms. Fifty-ohm coaxial lines connect between the pulse amplifiers and the cavities.

The cavities must be frequency modulated by only 2 kHz during acceleration from 150 to 1000 GeV to compensate for the change in the protons' velocity. Radial position changes of 2 cm average can be made by frequency changes of 1 kHz. Because the required modulation is small, the normal frequency program is a completely dead-reckoned digitally gene-rated modulation and runs off the control system Magnet Data Clock.

Though this system will run and accelerate beam without any feedback at low beam intensities, coherent phase oscillations are generally damped with a single-phase feedback loop. The phase detector itself must be capable of obtaining information from only one bucket (out of 1113 rf buckets) in the collider mode of operation.

The cavities have a 7-kHz bandwidth ($-3$ dB points) and are tuned to resonance (1 kHz per °C) by controlled changes in their operating temperature during the ramp cycle. In addition, it is necessary to compensate for the cavity heating that occurs during the long cycle time with the rf on for 40 out of 60 s. This is done by heating the water going to the cavities in proportion to the power not used when they are not at maximum power.

All eight cavities are used for acceleration of protons during fixed-target operation. However, the cavities have been positioned in pairs spaced by $3\lambda/4$ and phased in time by $90°$ so as to be in phase for beam going in one direction but $180°$ out of phase for beam traveling in the opposite direction. During collider operation, they operate as orthogonal sets of four for acceleration of protons and antiprotons independently. The colliding point of the two beams can be moved circumferentially around the accelerator and frozen at a particular point by independent adjustment of frequency of the two systems and by adjustment of the phase between the systems.

## 4. OPERATIONS

The Energy Doubler was commissioned between June and September of 1983 (19). Beam was first accelerated in July to 512 GeV. Fixed-target operation at 400 GeV began in October 1983 and the energy was increased to 800 GeV in February 1984. At this writing there have been three physics runs over a total of about 12 months duration, so one can begin to make some generalizations about the operational behavior of the accelerator. The three runs were at 400, 800, and 800 GeV with fast extraction. Intensity to date has not exceeded $1.1 \times 10^{13}$ ppp. Single-beam storage studies have been carried out (45).

The rapid commissioning was successful probably because of the many step-by-step tests of components and subsystems that were made over the years prior to the final beam test. Though the major systems were not particularly reliable, they were all checked out and working. Interruptions to beam studies were at the level of trips of the ramp caused by refrigerator instabilities or spurious quenches caused by the quench protection system rather than any basic flaws.

As an accelerator the Energy Doubler was much more stable and reproducible than anyone had expected. (There had been concern that the coils would move during cooldown, quenches, or ramping.) For instance, steering adjustments once made did not have to be corrected from day to day. In addition, settings worked well over wide energy intervals and did not have to be changed appreciably as a function of excitation. The natural tune and chromaticity, though not exactly what was expected from the magnetic measurements, could be explained by differences in the average multipole coefficients $a_1$, $b_1$, $b_2$ of about one unit ($1 \times 10^{-4}$ at one inch). The momentum aperture was reassuringly large. Nevertheless continuous study and effort will be necessary to carry out all the detailed measurements required for a sufficient understanding of the accelerator (46) to permit both high-intensity, fixed-target operation and $\bar{p}p$ storage.

Operational reliability is a big issue in the Tevatron facility. There are now four accelerators in the chain, followed by the switchyard beam lines.

Both the Energy Doubler and the switchyard have superconducting magnets and the complications that go with them. It is obvious that the success of the Tevatron, as well as the credibility of larger, higher energy proton accelerators, lies in the ability to obtain high operational reliability. Improvements over the present reliability will have to be made in the Tevatron.

For initial operation with the Energy Doubler, actual uptime for high-energy physics has been at best of the order of 60% of the scheduled time. This amounts to about 300 hours per month as compared with 400 hours per month (or 80% actual to scheduled time) for Main Ring operation prior to Energy Doubler installation. At present about 50% of all downtime is due to the Energy Doubler systems. The major causes are the cryogenic system, the power supply and quench protection system, and quenches of the magnets themselves. These three categories account for 75% of the Energy Doubler downtime in a ratio of 2:1:1. As a reference, the Main Ring power supply system has about the same failure time as the Energy Doubler power supply and quench protection system and is the largest cause of downtime in the rest of the accelerator.

Cryogenic system downtime is of three types: trips due to unstable operation or oscillations in the system that indicate the magnets may be too warm; failures of expansion engines; and failure or reduced capacity of the central plant as a result of contamination, filter clogging, or inadequate supply of nitrogen or helium.

In the original design of the cryogenic system, the central helium liquefier played the role of a backup system that could be used to supplement the satellite refrigerators' capacity and provide continued operation if a refrigerator were down. In reality the central liquefier is required for ramped operation in addition to the above roles. This essential use of the central liquefier came about because the heat leak in the magnets is larger than originally expected. The increased heat leak resulted from a last-minute modification to the magnet-cryostat anchors, which prevent the collared coil assembly from rotating in the external iron.

Quenches of the superconducting magnets are obviously very disruptive to the accelerator operation. Not only does it take time to recover from a quench but the refrigeration and quench protection systems are exercised with each quench. During operation in the first quarter of 1985, there have been on the average of nine quenches per week. High-field quenches require about 70 minutes for recovery, injection-field quenches take about 20 minutes. The reasons for quenching can be put into three categories: typically 20% are directly due to failures of the quench protection system (like heaters firing without cause) or cooling problems in the cryogenic system; 25% are beam induced from failures of devices like kickers or correction coils; and the remaining 55% are beam related and have to do with tuning, unstable operation, or high intensity. The third kind of quench happens most often at high field, whereas beam quenches from device failures happen most often at injection. There are about equal numbers of high- and low-field quenches.

The expense of operating the superconducting accelerator can be assessed from the power and cryogens used. Figure 18 illustrates the power savings of the superconducting ring. The shaded area illustrates the power required to operate the main accelerator prior to the Energy Doubler operation versus that necessary to operate the Main Ring–Doubler and Central Helium Liquefier after Doubler turn on. There is of the order of a factor of two savings or approximately 15 GW hours per month (10 million dollars per year) for continuous operation. Cryogen usage (Figure 19) runs about 5.4 million dollars; thus if the program is operating 11 months of the year and the refrigeration operation is not cut back during the one month off, a savings of about $3\frac{1}{3}$ million dollars could be realized. Of course, this saving is offset by the additional operating cost of the cryogenic systems from manpower and materials that would not be needed in a conventional system. The main advantages are that power usage per se has been substantially reduced and that higher energy beam and collider physics is possible.

Cryogen usage is substantially greater than anticipated. Nitrogen usage is more than 50% higher than expected. This excess is due in part to the higher heat leak of the magnets and in part to a preference for operation in the most stable mode rather than the most economical mode. The large helium leakage rate is something of a mystery and probably is distributed over many parts of the plant. For instance, leakage of valve stems is a suspected cause and checking all possibilities is an arduous task. Work must be done to reduce both nitrogen usage and helium leakage as the operation stabilizes.

## 5.  HINDSIGHT AND A LOOK TO THE FUTURE

It works! This is by far the most significant thing that can be said in retrospect. One tends to forget that it was not obvious that the Energy Doubler would work and that years of effort with many failures and setbacks went into the magnet and cryogenic development.

There are no major flaws in the system that we know of to date, but one should keep in mind that it is very much a prototype accelerator. It is as much an accelerator research tool as a high-energy physics tool. It will take time for the performance to become as dependable as expected from conventional accelerators.

So, first of all it works. Now if we were starting over today what would we have done differently? The most natural question to ask is whether the

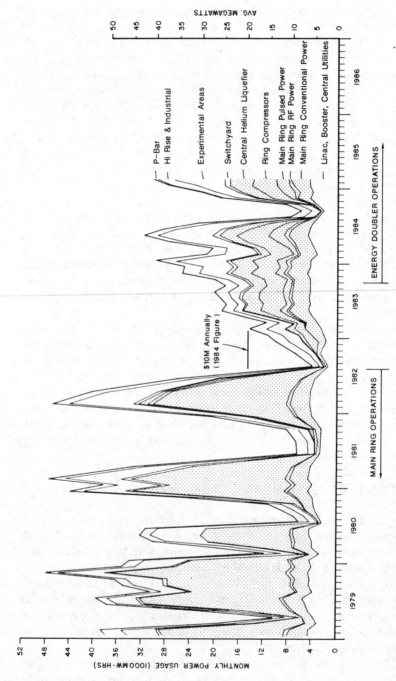

*Figure 18*   Power usage before and after Energy Doubler installation. Shaded area indicates Main Ring usage prior to Energy Doubler installation versus Main Ring and Energy Doubler usage after installation.

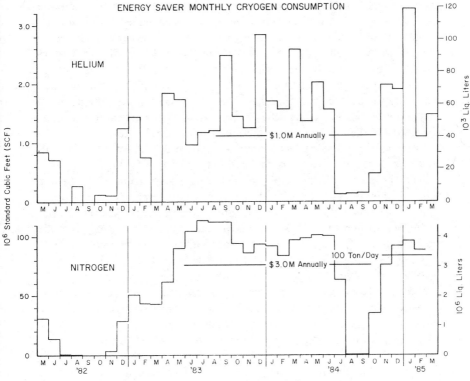

*Figure 19*   Energy Doubler helium and nitrogen usage.

Energy Doubler should have been built in the same tunnel as the Main Ring. Even at this point, one can only speculate as to the outcome if the same tunnel had not been used. It is quite possible that the whole project would not have happened at all. After all, with a risky research and development effort, one does not necessarily want to spend a lot on conventional construction in order to test the new idea. It is expedient having all the conventional utilities and buildings available and in good working order. By installing the Energy Doubler in the Main Ring tunnel and turning off all physics research during that time interval, one guaranteed total laboratory commitment to the project. That commitment would not have been available were the physics programs running simultaneously. On the other hand, the three long shutdowns for Doubler installation, p̄ injection extraction tunnel modifications, and colliding detector building construction have been very disruptive to physics research. Of the order of 2¾ years of operation will have been lost during the six years (1981–1986) of construction, installation, and commissioning of the Tevatron. But physics at higher energy has been made possible sooner.

Many design choices for the Energy Doubler magnets were precluded by the requirement that they fit below the Main Ring. This boundary condition may actually have helped focus the effort. Possibly the esthetically pleasing warm iron design with short suspension blocks and high heat leak might have been modified had there been more space. The compact cryostat cross section leaves little room for the cryogen flow and thus limits the length of the magnet strings serviced by one refrigerator.

Independent of the magnet geometry, better heat transfer from the coils to the two-phase system would have helped obtain higher energy and reduce ramp turn-on transients (resulting from expansion of the helium liquid with heating). Increasing the capacity of the satellite refrigerators to match better the magnet heat load or obtaining redundancy in the Central Helium Liquefier is the single, most important change that could presently be made to the Energy Doubler in order to improve its reliability.

The overriding issue for the next generation of large superconducting accelerators-colliders must be reliability. Some real assessment of the probable success in meeting operating goals must be made, and large prototype systems tested for fatal flaws. The reliability issue must be treated with a care it never has received in past accelerators.

A second difficult issue is that of magnet aperture. By this one does not mean the physical aperture of the beam tube but rather the usable dynamic aperture within which the beam will survive. There are at least two levels here: over what region will the beam behave reasonably linearly so one can analyze it and so that it will store with very long lifetimes; and over what region will it survive long enough so that it can be crudely measured and adjusted? As indicated in the introduction, evaluation of the problem is very controversial. Computer simulations are not necessarily sufficient to answer these questions and must be checked against measured behavior in existing accelerators such as the SP$\bar{\text{P}}$S and the Energy Doubler.

The field quality in the magnets is affected by the choice of coil radius in two ways. First, as the radius is reduced conductor placement errors become more important and the random variations of the multipole coefficients of the field will be increased. It is believed that these variations in the $b_n$, $a_n$ coefficients scale like $1/r_c^{n+\frac{1}{2}}$ all else being equal, where $r_c$ is the coil radius. This possibly simplistic approximation comes from assuming dominance of azimuthal errors that scale like the square root of the number of turns in the coil and consequently like the $1/r_c^{1/2}$ for constant cable thickness (47). The second effect is produced by "persistent currents"; it changes the average multipole coefficients at low excitation, as well as producing some random behavior. The average multipole coefficients of the dc residual field scale like $D/r_c^{n+1}$, where $D$ is the superconductor filament diameter. [They also scale like $J_c(B)$ and $\Delta r_c \varepsilon$, where $\Delta r_c$ is the coil thickness and $\varepsilon$ is the fractional area of the coil that is composed of superconductor (24).] The effect is dominant in the $b_0, b_2, b_4, \ldots$ coefficients; however, random behavior can appear in both normal and skew coefficients from variations in filament diameter, critical current $J_c(B)$, and temperature of the coils. Because persistent currents produce a field that is only slowly dependent on magnet excitation (Figure 7), the effect on the beam at injection varies inversely with the injection field.

Persistent currents lead to large chromaticity, $\xi$, in the accelerator if uncompensated $[\Delta v(\Delta p/p) = \xi \Delta p/p, \ \xi \simeq b_2 \langle \eta \beta \rangle]$. Typically, $\eta$ is unchanged as the radius of the accelerator is increased and $\beta$ increases proportionally to $\sqrt{R}$, where $R$ is the accelerator radius. Uncorrected chromaticities at injection 50 to 100 times larger than in the Energy Doubler can result in 20-TeV accelerators at 1 TeV injection, with coil radii reduced from 3.8 cm (the Energy Doubler coil radius) to 2.5 or 2.0 cm. Fortunately it appears that filament diameters of one half to one quarter that of the conductor used in the Energy Doubler (8 $\mu$m) may be possible. In any case, sextupole corrections will probably have to be made locally in each bend magnet (otherwise second-order terms may become important and misalignment between the correction and the bends will produce random quadrupole errors). Two such correction schemes are passive and require no external control. One relies upon a shorted superconducting sextupole coil driven by the unwanted sextupole flux (48). Another method relies on strips of superconductor mounted on the outside of the beam pipe (49). This scheme uses the magnetization induced by persistent currents in the passive strips to cancel the sextupole fields and does not involve any external (transport) current. Higher order, e.g. decapole, corrections may be required also.

The fundamental risk from the random multipole errors is the deterioration from rational linear behavior of the accelerator beam (50). This deterioration is roughly proportional to $\sqrt{L} x^{n-1} \langle a_n^2, b_n^2 \rangle^{1/2}$ in the approximation that the amplitude function $\beta$ is proportional to $R^{1/2}$. Here $L$ is the length of a bend magnet (or the coherent length over which multipole coefficients remain constant if shorter than a magnet length); $x$ is the aperture dimension of interest, either horizontal or vertical; and $\langle a_n^2, b_n^2 \rangle^{1/2}$ are the rms fluctuations of the multipole coefficients. The above expression can be written $\sqrt{L} r_c^{-3/2} (x/r_c)^{n-1}$ using the expected scaling of the multipoles. Thus, for sextupole-dominated effects the aperture is expected to be about five times smaller for a 2-cm coil radius than for the 3.8-cm Energy Doubler coil radius.

Recently there have been pleasant surprises in superconductor development. Not only does it appear possible to produce conductor with 4-$\mu$m filaments as mentioned, but the current-carrying capacity of the conductor has already improved markedly since production of Energy Doubler magnets. At present (1985) a $J_c$ of 2150 amp mm$^{-2}$ at 5 T and 4.2 K is typical in short sample tests of Energy Doubler–type cable, an improvement of 20% since Doubler production. Moreover, there are preliminary indications that both small filaments and high current density can be achieved simultaneously, and strand with 2900 amp mm$^{-2}$ has been produced.

Cryogenic requirements and system design will probably be quite different in a Superconducting Super Collider (SSC) from those for the Energy Doubler (51, 52). A large fraction of the magnet heat leak comes from synchrotron radiation heating by the beam and some comes from static or ramped heating loads. These must be substantially reduced. Static loads can be reduced through longer magnet supports; excitation ramping will be slow and only occasional. Total system cryogenic capacity may need to be only three times that required for the Energy Doubler. Heat leak estimates, possibly optimistic, are as little as 1–2 W at 5 K and 2 W at 10 K per (12-meter) dipole, as compared with 9 W per 6-meter Energy Doubler dipole.

The refrigerator system will probably consist of stand-alone refrigerators located around the ring at each of the dozen or so sectors. Each refrigerator might have a number of compressors, so there are backup spares. In addition, each refrigerator could be sized for 150% of nominal load so that if any one goes down its cooling load can be handled by its two neighbors.

It is hoped that transfer lines and most headers can be eliminated. By having more space for the magnet cryostat, the magnets themselves will serve as their own transfer lines. By designing the magnet cryostats to withstand high pressures, venting of helium during quenches and power failures can be substantially reduced.

The magnet quench protection system can potentially be "passive" for a slowly ramping collider in which radiation damage is not likely to be a problem for semiconductors (53). In such a system diodes are put across the coils of the individual magnets (or the coil is split in fractions and additional diodes are used). These conduct when a quench occurs. $L \, di/dt$ voltage across the magnet must be limited to less than the forward voltage drop across the diode, and the quench propagation velocity must be sufficient to allow the magnet to absorb its own energy without being destructively overheated. This scheme favors magnets with low stored energy or small coil radius. The quenches must still be detected in order to dump the energy from the rest of the magnet string, but heaters and very fast detection response have been eliminated. (There is a spectrum of possible solutions to the quench protection that range between this one and that used in the Energy Doubler.) The passive diode protection scheme does require a large number of diodes mounted in the magnet cryostats. Of the order of 25,000 would be required in a 20-TeV collider. Even if the probable number of times each one is used in quenching is less than one, there still are serious reliability questions with this scheme. (Magnets would have to be warmed up to change shorted diodes.) The HERA accelerator will probably try this method first, and its experience will be extremely valuable.

ACKNOWLEDGMENTS

The successful completion and operation of the first large superconducting accelerator are a result of the dedicated work and perseverance of the people at Fermilab. Special acknowledgment should be given to the following individuals for supplying material used in this article: D. Bogert, D. Edwards, E. Fisk, W. Fowler, M. Harrison, H. Jostlein, K. Koepke, P. Martin, C. Rode, and G. Tool.

*Literature Cited*

1. Bjorken, J. D. "Fixed-target physics at Fermilab," p. 144; and Lederman, L. M. "Physics at the Tevatron," p. 33; both in *Am. Phys. Soc., Proc. Santa Fe Meet., Oct. 31–Nov. 3, 1984.* Philadelphia/ Singapore: World Sci. (1985)
2. Sanford, J. R. *Ann. Rev. Nucl. Sci.* 26: 151–98 (1976)
3. Design Report Superconducting Accelerator. *Fermilab Rep.* (May 1979)
4. Design Report Tevatron I Project. *Fermilab Rep.* (Sept. 1984)
5. Peoples, J. *IEEE Trans. Nucl. Sci.* NS-30(4): 1970 (1983)
6. Wilson, R. R. *Phys. Today*, Oct. 23 pp. (1977)
7. Vander Arend, P. C., Fowler, W. B. *IEEE Trans. Nucl. Sci.* NS-20(3): 119 (1973)
8. Palmer, R., Tollestrup, A. V. *Ann. Rev. Nucl. Part. Sci.* 34: 247 (1984)
9. Wilson, M. N. *Superconducting Magnets*, pp. 131 ff. Oxford: Clarendon (1983)
10. Turkot, F., Cooper, W. E., Hanft, R., McInturff, A. *IEEE Trans. Nucl. Sci.* NS-30: 3387 (1983)
11. Hanft, R., Brown, B. C., Cooper, W. E., Gross, D. A., Michelotti, L. et al. *IEEE Trans. Nucl. Sci.* NS-30: 3381 (1983)
12. Cooper, W. E., Fisk, H. E., Gross, D. A., Lundy, R. A., Schmidt, E. E., et al. *IEEE Trans. Magn.* MAG-19: 1372 (1983)
13. McInturff, A. D., Gross, D. *IEEE Trans. Nucl. Sci.* NS-28: 3211 (1981); Wake, M., Gross, D., Yamada, R., Blatchley, D. *IEEE Trans. Magn.* MAG-15: 141 (1979); McInturff, A. D., Carson, J., Engler, N., Fisk, H., Hanft, R., et al. *IEEE Trans. Magn.* MAG-21: 478 (1985)
14. Brown, B. C., Fisk, H. E., Hanft, R. *IEEE Trans. Magn.* MAG-21: 979 (1985)
15. Edwards, D. A. In *Proc. Int. Conf. High Energy Accelerators, 12th, Fermilab*, ed. F. T. Cole, R. Donaldson. Batavia: Fermilab. 90 pp. (1983). This reference gives a summary of dipole multipole data
16. Michelotti, L. P., Ohnuma, S. *IEEE Trans. Nucl. Sci.* NS-30: 2472 (1983)
17. Gelfand, N. M. In Accelerator Physics Issues for a Superconducting Super Collider, 12–17 Dec., ed. M. Tigner. Ann Arbor: Univ. Mich. 124 pp. (1983)
18. Collins, T. L., Edwards, D. A. *Fermilab Tech. Memo TM-614* (1975)
19. Edwards, H. T. See Ref. 15, p. 1
20. Rode, C. H. See Ref. 15, p. 529
21. Rode, C. H., Theilacker, J. C. "What % of Carnot is Realistic?" *Fermilab Intern. Memo* (1983)
22. McInturff, A. D. Private communication
23. Martin, P., et al. *Fermilab Rep. TM-1134.* 57 pp. (1982)
24. Kuchnir, M. "Second SPTF Test." *Fermilab Intern. Memo* (1981)
25. Bartelson, C. L., et al. *J. Vac. Sci. Technol. A* 1(2): 187 (1983)
26. Tool, G., Flora, R., Martin, P., Wolff, D.

*IEEE Trans. Nucl. Sci.* NS-30:2889 (1983)
27. Koepke, K., Martin, P. *Fermilab Intern. Memo UPC-155* (1982)
28. This is an experimental observation; private communication, K. Koepke
29. Bogert, D., Segler, S. In *Europhysics Conf.—Computing in Accelerator Design and Operations, Berlin*, ed. W. Burse, R. Zelazny. Berlin: Springer-Verlag. 338 pp. (1983)
30. Bonner, M. K. In *Research and Development*, ed. R. R. Jones. Barrington, Ill. 106 pp. (1984)
31. Cox, B., Mazur, P. O., VanGinneken, A. *Fermilab Tech. Memo TM-828-A* (1978)
32. VanGinneken, A. *Fermilab Rep. FN-272* (1975)
33. VanGinneken, A., et al. *Fermilab Publ. 85/64* (1985)
34. Drozhdin, A. I., Harrison, M., Mokhov, N. V. *Fermilab Rep. FN-418* (1985)
35. Johnson, D. E. *Fermilab Tech. Memo TM-1106* (1982); Johnson, D. E., Koepke, K., Willeke, F., et al. In *Part. Accel. Conf., Vancouver, IEEE Trans. Nucl. Sci.* In press (1985)
36. Gerig, R., et al. See Ref. 35
37. Johnson, M., McInturff, A., Raja, R., Mantsch, P. M. See Ref. 15, p. 536
38. Pfeffer, H., Yarema, R. J. *IEEE Trans. Nucl. Sci.* NS-30:2870 (1983)
39. Rotolo, C. J. *IEEE Trans. Nucl. Sci.* NS-

30:2959 (1983)
40. Goodwin, R. W., Johnson, R. P. See Ref. 15, p. 594
41. Shafer, R. E., Gerig, R., Baumbaugh, A., Wegner, C. See Ref. 15, p. 609
42. Crisp, J., et al. See Ref. 35
43. Elseth, J., et al. See Ref. 35
44. Kerns, Q., Meisner, K., et al. See Ref. 35
45. Johnson, R. See Ref. 17, p. 45
46. Edwards, D. A., Johnson, R. P., Willeke, F. *Fermilab Publ. 85/59. Part. Accel.* Submitted (1985)
47. Edwards, D. A. Private communication
48. Gilbert, W., Borden, A., Hassenzahl, W., Moritz, G., Taylor, C. *IEEE Trans. Magn.* MAG-21:486 (1985)
49. Brown, B. C., Fisk, H. E. In *Proc. Summer Study on the Design and Utilization of the Superconducting Super Collider*, ed. R. Donaldson, J. Monfin. New York: Am. Phys. Soc. (1984)
50. Collins, T. *Fermilab Publ. 84/114.* 18 pp. (1984)
51. The Superconducting Super Collider. A Reference Design Study for the US Dept. Energy. Berkeley, Calif. (1984)
52. Rode, C. H. In *10th Int. Cryogenic Eng. Conf., Helsinki*, ed. H. Collan, P. Berglund, M. Knusius. Guildford, Surrey: Butterworth (1984)
53. Koepke, K., Tool, G. Workshop on SSC Commissioning and Operation. Unpublished (1985)

# CUMULATIVE INDEXES

## CONTRIBUTING AUTHORS VOLUMES 1–35

# CHAPTER TITLES, VOLUMES 1–35

## WEAK AND ELECTROMAGNETIC INTERACTIONS